CNC Programming Handbook

Second Edition

CNC Programming Handbook

Second Edition

A Comprehensive Guide to Practical CNC Programming

Peter Smid

Industrial Press, Inc.

200 Madison Avenue
New York, New York 10016-4078
http://www.industrialpress.com

TJ1189,S592 2003

Library of Congress Cataloging-in-Publication Data

Smid, Peter.
    CNC programming handbook: comprehensive guide to practical CNC programming/
Peter Smid.
        p. cm.
    ISBN 0-8311-3158-6
      1. Machine-tools--Numerical control--Programming --Handbooks, manuals,etc..I.
Title.

TJ1189 .S592 2000
621.9'023--dc21

00-023974

Second Edition

CNC Programming Handbook

Industrial Press Inc.
200 Madison Avenue
New York, NY 10016-4078

4 5 6 7 8 9 10

Dedication

To my father František and my mother Ludmila,
who taught me never to give up.

Acknowledgments

In this second edition of the *CNC Programming Handbook*, I would like to express my thanks and appreciation to Peter Eigler for being the bottomless source of new ideas, knowledge and inspiration - all that in more ways than one. My thanks also go to Eugene Chishow, for his always quick thinking and his ability to point out the elusive detail or two that I might have missed otherwise. To Ed Janzen, I thank for the many suggestions he offered and for always being able to see the bigger picture. To Greg Prentice, the President of GLP Technologies, Inc., - and my early mentor - you will always be my very good friend.

Even after three years of improving the *CNC Programming Handbook* and developing the enclosed compact disc, my wife Joan will always deserve my thanks and my gratitude. To my son Michael and my daughter Michelle - you guys have contributed to this handbook in more ways than you can ever imagine.

I have also made a reference to several manufacturers and software developers in the book. It is only fair to acknowledge their names:

- FANUC and CUSTOM MACRO or USER MACRO or MACRO B
  are registered trademarks of Fujitsu-Fanuc, Japan

- GE FANUC is a registered trademark of GE Fanuc Automation, Inc.,
  Charlottesville, VA, USA

- MASTERCAM is the registered trademark of CNC Software Inc.,
  Tolland, CT, USA

- AUTOCAD is a registered trademark of Autodesk, Inc.,
  San Rafael, CA, USA

- HP and HPGL are registered trademarks of Hewlett-Packard, Inc.,
  Palo Alto, CA, USA

- IBM is a registered trademark of International Business Machines, Inc.,
  Armonk, NY, USA

- WINDOWS is a registered trademarks of Microsoft, Inc.,
  Redmond, WA, USA

About the Author

Peter Smid is a professional consultant, educator and speaker, with many years of practical, hands-on experience, in the industrial and educational fields. During his career, he has gathered an extensive experience with CNC and CAD/CAM applications on all levels. He consults to manufacturing industry and educational institutions on practical use of Computerized Numerical Control technology, part programming, CAD/CAM, advanced machining, tooling, setup, and many other related fields. His comprehensive industrial background in CNC programming, machining and company oriented training has assisted several hundred companies to benefit from his wide-ranging knowledge.

Mr. Smid's long time association with advanced manufacturing companies and CNC machinery vendors, as well as his affiliation with a number of Community and Technical College industrial technology programs and machine shop skills training, have enabled him to broaden his professional and consulting skills in the areas of CNC and CAD/CAM training, computer applications and needs analysis, software evaluation, system benchmarking, programming, hardware selection, software customization, and operations management.

Over the years, Mr. Smid has developed and delivered hundreds of customized educational programs to thousands of instructors and students at colleges and universities across United States, Canada and Europe, as well as to a large number of manufacturing companies and private sector organizations and individuals.

He has actively participated in many industrial trade shows, conferences, workshops and various seminars, including submission of papers, delivering presentations and a number of speaking engagements to professional organizations. He is also the author of articles and many in-house publications on the subject of CNC and CAD/CAM. During his many years as a professional in the CNC industrial and educational field, he has developed tens of thousands of pages of high quality training materials.

The author welcomes comments, suggestions and other input from educators, students and industrial users. You can e-mail him through the publisher of this handbook from the *Main Menu* of the enclosed CD.

You can also e-mail him from the **CNC Programming Handbook** page at **www.industrialpress.com**

TABLE OF CONTENTS

7 - PART PROGRAM STRUCTURE   41

8 - PREPARATORY COMMANDS   47

9 - MISCELLANEOUS FUNCTIONS   53

10 - SEQUENCE BLOCK   61

35 - LATHE CYCLES   307

36 - GROOVING ON LATHES   323

37 - PART-OFF   335

38 - SINGLE POINT THREADING   339

53 - CNC AND CAD/CAM   483

A - REFERENCE TABLES   491

NUMERICAL CONTROL

Numerical Control technology as it is known today, emerged in the mid 20th century. It can be traced to the year of 1952, the U.S. Air Force, and the names of John Parsons and the Massachusetts Institute of Technology in Cambridge, MA, USA. It was not applied in production manufacturing until the early 1960's. The real boom came in the form of CNC, around the year of 1972, and a decade later with the introduction of affordable micro computers. The history and development of this fascinating technology has been well documented in many publications.

In the manufacturing field, and particularly in the area of metal working, Numerical Control technology has caused something of a revolution. Even in the days before computers became standard fixtures in every company and in many homes, the machine tools equipped with Numerical Control system found their special place in the machine shops. The recent evolution of micro electronics and the never ceasing computer development, including its impact on Numerical Control, has brought significant changes to the manufacturing sector in general and metalworking industry in particular.

DEFINITION OF NUMERICAL CONTROL

In various publications and articles, many descriptions have been used during the years, to define what Numerical Control is. It would be pointless to try to find yet another definition, just for the purpose of this handbook. Many of these definitions share the same idea, same basic concept, just use different wording.

The majority of all the known definitions can be summed up into a relatively simple statement:

> Numerical Control can be defined as an operation of machine tools by the means of specifically coded instructions to the machine control system

The instructions are combinations of the letters of alphabet, digits and selected symbols, for example, a decimal point, the percent sign or the parenthesis symbols. All instructions are written in a logical order and a predetermined form. The collection of *all* instructions necessary to machine a part is called an *NC Program, CNC Program,* or a *Part Program.* Such a program can be stored for a future use and used repeatedly to achieve identical machining results at any time.

◆ NC and CNC Technology

In strict adherence to the terminology, there is a difference in the meaning of the abbreviations *NC* and *CNC.* The *NC* stands for the older and original *Numerical Control* technology, whereby the abbreviation *CNC* stands for the newer *Computerized Numerical Control* technology, a modern spin-off of its older relative. However, in practice, *CNC* is the preferred abbreviation. To clarify the proper usage of each term, look at the major differences between the NC and the CNC systems.

Both systems perform the same tasks, namely manipulation of data for the purpose of machining a part. In both cases, the internal design of the control system contains the logical instructions that process the data. At this point the similarity ends.

The NC system (as opposed to the CNC system) uses a fixed logical functions, those that are built-in and permanently wired within the control unit. These functions cannot be changed by the programmer or the machine operator. Because of the fixed wiring of the control logic, the NC control system is synonymous with the term *'hardwired'.* The system can interpret a part program, but it does not allow any changes to the program, using the control features. All required changes must be made away from the control, typically in an office environment. Also, the NC system requires the compulsory use of punched tapes for input of the program information.

The modern CNC system, but not the old NC system, uses an internal micro processor (*i.e.,* a computer). This computer contains memory registers storing a variety of routines that are capable of manipulating logical functions. That means the part programmer or the machine operator can change the program on the control itself (at the machine), with instantaneous results. This flexibility is the greatest advantage of the CNC systems and probably the key element that contributed to such a wide use of the technology in modern manufacturing. The CNC programs and the logical functions are stored on special computer chips, as *software instructions,* rather than used by the hardware connections, such as wires, that control the logical functions. In contrast to the NC system, the CNC system is synonymous with the term *'softwired'.*

When describing a particular subject that relates to the numerical control technology, it is customary to use either the term NC or CNC. Keep in mind that NC can also mean CNC in everyday talk, but CNC can never refer to the older technology, described in this handbook under the abbreviation of NC. The letter *'C'* stands for *Computerized*, and it is not applicable to the hardwired system. All control systems manufactured today are of the CNC design. Abbreviations such as *C&C* or *C'n'C* are not correct and reflect poorly on anybody that uses them.

CONVENTIONAL AND CNC MACHINING

What makes the CNC machining superior to the conventional methods? Is it superior at all? Where are the main benefits? If the CNC and the conventional machining processes are compared, a common general approach to machining a part will emerge:

1. Obtain and study the drawing
2. Select the most suitable machining method
3. Decide on the setup method (work holding)
4. Select the cutting tools
5. Establish speeds and feeds
6. Machine the part

This basic approach is the same for both types of machining. The major difference is in the way *how* various data are input. A feedrate of 10 inches per minute (10 in/min) is the same in manual or CNC applications, but the method of applying it is not. The same can be said about a coolant - it can be activated by turning a knob, pushing a switch or programming a special code. All these actions will result in a coolant rushing out of a nozzle. In both kinds of machining, a certain amount of knowledge on the part of the user is required. After all, metal working, particularly metal cutting, is mainly a skill, but it is also, to a great degree, an art and a profession of large number of people. So is the application of *Computerized Numerical Control*. Like any skill or art or profession, mastering it to the last detail is necessary to be successful. It takes more than technical knowledge to be a CNC machinist or a CNC programmer. Work experience and intuition, and what is sometimes called a *'gut-feel'*, is a much needed supplement to any skill.

In a conventional machining, the machine operator sets up the machine and moves each cutting tool, using one or both hands, to produce the required part. The design of a manual machine tool offers many features that help the process of machining a part - levers, handles, gears and dials, to name just a few. The same body motions are repeated by the operator for every part in the batch. However, the word *'same'* in this context really means *'similar'* rather than *'identical'*. Humans are not capable to repeat every process exactly the same at all times - that is the job of machines. People cannot work at the same performance level all the time, without a rest. All of us have some good and some bad moments. The results of these moments, when applied to machining a part, are difficult to predict. There will be some differences and inconsistencies within each batch of parts. The parts will not always be *exactly* the same. Maintaining dimensional tolerances and surface fin-

ish quality are the most typical problems in conventional machining. Individual machinists may have their own time 'proven' methods, different from those of their fellow colleagues. Combination of these and other factors create a great amount of inconsistency.

The machining under numerical control does away with the majority of inconsistencies. It does not require the same physical involvement as manual machining. Numerically controlled machining does not need any levers or dials or handles, at least not in the same sense as conventional machining does. Once the part program has been proven, it can be used any number of times over, always returning consistent results. That does not mean there are no limiting factors. The cutting tools do wear out, the material blank in one batch is not identical to the material blank in another batch, the setups may vary, etc. These factors should be considered and compensated for, whenever necessary.

The emergence of the numerical control technology does not mean an instant, or even a long term, demise of all manual machines. There are times when a traditional machining method is preferable to a computerized method. For example, a simple one time job may be done more efficiently on a manual machine than a CNC machine. Certain types of machining jobs will benefit from manual or semiautomatic machining, rather than numerically controlled machining. The CNC machine tools are not meant to replace every manual machine, only to supplement them.

In many instances, the decision whether certain machining will be done on a CNC machine or not is based on the number of required parts and nothing else. Although the volume of parts machined as a batch is always an important criteria, it should never be the only factor. Consideration should also be given to the part complexity, its tolerances, the required quality of surface finish, etc. Often, a single complex part will benefit from CNC machining, while fifty relatively simple parts will not.

Keep in mind that numerical control has never machined a single part by itself. Numerical control is only a *process* or a *method* that enables a machine tool to be used in a productive, accurate and consistent way.

NUMERICAL CONTROL ADVANTAGES

What are the main advantages of numerical control?

It is important to know which areas of machining will benefit from it and which are better done the conventional way. It is absurd to think that a two horse power CNC mill will win over jobs that are currently done on a twenty times more powerful manual mill. Equally unreasonable are expectations of great improvements in cutting speeds and feedrates over a conventional machine. If the machining and tooling conditions are the same, the cutting time will be very close in both cases.

Some of the major areas where the CNC user can and should expect improvement:

❏ Setup time reduction

❏ Lead time reduction

❏ Accuracy and repeatability

❏ Contouring of complex shapes

❏ Simplified tooling and work holding

❏ Consistent cutting time

❏ General productivity increase

Each area offers only a *potential* improvement. Individual users will experience different levels of actual improvement, depending on the product manufactured on-site, the CNC machine used, the setup methods, complexity of fixturing, quality of cutting tools, management philosophy and engineering design, experience level of the workforce, individual attitudes, etc.

◆ Setup Time Reduction

In many cases, the setup time for a CNC machine can be reduced, sometimes quite dramatically. It is important to realize that setup is a manual operation, greatly dependent on the performance of CNC operator, the type of fixturing and general practices of the machine shop. Setup time is unproductive, but necessary - it is a part of the overhead costs of doing business. To keep the setup time to a minimum should be one of the primary considerations of any machine shop supervisor, programmer and operator.

Because of the design of CNC machines, the setup time should not be a major problem. Modular fixturing, standard tooling, fixed locators, automatic tool changing, pallets and other advanced features, make the setup time more efficient than a comparable setup of a conventional machine. With a good knowledge of modern manufacturing, productivity can be increased significantly.

The number of parts machined under one setup is also important, in order to assess the cost of a setup time. If a great number of parts is machined in one setup, the setup cost per part can be very insignificant. A very similar reduction can be achieved by grouping several different operations into a single setup. Even if the setup time is longer, it may be justified when compared to the time required to setup several conventional machines.

◆ Lead Time Reduction

Once a part program is written and proven, it is ready to be used again in the future, even at a short notice. Although the lead time for the first run is usually longer, it is virtually nil for any subsequent run. Even if an engineering change of the part design requires the program to be modified, it can be done usually quickly, reducing the lead time.

Long lead time, required to design and manufacture several special fixtures for conventional machines, can often be reduced by preparing a part program and the use of simplified fixturing.

◆ Accuracy and Repeatability

The high degree of accuracy and repeatability of modern CNC machines has been the single major benefit to many users. Whether the part program is stored on a disk or in the computer memory, or even on a tape (the original method), it always remains the same. Any program can be changed at will, but once proven, no changes are usually required any more. A given program can be reused as many times as needed, without losing a single bit of data it contains. True, program has to allow for such changeable factors as tool wear and operating temperatures, it has to be stored safely, but generally very little interference from the CNC programmer or operator will be required. The high accuracy of CNC machines and their repeatability allows high quality parts to be produced consistently time after time.

◆ Contouring of Complex Shapes

CNC lathes and machining centers are capable of contouring a variety of shapes. Many CNC users acquired their machines only to be able to handle complex parts. A good examples are CNC applications in the aircraft and automotive industries. The use of some form of computerized programming is virtually mandatory for any three dimensional tool path generation.

Complex shapes, such as molds, can be manufactured without the additional expense of making a model for tracing. Mirrored parts can be achieved literally at the switch of a button. Storage of programs is a lot simpler than storage of patterns, templates, wooden models, and other pattern making tools.

◆ Simplified Tooling and Work Holding

Nonstandard and 'homemade' tooling that clutters the benches and drawers around a conventional machine can be eliminated by using standard tooling, specially designed for numerical control applications. Multi-step tools such as pilot drills, step drills, combination tools, counter borers and others are replaced with several individual standard tools. These tools are often cheaper and easier to replace than special and nonstandard tools. Cost-cutting measures have forced many tool suppliers to keep a low or even a nonexistent inventory, increasing the delivery time to the customer. Standard, off-the-shelf tooling can usually be obtained faster then nonstandard tooling.

Fixturing and work holding for CNC machines have only one major purpose - to hold the part rigidly and in the same position for all parts within a batch. Fixtures designed for CNC work do not normally require jigs, pilot holes and other hole locating aids.

◆ Cutting Time and Productivity Increase

The cutting time on the CNC machine is commonly known as the *cycle time* - and is always consistent. Unlike a conventional machining, where the operator's skill, experience and personal fatigue are subject to changes, the CNC machining is under the control of a computer. The small amount of manual work is restricted to the setup and loading and unloading the part. For large batch runs, the high cost of the unproductive time is spread among many parts, making it less significant. The main benefit of a consistent cutting time is for repetitive jobs, where the production scheduling and work allocation to individual machine tools can be done very accurately.

The main reason companies often purchase CNC machines is strictly economic - it is a serious investment. Also, having a competitive edge is always on the mind of every plant manager. The numerical control technology offers excellent means to achieve a significant improvement in the manufacturing productivity and increasing the overall quality of the manufactured parts. Like any means, it has to be used wisely and knowledgeably. When more and more companies use the CNC technology, just having a CNC machine does not offer the extra edge anymore. The companies that get forward are those who know how to use the technology efficiently and practice it to be competitive in the global economy.

To reach the goal of a major increase in productivity, it is essential that users understand the fundamental principles on which CNC technology is based. These principles take many forms, for example, understanding the electronic circuitry, complex ladder diagrams, computer logic, metrology, machine design, machining principles and practices and many others. Each one has to be studied and mastered by the person in charge. In this handbook, the emphasis is on the topics that relate directly to the CNC programming and understanding the most common CNC machine tools, the *Machining Centers* and the lathes (sometimes also called the *Turning Centers*). The part quality consideration should be very important to every programmer and machine tool operator and this goal is also reflected in the handbook approach as well as in the numerous examples.

TYPES OF CNC MACHINE TOOLS

Different kinds of CNC machines cover an extremely large variety. Their numbers are rapidly increasing, as the technology development advances. It is impossible to identify all the applications, they would make a long list. Here is a brief list of some of the groups CNC machines can be part of:

❏ Mills and Machining centers

❏ Lathes and Turning Centers

❏ Drilling machines

❏ Boring mills and Profilers

❏ EDM machines

❏ Punch presses and Shears

❏ Flame cutting machines

❏ Routers

❏ Water jet and Laser profilers

❏ Cylindrical grinders

❏ Welding machines

❏ Benders, Winding and Spinning machines, etc.

CNC machining centers and lathes dominate the number of installations in industry. These two groups share the market just about equally. Some industries may have a higher need for one group of machines, depending on their needs. One must remember that there are many different kinds of lathes and equally many different kinds of machining centers. However, the programming process for a vertical machine is similar to the one for a horizontal machine or a simple CNC mill. Even between different machine groups, there is a great amount of general applications and the programming process is generally the same. For example, a contour milled with an end mill has a lot in common with a contour cut with a wire.

◆ Mills and Machining Centers

Standard number of axes on a milling machine is three - the X, Y and Z axes. The part set on a milling system is always stationary, mounted on a moving machine table. The cutting tool rotates, it can move up and down (or in and out), but it does not physically follow the tool path.

CNC mills - sometimes called CNC milling machines - are usually small, simple machines, without a tool changer or other automatic features. Their power rating is often quite low. In industry, they are used for toolroom work, maintenance purposes, or small part production. They are usually designed for contouring, unlike CNC drills.

CNC machining centers are far more popular and efficient than drills and mills, mainly for their flexibility. The main benefit the user gets out of a CNC machining center is the ability to group several diverse operations into a single setup. For example, drilling, boring, counter boring, tapping, spot facing and contour milling can be incorporated into a single CNC program. In addition, the flexibility is enhanced by automatic tool changing, using pallets to minimize idle time, indexing to a different side of the part, using a rotary movement of additional axes, and a number of other features. CNC machining centers can be equipped with special software that controls the speeds and feeds, the life of the cutting tool, automatic in-process gauging and offset adjustment and other production enhancing and time saving devices.

There are two basic designs of a typical CNC machining center. They are the *vertical* and the *horizontal* machining centers. The major difference between the two types is the nature of work that can be done on them efficiently. For a vertical CNC machining center, the most suitable type of work are flat parts, either mounted to the fixture on the table, or held in a vise or a chuck. The work that requires machining on two or more faces in a single setup is more desirable to be done on a CNC horizontal machining center. An good example is a pump housing and other cubic-like shapes. Some multi-face machining of small parts can also be done on a CNC vertical machining center equipped with a rotary table.

The programming process is the same for both designs, but an additional axis (usually a B axis) is added to the horizontal design. This axis is either a simple positioning axis (indexing axis) for the table, or a fully rotary axis for simultaneous contouring.

This handbook concentrates on the CNC vertical machining centers applications, with a special section dealing with the horizontal setup and machining. The programming methods are also applicable to the small CNC mills or drilling and/or tapping machines, but the programmer has to consider their restrictions.

◆ Lathes and Turning Centers

A CNC lathe is usually a machine tool with two axes, the vertical X axis and the horizontal Z axis. The main feature of a lathe that distinguishes it from a mill is that the part is rotating about the machine center line. In addition, the cutting tool is normally stationary, mounted in a sliding turret. The cutting tool follows the contour of the programmed tool path. For the CNC lathes with a milling attachment, so called *live tooling,* the milling tool has its own motor and rotates while the spindle is stationary.

The modern lathe design can be horizontal or vertical. Horizontal type is far more common than the vertical type, but both designs have their purpose in manufacturing. Several different designs exist for either group. For example, a typical CNC lathe of the horizontal group can be designed with a flat bed or a slant bed, as a bar type, chucker type or a universal type. Added to these combinations are many accessories that make a CNC lathe an extremely flexible machine tool. Typically, accessories such as a tailstock, steady rests or follow-up rests, part catchers, pullout-fingers and even a third axis milling attachment are popular components of the CNC lathe. A CNC lathe can be very versatile - so versatile in fact, that it is often called a *CNC Turning Center.* All text and program examples in this handbook use the more traditional term *CNC lathe*, yet still recognizing all its modern functions.

PERSONNEL FOR CNC

Computers and machine tools have no intelligence. They cannot think, they cannot evaluate a situation in a rational way. Only people with certain skills and knowledge can do that. In the field of numerical control, the skills are usually in the hands of two key people - one doing the *programming*, the other doing the *machining*. Their respective numbers and duties typically depend on the company preference, its size, as well as the product manufactured there. However, each position is quite distinct, although many companies combine the two functions into a one, often called a *CNC Programmer/Operator.*

◆ CNC Programmer

The CNC programmer is usually the person who has the most responsibility in the CNC machine shop. This person is often responsible for the success of numerical control technology in the plant. Equally, this person is held responsible for problems related to the CNC operations. Although duties may vary, the programmer is also responsible for a variety of tasks relating to the effective usage of the CNC machines. In fact, this person is often accountable for the production and quality of all CNC operations.

Many CNC programmers are experienced machinists, who have had a practical, hands-on experience as machine tool operators. They know how to read technical drawings and they can comprehend the engineering intent behind the design. This practical experience is the foundation for the ability to 'machine' a part in an office environment. A good CNC programmer must be able to visualize all the tool motions and recognize all restricting factors that may be involved. The programmer must be able to collect, analyze, process and logically integrate all the collected data into a single, cohesive program. In simple terms, the CNC programmer must be able to decide upon the best manufacturing methodology in all respects.

In addition to the machining skills, the CNC programmer has to have an understanding of mathematical principles, mainly application of equations, solution of arcs and angles. Equally important is the knowledge of trigonometry. Even with computerized programming, the knowledge of manual programming methods is absolutely essential to the thorough understanding of the computer output and the control of this output.

The last important quality of a truly professional CNC programmer is his or her ability to listen to the other people - the engineers, the CNC operators, the managers. Good listening skills are the first prerequisite to become flexible. A good CNC programmer must be flexible in order to offer high programming quality.

◆ CNC Machine Operator

The CNC machine tool operator is a complementary position to the CNC programmer. The programmer and the operator may exist in a single person, as is the case in many small shops. Although the majority of duties performed by a conventional machine operator has been transferred to the CNC programmer, the CNC operator has many unique responsibilities. In typical cases, the operator is responsible for the tool and machine setup, for the changing of the parts, often even for some in-process inspection. Many companies expect quality control at the machine - and the operator of any machine tool, manual or computerized, is also responsible for the quality of the work done on that machine. One of the very important responsibilities of the CNC machine operator is to report findings about each program to the programmer. Even with the best knowledge, skills, attitudes and intentions, the 'final' program can always be improved. The CNC operator, being the one who is the closest to the actual machining, knows precisely what extent such improvements can be.

SAFETY RELATED TO CNC WORK

On the wall of many companies is a safety poster with a simple, yet powerful message:

The first rule of safety is to follow all safety rules

The heading of this section does not indicate whether the safety is oriented at the programming or the machining level. The reason is that the safety is totally *independent*. It stands on its own and it governs behavior of everybody in a machine shop and outside of it. At first sight, it may appear that safety is something related to the machining and the machine operation, perhaps to the setup as well. That is definitely true but hardly presents a complete picture.

Safety is the most important element in programming, setup, machining, tooling, fixturing, inspection, shipping, and *you-name-it* operation within a typical machine shop daily work. Safety can never be overemphasized. Com-

panies talk about safety, conduct safety meetings, display posters, make speeches, call experts. This mass of information and instructions is presented to all of us for some very good reasons. Quite a few are based on past tragic occurrences - many laws, rules and regulations have been written as a result of inquests and inquiries into serious accidents.

At first sight, it may seem that in CNC work, the safety is a secondary issue. There is a lot of automation, a part program that runs over and over again, tooling that has ben used in the past, a simple setup, etc. All this can lead to complacency and false assumption that safety is taken care of. This is a view that can have serious consequences.

Safety is a large subject but a few points that relate to the CNC work are important. Every machinist should know the hazards of mechanical and electrical devices. The first step towards a safe work place is with a clean work area, where no chips, oil spills and other debris are allowed to accumulate on the floor. Taking care of personal safety is equally important. Loose clothing, jewelry, ties, scarfs, unprotected long hair, improper use of gloves and similar infractions, is dangerous in machining environment. Protection of eyes, ears, hands and feet is strongly recommended.

While a machine is operating, protective devices should be in place and no moving parts should be exposed. Special care should be taken around rotating spindles and automatic tool changers. Other devices that could pose a hazard are pallet changers, chip conveyors, high voltage areas, hoists, etc. Disconnecting any interlocks or other safety features is dangerous - and also illegal, without appropriate skills and authorization.

In programming, observation of safety rules is also important. A tool motion can be programmed in many ways. Speeds and feeds have to be realistic, not just mathematically 'correct'. Depth of cut, width of cut, the tool characteristics, all have a profound effect on overall safety.

All these ideas are just a very short summary and a reminder that safety should always be taken seriously.

Many different types of CNC machines are used in industry, and the majority of them are *CNC machining centers* and *CNC lathes*. They are followed by wire EDM, fabricating machines and machines of special designs. Although the focus of this handbook is on the two types that dominate the market, many general ideas can be applied to other CNC equipment.

CNC MACHINES - MILLING

The description of CNC milling machines is so large, it can fill a thick book all by itself. All machine tools from a simple knee type milling machine up to a five axis profiler can be included in this category. They vary in size, features, suitability for certain work, etc., but they do all have one common denominator - *their primary axes are the X and Y axes* - and for this reason, they are called the *XY machines*.

In the category of the XY machines are also wire EDM machine tools, laser and water jet cutting machines, flame cutters, burners, routers, etc. Although they do not qualify as milling type machine tools, we mention them because the majority of programming techniques applicable to the mills is identical to these machines types as well. The best example is a *contouring* operation, a process common to many CNC machines.

For the purpose of this handbook, a milling machine can be defined:

> Milling machine is a machine capable of a simultaneous cutting motion, using an end mill as the primary cutting tool, along at least two axes at the same time

This definition eliminates all CNC drill presses, since their design covers positioning but not profiling. The definition also eliminates wire EDM machines and a variety of burners, since they are capable of a profiling action but not with an end mill. Users of these machine tools will still benefit from many subjects covered here. The general principles are adaptable to the majority of CNC machine tools. For example, a wire EDM uses a very small cutter diameter, in the form of a wire. A laser cutting machine uses the laser beam as its cutter, also having a known diameter but the term *kerf* is used instead. The focus will be concentrated on metal cutting machine tools, using various styles of end mills as the primary tool for contouring. Since an end mill can be used in many ways, first look will be at the various types of available milling machines.

◆ Types of Milling Machines

Milling machines can divided into three categories:

❑ By the number of axes - two, three or more

❑ By the orientation of axes - vertical or horizontal

❑ By the presence or absence of a tool changer

Milling machines where the spindle motion is *up* and *down*, are categorized as *vertical* machines. Milling machines where the spindle motion is *in* and *out*, are categorized as *horizontal* machines - see *Figure 2-1* and *2-2*.

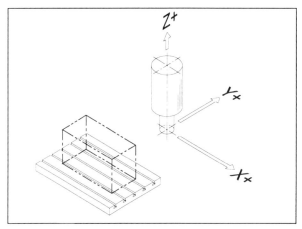

Figure 2-1
Schematic representation of a CNC vertical machining center

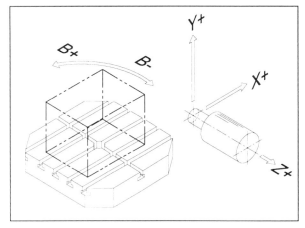

Figure 2-2
Schematic representation of a CNC horizontal machining center

These simplified definitions do not really reflect reality of the current state of art in machine tool manufacturing. The machine tool industry is constantly changing. New and more powerful machines are designed and produced by many manufacturers worldwide, with more features.

The majority of modern machines designed for milling are capable of doing a multitude of machining tasks, not only the traditional milling. These machines are also capable of many other metal removing operations, mainly drilling, reaming, boring, tapping, profiling, thread cutting and many others. They may be equipped with a multi-tool magazine (also known as a carousel), a fully automatic tool changer (abbreviated as ATC) and a pallet changer (abbreviated as APC), a powerful computerized control unit (abbreviated as CNC), and so on. Some machine models may have additional features, such as adaptive control, robot interface, automatic loading and unloading, probing system, high speed machining features and other marvels of modern technology. The question is - can machine tools of these capabilities be classified as simple CNC milling machines? In two words - certainly not. Milling machines that have at least *some* of the advanced features built-in, have become a new breed of machine tools - *CNC Machining Centers.* This term is strictly CNC related - a *manual machining center* is a description that does not exist.

◆ Machine Axes

Milling machines and machining centers have at least three axes - X, Y and Z. The machines become more flexible if they have the fourth axis, usually an indexing or a rotary axis (the A axis for vertical models or the B axis for horizontal models). Even higher level of flexibility can be found on machines with five or more axes. A simple machine with five axes may be a boring mill that has three major axes, plus a rotary axis (usually the B axis) and an axis parallel to the Z axis (usually the W axis). However, true complex and flexible five-axis profiling milling machine is the type used in aircraft industry, where a multi-axis, simultaneous cutting motion is necessary to machine complex shapes and reach cavities and various angles.

At times, the expression *two and a half axis* machine or a *three and a half axis* machine is used. These terms refer to the type of machines, where simultaneous cutting motion of all axes has certain limitations. For example, a four-axis vertical machine has X, Y and Z axis as primary axes, plus an indexing table, designated as an A axis. The indexing table is used for positioning, but it cannot rotate simultaneously with the motion of primary axes. That type of a machine is often called a *'three and a half axis'* machine. By contrast, a more complex but similar machine that is equipped with a fully rotating table, is designed as a four axis machine. The rotary table can move simultaneously with the cutting motion of the primary axes. This is a good example of a true *'four axis'* machine tool.

Each machining center is described by its specifications as provided by the machine tool manufacturer. The manufacturer lists many specifications as a quick method of comparison between one machine and another. It is not unusual to find a slightly biased information in the descriptive brochure - after all, it is a sales tool.

In the area of milling systems, three most common machine tools are available:

- ❏ CNC Vertical Machining Center - VMC
- ❏ CNC Horizontal Machining Center - HMC
- ❏ CNC Horizontal Boring Mill

Programming methods do not vary too much for either type, except for special accessories and options. Some of the major differences will be the orientation of machine axes, additional axis for indexing or full rotary motion, and the type of work suitable for individual models. Description of the most common type of a machining center - *the Vertical Machining Center (VMC)* - presents a fairly accurate sample of describing other machines of the group.

◆ Vertical Machining Centers

Vertical machining centers are mainly used for flat type of work, such as plates, where the majority of machining is done on only one face of the part in a single setup.

A vertical CNC machining center can also be used with an optional fourth axis, usually a rotary head mounted on the main table. The rotary head can be mounted either vertically or horizontally, depending on the desired results and the model type. This fourth axis can be used either for indexing or a full rotary motion. In the combination with a tailstock (usually supplied), the fourth axis in the vertical configuration can be used for machining long parts that need support at both ends.

The majority of vertical machining centers most operators work with are those with an empty table and three-axes configuration.

From the programming perspective, there are at least two items worth mentioning:

- ❏ *ONE* - programming always takes place *from the viewpoint of the spindle,* not the operator's. That means the view is as if looking straight down, at ninety degrees towards the machine table for development of the tool motion. Programmers always view the top of part!

- ❏ *TWO* - various markers located somewhere on the machine show the positive and the negative motion of the machine axes. For programming, these markers should be ignored! These indicate operating directions, not programming directions. As a matter of fact, typically the programming directions are exactly the opposite of the markers on the machine tool.

Vertical and Horizontal Machining Center - Typical Specifications		
Description	Vertical Machining Center	Horizontal Machining Center
Number of axes	3 axes (XYZ)	4 axes (XYZB)
Table dimensions	780 x 400 mm 31 x 16 inches	500 x 500 mm 20 x 20 inches
Number of tools	20	36
Maximum travel - X axis	575 mm 22.5 inches	725 mm 28.5 inches
Maximum travel - Y axis	380 mm 15 inches	560 mm 22 inches
Maximum travel - Z axis	470 mm 18.5 inches	560 mm 22 inches
Table indexing angle	N/A	0.001 degree
Spindle speed	60-8000 rpm	40 - 4000 rpm
Spindle output	AC 7.5/5.5 kW AC 10/7 HP	AC 11/8 kW AC 15/11 HP
Spindle nose-to-table distance - Z axis	150 - 625 mm 6 - 24.6 inches	150 - 710 mm 6 - 28 inches
Spindle center-to-column distance - Y axis	430 mm 17 inches	30 - 560 mm 1.2 - 22 inches
Spindle taper	No. 40	No. 50
Tool shank size	BT40	CAT50
Feedrate range	2 - 10000 mm/min 0.100 - 393 in/min	1 - 10000 mm/min 0.04 - 393 in/min
Rapid traverse rate	30000 mm/min (XY) - 24000 mm/min (Z) 1181 in/min (XY) - 945 in/min (Z)	30000 mm/min (XY) - 24000 mm/min (Z) 1181 in/min (XY) - 945 in/min (Z)
Tool selection	Random memory	Random memory
Maximum tool diameter	80 mm (150 w/empty pockets) 3.15 inches (5.9 w/empty pockets)	105 mm 4.1 inches
Maximum tool length	300 mm 11.8 inches	350 mm 13.75 inches
Maximum tool weight	6 kg 13 lbs	20 kg 44 lbs

◆ Horizontal Machining Centers

Horizontal CNC Machining Centers are also categorized as multi-tool and versatile machines, and are used for cubical parts, where the majority of machining has to be done on more than one face in a single setup.

There are many applications in this area. Common examples are large parts, such as pump housings, gear cases, manifolds, engine blocks and so on. Horizontal machining centers always include a special indexing table and are typically equipped with a pallet changer and other features.

Because of their flexibility and complexity, CNC horizontal machining centers are priced significantly higher than vertical CNC machining centers.

From the programming point of view, there are several unique differences, mainly relating to the Automatic Tool Changer, the indexing table, and - in some cases - to the additional accessories, for example, the pallet changer. All differences are relatively minor. Writing a program for the horizontal machining centers is no different than writing a program for vertical machining centers.

◆ Horizontal Boring Mill

Horizontal boring mill is just another CNC machine. It closely resembles a CNC horizontal machining center, but it does have its own differences. Generally, a horizontal boring mill is defined by the lack of some common features, such as the Automatic Tool Changer. As the name of the machine suggests, its primary purpose is boring operations, mainly lengthy bores. For that reason, the reach of the spindle is extended by a specially designed quill. Another typical feature is an axis parallel to the Z axis, called the W axis. Although this is, in effect, the fifth axis designation (X, Y, Z, B, W), a horizontal boring mill cannot be called a true five axis machine. The Z axis (quill) and the W axis (table) work in the opposite directions towards each other, so they can be used for large parts and *hard-to-reach* areas. It also means, that during drilling, the machine table moves against an extended quill. The quill is a physical part of the spindle. It is in the spindle where the cutting tool rotates - but the in-and-out motions are done by the table. Think of the alternate method offered on horizontal boring mills - if the quill were to be very long, it would lose its strength and rigidity. The better way was to split the traditional single Z axis movement into two - the quill extension along the Z axis will move only part of the way towards the table and the table itself, the new W axis, will move another

part of the way towards the spindle. They both meet in the area of the part that could be machined using all the machine tool resources.

Horizontal boring mill may be called a 3-1/2 axis CNC machine, but certainly not a 5-axis CNC machine, even if the count of the axes is five. Programming procedures for CNC boring mills are very similar to the horizontal and vertical CNC machining centers.

◆ Typical Specifications

On the preceding page is a comprehensive chart showing the typical specifications of a *CNC Vertical Machining Center* and a *CNC Horizontal Machining Center.* The specifications are side by side in two columns, strictly for convenience, not for any comparison purposes. These are two different machine types and comparison is not possible for all features. In order to compare individual machine tools within a certain category, machine tool specifications provided by the machine manufacturer often serve as the basis for comparison. These specifications are contained a list of verifiable data, mainly technical in nature, that describes the individual machine by its main features. Machine tool buyers frequently compare many brochures of several different machines as part of the pre purchase process. Managers and process planners compare individual machines in the machine shop and assign the available workload to the most suitable machine.

A fair and accurate comparison can be made between two vertical machining centers or between two horizontal machining centers, but cannot be done fairly to compare between two different machine types.

In a typical machine specification chart, additional data may be listed, not included in the earlier chart. In this handbook, the focus is on only those specifications that are of interest to the CNC programmer and the CNC operator.

CNC TURNING

CNC MACHINES - TURNING

A conventional engine lathe or a turret lathe is a common machine in just about every machine shop. A lathe is used for machining cylindrical or conical work, such as shafts, rings, wheels, bores, threads, etc. The most common lathe operation is removal of material from a round stock, using a turning tool for external cutting. A lathe can also be used for internal operations such as boring, as well as for grooving, threading, etc., if a proper cutting tool is used. Turret lathes are usually weaker in machining power than engine lathes, but they do have a special carousel that holds several mounted cutting tools. An engine lathe has often only one or two cutting tools mounted at a time, but has more machining power.

Typical lathe work controlled by a CNC system uses machines known in industry as the *CNC Turning Centers* - or more commonly - the *CNC lathes*.

The term *'turning center'* is rather unpopular, but an accurate overall description of a computerized lathe (a CNC lathe) that can be used for a great number of machining operations during a single setup. For example, in addition to the standard lathe operations such as turning and boring, a CNC lathe can be used for drilling, grooving, threading, knurling and even burnishing. It can also be used in different modes, such as chuck work, collet work, barfeeder, or between centers. Many other combinations also exist. CNC lathes are designed to hold several tools in special turrets, they can have a milling attachment, indexable chuck, a sub spindle, a tailstock, a steadyrest and many other features not always associated with a conventional lathe design. Lathes with more than four axes are also common. With constant advances in machine tool technologies, more CNC lathes appear on the market that are designed to do a number of operations in a single setup, many of them traditionally reserved for a mill or a machining center.

◆ Types of CNC Lathes

Basically, CNC lathes can be categorized by the *type of design* and by the *number of axes*. The two basic types are the *vertical* CNC lathe and the *horizontal* CNC lathe. Of the two, the horizontal type is by far the most common in manufacturing and machine shops. A vertical CNC lathe (incorrectly called a vertical boring mill) is somewhat less common but is irreplaceable for a large diameter work. For a CNC programmer, there are no significant differences in the programming approach between the two lathe types.

◆ Number of Axes

The most common distinction of different CNC lathes is by the number of programmable axes. Vertical CNC lathes have two axes in almost all designs available. The much more common CNC horizontal lathes, commonly designed with two programmable axes, are also available with three, four or six axes, thus adding extra flexibility to manufacturing of more complex parts.

A horizontal CNC lathe can further be described by the *type* of engineering design:

❑ FRONT lathe   ... an engine lathe type

❑ REAR lathe   ... a unique slant bed type

Slant bed type is very popular for general work, because its design allows cutting chips to fall away from the CNC operator and, in case of an accident, forces the part to fall down into a safe area, towards the chip conveyer.

Between the categories of flat bed and slant type lathes, front and rear lathes, horizontal and vertical lathe designs, there is another variety of a lathe. This category describes CNC lathes by the *number of axis*, which is probably the simplest and most common method of lathe identification.

AXES DESIGNATION

A typical CNC lathe is designed with two standard axes - one axis is the X axis, the other axis is the Z axis. Both axes are perpendicular to each other and represent the typical two-axis lathe motions. The X axis also represents the *cross travel* of the cutting tool, the Z axis represents its *longitudinal motion*. All varieties of cutting tools are mounted in a turret (a special tool magazine) and can be external or internal. Because of this design, the turret loaded with all cutting tools moves along the X and Z axes, which means all tools are in the work area.

Following the established standards of the milling machines and machining centers, the only machine axis capable of making a hole by methods of drilling, boring, piercing or punching, is the Z axis.

In CNC lathe work, the traditional axis orientation for a horizontal type of lathe is *upwards* and *downwards* motion for the X axis, and *left* and *right* motion for the Z axis, when looking from the machinist's position. This view is shown in the following three illustrations *Figure 3-1, Figure 3-2,* and *Figure 3-3.*

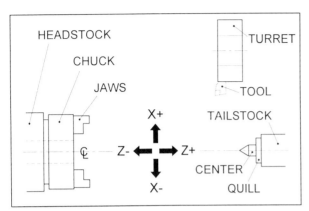

Figure 3-1

Typical configuration of a two axis slant bed CNC lathe - rear type

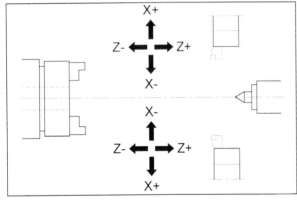

Figure 3-2

Typical configuration of a CNC lathe with two turrets

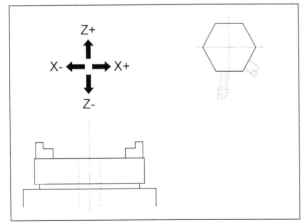

Figure 3-3

Schematic representation of a vertical CNC lathe

This is true for both the front and rear lathes and for lathes with three or more axes. The chuck face is oriented vertically to the horizontal spindle center line for all horizontal lathes. Vertical lathes, due to their design, are rotated by 90°, where the chuck face is oriented horizontally to the vertical spindle center line.

In addition to the X and Z primary axes, the multi-axis lathes have individual descriptions of each additional axis, for example, the C axis is usually the third axis, used for milling operations, using so called *live tooling*. More details on the subject of coordinate system and machine geometry are available in the next chapter.

◆ Two-axis Lathe

This is the most common type of CNC lathes. The work holding device, usually a chuck, is mounted on the left side of the machine (as viewed by the operator). The rear type, with the slant bed, is the most popular design for general work. For some special work, for example in the petroleum industry (where turning tube ends is a common work), a flat bed is usually more suitable. The cutting tools are held in a specially designed indexing turret that can hold four, six, eight, ten, twelve and more tools. Many such lathes also have two turrets.

Advanced machine tool designs incorporate tool storage away from the work area, similar to the design of machining centers. Tens, even hundreds, of cutting tools may be stored and used for a single CNC program. Many lathes also incorporate a quick changing tooling system.

◆ Three-axis Lathe

Three-axis lathe is essentially a two-axis lathe with an additional axis. This axis has its own designation, usually as a C axis in absolute mode (H axis in incremental mode), and is fully programmable. Normally, the third axis is used for cross-milling operations, slot cutting, bolt circle holes drilling, hex faces, side faces, helical slots, etc. This axis can replace some simple operations on a milling machine, reducing the setup time for the job. Some limitations do apply to many models, for example, the milling or drilling operations can take place only at positions projecting from the tool center line to the spindle center line (within a machining plane), although others offer off-center adjustments.

The third axis has its own power source but the power rating is relatively lower when compared with the majority of machining centers. Another limitation may be the smallest increment of the third axis, particularly on the early three axis lathes. Smallest increment of one degree is certainly more useful than an increment of two or five degrees. Even better is an increment of 0.1°, 0.01°, and commonly 0.001° on the latest models. Usually the lathes with three axes offer a very fine radial increment that allows a simultaneous rotary motion. Those with low increment values are usually designed with an oriented spindle stop only.

From the perspective of CNC part programming, the additional knowledge required is a subject not difficult to learn. General principles of milling apply and many programming features are also available, for example, fixed cycles and other shortcuts.

◆ Four-axis Lathe

By design, a four-axis CNC lathe is a totally different concept than a three-axis lathe. As a matter of fact, to program a four-axis lathe is nothing more than programming *two two-axis lathes at the same time*. That may sound strange at first, until the principle of a four-axis CNC lathe becomes clearer.

There are actually two controls (and two sets of the XZ axes), one for each pair (set) of axes. Only one program may be used to do the *external* - or *outside* - diameter roughing (OD) and another program to do the *inside* - or *internal* - roughing (ID). Since a four-axis lathe can work with each pair of axes *independently*, the OD and ID can be machined at the same time, doing two different operations simultaneously. The main keys to a successful 4-axis lathe programming is coordination of the tools and their operations, timing of the tool motions and a generous sense of healthy compromise.

For several reasons, both pairs of axes cannot work all the time. Because of this restriction, special programming features such as synchronized waiting codes (typically *Miscellaneous Function*), the ability to estimate how much time each tool requires to complete each operation, etc., are required. There is a level of compromise here, because only *one* spindle speed can be used for both active cutting tools, although feedrate is independent for both pairs of axes. This means that some machining operations simply cannot be done simultaneously.

Not every lathe job benefits from the 4-axis machining. There are cases when it is more costly to run a job on a 4-axis lathe inefficiently and it may be very efficient to run the same job on a 2-axis CNC lathe.

◆ Six-axis Lathe

Six-axis CNC lathes are specially designed lathes with a twin turret and a set of three axes per turret. This design incorporates many tool stations, many of them power driven, as well as back-machining capabilities. Programming these lathes is similar to programming a three-axis lathe twice. The control system automatically provides synchronization, when necessary.

A small to medium size six-axis CNC lathe is popular choice of screw machine shops and industries with similar small parts and large volume applications.

FEATURES AND SPECIFICATIONS

A look at a typical promotional brochure describing a CNC machine tool is very useful in many respects. In most cases, the artwork quality is impressive, the printing, photographs, paper selection and the use of colors is equally well done. It is the purpose of the brochure to make a good marketing tool and attract the potential buyer.

There is more in the promotional brochure than just attractive photographs - in fact, in a well designed brochure, there is a wealth of technical information, describing the machine tool. These are the *features* and *specifications* the CNC machine tool manufacturer considers important to the customer.

In the majority of brochures, there are practical data that can be used in programming a particular CNC machine, a lathe in the example.

◆ Typical Machine Specifications

A typical horizontal CNC lathe, with two axes and a slant bed design, may have the following specifications (taken from an actual brochure):

Description	Specification
Number of axes	Two (X, Z) or three (X, Z and C)
Maximum swing over bed	560 mm 22.05 inches
Maximum turning diameter	350 mm 13.76 inches
Maximum turning length	550 mm 21.65 inches
Spindle bore	85 mm 3.34 inches
Bar capacity	71 mm 2.79 inches
Number of tools	12
Tool size square	25 mm 1 inch
Tool size round	⌀40 mm ⌀1.57 inches
Indexing time	0.1 second
Axis travel in X axis	222 mm 8.75 inches
Axis travel in Z axis	635 mm 25 inches
Rapid traverse rate X axis	16000 mm/min 629 in/min
Rapid traverse rate Z axis	24000 mm/min 944 in/min
Cutting feedrate	0.01 - 500 mm/rev .0001 - 19.68 in/rev
Chuck size	254 mm 10 inches
Main spindle motor	AC 15/11kW AC 20/14.7HP
Spindle speed	35 - 3500 rpm
Minimum input increment	0.001 mm .0001 inch
Motorized head:	
Number of rotating tools	12
Rotating tool speed	30 - 3600 r/min
Milling motor	AC 3.7/2.2 kW AC 5/2.95 HP
Collet size	1 - 16 mm .04 - .63 inches
Tap size	M3 - M16 metric #5 - 5/8 inches

It is very important to understand the specifications and features of the CNC machine tools in the shop. Many features relate to the control system, many others to the machine tool itself. In CNC programming, many important decisions are based on one or several of these features, for example number of tool stations available, maximum spindle speed and others.

◆ Control Features

The last item in understanding the overall description of a CNC lathe is the look at some control features unique to lathes and how they differ form a typical milling control. The subject of control features is described in more detail in *Chapter 5*.

At this time, some features and codes may not make much sense - they are included for reference only. Common and typical features are listed:

- ❏ X axis represents a diameter, not a radius
- ❏ Constant surface speed (CSS) is standard control feature (G96 for CSS and G97 for r/min)
- ❏ Absolute programming mode is X or Z or C
- ❏ Incremental programming mode is U or W or H

- ❏ Thread cutting of various forms (including taper and circular) can be performed, depending on the control model
- ❏ Dwell can use the P, U or X address (G04)
- ❏ Tool selection uses 4-digit identification
- ❏ Feedrate selection (normal) in mm/rev or in/rev
- ❏ Feedrate selection (special) in m/min or in/min
- ❏ Rapid traverse rate different for X and Z axes
- ❏ Multiple repetitive cycles for turning, boring, facing, contour repeat, grooving, and threading are available
- ❏ Feedrate override is common from 0 to 200% in 10% increments (on some lathes only from 0 to 150%)
- ❏ X axis can be mirrored
- ❏ Tailstock can be programmable
- ❏ Automatic chamfering and corner rounding R and I / K in G01 mode
- ❏ Thread cutting feedrate available with six-decimal place accuracy (for inch units)
- ❏ Least input increment in X axis is 0.001 mm or .0001 inches on diameter - one half of that value per side

4 *COORDINATE GEOMETRY*

The major step towards the basic understanding of CNC principles and geometrical concepts is the understanding of a subject known in mathematics as the *system of coordinates*. System of coordinates is founded on a number of mathematical principles dating back over four hundred years. The most important of these principles are those that can be applied to the CNC technology of today. In various publications on mathematics and geometry, these principles are listed under the headings like the *real number system* and the *rectangular coordinates*.

REAL NUMBER SYSTEM

The key to understanding the rectangular coordinates is the knowledge of arithmetic, algebra and geometry. The key knowledge in this area is the knowledge of the *real number system*. Within the real number system, there are ten available numerals (digits), 0 to 9 (zero to nine), that can be used in any of the following groups:

- ❏ Zero integer ...   0

- ❏ Positive integers ...   1, 2, +3, 10, 12943, +45
  (with or without sign)

- ❏ Negative integers ...   -4, -381, -25, -77
  (minus sign required)

- ❏ Fractions ...   1/8, 3/16, 9/32, 35/64

- ❏ Decimal fractions ...   0.185, .2, .546875, 3.5

All groups are used almost daily. These groups represent the mainstream of just about all applications of numbers in modern life. In CNC programming, the primary goal is to use the numbers to 'translate' the drawing, based on its dimensions, into a cutter path.

Computerized Numerical Control means *control by the numbers using a computer*. All information in a drawing has to be translated into a CNC program, using primarily numbers. Numbers are also used to describe commands, functions, comments, and so on. The mathematical concept of a real number system can be expressed graphically on a straight line, called the *number scale*, where all divisions have the same length - *Figure 4-1*.

Figure 4-1
Graphical representation of the Number Scale

The length of each division on the scale represents the unit of measurement in a convenient and generally accepted scale. It may come as a surprise that this concept is used every day. For example, a simple ruler used in schools is based on the number scale concept, regardless of measuring units. Weight scales using tons, pounds, kilograms, grams and similar units of mass are other examples. A simple household thermometer uses the same principle. Other similar examples are available as well.

RECTANGULAR COORDINATE SYSTEM

Rectangular coordinate system is a concept used to define a planar 2D point, using the XY coordinates, or a spacial 3D point, using the XYZ coordinates. It was first defined in the 17th century by a French philosopher and mathematician Rene Descartes (1596-1650). His name is used as an alternative to the rectangular coordinate system, called the *Cartesian Coordinate System* - see *Figure 4-2*.

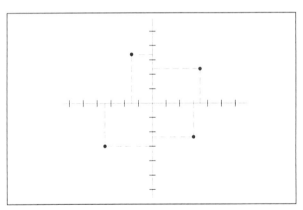

Figure 4-2
Rectangular coordinate system (Cartesian coordinate system)

The concepts used in design, drafting *and* in numerical control are over 400 years old. A given point can be mathematically defined on a plane (two coordinate values) or in space (three coordinate values). The definition of one point is relative to another point as a distance parallel with one of three axes that are perpendicular to each other. In a plane, only two axes are required, in the space, all three axes must be specified. In programming, point represents an exact location. If such a location is on a plane, the point is defined as a 2D point, along two axes. If the location is in a space, the point is defined as a 3D point, along three axes.

When *two* number scales that intersect at right angles are used, mathematical basis for a *rectangular coordinate system* is created. Several terms emerge from this representation, and all have an important role in CNC programming. Their understanding is very important for further progress.

◆ Axes and Planes

Each major line of the number scale is called an *axis*. This old principle, when applied to CNC programming, means that at least two axes - *two number scales* - will be used. This is the mathematical definition of an axis:

> An axis is a straight line passing through
> the center of a plane or a solid figure,
> around which the parts are symmetrically arranged

The definition can be enhanced by a statement that an axis can also be *a line of reference*. In CNC programming, an axis is used as a reference all the time. The definition contains the word *'plane'*. A *plane* is a term used in 2D applications, while a *solid object* is used in 3D applications. Mathematical definition of a plane is:

> A plane is a surface in which a straight line joining
> any two of its points will lie wholly in the surface

From the *top viewpoint* of the observer, looking straight down on the illustration *Figure 4-3*, a *viewing direction* is established. This is often called *viewing a plane*.

A plane is a 2D entity - the letter X identifies its horizon-

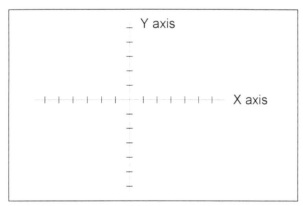

Figure 4-3
Axis designation - viewing plane
Mathematical designation is fully implemented in CNC

tal axis, the letter Y identifies its vertical axis. This plane is called the *XY plane*. Defined mathematically, the horizontal axis is *always* listed as the *first* letter of the pair. In drafting and CNC programming, this plane is also known as the *Top View* or a *Plan View*. Other planes are also used in CNC, but not to the same extent as in CAD/CAM work.

◆ Point of Origin

Another term that emerged from the rectangular coordinate system is called the *point of origin*, or just *origin*. It is the point where the two perpendicular axes intersect. This point has a zero coordinate value in each axis, specified as planar X0Y0 and spacial X0Y0Z0 - *Figure 4-4*.

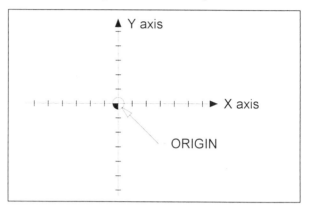

Figure 4-4
Point of origin - intersection of axes

This intersection has a special meaning in CNC programming. The origin acquires a new name, typically the *program reference point*. Other terms are also used: *program zero, part reference point, workpiece zero, part zero*, with the same meaning and purpose.

◆ Quadrants

Viewing the two intersecting axes and the new plane, four distinct areas can be clearly identified. Each area is bounded by two axes. These areas are called *quadrants*. Mathematically defined,

> A quadrant is any one of the four parts of the plane
> formed by the system of rectangular coordinates

The word *quadrant* (from the Latin word *quadrans* or *quadrantis*, meaning *the fourth part*), suggests *four* uniquely defined areas or *quadrants*. Looking down in the top view at the two intersecting axes, the following definitions apply to quadrants. They are mathematically correct and are used in CNC/CAD/CAM applications:

Quadrant I	UPPER RIGHT
Quadrant II	UPPER LEFT
Quadrant III	LOWER LEFT
Quadrant IV	LOWER RIGHT

The quadrants are defined in the *counterclockwise* direction from the horizontal X axis and the naming convention uses *Roman* numbers, not *Arabic* numbers normally used.

The counting starts at the *positive* side of the *horizontal* axis. *Figure 4-5* illustrates the definitions.

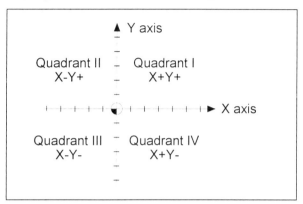

Figure 4-5

Quadrants in the XY plane and their identification

Any point coordinate value can be positive, negative or zero. Any coordinate value is determined solely by the *location* of the defined point in a particular quadrant and its distance along an axis, relative to the origin - *Figure 4-6*.

POINT LOCATION	COORDINATE	
	X AXIS	Y AXIS
QUADRANT I	+	+
QUADRANT II	–	+
QUADRANT III	–	–
QUADRANT IV	+	–

Figure 4-6

Algebraic signs for a point location in plane quadrants

❏ ***IMPORTANT:***
    … If the defined point lies exactly on the X axis,
      it has the *Y* value equal to zero (Y0).
    … If the point lies exactly on the Y axis,
      it has the *X* value equal to zero (X0).
    … If the point lies exactly on both X and Y axes,
      both X and *Y* values are zero (X0 Y0).

X0Y0Z0 is the point of origin. In part programming, positive values are written *without* the plus sign - *Figure 4-7*.

◆ Right Hand Coordinate System

In the illustrations of the *number scale*, *quadrants* and *axes*, the *origin* divides each axis into two portions. The zero point - *the point of origin* - separates the *positive* section of the axis from the *negative* section. In the right-hand coordinate system, the *positive* axis begins at the origin and is directed towards the *right* for the X axis, *upwards* for the Y axis and *towards the perpendicular viewpoint* for the Z axis. Opposite directions are negative.

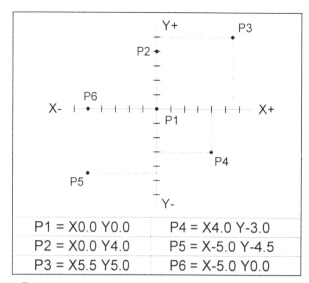

P1 = X0.0 Y0.0	P4 = X4.0 Y-3.0
P2 = X0.0 Y4.0	P5 = X-5.0 Y-4.5
P3 = X5.5 Y5.0	P6 = X-5.0 Y0.0

Figure 4-7

Coordinate definition of points within the rectangular coordinate system (point P1 = Origin = X0Y0)

If these directions were superimposed over a human right hand, they would correspond to the direction from the root of thumb or finger towards its tip. The thumb would point in the X direction, the index finger in the Y direction and the middle finger in the Z direction.

The majority of CNC machines are programmed using the so called *absolute* coordinate method, that is based on the point of origin X0Y0Z0. This absolute method of programming follows very strictly the rules of rectangular coordinate geometry and all concepts covered in this chapter.

MACHINE GEOMETRY

Machine geometry is the relationship of distances between the *fixed point of the machine* and the *selectable point of the part*. Typical geometry of a CNC machine uses the right hand coordinate system. The positive and negative axis direction is determined by an established viewing convention. The basic rule for the Z axis is that it is always the axis along which a simple hole can be machined with a single point tool, such as a drill, reamer, wire or a laser beam. The *Figure 4-8* illustrates the standard orientation of an XYZ type machine tools.

◆ Axis Orientation - Milling

A typical 3-axis machine uses three controlled axes of motion. They are defined as the *X axis*, the *Y axis*, and the *Z axis*. The X axis is parallel to the longest dimension of the machine table, Y axis is parallel to the shortest dimension of the table and the Z axis is the spindle movement. On a vertical machining center, the X axis is the table *longitudinal* direction, the Y axis is the saddle *cross* direction and

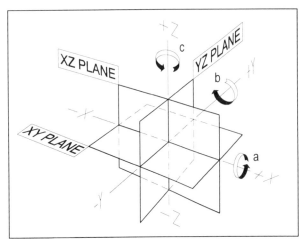

Figure 4-8
Standard orientation of planes and CNC machine tool axes

the Z axis is the *spindle* direction. For horizontal machining centers, the terminology is changed due to the design of these machines. The X axis is the table *longitudinal* direction, the Y axis is the column direction and the Z axis is the *spindle* direction. Horizontal machine can be viewed as a vertical machine rotated in space by ninety degrees. The additional feature of a horizontal machining center is the indexing B axis. Typical machine axes applied to CNC vertical machines are illustrated in *Figure 4-9*.

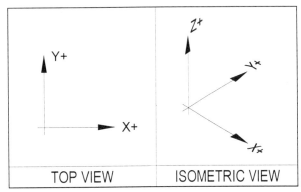

Figure 4-9
Typical machine axes of a vertical CNC machining center

◆ Axis Orientation - Turning

Most CNC lathes have two axes, X and Z. More axes are available, but they are not important at this point. A special third axis, the C axis, is designed for milling operations (live tooling) and is an option on the typical CNC lathe.

What is more common for CNC lathes in industry, is the double orientation of the XZ axes. Lathes are distinguished as *front* and a *rear* lathes. An example of a front lathe is similar to the conventional engine lathe. All the slant bed types of a lathe are of the rear kind. Identification of the axes have often not followed mathematical principles.

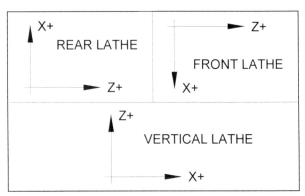

Figure 4-10
Typical machine axes of a CNC lathe (turning center)

Another variety, a vertical CNC lathe, is basically a horizontal lathe rotated 90°. Typical axes for the horizontal and vertical machine axes, as applied to turning, are illustrated in *Figure 4-10*.

◆ Additional Axes

A CNC machine of any type can be designed with one or more additional axes, normally designated as the secondary axes using the U, V and W letters. These axes are normally parallel to the primary X, Y and Z axes respectively. For a rotary or an indexing applications, the additional axes are defined as A, B and C axes, as being rotated about the X, Y and Z axes, again in their respective order. Positive direction of a rotary (or an indexing) axis is the direction required to advance a right handed screw in the positive X, Y or Z axis. The relationship of the primary and the secondary (or supplementary) axes is shown in *Figure 4-11*.

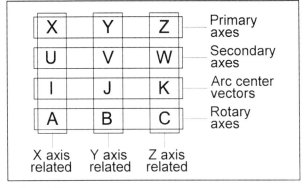

Figure 4-11
Relationship of the primary and the secondary machine axes

Arc center modifiers (sometimes called the arc center vectors) are not true axes, yet they are also related to the primary axes XYZ. This subject will be described in the section on *Circular Interpolation*, in *Chapter 29*.

5 CONTROL SYSTEM

A machine unit equipped with a computerized numerical control system is commonly known as a CNC machine. In an analogy of the machine tool as the *body* of a CNC machine system, the control unit is its *brain*, its nerve center. There are no levers, no knobs and no handles on a CNC machine the way they function on conventional milling machines and lathes. All the machine speeds, feeds, axes motions and hundreds of other tasks are programmed by a CNC programmer and controlled by a computer that is major part of the CNC unit. To make a program for a CNC machine tool means to make a program for the control system. True, the machine tool is a major consideration as well, but it is the *control unit* that determines the *format* of the program, its *structure* and its *syntax*.

In order to fully understand the CNC programming process, it is important to understand not only the intricacies of how to machine a part, what tools to select, what speeds and feeds to use, how to setup the job and many other features. It is equally important to know how the computer, the CNC unit, actually works without the need to be an expert in electronics or a computer scientist. *Figure 5-1* shows an actual Fanuc control panel.

The machine manufacturers add their own operation panel, with all the switches and button needed to operate the CNC machine and all its features. A typical operation panel is illustrated in *Figure 5-2*. Another item required for the system, the handle, will be described as well.

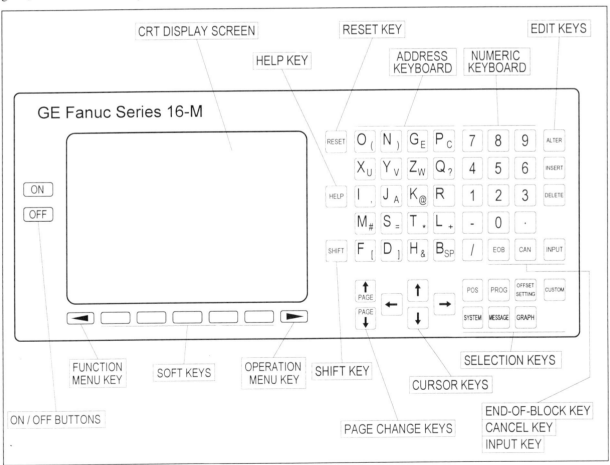

Figure 5-1
A typical example of a Fanuc control panel - actual layout and features will vary on different models (Fanuc 16M)

GENERAL DESCRIPTION

Even a brief look at any control unit reveals that there are two basic components - one is the *operation panel*, full of rotary switches, toggle switches and push buttons. The other component is the *display screen* with a keyboard or a keypad. The programmer who does not normally work on the CNC machine will seldom, if ever, have a reason to use either the operation panel or the display screen. They are available at the machine to the CNC machine operator, and used for the machine setup as well as to control the activities of the machine.

Should the CNC programmer be interested in the machine operation? Is it necessary for the programmer to *know* and *understand* all functions of the control system? There is only one answer to both questions - *definitely yes*.

The control unit - the CNC system - contains features that only work in conjunction *with* the program, it does not do anything useful on its own. Some features can be used *only* if the program itself supports them. All switches and buttons and keys are used by the machine operator, to exercise control over the program execution and machining process.

◆ Operation Panel

Depending on the type of the CNC machine, the following table covers the most typical and common features found on the modern operation panel. There are some small differences for the operation of a machining center and a lathe, but both operation panels are similar. As with any general reference book, it is always a good idea to double check with the manufacturer specifications and recommendations. It is common that many machines used in the shop have some special features.

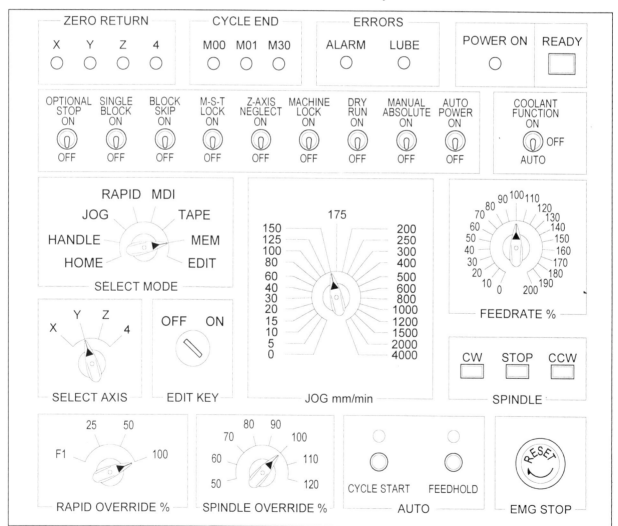

Figure 5-2

A typical operation panel of a CNC machining center - actual layout and features will vary on different models

Feature	Description
ON / OFF switch	Power and control switch for the main power and the control unit
Cycle Start	Starts program execution or MDI command
Emergency Stop	Stops all machine activity and turns off power to the control unit
Feedhold	Temporarily stops motion of all axes
Single Block	Allows program run one block at a time
Optional Stop	Temporarily stops the program execution (M01 required in program)
Block Skip	Ignores blocks preceded with a forward slash (/) in the program
Dry Run	Enables program testing at fast feedrates (without a mounted part)
Spindle Override	Overrides the programmed spindle speed, usually within 50-120% range
Feedrate Override	Overrides the programmed feedrate, usually within 0-200% range
Chuck Clamp	Shows current status of the chuck clamping (Outside / Inside clamping)
Table Clamp	Shows current status of table clamping
Coolant Switch	Coolant control ON / OFF / AUTO
Gear Selection	Shows current status of working gear range selection
Spindle Rotation	Indicates spindle rotation direction (clockwise or counterclockwise)
Spindle Orientation	Manual orientation of the spindle
Tool Change	Switch allowing a manual tool change
Reference Position	Switches and lights relating to setup of the machine from reference position
Handle	Manual Pulse Generator (MPG), used for Axis Select and Handle Increment switches
Tailstock Switch	Tailstock and/or quill switch to manually position the tailstock
Indexing Table Switch	Manually indexes machine table during setup
MDI Mode	Manual Data Input mode

Feature	Description
AUTO Mode	Allows automatic operations
MEMORY mode	Allows program execution from the memory of the CNC unit
TAPE mode	Allows program execution from an external device, such as a desktop computer or a punched tape
EDIT mode	Allows changes to be made to a program stored in the CNC memory
MANUAL Mode	Allows manual operations during setup
JOG Mode	Selects the jog mode for setup
RAPID Mode	Selects the rapid mode for setup
Memory Access	Key (switch) to allow program editing
Error Lights	Red light indicating an error

Even is some features may not be listed, virtually all of those in the table are somewhat related to the CNC program. Many control systems have unique features of their own. These features must be known to the CNC operator. The program supplied to the machine should be *flexible*, not *rigid* - it should be 'user friendly'.

◆ Screen Display and Keyboard

The screen display is the 'window' to the computer. Any active program can be viewed, including the status of the control, current tool position, various offsets, parameters, even a graphic representation of the Tool Path. On all CNC units, individual monochrome or color screens can be selected to have the desired display at any time, using the input keys (keyboard pads and soft keys). Setting for international languages is also possible.

The keyboard pads and soft keys are used to *input* instructions to the control. Existing programs can be modified or deleted, new programs can be added. Using the keyboard input, not only the machine axes motion can be controlled, but the spindle speed and feedrate as well. Changing the internal parameters and evaluating various diagnostics are more specific means of control, often restricted to service people. Keyboard and screen are used to set the program origin and to hook up to external devices, such as a connection with another computer. There are many other options. Every keyboard allows the use of *letters*, *digits* and *symbols* for data entry. Not every keyboard allows the use of *all* the alphabet letters or *all* available symbols. Some control panel keys have a description of an *operation*, rather than a letter, digit or symbol, for example, *Read* and *Punch* keys or the *Offset* key.

◆ Handle

For the setup purposes, each CNC machine has a rotary handle that can move one selected axis by as little as the least increment of the control system. The official Fanuc name for the handle is *Manual Pulse Generator*. Associated with the handle is the *Axis Select* switch (often duplicated on the operation panel as well as on the handle) and the range of increment (that is the least increment X1, X10 and X100). The letter X in this case is the multiplier and stands for *'X times'*. One handle division will move the selected axis by X times the minimum increment of the active units of measurement. In *Figure 5-3* and the following table are the details of a typical handle.

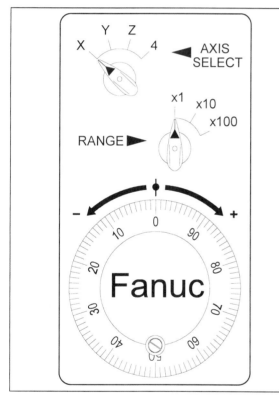

Figure 5-3

An example of a detached handle, called the Manual Pulse Generator (MPG), with a typical layout and features.
Layout and features may vary on different machine models.

Handle Multiplier	One handle division motion is ...	
	for Metric units	for English units
X1	0.001 mm	.0001 inch
X10	0.010 mm	.0010 inch
X100	0.100 mm	.0100 inch

SYSTEM FEATURES

The CNC unit is nothing more than a sophisticated special purpose computer. The 'special purpose' in this case is a *computer capable of controlling the activities of a machine tool*, such as a lathe or a machining center. It means the computer has to be designed by a company that has expertise in this type of special purpose computers. Unlike many business types of computers, each CNC unit is made for a particular customer. The customer is typically the machine manufacturer, *not* the end user. The manufacturer specifies certain requirements that the control system has to meet, requirements that reflect the uniqueness of the machines they build. The basic control does not change, but some customized features may be added (or taken away) for a specific machine. Once the control system is sold to the machine manufacturer, more features are added to the system. They mainly relate to the design and capabilities of the machine.

A good example is a CNC unit for two machines that are the same in all respects except one. One machine has a *manual* tool changer, the other has an *automatic* tool changer. In order to support the automatic tool changer, the CNC unit must have special features built-in, that are not required for a machine without the tool changer. The more complex the CNC system is, the more expensive it is. Users that do not require all the sophisticated features, do not pay a premium for features they do not need.

◆ Parameter Settings

The information that establishes the built-in connection between the CNC control and the machine tool is stored as special data in internal registers, called the *system parameters*. Some of the information in this handbook is quite specialized and listed for reference only. Programmers with limited experience do not need to know system parameters in a great depth. The original factory settings are sufficient for most machining jobs.

When the parameter screen is displayed, it shows the parameter number with some data in a row. Each row of numbers represents one *byte*, each digit in the byte is called a *bit*. The word *bit* is made from the words *Binary digIT* and is the smallest unit of a parameter input. Numbering of bits starts with 0, read from *the right* to *the left*:

Number	#7	#6	#5	#4	#3	#2	#1	#0
xxxx	0	1	1	0	1	0	0	1

The Fanuc control system parameters belong to one of three groups, specified within an allowed range:

❑   Binary codes

❑   Units inputs

❑   Setting values

The groups use different input values. The *binary input* can only have an input of a 0 or 1 for the *bit* data format, 0 to +127 for the byte type. *Units input* has a broader scope - the unit can be in mm, inches, mm/min, in/min, degrees, milliseconds, etc. A *value* can also be specified within a given range, for example, a number within the range of 0-99, or 0-99999, or +127 to -127, etc. .

A typical example of a *binary input* is a selection between *two options*. For instance, a feature called *dry run* can be set only as *effective* or *ineffective*. To select a preference, an arbitrary bit number of a parameter has be set to 0 to make the dry run effective and to 1 to make it ineffective.

Units input, for example, is used to set the increment system - the dimensional units. Computers in general do not distinguish between inch and metric, just numbers. It is up to the user *and* the parameter setting, whether the control will recognize 0.001 mm or .0001 inches as the least increment. Another example is a parameter setting that stores the maximum feedrate for each axis, the maximum spindle speed, etc. Such values must never be set higher than the machine can support. An indexing axis with a minimum increment of 1°, will *not* become a rotary axis with .001° increment, just because the parameter is set to a lower value, even if it is possible. **Such a setting is wrong and can cause serious damage!**

To better understand what the CNC system parameters can do, here is an abbreviated listing of parameter classification for a typical Fanuc control system (many of them are meaningful to the service technicians only):

Parameters related to Setting
Parameters related to Axis Control Data
Parameters related to Chopping
Parameters related to the Coordinate System
Parameters related to Feedrate
Parameters related to Acceleration/Deceleration Control
Parameters related to Servo
Parameters related to DI/DO
Parameters related to MDI, EDIT, and CRT
Parameters related to Programs
Parameters related to Serial Spindle Output
Parameters related to Graphic Display
Parameters related to I/O interface
Parameters related to Stroke Limit
Parameters related to Pitch Error Compensation
Parameters related to Inclination Compensation
Parameters related to Straightness Compensation
Parameters related to Spindle Control
Parameters related to Tool Offset
Parameters related to Canned Cycle
Parameters related to Scaling and Coordinate Rotation
Parameters related to Automatic Corner Override
Parameters related to Involute Interpolation
Parameters related to Uni-directional Positioning
Parameters related to Custom Macro (User Macro)
Parameters related to Program Restart

Parameters related to High-Speed Skip Signal Input
Parameters related to Automatic Tool Compensation
Parameters related to Tool Life Management
Parameters related to Turret Axis Control
Parameters related to High Precision Contour Control
Parameters related to Service   *... and other parameters*

Quite a few parameters have nothing to do with daily programming and are listed only as an actual example. All system parameters should be set or changed only by a qualified person, such as an experienced service technician. A programmer or operator should not modify any parameter settings. These changes require not only qualifications but authorization *as well*. Keep the list of original parameter settings away from the control, in a safe place, just in case.

Take care when changing control system parameters !

Many parameters are periodically updated during program processing. The CNC operator is usually not aware that this activity is going on at all. There is no real need to monitor this activity. The safest rule to observe is that once the parameters have been set by a qualified technician, any *temporary* changes required for a given work should be done through the CNC program. If *permanent* changes are required, an authorized person should be assigned to do them - *nobody else*.

◆ System Defaults

Many parameter settings stored in the control at the time of purchase have been entered by the manufacturer as either the *only choices*, the *most suitable choices*, or the *most common selections*. That does not mean they will be *the* preferred settings - it means they were selected on the basis of their common usage. Many settings are rather conservative in their values, for safety reasons.

The set of parameter values established at the time of installation are called the *default* settings. The English word *'default'* is a derivative of a French word *'defaut'*, that can be translated as *'assumed'*. When the main power to the control is turned on, there are no set values passed to parameters from a program, since no program has yet been used. However, certain settings become active automatically, *without* an external program. For instance, a cutter radius offset is automatically canceled at the startup of the control system. Also canceled are the fixed cycle mode and tool length offset. The control *'assumes'* that certain conditions are preferable to others. Many operators will agree with most of these initial settings, although not necessarily with all of them. Some settings are customizable by a change of a parameter settings. Such settings will become permanent and create a *new 'default'*.

Always document any changes to the parameters!

A computer is fast and accurate but has no intelligence. People are often slow and make errors, but have one unique ability - *they think*. A computer is just a machine that does not *assume* anything, does not *consider*, does not *feel* - computer *does not think*. A computer does not do anything that a human effort and ingenuity has not done during the design process, in form of hardware and software.

When the CNC machine is powered, the internal software sets *certain* existing parameters to their default condition, designed by engineers. Not *all* system parameters, only *certain* parameters can have an assumed condition - a condition that is known as the *default value* (condition).

For example, a tool motion has three basic modes - a *rapid* motion, a *linear* motion and a *circular* motion. The default motion setting is controlled by a parameter. Only one setting can be active at the startup. Which one? The answer *depends on the parameter setting*. Many parameters can be preset to a desired state. Only the *rapid* or the *linear* mode can be set as default in the example. Since the rapid motion is the first motion in the program, it seems to make sense to make it a default - *but wait!*

Most controls are set to the *linear* motion as the default (G01 command), to be in effect at the start - strictly *for safety reasons*. When the machine axes are moved manually, the parameter setting has no effect. If a manual input of an axis command value takes place, either through the program or from the control panel, a tool motion results. If the motion command is not specified, the system will use the command mode that had been preset as the default in the parameters. Since the default mode is a linear motion G01, the result is an *error condition,* faulting the system *for the lack of a feedrate!* There is no cutting feedrate in effect, which the G01 requires. Had the default setting been the rapid motion G00, a rapid motion would be performed, as it does not need programmed feedrate.

It is beneficial to know the default settings of all controls in the shop. Unless there is a good reason to do otherwise, defaults for similar controls should be the same.

◆ Memory Capacity

CNC programs can be stored in the control memory. The program size is only limited by the capacity of the control. This capacity is measured in a variety of ways, originally as the equivalent *length of tape* in meters or feet, lately as the *number of bytes* or the number of *screen pages*. A common minimum memory capacity of a CNC lathe control is 20 m of tape (66 ft). This is an old fashioned method that somehow persisted in staying with us. On CNC milling systems, the memory requirements based on the same criteria are generally larger and the typical minimum memory capacity is 80 m or about 263 ft. Optionally, larger memory capacity can be added to the control system. The minimum memory capacity of the control varies from one machine to another - always check control specifications carefully.

Modern methods of measuring memory capacity prefer to use *bytes* as the unit, rather that a length of an obsolete tape. A byte is the smallest unit of storage capacity and is very roughly equivalent to one character in the program.

The memory capacity of the control system should be large enough to store the longest CNC program expected on a regular basis. That requires some planning before the CNC machine is purchased. For example, in three dimensional mold work or high speed machining, the cost of additional memory capacity may be very high. Although any cost is a relative term, there are reliable and inexpensive alternatives, well worth looking into.

One alternative is running the CNC program from a personal computer. An inexpensive communication software and cabling is required to connect the computer with the CNC system. The simplest version is to *transfer* the CNC program from one computer to the other. More sophisticated possibility includes software and cables that can actually run the machine from the personal computer, without loading it to the memory of the CNC first. This method is often called '*drip-feeding*' or '*bitwise input*'. When operated from the personal computer, the CNC program can be as long as the capacity of the storage device, typically the hard drive.

Most CNC programs will fit into the internal memory of the control system. Many controls use the number of available characters or the equivalent length of tape. Here are some formulas that can be used to get at least the approximate memory capacity calculations:

➲ Formula 1 :

To find the program length in *meters*, when the capacity is known in *characters*, use the following formula:

$$S_m = N_c \times .00254$$

☞ where . . .

S_m = Storage capacity in meters
N_c = Memory capacity (number of characters)

➲ Formula 2 :

To find the length of program in *feet,* when the capacity is known in *characters*, use the following formula:

$$S_f = \frac{N_c}{120}$$

☞ where . . .

S_f = Storage capacity in feet
N_c = Memory capacity (number of characters)

➲ Formula 3 :

To find the number of *characters* in a given program, if the system memory capacity is known in *meters*:

$$C = \frac{m}{.00254}$$

☞ where . . .

C = Number of available characters
m = Memory capacity in meters

Virtually the same results can be achieved by a slightly restructured formula:

$$C = \frac{m \times 1000}{2.54}$$

➲ Formula 4:

To find the number of *characters*, if the system memory capacity is known in *feet*, use the following formula:

$$C = f \times 120$$

☞ where . . .

C = Number of available characters
f = Memory capacity in feet

Latest Fanuc controls show the available memory as the number of *free screen display pages*. This type of data is not easy to convert as the others.

In cases where the available memory capacity is too small to accept a large program, several techniques are available to minimize the problem, for example, the program length reduction methods, described in *Chapter 50*.

MANUAL PROGRAM INTERRUPTION

If a program needs to be interrupted in the middle of processing, the control system offers several ways to do that, using the machine operation panel. The most common features of this type are toggle switches or push buttons for a *single block* operation, *feedhold* and the *emergency stop*.

◆ Single Block Operation

The normal purpose of a program is to control the machine tool automatically and sequentially in a continuous mode. Every program is a series of formatted commands - or instructions - written as individual lines of code, called *blocks*. Blocks and their concepts will be described in the following chapters. All program commands in a single block are processed as a *single instruction*. The blocks are received by the control system in sequential order, from the top down and in the order they appear in the program. Normally, a CNC machine is run in a continuous mode, while the blocks are processed automatically, one after another. This continuity is important for production, but not practical when proving a new program, for example.

To disable the continuous program execution, a *Single Block* switch is provided on the operation panel. In the single block mode, only one block of the program will be processed each time the *Cycle Start* key is pressed. On the operation panel, the single block mode can be used separately or in combination with other settings that make program proving faster and more accurate.

◆ Feedhold

Feedhold is a special push button located on the operation panel, usually close to the *Cycle Start* button. When this button is pressed during a rapid, linear or circular axes motion, it will immediately stop the motion. The action applies to all axes active at the time. This feature is convenient for a machine setup or a first part run. Some types of motion restrict the function of the feedhold or disable it altogether. For example, threading or tapping modes make the switch inoperative.

Activating feedhold at the machine will not change any other program values - it will only affect the motion. The feedhold switch will be illuminated (in red light), as long as it is effective. The CNC programmer can override the feedhold from *within* the program, for special purposes.

◆ Emergency Stop

Every CNC machine has at least one special mushroom shaped push button, red in color, that is located in an accessible place on the machine. It is marked the *Emergency Stop* or *E-stop*. When this button is pressed, *all machine activities will cease immediately*. The main power supply will be interrupted and the machine will have to be restarted. The emergency stop switch is a mandatory safety feature on all CNC machines.

Pressing the emergency stop button is not always the *best* or even the *only* way to stop a machine operation. In fact, the latest controls offer other features, far less severe, designed to prevent a collision between a cutting tool and the part or fixture. Previously discussed feedhold button is only one option, along with other features. If the emergency stop must be used at all, it should be used as the *last resort*, when any other action would require unacceptably longer time. There is no need for panic, if something does go wrong.

For some machine actions, the effect of *Emergency Stop* is not always apparent. For example, the spindle requires a certain time for deceleration to stop.

## Chapter 5

MANUAL DATA INPUT - MDI

A CNC machine is not always operated by the means of a program. During a part setup, the CNC operator has to do a number of operations that require physical movements of the machine slides, rotation of spindle, tool change, etc. There are no mechanical devices on a CNC machine. The handle *(Manual Pulse Generator)* is an electronic, not a mechanical unit. In order to operate a CNC machine without conventional mechanical devices the control system offers a feature called the *Manual Data Input* - or MDI.

The *Manual Data Input* enables the input of a program data into the system *one program instruction* at a time. If too many instructions were to be input repeatedly, such as a long program, the procedure would be very inefficient. During a setup and for similar purposes, one or a few instructions at a time will benefit from the MDI.

To access the MDI mode, the MDI key on the operation panel must be selected. That opens the screen display with the current status of the system. Not all, but the majority of programming codes are allowed in the MDI mode. Their format is identical to the format of a CNC program in written form. This is one area where the CNC operator acts as a CNC programmer. It is very important that the operator is trained at least in the basics of CNC programming, certainly to the point of being able to handle the setup instructions for *Manual Data Input*.

PROGRAM DATA OVERRIDE

All CNC units are designed with a number of special rotary switches that share one common feature - they allow the CNC operator to *override* the programmed speed of the spindle or the programmed speed of the axis motion. For example, a 15 in/min feedrate in the program produces a slight chatter. A knowledgeable operator will know that by *increasing* the feedrate or *decreasing* the spindle speed, the chatter may be eliminated. It is possible to change the feedrate or the spindle speed by editing the program, but this method is not very efficient. A certain 'experimentation' may be necessary during the actual cut to find the optimum setting value. The manual override switches come to the rescue, because they can be used by trial during operation. There are four override switches found on most control panels:

❑ *Rapid feedrate override* (rapid traverse)
  (modifies the rapid motion of the machine tool)

❑ *Spindle speed override*
  (modifies the programmed spindle r/min)

❑ *Feedrate override* (cutting feedrate)
  (modifies the programmed feedrate)

❑ *Dry run mode*
  (changes cutting motions to a variable speed)

Override switches can be used individually or together. They are available on the control to make the work easier for both the operator and the programmer. The operator does not need to 'experiment' with speeds and feeds by constantly editing the program and the programmer has a certain latitude in setting reasonable values for the cutting feedrates and the spindle speed. The presence of the override switches is not a licence to program unreasonable cutting values. The overrides are fine tuning tools only - the program must always reflect the machining conditions of the work. The usage of override switches does not make any program changes, but gives the CNC operator the opportunity to edit the program later to reflect the optimum cutting conditions. Used properly, the override switches can save a great amount of valuable programming time as well as the setup time at the CNC machine.

◆ Rapid Motion Override

Rapid motions are selected in the CNC program by a preparatory command without a specified feedrate. If a machine is designed to move at 500 in/min (12700 mm/min) in the rapid mode, this rate will never appear in the program. Instead, you call the rapid motion mode by programming a special preparatory command G00. During the program execution, all motions in the G00 mode will be at the manufacturer's fixed rate. The same program will run faster on a machine with high rapid motion rating then on a machine with low rapid motion rating.

During setup, the rapid motion rate may require some control for program proving, when high rapid rates are uncomfortable to work with. After the program had been proven, rapid rate can be applied at its maximum. CNC machines are equipped with a *rapid override switch* to allow temporary rapid motion settings. Located on the control panel, this switch can be set to one of the four settings. Three of them are marked as the *percentage* of the maximum rate, typically as 100%, 50% and 25%. By switching to one of them, the rapid motion rate changes. For example, if the maximum rapid rate is 500 in/min or 12700 mm/min, the actual reduced rates are 250 in/min or 6350 mm/min at the 50% setting and 125 in/min or 3175 mm/min at the 25% setting. Each of the reduced rates is more comfortable to work with during setup.

The fourth position of the switch often has no percentage assigned and is identified as an F1 or by a small symbol. In this setting, the rapid motion rate is even slower than that of 25% setting. Why is it not identified as 10% or 15%, for example? The reason is simple - the control system allows a *customized* selection as to what the value will be. It may be a setting of between 0 and 100%. The default setting is also the most logical - usually 10% of the maximum rapid traverse rate. This setting should never be higher than 25% and can be done only through a setting of a system parameter. Make sure that all persons who work on such a machine are aware of the changes.

◆ Spindle Speed Override

The same logic used for the application of the rapid rate override can be used for the spindle speed override. The required change can be established during the actual cutting by using the *spindle speed override switch*, located on the control panel. For example, if the programmed spindle speed of 1000 r/min is too high or too low, it may be changed temporarily by the switch. During the actual cutting, the CNC operator may experiment with the spindle speed override switch to find the optimum speed for the given cutting conditions. This method is a much faster than 'experimenting' with the program values.

The spindle speed override switch can be continuous on some controls or selectable in increments of 10%, typically within the range of 50-120% of the programmed spindle speed. A spindle programmed at 1000 r/min can be overridden during machining to 500, 600, 700, 800, 900, 1000, 1100 and 1200 r/min. This large range allows the CNC operator the flexibility of optimizing the spindle rotation to suit the cutting conditions. There is a catch, however. The optimized spindle speed change may apply to only one tool of the many used in the program. No CNC operator can be expected to watch for that particular tool and switch the speed up or down when needed. A simple human oversight may ruin the part, the cutting tool or both. The recommended method is to find out the optimum speed for each tool, write it down, then change the program accordingly, so *all the tools* can be used at the 100% spindle override setting for production.

Comparison of the increments on the spindle override switch with the increments on switches for the rapid traverse override (described earlier) and the feedrate override (described next), offers much more limited range. The reason for the spindle speed range of 50% to 120% is safety. To illustrate with a rather exaggerated example, no operator would want to mill, drill or cut any material at 0 r/min (no spindle rotation), possibly combined with a heavy feedrate.

In order to change the selected override setting into 100% speed in the program, a new spindle speed has to be calculated. If a programmed spindle speed of 1200 r/min for a tool is *always* set to 80%, it should be edited in the program to 960 r/min, then used at 100%. The formula is quite simple:

$$S_n = S_p \times p \times 0.01$$

☞ where . . .

S_n = Optimized - or new - r/min
S_p = Originally programmed r/min
p = Percentage of spindle override

Overriding the programmed spindle speed on the CNC machine should have only one purpose - to establish the spindle speed rotation for the best cutting conditions.

◆ Feedrate Override

The most commonly used override switch is one that changes programmed feedrates. For the milling controls, the feedrate is programmed in *in/min* or *m/min*. For lathe controls, the feedrate is programmed in *in/rev* or in *mm/rev*. The feedrate per minute on lathes is used only in cases when the spindle is not rotating and the feedrate needs to be controlled.

The new feedrate calculation, based on the overridden feedrate setting, is similar to that for spindle speed:

$$F_n = F_p \times p \times 0.01$$

☞ where . . .

F_n = Optimized - or new - feedrate
F_p = Originally programmed feedrate
p = Percentage of feedrate override

Feedrate can be overridden within a large range, typically from 0% to 200% or at least 0% to 150%. When the feedrate override switch is set to 0%, the CNC machine will *stop* the cutting motion. Some CNC machines do not have the 0% percent setting and start at 10%. The maximum of 150% or 200% cutting feedrate will cut 1.5× or 2× faster than the programmed value.

There are situations, where the use of a feedrate override would damage the part or the cutting tool - or both. Typical examples are various tapping cycles and single point threading. These operations require spindle rotation synchronized with the feedrate. In such cases, the feedrate override will become *ineffective*. The feedrate override *will* be effective, if standard motion commands G00 and G01 are used to program any tapping or tread cutting motions. Single point threading command G32, tapping fixed cycles G74 and G84, as well as lathe threading cycles G92 and G76 have the feedrate override cancellation built into the software. All these and other related commands are described later in the handbook, in more detail.

◆ Dry Run Operation

Dry run is a special kind of override. It is activated from the control panel by the *Dry Run* switch. It only has a direct effect on the feedrate and allows much higher feedrate than that used for actual machining. In practice, it means the program can be executed much faster than using a feedrate override at the maximum setting. No actual machining takes place when the dry run switch is in effect.

What is the purpose of the dry run and what are its benefits? Its purpose is to test the integrity of the program *before* the CNC operator cuts the first part. The benefits are mainly in the time saved during program proving when no machining takes place. During a dry run, the part is normally *not* mounted in the machine. If the part is mounted in the holding device and the dry run is used as well, it is very important to provide sufficient clearances. Usually, it means *moving the tool away from the part*. The program is then executed 'dry', without actual cutting, without a coolant, just in the air. Because of the heavy feedrates in the dry run, the part cannot be machined safely. During a dry run, the program can be checked for all possible errors except those that relate to the actual contact of the cutting tool with the material.

The dry run is a very efficient setup aid to prove the overall integrity of the CNC program. Once the program is proven during a dry run, the CNC operator can concentrate on the sections of the program that contain actual machining. Dry run can be used in combination with several other features of the operation panel.

> Make sure to disable the dry run before machining!

◆ Z Axis Neglect

Another very useful tool for testing unproven programs on CNC machining centers (*not* lathes) is a toggle switch located on the operation panel called the *Z Axis Neglect* or *Z Axis Ignore*. As either name suggests, when this switch is activated, any motion programmed for the Z-axis will *not* be performed. Why the Z axis? Since the X and Y axes are used to profile a shape of the part (the most common contouring operations), it would make no sense to temporarily cancel either one of these axes. By neglecting (disabling) the Z axis temporarily, the CNC operator can concentrate on proving the accuracy of the part contour, without worrying about the depth. Needless to say, this method of program testing must take place without a mounted part (and normally without a coolant as well). *Be careful here!* It is important to *enable* or *disable* the switch *at the right time*. If the Z axis motion is disabled before the *Cycle Start* key is pressed, all following Z axis commands will be ignored. If the motion is enabled or disabled during program processing, the position of the Z axis may be inaccurate.

The Z axis neglect switch may be used in both the manual and automatic modes of operation. Just make sure that the motion along the Z axis is returned to the *enabled* mode, once the program proving is completed. Some CNC machines require resetting of the Z axis position settings.

◆ Manual Absolute Setting

If this feature is installed on the control (some controls use it automatically), it enables the CNC operator to resume a program in the middle of processing. Manual absolute can save time, particularly when processing long programs. *Manual Absolute* setting switch is not a typical option. To some extent, it is functionally related to the *Sequence Return* setting. Check the machine tool documentation before using either of these two features.

◆ Sequence Return

Sequence Return is a function controlled by a switch or a key on the control panel. Its purpose is to enable the CNC operator to start a program from the middle of an interrupted program. Certain programmed functions are memorized (usually the last speed and feed), others have to be input by the *Manual Data Input* key. The operation of this function is closely tied to the machine tool design. More information on the usage can be found in the machine tool manual. This function is very handy when a tool breaks during processing of long programs. It can save valuable production time, if used properly.

◆ Auxiliary Functions Lock

There are three functions available to the operation of a CNC machine that are part of the *'auxiliary functions'* group. These functions are:

Miscellaneous functions lock	Locks M functions
Spindle functions lock	Locks S functions
Tool functions lock	Locks T functions

As described later in this chapter, auxiliary functions generally relate to the *technological* aspects of the CNC programming. They control such machine functions as spindle rotation, spindle orientation, coolant selection, tool changing, indexing table, pallets and many others. To a lesser degree, they also control some program functions, such as compulsory or optional program stop, subprogram flow, program closing and others.

When the auxiliary functions are locked, all machine related miscellaneous functions M, all spindle functions S and all tool functions T will be suspended. Some machine tool manufacturers prefer the name *MST Lock* rather than *Auxiliary Functions Lock*. The MST is an acronym of the first letters from the words *Miscellaneous*, *Spindle* and *Tool*, referring to the program functions that will be locked.

The applications of these locking functions are limited to the job setup and program proving only and are not used for production machining.

◆ Machine Lock

Machine Lock function is yet another control feature for program proving. So far, we have looked at the *Z axis Neglect* function and the locking of the auxiliary functions. Remember that the *Z axis Neglect* function will disable the motion of the Z axis only and the *Auxiliary Functions Lock* (also known as the *MST lock*) locks the miscellaneous functions, the spindle functions and the tool functions. Another function, also available through the control panel, is called the *Machine Lock*. When this function is enabled, the motion of *all* axes is locked. It may seem strange to test a program by locking all the tool motions, but there is a good reason to use this feature. It gives the CNC operator the chance to test the program with virtually no chance of a collision.

When the machine lock is enabled, *only the axis motion is locked*. All other program functions are executed normally, including the tool change and spindle functions. This function can be used alone or in combination with other functions in order to discover possible program errors. Probably the most typical errors are syntax errors and the various tool offset functions.

◆ Practical Applications

Many of the control features described in this chapter, are used in *conjunction* with each other. A good example is *Dry Run* used in conjunction with the *Z axis Neglect* or the *Auxiliary Functions Lock*. By knowing what function are available, the CNC operator makes a choice to suit the needs of the moment. There are many areas of equal importance on which the CNC operator has to concentrate when setting up a new job or running a new program. Many features of the control unit are designed to make the operator's job easier. They allow concentration on one or two items at a time rather than the complexity of the whole program. These features have been covered in a reasonable detail, now is the time to look at some practical applications.

During the initialization of a new program run, a good CNC operator will take certain precautions as a matter of fact. For example, the first part of the job will most likely be tested with a rapid motion set to 25% or 50% of the available rapid rate. This relatively slow setting allows the operator to monitor the integrity of the program processing, as well as specific details. The details may include items such as a possibility of insufficient clearance between the tool and the material, checking if the Tool Path looks reasonable, and so on.

The CNC operator will have a number of tasks to perform simultaneously. Some of the tasks include monitoring the spindle speed, feedrate, tool motions, tool changes, coolant, etc. A careful and conscious approach results in building the confidence in the integrity of the CNC program. It may be the second or even the third part of the job when the CNC operator starts thinking of the optimization of the cutting values, such as the spindle speed and the cutting feedrates. This optimization will truly reflect the ideal speeds and feeds for a particular workpiece under given setup.

A production supervisor should not arbitrarily criticize an override setting less than 100%. Many managers consider the CNC program as an unchangeable document. They take the attitude that what is written is infallible - which is not always true. Often, the CNC operator may have no other choice *but to override* the programmed values. What is most important, is the modification of the program that reflects the optimized cutting conditions.

Once the machine operator finds what values must be changed in the program itself, the program must be edited to reflect these changes. Not only for the job currently worked on, but also for any repetition of the job in the future. After all, it should be the goal of every programmer and CNC operator to run any job at one hundred percent efficiency. This efficiency is most likely reached as a combined effort of the operator and the programmer. A good CNC programmer will always make the *effort* to reach 100% efficiency at the desk and then improve the program even further.

SYSTEM OPTIONS

Optional features on a CNC system are like options on a car. What is an option at one dealership, maybe a standard feature at another. Marketing strategies and corporate philosophies have a lot to do with this approach.

Here is a look at some control features that may or may not be classified as optional on a particular system. But some important disclaimer first:

> This handbook covers the subject matter relating to the majority of control features, regardless of whether they are sold as a standard or an optional feature of the system. It is up to the user to find out what exact options are installed on a particular control system.

◆ Graphic Display

Graphic representation of the tool path on the display screen is one of the most important, as well as sought after, control options. Do not confuse this option with any type of conversational programming, which also uses a graphic tool path interface. In the absence of a computer assisted programming (CAM), a graphic display on the control panel is a major benefit. Whether in monochrome or in color, the convenience of seeing the tool motions *before* actual machining is much appreciated by CNC operators and programmers alike.

A typical graphics option shows the machine axes and two cursors for zooming. When the tool path is tested, individual tools are distinguished by different colors, if available or different intensity. Rapid motions are represented by a dashed line type, cutting motions by a continuous line type. If the graphics function is applied during machining, the tool motions can be watched on the display screen - very helpful for those CNC machines that have dirty, oily and scratched safety shields.

Upwards or downwards scaling of the display allows for evaluation of a tool motion overall or for detail areas. Many controls also include actual tool path simulation, where the shape of the part and the cutting tool can be set first, then seen on the screen.

◆ In-Process Gauging

During many unattended machining operations, such as in manufacturing cells or Agile manufacturing, a periodic checking and adjusting dimensional tolerances of the part is imperative. As the cutting tool wears out, or perhaps because of other causes, the dimensions may fall into the 'out-of-tolerance' zone. Using a probe device and a suitable program, the *In-Process Gauging* option offers quite a satisfactory solution. The CNC part program for the *In-Process Gauging* option will contain some quite unique format features - it will be written parametrically, and will be using another option of the control system - the *Custom Macros* (sometimes called the *User Macros*), which offer the variable type programming.

If a company or a CNC machine shop is a user of the *In-Process Gauging* option, there are good chances that other control options are also installed and available to the CNC programmer. Some of the most typical options are probing software, tool life management, macros, etc. This technology goes a little too far beyond standard CNC programming, although it is closely related and frequently used. Companies that already use the numerical control technology, will be well advised to look into these options to remain competitive in their field.

◆ Stored Stroke Limits

Definition of an area on a CNC lathe or a cube on a CNC machining center that is safe to work within, can be stored as a control system parameter called *stored stroke limit*. These stored stroke limits are designed to prevent a collision between the cutting tool and a fixture, the machine tool or the part. The area (2D) or the cube (3D) can be defined as either *enabled* for the cutter entry or *disabled* for the cutter entry. It can be set manually on the machine or, if available, by a program input. Some controls allow only one area or cube to be defined, others allow more.

When this option is in effect and the CNC unit detects a motion in the program that takes place within the forbidden zone, an error condition results and the machining is interrupted. A typical applications may include zones occupied by a tailstock, a fixture, a chuck, a rotary table, and even an unusually shaped part.

◆ Drawing Dimensions Input

An option that seems somewhat neglected, is the programming method by using input of the dimensions from an engineering drawing. The ability to input known coordinates, radii, chamfers and given angles directly from the drawing makes it an attractive option. This ability is somewhat overshadowed by poor program portability. Such an option must be installed on all machines in the shop, in order to use the programmed features efficiently.

◆ Machining Cycles

Both the milling and the turning controls offer a variety of machining cycles. Typical machining cycles for milling operations are called *fixed cycles,* also known as the *canned cycles.* They simplify simple point-to-point machining operations such as drilling, reaming, boring, backboring and tapping. Some CNC systems also offer cycles for face milling, pocket milling, hole patterns, etc.

CNC lathes also have many machining cycles available to remove material by automatic roughing, profile finishing, facing, taper cutting, grooving and threading. Fanuc controls call these cycles *Multiple Repetitive Cycles.*

All these cycles are designed for easier programming and faster changes at the machine. They are *built in* the control and *cannot be changed.* Programmer supplies the cutting values during the program preparation by using the appropriate cycle call command. All the processing is done automatically, by the CNC system. Of course, there will always be special programming projects that cannot use any cycles and have to be programmed manually or with the use of an external computer.

◆ Cutting Tool Animation

Many of the graphic tool path displays defined earlier, are represented by simple lines and arcs. The current tool position is usually the location of the line or arc endpoint on the screen. Although this method of displaying the motion of the cutting tool graphically is certainly useful, there are two disadvantages to it. The shape of the cutting tool and the material being removed cannot be seen on the screen and a tool path simulation may help a bit. Many modern controls incorporate a graphic feature called *Cutting Tool Animation.* If available on the control, it shows the blank of the part, the mounting device and the tool shape. As the program is executed, the CNC operator has a very accurate visual aid in program proving. Each graphic element is identified by a different color, for even a better appearance. The blank size, the mounting device and the tool shape can be preset for exact proportions and a variety of tool shapes can be stored for repetitive use. This option is a good example of CAD/CAM-like features built into a stand-alone control system.

◆ Connection to External Devices

The CNC computer can be connected to an external device, usually another computer. Every CNC unit has one or more connectors, specifically designed for interfacing to peripheral devices. The most common device is called RS-232 (EIA standard), designed for communications between two computers. Setting up the connection with external devices is a specialized application. The CNC operator uses such a connection to transfer programs and other settings between two computers, usually for storage and backup purposes.

6 PROGRAM PLANNING

The development of any CNC program begins with a very carefully planned process. Such a process starts with the engineering *drawing* (technical print) of the required part released for production. Before the part is machined, several steps have to be considered and carefully evaluated. The more effort is put into the planning stage of the program, the better results may be expected at the end.

STEPS IN PROGRAM PLANNING

The steps required in program planning are decided by the nature of the work. There is no useful formula for all the jobs, but some basic steps should be considered:

- ❏ Initial information / Machine tools features
- ❏ Part complexity
- ❏ Manual programming / Computerized programming
- ❏ Typical programming procedure
- ❏ Part drawing / Engineering data
- ❏ Methods sheet / Material specifications
- ❏ Machining sequence
- ❏ Tooling selection
- ❏ Part setup
- ❏ Technological decisions
- ❏ Work sketch and calculations
- ❏ Quality considerations in CNC programming

The steps in the list are suggestions only - a guideline. They are quite flexible and should always be adapted for each job and to the specific conditions of the work.

INITIAL INFORMATION

Most drawings define only the shape and size of the completed part and normally do not specify data about the initial blank material. For programming, a good knowledge of the material is an essential start - mainly in terms of its size, type, shape, condition, hardness, etc. The drawing and material data are the primary information about the part. At this point, CNC program can be planned. The objective of such a plan is to use the initial information and establish the most efficient method of machining, with all related considerations - mainly part accuracy, productivity, safety and convenience.

The initial part information is not limited to the drawing and the material data - it also includes conditions *not covered* in the drawing, such as *pre-* and *post-* machining, grinding allowances, assembly features, requirements for hardening, next machine setup, and others. Collecting all this information provides enough material to start planning the CNC program.

MACHINE TOOLS FEATURES

No amount of initial information is useful if the CNC machine is not *suitable* for the job. During program planning, programmer concentrates on a *particular* machine tool, using a *particular* CNC system. Each part has to be setup in a fixture, the CNC machine has to be large enough to handle the size of the part, the part should not be heavier than the maximum weight allowed. The control system must be capable to provide the needed tool path, and so on.

In most cases, the CNC equipment is already available in the shop. Very few companies go and buy a new CNC machine just to suit a particular job. Such cases are rather rare and happen only if they make economic sense.

◆ Machine Type and Size

The most important considerations in program planning are the *type* and the *size* of the CNC machine, particularly its *work space* or *work area*. Other features, equally important, are the machine power rating, spindle speed and feedrate range, number of tool stations, tool changing system, available accessories, etc. Typically, small CNC machines have higher spindle speeds and lower power rating, large machines have lower spindle speeds available, but their power rating is higher.

◆ Control System

The control system is the heart of a CNC machine. Being familiar with all the standard and optional features available on all controls is a must. This knowledge allows the use of a variety of advanced programming methods, such as the machining cycles, subprograms, macros and other timesaving features of a modern CNC system.

A programmer does not have to physically run a CNC machine. Yet, the programs will become better and more creative with good understanding of the machine and its control system. Program development reflects programmer's knowledge of the CNC machine operation.

One of the main concerns in program planning should be the operator's *perception* of the program. To a large degree, such a perception is quite subjective, in the sense that different operators will express their personal preferences. On the other hand, every operator appreciates an error-free, concise, well documented and professionally prepared part program, consistently and one after another. A poorly designed program is disliked by any operator, regardless of personal preferences.

PART COMPLEXITY

At the time the drawing, material and the available CNC equipment are evaluated, the complexity of the programming task becomes much clearer. How difficult is to program the part manually? What are the capabilities of the machines? What are the costs? Many questions have to be answered before starting the program.

Simple programming jobs may be assigned to a less experienced programmer or the CNC operator. It makes sense from the management perspective and it is also a good way to gain experience.

Difficult or complex jobs will benefit from a computerized programming system. Technologies such as *Computer Aided Design* (CAD) and *Computer Aided Manufacturing* (CAM) have been a strong part of the manufacturing process for many years. The cost of a CAD/CAM system is only a fraction of what it used to be only a few years ago. Even small shops now find that the benefits offered by modern technology are too significant to be ignored. Several programming systems are available various computers and can handle virtually any job. For a typical machine shop, a Windows based programming software can be very beneficial. A typical example of this kind of application is the very popular and powerful *Mastercam*™, from CNC Software, Inc., Tolland, CT. There are several others.

MANUAL PROGRAMMING

Manual programming (without a computer) has been the most common method of preparing a part program for many years. The latest CNC controls make manual programming much easier than ever before by using fixed or repetitive machining cycles, variable type programming, graphic tool motion simulation, standard mathematical input and other time saving features. In manual programming, all calculations are done by hand, with the aid of a pocket calculator - no computer programming is used. Programmed data can be transferred to the CNC machine via a cable, using an inexpensive desktop or a laptop computer. This process is faster and more reliable than other methods. Short programs can also be entered manually, by keyboard entry, directly at the machine. A punched tape used to be the popular media of the past but has virtually disappeared from machine shops.

◆ Disadvantages

There are some disadvantages associated with manual programming. Perhaps the most common is the length of time required to actually develop a fully functioning CNC program. The manual calculations, verifications and other related activities in manual programming are very time consuming. Other disadvantages, also very high on the list, are a large percentage of errors, a lack of tool path verification, the difficulty in making changes to a program, and many others.

◆ Advantages

On the positive side, manual part programming does have quite a few unmatched qualities. Manual programming is so intense that it requires the total involvement of the CNC programmer and yet offers virtually unlimited freedom in the development of the program structure. Programming manually does have some disadvantages, but it teaches a tight discipline and organization in program development. It forces the programmer to understand programming techniques to the last detail. In fact, many useful skills learned in manual programming are directly applied to CAD/CAM programming. Programmer has to know *what* is happening at all times and *why* it is happening. Very important is the *in-depth understanding* of every detail during the program development.

Contrary to many beliefs, a thorough knowledge of manual programming methods is absolutely essential for efficient management of CAD/CAM programming.

CAD/CAM AND CNC

The need for improved efficiency and accuracy in CNC programming has been the major reason for development of a variety of methods that use a computer to prepare part programs. Computer assisted CNC programming has been around for many years. First, in the form of language based programming, such as APT™ or Compact II™. Since the late 1970's, CAD/CAM has played a significant role by adding the visual aspect to the programming process. The acronym *CAD/CAM* means *Computer Aided Design* and *Computer Aided Manufacturing*. The first three letters (CAD) cover the area of engineering design and drafting. The second three letters (CAM), cover the area of computerized manufacturing, where CNC programming is only a small part. The whole subject of CAD/CAM covers much more than just design, drafting and programming. It is a part of modern technology also known as *CIM - Computer Integrated Manufacturing*.

In the area of numerical control, computers have played a major role for a long time. Machine controls have become more sophisticated, incorporating the latest techniques of data processing, storage, tool path graphics, machining cycles, etc. Programs can now be prepared with the use of inexpensive computers, using graphical interface. Cost is no longer an issue, even small machine shops can afford a programming system in house. These systems are also popular because of their flexibility. A typical computerized programming system does not have to be dedicated only to programming - all related tasks, often done by the programmer, can be implemented on the same computer. For example, cutting tool inventory management, database of part programs, material information sheets, setup sheets and tooling sheets, etc. The same computer could also be used for uploading and downloading CNC programs.

◆ Integration

The keyword in the acronym CIM is - *integration*. It means putting all the elements of manufacturing together and work with them as a single unit and more efficiently. The main idea behind a successful integration is to avoid duplication. One of the most important rules of using a CAD/CAM computer software is:

> NEVER DO ANYTHING TWICE !

When a drawing is made in a CAD software (such as *AutoCAD*), then done again in a CAM software (such as *Mastercam*), there is a duplication. Duplication breeds errors. In order to avoid duplication, most of the CAD systems incorporate a transfer method of the design to the selected CAM system to be used for CNC programming. Typical transfers are achieved through special DXF or IGES files. The *DXF* stands for *Data Exchange Files* or *Drawing Exchange Files,* and the *IGES* abbreviation is a short form of *Initial Graphics Exchange Specification* files. Once the geometry is transferred from the CAD system to the CAM system, only the tool path related process is needed. Using a *post processor* (special kind of formatter), the computer software will prepare a part program, ready to be loaded directly to the CNC machine.

◆ Future of Manual Programming

It may seem that the manual programming is on the decline. In terms of actual use, this is probably true. However, it is necessary to keep in perspective that any computerized technology is based on the already well established methods of manual programming. Manual programming for CNC machines serves as the *source* of the new technology - it is the very elementary concept on which the computerized programming is based. This knowledge base opens the door for development of more powerful hardware and software applications.

The manual programming may be used somewhat less frequently today and eventually will be used even less - but knowing it well - *really understanding it* - is and always will be the key to control the power of CAM software. Even computers cannot do everything. There are some special programming projects that a CAM software, regardless of the price, may handle to an absolute satisfaction. If the control system can handle it, manual programming is the way to the ultimate control over such a project, when any other methods may not be suitable.

Even with a well customized and organized computerized programming system, how can the generated program output be exactly as intended? How can the CNC operator change any part of the program on the machine, without knowing its rules and structure?

> Successful use of computerized programming requires understanding of manual programming methods.

TYPICAL PROGRAMMING PROCEDURE

Planning of a CNC program is no different than any other planning - at home, at work, or elsewhere - it must be approached in a *logical and methodical* way. The first decisions relate to *what tasks* have to be done and *what goals* have to be reached. The other decisions relate to *how* to achieve the set goals in an efficient and safe manner. Such a progressive method not only isolates individual problems as they develop, it also forces their solution *before* the next step can be taken.

The following items form a fairly common and logical sequence of tasks done in CNC programming. The items are only in *a suggested order*, offered for further evaluation. This order may be changed to reflect special conditions or working habits. Some items may be missing or redundant:

1. Study of initial information (drawing and methods)
2. Material stock (blank) evaluation
3. Machine tool specifications
4. Control system features
5. Sequence of machining operations
6. Tooling selection and arrangement of cutting tools
7. Setup of the part
8. Technological data (speeds, feedrates, etc.)
9. Determination of the tool path
10. Working sketches and mathematical calculations
11. Program writing and preparation for transfer to CNC
12. Program testing and debugging
13. Program documentation

There is only one goal in CNC program planning and that is the completion of all instructions in the form of a program that will result in an error-free, safe and efficient CNC machining. The suggested procedures may require some changes - for example, should the tooling be selected *before* or *after* the part setup is determined? Can the manual part programming methods be used efficiently? Are the working sketches necessary? Do not be afraid to modify any so called *ideal* procedure - either temporarily, for a given job, or permanently, to reflect a particular CNC programming style. *Remember, there are no ideal procedures.*

PART DRAWING

The part drawing is the single most important document used in CNC programming. It visually identifies the shape, dimensions, tolerances, surface finish and many other requirements for the completed item. Drawings of complex parts often cover many sheets, with different views, details and sections. The programmer first evaluates *all* the drawing data first, then isolates those that are relevant for the development of a particular program. Unfortunately, many drafting methods do not reflect the actual CNC manufacturing process. They reflect the *designer's thinking*, rather than the method of manufacturing. Such drawings are generally correct in technical sense, but they are harder to study by the programmer and may need to be *'interpreted'* to be of any value in CNC programming. Typical examples are methods of applying dimensions, absence of a datum point that can be used as a program reference point and the view orientation in which the part is drawn. In the CAD/CAM environment, the traditional gap between the design, drafting and CNC programming must be eliminated. Just as it helps the programmer to understand designer's intentions, it helps the designer to understand the basics of CNC programming. Both, the designer and the programmer have to understand each other's methods and find common ground that makes the whole process of design and manufacturing coherent and efficient.

◆ Title Block

The title block - *Figure 6-1* - is typical to all professional drawings. Its purpose is to collect all the descriptive information related to the particular drawing.

No.	Date	Revision		By
Part Name:				
Scale:		Material:		
Dr.:		Date:		
Chk.:		Drawing number:		
App.:				

Figure 6-1

A title block example of an engineering drawing

The size and contents of a title block vary between companies, depending on the type of manufacturing and internal standards. It is usually a rectangular box, positioned in the corner of the drawing, divided into several small boxes. The contents of the title block include such items as the part name and part number, drawing number, material data, re-

visions, special instructions, etc. Data in the title block supply crucial information for CNC programming and can be used for program documentation to make easier cross referencing. Not all title block information is needed in programming, but may be used for program documentation.

Revision dates in a drawing are associated with the title block. They are important to the programmer, as they indicate *how current* is the drawing version. Only the latest version of the part design is important to manufacturing.

◆ Dimensioning

Dimensions on the part drawing are either in *English* or *metric* units. Individual dimensions can be referenced from a certain datum point or they can be consecutive, measured from the previous dimension. Often, both types of dimensions are mixed in the same drawing. When writing the program, it may be more convenient to translate all consecutive - or *incremental* - dimensions into datum - or *absolute* - dimensions. Most CNC programs benefit from drawings using datum, or absolute dimensioning. Similarly, when developing a subprogram for tool path translation, an incremental method of programming may be the right choice - and the choice depends on the application. The most common programming method for CNC machines uses the *absolute* dimensioning method *(Figure 6-2)*, mainly because of the editing ease within the CNC system.

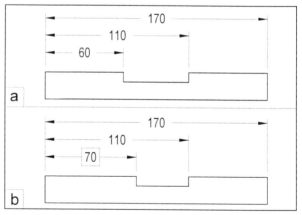

Figure 6-2

Program using ABSOLUTE dimensions
Only one change in the program is necessary

With the absolute system of dimensioning, many program changes can be done by a single modification. Incremental method requires at least two modifications. The differences between the two dimensioning systems can be compared in *Figure 6-2,* using the *absolute* dimensioning method, and in *Figure 6-3,* using the *incremental* dimensioning method. The word *incremental* is more common in CNC, in drafting the equivalent word would be *relative*. Both illustrations show the *a)* figure before revision, and the *b)* figure after revision.

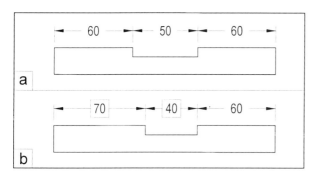

Figure 6-3

Program using INCREMENTAL dimensions
Two (or more) changes in the program are necessary

Fractions

Drawings in English units often contain fractions. A fractional dimension was sometimes used to identify a *less important* dimensional tolerances (such as ±.030 inches from the nominal size). The number of digits following the decimal point often indicated a tolerance (the more digits specified, the smaller the tolerance range). These methods are not an ISO standard and are of no use in programming. Fractional dimensions have to be changed into their decimal equivalents. The number of decimal places in the program is determined by the minimum increment of the control. A dimension of 3-3/4 is programmed as 3.75, and a dimension of 5-11/64 inches is programmed as 5.1719, its closest rounding. Many companies have upgraded their design standards to the ISO system and adhere to the principles of CNC dimensioning. In this respect, drawings using the metric units are much more practical.

Some dimensioning problems are related to an improper use of a CAD software, such as AutoCAD. Some designers do not change the default setting of the number of decimal places and every dimension ends up with four decimal places (inches) or three decimal places (metric). This is a poor practice and should be avoided. The best approach is to specify dimensional tolerances for all dimensions that require them, and even use *Geometric Dimensioning and Tolerancing* standards (GDT).

◆ Tolerances

For quality precision machining work, most part dimensions have a specified range of acceptable deviation from the nominal size, within its system of reference. For example, an English tolerance of +.001/-.000 inches will be different from a metric tolerance of +0.1/-0.0 mm. Dimensions of this type are usually critical dimensions and must be maintained during CNC machining. It may be true that the CNC operator is ultimately responsible for maintaining the part sizes within the tolerances (providing the program is correct) - but it is equally true, that the CNC programmer can make the machine operator's task easier. Consider the following example for a CNC lathe:

⊃ A drawing dimension specifies a hole as ∅75+0.00/-0.05 mm. What actual dimension should appear in the program?

There are some choices. The dimension on the high side may be programmed as X75.0 and X74.95 on the low side of the range. A middle value of X74.975 is also a choice. Each selection is mathematically correct. A creative CNC programmer looks not only for the mathematical points, but for the technical points as well. The cutting edge of a tool wears out with more parts machined. That means the machine operator has to *fine-tune* the machined size by using the *tool wear offsets*, available on most CNC systems. Such a manual interference during machining process is acceptable, but when done too often, it slows down the production and adds to the overall costs.

A particular programming approach can control the frequency of such manual adjustments to a great degree. Consider the ∅75 mm mentioned earlier. If it is an *external* diameter, the tool edge wear will cause the actual dimension during machining to become *larger*. In the case of an *internal* diameter, the actual dimension will become *smaller* as the cutting edge wears out. By programming X74.95 for the external diameter (the bottom limit) or X75.0 for the internal diameter (the top limit), the wear of the cutting edge will move *into* the tolerance range, rather than away from it. The manual tool offset adjustment by the machine operator may still be required, but less frequently. Another approach is to select the *middle size* of the tolerance range - this method will also have a positive effect but more manual adjustments may be necessary during machining.

◆ Surface Finish

Precision parts require a certain degree of surface finish quality. Technical drawing indicates the required finish for various features of the part. English drawings indicate the finish in *micro inches*, where *1 micro inch =. 000001"*. Metric drawings use specifications expressed in *microns,* where *1 micron = 0.001 mm*. Symbol for a micron is a Greek letter μ. Some drawings use symbols - *Figure 6-4*.

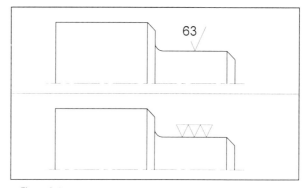

Figure 6-4

Surface finish marks in a drawing:
English (top) and metric (bottom)

The most important factors influencing the quality of surface finish are spindle speed, feedrate, cutting tool radius and the amount of material removed. Generally, a larger cutter radius and slower feedrates contribute towards finer surface finishes. The cycle time will be longer but can often be offset by elimination of any subsequent operations such as grinding, honing or lapping.

◆ Drawing Revisions

Another important section of the drawing, often overlooked by CNC programmers, shows the engineering changes (known as *revisions*) made on the drawing up to a certain date. Using reference numbers or letters, the designer identifies such changes, usually with both values - the previous and the new value - for example:

```
REV.3 / DIMENSION 5.75 WAS 5.65
```

Only the *latest* changes are important to the program development. Make sure the program not only reflects the current engineering design, but also is identified in some unique way to distinguish it from any previous program versions. Many programmers keep a copy of the part drawing corresponding to the program in the files, thus preventing a possible misunderstanding later.

◆ Special Instructions

Many drawings also include special instructions and comments that cannot be expressed with the traditional drafting symbols and are therefore spelled out independently, in words. Such instructions are very important for CNC program planning, as they may significantly influence the programming procedure. For example, an element of the part is identified as a *ground* surface or diameter. The drawing dimension always shows the *finished* size. In the program, this dimension must be adjusted for any grinding allowance necessary - an allowance selected by the programmer and written as a special instruction in the program. Another example of a special instruction required in the program relates to the machining performed during part assembly. For example, a certain hole on the drawing should be drilled and tapped and is dimensioned the same way as any other hole, but a special instruction indicates the drilling and tapping must be done when the part is handled during assembly. Operations relating to such a hole are *not* programmed and if any overlook of a small instruction such as this, may result in unusable part.

Many drawing instructions use a special pointer called a *leader*. Usually it is a line, with an arrow on the end, pointing towards the area that it relates to. For example, a leader may be pointing to a hole, with the caption:

```
φ12 - REAM 2 HOLES
```

This is a requirement to ream 2 holes with a reamer that has 12 mm diameter.

METHODS SHEET

Some companies have a staff of qualified manufacturing technologists or process planners responsible for determination of the manufacturing process. These people develop a series of machining instructions, detailing the route of each part through the manufacturing steps. They allocate the work to individual machines, develop machining sequences and setup methods, select tooling, etc. Their instructions are written in a *methods sheet* (*routing sheet*) that accompanies the part through all stages of manufacturing, typically in a plastic folder. If such a sheet is available, its copy should become a part of the documentation. One of the purposes of a methods sheet is to provide the CNC programmer with as much information as possible to shorten the turnover between programs. The greatest advantage of a methods sheet in programming is its comprehensive coverage of all required operations, both CNC and conventional, thus offering a complete overview of the manufacturing process. A good quality methods sheet will save a lot of decisions - it is made by a manufacturing engineer, who specializes in work detailing. The ideal methods sheet is one where the recommended manufacturing process closely matches established part programming methods.

For whatever reason, a large number of CNC machine shops does not use methods sheets, routing sheets or similar documentation. The CNC programmer acts as a process planner as well. Such an environment offers a certain degree of flexibility but demands a large degree of knowledge, skills and responsibility at the same time.

MATERIAL SPECIFICATIONS

Also important consideration in program planning is evaluation of the *material stock*. Typical material is raw and unmachined (a bar, billet, plate, forging, casting, etc). Some material may be already premachined, routed from another machine or operation. It may be solid or hollow, with a small or a large amount to be removed by CNC machining. The *size* and *shape* of the material determines the setup mounting method. The *type* of material (steel, cast iron, brass, etc.) will influence not only the selection of cutting tools, but the cutting conditions for machining as well.

> A program cannot be planned without knowing
> the type, size, shape and condition of the material.

◆ Material Uniformity

Another important consideration, often neglected by programmers and managers alike, is the *uniformity* of material specifications within a particular batch or from one batch to another. For example, a material ordered from two suppliers to be used for the *same part* may have slightly different sizes, hardness and even shape. A similar example is a material cut into single pieces on a saw, where the length of each piece varies beyond an acceptable range. This inconsistency between blank parts makes programming more difficult and time consuming. It also creates potentially unsafe machining conditions. If such problems are encountered, the best planning approach is to place emphasis on machining safety than on machining time. At worst, there will be some *air cutting* or slower than needed cutting feed, but no cuts will be too heavy for the tool to handle.

Another approach is to separate non-uniform material into groups and make separate programs for each group, properly identified. The best method is to cover all known and predictable inconsistencies under program control, for example, using the block skip function.

◆ Machinability Rating

Another important aspect of material specification is its *machinability*. Charts with suggested speeds and feeds for most common materials are available from major tooling companies. These charts are helpful in programming, particularly when an unknown material is used. The suggested values are a good starting point, and can be optimized later, when the material properties are better known.

Machinability rating in the English units is given in units called *feet per minute* (ft/min). Often the terms *surface feet per minute, constant surface speed* (CSS or CS), *peripheral speed* or just *surface speed* are used instead. For metric designation of the machinability rating, the meters per minute (m/min) are used. In both cases, the spindle speed (r/min) for a given tool diameter (for a mill) or a given part diameter (for a lathe) is calculated, using common formulas. For the English system, the spindle speed can be calculated in revolutions per minute (r/min):

$$ r / min = \frac{12 \times ft / min}{\pi \times D} $$

For a metric calculation, the formula is similar:

$$ r / min = \frac{1000 \times m / min}{\pi \times D} $$

☞ where ...

r/min	=	Revolutions per minute (spindle speed S)
12	=	Converts feet to inches
1000	=	Converts meters to millimeters
ft/min	=	Peripheral speed in feet per minute
m/min	=	Peripheral speed in meters per minute
π (pi)	=	Constant value of 3.141593...
D	=	Tool diameter (milling) or part diameter (turning) - in inches or mm

MACHINING SEQUENCE

Machining sequence defines the order of machining operations. Technical skill and machine shop experience does help in program planning, but some common sense approach is equally important. The sequence of machining must have a logical order - for example, drilling must be programmed before tapping, roughing operations before finishing, first operation before second, etc. Within this logical order, further specification of the order of individual tool motions is required for a particular tool. For example, in turning, a face cut may be programmed on the part first, then roughing all material on diameters will take place. Another method is to program a roughing pass for the first diameter, then face and continue with the remainder of the diameter roughing afterwards. In drilling, a center drill before drilling may be useful for some applications, but in another program a spot drill may be a better choice. There are no fixed rules on which method is better - each CNC programming assignment has to be considered individually, based on the criteria of safety and efficiency.

The basic approach for determining the machining sequence is the evaluation of all related operations. In general, program should be planned in such a way that the cutting tool, once selected, will do as much work as possible, before a tool change. On most CNC machines, less time is needed for positioning the tool than for a tool change. Another consideration is in benefits gained by programming all heavy operations first, then the lighter semifinishing or finishing operations. It may mean an extra tool change or two, but this method minimizes any shift of the material in the holding fixture while machining. Another important factor is the current position of a tool when a certain operation is completed. For example, when drilling a pattern of holes in the order of 1-2-3-4, the next tool (such as a boring bar, reamer or a tap) should be programmed in the order of 4-3-2-1 to minimize unnecessary tool motions - *Figure 6-5*.

T01 = Spot Drill	T02 = Drill	T03 = Tap
Hole 1	Hole 4	Hole 1
Hole 2	Hole 3	Hole 2
Hole 3	Hole 2	Hole 3
Hole 4	Hole 1	Hole 4

Figure 6-5

Typical machining sequence
(spot drill, drill and tap shown as an example)

This machining sequence may have to be changed after the final selection of tools and the setup method. The reverse sequence may not be practical in subprograms.

Program planning is not an independent execution of individual steps - it is a very interdependent and very logically coherent approach to achieve a certain goal.

TOOLING SELECTION

Selecting tool holders and cutting tools is another important step in planning a CNC program. The category of tooling covers a lot more than the cutting tools and tool holders - it includes an extensive line of accessories, including numerous vises, fixtures, chucks, indexing tables, clamps, collets and many other holding devices. Cutting tools require special attention, due to the large variety available and their direct effect in machining.

The cutting tool itself is usually the most important selection. It should be selected by two main criteria:

❑ Efficiency of usage

❑ Safety in operation

Many supervisors responsible for CNC programming try to make the existing tooling work at all times. Often they ignore the fact that a suitable new tool may do the job faster and more economically. A thorough knowledge of tooling and its applications is a separate technical profession - the programmer should know well all general principles of cutting tool applications. In many cases, a tooling representative may provide additional valuable assistance.

The arrangement of tools in the order of usage is also a subject of serious consideration in CNC program planning. On CNC lathes, each cutting tool is assigned to a certain turret station, making sure the *distribution of tools* is balanced between short and long tools (such as short turning tools versus long boring tools). This is important for the prevention of a possible interference during cutting or tool changing. Another concern should be the *order* in which each cutting tool is called, particularly for machines that do not have a bi-directional tool indexing. Most machining centers use a random type tool selection, where the order of tools is unimportant, only the diameter of the tool and its weight has to be considered.

All tool offset numbers and other program entries should be documented in a form known as the *tooling sheet*. Such a document serves as a guide to the operator during job setup. It should include at least the basic documentation relating to the selected tool. For example, the documentation may include the tool description, its length and diameter, the number of flutes, the tool and offset numbers, speed and feed selected for that tool and other relevant information.

PART SETUP

Another decision in program planning relates to the part setup - *how to mount* the raw or premachined material, what *supporting* tools and devices should be used, how many *operations* are required to complete as many machining sequences as possible, where to select a *program zero*, etc. Setup is necessary and it should be done efficiently.

Some machine types are designed to make the setup time more productive. Multispindle machining centers or lathes can handle two or more parts at the same time. Special features, such as barfeeder for a lathe, an automatic pallet changer or dual setup on the table, also help. Other solutions can be added as well.

◆ Setup Sheet

At this stage of program planning, once the setup is decided, making a *setup sheet* is a good idea. A setup sheet can be a simple sketch, designed mostly for the use at the machine, that shows the part orientation when mounted in a holding device, tool offset numbers used by the program, datum points and, of course, all the necessary identifications and descriptions. Other information in the setup sheet should relate to some unique requirements established during planning stages of the program (such as the position of clamps, bored jaws dimensions, limits of tool extension, etc.). Setup sheet and tooling sheet can be combined into a single source of information. Most programmers use their own various versions.

TECHNOLOGICAL DECISIONS

The next stage of CNC program planning involves the selection of spindle speeds, cutting feedrates, depth of cut, coolant application, etc. All of the already considered factors will have their influence. For example, the available range of spindle speeds is fixed for any CNC machine, the size of the cutter and the type of material will influence speeds and feeds, the power rating of the machine tool will help determine what amount of material can be removed safely, etc. Other factors that influence the program design include tool extensions, setup rigidity, cutting tool material and its condition. Not to be overlooked is the proper selection of cutting fluids and lubricants - they, too, are important for the part quality.

◆ Cutter Path

The core of any CNC programming is the determination of the cutter path - the *tool path*. This process involves individual cutter movements in its relationship to the part.

> In CNC programming, always look at the cutting
> tool as being moved around the work !!
> This principle applies to all CNC machine tools.

The key factor for understanding this principle is to visualize the *tool* motion, *not* the machine motion. The most noticeable difference between programming a machining center as compared to a lathe is the cutter rotation compared to the part rotation. In both cases, the programmer always must think in terms of the *cutter moving around the part - Figure 6-6*.

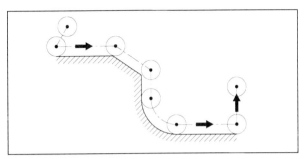

Figure 6-6
Contouring tool path motion - as intended (lathe or mill)

The tool path for all profiling tools has to take into consideration the cutter radius, either by programming the *equidistant path* for the center of the radius or by using cutter radius *offset*. CNC machines for milling and turning are provided with rapid motion, linear interpolation and circular interpolation, all as standard features. To generate more complex paths, such as a helical milling motion, a special option has to be available in the control unit. Two groups of typical tool paths exist:

- Point-to-point   *also called*   Positioning
- Continuous   *also called*   Contouring

Positioning is used for a point location operations, such as drilling, reaming, tapping and similar operations; *continuous path* generates a profile (contour). In either case, the programmed data refer to the position of the cutter when a certain motion is completed. This position is called the tool *target* position - *Figure 6-7.*

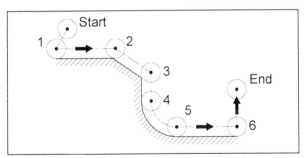

Figure 6-7
Contouring tool path motion with identified contour change points

The start and end positions of the profile are identified and so are the positions for each contour change. Each target position is called the *contour change point*, which has to be calculated. The order of target locations in the program is very important. That means the tool *position 1* is the *target* position commencing at the *Start* point, *position 2* is the *target* position beginning at point 1, *position 3* is the *target* from point 2 and so on, until the *End* position is reached. If the contour is for milling, the targets will be in X and Y axes. In turning, they will be in X and Z axes.

Most contouring operations require more than just one cutting motion, for example, roughing and finishing. Part of the programming process is to isolate the area that needs roughing. Can one cutting tool do both operations? Can all tolerances be maintained? Is the tool wear a problem? Can the surface finish be achieved? When programming non-cutting rapid motions, take the same care as with cutting motions. A particular focus should be to minimize rapid tool motions and ensure safe clearances.

◆ Machine Power Rating

Machine tools are rated by their power. Heavy cuts require more power than light cuts. A depth or width of a cut that is too large can break the tool and stall the machine. Such cases are unacceptable and must be prevented. The CNC machine specifications list the power rating of the motor at the machine spindle. The rating is in *kW* (kilowatts) or *HP* (horsepower). Formulas are available for power ratings, calculating metal removal rate, tool wear factors, etc. Useful is the comparison of *kW* and *HP* (based on 1 HP = 550 foot-pounds per second):

1 kW = 1.341 HP
1 HP = 0.746 kW

The topic of power and forces in machining can be complex and is not always needed in everyday programming. Work experience is often a better teacher than formulas.

◆ Coolants and Lubricants

When the tool contacts the material for an extended period of time, a great amount of heat is generated. The cutting edge gets overheated, becomes dull and may break. To prevent these possibilities, a suitable coolant must be used.

Water soluble oil is the most common coolant. A properly mixed coolant dissipates heat from the cutting edge and it also acts as a lubricant. The main purpose of lubrication is to reduce friction and make the metal removal easier. The flood of the coolant should aim at the tool cutting edge, with a flexible pipe or through a coolant hole in the tool.

Never use plain water as a coolant - it may severely damage the machine tool

The CNC operator is responsible for a suitable coolant in the machine. The coolant should be clean and mixed in recommended proportions. Water soluble oils should be biodegradable to preserve the environment and properly disposed of. The CNC programmer decides when to program the coolant and when not. Ceramic cutting tools are normally programmed dry, *without a coolant*. Some cast irons do not require flood coolant, but air blast or oil mist may be allowed. These coolant functions vary between machines, so check the machine reference manual for details.

Flood coolant may be used to cool down the part and gain better tolerances. It can also be used to flush away chips from congested areas, such as deep holes and cavities.

The benefits of cutting fluids far outweigh their inconveniences. Cutting fluids are often messy, the cutting edge cannot be seen, operator may get wet and sometimes old coolant smells. With proper management, all problems related to coolants can be controlled.

A coolant related programming issue is *when* to turn the coolant on in the program. As the coolant function M08 only turns on the pump motor, make sure the coolant actually reaches the tool edge *before* contact with work. Programming the coolant on early is better than late.

WORK SKETCH AND CALCULATIONS

Manually prepared programs require some mathematical calculations. This part of program preparation intimidates many programmers but is a necessary step. Many complex contours will require more calculations, but not more complex calculations. Almost any math problem in CNC programming can be solved by the use of arithmetic, algebra and trigonometry. Advanced fields of mathematics - analytic geometry, spherical trigonometry, calculus, surface calculations, etc. - are required for programming complex molds, dies and similar shapes. In such cases, a CAD/CAM programming system is necessary.

Those who can solve a right angle triangle can make calculations for almost any CNC program. At the end of the handbook is an overview of some common math problems. When working with more difficult contours, it is often not the solution itself that is difficult, it is the ability to arrive at the solution. The programmer must have the ability to *see* exactly what triangle has to be solved. It is not unusual to do several intermediate calculations before the required coordinate point can be established.

Calculations of any type often benefit from a pictorial representation. Such calculations usually need a working sketch. The sketch can be done by hand and should be done in an *approximate scale*. Larger sketch scales are easier to work with. Scaling the sketch has one great advantage - *you can immediately see the relationships* - what dimensions should be smaller or larger than the others, the relationship of individual elements, the shape of an extremely small detail, etc. However, there is one purpose you should never use the sketch for:

> Never use a scaled sketch to guess unknown dimensions !

Scaling a sketch is a poor and unprofessional practice, that creates more problems than it solves. It is a sign of laziness or incompetence.

◆ Identification Methods

A calculation sketch can be done directly in the drawing or on paper. Every sketch is associated with mathematical calculations. Using color coding or point numbering as identification methods offers benefits and better organization. Rather than writing coordinates at each contour change point in the drawing, use point reference numbers and create a separate coordinate sheet form using the reference numbers, as illustrated in *Figure 6-8*.

Position	X axis	Y axis	Z axis

Figure 6-8
Coordinate sheet example - blank form (no data)

Such a sheet can be used for milling or turning, by filling only the applicable columns. The aim is to develop a consistent programming style from one program to another. Fill-in all values, even those that do not change. A completed coordinate sheet is a better reference - *Figure 6-9*.

Position	X axis	Y axis	Z axis
START	X4.275	Y3.22	
P1	X4.155	Y3.01	
P2	X4.7878	Y3.01	
P3	X5.1668	Y2.755	
P4	X5.1668	Y2.456	
P5	X5.44	Y2.1833	
P6	X6.09	Y2.1833	
END	X6.09	Y2.575	

Figure 6-9
Coordinate sheet example - filled form for milling tool path

QUALITY IN CNC PROGRAMMING

An important consideration in program planning is a personal approach and attitudes. Our attitudes have a significant influence on the program development. Ask yourself some questions. Are you attentive to detail, well organized, concerned? Can a program be improved, is it safe, is it efficient? CNC program quality is more than writing an error free program. Part complexity is only related to *your* knowledge level and willingness to solve problems. It should be a personal goal to make a program that is the best program possible. *Set your standards high!*

7 · PART PROGRAM STRUCTURE

A CNC program is composed of a series of sequential instructions related to the machining of a part. Each instruction is specified in a format the CNC system can accept, interpret and process. Each instruction must also conform to the machine tool specifications. This input method of a program can be defined as *an arrangement of the machining and related instructions,* written in the format of the CNC system and aimed at a particular machine tool.

Various controls have a different format, but most are similar. Subtle differences exist among CNC machines from different manufacturers, even those equipped with the same control system. This is common, considering the specific demands individual machine builders place upon the control manufacturer to accommodate many original and unique machine design features. Such variations are usually minor but still important for programming.

BASIC PROGRAMMING TERMS

The field of CNC has its own terminology and special terms and its jargon. It has its own abbreviations and expressions that only the people in the field understand. CNC programming is only a small section of the computerized machining and it has a number of its own expressions. The majority of them relate to the *structure* of the program.

There are *four basic terms* used in CNC programming. They appear in professional articles, books, papers, lectures and so on. These words are the key to understanding the general CNC terminology:

Character → Word → Block → Program

Each term is very common and important in CNC programming and deserves its own detailed explanation.

◆ Character

A character is the smallest unit of CNC program. It can have one of three forms:

❑ Digit

❑ Letter

❑ Symbol

Characters are combined into meaningful words. This combination of digits, letters and symbols is called the *alpha-numerical* program input.

Digits

There are ten digits, 0 to 9, available for use in a program to create numbers. The digits are used in two modes - one for *integer* values (numbers without a decimal point), the other for *real numbers* (numbers with a decimal point). Numbers can have *positive* or *negative* values. On some controls, real numbers can be used with or without the decimal point. Numbers applied in either mode can only be entered within the range that is allowed by the control system.

Letters

The 26 letters of the English alphabet are all available for programming, at least in theory. Most control systems will accept only certain letters and reject others. For example, a CNC lathe control will reject the letter Y, as the Y axis is unique to milling operations (milling machines and machining centers). Capital letters are normal designation in CNC programming, but some controls accept low case letters with the same meaning as their upper case equivalent.

> If in doubt, use CAPITAL letters only!

Symbols

Several symbols are used for programming, in addition to the digits and letters. The most common symbols are the *decimal point, minus sign, percent sign, parenthesis* and others, depending on the control options.

◆ Word

A program *word* is a combination of *alpha-numerical* characters, creating a single instruction to the control system. Normally, each word begins with a capital letter that is followed by a number representing a program code or the actual value. Typical words indicate the *axes position, feedrate, speed, preparatory commands, miscellaneous functions* and many other definitions.

◆ Block

Just like the word is used as a *single* instruction to the CNC system, the block is used as a *multiple* instruction. A program entered into the control system consists of individual lines of instructions, sequenced in a logical order. Each line - called a *sequence block* or simply a *block* - is composed of one or several *words* and each word is composed of *two or more characters.*

In the control system, each block must be separated from all others. To separate blocks in the MDI *(Manual Data Input)* mode at the control, each block has to end with a special *End-Of-Block* code (symbol). This code is marked as EOB on the control panel. When preparing the program on a computer, the *Enter* key on the keyboard will terminate the block with the same result (similar to the old *Carriage Return* on typewriters). When writing a program on paper first, each program block should occupy only a single line on the paper. Each program block contains a series of single instructions that are executed together.

◆ **Program**

The part program structure varies for different controls, but the logical approach does not change from one control to another. A CNC program usually begins with a *program number* or similar identification, followed by the blocks of instructions in a logical order. The program ends with a *stop code* or a program termination symbol, such as the *percent sign - %*. Internal documentation and messages to the operator may be placed in strategic places within the program. The programming format has evolved significantly during the years and several formats have emerged.

PROGRAMMING FORMATS

Since the early days of numerical control, three formats had become significant in their time. They are listed in the order of their original introduction:

❏ Tab Sequential Format   NC only - no decimal point

❏ Fixed Format   NC only - no decimal point

❏ Word Address Format   NC or CNC - decimal point

Only the *very early* control systems use the *tab sequential* or *fixed* formats. Both of them disappeared in the early 1970's and are now obsolete. They have been replaced by the much more convenient *Word Address Format*.

WORD ADDRESS FORMAT

The *word address format* is based on a combination of one letter and one or more digits - *Figure 7-1*.

In some applications, such a combination can be supplemented by a symbol, such as a minus sign or a decimal point. Each letter, digit or symbol represents one character in the program and in the control memory. This unique alpha-numerical arrangement creates a *word*, where the letter is the *address*, followed by numerical data with or without symbols. The word *address* refers to a specific register of the control memory. Some typical words are:

```
G01  M30  D25  X5.75  N105  H01  Y0  S2500
Z-5.14  F12.0  T0505  T05  /M01  B180.0
```

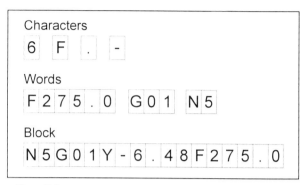

Figure 7-1

Typical word address programming format

The *address* - the letter - in the block defines the *meaning* of the word and must always be written first. For example, X5.75 is correct, 5.75X is not. No spaces (space characters) are allowed *within* a word but they are only allowed *before* the word, meaning before the letter.

Data indicates the word numerical assignment. This value varies greatly and depends on the preceding address. It may represent a sequence number N, a preparatory command G, a miscellaneous function M, an offset register number D or H, a coordinate word X, Y or Z, the feedrate function F, the spindle function S, the tool function T, etc.

Any one *word* is a series of characters (at least two) that define a single instruction to the control unit and the machine. The above examples of typical words have the following meaning in a CNC program:

G01	*Preparatory command*
M30	*Miscellaneous function*
D25	*Offset number selection - mills*
X5.75	*Coordinate word - positive value*
N105	*Sequence number (block number)*
H01	*Tool length offset number*
Y0	*Coordinate word - zero value*
S2500	*Spindle speed function*
Z-5.14	*Coordinate word - negative value*
F12.0	*Feedrate function*
T0505	*Tool function - lathes*
T05	*Tool function - mills*
/M01	*Miscellaneous function w/block skip symbol*
B180.0	*Indexing table function*

Individual words are instructions grouped together to form sequences of programming code. Each sequence that will process a series of instructions simultaneously, forms a unit called a *sequence block* or simply a *block*. The series of blocks arranged in a logical order that is required to machine a *complete* part or a *complete* operation is the *part program* also known as a *CNC program*.

The next block shows a rapid tool motion to the absolute position of X13.0Y4.6, with a coolant turned on:

```
N25 G90 G00 X13.0 Y4.6 M08
```

☞ where . . .

N25	Sequence or block number
G90	Absolute mode
G00	Rapid motion mode
X13.0 Y4.6	Coordinate location
M08	Coolant ON function

The control will process any one block as a complete unit - *never partially*. Most controls allow a random word order in a block, as long as the block number is specified first.

FORMAT NOTATION

Each word can only be written in a specific way. The number of digits allowed in a word, depending on the address and maximum number of decimal places, is set by the control manufacturer. Not all letters can be used. Only letters with an assigned meaning can be programmed, except in a comment. Symbols can be used in only some words, and their position in the word is fixed. Some symbols are used only in custom macros. Control limitations are important. Symbols supplement the *digits* and *letters* and provide them with an additional meaning. Typical programming symbols are the minus sign, decimal point, percent sign and a few others. All symbols are listed in a table below.

◆ Short Forms

Control manufacturers often specify the input format in an abbreviated form - *Figure 7-2*.

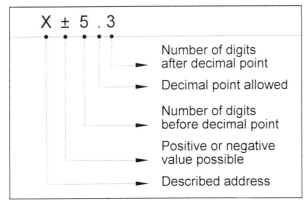

Figure 7-2

Word address format notation - X axis format in metric mode shown

The full format description for each meaning would be unnecessarily too long. Consider the following complete and not abbreviated description of the address X - as a coordinate word that is used in the metric system:

Address X accepts positive or negative data with the maximum of five digits in front of a decimal point and three digits maximum behind the decimal point - decimal point is allowed.

The absence of a decimal point in the notation means the decimal point is not used; the absence of a plus (+) sign in the notation means that the address value cannot be negative - *a lack of sign means a positive value by implication*. These samples of format notation explain the shorthand:

G2	Two digits maximum, no decimal point or sign
N5	Five digits maximum, no decimal point or sign
F5	Five digits maximum, no decimal point or sign
F3.2	Five digits maximum, three digits maximum in front of the decimal point, two digits maximum behind the decimal point, decimal point is allowed, no sign is used

Be careful when evaluating the shorthand notations from a manual. There are no industry standards and not all control manufacturers use the same methods, so the meaning of the short forms may vary significantly. The list of addresses, their format notation and description is listed in the following tables. They contain address notations based on a typical Fanuc control system.

◆ Milling System Format

The address description varies for many addresses, depending on the input units. The table below lists English format descriptions (metric format is in parenthesis, if applicable). Listed are format notations for milling units. The first column is the address, the second column is the format notation and the third column is a description:

Address	Notation	Description
A	A+5.3	Rotary or Indexing axis - unit is degrees - used about the X axis
B	B+5.3	Rotary or Indexing axis - unit is degrees - used about the Y axis
D	D2	Cutter radius offset number (sometimes uses address H)
F	F5.3	Feedrate function - may vary
G	G2	Preparatory commands
H	H3	Offset number (tool position and/or tool length offset)
I	I+4.4 (I+5.3)	Arc center modifier for X axis Shift amount in fixed cycles (X) Corner vector selection for X axis (old type of controls)
J	J+4.4 (J+5.3)	Arc center modifier for Y axis Shift amount in fixed cycles (Y) Corner vector selection for Y axis (old type of controls)

Address	Notation	Description
K	K+4.4 (K+5.3)	Arc center modifier for Z axis
L	L4	Fixed cycle repetition count Subprogram repetition count
M	M2	Miscellaneous function
N	N5	Block number or sequence number
0	04	Program number (EIA) or (:4 for ISO)
P	P4	Subprogram number call Custom macro number call
P	P3	Work offset number - used with G10
P	P53	Dwell time in milliseconds
P	P5	Block number in main program when used with M99
Q	Q4.4 (Q5.3)	Depth of peck in fixed cycles G73 and G83
Q	Q+4.4 (Q+5.3)	Shift amount in fixed cycle G76 and G87
R	R+4.4 (R+5.3)	Retract point in fixed cycles Arc radius designation
S	S5	Spindle speed in r/min
T	T4	Tool function
X	X+4.4 (X+5.3)	X axis coordinate value designation
X	X5.3	Dwell function with G04
Y	Y+4.4 (Y+5.3)	Y axis coordinate value designation
Z	Z+4.4 (Z+5.3)	Z axis coordinate value designation

◆ Turning System Format

Similar chart as for milling, this one is for lathe systems. A number of definitions are the same and are included only for convenience. Notation is in the English format, the metric notation is in parenthesis, if applicable to the address.

Address	Notation	Description
A	A3	Angle of thread for G76 Angle for direct drawing input
C	C+4.4 (C+5.3)	Chamfer for direct drawing input

Address	Notation	Description
D	D4	Number of divisions in G73
D	D44 (D53)	Depth of cut in G71 and G72 Relief amount in G74 and G75 Depth of first thread in G76
E	E2.6	Precision feedrate for threading
F	F2.6	Feedrate function - may vary
G	G2	Preparatory commands
I	I+4.4 (I+5.3)	Arc center modifier for X axis Taper height in X for cycles X axis relief in G73 Direction of chamfering Motion amount in X in G74
K	K+4.4 (K+5.3)	Arc center modifier for Z axis Taper height in Z for cycles Z axis relief in G73 Direction of chamfering Motion amount in Z in G75 Thread depth in G76
L	L4	Subprogram repetition count
M	M2	Miscellaneous function
N	N5	Block number or sequence number
0	04	Program number (EIA) or (:4 for ISO)
P	P4	Subprogram number call Custom macro number call Offset number with G10
P	P53	Dwell time in milliseconds
Q	Q5	End block number in G71 and G72
R	R+4.4 (R+5.3)	Arc radius designation Arc radius for corners
S	S5	Spindle speed in r/min or ft/min
T	T4	Tool function
U	U+4.4 (U+5.3)	Incremental value in X axis Stock allowance in X axis
U	U5.3	Dwell function with G04
W	W+4.4 (W+5.3)	Incremental value in Z axis Stock allowance in Z axis
X	X+4.4 (X+5.3)	Absolute value in X axis
X	X5.3	Dwell function with G04
Z	Z+4.4 (Z+5.3)	Absolute value in Z axis
Z	P5	Block number in a subprogram Start block number in G71 and G72

◆ Multiple Word Addresses

One feature that is noticeable in both tables is the abundance of different meanings for some addresses. This is a necessary feature of a word address format. After all, there are only 26 letters in the English alphabet, but more than that number of commands and functions. As new control features are added, even more variations may be necessary. Some of the addresses have such an established meaning (for example, X, Y and Z are coordinate words), that giving them an additional meaning would be confusing. Many letters, on the other hand, are not used very often and a multiple meaning for them is quite acceptable (addresses I, J, K, P, for example). In addition, the meaning of addresses varies between the milling and turning systems.

The control system has to have some means of accepting a particular word with a precisely defined meaning in the program. In most cases, the preparatory command G will define the meaning, at other times it will be the M function or a setting of system parameters.

SYMBOLS IN PROGRAMMING

In addition to the basic symbols, Fanuc can accept other symbols for different applications. The following table describes all symbols available on the Fanuc controls:

Symbol	Description	Comment
.	Decimal point	Fractional part of a number
+	Plus sign	Positive value or *addition* sign in Fanuc macros
−	Minus sign	Negative value or *subtraction* sign in Fanuc macros
*	Multiplication sign	*Multiplication* sign in Fanuc macros
/	Slash (front slash)	Block skip function symbol or *division* sign in Fanuc macros
( )	Parenthesis	Program comments & messages
%	Percent sign	Stop code (end of program file)
:	Colon	Program number designation
,	Comma	Used only within comments
[ ]	Brackets	Arguments in Fanuc macros
;	Semicolon	Non programmable End-Of-Block symbol (screen display only)
#	Sharp sign	Variable definition or call in Fanuc macros
=	Equal sign	Equality in Fanuc macros

The table lists both standard and special symbols. Special symbols are used only with optional features, such as the custom macro option. These symbols cannot be used in standard programming, as they would cause an error. Typical standard symbols are found on the computer keyboard. *Ctrl, Shift* and *Alt* character combinations are not allowed.

◆ Plus and Minus Sign

One of the most common symbols in CNC programming is an algebraic sign - plus or minus. Any data in a motion command can be either positive or negative. For convenience, virtually all control systems allow for an omission of a plus sign for all positive values. This feature is sometimes called positive bias of the control system. Positive bias is a term indicating an assumed positive value, if no sign is programmed in a word:

> **X+125.0** *is the same as* **X125.0**

The minus sign must always be programmed. If the minus sign is missing, the number becomes positive, with an incorrect results value (in this case the tool position):

> **X-125.0** *Negative value*
> **X125.0** *Positive value*
> **X+125.0** *Positive value (+ sign is ignored)*

Symbols supplement the letters and digits and are an integral part of the program structure.

PROGRAM HEADER

Comments or messages may be placed in the program, providing they are enclosed in parentheses. This kind of internal documentation is useful to both the programmer and operator. A series of comments at the program top is defined as the *program header*, where various program features are identified. The next example is an exaggerated sample of items that may be used in the program header:

```
(--------------------------------------------)
(FILE NAME ........................ O1234.NC)
(LAST VERSION DATE ................ 07-DEC-01)
(LAST VERSION TIME ................... 19:43)
(PROGRAMMER .................... PETER SMID)
(MACHINE ....................... OKK - VMC)
(CONTROL ...................... FANUC 15M)
(UNITS ............................ INCHES)
(JOB NUMBER ......................... 4321)
(OPERATION ................. DRILL-BORE-TAP)
(STOCK MATERIAL ............... H.R.S. PLATE)
(MATERIAL SIZE ................... 8 X 6 X 2)
(PROGRAM ZERO ................ X0 - LEFT EDGE)
(                            Y0 - BOTT EDGE)
(                            Z0 - TOP FACE )
(STATUS ...................... NOT VERIFIED)
(--------------------------------------------)
```

Within the program, each tool may be identified as well:

(*** T03 - 1/4-20 PLUG TAP ***)

Other comments and messages to the operator can be added to the program as required.

TYPICAL PROGRAM STRUCTURE

Although it may be a bit early to show a complete program, it will do no harm to look at a typical program structure. Developing a solid program structure is absolutely essential - it is going to be used all the time. Each block of the program is identified with a comment.

Note - Program blocks use only sample block numbers. Blocks in parentheses are *not* required for fixed cycles. The XY value in the block N88 should be the *current position*

of the X and Y axes. If the absolute position is unknown, change the block to the incremental version:

N88 G91 G28 X0 Y0

If a tool has to be repeated, make sure not to include the tool change block for the current tool. Many CNC systems will generate an alarm if the tool change command cannot find the tool in the magazine. In the following program example, the tool repeat blocks will be N5, N38 and N67.

The program structure example is for a machine with random tool selection mode and a typical control system, with some minor changes to be expected. Study the flow of the program, rather than its exact contents. Note the repetitiveness of blocks for each tool and also note the addition of a blank line (empty block) between individual tools for easier orientation in the program.

```
O0701 (ID MAX 15 CHARS)                        (PROGRAM NUMBER AND ID)
(SAMPLE PROGRAM STRUCTURE)                      (BRIEF PROGRAM DESCRIPTION)
(PETER SMID - 07-DEC-01)                        (PROGRAMMER AND DATE OF LAST REVISION)
                                                (BLANK LINE)
N1 G20                                          (UNITS SETTING IN A SEPARATE BLOCK)
N2 G17 G40 G80 G49                              (INITIAL SETTINGS AND CANCELLATIONS)
N3 T01                                          (TOOL T01 INTO WAITING POSITION)
N4 M06                                          (T01 INTO SPINDLE)
N5 G90 G54 G00 X.. Y.. S.. M03 T02              (T01 RESTART BLOCK - T02 INTO WAITING POSITION)
N6 G43 Z2.0 H01 M08                             (TOOL LG OFFSET - CLEAR ABOVE WORK - COOLANT ON)
(N7 G01 Z-.. F..)                               (FEED TO Z DEPTH IF NOT A CYCLE)
(---- CUTTING MOTIONS WITH TOOL T01 ----)
...
N33 G00 G80 Z2.0 M09                            (CLEAR ABOVE PART - COOLANT OFF)
N34 G28 Z2.0 M05                                (HOME IN Z ONLY-SPINDLE OFF)
N35 M01                                         (OPTIONAL STOP)
                                                (-- BLANK LINE --)
N36 T02                                         (TOOL T02 INTO WAITING POSITION - CHECK ONLY)
N37 M06                                         (T02 INTO SPINDLE)
N38 G90 G54 G00 X.. Y.. S.. M03 T03             (T02 RESTART BLOCK - T03 INTO WAITING POSITION)
N39 G43 Z2.0 H02 M08                            (TOOL LG OFFSET - CLEAR ABOVE WORK - COOLANT ON)
(N40 G01 Z-.. F..)                              (FEED TO Z DEPTH IF NOT A CYCLE)
(---- CUTTING MOTIONS WITH TOOL T02 ----)
...
N62 G00 G80 Z2.0 M09                            (CLEAR ABOVE PART - COOLANT OFF)
N63 G28 Z2.0 M05                                (HOME IN Z ONLY - SPINDLE OFF)
N64 M01                                         (OPTIONAL STOP)
                                                (-- BLANK LINE --)
N65 T03                                         (TOOL T03 INTO WAITING POSITION - CHECK ONLY)
N66 M06                                         (T03 INTO SPINDLE)
N67 G90 G54 G00 X.. Y.. S.. M03 T01             (T03 RESTART BLOCK - T01 INTO WAITING POSITION)
N68 G43 Z2.0 H03 M08                            (TOOL LG OFFSET - CLEAR ABOVE WORK - COOLANT ON)
(N69 G01 Z-.. F..)                              (FEED TO Z DEPTH IF NOT A CYCLE)
(---- CUTTING MOTIONS WITH TOOL T03 ----)
...
N86 G00 G80 Z2.0 M09                            (CLEAR ABOVE PART - COOLANT OFF)
N87 G28 Z2.0 M05                                (HOME IN Z ONLY - SPINDLE OFF)
N88 G28 X.. Y..                                 (HOME IN XY ONLY)
N89 M30                                         (END OF PROGRAM)
%                                               (STOP CODE - END OF FILE TRANSFER)
```

PREPARATORY COMMANDS

The program address G identifies a *preparatory command,* often called the *G code.* This address has one and only objective - that is *to preset* or *to prepare* the control system to a certain desired *condition,* or to a certain *mode* or a *state* of operation. For example, the address G00 presets a rapid motion mode for the machine tool, the address G81 presets the drilling cycle, etc. The term *preparatory command* indicates its meaning - a G code will *prepare* the control to accept the programming instructions *following* the G code in a specific way.

DESCRIPTION AND PURPOSE

A one block example will illustrate the purpose of the preparatory commands in the following program entry:

```
N7 X13.0 Y10.0
```

Even a casual look at this block shows that the coordinates X13.0Y10.0 relate to the *end position* of the cutting tool, when the block N7 is executed (i.e., processed by the control). The block does not indicate whether the coordinates are in the absolute or the incremental mode. It does not indicate whether the values are in the English or the metric units. Neither it indicates whether the motion to this specified target position is a rapid motion or a linear motion. If a look at the block cannot establish the meaning of the block contents, neither can the control system. The supplied information in such a block is *incomplete,* therefore unusable by itself. Some additional instructions for the block are required.

For example, in order to make the block N7 a tool destination in a rapid mode using absolute dimensions, *all* these instructions - or commands - must be specified *before* the block or *within* the block:

➲ **Example A :**

```
N7 G90 G00 X13.0 Y10.0
```

➲ **Example B :**

```
N3  G90
N4  ...
N5  ...
N6  ...
N7  G00 X13.0 Y10.0
```

➲ **Example C :**

```
N3  G90 G00
N4  ...
N5  ...
N6  ...
N7  X13.0 Y10.0
```

➲ **Example D :**

```
N2  G90
N3  G00
N4  ...
N5  ...
N6  ...
N7  X13.0 Y10.0
```

All four examples have the same machining result, providing that there is no change of any G code mode between blocks N4 and N6 in the examples B, C and D.

> One G code in a modal group replaces
> another G code from the same group

Modal and non-modal G-codes will be described shortly. Each control system has its own list of available G codes. Many G codes are very common and can be found on virtually all controls, others are unique to the particular control system, even the machine tool. Because of the nature of machining applications, the list of typical G codes will be different for the milling systems and the turning systems. The same applies for other types of machines. Each group of G codes must be kept separate.

> Check machine documentation for available G codes !

APPLICATIONS FOR MILLING

The G code table on the next page is a considerably detailed list of the most common preparatory commands used for programming CNC milling machines and CNC machining centers. The listed G codes may not be applicable to a particular machine and control system, so consult the machine and control reference manual to make sure. Some G codes listed are a special option that must be available on the machine and in the control system.

G code	Description	G code	Description
G00	Rapid positioning	G52	Local coordinate system setting
G01	Linear interpolation	G53	Machine coordinate system
G02	Circular interpolation clockwise	G54	Work coordinate offset 1
G03	Circular interpolation counterclockwise	G55	Work coordinate offset 2
G04	Dwell (as a separate block)	G56	Work coordinate offset 3
G09	Exact stop check - one block only	G57	Work coordinate offset 4
G10	Programmable data input (Data Setting)	G58	Work coordinate offset 5
G11	Data Setting mode cancel	G59	Work coordinate offset 6
G15	Polar Coordinate Command cancel	G60	Single direction positioning
G16	Polar Coordinate Command	G61	Exact stop mode
G17	XY plane designation	G62	Automatic corner override mode
G18	ZX plane designation	G63	Tapping mode
G19	YZ plane designation	G64	Cutting mode
G20	English units of input	G65	Custom macro call
G21	Metric units of input	G66	Custom macro modal call
G22	Stored stroke check ON	G67	Custom macro modal call cancel
G23	Stored stroke check OFF	G68	Coordinate system rotation
G25	Spindle speed fluctuation detection ON	G69	Coordinate system rotation cancel
G26	Spindle speed fluctuation detection OFF	G73	High speed peck drilling cycle (deep hole)
G27	Machine zero position check	G74	Left hand threading cycle
G28	Machine zero return (reference point 1)	G76	Fine boring cycle
G29	Return from machine zero	G80	Fixed cycle cancel
G30	Machine zero return (reference point 2)	G81	Drilling cycle
G31	Skip function	G82	Spot-drilling cycle
G40	Cutter radius compensation cancel	G83	Peck-drilling cycle (deep hole drilling cycle)
G41	Cutter radius compensation - left	G84	Right hand threading cycle
G42	Cutter radius compensation - right	G85	Boring cycle
G43	Tool length compensation - positive	G86	Boring cycle
G44	Tool length compensation - negative	G87	Back boring cycle
G45	Position compensation - single increase	G88	Boring cycle
G46	Position compensation - single decrease	G89	Boring cycle
G47	Position compensation - double increase	G90	Absolute dimensioning mode
G48	Position compensation - double decrease	G91	Incremental dimensioning mode
G49	Tool length offset cancel	G92	Tool position register
G50	Scaling function cancel	G98	Return to initial level in a fixed cycle
G51	Scaling function	G99	Return to R level in a fixed cycle

In any inconsistency between the listed codes in this handbook and the control system manual, the G codes listed by the control manufacturer must be selected.

APPLICATIONS FOR TURNING

Fanuc lathe controls use three G code group types - A, B and C. The *Type A* is the most common; in this handbook, all examples and explanations are Type A group, including the table below. Only one type can be set at a time. Types A and B can be set by a control system parameter, but type C is optional. Generally, most G codes are identical, only a few are different in the A and B types. More details on the subject of G code groups is listed at the end of this chapter.

G code	Description
G00	Rapid positioning
G01	Linear interpolation
G02	Circular interpolation clockwise
G03	Circular interpolation counterclockwise
G04	Dwell (as a separate block)
G09	Exact stop check - one block only
G10	Programmable data input (Data Setting)
G11	Data Setting mode cancel
G20	English units of input
G21	Metric units of input
G22	Stored stroke check ON
G23	Stored stroke check OFF
G25	Spindle speed fluctuation detection ON
G26	Spindle speed fluctuation detection OFF
G27	Machine zero position check
G28	Machine zero return (reference point 1)
G29	Return from machine zero
G30	Machine zero return (reference point 2)
G31	Skip function
G32	Threading - constant lead
G35	Circular threading CW
G36	Circular threading CCW
G40	Tool nose radius offset cancel
G41	Tool nose radius offset left
G42	Tool nose radius compensation right

G code	Description	
G50	Tool position register / Maximum r/min preset	
G52	Local coordinate system setting	
G53	Machine coordinate system setting	
G54	Work coordinate offset 1	
G55	Work coordinate offset 2	
G56	Work coordinate offset 3	
G57	Work coordinate offset 4	
G58	Work coordinate offset 5	
G59	Work coordinate offset 6	
G61	Exact stop mode	
G62	Automatic corner override mode	
G64	Cutting mode	
G65	Custom macro call	
G66	Custom macro modal call	
G67	Custom macro modal call cancel	
G68	Mirror image for double turrets	
G69	Mirror image for double turrets cancel	
G70	Profile finishing cycle	
G71	Profile roughing cycle - Z axis direction	
G72	Profile roughing cycle - X axis direction	
G73	Pattern repetition cycle	
G74	Drilling cycle	
G75	Grooving cycle	
G76	Threading cycle	
G90	Cutting cycle A	*(Group type A)*
G90	Absolute command	*(Group type B)*
G91	Incremental command	*(Group type B)*
G92	Thread cutting cycle	*(Group type A)*
G92	Tool position register	*(Group type B)*
G94	Cutting cycle B	*(Group type A)*
G94	Feedrate per minute	*(Group type B)*
G95	Feedrate per revolution	*(Group type B)*
G96	Constant surface speed mode	*(CSS)*
G97	Direct r/min input	*(CSS mode cancel)*
G98	Feedrate per minute	*(Group type A)*
G99	Feedrate per revolution	*(Group type A)*

Most of the preparatory commands are discussed under the individual applications, for example G01 under *Linear Interpolation*, G02 and G03 under *Circular Interpolation*, etc. In this section, G codes are described in general, regardless of the type of machine or control unit.

G CODES IN A PROGRAM BLOCK

Unlike the miscellaneous functions, known as the *M functions* and described in the next chapter, several preparatory commands may be used in a single block, providing they are not in a logical conflict with each other:

```
N25 G90 G00 G54 X6.75 Y10.5
```

This method of program writing is several blocks shorter than the single block alternative:

```
N25 G90
N26 G00
N27 G54
N28 X6.75 Y10.5
```

Both methods will appear identical during a continuing processing. However, the second example, when executed in a *single block* mode, each block will require pressing the *Cycle Start* key to activate the block. The shorter method is more practical, not only for its length, but for the logical connection between individual commands within the block.

Some rules of application and general considerations apply to G codes used with *other* data in a block. The most important of them is the subject of *modality*.

◆ Modality of G-commands

Earlier, the following example C was used to demonstrate the general placement of G codes into a program block:

➲ Example C - original :

```
N3 G90 G00
N4 ...
N5 ...
N6 ...
N7 X13.0 Y10.0
```

If the structure is changed slightly and filled with realistic data, these five blocks may be the result:

➲ Example C - modified (as programmed) :

```
N3 G90 G00 X5.0 Y3.0
N4 X0
N5 Y20.0
N6 X15.0 Y22.0
N7 X13.0 Y10.0
```

Note the rapid motion command G00 - how many times does it appear in the program? Just once - in the block N30. In fact, so is the command for absolute mode, G90. The reason neither G00 nor G90 has been repeated is because both commands remain active from the moment of their first appearance in the program. The term *modal* is used to describe this characteristic.

> For a command to be modal, it means it has to remain in a certain mode until canceled by another mode.

As most G codes are modal, there is no need to repeat a modal command in every block. Using the earlier example C once more, the control will make the following interpretation during program execution:

➲ Example C - modified (as processed) :

```
N3 G90 G00 X50.0 Y30.0
N4 G90 G00 X0
N5 G90 G00 Y200.0
N6 G90 G00 X150.0 Y220.0
N7 G90 G00 X130.0 Y100.0
```

The program does not have any practical application by moving from one location to another at a rapid rate, but it demonstrates the modality of preparatory commands. The purpose of modal values is to avoid unnecessary duplication of programming modes. G codes are used so often, that writing them in the program can be tedious. Fortunately, the majority of G codes can be applied only once, *providing they are modal*. In the control system specifications, preparatory commands are identified as modal and unmodal.

◆ Conflicting Commands in a Block

The purpose of preparatory commands is to select from two or more modes of operation. If the rapid motion command G00 is selected, it is a specific command relating to a tool motion. As it is impossible to have a rapid motion and a cutting motion active *at the same time*, it is impossible to have G00 and G01 active simultaneously. Such a combination creates a *conflict* in a block. If conflicting G codes are used in the same block, the *latter* G code will be used.

```
N74 G01 G00 X3.5 Y6.125 F20.0
```

In the example, the two commands G01 and G00 are in conflict. As G00 is the *latter* one in the block, it will become effective. The feedrate is ignored in this block.

```
N74 G00 G01 X3.5 Y6.125 F20.0
```

This is the exact opposite of the previous example. Here, the G00 is in the front, therefore the G01 will take precedence and the motion will take place as a cutting motion at the specified feedrate of 20.0 in/min.

◆ Word Order in a Block

G codes are normally programmed *at the beginning of a block*, after the block number, before other significant data:

```
N40 G91 G01 Z-0.625 F8.5
```

This is a traditional order, based on the idea that if the purpose of the G codes is to *prepare* or *preset* the control system to a certain condition, the preparatory commands should always be placed first. Supporting this argument is the fact that only non-conflicting codes are allowed in a single block. Strictly speaking, there is nothing wrong with rearranging the order to:

```
N40 G91 Z-0.625 F8.5 G01
```

Perhaps unusual, but quite correct. That is not the case with the next method of positioning a G code in a block:

```
N40 Z-0.625 F8.5 G01 G91
```

Watch for situations like this! What happens in this case is that the cutting motion G01, the feedrate F and the depth Z will be combined and executed *using the current dimensional mode*. If the current mode is absolute, the Z axis motion will be executed as an absolute value, *not* an incremental value. The reason for this exception is that Fanuc allows to mix dimensional values in the same block. That can be a very useful feature, if used carefully. A typical *correct* application of this feature can be illustrated in this example:

```
(G20)
N45 G90 G00 G54 X1.0 Y1.0 S1500 M03      (G90)
N46 G43 Z0.1 H02
N47 G01 Z-0.25 F5.0
N48 X2.5 G91 Y1.5         (G90 MIXED WITH G91)
N49 ...
...
```

Blocks N45 through N47 are all in the absolute mode. Before the block N48 is executed, the absolute position of the axes X and Y is 1.0,1.0. From this starting position, the target location is the absolute position of X2.5 combined with the incremental motion of 1.5 inches along the Y axis. The resulting absolute position will be X2.5Y2.5, making a 45° motion. In this case, the G91 will remain in effect for *all* subsequent blocks, until the G90 is programmed. Most likely, the block N48 will be written in absolute mode:

```
...
N48 X2.5 Y2.5
...
```

Normally, there is no reason to switch between the two modes. It can result in some very unpleasant surprises. There are some occasions when this special technique brings benefits, for example, in subprograms.

GROUPING OF COMMANDS

The example of conflicting G codes in one block brings one issue to the forefront. It makes sense, for example, that motion commands such as G00, G01, G02 and G03 cannot coexist in the same block. The distinction is not so clear for other preparatory commands. For example, can the tool length offset command G43 be programmed in the same block as the cutter radius offset command G41 or G42? The answer is yes, but let's look at the reason why.

Fanuc control system recognizes preparatory commands by separating them into arbitrary *groups*. Each group, called the *G code group*, has a Fanuc assigned arbitrary two-digit number. The rule governing the coexistence of G codes in one block is very simple. If two or more G codes from *the same group* are in the same block, they are in conflict with each other.

◆ Group Numbers

The G code groups are typically numbered from 00 to 25. This range varies between different control models, depending on the features. It can even be higher for the newest controls or where more G codes are required. One of these groups - the most unique one and perhaps the most important as well - is the *Group 00*.

All preparatory commands in the 00 group are *not* modal, sometimes using the descriptions *unmodal* or *non-modal*. They are only active in the block in which they were programmed. If unmodal G codes are to be effective in several consecutive blocks, they must be programmed in *each* of those blocks. In majority of unmodal commands, this repetition will not be used very often.

For example, a dwell is a programmed pause measured in milliseconds. It is needed only for the duration within the specified time, no longer. There is no logical need to program dwell in two or more consecutive blocks. After all, what is the benefit of the next three blocks?

```
N56 G04 P2000
N57 G04 P3000
N58 G04 P1000
```

All three blocks contain the same function, a dwell, one after another. The program can be made much more efficient by simply entering the total dwell value into a single block:

```
N56 G04 P6000
```

The following groups are typical for the Fanuc control systems. Applications for milling and turning controls are specially distinguished by the M and T letters respectively, in the *Type* column of the table:

Group	Description	G codes	Type
00	Unmodal G codes	G04 G09 G10	M/T
		G11 G27 G28 G29	M/T
		G30 G31 G37	M/T
		G45 G46 G47 G48	M/T
		G52 G53 G65	M/T
		G51 G60 G92	M
		G50	T
		G70 G71 G72 G73	T
		G74 G75 G76	T
01	Motion Commands, Cutting Cycles	G00 G01 G02 G03	M/T
		G32 G35 G36	T
		G90 G92 G94	T
02	Plane Selection	G17 G18 G19	M
03	Dimensioning Mode	G90 G91 (U and W for lathes)	M T
04	Stored Strokes	G22 G23	M/T
05	Feedrate	G93 G94 G95	T
06	Units Input	G20 G21	M/T
07	Cutter Radius Offset	G40 G41 G42	M/T
08	Tool Length Offset	G43 G44 G49	M
09	Cycles	G73 G74 G76 G80	M
		G81 G82 G83 G84	M
		G85 G86 G87 G88	M
		G89	M
10	Return Mode	G98 G99	M
11	Scaling Cancel, Mirror Image	G50	M
		G68 G69	T
12	Coordinate System	G54 G55 G56 G57	M/T
		G58 G59	M/T
13	Cutting Modes	G61 G62 G64	M/T
		G63	M
14	Macro Mode	G66 G67	M/T
16	Coordinate Rotation	G68 G69	M
17	CSS	G96 G97	T
18	Polar Input	G15 G16	M
24	Spindle Speed Fluctuation	G25 G26	M/T

The group relationship makes a perfect sense in all cases. One possible exception is Group 01 for *Motion Commands* and Group 09 for *Cycles*. The relationship between these two groups is this - if a G code from Group 01 is specified in any of the fixed cycle Group 09, the cycle is immediately canceled, *but the opposite is not true*. In other words, an active motion command is *not* canceled by a fixed cycle.

Group 01 is *not* affected by G codes from *Group 09*. In a summary ...

> Any G code from a given group automatically replaces another G code from the same group

G CODE TYPES

Fanuc control system offers a flexible selection of preparatory commands. This fact distinguishes Fanuc from many other controls. Considering the fact that Fanuc controls are used worldwide, it only makes sense to allow the standard control configuration to follow established style of each country. A typical example is the selection of dimensional units. In Europe, Japan and many other countries, metric system is the standard. In North America, the common system of dimensioning still uses the English units. As both markets are substantial in the world trade, a clever control manufacturer tries to reach them both. Almost all control manufacturers offer a selection of the dimensional system. But Fanuc and similar controls also offer selection of programming codes that were in effect *before* Fanuc reached the worldwide market.

The method Fanuc controls use is a simple method of parameter setting. By selecting the specific system parameter, one of two or three G code types can be selected, the one that is typical for a particular geographical user. Although the majority of the G codes are the same for every type, the most typical illustration are G codes used for English and metric selection of units. Many earlier US controls used G70 for English units and G71 for metric units. Fanuc system has traditionally used G20 and G21 codes for English and metric input respectively.

Setting up a parameter, the G code type that is the most practical can be selected. Such a practice, if done at all, should be done once and only when the control is installed, before any programs have been written for it. Change of the G code type at random is a guaranteed way to create an organizational nightmare. Keep in mind that a change of one code meaning will affect the meaning of another code. Using the units example for a lathe, if G70 means an English input of dimensions, you cannot use it to program a roughing cycle. Fanuc provides a different code. Always stay with the standard G code type. All G codes in this handbook use the default group of the *Type A*, and also the most common group.

◆ G Codes and Decimal Point

Many latest Fanuc controls include a G code with a decimal point, for example, G72.1 (Rotation copy) or G72.2 (Parallel copy). Several preparatory commands in this group are related to a particular machine tool or are not typical enough to be described in this handbook.

9

MISCELLANEOUS FUNCTIONS

The address *M* in a CNC program identifies a *miscellaneous function*, sometimes called a *machine function*. Not all miscellaneous functions are related to the operation of a CNC machine - quite a few are related to the processing of the program itself. The more suitable term *miscellaneous functions* is used throughout this handbook.

DESCRIPTION AND PURPOSE

Within the structure of a CNC program, programmers often need some means of activating certain aspects of the machine operation or controlling the program flow. Without the availability of such means, the program would be incomplete and impossible to run. First, let's look at the miscellaneous functions relating to the operation of the machine - the true *machine functions*.

◆ Machine Related Functions

Various physical operations of the CNC machine must be controlled by the program, to ensure fully automated machining. These functions generally use the M address and include the following operations:

❏	Spindle rotation	CW or CCW
❏	Gear range change	Low / Medium / High
❏	Automatic tool change	ATC
❏	Automatic pallet change	APC
❏	Coolant operation	ON or OFF
❏	Tailstock or quill motion	IN or OUT

These operations vary between machines, due to the different designs by various machine manufacturers. A machine design, from the engineering point of view, is based on a certain primary machining application. A CNC milling machine will require different functions related to the machine than a CNC machining center or a CNC lathe. A numerically controlled EDM wire cutting machine will have many unique functions, typical to that kind of machining and those found on no other machine.

Even two machines designed for the *same type* of work, for example, two kinds of a vertical machining center, will have functions different from each other, if they have a different CNC system or significantly different options. Different machine models from the same manufacturer will also have certain unique functions, even with the same model of the CNC system.

All machine tools designed for metal removal by cutting have certain common features and capabilities. For example, spindle rotation can have three - *and only three* - possible selections in a program:

- ❏ Spindle normal rotation
- ❏ Spindle reverse rotation
- ❏ Spindle stop

In addition to these three possibilities, there is a function called the spindle orientation, also a machine related function. Another example is a coolant. Coolant can only be controlled as being ON or being OFF.

These operations are typical to most CNC machines. All are programmed with an M function, followed by no more than two digits, although some control models allow the use of a three digit M function, Fanuc 16/18, for example.

Fanuc also uses the three digit M functions in several special applications, for example, for synchronization of two independent turrets on a 4-axis lathe. All these and other functions are related to the operation of the machine and belong to the group collectively known as *miscellaneous functions* or simply as the *M functions* or *M codes*.

◆ Program Related Functions

In addition to the machine functions, some M functions are used to control the execution of a CNC program. An interruption of a program execution requires an M function, for instance, during the change of a job setup, such as a part reversal. Another example is a situation, where one program calls one or more subprograms. In such a case, each program has to have a program call function, the number of repetitions, etc. M functions handle these requirements.

Based on the previous examples, the use of miscellaneous functions falls into two main groups, based on a particular application:

- ❏ Control of the machine functions
- ❏ Control of the program execution

This handbook covers only the most common miscellaneous functions, used by the majority of controls. Unfortunately, there are many functions that vary between machines and the control system. These functions are called *machine specific functions*. For this reason, always consult the documentation for the particular machine model and its control system.

TYPICAL APPLICATIONS

Before learning the M functions, note the type of activity these functions do, regardless of whether such activity relates to the machine or the program. Also note the abundance of two way toggle modes, such as *ON* and *OFF*, *IN* and *OUT*, *Forward* and *Backward*, etc. Always check your manual first - for the reasons of consistency, all M functions in this handbook are based on the following table:

◆ **Applications for Milling**

M code	Description
M00	Compulsory program stop
M01	Optional program stop
M02	End of program (usually with reset, no rewind)
M03	Spindle rotation normal
M04	Spindle rotation reverse
M05	Spindle stop
M06	Automatic tool change (ATC)
M07	Coolant mist ON
M08	Coolant ON (coolant pump motor ON)
M09	Coolant OFF (coolant pump motor OFF)
M19	Spindle orientation
M30	Program end (always with reset and rewind)
M48	Feedrate override cancel OFF *(deactivated)*
M49	Feedrate override cancel ON *(activated)*
M60	Automatic pallet change (APC)
M78	B axis clamp *(nonstandard)*
M79	B axis unclamp *(nonstandard)*
M98	Subprogram call
M99	Subprogram end

◆ **Applications for Turning**

M code	Description
M00	Compulsory program stop
M01	Optional program stop
M02	End of program (usually with reset, no rewind)
M03	Spindle rotation normal
M04	Spindle rotation reverse
M05	Spindle stop
M07	Coolant mist ON
M08	Coolant ON (coolant pump motor ON)
M09	Coolant OFF (coolant pump motor OFF)
M10	Chuck open
M11	Chuck close
M12	Tailstock quill IN
M13	Tailstock quill OUT
M17	Turret indexing forward
M18	Turret indexing reverse
M19	Spindle orientation (optional)
M21	Tailstock forward
M22	Tailstock backward
M23	Thread gradual pull-out ON
M24	Thread gradual pull-out OFF
M30	Program end (always with reset and rewind)
M41	Low gear selection
M42	Medium gear selection 1
M43	Medium gear selection 2
M44	High gear selection
M48	Feedrate override cancel OFF *(deactivated)*
M49	Feedrate override cancel ON *(activated)*
M98	Subprogram call
M99	Subprogram end

◆ **Special MDI Functions**

Several M functions cannot be used in the CNC program at all. This group is used in the *Manual Data Input* mode exclusively (MDI). An example of such a function is a step by step tool change for machining centers, used for service purposes only, never in the program. These functions are outside of the scope of this handbook.

◆ **Application Groups**

The two major categories, described earlier, can further be divided into several groups, based on the specific application of the miscellaneous functions within each group. A typical distribution list is contained in the following table:

Group	Typical M-functions
Program	M00 M01 M02 M30
Spindle	M03 M04 M05 M19
Tool change	M06
Coolant	M07 M08 M09
Accessories	M10 M11 M12 M13 M17 M18 M21 M22 M78 M79
Threading	M23 M24
Gear ranges	M41 M42 M43 M44
Feedrate override	M48 M49
Subprograms	M98 M99
Pallets	M60

The table does not cover all M functions or even all possible groups. Neither does it distinguish between machines. On the other hand, it does indicate the *types* of applications the miscellaneous functions are used for in everyday CNC programming.

The miscellaneous functions listed in this chapter are used throughout the book. Some of them appear more often than others, reflecting their general use in programming. The functions that do not correspond to a particular machine control system are either not used or not needed. However, the concepts for their applications are always similar for most control systems and CNC machines.

In this chapter, only the more general functions are covered in significant detail. Remaining miscellaneous functions are described in the sections covering individual applications. At this stage, the stress is on the usage and behavior of the most common miscellaneous functions.

M FUNCTIONS IN A BLOCK

If a miscellaneous function is programmed in a block by itself, with no other data supplementing it, only the function itself will be executed. For example,

```
N45 M01
```

is an optional stop. This block is correct - an M function *can* be the only block entry. Unlike the preparatory commands (G codes), only *one* M function is allowed in a block - unless the control allows multiple M functions in the same block, a program error will occur (latest controls only).

A practical method of programming certain miscellaneous functions is in a block that contains a *tool motion*. For example, turning the coolant on and - *at the same time* - moving the cutting tool to a certain part location may be required. As there is no conflict between the instructions, the block may look something like this:

```
N56 G00 X12.9854 Y9.474 M08
```

In this example, block N56, the precise *time* the M08 function will be activated is not very important. In other cases, the timing may be very important. Some M functions *must* be in effect before or after certain action takes place. For example, look at this combination - a Z axis motion is applied together with the *program stop* function M00 in the same block:

```
N319 G01 Z-12.8456 F20.0 M00
```

This is a far more serious situation and two answers are needed. One is *what exactly* will happen, the other is *when exactly* it will happen, when the M00 function is activated. There are three possibilities and three questions to ask:

1. Will the program stop take place immediately, when the motion is activated - at the start of the block?

2. Will the program stop take place while the tool is on the way - during a motion?

3. Will the program stop take place when the motion command is completed - at the end of the block?

One of the three options *will* happen - but which one? Even if a practical purpose of these examples may not be apparent at this stage, it is useful to know how the control system interprets blocks containing a tool motion and a miscellaneous function.

Each M function is designed logically - it is also designed to make a *common sense*.

The actual startup of a M function is divided into *two* groups - not three:

❑ M function activates at the start of a block (simultaneously with the tool motion)

❑ M function activates at the end of a block (when the tool motion has been completed)

No M function will be activated *during* the block execution, there is no logic to it. What is the logical startup of the coolant ON function M08 in the block N56 above? The correct answer is that the coolant will be activated at the *same time* as the tool motion *begins*. The correct answer for the example block N319 is that the M00 program stop function will be activated *after* the tool motion has been completed. Makes sense? Yes, but what about the other functions, how do they behave in a block? Let's look at them next.

◆ Startup of M Functions

Take a look at the list of typical M functions. Add a tool motion to each and try to determine the way the function is going to behave, based on the previous notes. A bit of logical thinking provides a good chance to arrive at the right conclusion. Compare the two following groups to confirm:

M functions activated at the START OF A BLOCK	
M03	Spindle rotation normal
M04	Spindle rotation reverse
M06	Automatic tool change (ATC)
M07	Coolant mist ON
M08	Coolant ON (coolant pump motor ON)

M functions activated at the END OF A BLOCK	
M00	Compulsory program stop
M01	Optional program stop
M02	End of program (usually with reset, no rewind)
M05	Spindle stop
M09	Coolant OFF (coolant pump motor OFF)
M30	Program end (always with reset and rewind)
M60	Automatic pallet change (APC)

If there is an uncertainty about how the function will interact with the tool motion, the safest choice is to program the M function as a *separate block*. That way the function will always be processed *before* or *after* the relevant program block. In the majority of applications this will be a safe solution.

◆ Duration of M Functions

Knowledge of *when* the M function takes effect is logically followed by the question about *how long* the function will be active. Some miscellaneous functions are active only in the block they appear. Others will continue to be in effect until canceled by another miscellaneous function. This is similar to the modality of the preparatory G commands, however the word *modal* is not usually used with M functions. As an example of a function duration, take miscellaneous functions M00 or M01. Either one will be active for *one* block only. The coolant ON function M08, will be active until a *canceling* or an *altering* function is programmed. Remember, any one of the following functions will cancel the coolant ON mode - M00, M01, M02, M09 and M30. Compare these two tables:

M functions completed in ONE BLOCK	
M00	Compulsory program stop
M01	Optional program stop
M02	End of program (usually with reset, no rewind)
M06	Automatic tool change (ATC)
M30	Program end (always with reset and rewind)
M60	Automatic pallet change (APC)

M functions active UNTIL CANCELED or ALTERED	
M03	Spindle rotation normal
M04	Spindle rotation reverse
M05	Spindle stop
M07	Coolant mist ON
M08	Coolant ON (coolant pump motor ON)
M09	Coolant OFF (coolant pump motor OFF)

The classification is quite logical and shows some common sense. There is no need to remember individual M functions and their exact activities. The best place to find out for certain, is to study manuals supplied with the CNC machine and watch the program run right on the machine.

PROGRAM FUNCTIONS

Miscellaneous functions that control program processing can be used either to interrupt the processing temporarily (in the middle of a program) or permanently (at the end of a program). Several functions are available for this purpose.

◆ Program Stop

The M00 function is defined as an *unconditional* or *compulsory* program stop. Any time the control system encounters this function during program processing, all automatic operations of the machine tool will stop:

❏ Motion of all axes

❏ Rotation of the spindle

❏ Coolant function

❏ Further program execution

The control will *not* be reset when the M00 function is processed. All significant program data currently active are retained (feedrate, coordinate setting, spindle speed, etc.). The program processing can only be resumed by activating the *Cycle Start* key. The M00 function cancels the spindle rotation and the coolant function - they have to be reprogrammed in subsequent blocks.

The M00 function can be programmed as an individual block or in a block containing other commands, usually axis motion. If the M00 function is programmed together with a motion command, the motion will be completed first, *then* the program stop will become effective:

➲ M00 programmed *after a motion command :*

```
N38 G00 X13.5682
N39 M00
```

➲ M00 programmed *with a motion command :*

```
N39 G00 X13.5682 M00
```

In both cases, the motion command will be completed first, *before* the program stop is executed. The difference between the two examples is apparent only in a single block processing mode (for example, during a trial cut). There will be no practical difference in auto mode of processing (Single Block switch set to OFF).

Practical Usage

The program stop function used in a program makes the CNC operator's job much easier. It is useful for many jobs. One common use is a part inspection on the machine, while the part is still mounted. During the stop, the part dimensions or the tool condition can be checked. Chips accumulated in a bored or drilled hole can be removed, for example, before another machining operation can start, such as blind hole tapping. The program stop function is also necessary to change the current setup in the middle of a program, for example, to reverse a part. A manual tool change also requires the M00 function in the program.

> The program stop function M00 is used only
> for a manual intervention during program processing

The control also offers an *optional* program stop M01, described next. The main rule of using M00 is the need of a *manual intervention* for *every* part machined. Manual tool change in a program qualifies for M00, because every part needs it. A dimensional check may not qualify, if is infrequent. M01 will be a better choice. Although the difference between the two functions is slight, the actual difference in cycle time can be significant for large number of parts.

When using the M00 function, always inform the operator *why* the function has been used and what its purpose is. Make the intent known to avoid a confusion. This intent can be made available to the operator in two ways:

❏ In the setup sheet, refer to the block number that contains the miscellaneous function M00 and describe the manual operation that has to be performed:

```
BLOCK N39 ...... REMOVE CHIPS
```

❏ In the program itself, issue a comment section with the necessary information. The comment section must be enclosed in parentheses (three versions shown):

```
[A]    N39 M00 (REMOVE CHIPS)

[B]    N39 X13.5682 M00 (REMOVE CHIPS)

[C]    N38 X13.5682 M00
       (REMOVE CHIPS)
```

Any one of the three methods will give the CNC operator the necessary information. From the two options, the second one [B], the comment section in the program, is preferable. The built-in instructions can be read directly from the display screen of the control panel.

◆ Optional Program Stop

The miscellaneous function M01 is an *optional* or a *conditional* program stop. It is similar to M00 function, with one difference. Unlike the M00 function, when M01 function is encountered in the program, the program processing will *not* stop, unless the operator interferes via the control panel. The *Optional Stop* toggle switch or a button key located on the panel can be set to either ON or OFF position. When the M01 function in the program is processed, the setting of the switch will determine whether the program will temporarily stop or continues to be processed:

Optional Stop switch setting	Result of M01
ON	Processing *will* stop
OFF	Processing *will not* stop

In case there is no M01 function programmed, the setting of the *Optional Stop* switch is irrelevant. Normally, it should be in the OFF position for production work.

When active, the M01 function behaves the same way as the M00 function. The motion of all axes, spindle rotation, coolant function and any further program execution will be temporarily interrupted. Feedrate, coordinate settings, spindle speed setting, etc., are retained. The further processing of the program can only be reactivated by the *Cycle Start* key. All programming rules for the M00 function also apply to the M01 function.

A good idea is to program the M01 function at the end of each tool, followed by a blank line with no data. If the program processing can continue without stopping, the *Optional Stop* switch will be set to OFF and no production time is lost. If there is a need to stop the program temporarily at the end of a tool, the switch will be set to ON and the processing stops at the end of the tool. The time loss is usually justified under the circumstances, for example, to change the cutting insert or to inspect a dimension or the surface finish of the part.

◆ Program End

Every program must include a special function defining the *end* of current program. For this purpose, there are two M functions available - the M02 and M30. Both are similar, but each has a distinct purpose. The M02 function will terminate the program, but *will cause no return* to the first block at the program top. The function M30 will terminate the program as well but it *will cause a return* to the program top. The word *'return'* is often replaced by the word *'rewind'*. It is a leftover from the times when a reel-to-reel tape reader was common on NC machines. The tape had to be *rewound* when the program has been completed for each part. M30 function provided this *rewind* capability.

When the control reads the program end function M02 or M30, it cancels all axis motions, spindle rotation, coolant function and usually resets the system to the default conditions. *On some controls the reset may not be automatic and any programmer should be aware of it.*

If the program ends with the M02 function, the control remains at the program end, ready for the next *Cycle Start*. On modern CNC equipment there is no need for M02 at all, except for backward compatibility. This function was used in addition to M30 for those machines (mainly NC lathes) that had tape readers without reels, using a short loop tape. The trailer of the tape was spliced to the tape leader, creating a closed loop. When the program was finished, the start of the tape was next to the end, so no rewind was necessary. Long tapes could not use loops and required reels and M30. So much for the history of M02 - just ignore its existence.

Is M02 the Same as M30 ?

On most modern controls, a system parameter can be set to make the M02 function with the same meaning as that of M30. This setting can give it the rewind capabilities, useful in situations where an old program can be used on a machine with a new control without changes.

In a summary, if the end of program is terminated by the M30 function, the rewind *will* be performed; if the M02 function is used, the rewind *will not* be performed.

When writing the program, make sure the last block in the program contains nothing else but M30 as the preferred end (sequence block *is* allowed to start the block):

```
N65 . . .
N66 G91 G28 X0 Y0
N67 M30                         (END OF PROGRAM)
%
```

On some controls, the M30 function can be used together with the axes motion - *NOT recommended !:*

```
N65 . . .
N66 G91 G28 X0 Y0 M30           (END OF PROGRAM)
%
```

Percent Sign

The percent sign (%) after M30 is a special *stop code*. This symbol terminates the loading of a program from an external device. It is also called the *end-of-file marker.*

◆ Subprogram End

The last M function for a program end is M99. Its primary usage is in the subprograms. Typically, the M99 function will terminate a subprogram and return to the processing of the previous program. If M99 is used in a standard program, it creates a program with *no end* - such a situation is called an *endless loop*. M99 should be used only in subprograms, not in the standard programs.

MACHINE FUNCTIONS

Miscellaneous functions relating to the operation of the machine tool are part of another group. This section describes the most important of them in detail.

◆ Coolant Functions

Most metal removal operations require that the cutting tool is flooded with a suitable coolant. In order to control the flow of coolant in the program, there are three miscellaneous functions usually provided for this purpose:

M07	Mist ON
M08	Flood ON
M09	Mist or Flood OFF

Mist is the combination of a small amount of cutting oil mixed with compressed air. It depends on the machine tool manufacturer whether this function is standard for a particular CNC machine tool or not. Some manufacturers replace the mixture of oil and air with air only, or with oil shot only, etc. In these cases, it is typical that an additional equipment is built into the machine. If this option exists on the machine, the most common miscellaneous function to activate the oil mist or air is M07.

The function similar to M07 is M08 - coolant *flooding*. This is by far the most common coolant application in CNC programming. It is standard for virtually all CNC machine. The coolant, usually a suitable mixture of soluble oil and water, is premixed and stored in the coolant tank of the machine tool. Flooding the cutting edge of the tool is important for three reasons:

❑ Heat dissipation

❑ Chip removal

❑ Lubrication

The primary reason to use a coolant flood aimed at the cutting edge is to dissipate the heat generated there during cutting. The secondary reason is to remove chips from the cutting area, using coolant pressure, Finally, the coolant also acts as a lubricant to ease the friction between the cutting tool and material. Lubrication helps to extend tool life and improves the surface finish.

During initial tool approach towards the part or during final return to the tool change position, the coolant is normally not required. To turn off the coolant function, use the M09 function - *coolant off*. M09 will turn off the oil mist or flood supply and nothing else. In reality, the M09 function will shut off the coolant pump motor.

Each of the three coolant related functions may be programmed in separate blocks or together with an axis motion. There are subtle but important differences in the order and timing of the program processing. The following examples explain the differences:

● Example A - oil mist is turned ON, if available :

N110 M07

● Example B - coolant is turned ON :

N340 M08

● Example C - coolant is turned OFF :

N500 M09

● Example D - axis motion and coolant ON :

N230 G00 X11.5 Y10.0 M08

● Example E - axis motion and coolant OFF :

N400 G00 Z1.0 M09

The examples show the differences in the program processing. The general rules of coolant programming are:

❑ Coolant ON or OFF in a *separate* block becomes active in the block in which it is programmed (Examples A, B and C)

❑ Coolant ON, when programmed *with the axes motion*, becomes active simultaneously with the axes motion (Example D)

❑ Coolant OFF, programmed *with the axes motion*, becomes effective only upon completion of the axes motion (Example E)

The main purpose of M08 function is *to turn the coolant pump motor on*. It does not guarantee that the cutting edge receives any coolant immediately. On large machines with long coolant pipes, or machines with low coolant pump pressure, some delay is to be expected before the coolant covers the distance between the pump and cutting tool.

Coolant should always be programmed with two important considerations in mind:

❑ There will be no coolant splashing outside of the work area (outside of the machine)

❑ There will never be a situation when the coolant reaches a hot edge of the tool

The first consideration is relatively minor. If the coolant function is programmed in the 'wrong' place, the result may be just an inconvenience. The wet area around the machine may present unsafe working conditions and should be quickly corrected. Even more serious situation happens when the coolant suddenly starts flooding a cutting tool that has already entered the material. The change in temperature at the cutting edge may cause the tool to break and damage the part. Carbide tools are far more easily affected by temperature changes than high speed steel tools. Such a possibility can be prevented during programming, by using the M08 function *a few blocks ahead* of the actual cutting block. Long pipes or insufficient coolant pressure on the machine may delay the start of the actual flooding.

◆ Spindle Functions

Chapter 12 - Spindle Control, details all aspects of controlling the machine spindle in a CNC program. Miscellaneous functions that are available for the spindle control its *rotation* and *orientation*.

Most spindles can rotate in both directions, *clockwise* (CW) and *counterclockwise* (CCW). The direction of rotation is always relative to a standard point of view. The viewpoint is established from the spindle side as the *direction along the spindle center line towards its face*. CW rotation in such a view is programmed as M03, CCW direction as M04, assuming the spindle can be rotated either way.

The drilling and milling types of machines use this established convention quite commonly. The same convention is also applied to CNC lathes. On a CNC milling machine or a machining center, it is more practical to look towards the part from the spindle side rather than from the table side. On a lathe (slant bed horizontal type), the more practical view is *from the tailstock towards the spindle*, because that is the closest to how the CNC machine operator stands in front of the lathe. However, M03 and M04 spindle directions are established the same way as for machining centers. A further complication is the fact that left hand tools are used in lathe work more frequently than in the work for milling applications. Make an effort to study the instruction manual for a specific machine carefully - also see details described in *Chapter 12*.

Spindle function to program a spindle stop is M05. This function will stop the spindle from rotating, *regardless of the rotation direction*. On many machines, the miscellaneous function M05 must also be programmed *before* reversing the spindle rotation:

```
M03                           (SPINDLE CW)
...
<... Machining at the current location ...>
...
M05                           (SPINDLE STOP)
<... Usually a tool change ... >
M04                           (SPINDLE CCW)
...
<... Machining at the current location ...>
...
```

The M05 function may also be required when changing gear ranges on CNC lathes. A spindle stop programmed in a block containing an axis motion, will take place *after* the motion has been completed.

The last spindle control function is the function M19, called the *spindle orientation*. Some control manufacturers call it the spindle key lock function. Regardless of the description, the M19 function will cause the spindle to stop in an oriented position. This function is used mostly during machine setup, seldom in the program. The spindle must be oriented in two main situations:

❏ Automatic tool change (ATC)

❏ Tool shift during a boring operation
   (G76 and G87 boring cycles only)

When the *Automatic Tool Change* (ATC) function M06 is used in the program, there is no need to program the spindle orientation for most CNC machining centers. The orientation is built into the automatic tool changing sequence and guarantees the correct positioning of all cutting tool holders. Some programmers like to program the M19 with the machine zero return for the tool change position, to save a second or two of the cycle time.

The spindle orientation is necessary for certain boring operations on milling systems. To exit a bored hole with a boring tool away from the finished cylindrical wall, the spindle must be stopped first, the tool cutting bit must be oriented, and then the tool can be retracted from the hole. A similar approach is used for backboring operations. However, these special cutting operations use fixed cycles in the program, where the spindle orientation is built in. For more details, *Chapter 25* describes *Fixed Cycles*.

In conclusion, the M19 function is rarely used in the program. It is available as a programming aid and to the machine operator for setup work, using MDI operations.

◆ Gear Range Selection

Virtually all programmable gear range selections apply to a CNC lathe. On machining centers, the spindle gear range is automatically changed. Most CNC lathes have two or more gear ranges available, some more powerful lathes are equipped with up to four selections. The basic programming rule is to select the gear range based on the machining application.

For example, most roughing operations require the *power* of the spindle more than the spindle *speed*. In this case, a low range is usually a better selection. For finishing work, a medium or high range is better, because high spindle rotation can be more beneficial to the metal removing process.

The distribution of the miscellaneous functions depends entirely on the number of gear ranges the CNC lathe has available. Number of ranges is 1, 2, 3 or 4. The following table shows typical distribution of the M functions, but check the actual commands in a machine tool manual.

Ranges	M function	Gear Range
1 available	N/A	None programmed
2 available	M41 M42	Low range High range
3 available	M41 M42 M43	Low range Medium range High range
4 available	M41 M42 M43 M44	Low range Medium range 1 Medium range 2 High range

The rule of thumb is that the higher the gear range, the more spindle speed is possible and less spindle power is required. The opposite is also true. Normally, the spindle rotation does not have to be stopped to change a gear, but consult the lathe manual anyway. In doubt, stop the spindle first, change the gear range, then restart the spindle.

◆ Machine Accessories

The majority of miscellaneous functions is used for some physical operation of the machine tool accessories. From this group, the more common applications have been already covered, specifically the coolant control and the gear changes. The remaining M functions in this group are described in detail elsewhere in this handbook, so only a short description is offered here. The most notable of the machine related M functions are:

M function	Description	Type
M06	Automatic tool change (ATC)	M
M60	Automatic pallet change (APC)	M
M23 M24	Thread gradual pull-out *ON / OFF*	T
M98 M99	Subprogram *call* / Subprogram *end*	M/T

Each line of in a CNC program is called *a block*. In the terminology established earlier, a block was defined as a single instruction processed by the CNC system.

. A sequence block, a program block - or simply a block - is normally one hand written line in the program copy, or a line typed in a text editor and terminated by the *Enter* key. This line can contain one or more program words - words that result in the definition of a single instruction to the CNC machine. Such a program instruction may contain a combination of preparatory commands, coordinate words, tool functions and commands, coolant function, speeds and feeds commands, position registration, offsets of different kinds, etc. In plain English, the contents of one block will be processed as a single unit before the control processes any following block. When the whole CNC program is processed, the system will evaluate individual instructions (blocks) *as one complete machine operation step*. Each program consists of a series of blocks necessary to complete a certain machining process. The overall program length will always depend on the total number of blocks and their size.

BLOCK STRUCTURE

As many program words as necessary are allowed in a single block. Some controls impose a limit on the number of characters in one block. There is only a theoretical maximum for Fanuc and similar controls, irrelevant in practice. The only restriction is that *two* or *more duplicated* words (functions or commands) cannot be used in the same block (with the exception of G codes). For example, only one miscellaneous M function (exceptions do exist) or only one coordinate word for the X axis in a single block are allowed. The order of individual words within a block follows a fairly free format - that means the required words may be in any order, providing that the sequence block (the N address) is written as the *first* address. Although the order of individual words in a block *is* allowed to be in any order, it is a standard practice to place words in a *logical order* within a block. It makes the CNC program easier to read and understand.

A typical program block structure is very dependent on the control system and the type of the CNC machine. A typical block may contain the following instructions, in the order suggested. Not all program data are necessary to be specified every time, only when required.

❏	Block number	N
❏	Preparatory commands	G
❏	Auxiliary functions	M
❏	Axis motion commands	X Y Z A B C U V W ...
❏	Words related to axes	I J K R Q ...
❏	Speed, feed or tool function	S F T

The contents of the program block will vary between machine tools of different kinds, but logically, the majority of general rules will always be followed, regardless of the CNC system or the machine tool.

◆ Building the Block Structure

Each block of a CNC program has to be built with the same thoughts and the same care as any other important structure, for example a building, a car, or an aircraft. It starts with good planning. Decisions have to be made as to what will and what will not be part of the program block, similar to a building, car, aircraft or other structure. Also, decisions have to be made as to what order of commands - *instructions* - are going to be established within the block and many other considerations.

The next few examples compare a typical structure of blocks for milling operations and blocks for turning operations. Each block is presented as a separate example.

◆ Block Structure for Milling

In milling operations, the structure of a typical program block will reflect the realities of a CNC machining center or a similar machine.

➲ Milling block examples:

```
N11 G43 Z2.0 S780 M03 H01          (EXAMPLE 1)

N98 G01 X2.15 Y4.575 F13.0         (EXAMPLE 2)
```

The first milling example in block N11, is an illustration of a tool length offset block, applied together with the spindle speed and the spindle rotation direction.

The second example in block N98, shows a typical programming instruction for a simple linear cutting motion, using the linear interpolation method and a suitable cutting feedrate.

◐ Turning block examples:

`N67 G00 G42 X2.5 Z0.1 T0202 M08      (EXAMPLE 1)`

`N23 G02 X7.5 Z-2.8 R0.5 F0.012       (EXAMPLE 2)`

In the lathe examples, block N67 illustrates a rapid motion to an XZ position, as well as a few other commands - the tool nose radius offset startup G42, activation of the tool offset (T0202), and the coolant ON function M08. The example in block N23 is a typical circular interpolation block with a feedrate.

PROGRAM IDENTIFICATION

A CNC program can be identified by its number and, on some controls, also by its name. The identification by the program number is necessary in order to store more than one program in the CNC memory. Program name, if available, can be used to make a brief description of the program, readable on the control screen display.

◆ Program Number

The first block used in any part program is commonly a program number, if required by the control system from the program. Two addresses are available for the program number - the standard letter O for EIA format, and the colon [:] for the ASCII (ISO) format. In memory operation, the control system always displays program number with the letter O. The block containing the program number is *not always* necessary to include in the CNC program.

If the program uses program numbers, they must be specified within an allowed range. Programs for typical Fanuc controls must be within the range of 1 - 9999, program number zero (O0 or O0000) is not allowed. Some controls allow a 5-digit program number. Also not allowed are decimal point or a negative sign in the program number. Suppression of leading zeros is allowed - for example, O1, O01, O001, and O0001 are all legitimate entries, in this case for a program number *one*.

◆ Program Name

On the latest Fanuc control systems, the name of the program can be included *in addition* to the program number, *not instead* of the program number. The program name (or a brief description of the program) can be up to sixteen characters long (spaces and symbols are counted). The program name must be *on the same line* (in the same block) as the program number:

`O1001 (DWG. A-124D IT. 2)`

This feature has the advantage that when the directory of the memory is displayed on the screen, the name of the program appears next to the program number, making the di-

rectory more descriptive and useful. The program description can be read on the display screen and provides an easier identification of each program stored.

If the program name is longer than the sixteen characters recommended, no error is generated, but only the first sixteen characters will be displayed. Make sure to avoid program names that can be ambiguous when displayed. Consider these two program names, they appear to be good:

`O1005 (LOWER SUPPORT ARM - OP 1)`
`O1006 (LOWER SUPPORT ARM - OP 2)`

Since the control screen display can show only the first *sixteen* characters of the program name, the program names will be ambiguous when displayed:

`O1005 (LOWER SUPPORT AR)`
`O1006 (LOWER SUPPORT AR)`

To eliminate this problem, use an abbreviated description that is within the sixteen characters and contains all the significant data:

`O1005 (LWR SUPP ARM OP1)`
`O1006 (LWR SUPP ARM OP2)`

If a more detailed description is required, the description has to be split over one or more comment lines:

`O1005 (LWR SUPP ARM OP1)`
`(OPERATION 1 - ROUGHING)`

The comments in the block or blocks following the program number will *not* appear on the directory screen listing, but still will be a useful aid to the CNC operator. They will be displayed during the program execution and, of course, in a hard copy printout.

Keep the program names short and descriptive - their purpose is to assist the CNC operator in the search of programs stored in the control memory. The suitable data to include in the program name are the drawing number or the part number, shortened part name, operation, etc. Data not suitable are the machine name, control model, programmer's name, date or time, company or customer's name and similar descriptions.

On many controls, when loading the program into the memory, the CNC operator must specify the program number on the control panel, regardless of the number in the CNC program. It can be a number that just happens to be available in the system, or it can be a number that has a unique meaning, perhaps indicating a unique group (for example, all programs that begin with the O10xx belong to the group associated with a single customer). Subprograms must always be stored under the number specified by the CNC programmer. Innovative use of program numbers may also serve to keep track of programs developed for each machine or part.

SEQUENCE NUMBERS

Individual sequence blocks in the CNC program can be referenced with a number for easier orientation within the program. The program address for a block number is the letter N, followed by up to five digits - from sequence 1 to sequence 9999 or 99999, depending on the control system. The block number range will be N1 to N9999 for the older controls and N1 to N99999 for the newer controls. Some rather old controls accept block number in the three digit range only, N1 - N999.

The N address must always be the *first word* in the block. For an easier orientation in programs that use subprograms, there should be no duplication of the numbers between the two types of programs. For example, a main program starting with N1 and a subprogram also starting with N1 may cause a confusing situation. Technically, there is nothing wrong with such a designation. Refer to *Chapter 39* for suggestions on block numbering in subprograms.

◆ Sequence Number Command

In the following table, the first column represents sequence numbers the way they are used normally, the second column shows the sequence numbers required in a format acceptable to the machine control system, as applied to a CNC program:

Increment	First block number
1	N1
2	N2
5	N5
10	N10
50	N50
100	N100
99999	N99999

Using sequence numbers (block numbers) in a CNC program offer several advantages and at least one likely disadvantage.

On the positive side, the block numbers will make the program search greatly simplified during editing or tool repetition on the machine. They also make the program much easier to read on the CNC display screen during processing or on the printed hard copy. That means both *the programmer* and *the operator* benefit.

On the negative side, block numbers will *reduce* the available computer memory of the CNC. That means a fewer number of programs can be stored in the memory, and long programs may not fit in their entirety.

◆ Sequence Block Format

The program input format notation for a block number, using the address N, is N5 for the more advanced controls, and N4 or even N3 for older controls. Block number N0 is not allowed, neither is a minus sign, a fractional number or a block number using a decimal point. Minimum block increment number must always be an integer - smallest integer allowed is one (N1, N2, N3, N4, N5, etc.). A larger increment is allowed and its selection depends on the personal programming style or the standard established within the company. The typical sequence block increments other then one are:

Increment	Program example
2	N2, N4, N6, N8, ...
5	N5, N10, N15, N20, ...
10	N10, N20, N30, N40, ...
100	N100, N200, N300, N400, ...

Some programmers like to start with the default of the last example - N100, usually programmed in the increments of 100, 10, or less. There is nothing wrong with this rather a large start and increment, but the CNC program will become unnecessarily too long, too soon, and possibly difficult to manage.

In all cases of block increments other than one, the purpose of the program is the same - to allow for additional blocks to be filled-in between the existing blocks, if such a need comes. The need may arise while proving or optimizing the program on the CNC machine, where an addition to the existing program will be required. Although the new blocks (the ones inserted) will *not* be in the order of an equal increment, at least they will be numerically ascending. For example, a face cut on a lathe with one cut (Example A) was modified by the machine operator for two cuts (Example B):

⮞ Example A - one face cut :

```
N40 G00 G41 X3.5 Z0 T0303 M08
N50 G01 X-0.07 F0.01
N60 G00 W0.1 M09
N70 G40 X3.5
```

⮞ Example B - two face cuts :

```
N40 G00 G41 X3.5 Z0.05 T0303 M08
N50 G01 X-0.07 F0.01
N60 G00 W0.1
N61 X3.5
N62 Z0
N63 G01 X-0.07
N64 G00 W0.1 M09
N70 G40 X3.5
```

Note the change in block N40 and added blocks N61 to N64. Preference in this handbook is to program in increments of one and if an addition is needed, the added blocks will have *no block numbers at all* (check if the control system allows block numbers to be omitted, most do).

➲ Example A - one face cut :

```
N40 G00 G41 X3.5 Z0 T0303 M08
N41 G01 X-0.07 F0.01
N42 G00 W0.1
N43 G40 X3.5
```

➲ Example B - two face cuts :

```
N40 G00 G41 X3.5 Z0.05 T0303 M08
N41 G01 X-0.07 F0.01
N42 G00 W0.1
X3.5
Z0
G01 X-0.07
G00 W0.1
N43 G40 X3.5
```

Note that the program is a little smaller and the additional blocks are quite visual and noticeable when printed or displayed on the screen.

Leading zeros may (and should) be omitted in the block number - for example, N00008 can be written as N8. Omitting the leading zeros will reduce the overall program length. The trailing zeros must always be written, to distinguish for such similarities as N08 and N80.

The use of block numbers in a program is optional, as shown in the earlier example. A program containing block numbers is easier to read. For the CNC operator, search and edit functions in program editing can be used easily. *Note* - some programming applications depend on the block numbers, for example, lathe multiple repetitive cycles G70, G71, G72, G73. In this case, *at least* the significant blocks have to be numbered (see *Chapter 35*).

◆ **Numbering Increment**

Block numbers in a program can be in any physical order - ascending, descending or mixed - they can also be duplicated or missing altogether. Some programming practices are established as preferable, because they are logical and make sense. Having a mixed order of sequence numbers in the program serves no useful purpose and neither do duplicated sequence numbers. If the program contains duplicate block numbers and a block number search is initiated at the machine, the control system will only search for the *first* occurrence of the particular block number, which may or may not be the block required. Any further search will have to be repeated from the string found last. The reason for the generous latitude in the sequence block numbering is to offer flexibility to the CNC operator *after* the program has been done and loaded into the control.

The block sequence number does not affect the order of program processing, regardless of the increment. Even if the blocks are numbered in a descending or mixed order, the part program will always be processed sequentially, on the basis of the block *contents*, not its number. The increment of 5 or 10 is the most practical, since it allows for insertion of up to 4 to 9 blocks respectively between any two original blocks. That should be more than sufficient for the majority of program modifications.

For those CNC programmers who use a computer based programming system, just a few words relating to the programming of sequence numbers. Although the computer programming system allows the start number of the block and its increment to almost any combination, adhere to *the start and increment numbers of one* (N1, N2, N3, ...). The purpose of a computer based programming is to keep an accurate database of the part geometry *and* the cutting tool path. If the CNC program is modified manually, the part computer database is not accurate any more. Any CNC program change should always be reflected in the *source* of the program, as well as its result - never in the result alone.

◆ **Long Programs and Block Numbers**

Long programs are always difficult to load into a CNC memory with limited capacity. In such cases, the program length may be shortened by omitting the block numbers altogether or - even better - by programming them only in the *significant blocks*. The significant blocks are those that have to be numbered for the purpose of program search, a tool repetition, or other procedure that depends on program numbers, such as a machining cycle or tool change. In these cases, select increments of two or five, for the operator's convenience. Even limited use of sequence numbers will increase the program length, but for a justifiable reason.

If *all* sequence block numbers have been omitted in the program, the search on the machine control will become rather difficult. The CNC operator will have no other option but to search for the next occurrence of a particular address within a block, such as X, Y, Z, etc., rather than a sequence block number. This method of search may unnecessarily prolong the search time.

END OF BLOCK CHARACTER

Because of the control system specifications, the individual sequence blocks must be separated by a special character, known as the *end-of-block* character or by its abbreviation EOB or E-O-B. On most computer systems, the EOB character is generated by the *Enter* key on the keyboard. When the program is input to the control by MDI (Manual Data Input), the EOB character key on the control panel terminates the block. The end-of-block symbol on Fanuc controls appears as a semicolon [;].

The semicolon symbol on the screen is only a *graphic representation* of the end-of-block character and is never entered literally in the CNC program. Under no circumstances it should be included in the program itself. Some older control systems have an asterisk [*] as the display symbol for the end-of-block, rather then the semicolon [;]. Many controls use other symbols, that also represent the end of block, for example, some use the dollar sign [$]. In any case, remember the symbol is only the *representation* of the end-of-block character, not its actual character.

STARTUP BLOCK OR SAFE BLOCK

A *startup block* (sometimes called a *safe block* or a *status block*) is a special sequence block. It contains one or more modal words (usually preparatory commands of several G groups) that preset the control system into a desired initial or default state. This block is placed at the beginning of each program or even at the beginning of each tool and it is the *first block* processed during a repetition of a program (or a tool within a program). In the CNC program, the startup block usually precedes any motion block or axis setting block, as well as the tool change or tool index block. This is the block to be searched for, if the program or the desired cutting tool is to be repeated during a machine operation. Such a block will be slightly different for the milling and turning systems, due to the unique requirements of each control system.

Earlier in this handbook, in the *Chapter 5*, one topic covered the state of the control system when the main power has been turned on, which sets the system default conditions. A CNC programmer should never count on these default conditions, since they can be easily changed by the machine tool operator, without the programmer's knowledge. If such a change does happen, the programmed settings will *not* correspond to those suggested by the machine tool manufacturer or the engineers who designed the control system.

A professional CNC programmer should always assume the attitude of a *safe* programming approach and will not leave anything to chance. The programmer will try to preset all the required conditions under the program control, rather that counting on the defaults of the CNC system. Such an approach is not only much safer, it will also result in the programs that are easy to use during the setup, the tool path proving and the tool repetition due to the tool breakage, dimensional adjustments, etc. It is also very beneficial to the CNC machine operators, particularly to those with limited experience. In all the applications listed, the startup block will not increase the machining cycle time at all. Another benefit of the startup block is that the program is more transportable from one machine tool to another, since it does not count on the default setting of a particular machine-control combination.

The name *safe block* - which is another name used for the startup block - does not become *safe* on its own - it must be *made* safe. Regardless of the name, this block should contain control settings for the program or the cutting tool that start the program in a 'clean' state. The most common entries that set the initial status are the dimensioning system (English/metric and absolute/incremental), cancellation of any active cycle, cancellation of the active cutter radius offset mode, the plane selection for milling, the feedrate default selection for lathes, etc. The presented examples show some startup blocks for both milling and turning controls.

At the beginning of the program for milling, a startup block may be programmed with the following contents:

```
N1 G00 G17 G20 G40 G54 G64 G80 G90 G98
```

N1 block is the first sequence number, G00 selects the rapid mode, G17 establishes the XY plane selection, G20 selects the English units, G40 cancels any active cutter radius offset, G64 sets a continuous cutting mode, G80 cancels any active fixed cycle, G90 selects the absolute mode, and G98 will retract to the initial level in a fixed cycle. These conditions apply only when the startup block is processed as the *first major block* in the CNC program - any subsequent program changes will become effective only with the block in which the change is applied. For example, if a G01 command is effective by default, any subsequent usage of G00, G02, or G03 will cancel the G01 command.

At the beginning of a CNC lathe program, the startup block may contain these G codes:

```
N1 G20 G00 G40 G99
```

N1 is the first block number, G20 selects the English units, G00 selects the rapid mode, G40 cancels any active tool nose radius offset, and the G99 selects feedrate per revolution mode. Reference to the absolute or incremental system is usually not required, since the lathe controls use addresses X and Z for the *absolute* dimensioning and the addresses U and W for the *incremental* dimensioning. For lathe controls that do not support the U and W addresses, the standard G91 code is used for incremental values in X and Z axes. As in the milling example, any of the words programmed in the safe block can be overridden by subsequent change of the G commands.

Some controls systems do not allow certain G codes on the same line. For example, G20 or G21 may not be programmed with other G codes. If you are not sure, place the G codes in separate blocks. Instead of

```
N1 G20 G17 G40 G49 G80
```

two or more blocks can be safely used:

```
N1 G20
N2 G17 G40 G49 G80
```

PROGRAM COMMENTS

Various comments and messages in the program can be included within the program body as separate blocks, or as parts of an existing block, mostly in cases when the message is short. In either case, the message must be enclosed in parenthesis (for ASCII/ISO format):

➲ Example A :

N330 M00 (REVERSE PART)

➲ Example B :

N330 M00 (REVERSE PART / CHECK TOOL)

➲ Example C :

N330 M00
(REVERSE PART / CHECK TOOL)

The purpose of a message or comment in the program is to inform the machine operator of a specific task that must be performed *every time* the program reaches the stage of processing where such message appears. Comments are also useful for understanding the program at a later date and can be used for documenting the program.

Typical messages and comments relate to information about setup changes, chip removal from a hole, dimensional check, cutting tool condition check and many others. A message or a comment block should be included only if the required task is *not clear* from the program itself - no need to describe what happens in each block. Messages and comments should be brief and focused, as they occupy a memory space in the CNC memory.

From the practical perspective, a series of messages and comment blocks can be provided at the beginning of each program, to list all significant drawing information and cutting tools required for the job. This subject has been covered in *Chapter 7* - here is just a reminder:

```
O1001 (SHAFT - DWG B451)
(SHAFT TOOLING - OP 1 - 3 JAW CHUCK)

(T01 - ROUGH TOOL - 1/32R - 80 DEG)
(T02 - FINISH TOOL - 1/32R - 55 DEG)
(T03 - OD GROOVING TOOL - 0.125 WIDE)
(T04 - OD THREADING TOOL - 60 DEG)

N1 G20 G99
N2 ...
```

If the available memory space of the CNC unit is limited, using comment blocks in this manner may prove impractical. It will be better if the messages and comments are listed in proper setup and tooling sheets, with all the required details.

CONFLICTING WORDS IN A BLOCK

The instructions in a program block must be logical and reasonable - not impossible. For instance, the first block of the program contains the following words:

N1 G20 G21 G17

What the block contains is simply not logically possible. It instructs the control to:

'Set the English system of dimensions, also set the metric system of dimensions and set the XY plane'.

Definitely not possible, but also not realistic - what will actually happen and how does the control interpret such a statement? The XY plane is all right, but what about the selection of dimensions? Obviously, both selection are not possible, the block contains *conflicting* words, opposite dimensional units. Some controls may give an error message, Fanuc systems will not. What will happen? The control unit will evaluate the sequence block and check for any words within the same group. The distribution of command groups have been described in the section dealing with the preparatory commands - G codes, in *Chapter 8*.

If the computer system finds two or more words that belong to the same group, it will not return an error message, it will automatically activate the *last* word of the group. In the example of conflicting dimensional selection, it will be the preparatory function G21 - selection of metric dimensions - that becomes active. That may or may not be the selection required. Rather than counting on some kind of elusive luck, make sure there are no conflicting words in any program block.

In the example illustrating the English and metric selection, the preparatory command G was used. What would happen if, for example, the address X was used? Consider the following example:

N120 G01 X11.774 X10.994 Y7.056 F15.0

There are two X addresses in the same block. The control system *will not* accept the second X value, but it will issue an alarm (error). Why? Because there is a great difference between the programming rules for a G code as such and for the coordinate system words. Fanuc controls allow to place as many G codes in the same block as needed, providing they are not in conflict with each other. But the same control system will not allow to program more than one coordinate word of the same address for each sequence block. Some other rules may also apply. For example, the words in a block may be programmed in any order, providing the N address is the first one listed. For example, the following block is legal (but very nontraditional in its order):

N340 Z-0.75 Y11.56 F10.0 X6.845 G01

As a matter of good programming practices, be sure to write the entries for each sequence block in a logical order. The block number must be the first word and is usually followed by G code(-s), primary axes in their alphabetical order X.., Y.., Z..), auxiliary axes or modifiers (I.., J.., K..), miscellaneous functions and words, and the feedrate word as the last item. Select only those words needed for the individual block:

```
N340 G01 X6.845 Y11.56 Z-0.75 F10.0
```

Two other possibilities exist that may require a special attention in programming approach. For example, how will the following block be interpreted?

```
N150 G01 G90 X5.5 G91 Y7.7 F12.0
```

There is an apparent conflict between the absolute and incremental modes. Most Fanuc controls will process this block exactly the way it is written. The X axis target position will be reached in absolute values, but the Y axis will be an incremental distance, measured from the current position of the cutter. It may not be a typical approach, but it offers advantages in some cases. Remember - the sequence block following the block N150 will be in the *incremental* mode, since G91 is specified *after* the G90 command!

The other programming application to watch for, is in a block programmed in the circular interpolation mode. The section dealing with this subject (*Chapter 29*), specifies that an arc or a circle can be programmed either with arc modifiers I, J and K (depending whether a milling or a turning control system is used). It also specifies that a direct radius input, using the address R, can be used. Both of the following examples are correct, resulting in a 90° arc with a 1.5 inch radius:

⬤ With I and J arc modifiers :

```
N21 G01 X15.35 Y11.348
N22 G02 X16.85 Y12.848 I1.5 J0
N23 G01 ...
```

⬤ With the direct radius R address :

```
N21 G01 X15.35 Y11.348
N22 G02 X16.85 Y12.848 R1.5
N23 G01 ...
```

Now, consider how the control system will process the block N22, if it contains *both,* the I and J modifiers as well as the radius input:

```
N22 G02 X16.85 Y12.848 I1.5 J0 R1.5
```

or

```
N22 G02 X16.85 Y12.848 R1.5 I1.5 J0
```

The answer may be surprising - in *both cases*, the control will ignore the I and J values and *will only process the value of radius R*. The order of address definition is irrelevant in this special case. The address R has a higher control priority than the I and J addresses, if programmed in the same block. All examples assume that the control system supports the R radius input.

MODAL PROGRAMMING VALUES

Many program words are modal. The word *modal* is based on the word *'mode'* and means that the specific command remains in this mode after it has been used in the program once. It can only be canceled by another modal command of the same group. Without this feature, a program using linear interpolation in absolute mode with a feedrate of 18.0 in/min, would contain the absolute command G90, the linear motion command G01 and the feedrate F18.0 in every block. With the modal values, the programming output is much shorter. Virtually all controls accept modal commands. The following two examples illustrate the differences:

⬤ Example A - *without* modal values :

```
N12 G90 G01 X1.5 Y3.4 F18.0
N13 G90 G01 X5.0 Y3.4 F18.0
N14 G90 G01 X5.0 Y6.5 F18.0
N15 G90 G01 X1.5 Y6.5 F18.0
N16 G90 G01 X1.5 Y3.4 F18.0
N17 G90 G00 X1.5 Y3.4 Z1.0
```

⬤ Example B - *with* modal values :

```
N12 G90 G01 X1.5 Y3.4 F18.0
N13 X5.0
N14 Y6.5
N15 X1.5
N16 Y3.4
N17 G00 Z1.0
```

Both examples will produce identical results. Compare each block of the Example A with the corresponding block of the Example B. Observe that the modal commands are *not* necessary to be repeated in the CNC program. In fact, in everyday programming, many program commands used are modal. The exceptions are those program instructions, whose functionality starts and ends in the same block (for example dwell, machine zero return, certain machining instructions, such as tool change, indexing table, etc.). The M functions behave in a similar fashion. For example, if the program contains a machine zero return in two consecutive blocks (usually for safety reasons), it may look like this:

```
N83 G28 Z1.0 M09
N84 G28 X5.375 Y4.0 M05
```

G28 cannot be removed from block N84, because the G28 command is *not* modal and must be repeated.

EXECUTION PRIORITY

There are special cases, mentioned earlier, where the *order* of commands in the block determines the priority in which the commands are executed. To complete the subject of a block, let's look at another situation.

Here are two unrelated blocks used as examples:

```
N410 G00 X22.0 Y34.6 S850 M03
```

and

```
N560 G00 Z5.0 M05
```

In the block N410, the rapid motion is programmed together with two spindle commands. What will *actually happen* during the program execution? It is very important to know *when* the spindle will be activated in relationship to the cutting tool motion. On Fanuc and many other controls, the spindle function will take effect *simultaneously* with the tool motion.

In the block N560, a Z axis tool motion is programmed (Z5.0), this time together with the spindle stop function (M05). Here, the result will be different. The spindle will be stopped only when the motion is *one hundred percent completed*. *Chapter 9* covering *Miscellaneous Functions* explains this subject.

Similar situations exist with a number of miscellaneous functions (M codes), and any programmer should find out exactly how a particular machine and control system handle a motion combined with an M function address in the same block. Here is a refresher in the form of a list of the most common results:

Functions that will be executed *simultaneously* with the cutting tool motion:

```
M03  M04  M07  M08
```

Functions that will be executed *after* the cutting tool motion has been completed:

```
M00  M01  M05  M09  M98
```

Be careful here - if in doubt, program it safe. Some miscellaneous functions require an *additional* condition, such as another command or function to be active For example, M03 and M04 will only work if the spindle function S is in effect (spindle is rotating). Other miscellaneous functions should be programmed in separate blocks, many of them for logical or safety reasons:

```
M02  M06  M10  M11  M19  M30  M60  M99
```

Functions indicating the end of a program or a subprogram (M02, M30, M99) should stand on their own and not combined with other commands in the same block, except in special cases. Functions relating to a mechanical activity of the machine tool (M06, M10, M11, M19, M60) *should* be programmed without any motion in effect, for safety. In the case of M19 (spindle orientation), the spindle rotation must be stopped first, otherwise machine may get damaged. Not all M functions are listed in the examples, but they should provide a good understanding of how they may work, when programmed together with a motion. The chapter describing the miscellaneous functions also covers the duration of typical functions within a program block.

It never hurts to play it safe and always program these possible troublemakers in a sequence block containing no tool motion. For the mechanical functions, make sure the program is structured in such a way that it provides safe working conditions - these functions are oriented mainly towards the machine setup.

Addresses in a CNC program that relate to the tool position at a given moment are called *the coordinate words*. Coordinate words always take a dimensional value, using the currently selected units, English or metric. Typical coordinate words are X ,Y, Z, I, J, K, R, etc. They are the basis of all dimensions in CNC programs. Tens, hundreds, even thousands of values may have to be calculated to make the program do what it is intended to do - to accurately machine a complete part.

The dimensions in a program assume two attributes:

❑ Dimensional *units*   … English *or* Metric

❑ Dimensional *references*   … Absolute *or* Incremental

The units of dimensions in a program can be of two kinds - *metric* or *English*. The reference of dimensions can be either *absolute* or *incremental*.

Fractional values, for example 1/8, are not allowed in a CNC program. In the metric format, *millimeters* and *meters* are used as units, in the English format it is *inches* and *feet* that are used as units. Regardless of the format selected, the number of decimal places can be controlled, the suppression of leading and trailing zeros can be set and the decimal point can be programed or omitted, as applicable to a particular CNC system.

ENGLISH AND METRIC UNITS

Drawing dimensions can be used in the program in either *English* or *metric* units. This handbook uses the combined examples of both the English system, common in the USA, to some extent in Canada and one or two other countries. The metric system is common in Europe, Japan and the rest of the world. With the economy reaching global markets, it is important to understand both systems. The use of metric system is on the increase even in countries that still use the English units of measurement, mainly the United States.

Machines that come equipped with Fanuc controls can be programmed in either mode. The initial CNC system selection (known as the *default* condition) is controlled by a parameter setting of the control system, but can be overridden by a preparatory command written in the part program. The default condition is usually set by the machine tool manufacturers or distributors (sometimes even by the CNC dealers) and is based on the engineering decisions of the manufacturer, as well as the demands of their customers.

During the program development, it is imperative to consider the impact of default conditions of the control system on program execution. The default conditions come into effect the moment the CNC machine tool has been *turned on*. Once a command is issued in the MDI mode or in a program, the default value may be overwritten and will remain changed from that point on. The dimensional unit selection in the CNC program will change the *default* value (that is the internal control setting). In other words, if the English unit selection is made, the control system will remain in that mode until a metric selection command is entered. That can be done either through the MDI mode, a program block, or a system parameter. This applies even for situations when the power has been turned *off* and then *on* again!

To select a specific dimensional input, regardless of the default conditions, a preparatory G command is required at the beginning of the CNC program:

G20	Selects English units (inches and feet)
G21	Selects metric units (millimeters and meters)

Without specifying the preparatory command in the program, control system will default to the status of current parameter setting. Both preparatory command selections are modal, which means the selected G code remains active until the opposite G code is programmed - so the metric system is active until the English system replaces it and vice versa.

This reality may suggest a certain freedom of switching between the two units anywhere in the program, almost at random and indiscriminately. *This is not true*. All controls, including Fanuc, are based on the *metric* system, partially because of the Japanese influence, but mainly because the metric system is more accurate. Any 'switching' by the use of the G20 or G21 command does not necessarily produce any real conversion of one unit into the other, but merely *shifts* the decimal point, not the actual digits. At best, only some conversions take place, not all. For example, G20 or G21 selection will convert one measuring unit to another on *some - but not all* - offset screens.

The following two examples will illustrate the *incorrect* result of changing G21 to G20 and G20 to G21 within the same program. Read the comments for each block - you may find a few surprises:

● Example 1 - from metric to English units :

G21   *Initial unit selection (metric)*

G00 X60.0   *X value is accepted as 60 mm*

G20   *Previous value will change into 6.0 inches*
             (real translation is 60 mm = 2.3622047 inches)

● Example 2 - from English to metric units :

G20   *Initial unit selection (English)*

G00 X6.0   *X value is accepted as 6.0 inches*

G21   *Previous value will change into 60 mm*
             (real translation is 6.0 inches = 152.4 mm)

Both examples illustrate the possible problem caused by switching between the two dimensional units in the same program. For this reason, always use only one unit of dimensioning in a part program. If the program calls a subprogram, the rule extends to subprograms as well:

> Never mix metric and English units in the same program

In fact, it is unwise to mix them, even if the results for the control system are predictable. The selection of the dimensional system will make a great difference how some control functions will work. The following functions *will be* affected by the change from one system of units to the other:

❑   Dimensional words (X, Y, Z axes, I, J, K modifiers, etc.)

❑   Constant Surface Speed (CSS - for CNC lathes)

❑   Feedrate function (the F address)

❑   Offset values (the H and D offsets for milling and tool preset values for a lathe)

❑   Screen position display (the number of decimal places)

❑   Manual pulse generator - the HANDLE (value of divisions)

❑   Some control system parameters

The initial selection of dimensional units can also be done by a system parameter setting. The control status when the power has been turned on is the same as is was at the time of the last power shut off. If neither G20 nor G21 is programmed, the control accepts the dimensional units selected by a parameter setting. If G20 or G21 is included in the program, *the program command will always take a priority* over any control system parameter setting. Programmer makes the decisions - the control system is only interpreting them, but it does not mean it is always 'right'.

Always program the units setting in a separate block, *before* any axis motion, offset selection, or setting of a coordinate system (G92, G50 and G54 to G59). The failure to follow this rule may produce incorrect results, particularly when frequently changing units for different jobs.

♦ **Comparable Unit Values**

There are many units available in both the metric and English systems. In CNC programming, only a very small portion of them is used. The metric units are based on a *millimeter* or a *meter*, depending on the application. The English units are based on *inches* and *feet*, again depending on the application. Common abbreviations for the different units are:

Millimeter	mm
Meter	m
Inch	in
Foot	ft

Many programming terms use these abbreviations. The next table shows the comparable terms between the two dimensional systems (older terms are in parentheses):

Metric	English
m/min (also MPM)	ft/min (also FPM or SFPM)
mm/min	in/min (also IPM or ipm)
mm/rev	in/rev (also IPR or ipr)
mm/tooth	in/tooth (also IPT or ipt)
kW	HP

ABSOLUTE AND INCREMENTAL MODES

A dimension in either input units must have *a specified point of reference*. For example, if X35.0 appears in the program and the selected units are millimeters, the statement does not indicate where the dimension of 35 mm has its origin. The control system needs more information to program dimensional values correctly.

There are two types of references in programming:

❑   Reference to a common point on the part
    ... known as the origin for ABSOLUTE input

❑   Reference to a previous point on the part
    ... known as the last tool position for INCREMENTAL input

In the example, the dimension X35.0 (and any other as well) can be measured from a selected fixed point on the part, called *origin*, or *program zero*, or *program reference point* - all these terms have the same meaning. The same value of X35.0 can also be measured from the *previous* position, which is always the *last* tool position. This position then becomes the *current* position for the next tool motion. The control system cannot distinguish one of the two possibilities from the X35.0 statement alone, so some other description must be added to the program.

All dimensions in a CNC program measured from the *common point* (origin) are *absolute* dimensions, as illustrated in *Figure 11-1*, and all dimensions in a program measured from the *current position* (last point) are *incremental* dimensions, as illustrated in *Figure 11-2*.

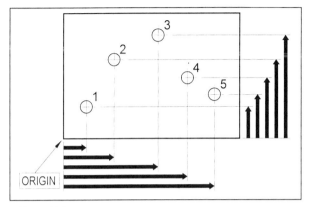

Figure 11-1

Absolute dimensioning - measured from part origin
G90 command will be used in the program

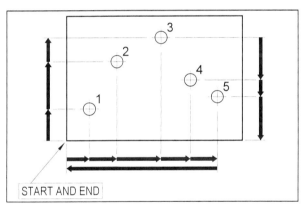

Figure 11-2

Incremental dimensioning - measured from the current tool location
G91 command will be used in the program

> Absolute dimensions in the program represent
> the target locations of the cutting tool from origin

> Incremental dimensions in the program represent
> the actual amount and direction of the cutting tool
> motion from the current location

Since the dimensional address X in the example, written as X35.0, is programmed the same way for either point of reference, some additional means must be available to the programmer. Without them, the control system would use a default setting of a system parameter, not always reflecting the programmer's intentions. The selection of the dimensioning mode is controlled by two modal G commands.

◆ Preparatory Commands G90 and G91

There are two preparatory commands available for the input of dimensional values, G90 and G91, to distinguish between two available modes:

G90	Absolute mode of dimensioning
G91	Incremental mode of dimensioning

Both commands are modal, therefore they will cancel each other. The control system uses an initial default setting when powered on, which is usually the *incremental mode*. This setting can be changed by a system parameter that presets the computer at the power startup or a reset. For individual CNC programs, the system setting can be controlled by including the proper preparatory command in the program, using either one of two available commands - the G90 or G91.

It is a good programming practice to always include the required setting in the CNC program, not to count on any default setting in the control system. It may come as a surprise that the common default setting of the control system is the *incremental mode,* rather than the absolute mode. After all, absolute programming has a lot more advantages than incremental programming and is far more popular. In addition, even if the incremental programming is used frequently, the program still starts up in the absolute mode. The question is *why* the incremental default? The reason is - as in many cases of defaults - the machining safety. Follow this reasoning:

Consider a typical start of a new program loaded into the machine control unit. The control had just been turned on, the part is safely mounted, the cutting tool is at the home position, offsets are set and the program is ready to start. Such a program is most likely written in the more practical absolute mode. Everything seems fine, except that the absolute G90 command is missing in the program. *What will happen at the machine?* Think before an answer and think logically.

When the first tool motion command is processed, the chances are that the tool target values will be positive or have small negative values. Because the dimensional input mode is missing in the program, the control system *'assumes'* the mode as incremental, which is the default value of the system parameter. The tool motion, generally in X and Y axes only, will take place to either the overtravel area, in the case of positive target values, or by a small amount, in the case of negative target values. In either case, the chances are that no damage will be done to the machine or the part. Of course, there is no guarantee, so always program with safety in mind.

> G91 is the standard default mode for input of dimensions.

◆ Absolute Data Input - G90

In the absolute programming mode, all dimensions are measured from the *point of origin.* The origin is the *program reference point* also known as *program zero.* The actual motion of the machine is the difference between the current absolute position of the tool and the previous absolute position. The algebraic signs [+] plus or [-] minus refer to the quadrant of rectangular coordinates, *not* the direction of motion. Positive sign does not have to be written for any address. All the zero values, such as X0, Y0 or Z0 refer to the tool position at program reference point, not to the tool motion itself. The zero value of any axis must be written when necessary.

The preparatory command G90 selected for absolute mode remains modal until the incremental command G91 is programmed. In the absolute mode, there will be no motion for any axis that is omitted in the program.

The main advantage of absolute programming is the ease of modification by the programmer or by the CNC operator. A change of one dimension does not effect any other dimensions in the program.

For CNC lathes with Fanuc controls, the common representation of the absolute mode is the axis designation as X and Z, *without* the G90 command. Some lathes may use the G90, but not those with Fanuc controls.

◆ Incremental Data Input - G91

In the incremental mode of programming, also called a *relative* mode, all program dimensions are measured as departure distances into a specified direction (equivalent to the *'distance-to-go'* on the control system). The actual motion of the machine is the specified amount along each axis, with the direction indicated as positive or negative.

The signs + or - specify *direction* of the tool motion, *not* the quadrant of rectangular coordinates. Plus sign for positive values does not have to be written, but minus sign must be used. All zero input values, such as X0, Y0 or Z0 mean there will be *no tool motion along that axis,* and *do not* have to be written at all. If a zero axis value is programmed in incremental mode, it will be ignored. The preparatory command for incremental mode is G91 and remains modal until the absolute command is programmed. There will be no motion for any axis omitted in the program block.

The main advantage of incremental programs is their portability between individual sections of a program. An incremental program can be called at different locations of the part, even in different programs. It is mostly used when developing subprograms or repeating an equal distance.

For Fanuc controlled CNC lathes, the common representation of the incremental mode is the axis designation as U and W, without the G91 command. Some lathes may use the G91, but not those with Fanuc controls.

◆ Combinations in a Single Block

On many Fanuc controls, the absolute and incremental modes can be combined in a single program block for special programming purposes. This may sound rather unusual, but there are significant benefits in this advanced application. Normally, the program is in one mode only - *either* in the absolute mode *or* in the incremental mode. On many controls, for any switch to the opposite mode, the motion command must be programmed in a separate block. Such controls, for example, do not allow to program an incremental motion along one axis and an absolute motion along the other axis in the same block.

Most Fanuc control systems do allow to program both modes *in the same block.* All that needs to be done is to specify the G90 or the G91 preparatory command *before* the significant dimensional address.

For lathe work, where G90 and G91 are not used, the switch is between the X and U axes and the Z and W axes. The X and Z contain the absolute values, U and W are the incremental values. Both types can be written in the same block without a problem. Here are some typical examples for both applications:

➡ Milling example :

```
N68 G01 G90 X12.5037 G91 Y4.5177 F18.5
```

The milling example shows a motion where the cutter has to reach the absolute position of X12.5037 inches and - *at the same time* - has to move along the Y axis by the distance of 4.5177 inches in the positive direction. Note the position of the commands G90 and G91 in the block - it is very important, but it may not work on all controls.

➡ Turning example :

```
N60 G01 X13.56 W-2.5 F0.013
```

This example for a CNC lathe shows a tool path motion, where the cutting tool has to reach the diameter of 13.56 inches and - *at the same time* - has to move 2.5 inches *into* the negative Z axis direction, represented by the incremental designation address W. G90 or G91 is not normally used, since the *Group A* of G codes is the most common one and does not support G code designation of dimensional mode selection.

Anytime there is a switch between the absolute and the incremental mode in a CNC program, the programmer must be careful not to remain in the *'wrong'* mode longer than needed. The switch between the modes is usually temporary, for a specific purpose. It may affect one block or several blocks. Ensure that the original setting for the program is reinstated. Remember that both the absolute and the incremental modes are *modal* - they remain in effect until canceled by the opposite mode.

DIAMETER PROGRAMMING

All dimensions along the X axis on a CNC lathe can be programmed as diameter values. This approach simplifies lathe programming and makes the program easier to read. Normally, the default of most Fanuc controls is the *diameter programming*. The control system parameter can be changed to interpret the X axis as a radius input:

```
G00 X4.0    Diameter dimension    ... when set by a parameter
G00 X2.0    Radius dimension      ... when set by a parameter
```

Either value is correct, with the appropriate parameter setting. The diameter programming is easier to understand by both the programmer and operator, because drawings use the diameter dimensions for cylindrical parts and measuring diameters at the machine is common. Exercise certain caution - if the diameter programming is used, all tool wear offsets for the X axis must be treated as applicable to the *diameter of the part*, not to its single side (radius value).

Another programming consideration, also very important, is the selection of the absolute or the incremental mode of dimensional input. The diameter programming, where the X axis value represents the part diameter, is much more common in the absolute mode. In those cases, when an incremental value is required, remember that all incremental dimensions in the program must also be specified *per diameter, not* per radius.

In the incremental mode, the intended X axis motion will be programmed as the U axis, specified as a distance and direction to travel on a *diameter*.

For example, the two sections of the following metric programs are identical - note that they both *start* in the absolute mode and only the diameters appear different:

➲ Example 1 - Absolute diameters :

```
G00 G42 X85.0 Z2.0 T0404 M08    (ABSOLUTE START)
G01 Z-24.0 F0.3
X95.0
Z-40.0
X112.0
Z-120.0
X116.0
G00 ..
```

➲ Example 2 - Incremental diameters :

```
G00 G42 X85.0 Z2.0 T0404 M08    (ABSOLUTE START)
G01 Z-24.0 F0.3
U10.0                              (X95.0)
Z-40.0
U17.0                              (X112.0)
Z-120.0
U4.0                               (X116.0)
G00 ..
```

MINIMUM MOTION INCREMENT

Minimum increment (also called the *least* increment) is the *smallest amount of an axis movement* the control system is capable of supporting. The minimum increment is the smallest amount that can be programmed within the selected dimensional input. Depending on the dimensional input selection, the minimum axis motion increment is expressed either in *millimeters* for the metric system or in *inches* for the English system.

Units system	Minimum increment
Metric	0.001 mm
English	.0001 inch

In the definition of minimum increment, the most common increments are 0.001 mm and 0.0001 inches for metric and English units respectively. For a typical CNC lathe, the minimum increment for the X axis is also 0.001 mm or .0001 inches but is measured on the *diameter* - that means a 0.0005 mm or .00005 inches minimum increment per side. Fine tuning for machining precision is much more flexible and precise in the metric system than in the English system:

Minimum increment	Converted equivalent
0.001 mm	.00003947 inches
.0001 inches	0.00254 millimeters

For high precision work, using metric units should be the preferred dimensional system for part programming. In fact, the metric system is *154% more accurate* than the English system, which makes the English system almost *60.63% less accurate* than the metric system.

FORMAT OF DIMENSIONAL INPUT

The year of 1959 is often considered to be the first year of practical numerical control. Since that time, several major changes have taken place that influenced the programming format of dimensional input.

Even to this day, dimensional data can be programmed in one of the four possible ways:

❏ Full address format

❏ Leading zeros suppression

❏ Trailing zeros suppression

❏ Decimal point

In order to understand the format differences, looking back some years may be beneficial. Older control systems (mainly the old NC systems as compared to the more modern CNC systems) were not able to accept the highest input level of dimensions - the decimal point format - but the newest controls accept *all* the earlier program formats, even when the decimal format is most common. The reason is compatibility with the existing programs (old programs). Since decimal point programming method is the latest of the four available, control systems that allow decimal point programming can also accept programs written many years earlier (assumed that the control and machine tool are also compatible). *The reverse is not true.*

This is a very important issue, because knowing how the control interprets a *number that has no decimal point* is critical for all tool motion commands and feedrates.

◆ Full Address Format

The full format of a dimensional address is described by the notation of +44 in the English units and +53 in the metric units. That means all eight available digits have to written for the axis words X, Y, Z, I, J, K, etc. For example, the English dimension of .625, when applied to the X axis, will be written as:

X00006250

The metric dimension of 0.42 mm, also when applied to the X axis, will be written as:

X00000420

The full format programming is applicable only to the *very early* control units, but is correct even today. The programmed axis was usually written *without* the axis designation, which is determined by *position* of the dimension within the block. For modern CNC programming, the full format is obsolete and is used here only for reference and comparison. Yes, this format will work quite nicely in modern programs, but don't used it as a standard.

◆ Zero Suppression

Zero suppression concept is a great improvement over the full programming format. It was the adaptation of a new format that reduced the number of zeros in the dimensional input. Many modern controls still support the method of zero suppression, but only for reasons of compatibility with old and proven programs.

Zero suppression means that either the *leading* or the *trailing* zeros of the maximum dimensional input do not have to be written in the CNC program. The result is a great reduction in the program length. The default setting has been done by the control manufacturer, although the default mode can be optionally set by a system parameter. *Don't make any changes without a valid reason!*

Since the *leading* zeros suppression and the *trailing* zeros suppression are mutually exclusive, which one should be programmed for addresses without a decimal point? As it depends on the parameter setting of the control system *or* the designation of the status by the control manufacturer, the actual control status must be known. The status determines which zeros can be suppressed. It may be the zeroes at the *beginning* or the zeros at the *end* of a dimension without a decimal point. In the extremely unlikely event that the CNC system is equipped with the zero suppression feature as the only mode, programming the decimal point will not be possible. To illustrate the results of zero suppression, earlier examples will be used.

If the English input of .625 inches is to be programmed in the *leading zero suppression* format and applied to the X axis, it will appear in the program as:

X6250

The same dimension of .625 inches with the *trailing zeros suppressed*, will appear in the program as:

X0000625

The metric units input of 0.42 mm, also applied to the X axis, is written with the leading zeros suppressed as:

X420

The same dimension of 0.42 mm with the trailing zeros suppressed will appear in the program as:

X0000042

Although the examples above illustrate only one small application, the impression that the leading zero suppression is more practical than the trailing zero suppression is quite right. Many older control systems are indeed set arbitrarily to accept the *leading zero suppression as the default*, because of its practicality. Here is the reason why - study it carefully, although today the subject is more trivial than practical. On the other hand, if even one decimal point is omitted (forgotten) in the program, this knowledge becomes very useful and the subject is not trivial any more.

Preference for Leading Zero Suppression

The minimum and the maximum dimensional input the control system can accept consists of eight digits, without a decimal point, ranging from 00000001 to 99999999:

❑ Minimum: 0000.0001 inches   *or*   00000.001 mm
❑ Maximum: 9999.9999 inches   *or*   99999.999 mm

The decimal point is not written. If the program uses zero suppression of either type, a comparison of input values should be useful:

Input value comparison - inches		
Decimal point	Leading zeros suppression	Trailing zeros suppression
X0.0001	X1	X00000001
X0.001	X10	X0000001
X0.01	X100	X000001
X0.1	X1000	X00001
X1.0	X10000	X0001
X10.0	X100000	X001
X100.0	X1000000	X01
X1000.0	X10000000	X1

The leading zero suppression is much more common, because it benefits numbers with a small fractional part rather than a large integer part.

For the metric input the results will be similar:

Input value comparison - millimeters		
Decimal point	Leading zeros suppression	Trailing zeros suppression
X0.001	X1	X00000001
X0.01	X10	X0000001
X0.1	X100	X000001
X1.0	X1000	X00001
X10.0	X10000	X0001
X100.0	X100000	X001
X1000.0	X1000000	X01
X10000.0	X10000000	X1

Even if programs use decimal point all the time, knowing the effect of zero suppression is important. For example, what will happen if the programmer forgets to program the decimal point or the CNC operator forgets to punch it in? These are serious - *and common* - errors that can be avoided with some care and good knowledge.

To complete the section on zero suppression, let's look at a program input that uses an axis letter but *not as a coordinate word*. A dwell command will be used to explain. *Chapter 24* covers all the details relating to the dwell programming. For now, just use the basic format and one second dwell time as the unit. The dwell format is specified by the dwelling axis X, as being X5.3. This format tells us that

the dwell can be programmed with the X axis, followed by the maximum of eight digits, always positive. If the control system allows the decimal point, there is no confusion. If the leading or the trailing zeros have to be suppressed, the programmed input is very important.

For example, a program requires dwell lasting 0.5 of a second (one half of a second). In the various formats the program block containing the *½ sec.* dwell will be:

- ❏ Full format   X0000050
- ❏ No leading zeros   X500
- ❏ No trailing zeros   X000005
- ❏ Decimal point   X0.5 *or* X.5

Note that the logic behind the format is the same for dwell as for the coordinate words. The programmed format will always adhere to the notation of the address. Incidentally, in some fixed cycles, the dwell is expressed by the P address, which does not take a decimal point at all and must be programmed with the leading zero suppression mode in effect. Half a second will be equal to P500.

◆ Decimal Point Programming

All modern programming will use the decimal point for dimensional input. Programming the decimal point, particularly for program data requiring a fractional portion, makes the CNC program much easier to develop and to read at a later date.

From all the available program addresses that can be used, not all can be programmed with the decimal point. The ones that can are those that specify the data in inches, millimeters or seconds (some exceptions exist).

The following two lists contain addresses where the decimal point is allowed in programs for both milling and turning controls:

➲ Milling control programs :

     X, Y, Z, I, J, K, A, B, C, Q, R

➲ Turning control programs :

     X, Z, U, W, I, K, R, C, E, F

The control system that supports the option of programming the decimal point, can also accept dimensional values *without* a decimal point, to allow compatibility with older programs. In such cases, it is important to understand the principles of programming format using the leading and the trailing zeros. If they are used correctly (see earlier explanations), there will be no problem to apply the various dimensional formats to any other control system, old or new. If possible, program the decimal point as a standard approach.

This compatibility enables many long time users to load their old programs (usually in the tape format), into the new CNC controls - but *not* the other way around - usually with minor modifications, or no modifications at all.

Some modern CNC units do not have the ability to accept an paper tape because they have no tape reader. To convert any tapes that contain good programs, there are two options - one, have someone to install a tape reader in the control, if possible and justified (probably not). The other method is to store the contents of a tape in the memory of a desktop computer. This method is very inexpensive and offers much better storage options than a paper tape. With suitable software and a portable tape reader, the task is not impossible. Keep also in mind that there are companies specializing in this kind of work.

The dimensional data in the metric system assume 0.001 mm minimum increment, while in the English system the increment is .0001 of an inch (leading zero suppression mode is in effect as a default):

Y12.56 *is*   Y125600 *... for English units*

Y12.56 *is*   Y12560   *... for metric units*

The programmed values with and without the decimal point can be mixed together in the same block:

N230 X4.0 Y-10

This may be beneficial for extreme conservation of the system memory. For example, the X4.0 word will require fewer characters than the word X40000 - on the other hand, the Y-10 is shorter than the decimal point equivalent of Y-0.001 (both examples are in English units). If all digits *before* or *after* the decimal point are zeros, they do not have to be written:

X0.5 = X.5
Y40.0 = X40.
Z-0.1 = Z-.1
F12.0 = F12.
R0.125 = R.125   *... etc.*

Any zero value *must* be written - for example, X0 cannot be written as X only. In this handbook, all the program examples use the decimal point format, whenever possible. Many programmers prefer to program zeros as in the left column of the example. They add a few characters into the system memory, but they are much easier to read. They are also easier for learning.

♦ **Input Comparison**

Differences in the input format for both the English and metric dimensioning can be seen clearly. One more time, the same examples will be shown, as before:

➲ English example - input of .625 inches :

Full format	X00006250
No leading zeros	X6250
No trailing zeros	X0000625
Decimal point	X0.625 *or* X.625

➲ Metric example - input of 0.42 mm :

Full format	X00000420
No leading zeros	X420
No trailing zeros	X0000042
Decimal point	X0.42 *or* X.42

CALCULATOR TYPE INPUT

In some specialized industries, such as woodworking or fabricating, the majority of dimensions (especially metric) do not require decimal parts, only whole numbers. In these cases, the decimal point would always be followed with a zero. Fanuc provides a solution to such situations by the feature called *calculator input*. Using this feature can shorten program size.

The calculator type input requires the setting of a system parameter. Once the parameter is set, the decimal point and the trailing zeros do not have to be written - they will be assumed. For example, X25 will be interpreted as X25.0, *not* the normally expected X0.0025.

In case the input value does require the decimal point, it can be written as usually. This means the values with a decimal point will be interpreted correctly and numbers without the decimal point will be treated as major units only (inches or millimeters). Here are some examples:

Standard Input	Calculator Input
X345.0	X345
X1.0	X1
Y0.67	Y0.67
Z7.48	Z7.48

Normally, the control system is set to the leading zero suppression mode and the non-decimal values are interpreted as the number of the smallest units. For example, Z1000 in G21 mode will be equivalent to Z1.0 (mm).

Both types of CNC machines, machining centers and lathes, use spindle rotation when removing excessive material from a part. The rotation may be that of the cutting tool (milling) or the part itself (lathes). In both cases, the activities of the machine spindle and the working feedrate of the cutting tool need to be strictly controlled by the program. These CNC machines require instructions that relate to the selection of a suitable *speed* of the machine spindle and a cutting *feedrate* for a given job.

There are several methods to control the spindle and cutting feedrate and they all depend mainly on the type of the CNC machine and the current machining application. In this chapter, we look at the *spindle control* and its programming applications.

SPINDLE FUNCTION

The program command relative to spindle speed is controlled in the CNC system by the address S. The programming format of the S address is usually within the range of 1 to 9999 and no decimal point is allowed:

S1 *to* S9999

For many high speed CNC machines is not unusual to have spindle speed available up to five digits, in the range of 1 to 99999, within the S address range:

S1 *to* S99999

The maximum spindle speed range available in the control must always be *greater* than the maximum spindle speed range of the machine itself. It is quite typical that virtually all control systems support a much greater range of spindle speeds than the CNC machine allows. In programming spindle speeds, the limitation is always caused by the machine unit, *not* by the control system.

◆ Spindle Speed Input

The address S relates to the machine spindle function, and must always be assigned a specific numeric value in the CNC program. There are three alternatives as to what the numeric value (input) of the spindle function may be:

❑ Spindle speed code number .. old controls - obsolete

❑ Direct spindle speed .. r/min

❑ Peripheral spindle speed .. ft/min or m/min

On the CNC lathes, all three alternatives may exist, depending on the control system. For the CNC milling systems, peripheral spindle speed is not applicable, but the spindle speed code number and the direct spindle speed are. The spindle speed selection by special code number is an obsolete concept, not required on modern controls.

The spindle speed designation S is not sufficient to be programmed by itself. In addition to the selected spindle speed address, certain additional attributes are necessary as well. These are attributes that control the spindle function environment. For example, if the spindle speed is specified as S400 in the program, the programming instruction is not complete, because the spindle function stands by itself in the program. It does not include *all* information the control system requires for the spindle data. A spindle speed value that is set, for example, to 400 r/min or 400 m/min or 400 in/min (depending on the machining application), does not contain all necessary information, namely, the spindle rotation *direction*.

Most machine spindles can be rotated in two directions - *clockwise* or *counterclockwise*, depending on the type and setup of the cutting tool used. The spindle rotation has to be specified in program, in addition to the spindle speed function. There are two miscellaneous functions provided by the control system that control the direction of the spindle - M03 and M04.

DIRECTION OF SPINDLE ROTATION

Thinking in terms of *right* and *left, up* and *down, clockwise* and *counterclockwise,* and similar directional terms, is thinking in terms that are *relative* to some known reference. To describe a spindle rotation as clockwise (CW), or as counterclockwise (CCW), some established and standard reference method is needed, in this case a reference point of view (reference viewpoint).

The direction of spindle rotation is always relative to the point of view that is established from the spindle side of the machine. This part of a machine that contains the spindle, and is generally called the machine headstock. Looking from the machine headstock area into *the direction along spindle center line and towards its face,* establishes the correct viewpoint for defining CW and CCW rotation of the spindle. For CNC drills, milling machines and CNC machining centers, the reference point of view is quite simple to understand. For CNC lathes, the rules are exactly the same, and will be described shortly.

◆ Direction for Milling

It may be rather impractical to look down along the center line of the spindle, perpendicularly towards the part. The common standard view is from the operator's position, facing the front of a vertical machine. Based on this view, the terms *clockwise* and *counterclockwise* can be used accurately, as they relate to the spindle rotation - *Figure 12-1*.

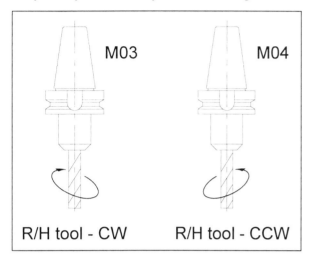

Figure 12-1
Direction of spindle rotation.
Front view of a vertical machining center is shown

◆ Direction for Turning

A comparable approach would seem logical for the CNC lathes as well. After all, the operator also faces the front of a machine, same as when facing a vertical machining center. *Figure 12-2* shows a front view of a typical CNC lathe.

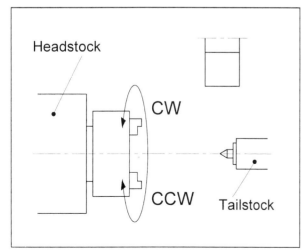

Figure 12-2
Typical view of a slant bed two axis CNC lathe.
CW and CCW directions only appear to be reversed

Although the descriptions CW and CCW in the illustration appear to be opposite to the direction of arrows, they are correct. The reason is that there are two possible points of view, and they are both using the spindle center line as the viewing axis. Only one of the viewpoints matches the standard definition and is, therefore, correct. The definition of spindle rotation for lathes is *exactly* the same as for machining centers.

> To establish spindle rotation as CW and CCW,
> look from the headstock towards the spindle face.

The first and proper method will establish the relative viewpoint starting *at the headstock* area of the lathe. From this position, looking towards the tailstock area, or into the same general area, the clockwise and counterclockwise directions are established correctly.

The second method of viewing establishes the relative viewpoint starting *at the tailstock* area, facing the chuck. This is an *incorrect* view!

Compare the following two illustrations - *Figure 12-3* shows the view from the headstock, *Figure 12-4* shows the view from the tailstock and arrows must be reversed.

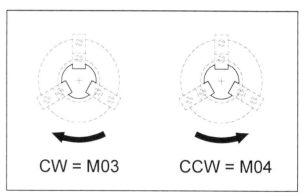

Figure 12-3
Spindle rotation direction as viewed from the headstock

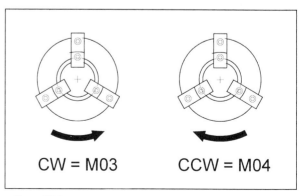

Figure 12-4
Spindle rotation direction as viewed from the tailstock

◆ Direction Specification

If the spindle rotation is clockwise, M03 function is used in the program - if the rotation is counterclockwise, M04 function is used in the program.

Since the spindle speed S in the program is dependent on the spindle rotation function M03 or M04, their relationship in a CNC program is important.

The spindle speed address S and spindle rotation function M03 or M04 must always be accepted by the control system together. One without the other will not mean anything to the control, particularly when the machine is switched on. There are at least two correct ways to program the spindle speed and spindle rotation:

❑ If the spindle speed and rotation are programmed together in the same block, the spindle speed and the spindle rotation will start simultaneously

❑ If the spindle speed and rotation are programmed in separate blocks, the spindle will not start rotating until both the speed and the rotation commands have been processed

◆ Spindle Startup

The following examples demonstrate a number of correct starts for the spindle speed and spindle rotation in the program. All examples assume that there is *no active setting* of the spindle speed S, either through a previous program or through the *Manual Data Input* (MDI). On CNC machines, there is no registered or default spindle speed when the machine power is turned on.

➲ Example A - Milling application :

```
N1 G20
N2 G17 G40 G80
N3 G90 G00 G54 X14.0 Y9.5
N4 G43 Z1.0 H01 S600 M03 (SPEED WITH ROTATION)
N5 ...
```

This example is one of the preferred formats for milling applications. Both the spindle speed and spindle rotation are set with the Z axis motion towards the part. Equally popular method is to start the spindle with the XY motion - N3 in the example:

```
N3 G90 G00 G54 X14.0 Y9.5 S600 M03
```

Selection is a matter of personal preference. The G20 in a separate block in not necessary for Fanuc controls.

➲ Example B - Milling application :

```
N1 G20
N2 G17 G40 G80
N3 G90 G00 G54 X14.0 Y9.5 S600      (SPEED ONLY)
N4 G43 Z1.0 H01 M03          (ROTATION STARTS)
N5 ...
```

This second example B is technically correct, but logically flawed. There is no benefit in splitting the spindle speed and spindle rotation into two blocks. This method makes the program harder to interpret.

➲ Example C - Milling application :

```
N1 G20
N2 G17 G40 G80
N3 G00 G90 G54 X14.0 Y9.5 M03    (ROTATION SET)
N4 G43 Z1.0 H01              (NO ROTATION)
N5 G01 Z0.1 F50.0 S600       (ROTATION STARTS)
N6 ...
```

Again, the C example is not wrong, but it is not very practical either. There is no danger, if the machine power has been switched on just prior to running this program. On the other hand, M03 *will* activate the spindle rotation, if another program was processed earlier. This could create a possibly dangerous situation, so follow a simple rule:

> Program M03 or M04 together with or
> after the S address, not before.

➲ Example D - Turning application with G50 :

```
N1 G20
N2 G50 X13.625 Z4.0 T0100
N3 G96 S420 M03  (SPEED SET - ROTATION STARTS)
N4 ...
```

This is the preferred example for CNC lathes, if the older G50 setting method is used. Because the spindle speed is set as CSS - *Constant Surface Speed*, the control system will calculate the actual revolutions per minute (r/min) based on the CSS value of 420 (ft/min) and the current part diameter at X13.625. The next example E is correct but not recommended (see caution box above).

➲ Example E - Turning application with G50 :

```
N1 G20
N2 G50 X13.625 Z4.0 T0100 M03     (ROTATION SET)
N3 G00 X6.0 Z0.1              (NO ROTATION)
N4 G96 G01 Z0 F0.04 T0101 S420 (ROTAT. STARTS)
N5 ...
```

➲ Example F - Turning application without G50 :

```
N1 G20 T0100
N2 G96 S420 M03   (SPEED SET - ROTATION STARTS)
N3 G00 ...
```

In this more contemporary example (G50 is not used as a position register command anymore), the machine spindle speed will be calculated for a tool offset value stored in the *Work Geometry Offset* register of the control system. The system will perform the calculation of actual r/min when the block N2 is executed.

These examples are only *technically correct* methods for a spindle start. All contain selected rotation at the beginning of a program and cover both, milling and turning applications. The example for the *beginning* of a program has been selected intentionally, because for any first tool in the program, there is no active speed or rotation in effect (normally carried on from a previous tool). However, the control unit may still store spindle speed and rotation from the last tool of the *previous* job!

Any tool *following* the first tool will normally assume the programmed speed selection and rotation for the previous tool. If only the spindle speed command S is programmed for the next tool, *without* the rotation direction, the tool will assume the *last* programmed rotation direction. If only the rotation direction code M03 or M04 is programmed, the spindle speed S will be the same as for the previous tool.

Be careful if a program contains the program stop functions M00 or M01, or the spindle stop function M05. Any one of them will automatically stop the spindle. It means to be absolutely sure as to when the spindle rotation will take place and what it will be. Always program the spindle speed selection and its rotation in the same block and for *each* tool. Both functions are logically connected and placing them within a single block will result in a coherent and logical program structure.

SPINDLE STOP

Normally, most work requires a spindle that rotates at a certain speed. In some cases, a rotating spindle is not always desirable. For example, before programming a tool change or reverse a part in the middle of a program, the spindle must be stopped first. The spindle must also be stopped during a tapping operation and at the end of program. Some miscellaneous functions will stop the spindle rotation automatically (for example, the functions M00, M01, M02 and M30). Spindle rotation will also stop automatically during certain fixed cycles. For a total control of program, the spindle stop should always be specified in the program. Counting on other functions to stop the spindle is not a good programming practice. There is a special function available in programming, to stop the spindle.

In order to stop the spindle rotation, use function M05. M05 will stop either the clockwise or the counterclockwise spindle rotation. Because M05 does not do anything else (unlike other functions that also stop the spindle, such as M00, M01, M02, M30 and others), it is used for situations, where the spindle must be stopped *without* affecting any other programmed activities. Some typical examples include reversal in tapping, tool motion to the indexing position, turret change position, or after machine zero return, depending on the application. Using one of the other miscellaneous functions that automatically stop the spindle, the M05 function is not required. On the other hand, it does no harm to program exactly what is required, in a particular order. This method may result in a slightly longer program, but it will be easier to read and maintain it, mainly by CNC operators with limited experience.

Spindle stop function can be programmed as a separate block, for example:

`N120 M05`

or in a program block containing the tool motion, such as in the next example:

`N120 Z1.0 M05`

The motion will always be completed first, *then* the spindle will be stopped. This is a safety feature built into the control system. Always remember to program M03 or M04 to reinstate the spindle rotation.

SPINDLE ORIENTATION

The last M function that also relates to a spindle activity, is M19. This function is most commonly used to set a machine spindle into an oriented position. Other M codes may be valid, depending on the control system, for example M20 on some controls. The spindle orientation function is a very specialized function, seldom appearing in the program itself. When M19 function is used, it is mainly during setup, in the *Manual Data Input* mode (MDI). This function is exclusive to the milling systems, because only specially equipped CNC lathes may require it. The function can only be used when the spindle is stationary, usually after the spindle has stopped. When the control system executes the M19 function, the following action will result:

The spindle will make a slight turn in both directions, clockwise and counterclockwise, and after a short period, the internal locking mechanism will be activated. In some cases, the locking that takes place is audible. The spindle will be locked in a *precise* position, and rotating it by hand, will not be successful. The exact locking position is determined by the machine tool manufacturer, indicated by the setting angle - *Figure 12-5*.

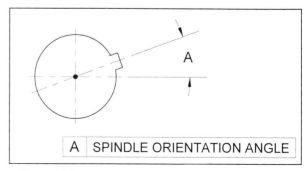

Figure 12-5

Spindle orientation angle is defined by the machine manufacturer and cannot be changed

In CNC machine tool operation, the M19 function enables the machine operator to place a tool into the spindle manually and guarantees a proper tool holder orientation. Later chapters will provide more details about spindle orientation and its applications, for example, in single point boring cycles.

> **WARNING** - An incorrect tool holder orientation may result in a damage to the part or machine.

Many CNC machining centers (not all) use tool holders that can be placed into the tool magazine only one way. To achieve this goal, the tool holder has a special notch built-in, that matches the internal design of the spindle - *Figure 12-6*. In order to find the side of the holder that has the notch, there is a small dimple on the notch side. This design is intentional.

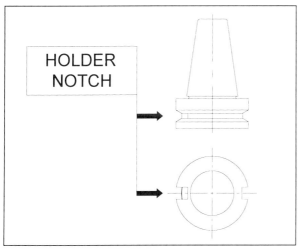

HOLDER NOTCH

Figure 12-6

Built-in notch in a tool holder used for correct tool orientation in the spindle - not all machines require this feature

For tools with several flutes (cutting edges), such as drills, end mills, reamers, face mills, etc., the orientation of cutting edge relative to the stopped spindle position is not that important. However, for single point tools such as boring bars, orientation of the cutting edge during setup is extremely important, particularly when certain fixed cycles are used. The two fixed cycles that use the *built-in* spindle orientation, G76 and G87, the tool retracts from the machined hole *without rotating*. In order to prevent damage to the finished bore, the tool retraction must be controlled. Spindle orientation guarantees that the tool will shift away from the finished bore into a clear direction. An accurate initial setup is necessary!

Those machines that allow placing tool holder into the spindle either way still require proper setting of tools that shift when G76 or G87 fixed cycles are programmed.

SPINDLE SPEED - R/MIN

When programming CNC machining centers, designate the spindle speed directly in *revolutions per minute* (r/min). A basic block that contains spindle speed of 200 r/min, for example, will require this data entry:

`N230 S200 M03`

Such format is typical to milling controls, where no peripheral speed is used. There is no need to use a special preparatory command to indicate the r/min setting, it is the control default. The r/min value must have a minimum increment of one. Fractional or decimal values are not allowed and the r/min must always be within the range of any machine specifications.

A few machining centers may be equipped with the option of a *dual* spindle speed selection - *direct r/min* and a *peripheral speed*. In this case, as well as for all lathe programming, a proper preparatory command is used to distinguish *which* selection is active. G96 is used for peripheral speeds, G97 for direct specification of r/min. The distinction between them is discussed next.

SPINDLE SPEED - SURFACE

Programmed spindle speed should be based on the machined *material* and the *cutting tool diameter* (machining centers), or the *part diameter* (lathes). The general rule is that the larger the diameter, the slower the spindle r/min must be. Spindle speed should never be guessed - *it should always be calculated*. Such a calculation will guarantee that the spindle speed is *directly proportional to the programmed diameter*. An incorrect spindle speed will have a negative effect on both the tool and the part.

◆ Material Machinability

To calculate spindle speed, each work material has a suggested machinability rating for a given tool material. This rating is either a percentage of some common material, such as mild steel, or a direct rating in terms of its *peripheral speed* or *surface speed*. Surface speed is specified in *feet per minute (ft/min)* in English units, and in *meters per minute (m/min)* in the metric system. An older abbreviation used for *ft/min* is *FPM*, meaning *Feet Per Minute*. The amounts of surface speeds indicate the level of machining difficulty with a *given tool material*. The lower the surface speed, the more difficult it is to machine the work material.

Note the emphasis on the words *'given tool material'*. To make all comparisons meaningful and fair, they must be made with the same type of cutting tool, for example, surface speeds for high speed steel tools will be much lower then those for cobalt based tools and, of course, for carbide tools.

Based on the surface speed and the cutter diameter (or the part diameter for lathes), machine spindle speed can be calculated in *revolutions per minute*, using one mathematical formula for English units system and another when metric units are programmed.

> Surface speed will increase for soft materials
> and decrease for hard materials.

> High speed steel tools will run slower
> than carbide tools in the same material.

◆ Spindle Speed - English Units

To calculate the spindle speed in *r/min,* the *peripheral* speed of the material for the cutting tool type must be known, as well as the diameter of the tool or the part:

$$r / min \ = \ \frac{12 \times ft / min}{\pi \times D}$$

☞ where . . .

r/min	=	Spindle speed in revolutions per minute
12	=	Multiplying factor - feet to inches
ft/min	=	Peripheral speed in *feet per minute*
π	=	Constant 3.1415927
D	=	Diameter in *inches* (cutter diameter for milling, or part diameter for turning)

➲ Example :

Peripheral speed for the selected material is 150 ft/min, and the cutting tool diameter is 1.75 inches:

```
r/min   =    (12 × 150) / (3.1415 × 1.75)
        =    327.4
        =    327 r/min
```

Many programming applications can use a shorter formula, without losing any significant accuracy:

$$r / min \ = \ \frac{3.82 \times ft / min}{D}$$

For less demanding calculations, the 3.82 constant may be rounded to 4 and used as an easier calculation without a calculator. The measuring units must be applied properly, otherwise the results will not be correct.

> Never mix English and metric units in the same program !

◆ Spindle Speed - Metric Units

When metric system is used in the program, the logic of previous formula is the same, but the units are different:

$$r / min \ = \ \frac{1000 \times m / min}{\pi \times D}$$

☞ where . . .

r/min	=	Spindle speed in revolutions per minute
1000	=	Multiplying factor - meters to mm
m/min	=	Peripheral speed in *m/min*
π	=	Constant 3.1415927
D	=	Diameter in *mm* (cutter diameter for milling, or part diameter for turning)

➲ Example:

Given surface speed is 30 m/min and the cutting tool diameter is 15 mm:

```
r/min   =    (1000 × 30) / (3.1415 × 15)
        =    636.6
        =    637 r/min
```

A shorter version of the formula is an acceptable alternative and almost as accurate as the precise formula:

$$r / min \ = \ \frac{318.3 \times m / min}{D}$$

Again, by replacing the constant 318.3 with constant 320 (or even 300), the r/min will be somewhat inaccurate, but most likely within an acceptable range.

CONSTANT SURFACE SPEED

On the CNC lathes, the machining process is different from the milling process. The turning tool has no diameter and the diameter of a boring bar has no relationship to the spindle speed. It is the *part diameter* that is the diameter used for spindle speed calculations. As the part is being machined, the diameter changes constantly. For example, during a facing cut or during roughing operations the diameter changes - see illustration in *Figure 12-7*. Programming the spindle speed in r/min is not practical - after all, which of the many diameters should be selected to calculate the r/min? The solution is to use the *surface speed* directly in the lathe program.

To select a surface speed is only a half of the procedure. The other half is to communicate this selection to the control system. The control has to be set to the *surface speed* mode, not the *r/min* mode. Operations as drilling, reaming, tapping, etc., are common on a lathe and they require the direct r/min in the program. To distinguish between the two alternatives in a lathe programming, the choice of the *surface speed* or the *revolutions per minute* must be specified. This is done with preparatory commands G96 and G97, *prior to* the spindle function:

```
G96 S.. M03    Surface speed selected
G97 S.. M03    Revolutions per minute selected
```

For milling, this distinction normally does not exist and spindle speed in *r/min* is always assumed.

By programming the surface speed command G96 for turning and boring, the control enters into a special mode, known as the *Constant Surface Speed* or *CSS*. In this mode, the actual spindle revolutions will *increase* and *decrease* automatically, depending on the diameter being cut (current diameter). The automatic *Constant Surface Speed* is built in the control systems available for most CNC lathes. It is a feature that not only saves programming time, it also allows the tool to remove *constant amount of material* at all times, thus saving the cutting tool from excessive wear and creating a better surface finish.

Figure 12-7 shows a typical example, when a facing cut starts at X6.2 (⌀6.2), and faces the part to the centerline (or slightly below). G96 S375 was used in the program, and 6000 r/min was the maximum spindle speed of the lathe.

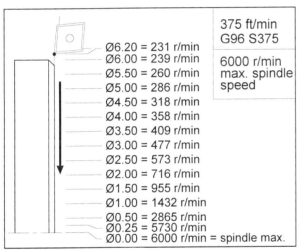

Figure 12-7

Example of a facing cut using constant surface speed mode G96

Although only selected diameters are shown in the illustration, along with their corresponding revolutions per minute, the updating process is constant. Note the sharp increase in r/min as the tool moves closer to machine center line. When the tool reaches X0 (⌀0.0), the speed will be at its maximum, within the current gear range. As this speed may be too high in some cases, the control system allows setting of a certain maximum, described later.

To program a peripheral speed for a CNC lathe, there are several options. In the following three examples, the most important ones will be examined. The gear change functions are omitted for all examples.

◆ Example 1 :

The surface speed is set right after the coordinate setting, using the G50 (or G92) command:

```
N1 G20
N2 G50 X16.0 Z5.0 T0100
N3 G96 S400 M03
...
```

In this quite common application, the actual spindle speed will be based on the current diameter of 16 inches, resulting in 95 r/min in block N3. In some cases, this will be too low. Consider another example:

◆ Example 2 :

On large CNC lathes, G50 setting of the X axis diameter is quite large, say ⌀24.0 inches. In the previous example, the target diameter of the next tool motion was not important, but in this case it is. For example:

```
N1 G20
N2 G50 X24.0 Z5.0 T0100
N3 G96 S400 M03
N4 G00 X20.0 T0101 M08
...
```

In the Example 2, the initial tool position is at X24.0 and the tool motion terminates at X20.0, both values are diameters. This translates to an actual motion of only 2.0 inches. At the X24.0, the spindle will rotate at 64 r/min, at X20.0 it will rotate at 76 r/min. The difference is very small to warrant any special programming. It is different, however, if the starting position is at a large diameter, but a tool moves to a much smaller target diameter.

◆ Example 3a :

From the initial position of ⌀24.0 inches, the tool will move to a rather small diameter of 2.0 inches:

```
N1 G20
N2 G50 X24.0 Z5.0 T0100
N3 G96 S400 M03
N4 G00 X2.0 T0101 M08
...
```

Spindle speed at the start of program (block N3) will be the same as in previous example, at 64 r/min. In the next block (N4), the speed calculated for ⌀2.0 inch will be 764 r/min, automatically calculated by the control. This rather large change in spindle speeds may have an adverse effect on some CNC lathes. What may happen is that the cutting tool will reach the ⌀2.0 inch *before* the spindle speed fully accelerates to the required 764 r/min. The tool may start removing material at a speed much slower than intended. In order to correct the problem, the CNC program needs to be modified:

● Example 3b :

The modification takes place in block N3. Instead of programming a constant surface speed mode, program the direct *r/min* for the target of ⌀2.0 inches, based on 400 ft/min surface speed. The r/min has to be calculated first, then the CSS setting will be programmed in a subsequent block:

```
N1 G20
N2 G50 X24.0 Z5.0 T0100
N3 G97 S764 M03
N4 G00 X2.0 T0101 M08
N5 G96 S400
. . .
```

In the example, at the ⌀24.0 (X24.0 in N2), the actual r/min would be only 64 r/min. At the ⌀2.0 (X2.0 in N4), the r/min will be 764. The cutting tool *may* reach X2.0 position before the spindle speed has accelerated to full 764 r/min, if it is not calculated and programmed earlier.

This technique is only useful if the CNC lathe does not support automatic time delay. Many modern lathes have a built-in timer, that forces the cutting tool to wait before actual cutting, until the spindle speed has fully accelerated.

Modern CNC lathes today *do not* use the G50 setting and use the *Geometry Offset* setting instead. In this case, the actual starting diameter at machine zero position is normally not known. Some experience can be useful in this case, otherwise program a short dwell before the actual cutting.

♦ **Maximum Spindle Speed Setting**

When the CNC lathe operates in the *Constant Surface Speed* mode, the spindle speed is directly related to the current part diameter. The smaller the work diameter is, the greater the spindle speed will be. So the natural question is - what will happen if the tool diameter is *zero*? It may seem impossible to ever program a zero diameter, but there are at least two cases when that is the case.

In the first case, zero diameter is programmed for all *center line operations*. All drilling, center drilling, tapping and similar operations are programmed at the zero diameter (X0). These operations are always programmed in the direct r/min mode, using G97 command. In G97 mode, the spindle speed is controlled directly, r/min does not change.

The second case of a zero diameter is when *facing off* a solid part all the way to the center line. This is a different situation. For all operations at X0, the cutting diameter does not change, because a direct r/min is programmed. During a face cutting operation, the diameter changes all the time while material removal continues until the tool reaches the spindle center line. No, don't reach for the formulas explained earlier. *Any calculation with a diameter in the formula being zero, will result in error!* Rest assured, there will *not* be 0 r/min at the center line of the spindle in the G96 mode. Return to *Figure 12-7* for illustration.

Whenever the CSS spindle mode is active and the tool reaches spindle center line at X0, the result will normally be the *highest* spindle rotation possible, *within the active gear range*. It is paradoxical, but that is exactly what will happen. Such situation is acceptable when the part is well mounted, does not extend from the chuck or fixture too far out, the tool is strong and robust, and so on. When the part is mounted in a special fixture, or an eccentric setup is used, when the part has a long overhang, or when some other adverse conditions are present, the maximum spindle speed at the center line may be *too high* for operating safety.

There is a simple solution to this problem, using a programming feature available for Fanuc and other controls. The CSS mode can be used with a *preset highest limit*, specified in revolutions per minute. The program function for the maximum spindle speed setting is normally G50. This maximum setting is sometimes called *maximum spindle speed clamping*. Do not confuse this G50 with its other meaning, position register preset. Here is an example:

```
O1201 (SPINDLE SPEED CLAMP)
N1 G20 T0100
N2 G50 X9.0 Z5.0 S1500        (1500 R/MIN MAX)
N3 M42                        (HIGH SPINDLE RANGE)
N4 G96 S400 M03               (CSS AND 400 FT/MIN)
N5 G00 G41 X5.5 Z0 T0101 M08
N6 G01 X-0.07 F0.012          (BELOW CENTER LINE)
N7 G00 Z0.1
N8 G40 X9.0 Z5.0 T0100
N9 M01
```

What actually happens in program O1201? Block N1 selects English units of measurement. The critical block N2 has *two meanings*:

❑ Sets only the tool coordinate position, as in:

    G50 X9.0 Z5.0

❑ *Also* sets the maximum r/min to 1500 as in:

    G50 X9.0 Z5.0 S1500

Block N3 selects the spindle gear range; block N4 sets the CSS mode, using 400 ft/min surface speed. Spindle rotation M03 is called in the same block. In block N5, the tool makes a rapid motion towards ⌀5.5 and the part front face. During rapid motion, tool nose radius offset and the coolant function are activated. The spindle speed at ⌀5.5 will be 278 r/min, using a formula described earlier in this chapter. Next block N6 is the actual facing cut. At the cutting feedrate of 0.012 in/rev, the tool tip faces off the blank to the center line. In reality, the end point is programmed on the other side of spindle center line. The tool point radius size must be taken into consideration when programming with the tool nose radius offset and to the machine center line. A special section later explains what exactly will happen during this cut.

Block N7 moves the tool tip .100 inches away from the face, at a rapid rate. In the remaining two blocks, the tool will rapid to the indexing position with a cancellation of radius offset in N8 and an optional program stop is provided in block N9.

Now, think of what happens in blocks N5 and N6. The spindle will rotate at the speed of 278 r/min at the ∅5.5. Since the CSS mode is in effect, as the tool tip faces off the part, the diameter is becoming smaller and smaller while the *r/min is constantly increasing*.

Without the maximum spindle speed limit in block N2, the spindle speed at the center line will be equivalent to the *maximum r/min* available within M42 gear range. A typical speed may be 3500 r/min or higher.

With the preset maximum spindle speed limit of 1500 r/min (G50 S1500), the spindle will be constantly increasing its speed, but only *until* it reaches the 1500 preset r/min, then it will *remain at that speed* for the rest of cut.

At the control, CNC operator can easily change the maximum limit value, to reflect true setup conditions or to optimize the cutting values.

Spindle speed is preset (or clamped) to the maximum *r/min* setting, by programming the S function together with the G50 preparatory command. If the S function is in a block not containing G50, the control will interpret it as a new spindle speed (CSS *or* r/min), active from that block on. *This error may be very costly!*

> Use caution when presetting maximum r/min of the spindle!

The maximum spindle speed can be clamped in a separate block or in a block that also includes the current tool coordinate setting. In the example O1201, block N2 contains both settings. Typically, the combined setting is useful at the beginning of a tool, the separate block setting is useful if the need arises to change the maximum spindle speed in the middle of a tool, for instance, between facing and turning cuts using the same tool.

To program the G50 command as a separate block, anywhere in the program, just issue the preparatory command combined with the spindle speed preset value. Such a block will have *no effect* whatsoever on any active coordinate setting, it represents just another meaning of G50 command. The following examples are all correct applications of G50 command for both, the coordinate setting and/or the maximum spindle speed preset:

`N12 G50 X20.0 Z3.0 S1500` *Double meaning*

`N38 G50 S1250` *Single meaning*

`N15 G50 X8.5 Z2.5` *Single meaning*

`N40 G50 Z4.75 S700` *Double meaning*

From these examples, G50 command should be easy to understand. There are two, completely independent, meanings of the G50 command. Either one can be programmed in a single block, or they can be separated into two individual blocks.

If the CNC lathe supports G92 instead of G50, keep in mind that they have exactly the same meaning and purpose. On lathes, the G50 command is more common than the G92 command but programming method is the same.

◆ Part Diameter Calculation in CSS

Often, knowing at *what diameter* the spindle will actually be clamped can be a useful information. Such knowledge may influence the preset value of spindle speed clamp. To find out at what diameter the *Constant Surface Speed* will remain fixed, the formula that finds the r/min at a given diameter must be reversed:

$$D = \frac{12 \times ft / min}{\pi \times r / min}$$

☞ where . . .

D	=	Diameter where CSS stops (in inches)
12	=	Multiplying factor - feet to inches
ft/min	=	Active surface speed
π	=	Constant 3.1415927
r/min	=	Preset maximum spindle speed

➲ Example - English units :

If the preset value in the program is G50 S1000 and the surface speed is selected as G96 S350, the CSS will be clamped when it reaches the ∅1.3369 inches:

```
D  =  (12 × 350) / (π × 1000)
   =  1.3369015
   =  ∅1.3369
```

The formula may be shortened:

$$D = \frac{3.82 \times ft / min}{r / min}$$

For completeness, the formulas based on the English system, can be adapted to a metric environment:

$$D = \frac{1000 \times m / min}{\pi \times r / min}$$

☞ where . . .

D	=	Diameter where CSS stops (in mm)
1000	=	Multiplying factor - meters to mm
m/min	=	Active surface speed
π	=	Constant 3.1415927
r/min	=	Preset maximum spindle speed

Just like in the English version, you may shorten the metric formula as well:

$$D \ = \ \frac{318.3 \times m/min}{r/min}$$

➡ Example - Metric units :

If the preset value in the program is G50 S1200 and the surface speed is selected as G96 S165, the CSS will be clamped when it reaches the ⌀43.768 mm:

```
D  =  (1000 × 165) / (π × 1200)
   =  43.767609
   =  ⌀43.768 mm
```

◆ CSS Calculation

The *Constant Surface Speed* (CSS) is required for most turning and boring operations on a CNC lathe. It is also the basic source of cutting data, from which the spindle speed is calculated for virtually all machining center operations. Now - consider a very common scenario - the CNC operator has optimized the current cutting conditions, including the spindle speed, so they are very favorable. Can these conditions be applied to subsequent jobs?

Yes, they can - provided that certain critical requirements will be satisfied:

❏ Machine and part setup are equivalent

❏ Cutting tools are equivalent

❏ Material conditions are equivalent

❏ Other common conditions are satisfied

If these requirements are met, the most important source data is the spindle speed actually used during machining. Once the optimum spindle speed is known, the cutting speed (CSS) can be calculated and used for any other tool diameter, providing the requirements above are met.

In a nutshell, the whole subject can be quickly summed up by categorizing it as a CSS calculation - that of *Constant Surface Speed*, also known as the *Cuting Speed* (CS), when the tool or part diameter and the spindle speed are known. From there on, it is a simple matter of formulas:

To calculate the *Cutting Speed* in English units:

$$ft/min \ = \ \frac{\pi \times D \times r/min}{12}$$

➡ EXAMPLE:

A ⌀5/8 inch drill works very well at 756 r/min - what is its cutting speed in ft/min?

```
ft/min = (3.14 × 0.625 × 756) / 12 = 123.64
```

To calculate the *Cutting Speed* in metric units:

$$m/min \ = \ \frac{\pi \times D \times r/min}{1000}$$

➡ EXAMPLE:

A ⌀7 mm end mill works very well at 1850 r/min - what is its cutting speed in m/min?

```
m/min = (3.14 × 7 × 1850) / 1000 = 40.66
```

The major benefit of using this method is a significant reduction of time spent at the CNC machine, usully required to find and 'fine-tune' the optimum spindle speed during setup or part optimization.

Feedrate is the closest programming companion to the spindle function. While the spindle function controls the spindle speed and the spindle rotation direction, feedrate controls *how fast the tool will move*, usually to remove excessive material (stock). In this handbook, the *rapid positioning*, sometimes called a *rapid motion* or *rapid traverse motion*, is not considered a true feedrate and will be described separately, in *Chapter 20*.

FEEDRATE CONTROL

> Cutting feedrate is the speed at which the
> cutting tool removes the material by cutting action.

The cutting action may be a rotary motion of the tool (drilling and milling, for example), the rotary motion of the part (lathe operations), or other action (flame cutting, laser cutting, water jet, electric discharge etc.). The feedrate function is used in the CNC program to select the feedrate value, suitable for the desired action.

Two feedrate types are used in CNC programming:

❑ Feedrate per minute

❑ Feedrate per revolution

The most common types of machines, CNC machining centers and lathes, can be programmed in either feedrate mode. In practice, it is much more common to use the *feedrate per minute* on machining centers and the *feedrate per revolution* on the lathes.

There is a significant difference in G codes used for machining centers and lathes.

FEEDRATE	Milling	Turning Group A	Turning Group B	Turning Group C
Per minute	G94	G98	G94	G94
Per revolution	G95	G99	G95	G95

> The G code *Group A* is the most commonly used
> on Fanuc controls and in this publication.

Another type of a special feedrate is called the *inverse time feedrate*. It is very seldom used and is not discussed in this handbook.

FEEDRATE FUNCTION

The calling address for a feedrate word in the program is the address F, followed by a number of digits. The number of digits following the address F depends on the feedrate mode and the machine tool application. Decimal place is usually allowed.

◆ Feedrate per Minute

For milling applications, all cutting feedrate in the linear and circular interpolation mode is programmed in *inches per minute (in/min)* or in *millimeters per minute (mm/min)*. The value of the feedrate is the distance a cutting tool will travel in one minute. This value is modal and is canceled only by another F address word. The main advantage of the feedrate per minute is that it is *not* dependent on the spindle speed. That makes it very useful in milling operations, using a large variety of tool diameters. Standard abbreviations for feedrate per minute are:

❑ Inches per minute in/min (or older ipm)

❑ Millimeters per minute mm/min

The most typical format for feedrate per minute is F3.1 for the English system and F4.1 for the metric system.

For example, the feedrate of 15.5 inches per minute, will be programmed as F15.5. In the metric system, the feedrate amount of 250 mm/min will appear in the program as F250.0. A slightly different programming format may be expected for special machine designs.

One important item to remember about feedrate is the *range* of the available feedrate values. The feedrate range of the control system *always exceeds* that of the machine servo system. For example, the feedrate range of a Fanuc CNC system is between .0001 and 24000.0 in/min or 0.0001 and 240000.0 mm/min. Note that the difference between the two units is only a decimal point shift, *not an actual translation*. In programming, only the feedrates that belong *within* the specified range can be used. Such a feedrate will be smaller than that for the control system.

In milling, the programming command (G code) for the *feedrate per minute* is G94. For most machines, it is set automatically, by the system default and does not have to be written in the program. For lathe operations, feedrate per minute is used very seldom. In *Group A*, the G code for feedrate per minute is G98, for *Groups B and C* it is G94. CNC lathes use primarily *feedrate per revolution* mode.

◆ Feedrate per Revolution

For the CNC lathe work, the feedrate is not measured in terms of time, but as the actual distance the tool travels in one spindle revolution (rotation). This *feedrate per revolution* is common on lathes (G99 for Group A). Its value is modal and another feedrate function cancels it (usually the G98). Lathes can also be programmed in *feedrate per minute* (G98), to control the feedrate when the spindle is stationary. Two standard abbreviations are used for *feedrate per revolution:*

- ❏ Inches per revolution   in/rev (or older ipr)
- ❏ Millimeters per revolution   mm/rev

The most typical format for feedrate per revolution is four decimal places in the English system and three decimal places in the metric system. This format means the feedrate of 0.083333 in/rev will be applied in the CNC program as F0.0833 on most controls. The metric feedrate example of 0.42937 mm/rev will be programmed as F0.429 on most controls. Many modern control systems accept feedrate of up to six decimal places for English units and five decimal for metric units.

Be careful when rounding feedrate values. For turning and boring operation, reasonably rounded feedrates are quite sufficient. Only in single point threading, the feedrate precision is critical for a proper thread lead, particularly for long or very fine threads. Some Fanuc controls can be programmed with up to six decimal places feedrate precision for threading only.

The programming command for the feedrate per revolution is G99. For most lathes, this is the system default, so it does not have to written in the program, unless the opposite command G98 is also used.

It is relatively more common to program a *feedrate per minute* (G98) for a CNC lathe program, than it is to program a *feedrate per revolution* (G95) in a milling program. The reason is that on a CNC lathe, this command controls the feedrate while the spindle is *not* rotating. For example, during a barfeed operation, a part stopper is used to 'push' the bar to a precise position in the chuck or a collet, or a pull-put finger to 'pull' the bar out. Rapid feed would be too fast and feedrate per revolution is not applicable. Feedrate per minute is used instead. In cases like these, the G98 and G99 commands are used in the lathe program as required. Both commands are modal and one cancels the other.

FEEDRATE SELECTION

To select the best feedrate, one that is most suitable for a given job, some general knowledge of machining is useful. This is an important part of programming process and should be done carefully. A feedrate selection depends on many factors, most notably on:

- ❏ Spindle speed - in rev/min
- ❏ Tool diameter [M] or the tool nose radius [T]
- ❏ Surface requirements of the part
- ❏ Cutting tool geometry
- ❏ Machining forces
- ❏ Setup of the part
- ❏ Tool overhang (extension)
- ❏ Length of the cutting motion
- ❏ Amount of material removal (depth or width of cut)
- ❏ Method of milling (climb or conventional)
- ❏ Number of flutes in the material (for milling cutters)
- ❏ Safety considerations

The last item is safety, always a programming responsibility number one, to assure the safety of the people and equipment. Safe speeds and feeds are only two aspects of safety awareness in CNC programming.

ACCELERATION AND DECELERATION

During a contouring operation, the direction of the cutting motion is changed quite often. There is nothing unusual about it, with all the intersections, tangency points and clearances. In contouring, it means that in order to program a sharp corner on a part, the tool motion along the X axis in one block will have to change into a motion along the Y axis in the next block. To make the change from one cutting motion to another, the control must *stop the X motion* first, then *start the Y motion*. Since it is impossible to *start* at a full feedrate instantly, without an *acceleration,* and equally impossible to *stop* a feedrate without a *deceleration,* a possible cutting error may occur. This error may cause the sharp corners on the profile to be cut with an undesirable overshoot, particularly during very high feedrates or extremely narrow angles. It only occurs during a cutting motion in G01, G02, G03 modes, not the rapid motion mode G00. During the rapid motion, the deceleration is automatic - and away from the part.

In a routine CNC machining, there is a small chance of ever encountering such an error. Even if the error is present, it will likely be within tolerances.

If the error does need correction, Fanuc controls provide two commands that will correct the problem:

G09	Exact stop	(one block only)
G61	Exact stop mode	(modal)

Exact stops increase the cycle time. For programs used on older machines, they may be required in some cases.

◆ Exact Stop Command

The first of two commands that control the feedrate when machining around corners is G09 command - *Exact Stop*. This is an *unmodal* command and has to be repeated in every block, whenever it is required.

In the program example O1301, there is no provision for acceleration and deceleration. That may cause uneven corners, due to the rather high feedrate of F90.0 (in/min):

```
O1301 (NORMAL CUTTING)
...
N13 G00 X15.0 Y12.0
N14 G01 X19.0 F90.0
N15 Y16.0
N16 X15.0
N17 Y12.0
...
```

By adding the G09 exact stop command in the program, the motion in that block will be fully completed *before* the motion in the other axis will start.

```
O1302 (G09 CUTTING)
...
N13 G00 X15.0 Y12.0
N14 G09 G01 X19.0 F90.0
N15 G09 Y16.0
N16 G09 X15.0
N17 Y12.0
...
```

Example O1302 guarantees a sharp corner at all three positions of the part. If only one corner is critical for sharpness, program the G09 command in the block that terminates at that corner (program O1303):

```
O1303 (G09 CUTTING)
...
N13 G00 X15.0 Y12.0
N14 G01 X19.0 F90.0
N15 G09 Y16.0
N16 X15.0
N17 Y12.0
...
```

The G09 command is useful only if a handful of blocks require the deceleration for a sharp corner. For a program where all corners must be precise, the constant repetition of the G09 is not very efficient.

◆ Exact Stop Mode Command

The second command that corrects an error at sharp corners is G61 - *Exact Stop Mode*. It is much more efficient than G09 and functions identically. The major difference is that G61 is a *modal* command that remains in effect until it is canceled by the G64 cutting mode command. G61 shortens the programming time, but not the cycle time. It is most useful when the G09 would be repeated too many times in the same program, making it unnecessarily too long.

```
O1304 (G61 CUTTING)
...
N13 G00 X15.0 Y12.0
N14 G61 G01 X19.0 F90.0
N15 Y16.0
N16 X15.0
N17 Y12.0
N18 G64
...
```

Note that the program example O1304 is identical in results to O1301. In both cases, the exact stop check applies to all cutting motions - unmodally in O1301, modally in O1304. Also note the additional block N18. It uses the G64 command - *normal cutting mode*. The normal cutting mode is the default setting when the machine power is turned on and is not usually programmed. *Figure 13-1* illustrates the tool motion with and without the G09/G61 command. The large overshoot amount is exaggerated only for the illustration, in reality it is very small.

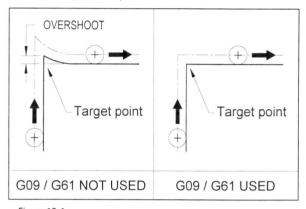

Figure 13-1

Feedrate control around corner - Exact Stop commands
The overshoot is exaggerated for clarity

◆ Automatic Corner Override

While a cutter radius offset is in effect for a milling cutter, the feedrate at the contour change points is normally *not* overridden. In a case like this, the preparatory command G62 can be used to automatically override the cutting feedrate at the corners of a part. This command is active until the G61 command (exact stop check mode), the G63 command (tapping mode), or the G64 command (cutting mode) is programmed.

◆ Tapping Mode

Programming in the tapping mode G63 will cause the control system to ignore any setting of the *feedrate override switch*, except the 100% setting. It will also cancel the function of the feedhold key, located on the control panel. The tapping mode will be canceled by programming the G61 command (exact stop check), or the G62 command (automatic corner override mode selection), or the G64 command (cutting mode selection).

◆ Cutting Mode

When the cutting mode G64 is programmed or is active by system default, it represents the *normal* cutting mode. When this command is active, the exact stop check G61 will *not* be performed, neither will the automatic corner override G62 or the tapping mode G63. That means the acceleration and deceleration will be done normally and the feedrate overrides will be effective. This is the most common default mode for the control system.

The *cutting mode* can be canceled by programming G61 command (exact stop mode), G62 command (automatic corner override mode) or G63 command (tapping mode).

The G64 command is not usually programmed, unless one or more of the other feedrate modes are used in the same program. To compare the G62 and the G64 modes, see illustration in *Figure 13-2*.

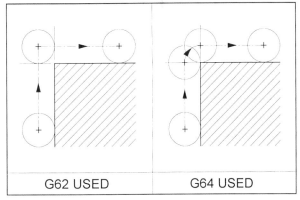

| G62 USED | G64 USED |

Figure 13-2
Corner override mode G62 and default G64 cutting mode

CONSTANT FEEDRATE

In *Chapter 29,* the topic is circular interpolation. In this chapter are detailed explanations and examples of maintaining a constant cutting feedrate for inside and outside arcs, from the *practical* point of view. At this point, the focus is on the *understanding* of the constant feedrate, rather than its *application*.

In programming, normal process is to calculate the coordinate values for all the contour change points, based on the part *drawing*. The cutter radius that produces the *center line* of the tool path is typically disregarded. When programming arcs to the drawing dimensions, rather than to the center line of the cutter, the feedrate applied to the programmed arc always relates to the *programmed* radius, *not* the actual radius cut at the tool center.

When the cutter radius offset is active and the tool path of the arc is offset by the cutter radius, the *actual* arc radius that is cut can be either *smaller* or *larger*, depending on the offset value for the cutting tool motion.

It is important to understand that the *effective* cutting radius will *decrease* in size for all internal arcs and it will *increase* in size for all external arcs. Since the cutting feedrate does not change automatically during the cutter radius offset mode, it must be adjusted in the program. Usually, this adjustment is not necessary, except in cases where the surface finish is of great importance or the cutter radius is very large. This consideration applies only to circular motions, not to linear cutting.

◆ Circular Motion Feedrates

Setting feedrates for circular motions is generally the same as for linear feedrates. In fact, most programs do not change feedrate for linear and circular tool motions. If the part surface finish is important, the 'normal' feedrate must be adjusted *higher* or *lower*, with consideration of the cutter radius, the type of radius cutting (outside or inside arc) and the cutting conditions. The larger the cutter radius, the more reason the cutting feedrate for programmed arcs will need some correction.

In case of arc cutting, the equidistant tool path (after applying cutter radius offset) may be *much larger* or *much smaller* than the arc programmed to drawing dimensions.

The feedrate for compensated arc motions is always based on the linear motion feedrate. Look for a more detailed explanation in *Chapter 29*, with an illustration and examples. First, here is the standard formula for calculating a linear feedrate:

$$F_l = r/min \times F_t \times n$$

☞ where . . .

$$
\begin{aligned}
F_l &= \text{Linear feedrate (in/min or mm/min)} \\
r/min &= \text{Spindle speed} \\
F_t &= \text{Feedrate per tooth (cutting edge)} \\
n &= \text{Number of cutting edges (flutes or inserts)}
\end{aligned}
$$

Based on the linear feedrate formula, the arc feedrate adjustments are influenced by the *side* of the machined arc - *outside* or *inside* arc. The linear feedrate should be *increased* for outside arcs and *decreased* for inside arcs.

For *outside* arcs, the feedrate is generally adjusted *upwards*, to a higher value:

$$F_o = \frac{F_l \times (R + r)}{R}$$

☞ where . . .

$$
\begin{aligned}
F_o &= \text{Feedrate for outside arc} \\
F_l &= \text{Linear feedrate} \\
R &= \text{Outside radius of the part} \\
r &= \text{Cutter radius}
\end{aligned}
$$

For *inside* arcs, the feedrate is generally adjusted *downwards*, to a lower value:

$$F_i = \frac{F_l \times (R - r)}{R}$$

☞ where . . .

F_i = Feedrate for inside arc
F_l = Linear feedrate
R = Inside radius of the part
r = Cutter radius

MAXIMUM FEEDRATE

The *maximum programmable feedrate* for the CNC machines is determined by the machine manufacturer, not the control manufacturer. For example, the maximum feedrate on a particular machine may be only 393 in/min, although the CNC system can support a feedrate several times greater. This is applicable to all controls, but there are additional programming considerations for CNC lathes, where the *feedrate per revolution* is the main method of programming a cutting tool.

◆ Maximum Feedrate Considerations

The maximum cutting feedrate per revolution is always restricted by the *programmed spindle speed (r/min)* and the *maximum rapid traverse rate* of the CNC lathe. It is quite easy to program the feedrate per revolution too high without even realizing it. This problem is most common in single point threading.

A CNC machine cannot deliver heavier feedrates than the maximum it was designed for, the results will not be accurate. Threading results could be unacceptable. When unusually heavy feedrates and fast spindle speeds are used in the same program, it is advisable to check whether the final feedrate does not exceed the maximum feedrate allowed on the given machine. It can be calculated the *maximum feedrate per revolution,* according to the following formula:

$$F_{max} = \frac{R_{max}}{r/min}$$

☞ where . . .

F_{max} = Max. allowed feedrate per revolution in/rev
R_{max} = Lower of the maximum feedrate,
selected from the X and the Z axis
r/min = Spindle speed in revolutions per minute

The R_{max} is either in *in/min* or *mm/min*, depending on the input units selected. In *Chapter 38* are details relating to the feedrate limits for threading.

FEEDHOLD AND OVERRIDE

While running a program, the programmed feedrate may be temporarily suspended or changed by using one of two available features of the control system. One is called a *feedhold* switch, the other is a *feedrate override* switch. Both switches are standard and allow the CNC operator to manually control the programmed feedrate during program execution. They are located on the operation panel.

◆ Feedhold Switch

Feedhold is a push button that can be toggled between *Feedhold ON* and *Feedhold OFF* modes. It can be used for both feedrate modes, the *feedrate per minute* or the *feedrate per revolution.* On many controls, the feedhold will stop not only a cutting feed with G01, G02, G03 in effect - it will also stop the rapid motion G00. Other program functions will remain active during a feedhold state.

For certain machining operations, the feedhold function is *automatically disabled* and becomes ineffective. This is typical for tapping and threading, using the G84 and G74 tapping cycles on machining centers and threading operations using the G32, G92 and G76 on lathes.

◆ Feedrate Override Switch

Feedrate override is normally controlled by means of a special rotary switch, located on the control panel of the CNC unit - *Figure 13-3*.

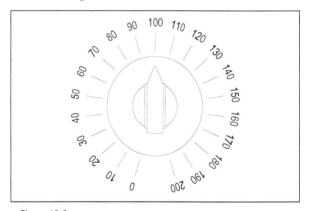

Figure 13-3
Typical feedrate override switch

This rotary switch has marked settings or divisions, indicating the *percentage of programmed feedrate.* A typical range of a feedrate override is 0 to 200%, where 0 may be no motion at all or the slowest motion, depending on the machine. The 200% setting *doubles* all programmed feedrates. A programmed feedrate of 12.0 in/min (F12.0) is the 100% feedrate. If the override switch is set to 80%, the actual cutting feedrate will be 9.6 in/min. If the switch is set to 110%, the actual cutting feedrate will be 13.2 in/min.

This simple logic applies to metric system as well. If the programmed feedrate is 300 mm/min, it becomes 100%. An 80% feedrate override results in 240 mm/min cutting feedrate and a 110% feedrate override setting is equivalent to 330 mm/min for the cutting tool.

The feedrate override switch works equally well for *feedrates per revolution*. For example, the programmed feedrate of .014 in/rev will result in actual feedrate of .0126 in/rev with 90% feedrate override and .0182 in/rev with 130% feedrate override. If a very precise feedrate per spindle revolution is required, be careful with the override settings. For example, programmed feedrate is F0.012, in *inches per revolution*. A change by one division on the override dial will either *increase* or *decrease* the programmed value by a full 10 percent. Therefore, the feedrate will be .0108 at 90%, .0120 at 100%, .0132 at 110%, etc. In most cases, the precise feedrate is not required, but keep in mind that some feedrates will not be accessible, for example, a feedrate of .0115 in/rev, because of the fixed 10% increments on the override switch.

In single point threading mode G32, the feedrate override switch is *disabled*. Feedrate override is also disabled for tapping cycles G84 and G74 on machining centers, and for single point threading cycles G92 and G76 on lathes. If the tapping mode is used for milling systems, with the command G63, both the *feedrate override* and the *feedhold* functions are disabled - *through the program !*

The control system offers two feedrate override functions for cutting motions *other than* tapping or threading cycles. They are M48 and M49. These are *programmable* functions, but may not be available for all controls.

◆ Feedrate Override Functions

Although the feedrate function uses the address F, two special miscellaneous functions M can be used in the program to set the *feedrate override ON* or *OFF*. On the operation panel, a switch is provided for feedrate override. If the CNC operator decides that programmed feedrate has to be temporarily *increased* or *decreased*, this switch is very handy. On the other hand, during machining operations, where the cutting feedrate *must* be used as programmed, the override switch has to be set to 100% only, not to any other setting.

A good example are special tapping operations *without* cycles, using G01 and G00 preparatory commands. Functions M48 and M49 are used precisely for such purposes:

M48	Feedrate override cancel function is OFF, which means feedrate override is *active*
M49	Feedrate override cancel function is ON, which means feedrate override is *inactive*

M48 function enables the CNC operator to use the feedrate override switch freely; the M49 function will cause feedrates to be executed *as programmed*, regardless of the feedrate override switch setting on the control panel. The most common usage of the two functions is for tapping or threading *without* a cycle, where the exact programmed feedrate *must* always be maintained. The following example shows the programming technique:

```
N10 S500 M03                    (USING TAP 12 TPI)
...
N14 G00 X5.0 Y4.0 M08
N15 Z0.25
N16 M49                 (DISABLE FEEDRATE OVERRIDE)
N17 G01 Z-0.625 F41.0 M05
N18 Z0.25 M04
N19 M48                 (ENABLE FEEDRATE OVERRIDE)
N20 G00 X.. Y.. M05
N21 M03
...
```

The tapping occurs between blocks N16 and N19 and the feedrate override is disabled for these blocks.

E ADDRESS IN THREADING

Some older CNC lathes use the feedrate address E for threading, rather than the more common address F.

The feedrate function E is similar to the F function. It also specifies the thread lead as *feedrate per revolution*, in in/rev or in mm/rev, but it has a *greater decimal place accuracy*. On older Fanuc control system model 6T, for example, the range for the thread lead is:

➲ English - Fanuc 6T control :

$$F = 0.0001 \quad to \quad 50.0000 \text{ in/rev}$$
$$E = 0.000001 \quad to \quad 50.000000 \text{ in/rev}$$

➲ Metric - Fanuc 6T control :

$$F = 0.001 \quad to \quad 500.000 \text{ mm/rev}$$
$$E = 0.0001 \quad to \quad 500.0000 \text{ mm/rev}$$

On the newest control models, FS-0/10/11/15/16T, the ranges are similar (there is no E address), but the safest way to find the available ranges is to lookup the specifications for your control system.

The E address is redundant on the newer controls and is retained only for compatibility with older programs that may be used on machines equipped with newer controls. The available threading feedrate ranges vary between different control systems, and depend on the type of feed screw and the input units used in the program.

Each numerically controlled machine using an automatic tool changer must have a special tool function (T function) that can be used in the program. This function controls the behavior of the *cutting tool*, depending on the type of machine tool. There are noticeable differences between T functions used on CNC machining centers and those used on CNC lathes. There are also differences between similar controls for the same machine type. The normal programming address for the tool function uses the address T.

For CNC machining centers, the T function typically controls the *tool number* only. For the CNC lathes, the function controls indexing to the *tool station number*, as well as the *tool offset number*.

T FUNCTION FOR MACHINING CENTERS

All vertical and horizontal CNC machining centers have a feature called the *Automatic Tool Changer*, abbreviated as *ATC*. In the program or MDI mode on the machine, this feature uses the function T, where the T address refers to a tool number selected by the programmer. The subsequent digits describe the tool number itself. On CNC machines with a manual tool change, the tool function may not be required at all.

Before programming for a particular CNC machining center begins, the *type of the tool selection* for that machine must be known. There are two major types of tool selection used in automatic tool change process:

❏   Fixed type

❏   Random memory type

To understand the difference between them, the first step is to understand the general principles of tool storage and tool selection, available for many modern CNC machining centers.

◆ Tool Storage Magazine

A typical CNC machining center (vertical or horizontal) is designed with a special *tool magazine* (sometimes called a *tool carousel*), that contains all tools required by the program. This magazine is not a permanent storage for the tools, but many machine operators keep the commonly used tools there at all times, if possible. A typical 20-tool magazine is illustrated in *Figure 14-1*.

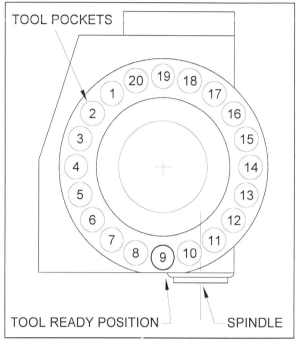

Figure 14-1

Typical side view of a 20-tool magazine

The capacity of such a magazine can be as small as ten or twelve tools and as high as several hundred tools on special machines. Typical medium size machining center may have between twenty and forty tools, larger machines will have more. The magazine is usually round or oval (larger capacity will be shaped in a zigzag form). It consists of a certain number of *pockets* - or *pots* - where the tool holder with a cutting tool is placed during setup. Each pocket is numbered in a consecutive order. It is important to know that the *pocket numbers are fixed* for each pocket. The magazine can be operated manually during setup and automatically, through the CNC program or MDI. The number of magazine pockets is the maximum number of tools that can be changed automatically on that machining center.

Within the travel of the tool magazine is one special position, used for the automatic tool change. This position is aligned with the tool changer and is commonly called the *waiting* position, the *stand-by* position, the *tool-ready* position, or just the *tool change* position.

◆ Fixed Tool Selection

A machining center that uses a fixed tool selection requires the CNC operator to place all tools into magazine pockets that *match the tool numbers*. For example, tool number 1 (called as T01 in the program) *must* be placed into the magazine pocket number 1, tool number 7 (called as T07 in the program) must be placed into the magazine pocket number 7, and so on.

The magazine pocket is mounted on a side of the CNC machine, usually away from the work area (work cube). With the fixed tool selection, the control system has no way of determining which tool number is in which magazine pocket number at any given time. The CNC operator has to match the tool numbers with the magazine pocket numbers during setup. This type of a tool selection is commonly found on many older CNC machining centers, or on some inexpensive machining centers.

Programming the tool is quite easy - whenever the T function number is used in the program, that will be the tool number selected during a tool change. For example,

```
N67 T04 M06
```

or

```
N67 M06 T04
```

or

```
N67 T04
N68 M06
```

simply means to bring tool number 4 into the spindle (the last method is preferred). What will happen to the tool that is in the spindle at that time? The M06 tool change function will cause the active tool to return to the magazine pocket it came from, before the new tool will be loaded. Usually, the tool changer takes the shortest way to select the new tool.

Today, this type of a tool selection is considered impractical and costly in a long run. There is a significant time loss during tool changes, because the machine tool has to wait until the selected tool is found in the magazine and placed into the spindle. The programmer can somewhat improve the efficiency by selecting tools and assigning tool numbers carefully, not necessarily in the order of usage. Examples in this handbook are based on a more modern type of the tool selection, called the *random memory*.

◆ Random Memory Tool Selection

This feature is the most common on modern machining centers. It also stores all tools required to machine a part in the tool magazine pockets, away from the machining area. The CNC programmer identifies each tool by a T number, usually in the order of usage. Calling the required tool number by the program will physically move the tool to the

waiting position within the tool magazine. This can happen simultaneously, while the machine is using another tool to cut a part. Actual tool change can take place anytime later. The is the concept of the *next tool waiting* - where the T function refers to the *next tool*, not the current tool. In the program, the next tool can be made ready by programming a few simple blocks:

```
T04                             (MAKE TOOL 4 READY)
...
```
   <... Machining with previous tool ...>
```
...
M06      (ACTUAL TOOL CHANGE - T04 IN SPINDLE)
T15                       (MAKE NEXT TOOL READY)
...
```
   <... Machining with tool 4 - T04 ...>
```
...
```

In the first block, the T04 tool was called into the waiting station of the tool magazine, while the previous tool was still cutting. When the machining has been completed, actual tool change will take place, where T04 will become the active tool. Immediately, the CNC system will search for the next tool (T15 in the example) and places it into the waiting position, while T04 is cutting.

This example illustrates that the T function will *not* make any physical tool change at all. For that, the *automatic tool change function* - M06 - also described later in this section, is needed and must be programmed.

Do not confuse the meaning of address T used with the fixed tool selection and the same address T used with the random tool selection. The former means the actual *number of the magazine pocket*, the latter means the *tool number of the next tool*. The tool call is programmed earlier than it is needed, so the control system can search for that tool while another tool is doing productive work.

◆ Registering Tool Numbers

Computers in general, and CNC systems in particular, can process given data very quickly and with the utmost precision. For the CNC work, the required data must be input first, to make the computer work in our favor. In the random tool selection method, the CNC operator is free to place *any tool* into *any magazine pocket*, as long as the actual setting is registered into the CNC unit, in the form of control system parameters. There is no need to worry too much about system parameters, just accept them as the collection of various system settings. Registering tool numbers has its own entry screen.

During the machine setup, the CNC operator will place the required tools into magazine pockets, writes down the numbers (which tool number is in which pocket number), and registers the information into the system. Such an operation is a normal part of the machine tool setup and various shortcuts can be used.

◆ Programming Format

Programming format for the T function used on milling systems depends on the maximum number of tools available for the CNC machine. Most machining centers have number of available tools under 100, although very large machines will have more tool magazines available (even several hundred). In the examples, two-digit tool function will used, covering tools within a range of T01 to T99.

In a typical program, the T01 tool command will call the tool identified in the setup sheet or a tooling sheet as tool number 1; T02 will call tool number 2, T20 will call tool number 20, etc. Leading zeros for tool number designation may be omitted, if desired - T01 can be written as T1, T02 as T2, etc. Trailing zeros *must* always be written, for example, T20 must be written as T20, otherwise the system will assume the leading zero and call the tool number 2 (T2 equals to T02, not T20).

◆ Empty Tool or Dummy Tool

Often, an empty spindle, free of any tool, is required. For this purpose, an empty tool station has to be assigned. Such a tool will also have to be identified by a unique number, even if no physical tool is used. If the magazine pocket or the spindle contains no tool, an *empty* tool number is necessary for maintaining the continuity of tool changes from one part to another. This nonexistent tool is often called the *dummy* tool or the *empty* tool.

The number of an empty tool should be selected as higher than the maximum number of tools. For example, if a machining center has 24 tool pockets, the *empty* tool should be identified as T25 or higher. It is a good practice to identify such a tool by the largest number within the T function format. For example, with a two digit format, the *empty* tool should be identified as T99, with a three digit format as T999. This number is easy to remember and is visible in the program.

As a rule, do not identify the *empty* tool as T00 - all tools *not assigned* may be registered as T00. There are, however, machine tools that do allow the use of T00, without possible complications.

TOOL CHANGE FUNCTION - M06

The tool function T, as applied to CNC machining centers, will *not* cause the actual tool change - the miscellaneous function M06 must be used in the program to do that. The purpose of tool change function is to exchange the tool in the spindle with the tool in the waiting position. The purpose of the T function for milling systems is to rotate the magazine and place the selected tool into the waiting position, where the actual tool change can take place. This *next tool search* happens while the control processes blocks following the T function call.

➲ Example :

```
N81  T01   ... makes T01 ready = loaded in the waiting position
N82  M06   ... brings T01 into the spindle
N83  T02   ... makes T02 ready = loaded in the waiting position
```

The three blocks appear to be simple enough, but let's explore them anyway. In block N81, the tool addressed as T01 in the program will be placed to the waiting position. The next block, N82, will activate the actual tool change - tool T01 will be placed into the spindle, ready to be used for machining. Immediately following the actual tool change is T02 in block N83. This block will cause the control system to search for the next tool, T02 in the example, to be placed into the waiting position. The search will take place *simultaneously* with the program data following block N83, usually a tool motion to the cutting position at the part. There will be no time lost, on the contrary, this method assures that the tool changing times will be always the same (the so called *chip-to-chip time*).

Some programmers prefer to shorten the program somewhat by programming the tool change command *together* with the next tool search in the same block. This method saves one block of program for each tool:

```
N81  T01
N82  M06  T02
```

The results will be identical - the choice is personal.

> Some machine tools will not accept the shortened two-block version and the three-block version must be programmed. If in doubt, always use the three-block version.

◆ Conditions for Tool Change

Before calling the M06 tool change function in the program, always create safe conditions. Most machines have a light located on the control panel for visual confirmation that the tool is at the tool change position.

The safe automatic tool change can take place only if these conditions are established:

❑ The machine axes had been zeroed

❑ The spindle must be fully retracted:

  (a) In Z axis at machine zero for vertical machines
  (b) In Y axis at machine zero for horizontal machines

❑ The X and Y axis positions of the tool must be selected in a clear area

❑ The next tool must be previously selected by a T function

A typical program sample illustrates the tool change between tools in the middle of the program (from T02 to T03) - graphically illustrated in *Figures 14-2* to *14-4*:

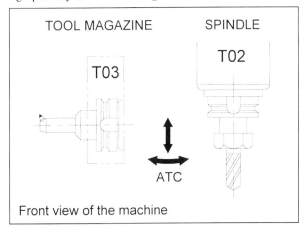

Figure 14-2
ATC example - Blocks N51 to N78 (current status)

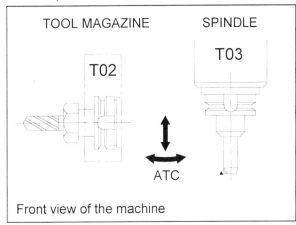

Figure 14-3
ATC example - Block N79 (actual tool change)

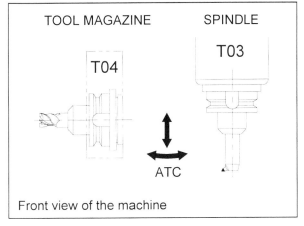

Figure 14-4
ATC example - Block N80 (new tool waiting = next tool)

⊃ Example for illustrations :

```
N51 ...                        (... T02 IN SPINDLE)
N52 ... T03    (... T03 READY FOR TOOL CHANGE)
. . .                       (MACHINING WITH T02)
N75 G00 Z1.0                  (RETRACT FROM DEPTH)
N76 G28 Z1.0 M05                  (T02 COMPLETED)
N77 M01                          (OPTIONAL STOP)
                      (BLANK LINE BETWEEN TOOLS)
N78 T03                    (T03 CALL REPEATED)
N79 M06          (T02 OUT - T03 IN THE SPINDLE)
N80 G90 G54 G00 X-18.56 Y14.43 S700 M03 T04
N81 . . .                  (MACHINING WITH T03)
```

In the example, block N76 represents the end of machining, using tool T02. It will cause tool T02 to move into the Z axis machine zero position, stopping the spindle at the same time. The optional program stop function M01 follows in the block N77.

In the following block N78, the call for T03 is repeated - this is not necessary, but may come very useful for repeating the tool later. Block N79 is the actual tool change. The T02 in the spindle will be replaced with T03 that is currently in the waiting position.

Finally, in block N80, the rapid motion in X and Y axes represents the first motion of T03, with spindle ON. Note the T04 at the block end. To save time, the next tool should be placed into the waiting position as soon as possible after the tool change.

Also note that when T02 is completed in block N77, it is *still in the spindle*! There are programmers who do not follow this method. If the tool change is included right *after* the G28 block (machine zero return) and *before* the M01 block, it will be more difficult for the operator to repeat the tool that just finished working, if it becomes necessary.

AUTOMATIC TOOL CHANGER - ATC

Several references to *Automatic Tool Changer* (ATC) were made in some examples. There are many designs of ATC's on various machines and they vary greatly from one machine manufacturer to another. Needless to say, the method of programming varies for different types, sometimes quite a bit. The machine tool changer, once it is setup, will automatically index the programmed cutting tool, in the proper order. Everything will be under program control. Programmer and operator should be thoroughly familiar with the type of ATC on all machining centers in the shop.

◆ Typical ATC System

A typical *Automatic Tool Change* system may have a double swing arm, one for the incoming tool, another for the outgoing tool. It will likely be based on *Random Memory* selection (described earlier), which means the next tool can be moved to a waiting position and be ready for a tool change, while the current tool works. This machine feature always guarantees the same tool change time. The typical time for the tool changing cycle can be very fast on modern CNC machines, often measured in fractions of a second.

The maximum number of tools that can be loaded into the tool magazine varies greatly, from as few as 10 to as many as 400 or more. A small CNC vertical machining center may have typically 10 to 30 tools. Larger machining centers will have a greater tool capacity.

Apart of the tool changer features, programmer and machine operator should be also aware of other technical considerations that may influence the tool change under program control. They relate to the physical characteristics of cutting tools when mounted in the tool holder:

❑ Maximum tool diameter

❑ Maximum tool length

❑ Maximum tool weight

◆ Maximum Tool Diameter

The maximum tool diameter that can be used without any special considerations is specified by the machine manufacturer. It assumes that a maximum diameter of a certain size may be used in *every* pocket of the tool magazine. Many machine manufacturers allow for a slightly larger tool diameter to be used, providing the *two adjacent* magazine pockets are empty *(Figure 14-5)*.

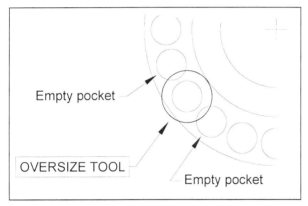

Figure 14-5
The adjacent pockets must be empty for a large tool diameter

For example, a machine description lists the maximum tool diameter *with* adjacent tools as 4 inches (100 mm). If both adjacent pockets are empty, the maximum tool diameter can be increased to 5.9 inches (150 mm), which may be quite a large increase. By using tools with a larger than recommended diameter, there is a *decrease* in the actual number of tools that can be placed in the tool magazine.

Adjacent pockets must be empty for oversize tools !

◆ Maximum Tool Length

The tool length in relation to the ATC, is the projection of a cutting tool from the spindle gauge line towards the part. The longer the tool length, the more important it is to pay attention to the Z axis clearance during the tool change. Any physical contact of the tool with the machine, the fixture or the part is extremely undesirable. Such a condition could be very dangerous - there is not much that can be done to interrupt the ATC cycle, except pressing the *Emergency Switch*, which is usually too late. *Figure 14-6* illustrates the concept of the tool length.

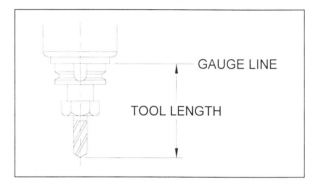

Figure 14-6
The concept of tool length

◆ Maximum Tool Weight

Most programmers will usually consider the tool diameter and the tool length, when developing a new program. However, some programmers will easily forget to consider the tool overall *weight*. Weight of the cutting tool does not generally makes a difference in programming, because the majority of tools are lighter than the maximum recommended weight. Keep in mind that the ATC is largely a mechanical device, and as such has certain load limitations. The weight of the tool is always the *combined weight* of the cutting tool *and* the tool holder, including collets, screws, pull studs and similar parts.

Do not exceed the recommended tool weight during setup !

For example, a given CNC machining center may have the maximum recommended tool weight specified as 22 pounds or about 10 kg. If even a *slightly* heavier tool is used, for example 24 lb. (10.8 Kg), the ATC should not be used at all - use a manual tool change for that tool only. The machine spindle may be able to withstand a slight weight increase but the tool changer may not. Since the word *'slight'* is only relative, the best advice in this case is - *do not overdo it!* If in doubt, always consult the manufacturer's recommendations. Examples in this chapter illustrate how to program such a unusual tool change, providing the tool weight is safe.

◆ ATC Cycle

A programmer does not have to know every detail related to the automatic tool changer actual operation. It is not a vital knowledge, although it may be quite a useful knowledge in many applications. On the other hand, a CNC machine operator should know *each and every step* of the ATC cycle inside out.

As an example, the following description is relevant to a typical CNC vertical machining center and may be a little different for some machines. Always study individual steps of the ATC operation - often, that knowledge will resolve a problem of a tool jam during the tool changing. This is a possible time loss that can be avoided. Some machines have a step-by-step cycle available with a special rotary switch, usually located near the tool magazine.

In the following example, a tool changer with a double arm swing system is used. It will take the cutting tool from the waiting position and exchange it with the tool currently in the machine spindle.

The ATC is a process that will execute the following order of steps when the tool change function M06 is programmed. All steps described are quite typical, but not necessarily standard for every CNC machining center, so take them only as a close example:

1. Spindle orients
2. Tool pot moves down
3. Arm rotates 60 degrees CCW
4. Tool is unclamped (in the magazine and spindle)
5. Arm moves down
6. Arms rotates 180 degrees CW
7. Arm moves up
8. Tool is clamped
9. Arm rotates 60 degrees CW
10. The rack returns
11. Tool pot moves up

The example is only presented as general information - its logic has to be adapted to each machine tool. The instruction manual for the machine usually lists relevant details about the ATC.

Regardless of the machine tool used, two conditions are always necessary to perform the ATC correctly:

❏ The spindle must be stopped (with the M05 function)

❏ The tool changing axis must be at the home position (machine reference position)

For CNC vertical machining centers, the tool changing axis is the Z axis, for the horizontal machining centers it is the Y axis. The M06 function will also stop the spindle, but never count on it. It is strongly recommended to stop the spindle with the M05 function (spindle stop) *before* the tool change cycle is executed.

◆ MDI Operation

Incidentally, each step of the tool change cycle can usually be executed through the MDI *(Manual Data Input)*, using special M functions. These functions are *only* used for service purposes, via the MDI operation and *cannot* be used in a CNC program. The benefit of this feature is that a tool changing problem can be traced to its cause and corrected from there. Check instructions for each machine to get details about these functions.

PROGRAMMING THE ATC

A number of possibilities exists in relation to the automatic tool changer. Some of the important ones are the number of tools used, what tool number is registered to the spindle (if any) at the start of a job, whether a manual tool change is required, whether an extra large tool is used, etc.

In the next several examples, some typical options will be presented - these examples can be used directly, if the CNC machine tool uses exactly the same format, or they can be *adapted* to a particular working environment. For the following examples, some conditions must be established that will help to understand the subject of programming a tool change much better.

To program the ATC successfully, all that is needed is the programming format for *three tools* - the *first* tool used, the tools used in the *middle* of the program and the *last* tool used in the program. To make the whole concept even easier to understand, the examples will use only four tool numbers - each tool number will *represent* one of the four available programming formats:

❏ T01 ... tool designation represents the first tool used in the CNC program

❏ T02 ... tool designation represents any tool in the CNC program between the first and the last tool

❏ T03 ... tool designation represents the last tool used in the CNC program

❏ T99 ... tool designation represents an empty tool (dummy tool) as an empty tool pocket identification

In all examples, the first three tools will always be used, the empty tool only if required. Hopefully, these examples will illustrate the concept of many possible ATC applications. Another possible situation is in situations where only *one tool* is used in the CNC program.

◆ Single Tool Work

Certain jobs or special operations may require only one tool to do the job. In this case, the tool is generally mounted in the spindle during setup and no tool calls or tool changes are required in the program:

```
O1401       (FIRST TOOL IN THE SPINDLE AT START)
N1 G20                            (INCH MODE)
N2 G17 G40 G80                    (SAFE BLOCK)
N3 G90 G54 G00 X.. Y.. S.. M03    (TOOL MOTION)
N4 G43 Z.. H01 M08               (APPROACH WORK)
...
       <...T01 working...>

...
N26 G00 Z.. M09             (T01 MACHINING DONE)
N27 G28 Z.. M05                  (T01 TO Z-HOME)
N28 G00 X.. Y..             (SAFE XY POSITION)
N29 M30                      (END OF PROGRAM)
%
```

Unless the tool is in the way of part changing, it remains in the spindle permanently for the job.

◆ Programming Several Tools

Machining a part using several tools is the most typical method of CNC work. Each tool is loaded into the spindle when required, using various ATC processes. From the programming viewpoint, the various tool changing methods do not affect the cutting section of the program, only the *start* of the tool (before machining) or the *end* of the tool (after machining).

As already discussed, the required tool can be changed automatically, only if the Z axis is at machine zero (for vertical machining centers) or the Y axis is at machine zero (for horizontal machining centers). The tool position in the remaining axes is only important to the safety of the tool change, so there is no tool contact with the machine, the fixture, or the part. All following examples are formatted for vertical machine models. Some programs use machine zero return for *all* axes at the end of last tool, for example:

```
...
N393 G00 Z.. M09        (CURRENT TOOL WORK DONE)
N394 G28 Z.. M05        (CURRENT TOOL TO Z HOME)
N395 G28 X.. Y..        (CURRENT TOOL TO XY HOME)
N396 M30                     (END OF PROGRAM)
%
```

Technically, there is nothing wrong with this practice, but it may cause a significant time loss for a large volume of parts. A preferred method is to either make the tool change *above* the last tool location, or to move the tool *away* from the part, to a *safe* location. This last method is illustrated in the examples that present various methods of program startup, as it relates to different methods of tool changing.

◆ Keeping Track of Tools

If the tool changing operation is simple, it should be easy to keep a track of where each tool is at any given moment. In later examples, more complex tool changes will take place. Keeping a visual track of which tool is waiting and which tool is in the spindle can be done with a 3 column table with block number, tool waiting and tool in the spindle.

Block Number	Tool Waiting	Tool in Spindle

To fill the table, start from the program top and find every occurrence of the T address and M06 function. All other data are irrelevant. In the example O1402, the table will be filled as a practical sample of usage.

◆ Any Tool in Spindle - Not the First

This is the most common method of programming an ATC. The operator sets all tools in the magazine, registers the settings but leaves the last tool measured in the spindle. On most machines, this tool should *not* be the first tool. The programmer matches this tool changing method within the program. The following example is probably the one that may be the most useful for everyday work. All activities are listed in the comments.

```
O1402              (ANY TOOL IN SPINDLE AT START)
                   (**** NOT THE FIRST TOOL ****)
N1 G20                            (INCH MODE)
N2 G17 G40 G80 T01            (GET T01 READY)
N3 M06                        (T01 TO SPINDLE)
N4 G90 G54 G00 X.. Y.. S.. M03 T02 (T02 READY)
N5 G43 Z.. H01 M08           (APPROACH WORK)
...
       <...T01 working...>

...
N26 G00 Z.. M09             (T01 MACHINING DONE)
N27 G28 Z.. M05                  (T01 TO Z HOME)
N28 G00 X.. Y..             (SAFE XY POSITION)
N29 M01                       (OPTIONAL STOP)

N30 T02                    (T02 CALL REPEATED)
N31 M06                       (T02 TO SPINDLE)
N32 G90 G00 G54 X.. Y.. S.. M03 T03 (T03 READY)
N33 G43 Z.. H02 M08          (APPROACH WORK)
...
       <...T02 working...>

...
N46 G00 Z.. M09             (T02 MACHINING DONE)
N57 G28 Z.. M05                  (T02 TO Z HOME)
N48 G00 X.. Y..             (SAFE XY POSITION)
N49 M01                       (OPTIONAL STOP)

N50 T03                    (T03 CALL REPEATED)
N51 M06                       (T03 TO SPINDLE)
N52 G90 G00 G54 X.. Y.. S.. M03 T01 (T01 READY)
N53 G43 Z.. H03 M08          (APPROACH WORK)
...
       <...T03 working...>

...
N66 G00 Z.. M09             (T03 MACHINING DONE)
N67 G28 Z.. M05                  (T03 TO Z HOME)
N68 G00 X.. Y..             (SAFE XY POSITION)
N69 M30                      (END OF PROGRAM)
%
```

The filled-in table below shows the status of tools for the *first part only*. '?' represents *any* tool number.

Block Number	Tool Waiting	Tool in Spindle
N1	?	?
N2	T01	?
N3	?	T01
N4	T02	T01
T01 WORKING		
N30	T02	T01
N31	T01	T02
N32	T03	T02
T02 WORKING		
N50	T03	T02
N51	T02	T03
N52	T01	T03
T03 WORKING		

When the *second part* is machined and any other part after that, the tools tracking is simplified and consistent. Compare the next table with the previous one - there are no question marks. The table shows where each tool is.

Block Number	Tool Waiting	Tool in Spindle
N1	T01	T03
N2	T01	T03
N3	T03	T01
N4	T02	T01
T01 WORKING		
N30	T02	T01
N31	T01	T02
N32	T03	T02
T02 WORKING		
N50	T03	T02
N51	T02	T03
N52	T01	T03
T03 WORKING		

Examples shown here use this method as is or slightly modified. For most jobs, there is no need to make a tool change at XY safe position, if the work area is clear of obstacles. Study this method before the others. It will help to see the logic of some more advanced methods a lot easier.

A few comments to the O1402 example. Always program M01 optional stop *before* a tool change - it will be easier to repeat the tool, if necessary. Also note beginning of each tool, containing the next tool search. The tool in the block containing the first motion has already been called - compare block N4 with N30 and block N32 with N50. The repetition of the tool search at the start of each tool has two reasons. It makes the program easier to read (tool is coming into the spindle will be known) and it allows a *repetition* of the tool, regardless of which tool is currently in the spindle.

◆ First Tool in the Spindle

Program may also start with the first tool in the spindle. This is a common practice for the ATC programming. The first tool in the program must be loaded into the spindle during setup. In the program, the first tool is called to the waiting station (ready position) during the *last* tool - *not the first tool*. Then, a tool change will be required in one of the last blocks in the program. The first tool in the program must be *first* for *all parts* within the job batch.

```
O1403              (FIRST TOOL IN SPINDLE AT START)
N1 G20                                    (INCH MODE)
N2 G17 G40 G80 T02                  (GET T02 READY)
N3 G90 G54 G00 X.. Y.. S.. M03
N4 G43 Z.. H01 M08                 (APPROACH WORK)
...
    <...T01 working...>
...
N26 G00 Z.. M09                (T01 MACHINING DONE)
N27 G28 Z.. M05                   (T01 TO Z HOME)
N28 G00 X.. Y..              (SAFE XY POSITION)
N29 M01                          (OPTIONAL STOP)

N30 T02                       (T02 CALL REPEATED)
N31 M06                          (T02 TO SPINDLE)
N32 G90 G54 G00 X.. Y.. S.. M03 T03 (T03 READY)
N33 G43 Z.. H02 M08                (APPROACH WORK)
...
    <...T02 working...>
...
N46 G00 Z.. M09                (T02 MACHINING DONE)
N47 G28 Z.. M05                   (T02 TO Z HOME)
N48 G00 X.. Y..              (SAFE XY POSITION)
N49 M01                          (OPTIONAL STOP)

N50 T03                       (T03 CALL REPEATED)
N51 M06                          (T03 TO SPINDLE)
N52 G90 G54 G00 X.. Y.. S.. M03 T01 (T01 READY)
N53 G43 Z.. H03 M03                (APPROACH WORK)
...
    <...T03 working...>
...
N66 G00 Z.. M09                (T03 MACHINING DONE)
N67 G28 Z.. M05                   (T03 TO Z HOME)
N68 G00 X.. Y..              (SAFE XY POSITION)
N69 M06                          (T01 TO SPINDLE)
N70 M30                         (END OF PROGRAM)
%
```

This method is not without a disadvantage. Since there is always a tool in the spindle, it may become an obstacle during setup or part changing. The solution is to program the tool change in such a way that there is *no tool* in the spindle during the part setup (*spindle empty* condition).

◆ No Tool in the Spindle

An empty spindle at the start and end of each machined part is less productive than starting with the first tool in the spindle. One extra tool change increases the cycle time. An empty spindle at start should only be used if the programmer has a valid reason, for example, to recover space above the part that would otherwise be occupied by the tool. The recovered space may be useful for removing the part, for example, with a crane or a hoist. The programming format for this situation is not much different from the previous example - except that there is an *extra tool change* at the end of the program. This tool change brings the first tool back into the spindle, for consistent startup of each program run.

```
O1404            (NO TOOL IN SPINDLE AT START)
N1 G20                            (INCH MODE)
N2 G17 G40 G80 T01             (GET T01 READY)
N3 M06                        (T01 TO SPINDLE)
N4 G90 G54 G00 X.. Y.. S.. M03 T02 (T02 READY)
N5 G43 Z.. H01 M08            (APPROACH WORK)
...
   <... T01 working...>

...
N26 G00 Z.. M09          (T01 MACHINING DONE)
N27 G28 Z.. M05               (T01 TO Z HOME)
N28 G00 X.. Y..            (SAFE XY POSITION)
N29 M01                        (OPTIONAL STOP)

N30 T02                     (T02 CALL REPEATED)
N31 M06                       (T02 TO SPINDLE)
N32 G90 G54 G00 X.. Y.. S.. M03 T03 (T03 READY)
N33 G43 Z.. H02 M08           (APPROACH WORK)
...
   <... T02 working...>

...
N46 G00 Z.. M09          (T02 MACHINING DONE)
N47 G28 Z.. M05               (T02 TO Z HOME)
N48 G00 X.. Y..            (SAFE XY POSITION)
N49 M01                        (OPTIONAL STOP)

N50 T03                     (T03 CALL REPEATED)
N51 M06                       (T03 TO SPINDLE)
N52 G90 G54 G00 X.. Y.. S.. M03 T99 (T99 READY)
N53 G43 Z.. H03 M08           (APPROACH WORK)
...
   <... T03 working...>

...
N66 G00 Z.. M09          (T03 MACHINING DONE)
N67 G28 Z.. M05               (T03 TO Z-HOME)
N68 G00 X.. Y..            (SAFE XY POSITION)
N69 M06                       (T99 TO SPINDLE)
N70 M30                       (END OF PROGRAM)
%
```

◆ First Tool in the Spindle with Manual Change

In the next example, the second tool represents *any middle* tool in the program using three or more tools. This tool may be too heavy or too long and cannot be indexed through the ATC cycle and must be loaded manually. This tool change can be done by the operator, but only if the *program supports manual tool change*. To achieve this goal is to use M00 program stop with an appropriate comment, describing the reason for the stop. Optional stop M01 is *not* a good selection - M00 is a much safer choice - it will *always* stop the machine without interference from the operator.

Follow the next example carefully, to understand how a manual tool change can be performed when the first tool is in the spindle. T02 in the example will be changed manually by the CNC operator.

```
O1405            (FIRST TOOL IN SPINDLE AT START)
N1 G20                            (INCH MODE)
N2 G17 G40 G80 T99             (GET T99 READY)
N3 G90 G54 G00 X.. Y.. S.. M03
N4 G43 Z.. H01 M08            (APPROACH WORK)
...
   <... T01 working...>

...
N26 G00 Z.. M09          (T01 MACHINING DONE)
N27 G28 Z.. M05               (T01 TO Z HOME)
N28 G00 X.. Y..            (SAFE XY POSITION)
N29 M01                        (OPTIONAL STOP)

N30 T99                     (T99 CALL REPEATED)
N31 M06                       (T99 TO SPINDLE)
N32 T03                          (T03 READY)
N33 M00          (STOP AND LOAD T02 MANUALLY)

N34 G90 G54 G00 X.. Y.. S.. M03 (NO NEXT TOOL)
N35 G43 Z.. H02 M08           (APPROACH WORK)
...
   <... T02 working...>

...
N46 G00 Z.. M09          (T02 MACHINING DONE)
N47 G28 Z.. M05               (T02 TO Z HOME)
N48 G00 X.. Y..            (SAFE XY POSITION)
N49 M19                   (SPINDLE ORIENTATION)
N50 M00        (STOP AND UNLOAD T02 MANUALLY)

N51 T03                     (T03 CALL REPEATED)
N52 M06                       (T03 TO SPINDLE)
N53 G90 G54 G00 X.. Y.. S.. M03 T01 (T01 READY)
N54 G43 Z.. H03 M08           (APPROACH WORK)
...
   <... T03 working...>

...
N66 G00 Z.. M09          (T01 MACHINING DONE)
N67 G28 Z.. M05               (T03 TO Z HOME)
N68 G00 X.. Y..            (SAFE XY POSITION)
N69 M01                        (OPTIONAL STOP)
N70 M06                       (T01 TO SPINDLE)
N71 M30                       (END OF PROGRAM)
%
```

Note the M19 function in the block N49. This miscellaneous function will orient the spindle to exactly the same position as if the automatic tool changing cycle were used. The CNC operator can then replace the current tool with the next tool and still maintain the tool position orientation. This consideration is mostly important for certain boring cycles, where the tool bit cutting edge has to be positioned away from the machined surface. If a boring bar is used, it is necessary to align its cutting tip.

◆ No Tool in the Spindle with Manual Change

The following program is a variation on the previous example, except that there is no tool in the spindle when the program starts.

```
O1406              (NO TOOL IN SPINDLE AT START)
N1 G20                                (INCH MODE)
N2 G17 G40 G80 T01                 (GET T01 READY)
N3 M06                            (T01 TO SPINDLE)
N4 G90 G54 G00 X.. Y.. S.. M03 T99 (T99 READY)
N5 G43 Z.. H01 M08                (APPROACH WORK)
...
    <...T01 working...>
...
N26 G00 Z.. M09              (T01 MACHINING DONE)
N27 G28 Z.. M05                   (T01 TO Z HOME)
N28 G00 X.. Y..               (SAFE XY POSITION)
N29 M01                          (OPTIONAL STOP)

N30 T99                       (T99 CALL REPEATED)
N31 M06                          (T99 TO SPINDLE)
N32 T03                               (T03 READY)
N33 M00            (STOP AND LOAD T02 MANUALLY)
N34 G90 G54 G00 X.. Y.. S.. M03 (NO NEXT TOOL)
N35 G43 Z.. H02 M08               (APPROACH WORK)
...
    <...T02 working...>
...
N46 G00 Z.. M09              (T02 MACHINING DONE)
N47 G28 Z.. M05                   (T02 TO Z HOME)
N48 G00 X.. Y..               (SAFE XY POSITION)
N49 M19                      (SPINDLE ORIENTATION)
N50 M00          (STOP AND UNLOAD T02 MANUALLY)

N51 T03                       (T03 CALL REPEATED)
N52 M06                          (T03 TO SPINDLE)
N53 G90 G54 G00 X.. Y.. S.. M03 T99(T99 READY)
N54 G43 Z.. H03 M08               (APPROACH WORK)
...
    <...T03 working...>
...
N66 G00 Z.. M09              (T03 MACHINING DONE)
N67 G28 Z.. M05                   (T03 TO Z HOME)
N68 G00 X.. Y..               (SAFE XY POSITION)
N69 M01                          (OPTIONAL STOP)
N70 M06                          (T99 TO SPINDLE)
N71 M30                          (END OF PROGRAM)
%
```

◆ First Tool in the Spindle and an Oversize Tool

Sometimes it is necessary to use a little larger diameter tool than the machine specifications allow. In that case, the oversize tool *must* return to the same pocket in the tool magazine it came from and the two adjacent magazine pockets must be empty. *Do not use a tool that is too heavy!* In the example O1407, the large tool is T02.

```
O1407           (FIRST TOOL IN SPINDLE AT START)
N1 G20                                (INCH MODE)
N2 G17 G40 G80 T99                 (GET T99 READY)
N3 G90 G54 G00 X.. Y.. S.. M03
N4 G43 Z.. H01 M08                (APPROACH WORK)
...
    <...T01 working...>
...
N26 G00 Z.. M09              (T01 MACHINING DONE)
N27 G28 Z.. M05                   (T01 TO Z HOME)
N28 G00 X.. Y..               (SAFE XY POSITION)
N29 M01                          (OPTIONAL STOP)

N30 T99                       (T99 CALL REPEATED)
N31 M06                          (T99 TO SPINDLE)
N32 T02                               (T02 READY)
N33 M06                          (T02 TO SPINDLE)
N34 G90 G54 G00 X.. Y.. S.. M03 (NO NEXT TOOL)
N35 G43 Z.. H02 M08               (APPROACH WORK)
...
    <...T02 working...>
...
N46 G00 Z.. M09              (T02 MACHINING DONE)
N47 G28 Z.. M05                   (T02 TO Z HOME)
N48 G00 X.. Y..               (SAFE XY POSITION)
N49 M01                          (OPTIONAL STOP)

N50 M06    (T02 OUT OF SPINDLE TO THE SAME POT)
N51 T03                               (T03 READY)
N52 M06                          (T03 TO SPINDLE)
N53 G90 G54 G00 X.. Y.. S.. M03 T01(T01 READY)
N54 G43 Z.. H03 M08               (APPROACH WORK)
...
    <...T03 working...>
...
N66 G00 Z.. M09              (T03 MACHINING DONE)
N67 G28 Z.. M05                   (T03 TO Z HOME)
N68 G00 X.. Y..               (SAFE XY POSITION)
N69 M01                          (OPTIONAL STOP)
N70 M06                          (T01 TO SPINDLE)
N71 M30                          (END OF PROGRAM)
%
```

◆ No Tool in the Spindle and an Oversize Tool

This is another tool change version. It assumes no tool in the spindle at the program start. It also assumes the next tool is *larger* than the maximum recommended diameter, within reason. In this case, the oversize tool must return to *exactly* the same pocket it came from. It is important that the adjacent pockets are both empty.

> Both adjacent magazines pockets
> must be empty for an oversize tool diameter !

In the O1408 example, T02 represents the large tool.

```
O1408           (NO TOOL IN SPINDLE AT START)
N1 G20                          (INCH MODE)
N2 G17 G40 G80 T01           (GET T01 READY)
N3 M06                    (T01 TO SPINDLE)
N4 G90 G54 G00 X.. Y.. S.. M03 T99 (T99 READY)
N5 G43 Z.. H01 M08          (APPROACH WORK)
...
    <...T01 working...>
...
N26 G00 Z.. M09        (T01 MACHINING DONE)
N27 G28 Z.. M05            (T01 TO Z HOME)
N28 G00 X.. Y..         (SAFE XY POSITION)
N29 M01                      (OPTIONAL STOP)

N30 T99                 (T99 CALL REPEATED)
N31 M06                    (T99 TO SPINDLE)
N32 T02                          (T02 READY)
N33 M06                    (T02 TO SPINDLE)
N34 G90 G54 G00 X.. Y.. S.. M03 (NO NEXT TOOL)
N35 G43 Z.. H02 M08         (APPROACH WORK)
...
    <...T02 working...>
...
N46 G00 Z.. M09        (T02 MACHINING DONE)
N47 G28 Z.. M05            (T02 TO Z HOME)
N48 G00 X.. Y..         (SAFE XY POSITION)
N49 M01                      (OPTIONAL STOP)

N50 M06    (T02 OUT OF SPINDLE TO THE SAME POT)
N51 T03                          (T03 READY)
N52 M06                    (T03 TO SPINDLE)
N53 G90 G54 G00 X.. Y.. S.. M03 T99 (T99 READY)
N54 G43 Z.. H03 M08         (APPROACH WORK)
...
    <...T03 working...>
...
N66 G00 Z.. M09        (T03 MACHINING DONE)
N67 G28 Z.. M05            (T03 TO Z HOME)
N68 G00 X.. Y..         (SAFE XY POSITION)
N69 M01                      (OPTIONAL STOP)
N70 M06                    (T99 TO SPINDLE)
N71 M30                    (END OF PROGRAM)
%
```

These examples illustrate some of the ATC programming methods. The task is not difficult once the tool changing mechanics of the machining center are known.

T FUNCTION FOR LATHES

So far, the tool function was covered as it applied to the CNC machining centers. CNC lathes also use the tool function T, but with a completely different structure.

◆ Lathe Tool Station

A typical slant bed lathe uses a polygonal turret holding all external and internal cutting tools in special holders. These *tool stations* are similar to a tool magazine on a machining center. Their design accepts 8, 10, 12 or more cutting tools - *Figure 14-7*.

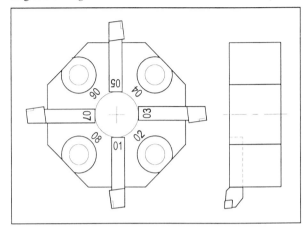

Figure 14-7
Typical view of an octagonal lathe turret

Many latest CNC lathe models start adopting the tool changer type similar to machining centers, with many more tools available and away from the work area.

Since all tools are normally held in a single turret, the one selected for cutting will always *carry along* all other tools into the work area. This may be a design whose time has passed but it is still very commonly used in industry. Because of a possible interference between a tool and the machine or part, care must be taken not only of the *active* cutting tool, but also of *all other tools* mounted in the turret, for all possible collision situations.

◆ Tool Indexing

To program a tool change, or rather to *index* the cutting tool into the active position, the T function must be programmed according to its proper format. For the CNC lathe, this format calls for the address T, followed by four digits - *Figure 14-8*.

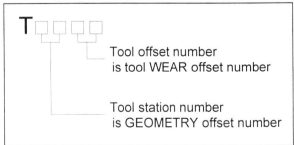

Figure 14-8
Structure of a 4-digit tool number for CNC lathes

It is important to understand this function well. Think about the four digits as *two pairs* of digits, rather than four single digits. Leading zeros within each pair may be omitted. Each pair has its own meaning:

The *first* pair (the first and the second digits), control the tool index station and the geometry offset.

➲ Example :

> T01xx - selects the tool mounted in position one
> and activates geometry offset number one

The *second* pair (the third and the fourth digits), control the tool wear offset number used with the selected tool.

➲ Example :

> Txx01 - selects the wear offset register number one

It is customary, not arbitrary, to match the pairs, if possible. For example, tool function T0101 will select tool station number one, geometry offset number one and the associated tool wear offset register number one. This format is easy to remember and should be used every time, if only one offset number is assigned to the tool number.

If *two or more different wear offsets* are used for the same tool, it is not possible to match the pairs. In such a case, *two or more different* wear offset numbers must be programmed for *the same* tool station number:

➲ Example :

> T0101   for turret station 01,
>   geometry offset 01 and wear offset 01

➲ Example :

> T0111   for turret station 01,
>   geometry offset 01 and wear offset 11

The first pair is always the tool station number *and* the geometry offset number. The examples assumed that tool wear offset 11 is not used by another tool. If tool 11 is used with the offset 11, another suitable wear offset number must be selected, for example 21, and program it as T0121. Most controls have 32 or more offset registers for geometry offsets and another 32+ wear offsets registers.

The offset values can be applied to the CNC lathe program by registering their value into the *offset registers*.

TOOL OFFSET REGISTERS

The word *offset* has been mentioned already several times with two adjectives - with the expression *geometry offset* and the expression *wear offset*. What exactly is an *offset*? What is the difference between one offset and the other?

On the OFFSET display of a typical Fanuc control, there is a choice of two screens, both *very* similar in appearance. One is called the *Geometry Offset* screen, the other is called the *Wear Offset* screen. *Figure 14-9* and *Figure 14-10* show examples of both screens, with typical (*i.e.*, reasonable) sample entries.

OFFSET - GEOMETRY				
No.	X-OFFSET	Z-OFFSET	Radius	Tip
01	-8.4290	-16.4820	0.0313	3
02	-8.4570	-14.7690	0.0000	0
03	-8.4063	-16.3960	0.0156	3
04	-8.4570	-12.6280	0.0000	0
05	-8.4350	-16.4127	0.0000	0
06	-9.8260	-13.2135	0.0313	2
07	0.0000	0.0000	0.0000	0

Figure 14-9

Example of the GEOMETRY offset screen display

OFFSET - WEAR				
No.	X-OFFSET	Z-OFFSET	Radius	Tip
01	0.0000	0.0000	0.0313	3
02	0.0000	0.0150	0.0000	0
03	0.0036	0.0000	0.0156	3
04	0.0000	-0.0250	0.0000	0
05	0.0010	-0.0022	0.0000	0
06	-0.0013	0.0000	0.0313	2
07	0.0000	0.0000	0.0000	0

Figure 14-10

Example of the WEAR offset screen display

◆ Geometry Offset

Geometry offset number is *always* the same as the turret station number. The operator measures and fills-in the geometry offsets for all tools used in the program.

> The GEOMETRY offset value is always measured
> from the machine zero position.

The distance from the machine zero position will reflect the distance from the tool reference point to the part reference point. *Figure 14-11* shows a typical measurement of the geometry offset applied to a common *external* tool.

All X values will normally have *diameter* values and are stored as *negative* for a typical rear lathe of the slant bed type. The Z axis values will normally be also negative (positive values are possible but impractical). How to actually measure the geometry offset is a subject of CNC machine tool operation training, not programming.

Figure 14-12 shows a typical measurement of the geometry offset applied to a common *internal* tool.

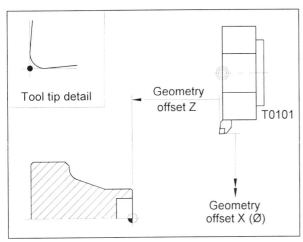

Figure 14-11
Typical geometry offset for external (turning) tools

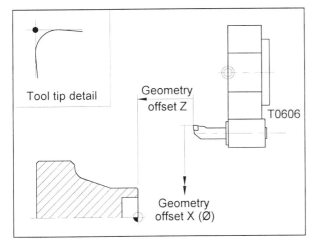

Figure 14-12
Typical geometry offset for internal (boring) tools

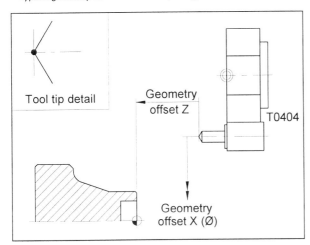

Figure 14-13
Typical geometry offset for center line (drilling) tools

The third and last possibility relating to the geometry offset is illustrated in *Figure 14-13*. It shows geometry offset applied to any tool used on the spindle center line (at X0 position). These tools include center drills, drills, taps, reamers, etc. Their X offset value will always be the same.

◆ Wear Offset

In writing a CNC program, the same dimensions are used in the program as those in the finished drawing. For example, a dimension of 3.0000 inches, is programmed as X3.0. This figure does not reflect any implied dimensional tolerances. Program entries X3.0, X3.00, X3.000 and X3.0000 have *exactly* the same result. What is needed to maintain dimensional tolerances, particularly when they are tight? What has to be done with a worn out tool that is still good enough to cut a few more parts? The answer is that the programmed tool path must be adjusted, *fine-tuned*, to match the machining conditions. The program itself will not be changed, but a *wear offset* for the selected tool is applied.

> The WEAR offset value is the difference between the *programmed* value and the *actual* measured size of the part.

Figure 14-14 illustrates the principle of the tool wear offset, although the scale is exaggerated for emphasis.

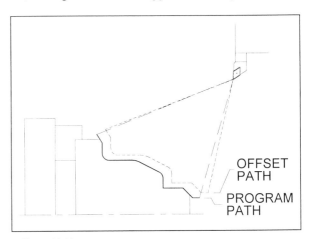

Figure 14-14
Programmed tool path and tool path with wear offset

The wear offset value has only one purpose - it compensates between the programmed value, for example of the 3.0 inch diameter, and the actual dimension *as measured* during inspection, for example 3.004. The differential value of -.004 is entered into the wear offset register. This is the offset number specified as *the second pair* of the tool function in the program. Since the program uses diameters for the X axis, the offsets will also be entered for a diameter. Details on this kind of adjustment are more useful to the CNC machine operator, but any programmer will benefit from them as well.

◆ Wear Offset Adjustment

To illustrate the concept of offset adjustment on a rear type lathe, T0404 in the program will be used as an example. The goal is to achieve an outside diameter of 3.0 inches and tolerance of ±.0005. The starting value of the wear offset in the register *Txx04* will be zero. The relevant section of the program may look something like this:

```
N31 M01
...
N32 T0400 M42
N33 G96 S450 M03
N34 G00 G42 X3.0 Z0.1 T0404 M08
N35 G01 Z-1.5 F0.012
N36 ...
```

When the machined part is inspected (measured), it can have only one of three possible inspection results:

- ❑   On-size dimension
- ❑   Oversize dimension
- ❑   Undersize dimension

If the part is measured *on size*, there is no need to interfere. The tool setup and the program are working correctly. If the part is *oversize*, it can usually be recut for machining an *outside* diameter. For an *inside diameter*, the exact opposite will apply. The recut may damage the surface finish, which could be a concern. If the part is *undersize*, it becomes a scrap. The aim is to prevent all *subsequent* parts from being undersize as well. The following table shows the inspection results for all existing possibilities:

Measurement	External diameter	Internal diameter
ON size	Size OK	Size OK
OVER size	Recut possible	SCRAP
UNDER size	SCRAP	Recut possible

Let's go a little further. Whether the part will be oversized or undersized, something has to be done to prevent this from happening again. The action to take is *adjusting the wear offset value*. Again, the emphasis here is that this is an example of an *outside* diameter.

The external diameter X3.0 in the example may result in 3.004 diameter measured size. That means it is 0.004 oversize - on diameter. The operator, who is in charge of the offset adjustments, will change the current 0.0000 value in the X register of the wear offset 04 to -0.0040. The subsequent cut should result in the part that will be measured within specified tolerances.

If the part in the example is undersize, say at 2.9990 inches, the wear offset must be adjusted by +.0010 in the X positive direction. The measured part is a scrap.

The principle of the wear offset adjustment is logical. If the machined diameter is *larger* then the drawing dimension allows, the wear offset is changed into the *minus* direction, towards the spindle center line, and vice versa. This principle applies equally to external and internal diameters. The only practical difference is that an oversized external diameter and undersize internal diameter *can* be recut (see the table above). *Chapter 34* presents several practical examples using the wear offset creatively.

◆ The R and T Settings

The last items are the R and T columns (*Geometry* and *Wear*). The offset screen columns are only useful during setup. The *R column* is the radius column, the *T column* is the tool tip orientation column (*Figure 14-15*).

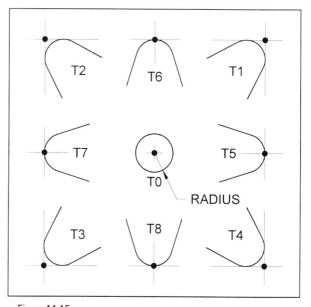

Figure 14-15

Arbitrary tool tip orientation numbers used with tool nose radius compensation (G41 or G42 mode)

The main rule of using R and T columns is that they are only effective in a tool nose radius offset mode. If no G41 or G42 is programmed, values in these columns are irrelevant. If G41/G42 command is used, non-zero values for that tool must be set in both columns. The R column requires the *tool nose radius* of the cutting tool, the T column requires the *tool tip orientation number* of the cutting tool. Both are described in *Chapter 30,* in more detail. The most common tool nose radii for turning and boring are:

1/64 of an inch = .0156 or 0,4 mm
1/32 of an inch = .0313 or 0,8 mm
3/64 of an inch = .0469 or 1,2 mm

The tool tip numbers are arbitrary and indicate the tool orientation number used to calculate the nose radius offset, regardless of the tool setting in the turret.

REFERENCE POINTS

In the previous chapters, the basic relationship between the machine geometry and the part setup was discussed. CNC programmers work in a fairly precise environment, and mathematical relationships are of extreme importance.

There are three major environments in programming that require an established mathematical relationship:

Environment	Relationship consists of ...
Machine	Machine tool + Control system (CNC unit)
Part	Workpiece + Drawing + Material
Tool	Holder + Cutting tool

Each environment by itself is independent of the other two. If the relationship is not apparent right away, consider the *sources* of each environment:

❑ *MACHINE TOOL* is made by a company specializing in machine tools, usually not controls or cutting tools

   ... this environment is combined with ...

❑ *CONTROL SYSTEM* is made by a company specializing in the application of electronics to machine tools. They do not normally manufacture machine tools or cutting tools.

❑ *PART* (workpiece) is a unique engineering design developed in a company that does not manufacture machine tools, control systems, or cutting tools and holders.

❑ *CUTTING TOOLS* are a specialty of tooling companies, which may or may not make cutting tool holders. These companies do not manufacture machine tools or CNC systems.

These sources inevitably meet when a customer buys a CNC machine. A certain engineering design (part), must be machined on a machine tool from one manufacturer, using a control system of another manufacturer, cutting tools from yet another manufacturer, and tool holders from a fourth source. These sources are similar to a musical quartet of first class musicians who never played together. In both cases there is a need to create a harmony.

By itself, each environment is not very useful. A machine without tools will not yield any profit; a tool that cannot be used on any machine is not going to benefit the manufacturing either. A part cannot be machined without tools.

The common point here is that all three environments cannot be useful without some 'team work'. They have to work together, *they have to interact.*

For programming purposes, these relationships and interactions are based on one common denominator of *each* environment - *a reference point.*

A reference point is a fixed or selected arbitrary location on the machine, on the tool and on the part. A fixed reference point is a precise location along two or more axes, designed during manufacturing or setup. Some reference points are established by the programmer, during the programming process. In these three environments, three reference points are needed - one reference point for each of the available groups:

❑ *Machine* reference point .. Machine zero or Home

❑ *Part* reference point .. Program zero or Part zero

❑ *Tool* reference point .. Tool tip or Command point

In a typical language of a machine shop, these reference points have somewhat more practical meaning. *Home position* or a *machine zero* are synonymous terms for *machine reference point.* A *program zero*, or *part zero*, or *part origin* are terms commonly used instead of the more official term *part reference point.* And the name *tool tip* or a *tool command point* are commonly used for the *tool reference point.*

REFERENCE POINT GROUPS

The first group is the *CNC machine tool* or *CNC machine* for short, which is the combination of machine proper and the control system. The numeric values that relate to the CNC machine tool include a variety of dimensions, specifications, parameters, ranges, ratings, etc. When a part is set in a fixture on the machine table or mounted into a lathe chuck, collet, face plate, or other work holding device, there is a second group of numbers to consider. The *part considerations*, such as its size, its height, diameter, shape, etc., are unique to each job. Finally, the third group of numbers relate to the *cutting tools.* Each cutting tool has its individual features, as well as features that are shared with the other cutting tools.

All available numeric values have a meaning - they are not merely numbers - they are actual *values* that programmers and operators have to work with individually as well as together.

◆ Reference Point Groups Relationship

The key to any successful CNC program is to make all three groups to work in a coordinated way. This goal can only be achieved by understanding the principles of reference points and how they work. Each reference point can have two characteristics:

❏ *Fixed* reference point

❏ *Flexible* or *floating* reference point

A fixed reference point is set by the machine manufacturer as part of the hardware design and cannot be physically changed by the user. A CNC machine has at least one fixed reference point. When it comes to deciding the reference points for the part or the cutting tool, the programmer has certain degree of freedom. A part reference point (program zero) is always a flexible point, meaning its actual position is in programmer's hands. The reference point for the mounted cutting tool can be either fixed or flexible, depending on machine design.

MACHINE REFERENCE POINT

The *machine zero point,* often called the *machine zero, home position* or just a *machine reference position,* is the origin of machine coordinate system. The location of this point may vary between the machine manufacturers, but the most obvious difference is between individual machine types, namely the vertical and horizontal models.

In general terms, a CNC machine has two, three, or more axes, depending on the type and model. Each axis has a maximum range of travel that is fixed by the manufacturer. This range is usually different for each axis. If the CNC operator exceeds the range on either end, an error condition known as *overtravel* will occur. Not a serious problem, but one that could be annoying. During machine setup, particularly after the power has been turned on, the position of all axes has to be preset to be always the same, from day to day, from one part to another. On older machines, this procedure is done by setting a grid, on modern machines, by performing a *machine zero return command.* Fanuc and many other control systems prevent automatic operation of a machine tool, unless the machine zero return command has been performed at least once - when the power to the machine has been turned on. A good safety feature.

On all CNC machines that use typical coordinate system, the machine zero is located at the *positive* end of each axis travel range. For a typical three-axis vertical machining center, look at the part in the XY plane, that is straight down from the tool position (tool tip). Also look into the XZ plane (operator's front view of the machine), or into the YZ plane (operator's right-side view of the machine). These three planes are perpendicular to each other and together create so called *work cube* or *work space* - *Figure 15-1.*

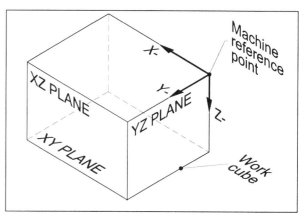

Figure 15-1

Machine reference point and axes orientation for a vertical machine

The cubical shape shown is useful only for overall understanding of the machine work area. For programming and setup, the majority of work is done with one or two axes at a time. To understand the work area and machine zero point in a plane, look at the machine from the top (XZ machine plane) and from the front (YZ machine plane). *Figures 15-2* and *15-3* illustrate both views.

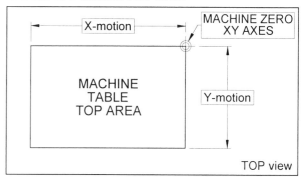

Figure 15-2

Top view of a vertical machine as viewed towards the table

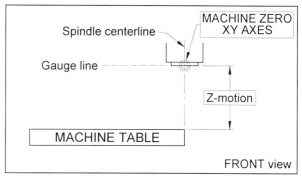

Figure 15-3

Front view of a vertical machine as viewed from the front

Compare the two views. In top view, the upper right corner is also the spindle center line shown in the front view.

Also note that in front view, there is a dashed line identified as the *gauge line*. This is an imaginary location for the proper fit of the tool holder tapered body and is set by the machine manufacturer. The inside of the spindle is a precision machined taper that accepts the tool holder with the cutting tool. Any tool holder mounted in the spindle will be in exactly the same position. The Z motion illustrated will be shortened by the cutting tool projection. This subject of tool referencing is discussed later in this chapter.

◆ Return to Machine Zero

In manual mode, the CNC operator physically moves the axes to the machine zero position. The operator is also responsible to register this position into the control system, if necessary. Never turn off power to the machine, while the machine slides are *at* or very *close to* the machine zero position. Being too close will make the manual machine zero return more difficult later, after the power had been restored. A clearance of 1.0 inch (25.0 mm) or more for each axis *from* machine zero is usually sufficient. A typical procedure to physically reach the machine zero position will follow these steps:

1. Turn the power on (machine and control)
2. Select machine zero return mode
3. Select the first axis to move (usually Z axis)
4. Repeat for the all other axes
5. Check the lighted *in-position* indicators
6. Check the position screen display
7. Set display to zero, if necessary

Mainly for safety reasons, the first selected axis should be the Z axis for machining centers and the X axis for lathes. In both cases, either axis will be moving *away* from work, into the clear area. When the axis has reached machine zero position, a small indicator light on the control panel turns on to confirm that the axis actually reached machine zero. The machine is now at its reference position, at the machine zero, or at the machine reference point, or at *home* - whichever term is used in the shop. The indicator light is the confirmation for each axis. Although the machine is ready for use, a good operator will go one step further. On the *position* display screen, the actual relative position should be set to zero readout for each axis, as a standard practice, if it is not set to zero automatically by the control. The POS button on the control panel selects the position screen display.

PART REFERENCE POINT

A part ready for machining is located within the machine motion limits. Every part must be mounted in a device that is safe, suitable for the required operation and does not change position for any other part of the job run. The fixed location of the device is very important for consistent results and precision. It is also very important to guarantee that each part of the job is set the same way as the first part. Once the setup is established, the part reference point can be selected.

This vital reference point will be used in a program to establish the relationship with machine reference point, reference point of the cutting tool and the drawing dimensions.

The part reference point is commonly known as a *program zero* or a *part zero*. Because the coordinate point that represents program zero can be selected by the programmer almost anywhere, it is not a fixed point, but a *floating* point. As this point is selectable, more details can be covered - after all, it is the programmer who selects part zero.

◆ Program Zero Selection

When selecting the program zero, often in the comfort of programmer's office, a major decision is made that will influence the efficiency of the part setup and its machining in the shop. Always be very attentive to all factors that are for and against a program zero selection in a certain position.

In theory, the program zero point may be selected literally anywhere. That is not much of an advice, although true in mathematical terms. Within the practical restrictions of the machine operations, only the most advantageous possibilities should be considered. Three such considerations should govern the selection of program zero:

❏ Accuracy of machining

❏ Convenience of setup and operation

❏ Safety of working conditions

Machining Accuracy

Machining accuracy is paramount - all parts must be machined exactly to the same drawing specifications. Accuracy is also important consideration in repeatability. All the parts in the batch must be the same and all subsequent jobs must be the same as well.

Convenience of Setup and Operation

Operating and setup convenience can only be considered once the machining accuracy is assured. Working easier is everyone's desire. An experienced CNC programmer will always think of the effect the program has in the machine shop. Defining program zero that is difficult to set on the machine or difficult to check is not very convenient. It slows down the setup process even more.

Working Safety

Safety is always important to whatever we do - machine and part setup are no different. Program zero selection has a lot to do with safety of the machining operation.

We look at the typical considerations of program zero selection for vertical machining centers and lathes individually. Differences in part design influence the program zero selections as well.

◆ Program Zero - Machining Centers

CNC machining centers allow a variety of setup methods. Depending on the type of work, some most common setup methods use vises, chucks, subplates and hundreds of special fixtures. In addition, CNC milling systems allow a multi-part setup, further increasing the available options. In order to select a program zero, all three machine axes must be considered. Machining centers with additional axes require zero point for each of these axes as well, for example, the indexing or rotary axes.

What are the most common setup methods? Most machining is done while clamped on machine table, in a vise or a fixture mounted on the table. These basic methods can be adapted to more complex applications.

CNC programmer determines the setup method for any given job, perhaps in cooperation with the machine operator. CNC programmer also selects the program zero position for each program. The process of selecting the program zero starts with drawing evaluation, but two steps have to be completed first:

Step 1.   Study how the drawing is dimensioned, which dimensions are critical and which are not

Step 2.   Decide on the method of part setup and holding

Program zero almost presents itself in the drawing. In any setup, make sure all critical dimensions and tolerances are maintained from one part to another. Drawing dimensions not specified are usually not critical.

The simplest setup on a machine table involves support for the part, some clamps and locating surfaces. The locating surfaces must be *fixed* during the job run and easy to be measured from. The most typical setup of this kind is based on the *three pin* concept. Two pins form a single row and the third pin is offset away at a right angle, creating a 90° angle setup corner as two locating surfaces - *Figure 15-4.*

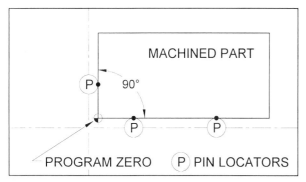

Figure 15-4

Three-pin concept of a part setup (all pins have the same diameter)

Since the part touches only one point on each pin, the setup is very accurate. Clamping is usually done with top clamps and parallels. The left and the bottom edge of the

part are both parallel to the machine axes and perpendicular to each other. Program zero (part zero) is at the intersection of the two locating edges.

The three-pin concept is common for virtually all setups, without using actual pins. If a part is mounted in a vise, there are similarities. The vise jaws must be parallel to or perpendicular with the machine axes and the fixed location must be established with a stopper or other fixed method.

Since a machine vise is the most common work holding device for small parts, let's use it as a practical example of how to select program zero. *Figure 15-5* illustrates a typical simple engineering drawing, with all the expected dimensions, descriptions and material specifications.

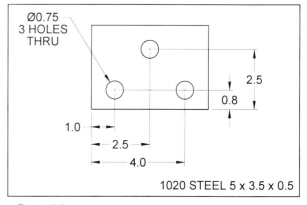

Figure 15-5

Sample drawing used for selecting program zero example

When selecting a program zero, first study the drawing dimensions. The designer's dimensioning style may have flaws, but it still is the engineering drawing. In the example, dimensioning for all holes is from the lower left corner of the work. Does the program zero of the part suggest itself?

For this example, there should be no question about programming the reference point anywhere else except at the lower left corner of the part. This is the drawing origin and it will become the part origin as well. It also satisfies *Step 1* of the program zero selection process. The *Step 2*, dealing with work holding device selection is next. A typical setup in a special CNC machine vise could be the one illustrated in *Figure 15-6.*

In the setup identified as *Version 1*, the part has been positioned between the vise jaws and a left part stopper. The part orientation is the same as per drawing, so all drawing dimensions will appear in the program using these drawing dimensions. It seems that this is a winning setup - yet, this setup is actually quite poor.

What is missing in the decision is any consideration of the actual overall size of material. The drawing specifies a rectangular stock of 5.00×3.50. These are *open* dimensions - they can vary $\pm.010$ or more and still be acceptable.

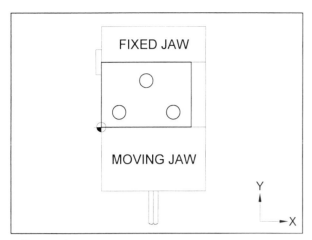

Figure 15-6

A sample part mounted in a machine vise - Version 1

Combine any acceptable tolerance with the vise design, where one jaw is a *fixed jaw* and the other one is a *moving jaw*, and the problem can be seen easily. *The critical Y axis reference is against a moving jaw!*

The program zero edge should be the *fixed* jaw - a jaw that does not move. Many programmers incorrectly use a moving jaw as the reference edge. The benefit of programming in the first quadrant (all absolute values are positive) is attractive, but can produce inaccurate machining results, unless the blank material is 100% percent identical for all parts (usually not a normal case). *Version 1* setup can be improved significantly by rotating the part 180° and aligning the part stopper to the opposite side - *Figure 15-7*.

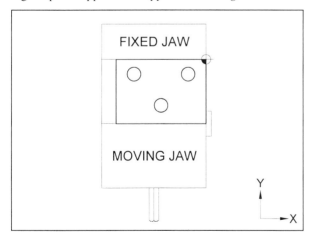

Figure 15-7

A sample part mounted in a machine vise - Version 2

In *Version 2*, results are consistent with the drawing. Part orientation by 180° has introduced another problem - *the part is located in the third quadrant!* All X and Y values will be *negative*. Drawing dimensions can be used in the program, but as negative. Just don't forget the minus signs.

If the choice is between *Version 1* and 2, select *Version 2* and make sure all negative signs are programmed correctly.

Is there another method? In most cases there is. The final *Version 3* will offer the best of both worlds. Part program will have all dimensions in the first quadrant, as per drawing. Also, the part reference edge will be against the fixed jaw! What is the solution? Rotate the *vise* 90° and position the part as shown - *Figure 15-8*, if possible.

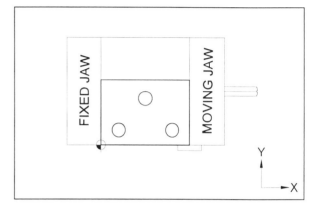

Figure 15-8

A sample part mounted in a machine vise - Version 3

To select a program zero for the Z axis, the common practice is to select the top face of the finished part. That will make the Z axis positive *above* the face and negative *below* the face. Another method is to select the bottom face of the part, where it is located in the fixture.

Special fixtures can also be used for a part setup. In order to hold a complex part, a fixture can be custom made. In many applications of special fixtures, the program zero position may be built into the fixture, away from the part.

Selecting a program zero for round parts or patterns (bolt circles, circular pockets), the most useful program zero is at the center of the circle - *Figure 15-9*.

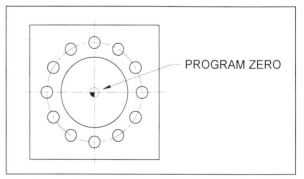

Figure 15-9

Common program zero for round objects is the center point

Chapter 40 describes the G52 command that may solve many problems associated with program zero at the center.

◆ Program Zero - Lathes

On CNC lathes, program zero selection is simple. There are only two axes to consider - the vertical X axis and the horizontal Z axis. Because of the lathe design, the X axis program zero selection is always the spindle center line.

> On CNC lathes, the program zero for the X axis
> *MUST* be on the center line of the spindle

For the Z axis, three popular methods are used:

❑ Chuck face   . . . main face of the chuck

❑ Jaw face   . . . locating face of the jaws

❑ Part face   . . . front of the finished part

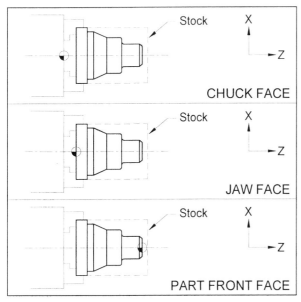

Figure 15-10

Common program zero options for a CNC lathe - center line is X0

Figure 15-10 illustrates the options. In setup, a chuck face offers only one benefit - it can be easily touched with the tool edge, using feelers to prevent tool chipping. On a negative side, unless the part rests against chuck face, additional calculations are needed for the coordinate data and drawing dimensions cannot be used easily.

Jaw or fixture face presents more favorable situation. The face can also be touched with the tool and is consistent for all parts. This location may benefit machining irregular shapes, such as castings, forgings and similar parts.

Many lathe parts require machining at both ends. During the first operation, material stock for the second operation must always be added to every Z value. This is the main reason why CNC programmers keep away from program zero located on jaw or fixture face, except in special cases.

The most popular method is setting program zero on the *front face* of the *finished* part. This is not a perfect selection either but has many other advantages. The only disadvantage is that during setup, there is no finished face. Many operators add the width of the rough face to the setup or cut a small face for the tool to touch.

What are the benefits of program zero at the front face? One benefit is that many drawing dimensions along Z axis can be transferred directly into the program, normally with a *negative* value. A lot depends on the dimensioning method but in majority of cases, the CNC programmer benefits. Another benefit, probably the most important, is that a negative Z value of a tool motion indicates the work area, a positive Z value is in the clear area. During program development, it is easy to forget a minus sign for the Z axis cutting motions. Such an error, if not caught in time, will position the tool *away* from part, with the tailstock as a possible obstacle. It is a wrong position, but a better one than hitting the part. Examples in this handbook use program zero at the *front finished face,* unless otherwise specified.

TOOL REFERENCE POINT

The last reference point is related to the tool. In milling and related operations, the reference point of the tool is usually the intersection of the tool centerline and the lowest positioned cutting tip (edge).

In turning and boring, the most common tool reference point is an imaginary tool point of the cutting insert, because most tools have a cutting edge with a built-in radius.

For tools such as drills and other point-to-point tools used in milling or turning, the reference point is always the extreme tip of the tool, as measured along Z axis. *Figure 15-11* shows some common tool tip points.

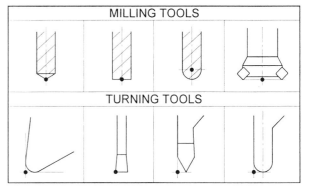

Figure 15-11

Typical tool reference points for various cutting tools

All three reference point groups are connected. An error in one setting will have an adverse effect on another. The knowledge of reference points is important to understand register commands, offsets and machine geometry.

REGISTER COMMANDS

The three reference points available for CNC programming must be harmonized to work together correctly. Having the reference points for the *part* (*i.e.,* program zero) and the cutting *tool* (*i.e.,* tool tip) available, there has to be some means to associate them together, to fit them together. There must be some means to *'tell'* the control system exactly where each tool is physically located, within the machine work area, *before* it can be used. The oldest method to do all this is to *register* the current position of the tool into the control system memory, through the program. This method required a command called the *Position Register.*

POSITION REGISTER COMMAND

The preparatory command for the tool position register is G92 for machining centers and G50 for lathes:

G92	Position register command	(used in milling)
G50	Position register command	(used in turning)

Some CNC lathes also use G92 command, but lathes supplied with Fanuc and similar controls normally use G50 command instead. In practical applications, both G92 and G50 commands have identical meaning and the following discussion applies to both commands equally. In the first part of this chapter, the focus will be on milling applications using G92 command, lathe applications using G50 command will be explained later.

In modern CNC programming, both position register commands were replaced by a much more sophisticated and flexible feature called the *Work Offsets* (G54 to G59), described in *Chapter 18*, and the *Tool Length Offset* (G43), described in *Chapter 19*. However, there are still quite a few older machine tools in shops that do not have the luxury of the G54 series of commands. There are also many companies using programs developed years ago, but still running on modern CNC equipment. In those cases, understanding the position registration command is an essential skill. This command has always been one that some programmers and operators found a little difficult to understand. In reality, it is a very simple command.

First, a look at some more detailed definition of this command. A typical description only specifies *Position Register Command*, which by itself is not very conclusive.

◆ Position Register Definition

A little more verbose definition of the position register command could be expressed this way:

> Position register command sets the tool location
> as the distance and direction
> FROM ... the program zero,
> TO ... the tool current position,
> measured along the axes

Note that the definition does not mention the *machine zero* at all - instead, it mentions the *current tool position*. This is a very important distinction. The current tool position *may be* at machine zero, but it also *may be* somewhere else, within travel limits of the machine axes.

Also note the emphasis on *from-to* direction. By definition, the distance is unidirectional, between the program zero and the current tool location. The direction is always *from* program zero, *to* tool position, never reversed. In a program, the correct sign of each axis value (positive, negative, or zero) is always required.

The position register is only applicable in the absolute mode of programming, while G90 command is in effect. It has no use in the incremental programming mode G91. In practical programming, virtually all programs written in the incremental mode do begin in the absolute mode, in order to reach the first tool location.

◆ Programming Format

As the name of the command suggests, the tool position data associated with the G92 command will be *registered* (*i.e.,* stored) into the control system memory.

The format for G92 command is as follows:

> G92 X.. Y.. Z..

In all cases, the address of each axis specifies the distance *from the program zero to the tool reference point* (tool tip). Programmer provides all coordinates based on the program reference point (program zero), discussed earlier. Any additional axis will also have to be registered with G92, for example the B axis for the indexing table on horizontal machining centers.

◆ Tool Position Setting

The only purpose of G92 command is to *register the current tool position into the control memory* - nothing more!

> No machine motion will ever occur
> in a block containing the G92 command !

The effect of G92 can be seen on the *absolute position* screen display. At all times, the absolute position display has some values for each axis. They could be zero or any other values. When G92 command is executed, all current values of the display will be *replaced* with the values specified with G92. If an axis was not specified with G92, there will be no change of display for that axis. At the machine, the operator has a major responsibility - to match the actual tool setting with the values specified in the G92 command.

MACHINING CENTERS APPLICATION

In programming for CNC machining centers *without* the *Work Coordinate System* feature (also known as *Work Offsets*), the *Position Register* must be established for each axis and each tool. There are two methods:

❏ The tool position is set *at machine zero*

❏ The tool position is set *away from machine zero*

Which method is better? We look at both of them.

◆ Tool Set at Machine Zero

The first method requires that the machine zero position will also be the tool change position for all axes. This is not necessary and definitely *very* impractical. Consider it for a moment and think why it is impractical.

A program is usually done away from the machine, but the part position on the table must be specified:

```
G92 X12.0 Y7.5 Z8.375
```

Numbers in the example look innocent enough. But consider the CNC operator at the machine, trying to setup the part (without a special fixture), to be *exactly* 12.0 inches away from machine zero in the X axis. At the same time, the operator must setup the same part *exactly* 7.5 inches away from machine zero for the Y axis. The same effort has to be done for the Z axis as well.

It is an almost impossible task, at least without some special fixtures. It is definitely an extremely *unproductive* task. There is no *need* for those numbers, they are strictly arbitrary - the X12.0 could have easily been X12.5, with no added benefit whatsoever. All this difficulty is encountered only because the programmer has chosen the machine zero reference point for tool change position (mainly in the X and Y axes).

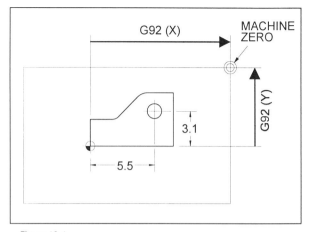

Figure 16-1

Current tool position register set at machine zero (only XY axes shown)

Figure 16-1 shows a G92 setup based on the tool set at machine zero position. This method of starting program at machine zero is useful. There could be an advantage, for example, if a special fixture is permanently attached to the machine table. A subplate with a locator grid is a common example. Permanently set one or more vises may also benefit. There are numerous variations on this type of setup.

◆ Tool Set Away from Machine Zero

The second method eliminates the difficulty of the previous setup. It allows the programmer to set XY tool position anywhere within the machine travel limits (considering safety first) and use that position as the tool change position for XY axes. As there is no need for the machine zero itself, the CNC operator can setup the part anywhere on the table, in any reasonable position, within limits of the machine axes. *Figure 16-2* shows an example of a tool set at a negative X axis and a positive Y axis.

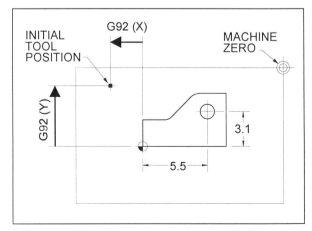

Figure 16-2

Current tool position register set away from machine zero (only XY axes shown)

In order to place tool into the starting tool change position, the operator physically moves the tool *from the program zero* by amounts specified in the G92 statement. This is a lot easier job and also much more efficient that restricting setup to the machine zero.

Once the tool change position is established, all tools in the program will return to this position for a tool change. The Z axis automatic tool change position on vertical machining centers must be programmed at machine zero as *the only* automatic tool change position. So the discussion really applies to XY axes only. Regardless of a tool position, the G92 setting will be the same for all tools, unless there is a good reason to change it.

The only major disadvantage of this method is that the new tool change position is only memorized by the control system while the power is on. When the power to the machine is turned off, the tool change position is lost. Many experienced CNC operators solve this problem by simply finding the actual distance from the machine zero to the tool change position, register it once for each particular setup and then move the tool by that distance after restoring power, for example, at the start of a new day.

◆ Position Register in Z Axis

For a typical vertical machine, the Z axis must be fully retracted to the machine zero, in order to make the automatic tool change. The position register value is measured from the program zero of the Z axis (usually the top of finished face), to the tool reference tip, while the Z axis is at machine zero position. There is no other option.

Normally, each tool will have a *different* Z value of the G92 command, assuming the tool length is different for each tool. As a rule, the XY settings will not change. *Figure 16-3* shows a typical setting for G92 command along the Z axis. Example O1601 illustrates the concept.

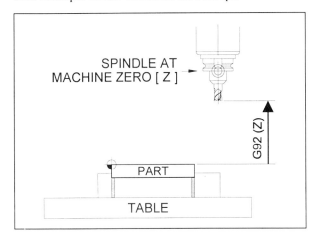

SPINDLE AT
MACHINE ZERO [Z]

G92 (Z)

PART

TABLE

Figure 16-3

Current tool position register set at machine zero for the Z axis (each tool will normally have a different setting)

◆ Programming Example

To illustrate how to use the position register command in a part program for vertical machining centers, certain rules have to be followed:

❏   The cutting tool should be changed first

❏   G92 must be established *before* any tool motions

❏   Tool *must* return to the G92 position when all the cutting is completed

All three rules are followed in a sample program:

```
O1601                        (PROGRAM NUMBER)
N1 G20                       (SET ENGLISH UNITS)
N2 G17 G40 G80 G90 T01       (GET TOOL 1 READY)
N3 M06                       (TOOL 1 TO SPINDLE)
N4 G92 X9.75 Y6.5 Z11.0      (SET CURRENT XY)
N5 G00 X1.0 Y0.5 S800 M03    (MOVE TO POSITION)
N6 Z0.1 M08               (MOVE TO CLEAR ABOVE)
N7 G01 Z-0.55 F5.0            (FEED TO DEPTH)
N8 X3.0 Y4.0 F7.0              (CUT A SLOT)
N9 G00 Z11.0 M09      (RAPID TO Z MACHINE ZERO)
N10 X9.75 Y6.5 M05    (RAPID TO XY SET POSITION)
N11 M01              (OPTIONAL STOP FOR TOOL 1)
...
```

This is a simple example to write but more difficult to set-up on the machine. Don't worry about unknown program entries at the moment, the explanations should be clear.

Note - the Z axis setting position must always be known - *at machine zero!* It does not matter whether the tool change in XY is made, at machine zero or away from it - the program format will be the same, just *meaning* of the values will be different. Only one tool is used, but normally, each tool will have a different Z value as the position register, since each tool has a different length.

LATHE APPLICATION

For the CNC lathes with Fanuc and similar controls, G50 command is used instead of G92 command:

> G50 X.. Z..

If G92 is used for a lathe, the command is similar:

> G92 X.. Z..

Either command has exactly the same definition and rules as for milling - it indicates the distance *from* program zero, *to* the current tool position along axes.

> No machine motion will occur
> in a block containing G50 or G92 command !

Commands G50 and G92 are identical, except that they belong to two different G code groups. Fanuc actually offers three G code groups for lathe controls. Based on history, typical Japanese made controls use G50, whereby typical US made controls used G92. A cooperative US and Japanese venture known as *GE Fanuc* (*General Electric* and *Fanuc*) produces controls that are the most common in North American industry, and using the G50 command.

To program the position register for lathe applications is very similar to that of G92 for the mills. However, due to design of CNC lathes, where all tools are mounted in the turret, the *projection* of each tool (for both axes) from the turret holder must be considered. Not only that, possible interference must be prevented, because all mounted inactive tools move simultaneously along the one that is used for cutting. In milling, all non-active tools are safely out of way, placed in a tool magazine. Several new designs of CNC lathes are available, where tool changer on the lathe resembles the milling type.

◆ Tool Setup

The most important programming decision for lathe work relates to the setup. Although there are several options to select from, some are preferable to others.

Probably the most practical approach for lathe setup will be to have the tool change position for all tools corresponding to the machine zero position. This is a very easy position to move the turret to, just using the control panel switches. The position register measured to machine zero does have one major disadvantage - it may be *too far* for most jobs, particularly on larger lathes along the Z axis. Just imagine a tool motion of 30 inches or more along the Z axis only to index the turret and than the same 30 inch motion back to continue the cutting cycle. It is not efficient at all. There is a solution, however.

Much more efficient method is to select the tool indexing position as *close to the part* as possible. This position should always be based on the *longest* tool mounted in the turret (usually internal tools), whether the tool is used in the program or not. If there is enough clearance for the longest tool, there will also be enough clearance for all remaining shorter tools.

A possible compromise of the two methods described is to keep tool indexing position at the X axis machine zero only (which is usually not too distant) and just establish the Z axis position.

On a CNC lathe, do not forget to keep in mind the general layout of all tools in the turret, to prevent a collision with the part, the chuck, or the machine.

There are other, but less common, methods to setup a tool on the lathe using the G50 command.

◆ Three-Tool Setup Groups

On a typical slant bed CNC lathe, equipped with a polygonal turret (6 to 14 stations), all cutting tools reside in individual stations of the turret. During tool indexing, only the selected tool is in the active station. Upon evaluation of the type of tools used for CNC lathe operations, it will be clear that there are only three groups of cutting tools, based on the type of work they normally do:

❏ Tools working on the part *center line*

❏ Tools working *externally* on the part

❏ Tools working *internally* on the part

If the position register for each group is understood well, it will be easy to apply it to any tool within a group, regardless of the number of tools used.

◆ Center Line Tools Setup

Lathe tools classified as center line tools are typically center drills, spot drills, standard twist drills, indexable carbide drills, taps, reamers, and so on. Even an end mill can be used at the spindle center line. All tools in this group have a single common denominator, whereby the tool tip is always located on the spindle center line, while they cut. These tools must always be setup exactly at 90° to the work face (parallel to the Z axis).

The position register value in the X axis is from the spindle center line (X0) to the center line of the tool. For the Z axis, the position register value is measured from program zero to the tool tip. Typically, the center line tools will have a fairly large overhang - that means their G50 value along the Z axis will be relatively small, when compared to the external tools, which generally do not project too much.

Figure 16-4 illustrates a typical setup for center line tools, using an indexable drill as an example.

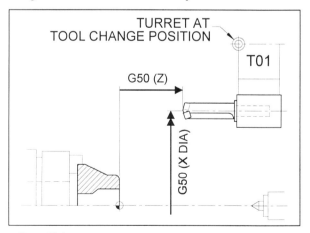

Figure 16-4
Typical G50 setting for center line lathe tools

◆ External Tools Setup

For external machining operations such as roughing and finishing outside diameters, taper cutting, grooving, knurling, single point threading, part-off and others, the cutting tool is rather small and approaches the part in an open space with generous clearances.

The position register value is measured from the program zero to the imaginary tool tip of the insert (see details at the end of this chapter). In case of tools like threading tool or a grooving tool, G50 amount is usually measured to the left side of the insert, for safety reasons.

Figure 16-5 illustrates a typical position register setup for an external tool (turning tool shown in the example).

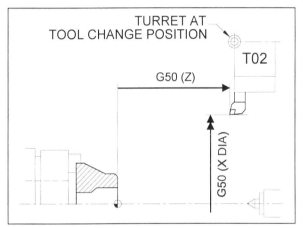

Figure 16-5
Typical G50 setting for external lathe tools

◆ Internal Tool Setup

Internal tools are all tools that do majority of their work inside of a part, in a premachined hole, core or other cavity. Typically, we may first think of a boring bar, but other tools can be used as well for various internal operations. For example, an internal grooving and internal threading are common operations on a CNC lathe. The setup rules along the Z axis apply in the same way for internal tools as for external tools of the same type.

Along the X axis, the tool position register setting must be made to the imaginary tip of the insert. *Figure 16-6* shows a typical position register setup for an internal tool (boring bar shown in the example).

All three illustrations *(Figures 16-4, 16-5* and *16-6)* show a possible order of the three operations (drill - turn - bore) for a typical CNC lathe job. Note that the turret position is identified as a *tool change position,* not necessarily as the machine zero position. That means G50 may be set anywhere within travel limits of the machine, even at the machine zero.

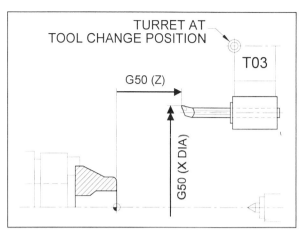

Figure 16-6
Typical G50 setting for internal lathe tools

For safety reasons, no tool should extend from a turret into the Z minus zone - that is to the left of part front face. Many CNC lathes have a fairly long travel *beyond* the Z axis machine zero (about 1-2 inches or 25-50 mm). Sometimes, this zone can be entered to make a safe tool change for very long tools. However, this is a more advanced programming method and requires strict safety considerations. There is virtually no extended zone for the X axis above machine zero position (only about .02 inches or 0.5 mm).

Another safety concern relating to long tools is clearance in the part holding area, including chuck and jaws. Make sure to extend only those tools where the work requires it.

◆ Corner Tip Detail

Typical turning tool contains an indexable insert, with a corner radius for strength and surface finish control. When the position register command is used for a tool that has a radius built-in, the programmer has to know (and also tell the CNC operator), *which edge* the G50 corresponds to. In many cases, the choice is simple. The G50 value is measured from program zero to the imaginary intersection of tangential X and Z axis. Depending on the tool shape and its orientation in the turret, G50 setting will vary. *Figure 16-7* on the next page shows several typical settings for the most common orientations of a tool with a corner radius, including two grooving tools.

◆ Programming Example

The example showing how to use a position register command G50 on a lathe will be very similar to that of a machining center. First, the tool change is made, followed with G50 setting for the selected tool. When the machining is done with that tool, it has to return to the same absolute position as specified in the G50 block. The following simplified example is for two tools - the first tool is programmed to cut a face, the second tool is programmed to cut a 2.5 inch diameter:

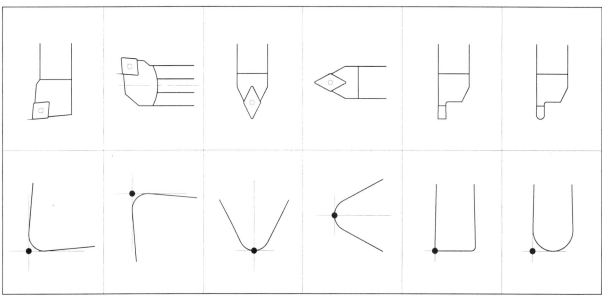

Figure 16-7

Position register setting G50 for common tool tip orientations - the heavy dot indicates XZ coordinates set by G50 X.. Z.. for the tool above

```
O1602
N1 T0100
N2 G50 X7.45 Z5.5
N3 G96 S400 M03
N4 G00 X2.7 Z0 T0101 M08
N5 G01 X-0.07 F0.007
N6 G00 Z0.1 M09
N7 X7.45 Z5.5 T0100
N8 M01

N9 T0200
N10 G50 X8.3 Z4.8
N11 G96 S425 M03
N12 G00 X2.5 Z0.1 T0202 M08
N13 Z-1.75 F0.008
N14 G00 X2.7 M09
N15 X8.3 Z4.8 T0200
N16 M30
%
```

Note blocks N2 and N7 for the first tool, and N10 and N15 for the second tool. For each tool, these XZ pairs of blocks are exactly the same. What the program is 'telling' the control system here is that block N2 only registers the current tool position, but block N7 actually returns that tool to the same position it came from. For the second tool, block N10 registers the current tool position, block N15 forces the tool to return there.

Other important blocks to consider together are the blocks N7 and N10. Block N7 is the *tool change position* for the first tool, block N10 is the *tool position register* for the second tool - *both tool are at the same physical position of the turret!* The difference in the XZ values reflects the difference in the projection (length) of each tool from the turret station. All that is done with G50 command is telling the control *where the current tool tip is from program zero -* always keep that in mind!

POSITION COMPENSATION

The relationships between various reference points in CNC programming are expressed as preset numeric values. More often than not, these numbers, these specific values, are required well *before* the actual machine setup takes place. During part programming, many dimensions are known exactly, others are known approximately and there are also many that are not known at all. Some known dimensions are subject to variations between jobs. Without any corrective facility available to the programmer, it will be almost impossible to setup the machine precisely and efficiently. Fortunately, modern controls offer many features to make both programming and machine setup an easier, faster and more precise activity. A number of coordinate systems, offsets and compensations are typical support tools used in programming for corrective purposes.

One of the oldest programming techniques available in programming is called a *position compensation*. As the name suggests, using position compensation functions, the *actual* tool position is compensated relative to its *theoretical* or *assumed* position.

It is only one of several corrective methods available to the programmer and machine operator. On modern CNC systems, this method is still available for compatibility with older programs. Today, this technique is not really needed. It has been replaced by the much more flexible *Work Offsets (Work Coordinate System)*, described in the next chapter of this handbook. The current chapter describes some typical programming applications that can benefit from using the old-fashioned position compensation method.

DESCRIPTION

The main purpose of position compensation is to correct any difference between machine zero and program zero tool positions. In practice, it is used in those cases, where the distance between the two reference points is subject to variations or is not known at all. For example, when working with castings, the program zero taken from the cast surface will be subject to frequent change. Using position compensation will eliminate the need to make constant program changes or realignment of the fixture setup. Normally, the part is mounted in a fixture on the table and the whole setup is compensated. For this reason, the position compensation is sometimes called *fixture offset* or the *table offset*. The difference between an offset and a compensation is often very subtle, and for any practical purposes, the two terms are same.

In this handbook, each term is used in the same meaning as the majority of users interpret it. Position compensation can also be used for a very limited replacement of the cutter radius offset - this usage is not covered at all for its obsolescence. Instead, the emphasis will be on positioning of the tool from machine zero towards the part.

Like several other functions, the position compensation is a programming method that requires input of the CNC machine operator. Programmer specifies the type of compensation and the memory register number, the operator enters the actual values at the machine, using appropriate display screens, during part setup.

◆ Programming Commands

On Fanuc and similar controls, there are four preparatory commands (G codes) available to program position compensation functions:

G45	*Single increase* in the programmed direction by the compensation amount
G46	*Single decrease* in the programmed direction by the compensation amount
G47	*Double increase* in the programmed direction by double the compensation amount
G48	*Double decrease* in the programmed direction by double the compensation amount

These definitions are based on *positive* compensation values stored in the control register. If the stored values are negative, the meaning of all definitions is valid only when the signs are inverted. None of these four preparatory commands is modal and all are valid only within the block in which they appear. If required in many blocks, they must be repeated in any subsequent block, if needed again.

◆ Programming Format

Each G code (G45 to G48) is associated with a unique position compensation number, programmed with the address H. The H address points to the memory area storage number of the control system. On most Fanuc control systems, the programmed letter can also be D, with exactly the same meaning. Whether the H or D address is used in the program, depends on the actual setting of a control system parameter.

A typical programming format for position compensation function is:

```
G91 G00 G45 X.. H..
```

or

```
G91 G00 G45 X.. D..
```

where the appropriate G code (G45 through G48), is followed by the target position and number of the memory storage area (using H or D address).

Note that the example uses *incremental* and *rapid motion* modes and only *one axis*. Normally, the compensation has to be applied to both X and Y axes. However, only a single measured amount can be stored under either H or D number. Since it is most probable that the compensation value will be different for each axis, it must be specified on separate blocks, with two *different* offset numbers H (or offset numbers D), for example:

```
G91 G00 G45 X.. H31     (H31 STORES THE X VALUE)
G45 Y.. H32             (H32 STORES THE Y VALUE)
```

or

```
G91 G00 G45 X.. D31     (D31 STORES THE X VALUE)
G45 Y.. D32             (D32 STORES THE Y VALUE)
```

For the record, the H address is also used with another type of compensation, known as the *tool length offset* (or *tool length compensation*), described in *Chapter 19*. The D address is also used with another type of compensation, known as the *cutter radius offset* (or *cutter radius compensation*), described in *Chapter 30*.

The applicable preparatory G code will determine how the address H or address D will be interpreted. In the examples, more common address H will be used - *Figure 17-1*.

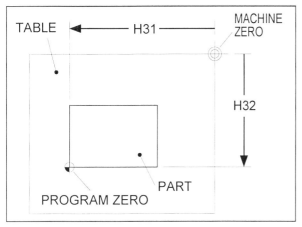

Figure 17-1

Position compensation - general concept

◆ Incremental Mode

The question may arise why the compensated motion is in the incremental mode. Remember that the main purpose of position compensation is to allow a correction of the distance between machine zero and program zero. The normal use is when starting the tool motion *from* machine zero position. By default, and without any offsets, coordinate settings or active compensations, the machine zero is the absolute zero, it is the *only* zero the machine control system 'knows' at the time.

Take the following example of several blocks, typically programmed at the beginning of a program with position compensation:

```
N1 G20
N2 G17 G80 T01
N3 M06
N4 G90 G00 G45 X0 H31        (NO X MOTION)
N5 G45 Y0 H32                (NO Y MOTION)
N6 . . .
```

This example illustrates a motion from machine zero (the current tool position), to program zero, which is the target position, along XY axes. Note the absolute mode setting G90 in block N4. Assume that the control system is set to H31=-12.0000 inches. The control will evaluate the block and interpret it as programmer's intention to go to the absolute zero, specified by G90. It checks the current position, finds it is at the absolute zero already and *does nothing*. There will be *no motion*, regardless of the compensation value setting, if the absolute motion is programmed to either X0 or Y0 target position. If the G90 is changed to G91, from absolute to incremental mode, there will be a motion along the *negative* direction of X axis, by the distance of exactly 12 inches and there will be a similar motion along Y axis, in block N5. The conclusion? *Use position compensation commands in the incremental mode G91 only.*

◆ Motion Length Calculation

Let's look a little closer at how the control system interprets a position compensation block. Interpreting the way how the control unit manipulates numbers is important for understanding how a particular offset or compensation works. Earlier definition has stated that a single increase is programmed with G45 command and a single decrease with G46 command. Both G47 and G48 commands are of no consequence at the moment. Since both commands are tied up with a particular axis and with a unique H address, all possible combinations available must be evaluated:

❑ Either an *increase* or a *decrease* is programmed
    (G45 or G46)

❑ Axis target can have a *zero* value, or a *positive* value,
    or a *negative* value

❑ Compensation amount may have a *zero* value,
    or a *positive* value, or a *negative* value

In programming, it is important to set certain standards and consistently abide by them. For example, on vertical machining centers, the compensation is measured *from* machine zero *to* program zero. That means a negative direction from the operator's viewpoint. The result is a logical decision to set negative compensation values as standard.

It is very crucial to understand *how* the control interprets information in a block. In position compensation, it evaluates the value stored in memory called by the address H (or D). If the value is zero, no compensation takes place. If the value of H is stored as a negative value, it adds this value to the value of the axis target position and the result is the motion length and direction. For example, assume the memory register H31 stores the value of -15.0 inches, and the machine current location is at its zero position and the axis setting on the control is also set to zero. Then the block

```
G91 G00 G45 X0 H31
```

will be interpreted as

```
-15.0 + 0 = -15.0000
```

resulting in the total motion of negative 15.0 inches along the X axis.

If the value of X axis target position is a non-zero and positive, the same formula applies:

```
G91 G00 G45 X1.5 H31
```

will be interpreted as

```
-15.0 + 1.5 = -13.5000
```

However, the next example is *not* correct:

```
G91 G00 G45 X-1.5 H31
```

Here, the motion will try to go into the *positive* X axis direction and the result will be X axis overtravel. Since the value of X is negative, G45 command cannot be used and G46 command must be *used* instead:

```
G91 G00 G46 X-1.5 H31
```

will be interpreted as

```
-15.0 + (-1.5) = -15.0000 - 1.5 = -16.5000
```

G45 could have been left in the program and the negative offset value could have been changed to a positive value. This could be quite confusing and definitely not consistent, but it would work quite well. To see the different possibilities, program O1701 is not doing very much, except moving from machine zero to different positions and back to machine zero (G28 command refers to a machine zero return and is explained separately in *Chapter 21*).

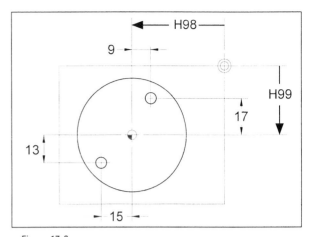

Figure 17-2

Position compensation applied to different target locations: zero, positive and negative - see O1701 program example

Figure 17-2 shows illustration for the following program example O1701. The logic applies to the X and Y axes exactly the same way. In is written in metric units and has been tested on Fanuc 11M, with the H address (D would work the same way). The position compensation values H98 and H99 were set to:

```
H98  =  -250.000
H99  =  -150.000
```

for the X and Y axes respectively. The modal commands were not repeated:

```
O1701 (G45 AND G46 TEST)
N1 G21 G17
N2 G92 X0 Y0 Z0
N3 G90 G00 G45 X0 H98          (ABS X0 TARGET)
N4 G46 Y0 H99                  (ABS Y0 TARGET)
N5 G28 X0 Y0

N6 G91 G00 G45 X0 H98          (INC X0 TARGET)
N7 G46 Y0 H99                  (INC Y0 TARGET)
N8 G28 X0 Y0

N9 G90 G00 G45 X9.0 H98        (ABS X+ TARGET)
N10 G46 Y17.0 H99             (ABS Y+ TARGET)
N11 G28 X0 Y0

N12 G91 G00 G45 X9.0 H98       (INC X+ TARGET)
N13 G46 Y17.0 H99            (INC Y+ TARGET)
N14 G28 X0 Y0

N15 G90 G00 G45 X-15.0 H98     (ABS X- TARGET)
N16 G46 Y-13.0 H99           (ABS Y- TARGET)
N17 G28 X0 Y0

N18 G91 G00 G45 X-15.0 H98     (INC X- TARGET)
N19 G46 Y-13.0 H99           (INC Y- TARGET)
N20 G28 X0 Y0
N21 M30
%
```

The control system will process each motion block separately - either the way it was intended or the wrong way (symbol O/T means an *overtravel* condition, preceded with the axis and direction of the overtravel):

```
N3    G90   ->   G45   ->   0    .   .   .    no motion
N4    G90   ->   G46   ->   0    .   .   .    no motion
N6    G91   ->   G45   ->   0    .   .   .    X-250.0
N7    G91   ->   G46   ->   0    .   .   .    Y+ O/T

N9    G90   ->   G45   ->   +    .   .   .    X-241.0
N10   G90   ->   G46   ->   +    .   .   .    Y+ O/T
N12   G91   ->   G45   ->   +    .   .   .    X-241.0
N13   G91   ->   G46   ->   +    .   .   .    Y+ O/T

N15   G90   ->   G45   ->   -    .   .   .    X+ O/T
N16   G90   ->   G46   ->   -    .   .   .    Y-163.0
N18   G91   ->   G45   ->   -    .   .   .    X+ O/T
N19   G91   ->   G46   ->   -    .   .   .    Y-163.0
```

◆ Position Compensation Along the Z axis

Position compensation feature usually applies only to the X and Y axes and will not normally be used with the Z axis. In most cases, the Z axis has to be controlled by another kind of compensation - known as the *tool length offset*. This method is described in *Chapter 19* of the handbook. If the Z axis is programmed with G45 or G46 commands, it will also be affected.

◆ Using G47 and G48

In the examples, position compensation feature was used only between the machine zero and program zero, as a method of determining where exactly is the part located on the table. The single increase using G45 and the single decrease using G46 commands were used, because they were the only commands needed.

Commands G47 (double increase) and G48 (double decrease) are only necessary for a *very* simplified cutter radius offset and are not covered in this handbook because of their obsolescence. However, they can still be used.

◆ Face Milling

In a later section *(Chapter 28)*, the principles of face milling will be explained in more detail. In that chapter is a very good example of how to apply position compensation to offset diameter of the face mill in a clear position, regardless of its size. This is probably the only possible use of G45 and G46 commands in contemporary programming.

WORK OFFSETS

Using the method of *Work Offsets* for tool positioning based on machine zero is much faster and more efficient than using the older methods of position compensation functions G45 and G46 described in the previous chapter. The work offsets are also known as *Work Coordinate System,* or even as *Fixture Offsets.* Work offsets are much more efficient than using the position register commands G92 (milling systems) or G50 (turning systems). CNC programmers who do not know the meaning of *position compensation* functions or the meaning of *position register* commands, are most likely working with the most modern CNC machines only. However, there are many machines in industry that still require these rather obsolete functions. Knowing them well will increase the number of available programming tools.

This chapter describes the most modern methods to coordinate the relationship between machine zero reference position and the program zero reference point. We will use the *Work Coordinate System* feature of any modern control system, whether it is called the *Work Coordinate System* or the *Work Offsets.* The latter term seems to be more popular because it is a little shorter. Think of the work offsets as an *alignment* between two or more coordinate systems.

WORK AREAS AVAILABLE

Before some more detailed descriptions can be covered, just what *is* a work coordinate system - or a *work offset?*

Work offset is a method that allows the CNC programmer to program a part away from the CNC machine, without knowing its exact position on the machine table. This is a very similar approach as in the position compensation method, but much more advanced and flexible. In the work offset system, up to six parts may be set up on the machine table, each having a different work offset number. Programmer can move the tool from one part to another with absolute ease. To achieve this goal, a special preparatory command for the active work offset is needed in the program and the control system will do the rest. The system will automatically make any adjustment for the difference between the two part locations.

Unlike the position compensation function, two, three, or more axes may be moved simultaneously with work offsets, although the Z axis for CNC machining centers is controlled independently, using G43 or G44 tool length offset commands. Commands relating to the Z axis offset are fully described in the next chapter, *Chapter 19.*

In position compensation, to switch machining from one part to another within the same setup, the program has to contain a different compensation number from program zero of the previous part. Using the work offset method, *all program zeros are measured from the machine zero position,* normally up to six, but more offsets are available.

The six work coordinate systems - or *work offsets* - that are available on Fanuc control systems are assigned the following preparatory commands:

G54	G55	G56	G57	G58	G59

When the control unit is turned on, the default coordinate system is normally G54.

Basically, the work offsets establish up to six independent work areas as a standard feature. The values input into the CNC unit are always distances measured *from the machine zero to the program zero.* As there are up to six work areas, up to six independent program zero positions can be defined. *Figure 18-1* shows the basic relationships, using the default G54 setting.

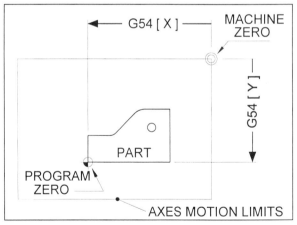

Figure 18-1

Basic relationships of the work offset method

The ⬚⬚⬚⬚ he default work offset ⬚⬚⬚⬚ other five available w⬚⬚⬚ tored in the control sy⬚⬚⬚ d from the machine ⬚⬚⬚ of the part, as determ⬚⬚⬚

The distance *from* machine zero *to* program zero of each work area is measured separately along the X and Y axes and input into the appropriate work offset register of the control unit. Note that the measurement direction is *from* machine zero *to* program zero, never the other way around. If the direction is negative, the minus sign must be entered in the offset screen.

For comparison with the position register command G92, *Figure 18-2* shows the same part set with the older method of G92 and machine zero as a start point. Note the *opposite* arrows designation, indicating the direction of measurement - *from* program zero *to* machine zero.

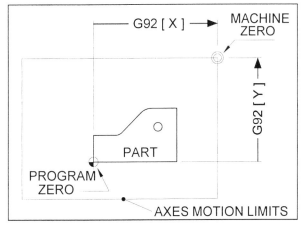

Figure 18-2
Basic relationships of the Position Register command G92

For work offsets G54 to G59, a typical entry into the co-ordinate offset position register will be the X axis as a negative value, the Y axis as a negative value and the Z axis as a zero value, for the majority of vertical machining centers. This is done by the CNC operator at the machine. *Figure 18-3* shows an example of a typical control system entry.

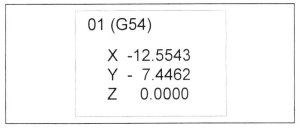

Figure 18-3
Typical data entry for the G54 work coordinate system

By using the G54 to G59 settings in the program, the control system selects the stored measured distances and the cutting tool may be moved to any position within the selected work offset simultaneously in both the X and Y axes, whenever desired.

Part position on the machine table is usually unknown during the programming process. The main purpose of work offset is to synchronize the actual position of the part as it relates to the machine zero position.

◆ Additional Work Offsets

The standard number of six work coordinate offsets is usually enough for most types of work. However, there are jobs that may require machining with more program reference points, for example, a multi-sided part on a horizontal machining table. What options do exist, if the job requires ten work coordinate systems, for example?

Fanuc offers - *as an option* - up to 48 additional work offsets, for the total of 54 (6+48). If this option is available on the CNC system, any one of the 48 work offsets can be accessed by programming a special G code:

G54.1 P..	Selection of additional work offset, where P = 1 to 48

● G54.1 P.. example :

G54.1 P1	Selection of additional work offset 1
G54.1 P2	Selection of additional work offset 2
G54.1 P3	Selection of additional work offset 3
G54 1 P*x*..	Selection of additional work offset *x*..
G54.1 P48	Selection of additional work offset 48

The utilization of additional work offsets in the program is exactly the same as that of the standard commands:

```
N2 G90 G00 G54.1 P1 X5.5 Y3.1 S1000 M03
```

Most Fanuc controls will allow omission of the decimal portion of the G54.1 command. There should be no problem programming:

```
N2 G90 G00 G54 P1 X5.5 Y3.1 S1000 M03
```

The presence of P1 to P48 function within a block will select an *additional* work offset. If the P1 to P48 parameter is missing, the default work offset command G54 will be selected by the control system.

WORK OFFSET DEFAULT AND STARTUP

If no work offset is specified in the program and the control system supports work offsets, the control will automatically select G54 - that is the *normal default selection*. In programming, it is always a good practice to program the work offset command and other default functions, even if the default G54 is used constantly from one program to another. The machine operator will have a better feel for the CNC program. Keep in mind that the control still has to have accurate work coordinates stored in the G54 register.

In the program, the work offset may be established in two ways - either as a separate block, with no additional information, as in this example:

`N1 G54`

The work offset can also be programmed as part of a startup block, usually at the head of program or at the beginning of each tool:

`N1 G17 G40 G80 G54`

The most common application is to program the appropriate work offset G code in the same block as the first cutting tool motion:

`N40 G00 G90 G54 X5.5 Y3.1 S1500 M03`

Figure 18-4 illustrates this concept. In the above block N40, the absolute position of the tool has been established as X5.5Y3.1, *within* the G54 work offset. What will *actually* happen when this block is processed?

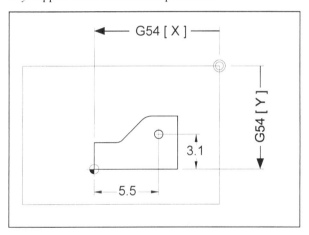

Figure 18-4
Direct tool motion to a given location using G54 work offset

Note that there are no X or Y values associated with the G54 command in the illustration. There is no need for them. The CNC operator places the part in *any suitable location* on the machine table, squares it up, finds how far is the program zero away from machine zero and enters these values into the control register, under the G54 heading. The entry could be either manual or automatic.

Assume for a moment, that after setup, the measured distances from machine zero to program zero were X-12.5543 and Y-7.4462. The computer will determine the actual motion by a simple calculation - it will always *add* the programmed target value X to the measured value X, and the programmed target value Y to the measured value Y.

The actual tool motion in the block N40 will be:

```
X = -12.5543 + 5.5 = -7.0543
Y =  -7.4462 + 3.1 = -4.3462
```

These calculations are absolutely unnecessary in everyday programming - they are only useful to the thorough understanding of how the control unit interprets given data.

The whole calculation is so consistent, it can be assigned into a simple formula. For simplicity, the settings of the EXT (external or common) offset are not included in the formula, but are explained separately, later in the chapter:

$$A \ = \ M \ + \ P$$

☞ where ...

   A = Actual motion length (distance-to-go displayed)
   M = Measured distance from machine zero
   P = Programmed absolute target position (axis value)

Be very careful when adding a negative value - mathematically, the double signs are handled according to the standard rules:

PLUS and PLUS *becomes* **PLUS**	`a + (+ b) = a + b`
PLUS and MINUS *becomes* **MINUS**	`a + (- b) = a - b`
MINUS and PLUS *becomes* **MINUS**	`a - (+ b) = a - b`
MINUS and MINUS *becomes* **PLUS**	`a - (- b) = a + b`

In the example, plus and minus combination creates a negative calculation:

`-10 + (-12) = -10 - 12 = -22`

If any other work offset is programmed, it will be automatically replaced by the new one, *before* the actual tool motion takes place.

◆ Work Offset Change

A single CNC program may use one, two, or all work offsets available. In all multi-offset cases, the work offset setting stores *the distance from the machine zero to the program zero of the each part in the setup.*

For example, if there are three parts mounted on the table, each individual part will have its own program zero position associated with one work offset G code.

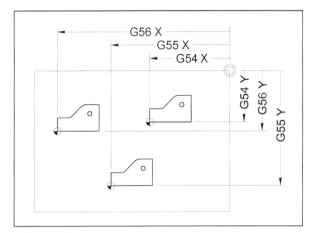

Figure 18-5

Using multiple work offsets in one setup and one program. Three parts shown in the example.

Compare all possible motions in *Figure 18-5*:

```
G90 G00 G54 X0 Y0
```

... will rapid from the *current* tool position, to the program zero position of the *first* part.

```
G90 G00 G55 X0 Y0
```

... will rapid from the *current* tool position, to the program zero position of the *second* part.

```
G90 G00 G56 X0 Y0
```

... will rapid from the *current* tool position, to the program zero position of the *third* part.

Of course, the target position *does not* have to be part zero (program zero) as shown in the example - normally, the tool will be moved to the first cutting position right away, to save the cycle time. The following program example will illustrate that concept.

In the example, a single hole will be spot drilled on each of the three parts to the calculated depth of Z-0.14 (program O1801). Study the simplicity of transition from one work offset to another - there are no cancellations - just a new G code, new work offset. The control will do the rest:

```
O1801
N1 G20
N2 G17 G40 G80
N3 G90 G54 G00 X5.5 Y3.1 S1000 M03    (G54 USED)
N4 G43 Z0.1 H01 M08
N5 G99 G82 R0.1 Z-0.14 P100 F8.0
N6 G55 X5.5 Y3.1                      (SWITCH TO G55)
```

```
N7 G56 X5.5 Y3.1                      (SWITCH TO G56)
N8 G80 Z1.0 M09
N9 G91 G54 G28 Z0 M05                 (SWITCH TO G54)
N10 M01
...
```

Blocks N3 through N5 relate to the first part, within the G54 work offset. The block N6 will spot drill the hole of the second part of the same setup, within the G55 work offset and the block N7 will spot drill the hole of the third part of the same setup, within the G56 work offset. Note the return to the G54 work offset in block N9. Return to the default coordinate system is not required - it is only a suggested good practice when the tool operation is completed. The work offset selection is modal - take care of the transitions between tools from one work offset to another. Bringing back the default offset G54 may always be helpful at the end of each tool.

If all these blocks are in the *same* program, the control unit will automatically determine the *difference* between the current tool position and the same tool position within the next work offset. This is the greatest advantage of using work offsets - an advantage over the position compensation and the position register alternatives. All mounted parts may be identical or different from each other, as long as they are in the same positions for the whole setup.

◆ Z Axis Application

So far, there was a conspicuous absence of the Z axis from all discussions relating to the work offset. That was no accident - it was intentional. Although any selected work offset can apply to the Z axis as well, and with exactly the same logic as for X and Y axes, there is a better way of controlling the Z axis. The method used for Z axis is in the form of G43 and G44 commands that relate specifically to the *tool length compensation*, more commonly known as the *tool length offset*. This important subject is discussed separately in the next chapter. In the majority of programming applications, the work offset is used only within the XY plane. This is a typical control system setting and may be represented by the following setup example of the stored values within the control register:

```
(G54) X-8.761 Y-7.819 Z0
(G55) X-15.387 Y-14.122 Z0
(G56) X-22.733 Y-8.352 Z0
(G57) ...
```

The Z0 offset entry is very important in the examples and in the machine control. The specified Z0 means that the coordinate setting for the Z amount (representing the height of the part) does not change from one part to another, even if the XY setting does.

The only time there is a need to consider Z axis within the work offset setting is in those cases, where the height of each part in the setup is *different*. So far, only the XY positions were considered, as they had been the ones changing.

If the Z amount changes as well, that change must be considered by modifying the coordinate register setting of the control. This is the responsibility of the CNC operator, but the programmer can learn an important lesson as well.

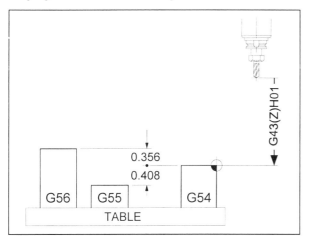

Figure 18-6

Setting of work offsets for a variable part height

Figure 18-6 shows some typical and common possibilities used for special parts that have a variable height within the same tool setup. The difference between part heights has to be always known, either from the part drawing specifications or from actual measurements at the machine.

If the previous multi-offset example for XY setting are also adapted for the Z axis, the work offset can be set up for parts within the same setup, but with variable heights. This variable height is controlled by the Z axis. The result of the setting will reflect the difference in height between the measured Z axis surface for one part and the measured Z axis surface for the other parts. Based on the data in the previous example, combined with the Z values shown in *Figure 18-6*, the control system settings may look like this:

```
(G54)  X-8.761  Y-7.819  Z0
(G55)  X-15.387 Y-14.122 Z-0.408
(G56)  X-22.733 Y-8.352  Z0.356
```

The important thing to know about the control of the Z axis within the selected work offset is that it works in very close *conjunction* with the tool length offset, discussed in the next chapter *(Chapter 19)*. Stored amount of the Z axis setting within a work offset will be applied to the actual tool motion and used to *adjust* this motion, according to the setting of the tool length offset. An example may help.

For instance, if the tool length offset of a particular cutting tool is measured as Z-10.0, the actual motion of such a tool to the program zero along Z axis will be -10.0 inches within the G54 work offset, -10.408 within the G55 work offset, and -9.644 within the G56 offset - all using the examples in the previous illustration, shown in *Figure 18-6*.

HORIZONTAL MACHINE APPLICATION

Machining several parts in a single setup is done quite frequently on CNC vertical machining centers. The multiple work offset concept is especially useful for CNC horizontal machining centers or boring mills, where many part faces may have to be machined during a single setup.

Machining two, three, four, or more faces of the part on a CNC horizontal machining center is a typical everyday work in many companies. For this purpose, the work offset selection is a welcome tool. For example, the *program zero* at the pivot point of the indexing table can be set for the X and Y axes. Program setting of the Z axis may be in the same position (the pivot point of the indexing table) or it can be on the face of each indexed position - either choice is acceptable. The work offset handles this application very nicely, up to six faces with a standard range of the G codes.

There is no significant difference in the programming approach - the switch from one work offset to another is programmed exactly the same way as for the vertical machining applications. The only change is that the Z axis will be retracted to a clear position and the table indexing will usually be programmed between the work offset change.

Figure 18-7 illustrates a typical setting for four faces of a part, where Z0 is at the top of each part face. There could be as many faces as there are table indexing positions. In either case, the programming approach would be similar if Z0 were at the center of indexing table, which is also quite a common setup application. See *Chapter 46* for more details relating to horizontal machining.

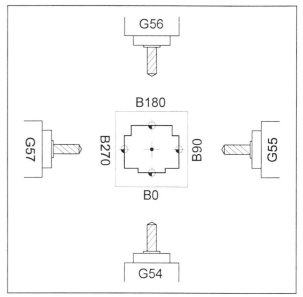

Figure 18-7

Example of work offsets applied to a horizontal machining center

EXTERNAL WORK OFFSETS

A careful look at a typical work offset screen display reveals one special offset that is identified by one of the following designations:

❏   00   (EXT)

❏   00   (COM)

The two zeros - 00 - indicate that this work offset is *not* one of the standard six offsets G54-G59. These offsets are identified by numbers 01 through 06. The 00 designation also implies that this is not a programmable offset, at least not by using the standard CNC programming methods. *Fanuc Macro B* option does allow programming this offset.

The abbreviation *EXT* means *External*, and the abbreviation *COM* means *Common*. The machine control will have one or the other designation, but not both. As a matter of curiosity, the *COM* designation is found on older controls, whereby the *EXT* designation is more recent. The reason? With the explosion of personal computer market, the *COM* abbreviation has become the de facto standard abbreviation for the word *communications*. As Fanuc controls also support several communication methods, including the connection with a personal computer, some time ago, the COM offset designation has been replaced with the designation EXT, to prevent possible confusion between the two abbreviations used in computing.

Either abbreviation refers to the same offset and has the same purpose. On the screen display, this special offset is usually located *before* or *above* the offset for G54, for example, as illustrated in *Figure 18-8*:

00 (EXT)	01 (G54)
X 0.0000	X -12.5543
Y 0.0000	Y - 7.4462
Z 0.0000	Z 0.0000

Figure 18-8
Example of an EXT (external) work offset display (EXT = COM)

The major difference between an external or common work offset is that it is not programmable with any particular G code. Its setting is normally set to zero for all axes. Any *non-zero* setting will activate this work offset in a very important way:

****** IMPORTANT ******
Any setting of the external work offset will always affect ALL work offsets used in the CNC program.

All six standard work offsets, as well as any additional work offsets, will be affected by the values set in the external work offset, based on the setting of each axis. Because all programmable coordinate systems will be affected, the name for this special offset is *Common Work Offset* or more often, the *External Work Offset*.

LATHE APPLICATIONS

Originally, the work coordinate system was designed for CNC machining centers only. It did not take long to apply it to CNC lathes as well. The operation, logically and physically, is identical to that for machining centers. Using work offsets for CNC lathes eliminates the awkward use of G50 or G92 and makes the CNC lathe setup and operation much faster and easier.

◆ Types of Offsets

The main difference in applying work offsets on a lathe is that seldom will there be a need for more than one work offset. Two work offsets are a possibility, three or more are used for some very special and complex setups. G54 to G59 commands are available on all modern CNC lathes and it is quite customary to ignore the work offset selection in the program, unless more than one offset is used. That means the CNC lathe programmer depends on the default G54 setting as a rule.

Two special offset features found on the latest control systems are the *Geometry* and *Wear* offsets, either on the same screen display, or on separate screens, depending on the control model.

◆ Geometry Offset

Geometry offset is the equivalent of a work offset as it is known from the milling controls. It represents the distance from tool reference point to program zero, measured from the machine zero along a selected axis. Typically, on a slant bed CNC lathes, with the tool turret *above* the spindle centerline, the geometry offset for both X and Z axes will be negative. *Figure 18-9* illustrates reasonable geometry values for a drill, turning tool and boring bar (T01, T02, T03).

GEOMETRY OFFSET				
No.	X OFFSET	Z OFFSET	RADIUS	TIP
01	-8.6330	-2.3630	0.0000	0
02	-8.6470	-6.6780	0.0469	3
03	-9.0720	-2.4950	0.0313	2
04	0.0000	0.0000	0.0000	0
05	0.0000	0.0000	0.0000	0

Figure 18-9
Typical data entries for a lathe tool GEOMETRY offset

◆ Wear Offset

The wear offset is also known and used on milling controls, but only for the tool length offset and the cutter radius offset, not for the work coordinate system (work offset).

On the CNC lathes, the purpose of the wear offset is identical to that for machining centers. This offset compensates for the tool wear and is also used to make fine adjustments to the geometry offsets. As a rule, once the geometry offset for a given tool is set, that setting should be left unchanged. Any adjustments and fine tuning of actual part dimensions should be done by the wear offset only.

WEAR OFFSET				
No.	X OFFSET	Z OFFSET	RADIUS	TIP
01	0.0000	0.0000	0.0000	0
02	-0.0060	0.0000	0.0469	3
03	0.0000	0.0040	0.0313	2
04	0.0000	0.0000	0.0000	0
05	0.0000	0.0000	0.0000	0

Figure 18-10

Typical data entries for a lathe tool WEAR offset

Figure 18-10 shows some reasonable sample entries in the wear offset registers. The tool radius and tip number settings appear in both displays and the display in both screens is automatic after the offset value input. The tool nose radius and the tool tip orientation number are unique to CNC lathe controls.

◆ Tool and Offset Numbers

Just like tools on CNC machining centers have numbers, they have numbers on CNC lathes as well. Usually, only one coordinate offset is used, but different tool numbers. Remember, the tool number for a lathe has *four digits,* for example, T0404:

❑ The first two digits select the tool indexing station (turret station) *and* the geometry offset number. There is no choice here. Tool in station 4, for example, will also use geometry offset number 4.

❑ The second two digits are for the wear offset register number only. They do not have to be the same as the tool number, but it makes sense to match the numbers, if possible.

Depending on the control model and the display screen size, the tool offset register may have a separate screen display (page) for the geometry and wear offsets, or both offset types may be shown on the same screen display. The work offset values (work coordinates) are always placed in the *Geometry* offset column.

TOOL SETUP

In the next three illustrations is a very similar layout as that shown in *Chapter 16*, describing the use of G50 register method (position register command used in the program). *Compare the two illustrations!*

The setup of the CNC lathe is identical in both cases, except for the method and purpose of the position measuring. All illustrations in the applications also match the reasonable data entered in the tool geometry and the tool wear offset screens of the control.

Typical values along the X axis are always negative (as shown in illustrations), typical values along the Z axis are usually negative. A positive value is also possible, but that means the tool is above work and tool changing can be very dangerous. *Watch out for such situations!*

The actual setting procedures are subject of a CNC machine operation training and not practical to cover in a programming handbook. There are additional methods, also part of machine training, that allow faster tool setting, using one tool as a master and setting all the remaining tools *relative to the master tool.*

◆ Center Line Tools

Tools that work on the spindle center line are tools that have their tool tip located on the center line during machining. This area covers all center drills, spot drills, various drills, reamers, taps, even end mills used for flat bottom holes. At the same time, it disqualifies all boring bars, since their tool tip does not normally lie on the spindle center line during machining. Center line tools are *always* measured from the center line of the tool to the center line of the spindle along the X axis and from the tool tip to the program zero along the Z axis. *Figure 18-11* illustrates a typical setting for center line tools.

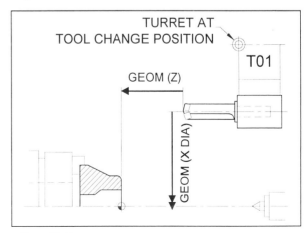

Figure 18-11

Typical geometry offset setting for CENTER LINE tools

◆ Turning Tools

Turning tools - or *external* tools - are measured from the imaginary tool tip to the program zero, along the X axis (as a negative diameter) and along the Z axis, usually as a negative value as well. Keep in mind that if the cutting tool insert (for turning or boring) is changed from one radius to another radius in the *same* tool holder, the entered setup amount must also change. The change may be marginal, but even a marginal change is enough to cause a scrap, so a good care is needed. For turning, be extra careful for a tool nose radius that changes from a larger size to a smaller size, for example, from 3/64 (R0.0469) to 1/32 (R0.0313).

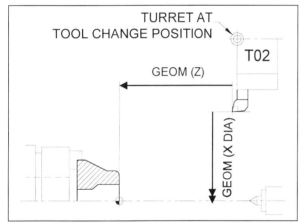

Figure 18-12
Typical geometry offset setting for EXTERNAL tools

Figure 18-12 illustrates a typical geometry setting for a turning (external) tool and *Figure 18-13* illustrates a typical geometry setting for a boring (internal) tool.

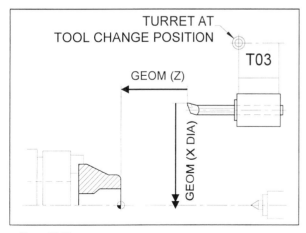

Figure 18-13
Typical geometry offset setting for INTERNAL tools

◆ Boring Tools

Boring tools - or *internal* tools - are always measured from the imaginary tool tip to program zero, along the X axis (typically as a negative diameter) and along the Z axis, typically as a negative value as well. In majority of cases, the X value of a boring tool will be noticeably larger than that for a turning or other external tool.

For boring operations, same as for turning operations, also be extra careful for a tool nose radius that changes from a larger size to a smaller size. It is the same as for a turning tool. The scrap can be made very easily.

◆ Command Point and Tool Work Offset

For various reasons, it is quite common to change a cutting insert in the middle of work, primarily to maintain favorable cutting conditions and to keep dimensional tolerances within drawing specifications. Cutting inserts are manufactured to very high standards, but a certain tolerance deviation should be expected between inserts obtained from different sources. If changing an insert, it is advisable to adjust the wear offset for precision work, in order to prevent scrapping the part.

Tool holders accept inserts of the same shape and size but with a different nose radius. Always be cautious when replacing an insert with an insert that has a larger or smaller tool nose radius. The offset has to be adjusted in both axes, by the proper amount.

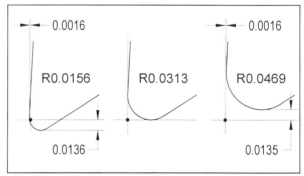

Figure 18-14
Setting error caused by a different insert radius in the same holder

The example in *Figure 18-14* shows standard setting for a 1/32 (.0313) nose radius (middle), and the setting error for a radius that is smaller (left) and one that is larger (right). The dimensions indicate the amount of error for the particular insert shown in the example.

When changing an insert, adjust the required offset(-s)

So far, we have looked at two methods of compensation for the actual position of the *cutting tool* in relation to the *machine reference point*. One method was the older type, using *position compensation*, the other was the contemporary *work coordinate system* method *(work offset)*. In both cases, the emphasis was only on the X and Y axes, not on the Z axis. Although the Z axis could have been included with either method, the results would not have been very practical. The main reason is the nature of the CNC work.

Normally, programmer decides on the setup of a part in the fixture and selects the appropriate location of the XYZ program zero (part reference point or part zero). When using work offsets, XY axes are always measured from the *machine reference point* to the *program zero* position. By a strict definition, the same rule applies to the Z axis. The major difference is that the measured XY values will *remain unchanged for all tools*, whether there is one tool used or one hundred tools. That is not the case with the Z axis.

The reason? Each tool has a *different length*.

GENERAL PRINCIPLES

The length of each cutting tool has to be accounted for in every program for a CNC machining center. Since the earliest applications of numerical control, various techniques of programming tool length have emerged. They all belong into one of two basic groups:

❏ Actual tool length is known

❏ Actual tool length is unknown

Needless to say, each group requires its own unique programming technique. To understand the concept of tool length in CNC programming, it is important to understand meaning of the phrase *actual tool length*. This length is sometimes known as the physical tool length or just tool length and has a very specific meaning in CNC programming and setup.

◆ Actual Tool Length

Let's evaluate a simple tool first. By holding a typical drill, we can determine its physical length with a measuring device. In human terms, a six inch long drill has a length of six inches, measured from one end to the other. In CNC programming that is still true, but not quite as relevant. A drill - or any other cutting tool - is normally mounted in a tool holder and only a *portion* of the actual tool projects

out, the rest is hidden in the holder. The tool holder is mounted in the spindle, by means of a standardized tooling system. Tool designations, such as the common sizes HSK63, HSK100, BT40 and CAT50, are examples of established European and American standards. Any tool holder within its category will fit any machine tool designed for that category. This is just one more precision feature built into the CNC machine.

The length of a tool for the purposes of CNC programming must always be associated with the tool holder and in relation to the machine design. For that purpose, manufacturers build a precision reference position into the spindle, called the *gauge line*.

◆ Gauge Line

When the tool holder with the cutting tool is mounted in the spindle of a CNC machine, its own taper is mounted against an opposite taper in the spindle and held in tightly by a pullbar. The precision of manufacturing allows for a constant location of the tool holder (any tool holder) in the spindle. This position is used for reference and is commonly called the *gauge line*. As the name suggests, it is an imaginary reference line used for gauging - or measuring - along the Z axis - *Figure 19-1*.

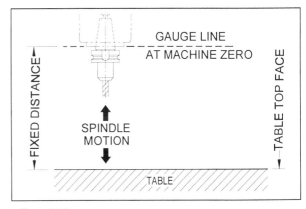

Figure 19-1

Typical front view of a CNC vertical machining center

We use the gauge line for accurate measuring of tool length and any tool motion along the Z axis. Gauge line is determined by the machine manufacturer and is closely related to another precision face, called the *machine table,* actually, the table top *face*. The gauge line is one side of a plane that is parallel with another plane - the *table top face*.

◆ Table Top Face

Every CNC machining center has a built-in machine table on which the fixture and part are mounted. Top of the machine table is precision ground to guarantee flatness and squareness for the located part.

In addition, the table is located a certain fixed distance from the gauge line. Just like the position of the tool holder in the spindle cannot be changed, the position of the table (even for a removable table using a palette system) cannot be changed either. The surface of the table creates another reference plane that is related to the gauge line and parallel to it as well. This arrangement allows to accurately program a tool motion along the Z axis.

The tool length offset (compensation) can be defined:

> The tool length offset is a procedure that corrects the difference between the programmed length of the tool and its actual length.

The most significant benefit of tool length offset in CNC programming is that it enables the programmer to design a complete program, using as many tools as necessary, without actually knowing the actual length of any tool.

TOOL LENGTH OFFSET COMMANDS

Fanuc systems and several other machine controls offer three commands relating to the tool length offset - all are preparatory G commands:

> G43   G44   G49

All three commands are only applicable to the Z axis. Unlike the work offset commands G54-G59, G43 or G44 cannot be used without a further specification. They can only be used with an offset number designated by the address H. The address H must be followed by up to three digits, depending on the number of offsets available within the control system:

G43	Positive tool length offset
G44	Negative tool length offset
G49	Tool length offset cancel
H00	Tool length offset cancel
H..	Tool length offset number selection

Tool length offset should always be programmed in the absolute mode G90. A typical program entry will be the G43 or G44 command, followed by the Z axis target position and the H offset number:

`N66 G43 Z1.0 H04`

This is also a convenient block to add coolant function M08 for the current tool:

`N66 G43 Z1.0 H04 M08`

The resulting motion in the example will be to 1.0 inch above the part zero. The control system will calculate the *distance to go*, based on the value of H offset stored by the operator during setup.

Figure 19-2 shows a typical screen for the tool length off-

TOOL OFFSET   (LENGTH)

No.	GEOMETRY	WEAR
001	-6.7430	0.0000
002	-8.8970	0.0000
003	-7.4700	0.0000
004	0.0000	0.0000
005	0.0000	0.0000
006	0.0000	0.0000
. . .	. . .	. . .

Figure 19-2

Typical tool length offset entry screen

set entry. Note that the actual display will vary from one control to another and the wear offset may not be available on some controls. The wear offset (if available) is only used for *adjustments* to the tool length as a separate screen entry.

The G44 command is hardly ever used in a program - in fact, it has the dubious distinction of being the least used commands of all Fanuc G codes. Its comparison with G43 is described later in this chapter.

Many CNC programmers and operators may not realize that the Z axis setting in a work offset (G54-G59) is also very important for the tool length offset. The reason why will be clear in the coming descriptions of different methods of tool length offset setting.

Some programming manuals suggest the older G45 or G46 commands can also be used for tool length offset. Although this is still true today and may have had some advantages in the early days, it is best to avoid them. First, the position commands are not used very much anymore and, second, they can also be used with the X and Y axes and do not truly represent the Z axis exclusively.

◆ Distance-To-Go in Z Axis

In order to interpret how the CNC system uses the tool length command, the programmer or operator should be able to calculate the *distance-to-go* of the cutting tool. The logic behind the tool length offset is simple:

❑ The value of the H offset will be *added* to the target Z position if G43 is used, because G43 is defined as the *positive* tool length offset

❑ The value of the H offset will be *subtracted* from the target Z position if G44 is used, because G44 is defined as the *negative* tool length offset

The target position in both cases is the absolute Z axis coordinate in the program. If the Z axis setting of the work offset (G54-G59), the H value, and the Z axis target are all known, the *distance-to-go.* can be accurately calculated. The control system will use this formula:

$$Z_d = W_z + Z_t + H$$

☞ where ...

Z_d = Distance-to-go along Z axis (actual travel)
W_z = Work coordinate value for Z axis
Z_t = Target position in Z axis (Z coordinate)
H = Value of the applied H offset number

➲ Example - W_z = 0:

`G43 Z0.1 H01` ..... *where:*

G54 in Z is set to Z0, Z axis target position is 0.1 and H01 is set to -6.743, then the distance-to-go Z_d will be:

```
Z_d  =    0 + (+0.1) + (-6.743)
     =    0 + 0.1 - 6.743
     =    -6.643
```

The displayed *distance-to-go* will be Z-6.673.

In order to make sure the formula is always correct, try to change a few values.

➲ Example - W_z = 0.0200:

In this example, the program contains block

`G43 Z1.0 H03` ..... *where:*

G54 in Z is set to 0.0200, Z axis target is Z1.0 and the value of H03 is -7.47:

```
Z_d  =    (+0.02) + (+1.0) + (-7.47)
     =    0.02 + 1.0 - 7.47
     =    -6.45
```

The result is correct, the tool will travel along the Z axis, towards the part and the *distance-to-go* will be Z-6.45.

In the last example, a negative target position is shown:

➲ Example - W_z = 0.0500:

The program block contains a negative Z coordinate:

`G43 Z-0.625 H07` ..... *where:*

G54 along Z is set to 0.0500, Z axis target is -0.625 and the H07 is -8.28. The *distance-to-go* calculation uses the same formula, but with different values:

```
Z_d  =    (+0.05) + (-0.625) + (-8.28)
     =    0.05 - 0.625 - 8.28
     =    -8.855
```

Again, the formula works correctly and can be used for *any* distance-to-go calculation along the Z axis. Experimenting with other settings may also be useful.

TOOL LENGTH SETUP

The length of a tool used for machining (consisting of the cutting tool and the tool holder), can be set directly on the CNC machine or away from it. These setup options are often called *on-machine* or *off-machine* tool length setups. Each option has an advantage and it has its corresponding disadvantage. They both share a certain relationship to the gauge line, as it applies to the length of the tool or its projection. These two setup options are directly opposite to each other and often cause philosophical divisions (or at least some friendly disagreements) among CNC programmers. Evaluate each setup option and compare its advantages with its disadvantages. Which one appears to be somewhat better will depend on many other factors as well.

Both options require involvement of two people, or at least two professional skills - the CNC programmer and the CNC operator. The question narrows down to *who is going to do what - and when.* To be fair, both sides have to do something. The programmer has to identify all selected tools by their number (the T address) and assign tool length offset numbers, for example, G43 or G44 with the H address for each tool. The operator has to physically set the tools into the holder and register the measured values of H addresses into the CNC system memory.

◆ On-Machine Tool Length Setting

In technical terms, the bulk of *on-machine* setting requires the work of a CNC operator. Typically, the operator places a tool into the spindle and measures the distance the tool travels *from machine zero* to *part zero* (program zero). This work can only be done between jobs and is definitely nonproductive. It can be justified under certain circumstances, particularly for jobbing shops and short-run jobs or for machine shops with very few people. Although the setting of a large number of tools will take much longer than setting of a few tools, there are setup methods available to the CNC operator that allow reasonably speedy *on-machine* tool length setup, namely using the *master tool* method, described later in this section. The one major benefit of this method is that it does not require the expense of additional equipment and a skilled person to operate it.

◆ Off-Machine Tool Length Setting

In technical terms, the *off-machine* setting requires the work of a skilled tool setter or a CNC operator. Since the setting is done away from the machine, a special equipment is required, adding to the overall cost of manufacturing. This equipment can be a simple fixture with a height gage (even made *in-house*), or a more expensive, commercially available digital display device.

◆ Tool Length Offset Value Register

Whichever method of the tool length setting is used, it produces a measured value that represents the length of the selected tool. This value is useless by itself and must be somehow supplied to the program, before the job is machined. The operator must register the measured value into the system, under the proper heading on the control panel.

The control system contains a special registry for the tool length offset, usually under the heading of *tool length setting*, *tool length offset*, *tool length compensation* or just *offset*. Regardless of the exact heading, the setting procedure is to make sure the measured length is entered into the control, so it can be used by the program. The measured length is always well within the Z axis travel limits of the machine, yet still allows for sufficient clearances for the part and the tool changes.

To understand the tool length offset, try to fully understand the Z axis motion and geometry of the CNC machine first. On vertical and horizontal machining centers, look at the XZ plane, which is the top of part for both. The principles are identical, but the focus will be on the vertical machining center layout.

Z AXIS RELATIONSHIPS

To understand the general principles of tool length offset, let's look at the schematic illustration of a typical setup for a vertical machining center - *Figure 19-3*.

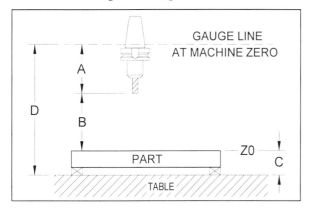

Figure 19-3

Z axis relationships of the machine, cutting tool, table top face, and the part height

The figure represents a common setup of a CNC vertical machining center, looking from the front of the machine, a typical operator's viewpoint. The spindle column is located at the machine zero position. This is the limit switch position for positive Z axis travel and is necessary for the automatic tool change on virtually all machining centers. All four illustrated dimensions are either known, can be found in various instruction or service manuals, or can be physically measured. They are always considered as *known* dimensions or *given* dimensions and used as *equally critical* for accurate machine setup:

❏ Distance between the tool gauge line and
    the tool cutting point

    ... dimension A in the illustration

❏ Distance between the tool cutting point and the Z0
    (program zero of the part)

    ... dimension B in the illustration

❏ Height of the part (distance between
    the table top and Z0 of the part)

    ... dimension C in the illustration

❏ Total of all three previous dimensions
    (distance between the tool gauge line and the table top)

    ... dimension D in the illustration

It is rather rare that the programmer or the operator would always know all four dimensions. Even if that were possible, some calculations would not be worthwhile doing. The reality is that only *some* dimensions are known or can be found out relatively easily.

In the illustration, the dimension D is always known, because it is the distance determined by the machine manufacturer. It may not be possible to know the C dimension (height of the part with clearances), but with planning and a common setup, this dimension can be known as well.

That leaves dimension A - the distance between the tool gauge line and the tool cutting point. There is no other method to find this dimension, but to actually measure it. In the earlier days of numerical control, this length A had to be always known and embedded in the program. Because of the inconveniences involved in finding this dimension, other methods have developed later.

Today, three methods are considered in programming the tool length setup, including the original method:

❏ Preset tool method is the original method
    ... it is based on an external tool setting device

❏ Touch-off method is the most common method
    ... it is based on the measurement at the machine

❏ Master tool method is the most efficient method
    ... it is based relative to the length of the longest tool

Each method has its benefits. The CNC programmer considers these benefits and chooses one method over the other. Applications of these methods and operations do not relate to the programming process directly - they are methods of *physical setup* on the machine only. For proper understanding of the subject by CNC programmers, they are included here as well. Regardless of which setting method is chosen, include a reference to the selected setting in the program, in the form of a comment or message.

◆ Preset Tool Length

Some users prefer to *preset* the length of cutting tools *away* from the machine, rather than during the machine setup. This has been the original method of setting tool lengths. There are some benefits in this approach - the most notable is the elimination of nonproductive time spent during setup. Another benefit applies to horizontal machining centers, where program zero is often preset to the center of the rotary or indexing table. There are disadvantages as well. Presetting tool length away from machine requires an external device, known as the *tool presetter*, which could be a relatively expensive addition to the CNC machine.

Using the tool presetter, all cutting tools are set at the external device, while the CNC machine runs a production job. There is no measuring on the machine when jobs do change. All the operator has to do, is to enter the measured values into the offset registers. Even that portion of the setup can be done through the program by using the optional G10 command (if available).

This method also requires a qualified person responsible for presetting the cutting tools. A large number of small and medium users with vertical machining centers cannot afford the additional expenses and do the setting of the cutting tools during the part setup, mostly using the touch-off method. This method may also be suitable choice when small job runs are machined. The touch-off method is described in the next section.

During the tool length measurement process, distance from the cutting tip of the tool to the gauge line is accurately determined - *Figure 19-4*. Preset tools will reach the machine already mounted in a tool holder, identified by the number of the tool and with the list of measured (preset) tool lengths. All the CNC operator has to do, is to set the required tools into the magazine and register each tool length in the offset register, using the proper offset number.

The preset dimensions have *positive* values, measured from the tool reference point to the gauge line of the holder. The gauge line of the machine is simulated in the presetting device to match. Each dimension will be entered as the H offset value in the tool length offset screen. For example, a tool length is preset to the value of 8.5 inches, with the programmed offset number for this tool as H05. On the offset screen, under number 05, the operator enters the measured length of 8.5000:

```
04  ...
05  8.5000
06  ...
```

◆ Tool Length by Touch Off

The tool length that uses the touch-off method is very common, in spite of some time loss during setup. As the illustration in *Figure 19-5* shows, each tool is assigned an H number (similar to the previous example), called the *tool length offset number.*

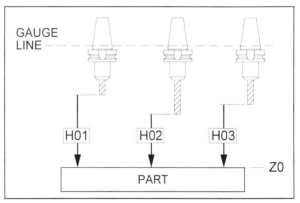

Figure 19-5

Touch-off method of the tool length offset setting

This number is programmed as the address H followed by the number itself. The H number usually corresponds to the tool number for convenience. The setup procedure is to measure the distance the tool travels *from machine zero position (home) to program zero position* (Z0). This distance is always negative and is entered into the corresponding H offset numbers under the tool length offset menu of the control system. The important notion here is that the Z axis settings for any work offset G54-G59 and the *common* offset are normally set to Z0.0000.

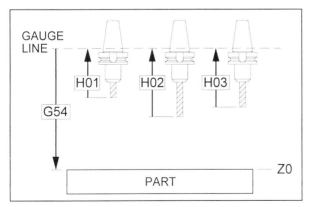

Figure 19-4

Tool length preset away from the machine (tool presetter method). Work offset (G54-G59) must be used

◆ Using a Master Tool Length

Using the touch-off method to measure tool length can be significantly speeded up by using a special method of a *master tool*, usually the longest tool. This tool can be a real tool or just a long bar with a rounded tip, permanently mounted in a tool holder. Within the Z axis travel, this new 'tool' would usually extend out *more* than any anticipated tool that may be used.

Offsets G54 to G59 and the *external* work offset normally contain the Z value set to 0.0, when the part touch-off method is used. This setting will change for the master tool length method. The master tool length measurement is very efficient and requires the following setup procedure. It provides suggested steps that may need some modification:

1. Take the master tool and place it in the spindle.

2. Zero the Z axis and make sure the read-out on the *relative* screen is Z0.000 or Z0.0000.

3. Measure the tool length for the master tool, using the touch-off method described previously. After touching the measured face, *leave the tool in that position !*

4. Instead of registering the measured value to the tool length offset number, register it into the common work offset or one of the G54-G59 work offsets under the Z setting ! *It will be a negative value.*

5. While the master tool is touching the measured face, set the relative Z axis read-out to *zero* !

6. Measure every other tool, using the touch-off method. The reading will be from the master tool tip, not from machine zero.

7. Enter the measured values under the H offset number, in the tool length offset screen. It will always be a negative value for any tool shorter than the master tool.

◎ *Note:*

> The master tool *does not* have to be the longest tool at all.
> The concept of the longest tool is strictly for safety.
> It means that every other tool will be shorter.

Choosing any other tool as master tool, the procedure is logically same, except the H offset entries will be *positive* for any tool that is *longer* than the master and they will be *negative* for any tool that is *shorter* than the master. In the rare case where the measured tool will have exactly the same length as the master tool, the offset entry for that tool will be zero. Illustration in *Figure 19-6* shows the concept of master tool setting.

After the master tool length is set and registered into the Z axis of work offset, enter the distance from the tool tip of the new tool to the tool tip of the master tool, and register it in the appropriate H offset number. If the longest tool is an actual tool, rather than a plain bar used for setup, its H offset value must be always set to 0.0.

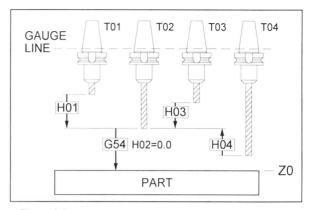

Figure 19-6

Tool length offset using the master tool length method. T02 is the master tool, with setting of H02=0.0

The greatest benefit of this setting method is shortened setup time. If certain tools are used for many of jobs, only the length of the master tool needs to be redefined for any new part height while all other tools remain unchanged. They are related to the master tool only.

◆ G43-G44 Difference

Initial description at the beginning of this chapter indicates that Fanuc and similar CNC systems offer *two* preparatory commands that activate the tool length offset. These two commands are G43 and G44. Most programmers use the G43 command exclusively in the program and may have some difficulty to interpret the meaning of G44 command, because they have never used it. There is a good reason why G44 is a dormant command - not quite dead but barely breathing. Programmers would like to know *how* - and *when* - or even *if* - to use one over the other. Here is an attempt at explanation.

First, take a look at the definitions found in various CNC reference books and manufacturers' specifications sheets. In different versions of these publications, the following typical definitions are used - all are quoted literally and all are correct:

```
G43     Plus offset
G44     Minus offset

G43     Tool length offset positive
G44     Tool length offset negative

G43     Plus direction
G44     Minus direction
```

These definitions are correct only if taken within the context of their meaning into consideration. That context is not really clear from any of these definitions. *Plus to where? Positive of what?* To find the context, think about the use of the tool length offset on a CNC machine. *What is the purpose of the tool length offset?*

The main and most important purpose of any tool length offset is to allow a CNC program to be developed *away* from the machine, *away* from tooling and fixturing, and *without* knowing the actual cutting tool length during program development.

The process has two parts - one is in the program, the other at the machine. In the program, either G43 command or G44 command is required, together with proper H offset number - that is done by the programmer. At the machine, the tool length offset can be set on or off the machine. Either way, the tool length is measured and the measured value is entered into the control - that is the job of the operator. It is the *measuring at the machine* that has a number of variations - programmer has a choice of only two G codes.

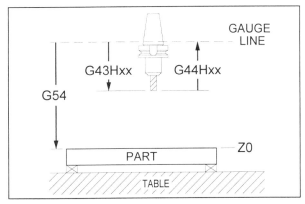

Figure 19-7

Less common method of using the tool length offset.
Work offset (typically G54) must be set as well.

Figure 19-7 illustrates one of two methods to set a tool length command - G54 or other work offset must be used.

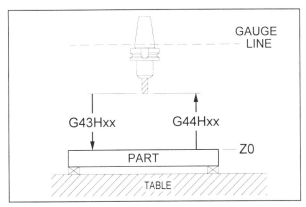

Figure 19-8

More common method of using the tool length offset.
No work offset setting is required and G43 is the preferred choice.

Figure 19-8 illustrates the other, and much more common, method. In this case, all work offset commands G54 to G59 will normally have a Z value set to 0.0.

In either case, written program will be exactly the same (it is only the setting method that changes, *not* the programming method). Program will contain the tool length offset command (G43 or G44), followed by the target position along the Z axis and the H offset number:

```
G43 Z1.0 H06      or      G44 Z1.0 H06
```

The control system cannot offer any benefits, until the measured value for H06 is stored in the offset registers. For example, if the H06 has been measured as 7.6385, it will be entered as a *negative* value, if G43 is used, and as a *positive* value, if G44 is used (tool motions will be identical):

```
G43 Z1.0 H06 ...... H06 = -7.6385
G44 Z1.0 H06 ...... H06 = +7.6385
```

It is clear that the 'secret' of G43 versus G44 difference is nothing more than a sign reversal. Either command instructs the control as to *how* the actual Z axis motion is calculated. Using G43, the H offset value will be *added* (+) in the calculation. Using G44, the H offset value will be *subtracted* (-). The actual Z travel motion will be:

```
G43:  Z + H06 = (1.0) + (-7.6385) = -6.6385
G44:  Z - H06 = (1.0) - (+7.6385) = -6.6385
```

The tool length measuring method done *on* the machine (touch-off) will result in offsets with *negative* values. The setup process can automatically input all measured values into the offset register, *as negative*. That is the reason why G43 is the standard command to program tool length offset. G44 is just not practical for everyday work.

PROGRAMMING FORMATS

Programming format for tool length offset is very simple and has been illustrated many times. In the following examples are some general applications of various methods. The first one will show programming method if no tool length offset is available. Understanding the development of tool length offset over the years makes it easier to apply it in the program. Other example shows a comparison of programming methods for the older G92 programming style and the modern G54 to G59 method. The last example shows the G54 to G59 method applied to a simple program using three tools, a typical way of programming today.

◆ Tool Length Offset not Available

In the early days of programming, tool length offset and work offsets were not available. G92 position register command was the only G code used for setting the current tool position. The programmer had to know *all and every* dimension specified by the machine manufacturer and *all and every* dimension of the job being setup, specifically the distance from the Z0 to the tool tip.

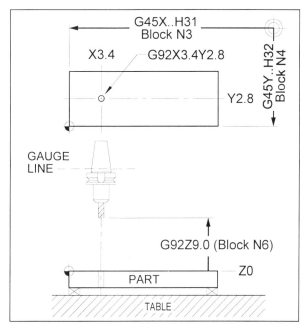

Figure 19-9
Setting tool length without tool length offset - program O1901

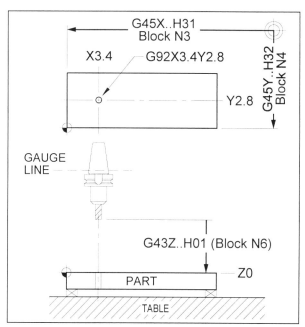

Figure 19-10
Setting tool length with G43 (Z) and G92 (XY) - program O1902

This early program required the position compensation command G45 or G46 in XY axes and the position register command G92 in XYZ axes. Each part must start at machine zero - *Figure 19-9:*

```
O1901
N1 G20                        (INCH MODE SELECTED)
N2 G92 X0 Y0 Z0          (MACHINE ZERO POSITION)
N3 G90 G00 G45 X3.4 H31       (X POSITION COMP)
N4 G45 Y2.8 H32               (Y POSITION COMP)
N5 G92 X3.4 Y2.8        (TOOL POS REGISTER XY)
N6 G92 Z9.0              (TOOL POS REGISTER Z)
N7 S850 M03                 (SPINDLE COMMANDS)
N8 G01 Z0.1 F15.0 M08        (Z APPROACH MOTION)
N9 Z-0.89 F7.0               (Z CUTTING MOTION)
N10 G00 Z0.1 M09             (Z RAPID RETRACT)
N11 Z9.0                (MACHINE ZERO RETURN Z)
N12 X-2.0 Y10.0         (CLEAR POSITION XY)
N13 M30                     (END OF PROGRAM)
%
```

◆ Tool Length Offset and G92

When the tool length offset became available, programming became easier. The position compensation G45/G46 was still in use at the time and G92 had to be set for both X and Y axes. However, G92 setting for the Z axis was replaced by G43 or G44 command, with an assigned H offset number - *Figure 19-10.*

Today, this method of combining the position compensation G45/G46 and tool length offset G43/G44 is considered obsolete, or at least quite old-fashioned. Only the G43H.. is used in modern programming, with the target position.

In an improved program, the tool length offset G43 is applied to the *first motion* command of the Z axis:

```
O1902
N1 G20                        (INCH MODE SELECTED)
N2 G92 X0 Y0 Z0          (MACHINE ZERO POSITION)
N3 G90 G00 G45 X3.4 H31       (X POSITION COMP)
N4 G45 Y2.8 H32               (Y POSITION COMP)
N5 G92 X3.4 Y2.8        (TOOL POSITION REGISTER)
N6 G43 Z1.0 H01         (TOOL LENGTH COMP Z)
N7 S850 M03                 (SPINDLE COMMANDS)
N8 G01 Z0.1 F15.0 M08        (Z APPROACH MOTION)
N9 Z-0.89 F7.0               (Z CUTTING MOTION)
N10 G00 Z0.1 M09             (Z RAPID RETRACT)
N11 G28 X3.4 Y2.8 Z1.0   (MACHINE ZERO RETURN)
N12 G49 D00 H00         (OFFSETS CANCELLATION)
N13 M30                     (END OF PROGRAM)
%
```

When a program is developed using G92, blocks N6 and N7 can be joined together for convenience, if preferred:

```
N6 G43 Z1.0 S850 M03 H01
N7 ...
```

This method has no effect on the tool length offset, only on the moment at which the spindle starts rotating. Position compensation and the tool length offset *cannot* be programmed in the same block.

Note that the position compensation is still in effect in this example, due to the lack of work coordinate offset of the G54 to G59 series.

◆ Tool Length Offset and G54-G59

The most modern programming has many commands and functions available and G54-G59 series is one of them. The G92 command has been replaced with work offset system G54-G59 and, optionally, more. Normally, G92 is not used in the same program that contains any work offset selection G54 through G59 or the extended series.

Here is a program example of using the tool length offset in a G54-G59 work offset environment:

```
O1903
N1 G20                      (INCH MODE SELECTED)
N2 G90 G00 G54 X3.4 Y2.8  (XY TARGET LOCATION)
N3 G43 Z1.0 H01           (TOOL LENGTH COMP Z)
N4 S850 M03               (SPINDLE COMMANDS)
N5 G01 Z0.1 F15.0 M08     (Z APPROACH MOTION)
N6 Z-0.89 F7.0            (Z CUTTING MOTION)
N7 G00 Z0.1 M09           (Z RAPID RETRACT)
N8 G28 X3.4 Y2.8 Z1.0     (MACHINE ZERO RETURN)
N9 G49 D00 H00            (OFFSETS CANCELLATION)
N10 M30                   (END OF PROGRAM)
%
```

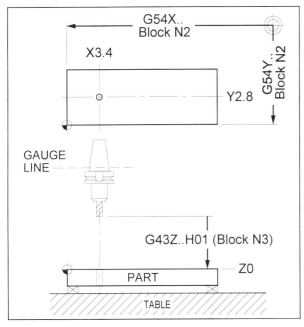

Figure 19-11

Setting tool length with G43 (Z) and G54-G59 (XY) - program O1903

In this example - *Figure 19-11*, using work offsets G54 through G59, the blocks N2, N3 and N4 can be joined together without a problem, perhaps to speed up processing:

```
N2 G90 G00 G54 G43 X3.4 Y2.8 Z1.0 S850 M3 H01
N3 ...
```

The command G54 will affect all axes, G43 with H01 will affect only the Z axis. Tool must move in the clear.

◆ Tool Length Offset and Multiple Tools

The majority of CNC programs include more than one tool; in fact, most jobs will require many different tools. Our next example (independent of the previous drawings) illustrates a common method how the programmer enters the tool length offset for three tools.

Three holes need to be spot-drilled, drilled and tapped. Drawing or explanation of the machining is not important at this time - just concentrate on the G43 tool length application. It is the program structure that is important now - note that there is *no change* in the program structure of any tool, only in the programmed values.

```
O1904
N1 G20
N2 G17 G40 G80 T01
N3 M06
N4 G90 G00 G54 X1.0 Y1.5 S1800 M03 T02
N5 G43 Z0.5 H01 M08    (TOOL LG OFFSET FOR T01)
N6 G99 G82 R0.1 Z-0.145 P200 F5.0
N7 X2.0 Y2.5
N8 X3.0 Y1.5
N9 G80 Z0.5 M09
N10 G28 Z0.5 M05
N11 M01

N12 T02
N13 M06
N14 G90 G00 G54 X3.0 Y1.5 S1600 M03 T03
N15 G43 Z0.5 H02 M08   (TOOL LG OFFSET FOR T02)
N16 G99 G81 R0.1 Z-0.89 F7.0
N17 X2.0 Y2.5
N18 X1.0 Y1.5
N19 G80 Z0.5 M09
N20 G28 Z0.5 M05
N21 M01

N22 T03
N23 M06
N24 G90 G00 G54 X1.0 Y1.5 S740 M03 T01
N25 G43 Z1.0 H03 M08   (TOOL LG OFFSET FOR T03)
N26 G99 G84 R0.5 Z-1.0 F37.0
N27 X2.0 Y2.5
N28 X3.0 Y1.5
N29 G80 Z1.0 M09
N30 G28 Z1.0 M05
N31 M30
%
```

This is a practical example of contemporary use of G43 tool length offset in a CNC program. In summary, G43 tool length offset requires target Z position and the address H for each tool. The actual offset value is set at the control, during the given job setup. Two or more length offsets may be used for the same tool if necessary, but that is a little more advanced subject, described separately in the next section of this chapter.

Also note that there is no tool length offset cancellation. Cancellation will also be explained later in this chapter.

CHANGING TOOL LENGTH OFFSET

The vast majority of programming jobs requires only a single tool length offset command per cutting tool. Based on this principle, we have identified *Tool 1* (T01) with tool length offset H01, *Tool 2* (T02) with tool length offset H02, etc. However, in some special circumstances, the tool length offset may have to be changed for the *same tool*. In those applications, there will be *two* or *more* tool length offsets for one tool.

An example of a single tool length offset change would be any part that uses two or more drawing references along the Z axis. *Figure 19-12* illustrates this concept with a groove dimensioned by its depth location for the top and bottom (groove width of .220 is implied).

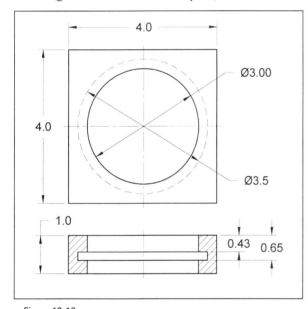

Figure 19-12

Example of programming more than one tool length offset for a single tool - program O1905

Based on the illustration, we have to decide on the cutting method first (premachining of the ∅3.000 hole is assumed). A .125 wide slot mill will be a good choice to profile the circle, using typical milling method for a full circle (see *Chapter 29*). The program can be shortened by using a subprogram method (see *Chapter 39*). Because the .220 groove width is larger than the cutter, more than one cut is needed - two in this case. For the first cut, the tool is positioned at the Z-0.65 depth (as per drawing) and makes the first cut at the bottom of the groove. The *bottom edge* of the tool will reach the Z-0.65 depth.

For the second cut, the *top edge* of the slotting mill is used and the tool makes the profile for the second groove (actually, it will widen the first groove) at the depth of Z-0.43 (again, as per drawing).

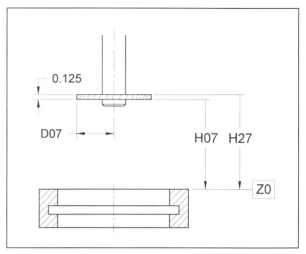

Figure 19-13

Setting of two length offsets for a single tool. The difference between H07 and H27 offsets is the width of slot (.125 shown).

Note the words - the *bottom* edge versus the *top* edge of the slot mill. Which edge is programmed as a reference for the tool length? The one at the bottom or the top?

Figure 19-13 shows that *two* reference positions are used for the same tool. For this reason, the program requires *two* tool length offsets, H07 and H27 in the illustration. D07 is the cutter radius offset, and .125 is the slotting mill width.

Other methods of programming can be used, for example, calculating the difference manually, but the method using multiple tool length offsets is very useful during machining to allow fine groove width adjustments. It is shown in the following example - program O1905:

```
O1905
(TWO TOOL LENGTH OFFSETS FOR ONE TOOL)
N1 G20
N2 G17 G40 G80
N3 G90 G00 G54 X0 Y0 S600 M03
N4 G43 Z1.0 H07 M08      (ABOVE JOB CLEARANCE)
N5 G01 Z-0.65 F20.0      (CUTTER EDGE - BOTTOM)
N6 M98 P7000            (CUTTING GROOVE AT Z-0.65)
N7 G43 Z-0.43 H27             (CUTTER EDGE - TOP)
N8 M98 P7000            (CUTTING GROOVE AT Z-0.43)
N9 G00 Z1.0 M09
N10 G28 Z1.0 M05
N11 M30
%

O7000
(SUBPROGRAM FOR GROOVE IN O1905)
N1 G01 G41 X0.875 Y-0.875 D07 F15.0
N2 G03 X1.75 Y0 R0.875 F10.0
N3 I-1.75
N4 X0.875 Y0.875 R0.875 F15.0
N5 G01 G40 X0 Y0
N6 M99
%
```

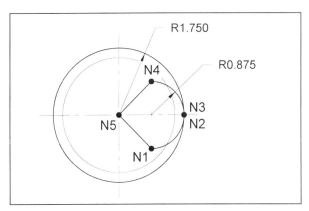

Figure 19-14
Full circle milling - subprogram O7000.
Start and finish of cutting is at the center of the groove.

In the example, tool length offset H07 is used for the bottom reference edge of the slotting mill and H27 is used for the top reference edge of the slotting mill. D07 is used for the cutter radius only. *Figure 19-14* shows the tool motions used in subprogram O7000.

HORIZONTAL MACHINE APPLICATION

So far, all presented examples were aimed towards a CNC vertical machining center. Although the logic of tool length offset applies equally to *any* machining center, regardless of the Z axis orientation, there are some noticeable differences in the practical applications on horizontal machining centers *(Chapter 46)*.

A horizontal machining center allows programming of a tool path on several faces of the part. Since each face has a different distance from the tool tip (along the Z axis), the tool length offset for each face will vary. It is common to program different work offsets and different tool length offsets for each face.

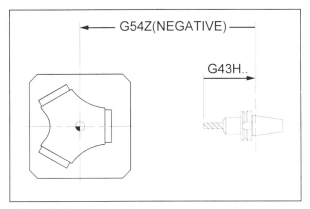

Figure 19-15
Typical tool length offset setting for a preset tool.
Program zero is at the center of the table.

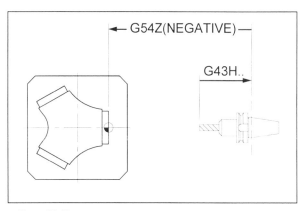

Figure 19-16
Typical tool length offset setting for a preset tool.
Program zero is at the face of the part.

The two related illustrations show typical setup of the tool length offset for preset tools on a horizontal machining center. *Figure 19-15* shows the program zero at the center of the table, *Figure 19-16* shows the program zero at the face of the part.

TOOL LENGTH OFFSET CANCEL

In programming, a well organized approach is always important. That means, a program command that is turned on when needed should also be turned off, when not needed anymore. Tool length offset commands are no exception.

The tool length offset cancellation may be included in the program. There is a special preparatory command available that cancels any selected method of the tool length offset, either G43 or G44. The command to cancel the tool length offset in the program (or via MDI) is G49:

G49	Tool length offset cancel

One method of using the G49 command is on its own - in a single block - just before returning to the machine zero in the Z axis, for example:

```
N176 G49
N177 G91 G28 Z0
...
```

A similar method also cancels the offset numbers:

```
N53 G91 G28 Z0 H00
```

In this case, the G28 command is coupled with an H offset number zero - H00. Note, there is no G49 in the block and H00 does the job of cancellation. There is no setting for H00 on the control. It just means cancellation of the tool length offset.

A program may also be *started* with the tool length offset command canceled (under program control), usually in the safety line (safety block or initial block):

```
N1 G20 G17 G40 G80 G49
```

... or a variation of the same block:

```
N1 G20
N2 G17 G40 G80 G49
```

There is one more way to cancel the tool length offset - *do not program it at all.*

A strange suggestion, perhaps, but well founded. Most examples in this handbook *do not* use G49 command at all. *Why not? What happens at the end of each tool?*

The Fanuc rule is quite explicit - any G28 or G30 commands (both execute the tool return to the machine zero) will cancel the tool length *automatically.*

The meaning is simple - programmer may take advantage of this rule and does not need to specifically cancel the tool length offset, if the machine returns to the tool change position. This is normal for all machines with an automatic tool changer. This approach is illustrated in many examples included in this handbook.

Any one of the methods will guarantee that the active tool length offset will be canceled. There may be some differences between machine manufacturers and consulting the machine manual will always be the responsible approach.

> There are machines that require
> the use of G49 for every tool.

A CNC machine tool does not always cut material and 'make' chips. From the moment the cutting tool becomes active in a part program, it goes through a number of motions - some are productive (cutting), others are nonproductive (positioning).

Positioning motions are necessary but nonproductive. Unfortunately, these motions cannot be totally eliminated and have to be managed as efficiently as possible. For this purpose, the CNC system provides a feature called the *rapid traverse* motion. Its main objective is to shorten the positioning time between non-cutting operations, where the cutting tool is not in contact with the part. Rapid motion operations usually involve four types of motion:

❑   From the tool change position towards the part

❑   From the part towards the tool change position

❑   Motions to bypass obstacles

❑   Motions between different positions on the part

RAPID TRAVERSE MOTION

Rapid traverse motion, sometimes called a *positioning motion,* is a method of moving the cutting tool from one position to another position at a *rapid rate of the machine.* The maximum rapid rate is determined by the CNC machine manufacturer and takes place within the travel limits of the machine.

The common rapid rate for many larger CNC machines is about 450 in/min (11430 mm/min). The modern machines offer a rapid motion up to 1500 in/min (38100 mm/min) or even more, particularly for smaller machines. The machine manufacturer determines the rate of rapid motion for each of the machine axes. The motion rate can be the same for each axis or it can be different. A different rapid rate is usually assigned to the Z axis, while the X and Y axes have the same rapid motion rate.

Rapid motion can be executed as a single axis motion, or as a compound motion of two or more axes simultaneously. It can be programmed in the absolute or incremental mode of dimensioning and it can be used whether the spindle is rotating or stationary. During program execution, the CNC operator may temporarily interrupt the rapid motion by pressing the feedhold key on the control panel, or even setting the feedrate override switch to zero or a decreased rate. Another kind of rapid rate control can be achieved by the *dry run* function, usually during setup.

◆ G00 Command

Preparatory command G00 is required in CNC program to initiate the rapid motion mode. Feedrate function F is *not required* with G00 and, if programmed, will be ignored during the rapid motion (in G00 mode). Such a feedrate will be stored in memory and becomes effective beginning with the first occurrence of any cutting motion (G01, G02, G03, etc.), unless a new F function is programmed with the cutting motion:

➲ Example A :

```
N21 G00 X24.5 F30.0
N22 Y12.0
N23 G01 X30.0
```

In block N21, only the rapid motion will be executed. The feedrate of 30.0 in/min will be ignored in this block, but stored for later use. The block N22 will also be in the rapid positioning mode, since G00 is a modal command. The last block, N23, is a linear motion (cutting motion), that requires a feedrate. As there is no feedrate assigned to this block, the *last programmed* feedrate will be used. That was specified in block N21 and it will become the *current* feedrate in block N23, as F30.0.

➲ Example B :

```
N21 G00 X24.5 F30.0
N22 Y12.0
N23 G01 X30.0 F20.0
```

In block N21, the G00 command becomes modal and remains in effect until it is canceled by another command of the same group. In the example (B), the G01 command in block N23 cancels the rapid motion mode and changes the rapid mode to a linear mode. Also, the feedrate is reprogrammed and will be at 20.0 in/min starting at block N23. The feedrate F30.0 in block N21 has never been used. It is redundant and should be removed.

The rapid traverse motion is measured as the *distance in current units traveled in one minute* (measured in *in/min* or *mm/min*). The maximum rate is always set by the machine manufacturer, *never* by the control system or the program. A typical limit set by the machine builder is a rate between 300 and 1500 in/min (7620 and 38100 mm/min), and even higher. Since motion per time is independent of the spindle rotation, it can be applied at any time, regardless of the last spindle rotation function mode (M03, M04, M05).

Depending on the CNC machine design, rapid motion rate can be the same for all axes, or each axis can have its own maximum rate. The maximum rapid rates for a typical machining center may be 1181 in/min (30000 mm/min) for the X and Y axes and about 945 in/min (24000 mm/min) for the Z axis. For a CNC lathe, the rates are somewhat slower, for example 197 in/min (5000 mm/min) for the X axis, and 394 in/min (10000 mm/min) for the Z axis. The rapid rates can be much higher for modern machines.

RAPID MOTION TOOL PATH

Every motion in the G00 mode is a rapid non-circular motion (circular or helical motion cannot normally be made at the rapid rate). The actual linear motion of the tool between two points is not necessarily the shortest tool path in the form of a straight line. Programmed tool path and the resulting actual tool path will be different, depending on several factors:

❑ The number of axes programmed simultaneously

❑ The actual length of motion for each axis

❑ The rapid traverse rate of each axis

Since the only purpose of the rapid motion is saving the unproductive time (motion from the current tool position to the target tool position), the tool path itself is irrelevant to the shape of the machined part. Always be aware of the actual rapid motion tool path for reasons of safety, particularly when two or more axes are programmed at the same time. No physical obstacles must be in the way of the tool path motion.

If there is an obstacle between any two points of the tool path, the obstacle *will not be* automatically bypassed by the control for one very simple reason - the control has no way of detecting such an obstacle. It is the programmer's responsibility to assure that any tool motion (rapid motion included) occurs without any obstacles in its way.

Some typical examples of physical obstacles that can interfere with the tool motion are:

❑ *FOR MACHINING CENTERS :*

Clamps, vises, fixtures, rotary or indexing table, machine table, part itself, etc.

❑ *FOR LATHES :*

Tailstock quill and body, chuck, steadyrest, live center, face plate, fixture, other tool, part itself, etc.

Additional obstacles during a tool motion may be caused by special types of setup, machine design, tool mounting method, etc.

> Always watch for obstacles during rapid motion.

Although an obstacle may be in the way of cutting motions in G01, G02 or G03 mode (for example a face turning towards a tailstock on a lathe), the most problems occur during rapid motions G00, G28, G29, G30 and with fixed cycles G81 to G89, G73, G74 and G76. During a rapid motion, the tool path is much less predictable than during cutting motions. Keep in mind that the only purpose of rapid motion is to get from one part location to another location *fast* - but not necessarily straight.

In order to bypass obstacles and still assure a safe rapid motion in the program at all times, let's take a closer look at the available options while programming a rapid motion.

◆ Single Axis Motion

Any tool motion programmed specifically for only one axis at a time is always a straight line along the selected axis. In other words, each rapid motion that is parallel to one of the available axes, must be programmed in a separate block. The resulting motion is always equivalent to the shortest distance between the start and end points of the motion - *Figure 20-1*.

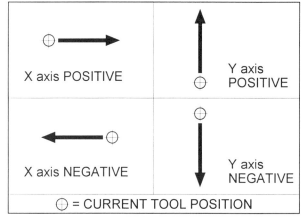

Figure 20-1

Single axis motion for a machining center application (XY shown)

Several consecutive program blocks, each containing only a single axis motion, can be included in the program to bypass obstacles to machining. This method of programming is preferable in cases where only the *exact* or *approximate* position of certain obstacles (such as clamps or fixtures) is known during the program preparation.

◆ Multiaxis Motion

We have already learned that the cutting tool is moved at a rapid rate using the G00 command. If this motion is a motion of two or more axes simultaneously, the programmed rapid path and the actual rapid path of the tool are not always the same. The resulting compound motion can be - and often is - much different from the theoretical programmed motion (the intended motion).

In theory, the motion along any two axes is equivalent to a *straight diagonal motion*. The real motion, however, may or may not be a straight diagonal tool path at all. Consider the following example in *Figure 20-2*.

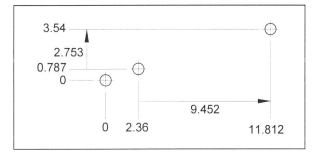

Figure 20-2

Drawing sketch for rapid motion examples

The current tool position (the start point) is at X2.36 Y0.787 coordinate location. The tool motion terminates at X11.812 Y3.54 location. In the terms of incremental motion, the cutting tool has to travel 9.452 inches along the X axis and 2.753 inches along the Y axis.

If the rapid rate for both axes is the same (XY rapid motion rates usually are), such as 394 in/min, it will take

`(9.452 × 60) / 394 = 1.44 seconds`

to complete the X axis motion - but only

`(2.753 × 60) / 394 = 0.42 seconds`

is required to complete the Y axis motion. Since the motion is not completed until *both* axes reach the end point, it is logical that the actual tool path will be *different* from the programmed tool path.

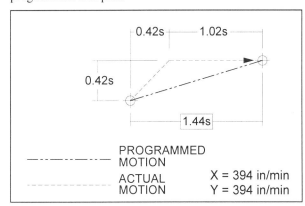

Figure 20-3

Rapid motion deviation - same rapid rate for each axes

Figure 20-3 shows a combination of an angular and a straight motion as the actual tool path. The tool departs at the rate of 394 in/min (10000 mm/min) simultaneously in

both axes, with a resulting 45° motion. The total time required to reach the end position is 1.44 seconds, which is the longest time required for *either* axis to reach its target position. After the elapsed time of 0.42 seconds, the Y axis target position has been reached, but there is still another 1.02 seconds left to complete the motion along the X axis. The target must be reached in both axes, so the tool then continues along the X axis only (for 1.02 seconds), in order to reach the final position.

Another example, also using the location coordinates in *Figure 20-2*, illustrates another situation, with the rapid rate *different* for each axis - *Figure 20-4*.

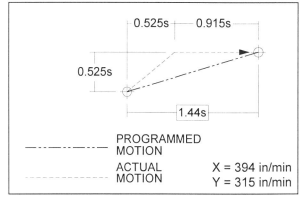

Figure 20-4

Rapid motion deviation - different rapid rate for each axis

In this not so common example, the X axis rate is set to 394 in/min (10000 mm/min) and the Y axis rate is set to 315 in/min (8000 mm/min). It will than take

`(9.452 × 60) / 394 = 1.44 seconds`

to complete the X axis motion - but only

`(2.753 × 60) / 315 = 0.525 seconds`

to complete the Y axis motion. In this case, the resulting motion will also include an angular departure, but *not* at 45°, because of the different rating of rapid traverse rate for each axis. During the 0.525 seconds (which is the common time to both axes), the X axis motion will travel

`0.525 / 60 × 394 = 3.448 inches`

but the Y axis motion will be only

`.525 / 60 × 315 = 2.753 inches`

The resulting motion is at 38.605° and a slight rounding have been applied. The actual departure angle is not always necessary to be known, but it helps to calculate it for rapid motions in some very tight areas of the part. It only takes a few simple trigonometric functions to make sure of the true rapid tool path, provided the rapid rate is known.

Both of the above examples illustrate an angular motion along two axes, followed by a straight single axis motion in the remaining axis. The graphical expression of these motions is a bent line, resembling a *hockey stick* or a *dog leg* shape, which are also very common terms applied to such a rapid motion.

Calculation of the actual motion shape, as we have done earlier, is only seldom necessary. Taking some basic precautions, the rapid motion can be programmed safely *without* any calculations. If no obstacle is within the work area (the imaginary rectangle created by the diagonally positioned start and end point), there is no danger of collision due to the diverted rapid tool path. On CNC milling systems, the third axis can also used. The rectangle of the above example will be enhanced by the third dimension and a three dimensional space must be considered. In this case, no obstacle should be *within this space*, otherwise the same rules apply for a rapid motion along three axes as for a two-axis simultaneous rapid motion. Note that the rapid rate for the Z axis on CNC machining centers is usually lower than the rapid rate for the X and Y axes.

◆ Straight Angular Motion

In some uncommon circumstances, the theoretical rapid tool path will correspond to the actual tool path (with no bent line as a result). This will happen if the simultaneous tool motion has the same length in each axis and the rapid rates of all axes are identical. Such an occurrence is rather rare, although not impossible. Some machine manufacturers provide this feature as a standard and the programmer should know whether the machining center does have that feature or not. Another situation where the resulting motion is a straight angle, is when the rapid rating varies for each axis, but the required length of motion just 'falls' into the range that results in a straight angular motion.

Both of these occurrences are rare (more or less a case of good luck) and in actual programming will seldom happen. To be on the safe side, never take any chances - it is always more practical to program the rapid motion without the actual calculation of the tool path but with safety as a primary consideration.

◆ Reverse Rapid Motion

Any rapid motion must be considered in terms of *approach* towards a part and the *return* back to the tool changing position. This is the way a cutting tool is normally programmed - we start at a certain position and then return there, when all cutting activity for the tool is completed. It is not a mandatory method, but it is an organized method, it is consistent, and it makes programming much easier.

So far, we have examined a rapid motion *before* an actual cut, starting from the tool change position. When the tool cutting function is completed, a rapid motion is required to return back to the tool change position.

This consideration is more important in turning applications than in milling, due to the nature of programming for the two types of machines. In turning, the approach motion may be along the Z axis first, to avoid a collision with the tailstock, and *then* along the X axis. The reverse motion should be along the X axis first, *then* along the Z axis motion, in order to achieve the same safety goal when returning to the tool change position.

A typical application of this programming technique may be useful after using a machining cycle (such as turning, boring, facing, threading, etc.), where the cycle starting point is also its end point.

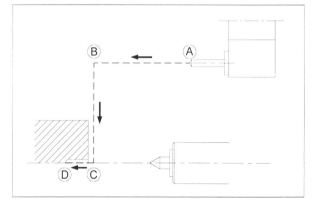

Figure 20-5

Typical example of a reversed rapid motion on a CNC lathe, used to bypass obstacles, for example, a tailstock

As *Figure 20-5* shows, rather than programming a direct motion from the turret position to the cutting position (which would be from point A to point C), the tool motion was split. The approach towards the part will be in the order of A to B to C, at a rapid rate. From point C to point D, the actual cutting takes place. When the cutting is completed, the tool will rapid in the reverse order, back to the starting position. Rapid motion will be from D to C to B to A. This is a necessary precaution to bypass a potential obstacle, for example, the tailstock.

TYPE OF MOTION & TIME COMPARISON

The technique of programming each axis separately in individual blocks of the program, is recommended only for the purpose of bypassing possible obstacles during the tool path - and strictly for safety. This method of programming requires a slightly longer cycle time than the simultaneous multiaxis rapid motion. To compare the difference, consider a three axis rapid motion, such as a typical tool approach in milling.

As an example, the rapid rate is at 394 in/min (10000 mm/min) for each axis. The motion takes place between the coordinate location of X2.36 Y0.787 Z0.2 (start point) and X11.812 Y3.54 Z1.0 (end point).

The required time for tool travel along each axis can be easily calculated:

❏ X axis time:

```
((11.812 - 2.36) × 60) / 394 = 1.440 sec.
```

❏ Y axis time:

```
((3.54 - .787) × 60) / 394 = 0.420 sec.
```

❏ Z axis time:

```
((1.0 - .2) × 60) / 394 = 0.121 sec.
```

If all three axes are moved simultaneously, the total time for positioning is 1.44 seconds, which is the longest time required for any axis to reach the end point. The program block will be:

```
G00 X11.812 Y3.54 Z1.0
```

If this motion were to be separated into three individual program blocks, the total time would be the result of individual times added together:

```
1.44 + 0.42 + 0.121 = 1.981 seconds
```

which is about 37.5% longer. The percentage will vary, depending on the rapid motion rate and the rapid travel length, measured along each machine axis. The program blocks will be written separately:

```
G00 X11.812
Y3.54
Z1.0
```

Note that the modality of G00 rapid motion command does not require its repetition in the subsequent blocks.

REDUCTION OF RAPID MOTION RATE

During a part setup or while proving a new program on the machine, the CNC operator has an option to select a *slower rapid traverse rate* than the maximum established by the machine manufacturer. This adjustment is done by the means of a special *rapid override switch,* located on the control system panel. This switch has typically four selectable positions, depending on the machine brand and the type of control system - *Figure 20-6.*

The second, third and fourth positions on the rapid motion override switch are rated as the *percentage of the actual rapid rate* - 25%, 50%,100% respectively. They are set by the machine manufacturer. The first setting, typically identified by F0 (or F1) is a rapid motion rate set through a *control system parameter.* The F0 (F1) setting should always be slower than any other setting, typically less than the lowest setting of 25%.

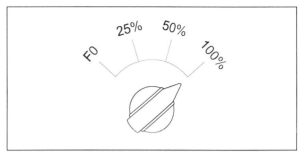

Figure 20-6
Rapid motion override switch set to 100% of rapid rate

The configuration of the rapid override switch varies between machines from different manufacturers. On some machines, the rapid motion may be stopped altogether, on others, the tool will move at the slowest percentage and cannot be stopped with the override switch alone.

During actual production, after the program has been verified and optimized for the best tool performance and productivity, the override switch should be set to the 100% pointer, to shorten the cycle time.

RAPID MOTION FORMULAS

The calculations relating to the rapid tool motion can be expressed as formulas and used quickly at any time by substituting the known parameters. Relationships between the rapid traverse rate, length of the motion and the elapsed time can be expressed in the following three formulas:

$$T = \frac{L \times 60}{R}$$

$$R = \frac{L \times 60}{T}$$

$$L = \frac{T \times R}{60}$$

☞ where ...

```
T  =  Required time in seconds
R  =  Rapid traverse rate per minute
      for the selected axis - in/min or mm/min
L  =  Length of motion - inches or mm
```

Units applied to the formulas must always be consistent within the selected system of measurement in the program. *Inches* and *inches per minute (in/min)* must be used with the English system. *Millimeters* and *millimeters per minute (mm/min)* must be used in the metric system. For any calculation relating to the rapid traverse time, the measuring units cannot be mixed.

APPROACH TO THE PART

The previous *Figure 20-5* had an illustration of a safe tool approach as applied to a CNC lathe. For CNC machining centers, the safety of part approach should be considered with equal care. Keep in mind that the general principles of rapid tool motion have to be considered for any machine. When approaching a part at a rapid rate, the cycle time can be somewhat shortened by keeping the part clearances to the smallest safe minimum. Let's have a look at some potential problems.

In the following example, an approach to the part is made along the Z axis, with a clearance of .05 inches (1.27 mm) in block N315:

```
N314 G90 G54 G00 X10.0 Y8.0 S1200 M03
N315 G43 Z0.05 H01
N316 G01 Z-1.5 F12.0
```

There is nothing wrong with such a method of programming, providing the cutting tool is properly set and the part height is consistent from one part to another, as it should be. The clearance of .05 inches (1.27 mm) allows very little amount of unproductive cutting. On the other hand, an inexperienced CNC operator may not feel quite comfortable with such a small clearance, particularly during the early training stages. If the operator's convenience is considered as a significant factor contributing to the overall productiv-

ity, it might be a reasonable compromise to split the Z axis motion into two separate motions:

```
N314 G90 G54 G00 X10.0 Y8.0 S1200 M03
N315 G43 Z0.5 H01
N316 G01 Z0.05 F100.0
N317 Z-1.5 F12.0
```

In this method, the rapid motion has been first programmed to a much more comfortable position of .500 inches above the part (N315). Then, the motion continued to the cutting start point, using the *linear* interpolation G01 in block N316. Since this is still a motion in the air, therefore not productive, a relatively heavy feedrate was used at the same time. As may be expected in such situations, there is a trade off.

Although the cutting time was slightly increased, at the same time, the CNC operator has been given an opportunity to use the feedrate override switch for testing the first part (used perhaps in a single block mode). Once the program is verified and debugged, the heavy feedrate in the non-cutting motion will speed up the operation and at the same time provide an extra safety clearance. The program with the split Z axis motion can always be optimized later, although this may not be the best approach for repetitive jobs, since the setup is always 'new' for any repetition at a later date. However, it may be very useful when running lots of large numbers (several thousands, for example).

21

MACHINE ZERO RETURN

The ability of a control system to return a cutting tool from any position to the machine reference position is a critical feature of all modern CNC systems. Programmers and operators understand the term *machine reference position* as synonymous with the *home* position or *machine zero* position. This is *the position of all machine slides at one of the extreme travel limits of each axis*. The exact position is determined by the machine manufacturer and is *not* normally changed during the machine working life. Return to that position is automatic, on request from the control panel, in MDI operation, or via the program.

MACHINE REFERENCE POSITION

The existence of machine reference position is for referencing purposes. In order that the CNC machine is accurate, we need more than just the high quality components, we need some unique location that can be considered the origin point of the machine - a zero position - a home position. Machine reference position is exactly such a point.

> Machine zero is a fixed position on a CNC machine that can be reached repeatedly, on request, through the control panel, MDI, or program code execution.

◆ Machining Centers

Although the design of CNC machining centers varies for different models, there are only four possible locations for the machine zero, within the XY view:

- ❑ Lower left corner of the machine
- ❑ Upper left corner of the machine
- ❑ Lower right corner of the machine
- ❑ Upper right corner of the machine

It is quite common, in fact normal, to start the first part of a new program from the machine zero position. Often, it is also necessary to make a tool change at machine zero position and return there when the program execution is completed. So, several of the four alternatives are not very convenient for setup of the part on the machine table and its removal when the machining is done.

The most common and standard machine reference position for vertical machining centers is at the *upper right corner* of the machine, looking perpendicularly towards the XY plane - *Figure 21-1*.

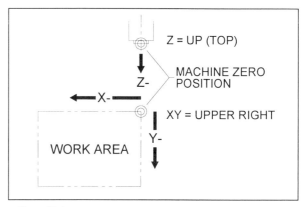

Figure 21-1

Machine zero position located at the upper right XY corner of a CNC vertical machining center

So far, any reference to the Z axis in the description was quite intentional. The Z axis machine zero position for a vertical machining center is always where the Automatic Tool Change (ATC) takes place. This is a built-in location, normally placed a safe distance from the machine table and the work area. For most machines, the standard machine zero of CNC machining centers is at the extreme travel limit of each axis in the *positive* direction. There are exceptions, as may be expected.

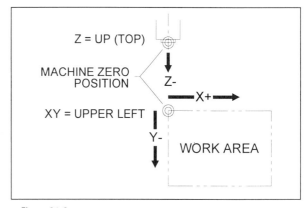

Figure 21-2

Machine zero position located at the upper left XY corner of a CNC vertical machining center

As *Figure 21-2* illustrates, some CNC vertical machining centers have the machine zero position at the *upper left* corner of the XY plane.

In both illustrations, the arrows indicate the tool motion direction *towards the work area*. Moving the tool from machine zero into the opposite direction will result in a condition known as *overtravel* - compare the two possibilities:

❏ Tool motion from machine zero, if machine zero is located at the upper *right* corner:

   X+ Y+ Z+   ... tool motion will overtravel

❏ Tool motion from machine zero, if machine zero is located at the upper *left* corner:

   X- Y+ Z+   ... tool motion will overtravel

The other two corners (lower left and lower right of the XY view) are not used as machine zero.

◆ Lathes

The machine reference position for two axis CNC lathes is logically no different from the reference position of the machining centers. An easy access by the CNC operator to the mounted part is the main determining factor. Both, the X and the Z axes have their machine reference position at the furthest distance from the rotating part, which means away from the headstock area, consisting of the chuck, collet, face plate, etc.

For the X axis, the machine zero reference position is always at the extreme limit of the travel *away from the spindle center line*. For the Z axis, the machine reference position is always at the extreme travel *away from the machine headstock*. In both cases, it normally means a positive direction towards the machine zero, the same as for the machining centers. The illustration in *Figure 21-3* shows a machine zero for a typical CNC lathe.

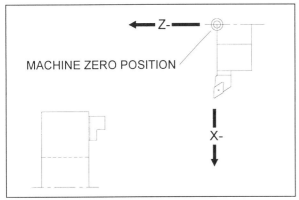

Figure 21-3
Machine zero position for a typical CNC lathe (rear type)

In the illustration, the arrows indicate the tool motion direction towards the work area. Moving the tool from the machine zero into the opposite direction will result in *overtravel* in the particular axis:

❏ Tool motion from machine zero of a typical rear lathe:

   X+ Z+   ... tool motion will overtravel

◆ Setting the Machine Axes

From the previous sections, remember that there is a direct relationship between the CNC machine, the cutting tool and the part itself. The work reference point (program zero or part zero) is always determined by the CNC programmer, the tool reference point is determined by the tool length at the cutting edge, also by the programmer.

Only the machine reference point (home position) is determined by the manufacturer of the machine and is located at a *fixed* position. This is a very important consideration.

Fixed machine zero means that all other references are dependent on this location.

In order to physically reach the machine reference position (home) and set the machine axes, for example, during the part or fixture setup, there are three methods available to the CNC operator:

❏ *Manually* - using the control panel of the system

   The machine operator will use the XYZ (machining centers) or the XZ (lathes) switches or buttons available for that purpose. One or more machine axes can be activated simultaneously, depending on the control unit.

❏ *Using the MDI* - Manual Data Input mode

   This method also uses the control panel. In this case, the machine operator sets the MDI mode and actually programs the tool motion, using the suitable program commands (G28, G30).

❏ *In the CNC program* - during a cycle operation

   Using the same program commands as for the MDI operation, the CNC programmer, not the machine operator, includes machine zero return command (or commands) in the program, at desired places.

When the operator has performed the actual machine zero return, it is always a good idea to set the relative and absolute positions to zero on the display screen. Keep in mind that the relative display can only be set to zero from the control panel and the absolute display can only be changed through a work offset, MDI mode, or the part program. This topic normally a part of CNC machine operation training, directly at the machine.

For the last two methods of a machine zero return, the CNC system offers specific preparatory commands.

◆ Program Commands

There are four preparatory commands relating to the machine zero reference position:

G27	Machine zero reference position return *check*
G28	Return *to* the *primary* machine zero reference position
G29	Return *from* the machine zero reference position
G30	Return *to* the *secondary* machine zero reference position (more than one is possible)

Of the four listed commands, G28 is used almost exclusively in two and three axis CNC programming. Its only purpose is to return the current tool to the machine zero position and do it along the one or more axes specified in the G28 program block.

◆ Command Group

All four preparatory commands G27 to G30 belong to the group 00 of the standard Fanuc designation that describes the *non modal* or *one-shot* G codes. In this designation, each G code of the 00 group must be repeated in every block it is used in. For example, when G28 command is used in one block for the Z axis and then it is used in the next block for the X and Y axes, it has to be repeated in *each* block as needed:

```
N230 G28 Z..    (MACHINE ZERO RETURN Z AXIS)
N231 G28 X.. Y.. (MACHINE ZERO RETURN XY AXES)
```

The G28 command in block N231 *must* be repeated. If the command is omitted, the *last motion* command programmed will be effective, for example, G00 or G01!

RETURN TO PRIMARY MACHINE ZERO

Any CNC machine may have more than one machine zero reference point (home position), depending on its design. For example, many machining centers with a pallet changer have a *secondary* machine reference position, that is often used to align both the left and right pallets during pallet changing. The most common machine tool design is the one that uses only a single home position. To reach this primary home position, the preparatory command G28 is used in the program and can also be used during the MDI control operation.

The G28 command moves the *specified* axis or axes to the home position, always at a rapid traverse rate. That means G00 command is assumed and does not have to be programmed. The axis or axes of the desired motion (with a value) must *always* be programmed. Only the programmed axes will be affected.

For example,

```
N67 G28
```

shows G28 programmed by itself in the block - this is an *incomplete* instruction. At least one axis must be specified with the G28 command, for example,

```
N67 G28 Y..
```

which will only send the Y axis to the machine zero reference position, or ...

```
N67 G28 Z..
```

will only send the Z axis to the machine zero reference position, and ...

```
N67 G28 X.. Y.. Z..
```

will send all three specified axes to the machine zero reference position. Any multiaxis motion requires caution - *watch for the infamous 'hockey stick' motion.*

◆ Intermediate Point

One of the elementary requirements of programming is the alpha numerical composition of a word. In the program, every letter must be followed by one or more digits. The question is what *values* will the axes in G28 have? They will be the *intermediate* point for machine zero return motion. The concept of the intermediate motion in G28 or G30 is one of the most misunderstood programming features.

Commands G28 and G30 must always contain the *intermediate point* (tool position). By Fanuc design and definition, the G28/G30 commands have a *built-in* motion to an intermediate point, *on the way* to machine zero. An analogy can be made to an airplane flight from Los Angeles, USA to Paris, France, that temporarily stops over in New York City. It may not be the most direct route, but it serves a certain specific purpose, for example, to refuel the aircraft.

> The coordinate values of the axes associated with G28 and G30 commands always indicate an intermediate point.

The purpose of the intermediate point, or position, is to shorten the program, normally by one block. This reduction is so marginal that the philosophy behind the design may be debated. Here is how the concept of the intermediate point (position) works.

When the G28 or G30 command is used in the program, *at least one axis must be specified* in the block. The value of that axis is the intermediate point, as interpreted by the control system. Absolute and incremental modes G90 and G91 make a great difference in interpretation of the G28 or G30 behavior, and will be described shortly.

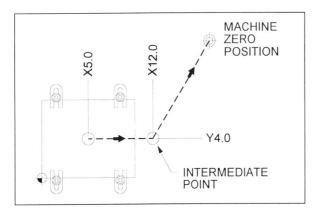

Figure 21-4
Intermediate point for machine zero return - XY axes shown

The tool motion in *Figure 21-4* is from the central hole of the part. During such a motion, the tool can collide with the upper right clamp on its way to machine zero, if the motion to the home position were programmed directly. Only the X and Y axes are considered in the illustration. An *intermediate point* can be programmed in a safe location, without making the program any longer. The program *without* an intermediate point can be constructed as:

```
G90
...
G00 X5.0 Y4.0                  (MACHINED HOLE)
G28 X5.0 Y4.0            (MACHINE ZERO MOTION)
...
```

The same program *with* an intermediate point at a safe location will change slightly:

```
G90
...
G00 X5.0 Y4.0                  (MACHINED HOLE)
G28 X12.0 Y4.0          (MACHINE ZERO MOTION)
...
```

Earlier examples have shown the reason behind this double motion. It is very simple - only to *save a single program block* - that is all. Its intended purpose is to use one block of program to achieve two motions, that would otherwise require two blocks. A safe program could also be:

```
G90
...
G00 X5.0 Y4.0                  (MACHINED HOLE)
X12.0                          (SAFE LOCATION)
G28 X12.0 Y4.0          (MACHINE ZERO RETURN)
...
```

to produce the same final result, but *with an extra block*.

For example, using the intermediate position, the tool can be programmed to avoid an obstacle on the way to the machine zero. If programmed with care, the intermediate position may be quite useful. Normally, it is more practical to

make the intermediate motion equal to zero and move the cutting tool to the machine zero directly. This is done by specifying the intermediate point as *identical* to the current tool position in the *absolute* mode - or - by specifying a zero tool motion in the *incremental* mode.

◆ Absolute and Incremental Mode

There is a major difference in programming the machine zero return command G28 or G30 in the *absolute* and *incremental* modes. Remember the basic difference between two similar statements:

```
G90 G00 X0 Y0 Z0      and      G91 G00 X0 Y0 Z0
```

Each coordinate statement X0Y0Z0 is interpreted by the control system differently. To review, an address followed by a zero, for example X0, means *position at the program reference point*, if the mode is *absolute*, using the G90 command. If the mode is incremental, using the G91 command, the X0 word means *no motion* for the specified axis.

Most CNC lathes use the U and W axes for incremental motion (based on absolute X and Z axes respectively), with the same logical applications. Absolute axes coordinates will be interpreted as the *programmed tool position,* incremental coordinates indicate the *programmed tool motion.*

Compare the two program examples below - they are the same - they are *identical* in terms of the actual tool motion:

```
(---> G28 USED IN THE ABSOLUTE MODE)
G90
...
N12 G01 Z-0.75 F4.0 M08
...
N25 G01 X9.5 Y4.874
N26 G28 Z-0.75 M09        (G28 IN ABSOLUTE MODE)
...

(---> G28 USED IN THE INCREMENTAL MODE)
G90
...
N12 G01 Z-0.75 F4.0 M08
...
N25 G01 X9.5 Y4.874
N26 G91 G28 Z0 M09    (G28 IN INCREMENTAL MODE)
...
```

Which method is better? Since both methods produce identical results, the choice is based on a given situation or personal preference. To switch to the incremental mode has its benefit, because the current tool location may not always be known. The disadvantage of this method is that G91 is most likely a temporary setting only and must be reset back to G90 mode, used by the majority of the program.

> A failure to reinstate the absolute mode may result
> in an expensive and possibly serious error.

Absolute mode of programming specifies the current tool position from program zero - *always* and *at all times*. Many examples presented here use the absolute programming mode - after all, this is - or it should be - the standard programming mode, for the majority of programs.

There is one time, where the incremental mode of machine zero return has some very practical advantages. It happens in those cases when the current tool position *is not known* to the programmer. Such a situation typically happens when using subprograms, where incremental mode is used repeatedly to move the tool incrementally to different XY locations. For instance - where *exactly* is the cutting tool located when the drilling cycle is completed in the N35 block of the following example?

```
G90
...
N32 G99 G81 X1.5 Y2.25 R0.1 Z-0.163 F12.0
N33 G91 X0.3874 Y0.6482 L7    (REPEAT 7 TIMES)
N34 G90 G80 Z1.0 M09          (CANCEL CYCLE)
N35 G28 (X???? Y????) Z1.0   (UNKNOWN POSITION)
...
```

Is it worth the extra effort to find the absolute location at all costs? Probably not. Let's look at some other examples. While in the absolute mode G90, the axis coordinate values define the intermediate point *location*. When incremental mode G91 is programmed, the coordinate values define the actual distance and direction of the intermediate *motion*. In both cases, the intermediate tool motion will be performed first. Then - and only then - the final return to the machine zero reference position will take place.

Take the current tool position as X5.0 and Y1.0 (absolute position). In the program, the XY values of the G28 command that follows the position block are very important:

```
G90
...
N12 G00 X5.0 Y1.0
N13 G28 X0 Y0
...
```

In this example, the G28 command specifies that the cutting tool should reach the machine zero position - identified as X0Y0 in the block N13. Since the G28 command relates to the machine zero only, it would be reasonable to assume that the X0Y0 relates to the machine zero, rather than the part zero. *That is not correct.*

The X0Y0 refers *to the point through which the tool will reach the machine zero position*. That is the defined point already known to be the *intermediate position* for the machine zero return command. This intermediate point is assigned the coordinates relating to the part (in absolute mode). In the example, the cutting tool will move to the program zero *before* continuing to the machine zero, resulting in a single block definition of two tool motions. This, of course, is not likely to be the intended motion.

The above example can be changed, so the intermediate motion is eliminated - or - *defined as the current tool position*. The intermediate motion can never be eliminated, but it can be programmed as a physical zero distance.

```
G90
...
N12 G00 X5.0 Y1.0
N13 G28 X5.0 Y1.0
...
```

By this modification, the intermediate point becomes the *current tool position*, which results in *direct* motion to the machine zero. The reason is that the intermediate tool position coincides with the current tool position. This programming format has nothing to do with modal values of axes. In the part program, X5.0Y1.0 in the block N13 must be repeated, while the absolute mode G90 is still in effect.

In cases when the current tool position is *not* known, the machine zero return has to be done in *incremental mode*. In this case, change temporarily to incremental mode and program a zero length motion for each specified axis:

```
G90
...
N12 G00 X5.0 Y1.0
N13 G91 G28 X0 Y0
N14 G90 ...
...
```

Again, an important remainder is in place here - always remember to switch back to the absolute mode as soon as possible, in order to avoid misinterpreting the consecutive program data.

In a brief overview, the intermediate point cannot be eliminated from the G28/G30 block. If situation demands a return to machine zero without going through a separate intermediate point, use a zero tool motion towards the intermediate point. The method depends on the active G90 or G91 mode at the time:

❑ In G90 absolute mode motion to machine zero, the current tool coordinate location must be repeated for each axis specified with G28 command.

❑ In G91 incremental motion to machine zero, the current tool motion must be equal to zero for each axis specified with the G28 command.

◆ Return from the Z Depth Position

One common example of using the intermediate tool position in a program block, is the return from a deep hole or a cavity to the machine zero. In the following example, and solely for the purpose of better explanation, regular tool motions are used rather than a drilling cycle, to retract the tool from the hole depth. In the example, the current XY position is X9.5Y4.874, and a peck drilling operation will be simulated in separate blocks:

```
...
N21 G90 G00 G54 X9.5 Y4.874 S900 M03
N22 G43 Z0.1 H01 M08
N23 G01 Z-0.45 F10.0
N24 G00 Z-0.43
N25 G01 Z-0.75
...
```

In block N25, the tool is at the bottom of the hole, at a current tool position of X9.5 Y4.874 Z-0.75 absolute coordinates. All the cutting is done and the tool has to be returned home in all three axes. For safety reasons, the Z axis must retract first. Several options can be selected, but three of them are the most common:

❏ Retract the Z axis above work in one block, then return XYZ axes to machine zero

❏ Retract the Z axis all the way to machine zero, then return the XY axes in the next block

❏ Return XYZ axes to machine zero directly from the current tool position (at the depth)

The *Figure 21-5* shows the available options.

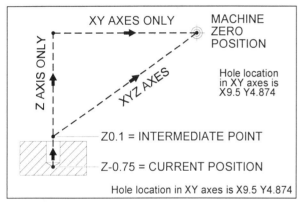

Figure 21-5

Machine zero return from a hole depth - milling

◉ Option 1

To retract the Z axis above work in one block first, then return the XYZ axes to the machine zero position, would be the 'normal' method, commonly used:

```
N26 G00 Z0.1 M09
```

This block must be followed by a return to the home position, along the Z axis:

```
N27 G28 Z0.1 M05
```

The complete program for *Option 1* will be:

```
...
N21 G90 G00 G54 X9.5 Y4.874 S900 M03
N22 G43 Z0.1 H01 M08
N23 G01 Z-0.45 F10.0
```

```
N24 G00 Z-0.43
N25 G01 Z-0.75
N26 G00 Z0.1 M09
N27 G28 Z0.1 M05
N28 G28 X9.5 Y4.874
N29 M01
```

◉ Option 2

To retract the Z axis all the way to machine zero first and then return the XY axes in the next block, is a variation on *Option 1*. First, return the Z axis to the machine zero:

```
N26 G28 Z-0.75 M09
```

Then, return the XY axes to machine zero as well:

```
N27 G28 X9.5 Y4.874
```

The complete program for *Option 2* will be:

```
...
N21 G90 G00 G54 X9.5 Y4.874 S900 M03
N22 G43 Z0.1 H01 M08
N23 G01 Z-0.45 F10.0
N24 G00 Z-0.43
N25 G01 Z-0.75
N26 G28 Z-0.75 M09
N27 G28 X9.5 Y4.874 M05
N28 M01
```

◉ Option 3

To return all three axes XYZ to machine zero directly from the current tool position (while the tool is still at the hole full depth), only *one* zero return block will be needed:

```
N26 G28 X9.5 Y4.874 Z0.1 M09
```

This is the intended method of programming, as Fanuc controls are designed. Some programmers may disagree with Fanuc on this issue, but that is how it works.

Here is the complete program for *Option 3*:

```
...
N21 G90 G00 G54 X9.5 Y4.874 S900 M03
N22 G43 Z0.1 H01 M08
N23 G01 Z-0.45 F10.0
N24 G00 Z-0.43
N25 G01 Z-0.75 M09
N26 G28 X9.5 Y4.874 Z0.1 M05
N27 M01
```

The motion to machine zero will take two steps:

Step 1:   Z axis will rapid to Z0.1 position

Step 2:   All axes will return to machine zero

Also note the rearrangements of M09 and M05 miscellaneous functions. Turning the coolant off first is more practical than stopping the spindle.

Although this is a matter of opinion, the choice of many programmers is to move the tool out of a cavity or hole first, then call the machine zero return command. If there is any justification for this preference, it is the perceived safety the CNC programmer puts into the program design. To be fair here, there is absolutely nothing wrong with the alternate method, if it is used with care. Comparing individual options with each other does offer some valuable conclusions:

❑ *OPTION 1 ...*

  ... is only reasonably safe, but quite efficient in terms of cycle time. There may be a possibility of an obstacle within the three-axis motion to machine zero.

❑ *OPTION 2 ...*

  ... is somewhat less efficient than the previous option, but definitely the safest one of all three.

❑ *OPTION 3 ...*

  ... is the most efficient in terms of program cycle time, but any error in position could result in a collision.

◆ Axes Return Required for the ATC

If the only purpose of machine zero return is to make an automatic tool change, only certain axes must be moved for that purpose. For a vertical machining center, only the Z axis is required to make the tool change:

```
G91 G28 Z0 M06
```

Horizontal machining centers require only the Y axis to reach its reference position for the automatic tool change. For safety and extra convenience, the Z axis is usually programmed as well, along with the Y axis, to prevent a collision with an adjacent tool in the magazine:

```
G91 G28 Y0 Z0 M06
```

In both examples, the tool change function M06 will *not* be effective, until the machine zero reference position has been physically reached. The M06 function can be programmed in a separate block later, if desired.

Indexing or rotary axes also have their own reference point and are used with G28 command the same way as linear axes. For example, a B axis will return to the machine zero reference position in the following block:

```
G91 G28 B0
```

If it is safe, the B axis may be programmed simultaneously with another axis:

```
G91 G28 X0 B0
```

Absolute mode designation follows the same rules for a rotary or indexing axis, as for the linear axes.

◆ Zero Return for CNC Lathes

For CNC lathe work, the G28 command may also be used, usually for setup. Common application of the machine zero return is also used, when at least one axis starts and ends at the machine zero position. This is quite often true of the X axis but not of the Z axis, which may be too far away on some larger lathe models.

Typically, a CNC lathe program will be designed in such a way, that machining of the first part will start from the machine zero, but any subsequent part will be machined from a safe tool change position. This method is only practical if the program uses geometry offset, rather than the older G50 setting. The most common method of machine zero return on the lathes is the direct method, without an intermediate point, because no G91 is required, therefore, an error is more difficult to make:

```
N78 G28 U0
N79 G28 W0
```

These two blocks will return the cutting tool to the machine zero in incremental mode, there is no intermediate motion applied. It is safer to move the X axis first, using the incremental mode U, then the Z axis, using the incremental mode W. If the work area is clear (watch for the tailstock), both X and Z axes can be returned to the machine zero at the same time:

```
N78 G28 U0 W0
```

Figure 21-6 illustrates a typical withdrawal of a boring bar from a hole, when the machining is completed.

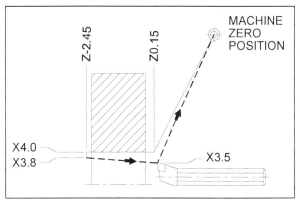

Figure 21-6

Machine zero return from a hole depth - turning application

When using position register command G50, the XZ setting must always be known for this command. In this case, the programming rules for machine zero return are very similar. Assuming that the machine zero position is at the coordinate position X10.0 Z3.0, the program for the boring tool can be written in two ways - one *without* using the G28 command, the other one *with* the G28 command.

➔ Example 1 :

The first example does not use G28 machine zero return command at all:

```
N1 G20 (EXAMPLE 1)
...
N58 G50 X10.0 Z3.0 S1000        (OLDER METHOD ONLY)
N59 G00 T0300 M42
N60 G96 S400 M03
N61 G00 G41 X4.0 Z0.15 T0303 M08
N62 G01 Z-2.45 F0.012
N63 X3.8 M09
N64 G00 G40 X3.5 Z0.15 M05
N65 X10.0 Z3.0 T0300
N66 M01
```

➔ Example 2 :

The second example will use G28 machine zero reference command, to achieve the same target position:

```
N1 G20 (EXAMPLE 2)
...
N58 G50 X10.0 Z3.0 S1000        (OLDER METHOD ONLY)
N59 G00 T0300 M42
N60 G96 S400 M03
N61 G00 G41 X4.0 Z0.15 T0303 M08
N62 G01 Z-2.45 F0.012
N63 G40 X3.8 M09
N64 G28 X3.5 Z0.15 M05 T0300
N65 M01
```

Most CNC programmers will likely feel more comfortable with the first example and saving one program block program will not likely be compelling enough to change their programming style. The second example *(Example 2)* can be programmed in the incremental mode as well, using the U and W addresses, but it would not be too practical.

RETURN POSITION CHECK COMMAND

The less common preparatory command G27 performs a checking function - and nothing else. Its only purpose is to check (which means to *confirm*), if the programmed position in the block containing G27 is at the machine zero reference point or not. If it is, the control panel indicator light for each axis that has reached the position will go on. If the reached position is not at the machine zero, the program processing is interrupted by an error condition displayed on the screen as an alarm.

If the tool starting position is programmed at the machine zero reference (home), it is a good practice to return there as well, when the machining with that cutting tool is completed. This is quite commonly done for CNC lathes, where the tool change (indexing) normally takes place in the same position, although this position does not always have to be the machine zero. Usually, it is a safe position near the machined part.

The format for G27 command is:

```
G27 X.. Y.. Z..
```

where at least one axis must be specified.

When used in the program, the cutting tool will automatically rapid (no G00 necessary) to the position as specified by the axes in the G27 block. The motion can be either in the absolute or incremental mode. Note that no G28 command is used.

```
N1 G20
N2 G50 X7.85 Z2.0               (OLDER METHOD ONLY)
N3 G00 T0400 M42
N4 G96 S350 M03
N5 G00 G42 X4.125 Z0.1 T0404 M08
N6 G01 Z-1.75 F0.012
N7 U0.2 F0.04
N8 G27 G40 X7.85 Z2.0 T0400 M09
N9 M01
```

In the example, block N8 contains G27, but no G00 or G28. This block instructs the CNC machine to return to the position X7.85 Z2.0 and check, upon arrival to the target position, if that position is the machine zero in *all* specified axes (two axes in the example). A confirmation light will turn on, if the machine zero position is confirmed. If the position is not confirmed, the program will not proceed any further until the cause (misposition) is eliminated.

Compare the starting position in block N2 and the return position in block N8. Assuming that this position is at machine zero reference point in both the X and Z axes, the above example will confirm OK position in the N8 block. Now, suppose that a small error has been made while writing block N8, and the X value was entered as X7.58 rather than the expected X7.85:

```
N8 G27 G40 X7.58 Z2.0 T0400 M09
```

In this case, the control system will return an error condition. The error is displayed automatically on the control screen (as an alarm). The system will *not* process the remainder of the program, until the error is corrected. The light indicating *Cycle Start* condition will turn off and the *source* of the problem has to be found. When looking for the source of the problem, always check *both positions*, the start position block, as well as the end position block. The error is quite easy to make in either block. Also note that any axis *not* specified in the block will not be checked for its actual position.

Another important point is the cancellation of the cutter radius offset and the tool offset. The G27 preparatory command should always be programmed with the G40 command and the *Txx00* in effect (G49 or H00). If the tool offset or the cutter radius offset is still in effect, the checking cannot be done properly, because the tool reference point is displaced by the offset value.

Here is how the first program *(Example 1)* listed earlier, can be modified to accept the G27 command. Note that the G27 will only move to the coordinates specified, *not* to any intermediate or other point. Block N65 will become the actual check block. The control system will move the machine axes to X10.0 Y3.0 and checks (confirms) whether this position is in fact the machine zero reference point. This is the reason *Example 1* could be modified, but not the second *Example 2*.

```
N1 G20
...
N58 G50 X10.0 Z3.0 S1000     (OLDER METHOD ONLY)
N59 G00 T0300 M42
N60 G96 S400 M03
N61 G00 G41 X4.0 Z0.15 T0303 M08
N62 G01 Z-2.45 F0.012
N63 X3.0 M09
N64 G00 G40 X3.5 Z0.15 M05
N65 G27 X10.0 Z3.0 T0300
N66 M01
```

The machine reference point return check can be done in either the absolute or incremental mode. The absolute statement in block N65 (in the last example) can be replaced with the incremental version:

```
N65 G27 U6.5 W2.85 T0300
```

There is a drawback to this command. A small price to pay when using this checking command is a slight cycle time loss. Because the deceleration of tool motion is built into the command by the control system, about one to three seconds may be lost when G27 command is implemented. This may be a significant loss if a large number of tools use G27 check in every program.

The G27 command is seldom used with *geometry offset* setting of the tools, which is the current modern method. The G50 command is older and not used anymore on the newest CNC lathes, but many lathes are still used in industry that do need the G50 setting.

RETURN FROM MACHINE ZERO POINT

The preparatory command G29 is the exact opposite of G28 or G30 command. While G28 will automatically return the cutting tool to machine zero position, G29 command will return the tool to its *original* position - again, *via an intermediate point*.

In normal programming usage, the command G29 usually follows G28 or G30 command. The rules relating to the absolute and incremental axis designation are valid for G29 in exactly the same respect as to the G28 and G30. All programmed axes are moved at the rapid traverse rate to the intermediate position first, defined by the preceding G28 or G30 command block. An example for a lathe application illustrates the concept:

```
(LATHE EXAMPLE)
...
T0303
...
G28 U5.0 W3.0
G29 U-4.0 W2.375
```

The G29 command should always be issued in the canceled mode of both the cutter radius offset (G40) and the fixed cycles (G80), if either is employed in the program. Use the standard cancellation G codes - G40 to cancel cutter radius offset and G80 to cancel a fixed cycle, before the G29 command is issued in the program.

A schematic sketch of the tool motion is illustrated in *Figure 21-7*.

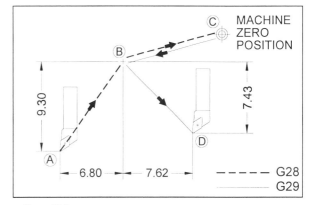

Figure 21-7
Automatic return from machine zero position

The illustration shows a tool motion from point A to point B first, then to point C, back to point B, and finally, to the point D. The point A is the starting point of the motion, point B is the intermediate point, point C is the machine zero reference point, and point D is the final point to reach, the actual target position.

The equivalent program commands, starting at the current tool position, which is point A, and resulting in the A to B to C to B to D tool path are quite simple:

```
G28 U18.6 W6.8
...
G29 U-14.86 W7.62
```

Of course, there would be some appropriate action programmed between the two blocks, for example, a tool change or some other machine activity.

Similar to G27 command, there is only a weak support for G29 among CNC programmers. It is one of the commands that can be very useful in some rare cases, but virtually unnecessary for everyday work. However, it is always an advantage to know what 'tools of trade' are available in CNC programming. They may come handy.

RETURN TO SECONDARY MACHINE ZERO

In addition to the G28 machine zero command, specific CNC machines also have the G30 command. In this chapter, and the handbook generally, many examples apply equally to G28 and G30 commands and were sometimes identified as G28/G30 to cover both. So what is different in G30 and why is this command needed it in the first place?

By definition, G30 preparatory command is a machine zero return command to the *secondary machine zero position*. That position must be available on the machine at the time of purchase. Note the descriptive word is *secondary*, not *second*. In virtually all respects, G30 is identical to the G28, except that it refers to a *secondary* program zero.

This secondary program zero can be the physical second, third, or even fourth reference point, as specified by the machine manufacturer. Not every CNC machine has a secondary machine zero reference position, and not every CNC machine even needs one. This secondary machine reference point serves only some very special purposes, mainly for horizontal machining centers.

The programming format for G30 command is similar to the G28 command, with an addition of the P address:

| G30 P.. X.. Y.. Z.. |

☞ where ...

G30	=	indicates the selection of a secondary reference position
P	=	can be P2, P3 and P4 to identify the secondary position (2-4)
XYZ	=	is the intermediate point definition (one axis minimum must be specified)

The most common use of a secondary machine zero reference point in CNC programming is for pallet changing. In the control unit parameter setting, the distance of the secondary reference point is set from the primary reference point and is not normally changed during the working life of the machine and the pallet changer.

To distinguish between multiple secondary machine zero positions, address P is added in the G30 block (there is no P address used for G28). If the CNC machine has only a single secondary machine reference position, the address P is usually *not* required in the program, and P1 is assumed in such a case:

`G30 X.. Y..`

is the same as

`G30 P1 X.. Y..`

In this case, the setting of the second reference point is within the parameters of the control system. In respect to other programming considerations, the G30 command is used in exactly the same way as the much more common G28 machine zero return command.

LINEAR INTERPOLATION

Linear interpolation is closely related to the rapid positioning motion. While the rapid tool motion is meant to be used from one position of the work area to another position *without cutting*, the linear interpolation mode is designed for actual *material removal*, such as contouring, pocketing, face milling and many other cutting motions.

Linear interpolation is used in part programming to make a straight cutting motion from the start position of the cut to its end position. It always uses the *shortest distance* the cutting tool path can take. The motion programmed in linear interpolation mode is always a straight line, connecting the contour start and end points. In this mode, the cutter moves from one position to another by the shortest distance between the end points. This is a very important programming feature, used mainly in contouring and profiling. Any angular motion (such as chamfers, bevels, angles, tapers, etc.) must be programmed in this mode to be accurate. Three types of motion can be generated in the linear interpolation mode:

❏ Horizontal motion   ... single axis only

❏ Vertical motion   ... single axis only

❏ Angular motion   ... multiple axes

The term *linear interpolation* means that the control system is capable to calculate thousands of intermediate coordinate points between the start point and end point of the cut. The result of this calculation is the shortest path between the two points. All calculations are automatic - the control system constantly coordinates and adjusts the feedrate for all cutting axes, normally two or three.

LINEAR COMMAND

G01	Linear interpolation

In G01 mode, the feedrate function F must be in effect. The first program block that starts the linear interpolation mode must have a feedrate in effect, otherwise an alarm will occur during the first run, after power on. Command G01 and feedrate F are modal, which means they may be omitted in all subsequent linear interpolation blocks, once they have been designated and providing the feedrate remains unchanged. Only a change of coordinate location is required for the axis designation in a program block. In addition to a single axis motion, a linear motion along two or three axes may be also programmed simultaneously.

◆ Start and End of the Linear Motion

Linear motion, like any other motion in CNC programming, is a motion between two end points of the contour. It has a *start* position and the *end* position. Any *start* position is often called the *departure* position, the *end* position is often called the *target* position. The start of a linear motion is defined by the current tool position, the end is defined by the target coordinates of the current block. It is easy to see that the end position of one motion will become the start position of the next motion, as the tool moves along the part, through all contour change points.

◆ Single Axis Linear Interpolation

The programmed tool motion along any single axis is always a motion parallel to that axis, regardless of the motion mode. Programming in either G00 or G01 mode will result in the same programmed end point, but at different feedrates and with different results. Evaluate *Figure 22-1* for comparison of the two motion modes.

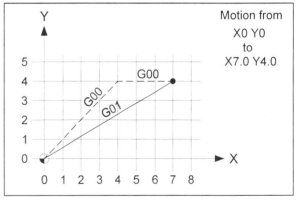

Figure 22-1

Comparison of the rapid mode and the linear interpolation mode

For CNC machining centers and the related machines, all tool motions that are parallel to the table edges are single axis motions. On the CNC lathes, many external and internal operations, such as facing, shoulder turning, diameter turning, drilling, tapping and others, are programmed as single axis motions. In all cases, a single axis motion can be along either the *vertical* or the *horizontal* axis, within the current (working) plane. A single axis motion can never be an angular motion, which requires two, three, or more axes. Another name for a motion that is parallel to a machine axis is *orthogonal* - horizontal or vertical only.

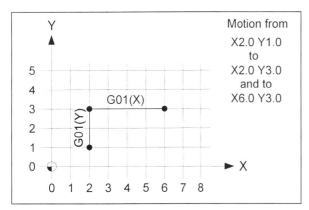

Figure 22-2
Single axis linear interpolation motion

Figure 22-2 illustrates a single axis linear interpolation motion, one along the X axis and the other along the Y axis.

◆ Two Axes Linear Interpolation

A linear motion can also be programmed along *two axes* simultaneously. This is a very common situation when the start point of the linear motion and its end point have at least two coordinates that are different from each other, while in the linear interpolation mode G01. The result of this two-axis motion is a straight tool motion at an angle. The motion will always be the shortest distance between the start point and the end point and results in a straight line at an angle calculated by the control - *Figure 22-3*.

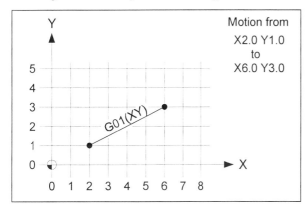

Figure 22-3
Two axes simultaneous linear interpolation motion

◆ Three Axis Linear Interpolation

A linear motion that takes place along three axes at the same time, is called the *three axis linear interpolation*. A simultaneous linear motion along three axes is possible on virtually all CNC machining centers. Programming a linear motion of this kind is not always easy, particularly when working with complex parts. Due to many difficult calculations involved in this type of tool motion, the manual pro-

gramming method is not efficient enough. Such programming projects more than justify an investment into a professional computer based programming system, such as the very powerful and widely used *Mastercam™*, that is based on modern computer technology combined with machining know-how. This type of programming is using desktop computers and is affordable by virtually all machine shops. Computer based programming is not a subject of this handbook, but its general concepts are discussed briefly in the last chapter of the handbook (*Chapter 53*).

The three-axis (XYZ) simultaneous linear motion is illustrated in *Figure 22-4*.

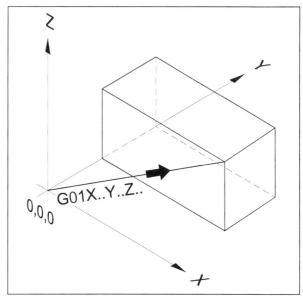

Figure 22-4
Three axes simultaneous linear interpolation motion

PROGRAMMING FORMAT

In order to program a tool motion in the linear interpolation mode, use preparatory command G01 along with one, two, or three axes of tool motion, as well as a cutting feedrate (F address) suitable for the job at hand:

```
G01 X.. Y.. Z.. F..
```

All entries in the linear motion block are modal and need to be programmed only if they are new or changed. Only the block instruction (word) that is affected by the change needs to be included in the program block.

Depending on which programming method is selected, the linear interpolation motion may be programmed in the absolute or incremental mode, using G90 and G91 preparatory commands for milling and incremental addresses U and W for turning.

LINEAR FEEDRATE

The actual cutting feedrate for a defined tool motion can be programmed in two modes:

- ❑ ... per time   mm/min *or* in/min
- ❑ ... per spindle revolution   mm/rev *or* in/rev

The selection depends on the machine type and dimensional units used. Typically, CNC machining centers, drills, mills, routers, flame cutters, laser profilers, wire EDM, etc., use feedrate per time. CNC lathes and turning centers typically use feedrate per revolution.

◆ Feedrate Range

Every CNC system supports cutting feedrate only within a certain range. For linear interpolation in milling applications, the typical lowest feedrate is 0.0001, either as *in/min*, *mm/min* or *deg/min*. The lowest feedrate for linear interpolation in turning is dependent on the minimum increment of the coordinate axes XZ. The following two tables point out typical ranges a normal CNC system can support. The first table is for milling, the second table is for turning. All units used in part programming are represented.

Minimum motion increment	MILLING
0.001 mm	0.0001 - 240000.00 mm/min
0.001 degree	0.0001 - 240000.00 deg/min
.0001 inch	.0001 - 240000.00 in/min

Minimum motion increment	TURNING
0.001 mm	0.00001 - 500.00000 mm/min
0.001 degree	0.00001 - 500.00000 deg/min
.0001 inch	.000001 - 50.000000 in/min

It may appear that the maximum feedrate that can be used is unusually high. For actual cutting, that is true. However, keep in mind that these ranges are relative to the control system, *not* to the machine. The machine manufacturer will always limit the maximum feedrate, according to the machine design and its capabilities. Control system only provides the theoretical range, that is more for the benefit of the machine manufacturer than the actual user. The intent in this case is to allow the machine manufacturers flexibility within current technological advances. As technology changes, the control system manufacturers will have to respond to the changes as well, by increasing the ranges.

◆ Individual Axis Feedrate

The subject of actual cutting feedrate per axis is not crucial in programming at all. It is included here for the mathematically oriented and interested individuals only. There is no need to know the following calculations at all - the CNC system will do them every time, all the time, accurately and automatically. On the other hand, here it is anyway.

In order to keep the linear motion as the *shortest* motion between two points, the CNC unit must always calculate the feedrate for *each axis individually*. Depending on the direction of the linear motion (its angular value), the computer will *'speed up'* one axis and *'hold back'* the other axis *at the same time*, and it will do it constantly during the cut. The result is a straight line between the start and end points of the linear contour. Strictly speaking, it is not a straight line but a jagged line, with edges so diminutive in size that they are virtually impossible to see, even under magnification. For all practical purposes, the result is a straight line.

The calculations are done by the CNC system, according to the following entries, as illustrated in *Figure 22-5*.

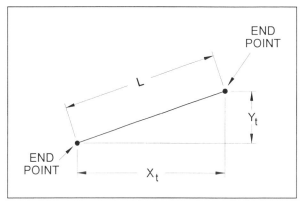

Figure 22-5
Data for the calculation of individual axis linear feedrate

Evaluate the following example of a linear motion and try to apply the formulas listed afterwards:

```
G00 X10.0 Y6.0                              (START POINT)
G01 X14.5 Y7.25 F12.0                       (END POINT)
```

The linear motion takes place between two end points, from the starting point at X10.0 Y6.0 to the end point at X14.5 Y7.25 - the feedrate is programmed at 12 in/min as F12.0. That means the actual travel motion along each axis is either known or it can be calculated:

$$X_t = 14.5 - 10.0 = 4.5$$
$$Y_t = 7.25 - 6.0 = 1.25$$
$$Z_t = 0$$

The length L of tool total motion (as illustrated) is the actual compound motion, and can be calculated by using the well known *Pythagorean Theorem*:

$$L = \sqrt{X_t^2 + Y_t^2 + Z_t^2}$$

The above formula is quite common, based on the square root of the total sum of squares of sides, that will result in the value of 4.6703854 as the travel length in the example:

$$L = \sqrt{4.5^2 + 1.25^2 + 0^2} = 4.6703854$$

The control system will *internally* apply the formulas and calculate the actual motion along the X axis (4.25), as well as along the Y axis (1.25), plus the length of the motion itself (4.6703854). From these values, the computer system will calculate the X and Y axis feedrate - there is no motion that takes place along the Z axis:

$$F_x = \frac{X_t}{L \times F}$$

$F_x = 4.5 / 4.6703854 \times 12 = 11.562215$

$$F_y = \frac{Y_t}{L \times F}$$

$F_x = 1.25 / 4.6703854 \times 12 = 3.2117263$

$$F_z = \frac{Z_t}{L \times F}$$

$F_x = 0 / 4.6703854 \times 12 = 0.0$

In this example, there is no Z axis motion. If the Z axis were part of the tool motion, for example, during a simultaneous three dimensional linear motion, the procedure will be logically identical, with the inclusion of Z axis in the calculations.

PROGRAMMING EXAMPLE

In order to illustrate the practical use of linear interpolation mode in a CNC program, here is a simple example, shown in *Figure 22-6*.

For even more comprehensive understanding, the example will be presented twice. One tool motion will start and end at the P1 location and will be programmed in the clockwise direction, the other program example will start at the same P1 location, but will continue in the counterclockwise direction.

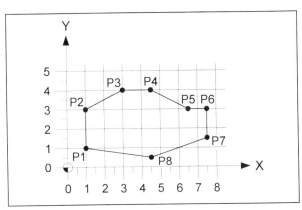

Figure 22-6
Example illustration for a simple linear interpolation

● Example 1 :

(CLOCKWISE DIRECTION FROM P1)

```
G90 ...                          (ABSOLUTE MODE)
G01 X1.0 Y3.0 F...                  (P1 TO P2)
X3.0 Y4.0                           (P2 TO P3)
X4.5                                (P3 TO P4)
X6.5 Y3.0                           (P4 TO P5)
X7.5                                (P5 TO P6)
Y1.5                                (P6 TO P7)
X4.5 Y0.5                           (P7 TO P8)
X1.0 Y1.0                           (P8 TO P1)
...
```

● Example 2 :

(COUNTERCLOCKWISE DIRECTION FROM P1)

```
G90 ...                          (ABSOLUTE MODE)
G01 X4.5 Y0.5 F...                  (P1 TO P8)
X7.5 Y1.5                           (P8 TO P7)
Y3.0                                (P7 TO P6)
X6.5                                (P6 TO P5)
X4.5 Y4.0                           (P5 TO P4)
X3.0                                (P4 TO P3)
X1.0 Y3.0                           (P3 TO P2)
Y1.0                                (P2 TO P1)
...
```

Linear interpolation provides means of programming all orthogonal (*i.e.*, vertical and horizontal) motions, as well as angular tool motions as the shortest linear distance between two points. Cutting feedrate must be programmed in this mode, for proper metal removal. Note that the coordinate location that has not changed from one point to the next - one block to the next - is not repeated in the subsequent block or blocks.

In many control manuals, the block skip function is also called the *block delete* function. The expression *'block delete'* offers rather a misleading description, since no program blocks will actually be deleted but only *skipped* during program processing. For this good reason, the more accurate description of the function is the *block skip* function, a term used in the handbook. This function is a standard feature of virtually all CNC controls. Its main purpose is to offer the programmer some additional flexibility in designing a program for no more than *two conflicting possibilities.* In the absence of a block skip function, the only alternative is to develop two individual part programs, each covering one unique possibility.

TYPICAL APPLICATIONS

To understand the idea of two conflicting possibilities, consider this programming application. The assignment is to write a program for a facing cut. The problem is that the blank material for parts delivered to the CNC machine is not consistent in size. Some blanks are slightly smaller in size and can be faced with a single cut. Others are larger and will require two facing cuts. This is not an uncommon occurrence in CNC shops and is not always handled efficiently. Making two inefficient programs is always an option, but a single program that covers both options is a better choice - but only if the *block skip function* is used in such a program.

This challenge illustrates a situation, where two conflicting options are required in a program at the *same* time. The most obvious solution would be to prepare two separate programs, each properly identified as to its purpose. Such a task can be done quite easily, but it will be a tedious, time consuming and definitely an inefficient process. The only other solution is to write a *single* program, with tool motions covering facing cuts for *both* possibilities. To avoid air cutting for those parts that require only one cut, a block skip function will be provided in the program and applied to all blocks relating to the *first* facing cut. The 'second' cut will always be needed!

Other common applications of the block skip function include a selective ON/OFF status toggle, such as the coolant function, optional program stop, program reset, etc. Also useful are applications for bypassing a certain program operation, applying or not applying a selected tool to a part contour and others. Any programming decision that requires a choice from *two* predetermined options is a good candidate for the block skip function.

BLOCK SKIP SYMBOL

To identify the block skip function in a program, a special programming symbol is required. This block skip function symbol is represented by a forward slash [/]. The system will recognize the slash as a code for the block skip. For most of CNC programming applications, the slash symbol is placed as the *first* character in a block:

➲ Example 1 :

```
N1 ...                              (ALWAYS PROCESSED)
N2 ...                              (ALWAYS PROCESSED)
N3 ...                              (ALWAYS PROCESSED)
/ N4 ...        (PROCESSED IF BLOCK SKIP IS OFF)
/ N5 ...        (PROCESSED IF BLOCK SKIP IS OFF)
/ N6 ...        (PROCESSED IF BLOCK SKIP IS OFF)
N7 ...                              (ALWAYS PROCESSED)
N8 ...                              (ALWAYS PROCESSED)
```

On *some* control systems, the block skip code can also be used selectively for certain addresses *within* a block, rather than at its beginning. Check the manual if such a technique can be used - it can be very powerful:

➲ Example 2 :

```
N6 ...
N7 G00 X50.0 / M08
N8 G01 ...
...
```

In those cases, when the control system does allow the block skip *within* a programmed block, all instructions *before* the slash code will be executed, regardless of the block skip toggle setting. If the block skip function is turned ON (block skip function is active), only the instructions *following* the slash code, will be skipped. In the *Example 2*, the coolant function M08 (block N7) will be skipped. If the block skip function is turned OFF (block skip function is not active), the whole block will be executed in *Example 2,* including the coolant function.

CONTROL UNIT SETTING

Regardless of the slash code position within a block, the program will be processed in two ways. Either in its entirety, or the instruction following the slash will be skipped (ignored). The final decision whether or not to use the block skip function is made during actual machining, by the operator, depending on the type of machining. For this purpose, a push button key, a toggle switch, or a menu item selection is provided on the control panel of the CNC unit. Selection of the block skip mode can be either as *active* (ON) - or *inactive* (OFF).

Most programs will not require any block skip codes. In such cases, the setting mode for the block skip function on the control panel is irrelevant, but OFF mode is strongly recommended. The switch setting becomes *very* important, if the program contains even a *single* block containing the slash symbol. The active setting ON will cause all instructions in a block following the slash code to be ignored during program processing. The inactive setting OFF will cause the control to ignore the slash code and process all instructions written in the program.

> Block skip function set to ON position means
> "Ignore all block instructions following the slash."

> Block skip function set to OFF position means
> "Process all block instructions."

In the *Example 1* listed earlier, the contents of blocks N4, N5 and N6 will be ignored, if the block skip function is ON. They will be processed, if the switch setting is OFF. The *Example 2*, also listed earlier, contains a slash in the block N7. The slash symbol is preceding the miscellaneous function M08 (coolant ON). If the skip function switch is ON, the coolant will be ignored; if it is OFF, the coolant function will be effective. This application may be useful in a dry run mode, to bypass the coolant flood during program verification, if no manual override is available.

> Not all controls allow the slash code in any other block
> position, except as the first character in the block: / N..

BLOCK SKIP AND MODAL COMMANDS

To understand the way how modal values work with skipped blocks, recall that modal commands can be specified only once in the program, in the block where they occur first. Modal commands are not repeated in the subsequent blocks, as long as they remain unchanged.

In programs where the block skip function is not used at all, there is nothing to do. When the block skip function is used, watch carefully all *modal* commands. Remember that a command established in a block using the slash code will not always be in effect. It depends on the setting of block skip switch. Any modal command that has to be carried over from a section with slash codes to the section without slash codes may be lost if the block skip function is used. Overlooking modal commands when programming block skip function can result in a program with serious errors.

A simple programming solution to avoid this potential problem is available. Just *repeat* all modal commands in the program section that will not be affected by the block skip function.

Compare the following two examples:

◯ Example A - Modal commands *are not* repeated :

```
N5 G00 X10.0 Y5.0 Z2.0
/ N6 G01 Z0.1 F30.0 M08
N7 Z-1.0 F12.0              (G01 AND M08 MISSING)
N8 ...
```

◯ Example B - Modal commands *are* repeated :

```
N5 G00 X10.0 Y5.0 Z2.0
/ N6 G01 Z0.1 F30.0 M08
N7 G01 Z-1.0 F12.0 M08
N8 ...
```

In both examples A and B, the program block containing the slash code indicates an intermediate Z axis position as Z0.1. This position may be required only in certain cases during machining and the operator will decide whether to use it or not, and also when to use it.

The critical block, identified in the examples as N6, contains several modal functions. The commands G01, Z0.1, F30.0 and M08 will all remain in effect, unless they are canceled or changed in any following block. From block N7 it is apparent that the Z coordinate position and the cutting feedrate value have changed. However, the G01 and M08 commands are not repeated in the example A and will *not* be in effect, if the block skip switch is set ON.

Both examples A and B will produce identical results, but only if the block skip function is in the inactive (OFF) mode. The control system will then execute the instructions in *all* blocks, in the order of programming sequence.

The processing result will be different for each programming example shown. If the block skip function is active (ON) - the block instructions following the slash code will *not* be processed. The next example A yields an unacceptable result, with a fairly possible collision. The example B uses careful and thoughtful approach with very little extra work. These are the results when block N6 is skipped:

◯ Example A - Modal commands *are not* repeated :

```
N5 G00 X10.0 Y5.0 Z2.0        (RAPID MOTION)
N7 Z-1.0 F12.0                (RAPID MOTION)
N8 ...
```

◯ Example B - Modal commands *are* repeated :

```
N5 G00 X10.0 Y5.0 Z2.0            (RAPID MOTION)
N7 G01 Z-1.0 F12.0 M08        (FEEDRATE MOTION)
N8 ...
```

Note that the linear motion G01, the feedrate F30.0 and the coolant M08 are all skipped in the example A. The X and Y axes have not been updated in either example and will remain unchanged. The conclusion is that the example A will result in a Z axis rapid motion in two consecutive blocks, causing a potentially dangerous situation. In the correct version, listed as example B, the programmed repetition of all commands - G01, F12.0 and M08 - assures the program will be run as intended. In the next section of this chapter we will look at the principles of program design for different practical applications.

In the summary, there is one basic rule for developing CNC programs with blocks using the block skip function:

> Always program *all* the instructions, even if it means repeating some program values and commands that have to be preserved.

The slash symbol can be placed into the program *after* the program has been designed for *both* options. Just place the slash in those blocks that define the optional skip of all selected program blocks. *Always check program!*

> Any CNC program containing block skip function should be checked at least twice.

The result of this double check must be *always* satisfactory, whether processed with the block skip in effect or without it. If an error is detected, *even a very minor error*, correct it first! After the correction, check the program at least *twice* again, covering both types of processing. The reason for the double check is that a correction made for one type of processing may cause a different error for the other type of processing.

PROGRAMMING EXAMPLES

The block skip function is very simple, often neglected, yet, it is a powerful programming tool. Many programs can benefit from a creative use of this feature. The type of work and some thinking ingenuity are the only criteria for its successful implementation. In the following examples, some practical applications of the block skip function are shown. Use the examples as start points for a general program design or when covering similar machining applications.

◆ Variable Stock Removal

Removal of the excessive stock material is typical during a rough cutting. When machining irregular shapes (castings, forgings, etc.) or rough facing on lathes, it may be difficult to determine the number of cuts. For example, some castings for a given job may have only the minimum excessive material, so one roughing or facing cut will be sufficient. Other castings for the same job may be larger and two roughing or facing cuts are needed.

If the program is designed in such a way that there is only *one* roughing or facing cut, problems may occur during machining of heavy stock. Programming *two* cuts for all parts produces a safer program, but will be inefficient for parts with a minimum stock. There will be too many tool motions known as 'cutting air', when the stock is minimal.

➲ Example - Variable stock face :

A face cutting of a stock that varies in size is a common problem in CNC work. A suitable solution is identical for turning and milling - the program should include tool motions for *two* cuts and the block skip function will be used on all blocks relating to the *first* cut.

Here is a lathe example of a typical face cut, when the facing stock varies between .08 (2 mm) and .275 (7 mm). After considering several machining options, the programmer decides that the reasonable maximum stock that can be faced in a single cut will be .135 (3.5 mm) - *Figure 23-1*.

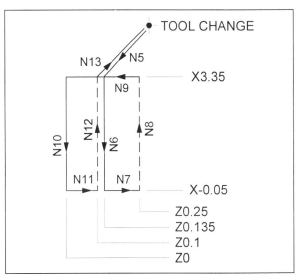

Figure 23-1

Variable stock for facing in a turning application - program O2301

```
O2301 (TURNING)
(VARIABLE FACE STOCK)
N1 G20 G40 G99
N2 G50 S2000
N3 G00 T0200 M42
N4 G96 S400 M03
N5 G41 X3.35 Z0.135 T0202 M08
/ N6 G01 X-0.05 F0.01
/ N7 G00 Z0.25
/ N8 X3.35
N9 G01 Z0 F0.05
N10 X-0.05 F0.01
N11 G00 Z0.1
N12 X3.5
N13 G40 X12.0 Z2.0 T0200
N14 M30
%
```

Block N5 contains the initial tool approach motion. The next three blocks are preceded by a slash. In N6, the tool cuts off the front face, at Z0.135; N7 moves the tool away from the face, block N8 is a rapid motion back to the initial diameter. There are no other blocks to be skipped after the block N8. N9 block contains a feedrate to the front face Z0, N10 is the front face cutting motion, N11 is the clearance motion, followed by standard final blocks.

Evaluate the example not once but at least *twice* - it shows what exactly happens. During the *first* evaluation, read *all* blocks and ignore the block skip function. During the *second* time, ignore all blocks containing the slash code. There will be identical results when compared with the first evaluation. The only difference will be the number of actual cuts - *one*, not two. In milling, the procedure is very similar.

An example for a milling application uses a ∅5 inch face mill. The excessive material stock to be faced varies between .120 and .315. The largest reasonable depth of cut selected will be .177 (4.5 mm) - *Figure 23-2*.

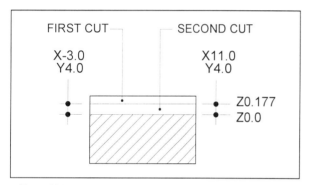

Figure 23-2

Variable stock for facing in a milling application - program O2302

```
O2302 (MILLING)
(VARIABLE FACE STOCK)
N1 G20
N2 G17 G40 G49 G80
N3 G90 G00 G54 X11.0 Y4.0
N4 G43 Z1.0 S550 M03 H01
N5 G01 Z0.177 F15.0 M08
/ N6 X-3.0 F18.0
/ N7 Z0.375
/ N8 G00 X11.0
N9 G01 Z0
N10 X-3.0 F18.0
N11 G00 Z1.0 M09
N12 G28 X-3.0 Y4.0 Z1.0
M13 M30
%
```

Block N5 in the example contains the Z axis approach to the first cut, at Z0.177 level. The next three blocks can be skipped if necessary. In the N6 block, the face mill actually cuts at Z0.177 position, N7 is the tool clearance motion after the cut, and N8 returns the tool to the initial X position. There are no other blocks to be skipped after block N8.

Block N9 does not need a feedrate for a good reason - it will be *either* F15.0 *or* F18.0, depending on whether blocks N6 to N8 were skipped or not. The feedrate is very important in block N10. Such a repetition guarantees the required feedrate in the critical block, when actual cutting takes place.

Both lathe and mill examples should offer at least some basic understanding of the *logic* used in program development, using the block skip function. Exactly the same logical approach can be used for more than two cuts and can also be applied to operations other than face cutting.

◆ Machining Pattern Change

Another application, where the block skip function may be used efficiently, is a simple family programming. The term *family programming* means a programming situation where there may be a slight difference in the design between two or more parts. Such a small variation between *similar* parts is often a good prospect for block skip function. A minor deviation in a machining pattern from one drawing to another can be adapted in a single program using the block skip function. Following two examples show typical possibilities of programming a *change of the tool path*. In one example, the emphasis is on a skipped machining location. In the other example, the emphasis is on the pattern change itself. Both examples are in metric and illustrate a simple grooving operation. In the lathe example, the *Figure 23-3* is related to program O2303.

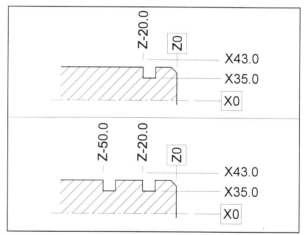

Figure 23-3

Variable machining pattern - turning application

The upper picture shows the result with block skip function set ON, the lower picture shows the result with block skip function set OFF, using the *same* program.

```
O2303 (LATHE EXAMPLE)
N1 G21
...
N12 G50 S1800
N13 G00 T0600 M42
N14 G96 S100 M03
```

```
N15 X43.0 Z-20.0 T0606 M08
N16 G01 X35.0 F0.13
N17 G00 X43.0
/ N18 Z-50.0
/ N19 G01 X35.0
/ N20 G00 X43.0
N21 X400.0 Z45.0 T0600 M01
```

Program O2303 demonstrates a single program for two parts with similar characteristics. One part requires a single groove, the other requires two grooves on the same diameter. In the example, both grooves are identical - they have the same width and depth and are machined with the same tool. The only difference between the two examples is the number of grooves and the second groove position. Machining the part will require the block skip function set ON or OFF, depending on the grove to be machined.

Evaluate the more important blocks in the program example. The N15 block is the initial tool motion to the start of the first groove at Z-20.0. In the next two blocks, N16 and N17, the groove will be cut and the tool returns to the clearance diameter. The following three blocks will cut the second groove, if it is required. That is the reason for the block skip code. In the block N18, the tool moves to the initial position of groove 2 at Z-50.0, in N19 the groove is cut. In the block N20, the tool retracts from the groove to a clearance position.

The milling example shown in *Figure 23-4*, also in metric, is represented in program O2304. The program handles two similar patterns that have four identical holes for both parts and two missing holes in the second part only. This is a good example of similar parts program, using block skip.

Both variations of program O2304 machine a hole pattern with 6 or 4 holes. Block skip function has been used to make a single program covering *both* patterns. The top of *Figure 23-4* shows the hole pattern when block skip function is set OFF, the bottom shows the hole pattern when block skip mode is set ON.

```
O2304  (MILLING EXAMPLE)
N1 G21
...
N16 G90 G00 G54 X30.0 Y25.0 M08
N17 G43 Z25.0 S1200 M03 H04
N18 G99 G81 R2.5 Z-4.0 F100.0          (HOLE 1)
N19 X105.0                             (HOLE 2)
N20 Y75.0                              (HOLE 3)
/ N21 X80.0 Y50.0                      (HOLE 4)
/ N22 X55.0                            (HOLE 5)
N23 G98 X30.0 Y75.0                    (HOLE 6)
N24 G80 G28 X30.0 Y75.0 Z25.0
N25 M01
```

Blocks N18 to N20 will drill holes 1, 2 and 3. Hole 4 in N21 and hole 5 in N22 will be drilled only if the block skip function is set to inactive mode (OFF), but neither one will not be drilled when the block skip setting is active (ON). Block N23 will always drill hole number 6.

A variation of this application is in the program O2305. There are *five* hole positions, but the block skip function is used *within* a block, to control only the Y position of the hole. Top of *Figure 23-5* shows the pattern when block skip function is OFF, the bottom shows the pattern when skip function has been set ON. The middle hole will have a different Y axis position, depending on the setting of the block skip function at the machine.

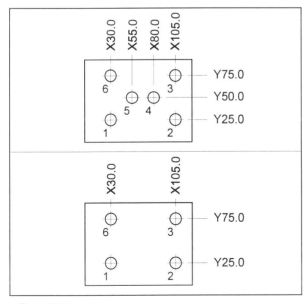

Figure 23-4

Program O2304 - variable machining pattern for a milling application - result with block skip OFF (top) and ON (bottom)

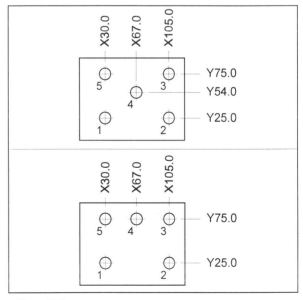

Figure 23-5

Program O2305 - variable machining pattern for a milling application - result with block skip OFF (top) and ON (bottom)

```
O2305 (MILLING EXAMPLE)
N1 G21
...
N16 G90 G00 G54 X30.0 Y25.0 M08
N17 G43 Z25.0 S1200 M03 H04
N18 G99 G81 R2.5 Z-4.0 F100.0         (HOLE 1)
N19 X105.0                           (HOLE 2)
N20 Y75.0                            (HOLE 3)
N21 X67.0 / Y54.0                    (HOLE 4)
N22 G98 X30.0 Y75.0                  (HOLE 5)
N23 G80 G28 X30.0 Y75.0 Z25.0
N24 M01
```

The hole 4 in block N21 will be drilled at the location of X67.0 Y75.0, if the block skip mode is ON. The address Y54.0 in block N21, will not be processed. If the block skip mode is OFF, the hole 4 will be drilled at coordinate location of X67.0 Y54.0. In that case, the Y75.0 position from the block N20 will be overridden. In order to guarantee the proper drilling at position 5, the coordinate Y75.0 in block N22 must be written. If it is omitted, the Y54.0 from block N22 will take precedence in block skip OFF mode.

Using the block skip feature is the simplest way of designing a family of similar parts. The applications are limited with the block skip function alone, but they offer the fundamentals of a powerful programming technique and an example of logical thinking. Many detailed explanations and examples of programming complex families of parts can be found in a special *Custom Macro* option Fanuc offers on most control systems.

◆ Trial Cut for Measuring

Another useful application of the block skip is to provide the machine operator with means of measuring the part *before* any final machining on the part has been done. Due to various dimensional imperfections of the cutting tool combined with other factors, the completed part may be slightly outside of the required tolerance range.

The following method of programming is very useful for programming parts requiring very close tolerances. It is also a useful method for those parts, where the part shape is difficult to measure after all machining is completed, for example conical shapes, such as tapers. The same method is also quite useful for parts where the cycle time of an individual tool is relatively long and all the tool offsets have to be fine tuned *before* production machining.

This approach to part programming is more efficient, as it eliminates a recut, increases surface finish, and can even prevent a scrap. In either case, a *trial cut* programming method that employs the block skip function is used. Setting the block skip mode OFF, the machine operator checks the trial dimension, adjusts the individual offset, if necessary, and continues the machining with block skip set ON.

The general concepts described in example O2306 are equally applicable to turning and milling - *Figure 23-6*.

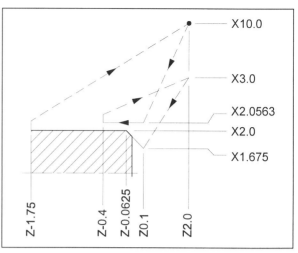

Figure 23-6

Application of a trial cut for measuring on a lathe - program O2306

```
O2306
(TRIAL CUT - LATHE)
N1 G20
...
N10 G50 S1400
N11 G00 T0600 M43
N12 G96 S600 M03
/ N13 G42 X2.0563 Z0.1 T0606 M08
/ N14 G01 Z-0.4 F0.008
/ N15 X2.3 F0.03
/ N16 G00 G40 X3.0 Z2.0 T0600 M00
/ (TRIAL DIA IS 2.0563 INCHES)

/ N17 G96 S600 M03
N18 G00 G42 X1.675 Z0.1 T0606 M08
N19 G01 X2.0 Z-0.0625 F0.007
N20 Z-1.75
N21 X3.5 F0.01
N22 G00 G40 X10.0 Z2.0 T0600
N23 M01
```

When program O2306 is processed with the block skip set OFF, all blocks will be executed, including the trial cut and finish profile. With the block skip set ON, the only operation executed will be the finishing to size, *without* the trial cut. In this case, all significant instructions are retained by repetition of the key commands (N18 and N19). Such a repetition is very crucial for successful processing in both modes of the block skip function. M00 function in N16 always stops the machine and enables a dimensional check.

Selecting trial diameter of 2.0563 in the example may be questioned. What is the logic for it? The trial diameter can be other reasonable size, say 2.05. That would leave a .025 stock per side for the finish cut. It is true that a different diameter *could* have been selected. The four decimal number was only selected for one reason - *to psychologically* encourage the operator to maintain accurate offset settings. Feel free to disagree - programmers may prefer a three or even two decimal number instead - the choices are open.

In the next example, another trial cut will also be programmed before the actual machining, but for a different reason - *Figure 23-7*.

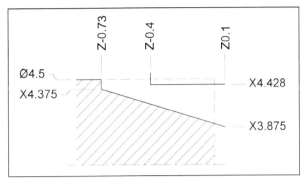

Figure 23-7

Trial cut for a taper cutting on a lathe - program O2307

In program O2307, the finished shape of the part is a taper, a feature difficult to measure when completed. Adjusting the tool offset in a trial and error way is not the right solution. Programming a trial cut within an area of the solid material, *along a straight diameter,* enables the operator to check the trial dimension comfortably and to adjust the offset *before* cutting the finished taper.

```
O2307
(TRIAL CUT FOR TAPER - ONE TOOL)
N1 G20 G99 G40
N2 G50 S1750 T0200 M42
N3 G96 S500 M03
/ N4 G00 G42 X4.428 Z0.1 T0202 M08
/ N5 G01 Z-0.4 F0.008
/ N6 U0.2 F0.03
/ N7 G00 G40 X10.0 Z5.0 T0200 M00
/ (TRIAL CUT DIA IS 4.428 INCHES)

/ N8 G96 S500 M03
N9 G00 G42 X4.6 Z0.1 T0202 M08
N10 G71 P11 Q13 U0.06 W0.005 D1500 F0.01
N11 G00 X3.875
N12 G01 X4.375 Z-0.73 F0.008
N13 X4.6 F0.012
N14 S550 M43
N15 G70 P11 Q13
N16 G00 G40 X10.0 Z5.0 T0200 M01
```

Program O2307 illustrates a common situation, where a single cutting tool is used for both roughing *and* finishing operations. It shows a logical way of using the block skip function, in a simple form. In most applications, separate tools for roughing and finishing may be needed, depending on the degree of required accuracy. When using two cutting tools, the trial cut dimension is usually more important for the *finishing* tool than for the roughing tool. In program O2308, the block skip function is illustrated using two cutting tools - T02 is for roughing, T04 is for finishing. Previous example in *Figure 23-7* is used.

```
O2308
(TRIAL CUT FOR TAPER - TWO TOOLS)
N1 G20 G99 G40
N2 G50 S1750 T0200 M42
N3 G96 S500 M03
/ N4 G00 G42 X4.46 Z0.1 T0202 M08
/ N5 G01 Z-0.4 F0.008
/ N6 U0.2 F0.03
/ N7 G00 G40 X10.0 Z5.0 T0200 M00
/ (T02 TRIAL CUT DIA IS 4.46 INCHES)

/ N8 G50 S1750 T0400 M43
/ N9 G96 S550 M03
/ N10 G00 G42 X4.428 Z0.1 T0404 M08
/ N11 G01 Z-0.4 F0.008
/ N12 U0.2 F0.03
/ N13 G00 G40 X10.0 Z5.0 T0400 M00
/ (T04 TRIAL CUT DIA IS 4.428 INCHES)

/ N14 G50 S1750 T0200 M42
/ N15 G96 S500 M03
N16 G00 G42 X4.6 Z0.1 T0202 M08
N17 G71 P18 Q20 U0.06 W0.005 D1500 F0.01
N18 G00 X3.875
N19 G01 X4.375 Z-0.73 F0.008
N20 X4.6 F0.012
N21 G00 G40 X10.0 Z5.0 T0200 M01

N22 G50 S1750 T0400 M43
N23 G96 S550 M03
N24 G00 G42 X122.0 Z3.0 T0404 M08
N25 G70 P18 Q20
N26 G00 G40 X10.0 Z5.0 T0400 M09
N27 M30
%
```

The example O2308 can be improved further by including the control of taper on the width, for example. Programming a trial cut is useful but often a neglected technique, although it does present many possible applications.

◆ Program Proving

The block skip function can also be useful to prove a new program on the machine, to check it against obvious errors. CNC operators with limited experience may be a little uneasy to run a program for the first time. One of the most common concerns of operators is the initial rapid motion towards a part, particularly when the clearances are small. The rapid motion rate of many modern CNC machines can be very high, well over 1500 in/min. At such high speeds, the rapid approach to the cutting position on the part may not add to the operator's confidence, particularly when the approach is programmed to the close proximity of the material. On most controls, the operator can set the rapid override rate to 100%, 50%, 25% and slower. On older controls, the rapid rate override cannot be done.

The next two examples, O2309 and O2310, show a typical programming method to eliminate the problem during setup and program proving, yet maintain the full rapid motion rate during repeated operations for productivity.

Block skip function in these examples takes a less usual role - it is used for a *section* of a block, rather than the whole block itself, if the control supports such a method.

```
O2309 (TURNING EXAMPLE)
N1 G20 G40 G99
N2 G50 S2000
N3 G00 T0200 M42
N4 G96 S400 M03
N5 G41 X2.75 Z0 T0202 M08 /G01 F0.1
N6 G01 X.. F0.004
N7 ...


O2310 (MILLING EXAMPLE)
N1 G20 G17 G40 G80
N2 G90 G00 G54 X219.0 Y75.0 M08
N3 G43 Z-1.0 S600 M03 H01 /G01 F30.0
N4 G01 X.. F12.0
N5 ...
```

In both examples, the block skip is used *within* a single block. The design of both programs takes advantage of *two conflicting commands within the same block*. If two conflicting commands exist in a single block, the *latter* command used in the block will become effective.

In both examples, the *first* command is G00, the *second* one is G01. Normally, the G01 motion will take a priority. Because of the slash code, the control will accept G00, if the block skip is set ON, but it will accept G01, if the block skip is set OFF. When the block skip mode is OFF, both motion commands will be read and the *second* command in that block becomes effective (G01 overrides G00). Watch for one possibility, already emphasized:

Block skip within a block may not work with all controls.

During the first machine run, the operator should set the block skip OFF, making the G01 command effective. The tool motion will be slower than in the rapid mode, but much safer. Also, the feedrate override switch of the control system will become effective, offering additional flexibility.

When the program proving is completed and the safe tool approach is confirmed, the block skip can be set ON, to prevent the G01 motion from being processed. Both O2309 and O2310 are typical examples of breaking with tradition to achieve a specific result.

◆ Barfeeder Application

On a CNC lathe, the block skip function can be used in barfeeding, for a continuously running machining. If the barfeeder allows it, the techniques is quite simple. The typical program will actually have *two ends* - one will use the M99 function, the other end will use the M30 function. The M99 block will be preceded by the block skip symbol and will be placed *before* the M30 code in the part program. This special technique is detailed in *Chapter 44*.

◆ Numbered Block Skip

For machining, the block skip function is set to either the ON or OFF position and remains in this mode for the *whole* program. If the ON setting is required for one section of the program, but not for another, the operator has to be informed, usually in the program comments. This practice of changing block skip mode in the middle of a program can be unsafe and possibly create problems.

An optional feature on some controls is a *selective* or a *numbered* block skip function. This option allows the operator to select which portions of the program required the ON setting and which portions require the OFF setting. The settings can be done *before* pressing the *Cycle Start* key to initialize the program. This method also uses the slash symbol, but followed by an integer, within the range of 1 to 9. The actual selection of the mode is done on the control screen *(Settings)*, under the matching switch number.

For example, a program may contain three groups, each expecting a different setting of the skip function. By using the switch number after the slash symbol, the groups are clearly defined and all the operator must do is to *match* the control settings with the required activity.

```
N1 ...
N2 ..
/1 N3 ...                    (BLOCK SKIP GROUP 1)
/1 N4 ...                    (BLOCK SKIP GROUP 1)
...
...
N16 ...
/2 N17 ...                   (BLOCK SKIP GROUP 2)
/2 N18 ...                   (BLOCK SKIP GROUP 2)
/2 N19 ...                   (BLOCK SKIP GROUP 2)
...
...
N29 ...
/3 N30 ...                   (BLOCK SKIP GROUP 3)
/3 N31 ...                   (BLOCK SKIP GROUP 3)
...
...
N45 ...
```

Identical rules apply for selective block skip function as for the normal version. Incidentally, the /1 selection is the same as a plain slash only, so blocks N3 and N4 above, could have also be written this way:

```
/ N3 ...
/ N4 ...
```

Numbered block skip function is not available on all controls.

Programs using the selective block skip function can be very clever and even efficient, but they may place quite a burden on the machine operator. For the majority of jobs, there will be a plenty of programming power available by using the standard block skip function.

24

DWELL COMMAND

Dwell is another name for a *pause* in the program - it is an intentional time delay applied during program processing. In this period of time - specified in a CNC program - any axis motion is stopped, while all other program commands and functions remain unaffected. When the designated time expires, the control system resumes processing the program with the block immediately *following* the block that contains the dwell.

PROGRAMMING APPLICATIONS

Programming a dwell is very easy and can be quite useful in two main applications:

❑ During actual cutting,
when the tool is in contact with material

❑ For operation of machine accessories,
when no cutting takes place

Each application is equally important to programmers, although the two are not used simultaneously.

◆ Applications for Cutting

When cutting tool is removing material, it is in contact with the machined part. A dwell can be applied during machining for a number of reasons. If the spindle is running, the spindle rotation *is* very important .

In practice, the application of a dwell during a cut is mainly used for breaking chips while drilling, counterboring, grooving or parting-off. Dwell may also be used while turning or boring, in order to eliminate the physical marks left on the part by end thrust of the cutting tool. This thrust is the result attributed to the tool pressures during cutting. In many other applications, the dwell function is useful to control deceleration of the cutting feed on a corner during fast feedrates, for example. This use of dwell could be particularly useful for older control systems. In both cases, the dwell command 'forces' the machining operation to be *fully completed* in one block, *before* the next block can be executed. The programmer still has to supply the exact period of time required for the pause. This time has to be sufficient - neither too short nor too long.

> Dwell command is always completed
> before the next operation begins.

◆ Applications for Accessories

The second common application of the dwell command is after certain miscellaneous functions - M functions. Several such functions are used to control a variety of CNC machine accessories, such as a barfeeder, tailstock, quill, part catcher, custom features, and others. The programmed dwell time will allow the full completion of a certain procedure, such as the operation of a tailstock. The machine spindle may be either stationary or rotating in these cases. Since there will be no contact of the cutting tool with part material in this category, it is *not* important whether the machine spindle rotates or not.

On some CNC machines, the dwell command may also be required when changing spindle speed, usually after a gear range change. This is used mainly on CNC lathes. In these cases, the best guidance as to *how* and *when* to program a dwell time is to follow the recommendation of the CNC machine manufacturer. Typical examples of a dwell used for lathe accessories are described in *Chapter 44*, covering the subject.

DWELL COMMAND

G04	Dwell command

The common preparatory command for dwell is G04. Like other G commands, G04 used by itself only will do nothing. It must always be used with another address, in this case specifying the amount of time to dwell (pause). The correct addresses for dwell are X, P or U (address U can only be used for a CNC lathe). The actual time duration specified by the selected address is either in *milliseconds*, or in *seconds,* depending on the selected address. Some control systems use a different address for programming dwell altogether, but the main purpose as well as the programming methods remain identical.

Some fixed cycles for machining centers also use dwell. This dwell is programmed together with the cycle data, not in a separate block. Only fixed cycles that require a dwell time can use it in the same block. For all other applications, the dwell command must be programmed as an *independent block*. It will remain active for that block only and does not carry over to the next block. Dwell is a only one block function and is *not* modal. During dwell execution, the current status of program processing is unchanged, but the overall cycle time will be affected.

◆ Dwell Command Structure

The structure - or format - for the dwell function is:

```
X5.3        ... All machines, excluding fixed cycles
U5.3        ... Lathes only
P53         ... All machines, including fixed cycles
```

In any case, the typical representation is five digits *before* and three digits *after* the decimal point, although that may vary on different control systems.

Since milliseconds or seconds can be used as units of dwell, the relationship can be established:

1s = 1000ms
1ms = 0.001s

☞ where ...

```
s  =  second
ms =  millisecond
```

Examples of practical application of the dwell format are:

```
G04 X2.0     ... preferred for long dwells
G04 P2000    ... preferred for short or medium dwells
G04 U2.0     ... lathe only - in seconds
```

In this example, the dwell is 2 seconds or 2000 milliseconds. All three formats are shown. The next example is similar:

```
G04 X0.5
G04 P500
G04 U0.5
```

This example illustrates a dwell of 500 milliseconds, or one half of a second. Again, all three formats are shown.

In a CNC program, the dwell function may appear in the following way - note the dwell as a separate block:

```
N21 G01 Z-1.5 F12.0
N22 G04 X0.3          (DWELL COMMAND 0.3 SEC)
N23 Z-2.7 F8.0
```

Programs using X or U addresses may cause a possible confusion, particularly to new programmers. The X and U addresses may incorrectly be interpreted as an axis motion. *This will never be the case.* By definition, the X axis and its lathe application, the U axis, is the *dwelling* axis. X axis is the only axis common to all CNC machines.

No axis motion will take place when the X, P or U address is used with the dwell command G04

The control unit interprets such a command as a dwell, *not as a axis motion.* This is because of the presence of the preparatory command G04, which establishes *meaning* of the address that follows it. If using the X or U address for dwell does not feel comfortable, use the third alternative - the address P. Keep in mind, that the address P does *not* accept the decimal point, so the dwell is programmed directly as the *number of milliseconds* to control the pause duration. One millisecond is 1/1000th of a second, therefore one second is equivalent to 1000 milliseconds.

The addresses X and U can also be programmed in milliseconds, without a decimal point - for example,

```
G04 X2.0     is equal to     G04 X2000
```

Leading zero suppression is assumed in the format without the decimal point (trailing zeros are required):

```
P1    =   P0001    ... 1 millisecond
P10   =   P0010    ... 10 milliseconds
P100  =   P0100    ... 100 milliseconds
```

Depending on the programming format for dwell, the range of programmable time varies. For the format using five digits in front of a decimal point and three digits following it, the range is between 0.001 of a second and up to 99999.999 seconds. That presents a range from the minimum of 1/1000th of a second, up to *27 hours, 46 minutes and 39.999 seconds.*

Dwell programming applications are identical to both machining centers and lathes, but the U address can only be used in lathe programs. The selection of either metric or English dimensional units has no effect on the dwell function whatsoever, as time is not dimensional.

DWELL TIME SELECTION

Seldom ever the dwell time will exceed more than just a few seconds, most often much less than only one second. Dwell is always a *nonproductive time* and it should be selected as the shortest time needed to accomplish the required action. The time delay for completion of a particular machine operation or a special machine accessory is usually recommended by the machine manufacturer. Selecting dwell time for cutting purposes is always programmer's responsibility. Unfortunately, some programmers often overprogram the dwell duration. After all, one second seems like a very short time, but think about this example:

In one block of the program, a dwell function is assigned for the duration of one second. The spindle speed is set to 480 r/min and the dwell is applied at 50 locations on the part, perhaps during a spot face operation. That means the cycle time for *each* part is 50 seconds longer *with* the dwell, then it would be *without* the dwell. Fifty seconds may not seem too unreasonable, but are they really necessary? Give it a little thought or - even better - calculate it. If the dwell must be used at all, make sure to calculate the *minimum* dwell that can do the job. It is easy to select the dwell arbitrarily, by guessing and without much thinking. In the example, the minimum dwell required is only 0.125 seconds:

```
60 / 480 = 0.125
```

This minimum dwell is *eight times less* than the programmed dwell of one second. If the minimum dwell is used rather than the estimated dwell, the cycle time will increase by only 6.25 seconds, rather than the original 50 seconds - a significant improvement in programming efficiency and productivity improvement on the machine.

Minimum dwell calculation and other issues related to it are described shortly.

SETTING MODE AND DWELL

Most programs for machining centers will use feedrate per time (programmed in inches per minute - *in/min* - or millimeters per minute - *mm/min*). Lathe applications are normally programmed in feedrate per revolution, as inches per revolution - *in/rev* - or millimeters per revolution - *mm/rev*. On many Fanuc controls, a parameter setting allows programming a dwell in either the elapsed time in seconds or milliseconds - or the *number of spindle revolutions*. Each application has its practical uses and benefits. Depending on the system parameter setting, the dwell command will assume a different meaning with each setting:

◆ Time Setting

This is the common and normal default setting for virtually all CNC units. For the *elapsed time* setting, the dwell is always programmed in seconds or milliseconds, within the range allowed by the control unit, for example, from 0.001 to 99999.999 seconds, a typical range for Fanuc and many similar control systems:

```
G04 P1000
```

... represents the dwell of one second, equivalent to 1000 milliseconds.

◆ Number of Revolutions Setting

For the number of *spindle revolutions* setting, the dwell is expressed as the number of times the spindle rotates, within the range of 0.001 to 99999.999 revolutions, for example:

```
G04 P1000
```

... represents the dwell for the duration of one revolution of the spindle.

MINIMUM DWELL

During a cut, that is for operations where cutting tool is in contact with the machined part, the minimum dwell definition is important, but the setting mode is unimportant (time or number of revolutions).

> Minimum dwell is the time required
> to complete one revolution of the spindle.

Minimum dwell, programmed in seconds, can be calculated, using a simple formula:

$$\text{Minimum dwell (sec)} = \frac{60}{r / min}$$

⬥ Example :

To calculate minimum dwell in seconds for spindle rotation of 420 r/min, divide the *r/min* into sixty (there are 60 seconds in one minute):

```
60 / 420 = 0.143  seconds dwell
```

The format selection of dwell block in the program will vary, depending on the machine type used and a particular programming style. All following examples represent the same dwell time of 0.143 of a second:

```
G04 X0.143
G04 P143
G04 U0.143
```

Regardless which format is used, all dwell values in the example specify dwell time of 143 milliseconds, which is 0.143 of a second. It is allowed to mix different formats in one program, but such a practice does not represent consistent programming style.

For *practical* dwell applications in a CNC program, the calculated minimum dwell is only mathematically correct and may *not* be the most practical value to use. It is always better to round off the calculated value of the minimum dwell slightly *upwards*. For example, the G04 X0.143 may become G04 X0.2, or - if a double value is used - then G04 X0.143 will become G04 X0.286, or even G04 X0.3 to round off the value.

The reasoning for this adjustment takes into consideration some machining realities. It is quite normal that the CNC operator may be running a certain job with the spindle speed in an override mode, perhaps even set at its lowest setting, at 50%. Since 50% spindle speed override is the usual minimum on most CNC controls, the *double* minimum dwell will *always guarantee at least one complete revolution*, without the loss of production time.

NUMBER OF REVOLUTIONS

In the other dwell mode (selected by a system parameter), the programming format will only *appear* to be the same, but its meaning will be much different. In some applications, it may be desirable to program a delay for a certain *number of spindle revolutions,* rather than duration for a specific time.

In a lathe grooving application, for example, a tool motion programmed to the bottom of a groove will require the grooving tool to remain in that position for a period of time to clean up the groove bottom. Of course, calculating the time in seconds (see below) would solve the problem, but an alternative approach could be very attractive. Many controls offer programming the required number of spindle revolutions directly, providing a system parameter is set. For example, to dwell as long as it takes to complete *three spindle revolutions* can be programmed directly, regardless of the spindle r/min.

◆ System Setting

If the control system is set to accept the dwell as the *number of spindle revolutions*, rather than as time in seconds or milliseconds, the programming is very straightforward. All that is needed is to call the dwell command G04, followed by the number of required revolutions:

```
G04 X3.0
G04 P3000
G04 U3.0
```

Each format represents the same result - a dwell in the duration of three spindle revolutions. How can we tell from the program whether the value means time or revolutions? We cannot. We *have to know* the control settings. The only clue may be the rather large input values of the dwell input. 3.0 revolutions are usually much shorter than 3.0 seconds of dwell. Note that the decimal point is still written, to allow fractions of a revolution, such as one half or one quarter of a revolution, for example.

◆ Time Equivalent

The two modes cannot be mixed in one program deliberately and even between programs, the mix is difficult. The CNC system parameter can be set to only one dwell mode at a time. Since control parameters are normally set for the dwell in seconds or milliseconds, rather than the dwell expressed by the number of spindle revolutions, the equivalent time must be calculated. The spindle speed (in *r/min*) must always be known in such a case.

The dwell time in seconds can be calculated to be equal to the required number of spindle revolutions. Use the following formula:

$$\text{Dwell}_{sec} = \frac{60 \times n}{r / min}$$

☞ where . . .

$$
\begin{array}{rcl}
60 & = & \text{Number of minutes (translation factor)} \\
n & = & \text{Required number of spindle revolutions} \\
r/min & = & \text{Current spindle speed (revolutions per minute)}
\end{array}
$$

➲ Example :

To calculate the dwell time in seconds for *full three* spindle revolutions, at spindle speed of 420 r/min, the formula can be applied:

```
Dwellsec = 60 × 3 / 420 = 0.429
```

The program block representing the required three spindle revolutions in terms of dwell *time* will take one of the following forms:

```
G04 X0.429
G04 P429
G04 U0.429
```

It may also be a good idea to work backwards and calculate the *equivalent of dwell time*, represented as the number of spindle revolutions. Usually, the result will not be an integer number and will require rounding to the nearest value upwards. The above formula can be easily reversed:

$$\text{Dwell}_{rev} = \frac{r / min \times \text{Dwell}_{sec}}{60}$$

➲ Example :

To confirm that the formula is correct, use the value of 0.429 of the previous example and calculate the number of revolutions for a dwell of 0.429 seconds at 420 r/min:

```
Dwellrev = 420 × 0.429 / 60 = 3.003 revolutions
```

The result confirms the formula is correct. It is more than likely, that the calculation will start with a dwell that is already rounded, for example, to one half of a second:

```
Dwellrev = 420 × 0.5 / 60 = 3.5 revolutions
```

The dwell time based on a required number of revolutions is mostly used for CNC lathe applications, especially when cutting with very slow spindle speeds. A slow spindle rotation does not have the latitude of faster speeds and does not allow for a large error in the dwell calculation. Keep in mind that the goal is to get *at least one complete* part rotation in order to achieve desired machining results. Otherwise, why program dwell at all? Consider another example:

Dwell is programmed for *one half of a second* duration, with the spindle rotation set to 80 r/min. The number of revolutions for one half of a second will be:

```
80 × 0.5 / 60 = 0.6666667
```

which is *less* than one complete spindle revolution. The reason for programming the dwell function in the first place is not honored here and the dwell time has to be increased. Dwell of 0.5 seconds is therefore not sufficient. The minimum dwell has to be *calculated*, using the formula presented earlier:

```
60 × 1 / 80 = 0.75 seconds
```

Generally, there is not much use for this type of calculations - most programming assignments can be handled very well with the standard *dwell per time* calculations.

LONG DWELL TIME

For machining purposes on CNC machines, an unusually long dwell time is neither required nor necessary. Does that mean long dwell times are not needed at all?

A long dwell time is the programmed time that is well *above the established average* for most normal applications. Seldom ever there is a need to program dwell time during a part machining in excess of one, two, three, or four seconds. The large range available on the control system *(over 27 hours)* is more important to the *maintenance personnel*, than to the CNC programmer. As an example of a typical application when a long dwell may be beneficial, is a program developed by the maintenance technicians for testing the spindle functionality.

Consider carefully the following actual situation common to machine service - a spindle of the CNC machine has been repaired and must be tested before the machine can be released back to production. The testing will mainly consist of running the spindle at various speeds, for a certain period of time of each speed selection.

In a typical example, the maintenance department requires a small CNC program, in which the machine spindle will rotate for 10 minutes at 100 r/min, then for another 20 minutes at 500 r/min, followed by the spindle rotation at the highest rate of 1500 r/min for additional 30 minutes. The program development is not an absolute necessity, since the maintenance technician may do the test by manual methods. The manual approach will not be very efficient but it will still serve the purpose of the maintenance test.

A better choice in these cases is to *store* the testing procedure as a small program, directly into the CNC memory. The maintenance (service) program will be a little different for machining centers than for lathes but the main objectives will remain the same.

⮕ Example - Machining Centers - Spindle test :

```
S100 M03              (100 R/MIN INITIAL SPEED)
G04 X600.0            (600 SECONDS IS 10 MINUTES)
S500              (SPEED INCREASED TO 500 R/MIN)
G04 X1200.0         (1200 SECONDS IS 20 MINUTES)
S1500            (SPEED INCREASED TO 1500 R/MIN)
G04 X1800.0         (1800 SECONDS IS 30 MINUTES)
M05                          (SPINDLE STOP)
```

The example for machining centers starts with the initial spindle rotation of 100 r/min. That selection is followed by the dwell of 600 seconds, which guarantees a 10 minute constant run. The spindle speed is then increased to 500 r/min and the dwell time to 1200 seconds, for another 20 minutes. The last selection is 1500 r/min spindle speed running for 1800 seconds, or another 30 minutes, before the spindle finally stops.

⮕ Example - Lathes - Spindle test :

```
M43                      (GEAR RANGE SELECTION)
G97 S100 M03          (100 R/MIN INITIAL SPEED)
G04 X600.0            (600 SECONDS IS 10 MINUTES)
S500              (SPEED INCREASED TO 500 R/MIN)
G04 X1200.0         (1200 SECONDS IS 20 MINUTES)
S1500            (SPEED INCREASED TO 1500 R/MIN)
G04 X1800.0         (1800 SECONDS IS 30 MINUTES)
M05                          (SPINDLE STOP)
```

This lathe example is very similar to the one for a machining center in the first example. The initial spindle speed setting includes gear range selection, for example, M43. The spindle rotation has been set to 100 r/min. The dwell of 600 seconds follows, leaving the spindle rotating for full 10 minutes. Then the speed is increased to 500 r/min and remains that way for another 20 minutes (1200 seconds). Before the spindle is stopped, one more change is done - the spindle speed increases to 1500 r/min and remains at that speed for another 30 minutes (1800 seconds).

> Maintain all safety rules when using long dwell times !

◆ **Machine Warm-Up**

A similar program (typically a subprogram) that uses a long dwell time is favored by many CNC programmers and CNC machine operators, to *'warm-up'* the machine before running a critical job. This machine warming activity takes place typically at the start of a morning shift during winter months or in a cold shop. This programming approach allows the machine to reach a certain ambient temperature before any precision components are machined. The same approach can also be used to gradually reach the maximum spindle speed for high-speed machining (5000 r/min and up). As usually, all safety considerations must have a high priority in all cases.

◆ X Axis is the Dwelling Axis

The control display screen shows how much time is still left before the dwell time expires. This can be viewed by looking at the X display of the *Distance-To-Go* indicator on the POS (position) screen of a typical Fanuc control. This display will always be displayed as X, since X axis is the only dwelling axis, regardless of control system, even if addresses P or U are programmed. Why the X axis has been selected as the dwelling axis and not any other axis? There is a simple reason - because *the X axis is the only axis common to all CNC machine tools - i.e.,* drilling machines, mills, machining centers, flame cutters, waterjets, lasers, and so on. They all use XYZ axes. Lathes use XZ axes (there is no Y axis) and wire EDM uses XY axes (there is no Z axis). Other machines are similar.

◆ Safety and Dwell

Several safety reminders have been mentioned already. Always exercise a great degree of caution when using programs with long dwell times, particularly for service or maintenance purposes. The CNC machine should *never* be left completely unattended. In case of long times needed for testing, proper warning signs should be prominently posted to prevent a potentially unsafe situation. If the signs are not practical, someone else should supervise the machine serviced.

Still on the subject of safety - the dwell function should never be programmed for the single purpose of allowing the machine operator a certain time to perform certain manual operation during program processing. Manual jobs such as polishing, filing, deburring, part reversal, tool or insert change, chip removal, inspection, lubrication, etc., must always be done - if absolutely necessary during program execution - as a manual operation, *never under the program control!*

> Never use dwell to perform manual operations on the machine !

FIXED CYCLES AND DWELL

Chapter 25 of this handbook covers the subject of fixed cycles for CNC machining centers and drills in a significant detail. In-depth descriptions of all cycles can be found in this chapter. For the purpose of the current topic, here are just some comments relative to the subject of dwell, this time, as the dwell relates to fixed cycles.

Several fixed cycles can be programmed with a dwell:

❏   Normally, cycles G76, G82, G88, G89

❏   Also cycles G74 and G84, only by parameter setting

The dwell address in fixed cycles is always P, to avoid duplication of the X address in the same block. The address U and the command G04 are *never* programmed in a fixed cycle - the dwell function is 'built' into all fixed cycles that allow the dwell parameter (technically all cycles do). The rules for calculating the dwell time remain the same for fixed cycles, as for any other machining application.

➲ Example :

```
N9 G82 X1.2 Y0.6 R0.2 Z-0.7 P300 F12.0
```

In the example, dwell of 300 milliseconds - 0.3 seconds - is specified by the address P. The dwell will become effective upon completion of the motion along the Z axis (actual cutting motion), but *before* the rapid return motion.

If a G04 P.. is programmed as a separate block in a fixed cycle mode, for example between the G82 block and the G80 block, *no cycle* will be executed in that block and the value of P in the fixed cycle definition is *not* updated. On the latest controls, a system parameter setting enables or disables this usage. If this method is used, the command G04 P.. will be active *before* the tool rapid motion from the location just completed. The dwell function will always be executed while the cutting tool is out of a hole, in the clear space. *This feature is seldom required.*

25
FIXED CYCLES

Machining holes is probably the most common operation, mainly done on CNC milling machines and machining centers. Even in the industries traditionally known for their complex parts, such as aircraft and aerospace components manufacturing, electronics, instrumentation, optical or mold making industries, machining holes is a vital part of the manufacturing process.

When we think of what machining holes means, we probably think first of such operations as center drilling, spot drilling and standard drilling, using common tools. However, this category is much wider. Other related operations also belong to the category of machining holes. The standard center drilling, spot drilling and drilling are used together with related operations such as reaming, tapping, single point boring, boring with block tools, countersinking and counterboring, spotfacing and even backboring.

Machining one simple hole may require only one tool but a precise and complex hole may require several tools to be completed. Number of holes required for a given job is important for selection of proper programming approach.

Even holes machined with the same tool may be different. Holes having the same diameter may have a variable depth, they may even be at different depths of the part. If all possible combinations are considered, it is easy to realize that making one hole may be a simple matter, but making a series of many different hole operations in one program requires a well planned and organized approach.

In the majority of programming applications, hole operations offer a great number of similarities from one job to another. Hole machining is a reasonably predictable operation and any operation that is predictable is an ideal subject to be handled very efficiently by a computer. For this reason, virtually all CNC control manufacturers have incorporated several ingenious programming methods for machining holes in their control systems. These methods use so the called *canned cycles* or - more commonly - the *fixed cycles*.

POINT-TO-POINT MACHINING

Machining holes is generally not a very sophisticated procedure. There is no contouring required and there is no multiaxis cutting motion. The only motion when actual cutting takes place is along a single axis - virtually always the Z axis. This type of machining is commonly known as *point-to-point machining*.

The method of point-to-point machining for holes is a method of controlling the motions of a cutting tool in the X and Y axes at a *rapid* machine rate, and in the Z axis at a *cutting* feedrate. Some motions along Z axis may also include rapid motions. All this means is that there is no cutting along XY axes for holes operations. When the cutting tool completes all motions along the Z axis and returns from the hole to the clearance position, motions along the X and Y axes resume and proceed to a new location of the part. There, the Z axis motions are repeated. Usually, this sequence of motions occurs at many locations. The hole shape and diameter is controlled by the cutting tool selection, the cutting depth is controlled by the part program. This method of machining is typical to fixed cycles for drilling, reaming, tapping, boring and related operations.

The elementary programming structure for point-to-point machining can be summed up into four general steps (typical drilling sequence shown in the example):

❏ Step 1: Rapid motion to the hole location
... along the X and/or Y axis

❏ Step 2: Rapid motion to the starting point of the cut
... along the Z axis

❏ Step 3: Feedrate motion to the specified depth
... along the Z axis

❏ Step 4: Return to a clear position
... along the Z axis

These four steps also represent the *minimum* number of blocks required to program a single hole, using manual programming method, *without* using fixed cycles. If there is only one or two holes in a part drawing and the machining operation is nothing more than a simple center drilling or drilling, the program length is of no significant importance. That is not the common case - normally, there are many holes in a part and several tools have to be used to complete each hole to engineering specifications. Such a program could be extremely long and very difficult to interpret and change. In fact, it may even be too long to fit into the standard CNC memory.

Possibly the most time consuming task in programming point-to-point operations is the amount of *repetitive* information that has to be written into the CNC program. This problem was solved by the introduction of fixed cycles. These cycles are also known as *canned cycles* because a lot of repetitive information is *canned* (or fitted) into a relatively small space of a computer chip.

◆ Single Tool Motions vs. Fixed Cycles

The following two examples compare the differences of programming a hole pattern in individual blocks (O2501), where each step of the tool path must be programmed as a single motion block, and the same pattern of holes using a fixed cycle (O2502). No explanations to the programs are given at this stage and the comparison is only a visual illustration between two distinct programming methods. It shows an application of a ∅3/16 standard drill that is used to cut a full blind depth of .625 inches. Only three holes are programmed in the example, illustrated in *Figure 25-1*.

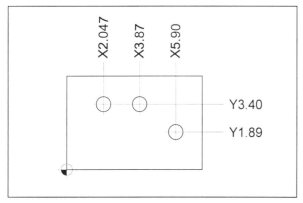

Figure 25-1

Simple hole pattern - programs O2501 and O2502

```
O2501 (EXAMPLE 1)
(PROGRAM USES INDIVIDUAL BLOCKS)
N1 G20
N2 G17 G40 G80
N3 G90 G54 G00 X5.9 Y1.89 S900 M03
N4 G43 Z1.0 H01 M08
N5 Z0.1 M08
N6 G01 Z-0.6813 F4.5
N7 G04 P200
N8 G00 Z0.1
N9 X3.87 Y3.4
N10 G01 Z-0.6813
N11 G04 P200
N12 G00 Z0.1
N13 X2.047
N14 G01 Z-0.6813
N15 G04 P200
N16 G00 Z0.1 M09
N17 G28 X2.047 Y3.4 Z1.0
N18 M30
%
```

The second example in program O2502 uses the same hole pattern, but fixed cycles are used for efficiency.

```
O2502 (EXAMPLE 2)
(PROGRAM USES FIXED CYCLE)
N1 G20
N2 G17 G40 G80
N3 G90 G54 G00 X5.9 Y1.89 S900 M03
N4 G43 Z1.0 H01 M08
```

```
N5 G99 G82 R0.1 Z-0.6813 P200 F4.5
N6 X3.87 Y3.4
N7 X2.047
N8 G80 G28 X2.047 Y3.4 Z1.0 M09
N9 M30
%
```

Program O2501 required the total of 18 blocks, even for three holes only. In program O2502, using fixed cycles, only nine blocks were needed. The shorter program O2502 is also easier to read, there are no repetitious blocks. The program modifications, updates and other changes can be done much easier, whenever required. Always use fixed cycles for machining holes, even if a single hole is machined.

FIXED CYCLE SELECTION

The fixed cycles have been designed by control manufacturers to eliminate the *repetition* in manual programming and allow an easy *program data changes* at the machine.

For example, a number of identical holes may share the same starting point, the same depth, the same feedrate, the same dwell, etc. Only the X and Y axes locations are different for each hole of the pattern. The purpose of the fixed cycles is to allow for programming necessary values only once - *for the first hole of the pattern*. The specified values become modal for the duration of the cycle and do not have to be repeated, unless and until one or more of them change. This change is usually for the XY location of a new hole, but other values may be changed for any hole at any time, particularly for more complex holes.

A fixed cycle is called in the program by a special preparatory G command. Fanuc and similar control systems support the following fixed cycles:

G73	High speed peck drilling cycle
G74	Left-hand tapping cycle
G76	Precision boring cycle
G80	Fixed cycle cancellation (any cycle)
G81	Drilling cycle
G82	Drilling cycle with dwell
G83	Peck drilling cycle
G84	Right-hand tapping cycle
G85	Boring cycle
G86	Boring cycle
G87	Back boring cycle
G88	Boring cycle
G89	Boring cycle

The list is only general and indicates the most *common* use of each cycle, not always the *only* use. For example, certain boring cycles may be quite suitable for reaming, although there is no reaming cycle directly specified. The next section describes programming format and details of each cycle and offers suggestions for their proper applications. Think of fixed cycles in terms of their *built-in* capabilities, not their general description.

PROGRAMMING FORMAT

General format for a fixed cycle is a series of parameter values specified by a unique address (not all parameters are available for every available cycle):

> N.. G.. G.. X.. Y.. R.. Z.. P.. Q.. I.. J.. F.. L.. (or K..)

Explanation of the addresses used in fixed cycles (in the order of the usual block appearance):

> ### N = Block number

Within the range of N1 to N9999 or N1 to N99999, depending on the control system

> ### G (first G command) = G98 or G99

❑ G98 returns tool to the initial Z position

❑ G99 returns tool to the point specified by the address R

> ### G (second G command) = Cycle number

❑ Only one of the following G commands can be selected:

G73   G74   G76   G81   G82   G83
G84   G85   G86   G87   G88   G89

> ### X = Hole position in X axis

X value can be an absolute or incremental value

> ### Y = Hole position in Y axis

Y value can be an absolute or incremental value

> ### R = Z axis start position = R level

❑ Position at which the cutting feedrate is activated

The R level position can have an absolute value or an incremental value.

> ### Z = Z axis end position = Z depth

❑ Position at which the feedrate ends

The Z depth position can have an absolute value or an incremental value.

> ### P = Dwell time

❑ Programmed in milliseconds (1 second = 1000 ms)

The dwell time is practically applicable only to G76, G82, G88 and G89 fixed cycles. It may also apply to G74, G84 and other fixed cycles, depending on the control parameter setting.

❑ Dwell time can be in the range of 0.001 to 99999.999 seconds, programmed as P1 to P99999999

> ### Q = Address Q has two meanings

❑ When used with cycles G73 or G83, it means a depth of each peck

❑ When used with cycles G76 or G87, it means the amount of shift for boring

The addresses I and J may be used instead of address Q, depending on the control parameter setting.

> ### I = Shift amount

❑ Must include the X axis shift direction for boring cycles G76 or G87

The I shift may be used instead of Q - see above.

> ### J = Shift amount

❑ Must include the Y axis shift direction for boring cycles G76 or G87

The J shift may be used instead of the Q - see above.

> ### F = Feedrate specification

❑ Applies to the cutting motion only

This value is expressed in *in/min* or *mm/min*, depending on the dimensional input selection.

> ### L (or K) = Number of cycle repetitions

❑ Must be within the range of L0 - L9999 (K0 - K9999) L1 (K1) is the default condition

GENERAL RULES

Programming is a very controlled discipline - it means there are rules, there are strict conditions, there are limitations, and there are restrictions. CNC programming is not a language programming but shares a lot with it. We talk about a *Fanuc* or *Siemens* programming, a *Cincinnati* programming, a *Mitsubishi* or *Mazatrol* programming, for example. Fixed cycles are miniature programs.

Consider fixed cycles as a set of small condensed modules - modules that contain a step-by-step series of *preprogrammed* machining instructions. The cycles are called 'fixed', because their internal format cannot be changed. These program instructions relate to the specific kind of predictable tool motion that repeats from job to job. The basic rules and restrictions relating to fixed cycles can be summed up in the following items:

❏ Absolute or incremental mode of dimensioning can be established *before* a fixed cycle is programmed or anytime *within* the fixed cycle mode.

❏ G90 must be programmed to select the absolute mode, the G91 command is required to select the incremental mode.

❏ Both G90 and G91 modes are modal !

❏ If one of the X and Y axes is omitted in the fixed cycle mode, the cycle will be executed at the specified location of one axis and the current location of the other axis.

❏ If both X and Y axes are omitted in the fixed cycle mode, the cycle will be executed at the current tool position.

❏ If neither G98 nor G99 command is programmed for a fixed cycle, the control system will select the default command as set by a system parameter (usually the G98 command).

❏ Address P for the dwell time designation cannot use a decimal point (G04 is not used) - dwell is always programmed in milliseconds.

❏ If L0 is programmed in a fixed cycle block, the control system will store the data of the block for a later use, but will not execute them at the current coordinate location.

❏ The command G80 will always cancel any active fixed cycle and will cause a rapid motion for any subsequent tool motion command. No fixed cycle will be processed in a block containing G80.

➲ Example :

```
G80 Z1.125      is the same as

G80 G00 Z1.125      or

G00 Z1.125
```

Preparatory G codes of the Group 01, namely G00, G01, G02, G03 and G32 commands, are the main motion commands and will also cancel any active fixed cycle.

Caution: In case of combining a fixed cycle command and a motion command of *Group 01* in the same block, the *order* of programming those commands is very important:

```
G00 G81 X.. Y.. R.. Z.. P.. Q.. L.. F..
```

fixed cycle is processed, while in

```
G81 G00 X.. Y.. R.. Z.. P.. Q.. L.. F..
```

fixed cycle is *not* processed, but the X and Y motions will be performed; other values will be ignored, with the exception of the F feedrate value, which is stored. *Avoid such situations at all costs!*

In this chapter, the individual fixed cycles are described in detail and each cycle has an illustration of its structure. The illustrations use shorthand graphic symbols, each with a specific meaning. In *Figure 25-2*, the meaning of all symbols used in the illustrations is described.

 – – – – ⇀	Rapid motion and direction
 ——▶	Cutting motion and direction
 ∿∿∿∿▶	Manual motion and direction
 ➡	Boring bar shift and direction
 ●—	Programmed coordinate
Q	Shift value / depth of peck
d	Clearance value
CW / CCW	Spindle rotation
OSS	Oriented spindle stop
DWELL	Dwell function executed

Figure 25-2
Symbols and abbreviations used in fixed cycles illustrations

ABSOLUTE AND INCREMENTAL VALUES

Like all machining processes, hole machining uses either the absolute programming mode G90 or the incremental mode G91 for programming fixed cycles. The selection will mainly affect the XY position of the hole, the R level and the Z depth - *Figure 25-3*.

In the absolute method of programming, all values are related to the point of origin - the *program zero*. In the incremental method, the XY position of one hole is the distance from the XY position of the previous hole. The R level is the distance from the *last* Z value, one established before calling the cycle, to the position where the feedrate is activated. The Z depth value is the distance between the R level and the termination of feedrate motion. At the start of any fixed cycle, tool motion to the R level will always be in rapid mode.

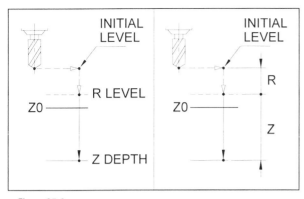

Figure 25-3
Absolute and incremental input values for fixed cycles

INITIAL LEVEL SELECTION

There are two preparatory commands controlling the Z axis tool return (retract) when a fixed cycle is *completed*.

G98	... will cause the cutting tool to retract to the *initial* position = *Z address designation*
G99	... will cause the cutting tool to retract to the *R level* position = *R address designation*

G98 and G99 codes are used for fixed cycles only. Their main function is to bypass obstacles between holes within a machined pattern. Obstacles may include clamps, holding fixtures, protruding sections of the part, unmachined areas, accessories, etc. Without these commands, the cycle would have to be canceled and the tool moved to a safe position. The cycle could then be resumed. With the G98 and G99 commands, such obstacles can be bypassed *without* canceling the fixed cycle, for more efficient programming.

Initial level is, by definition, the absolute value of the *last Z axis* coordinate in the program - before a fixed cycle is called - *Figure 25-4*.

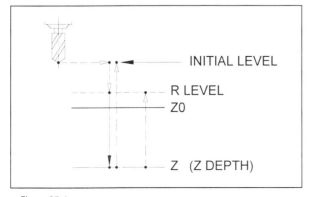

Figure 25-4
Initial level selection for fixed cycles

From the practical point of view, always select this position as the *safe* level - not just anywhere and not without some prior thoughts. It is important that the level to which the tool retracts when G98 command is in effect is *physically above all obstacles*. Use the initial level with other precautions, to prevent a collision of the cutting tool during rapid motions. A collision occurs when the cutting tool is in an undesirable contact with the part, the holding fixture, or the machine itself.

➲ Example of the initial level programming :

The following program segment is a typical example of programming the initial level position:

```
...
N11 G90 G54 G00 X10.0 Y4.5 S1200 M03
N12 G43 Z2.0 H01 M08   (INITIAL LEVEL AT Z2.0)
N13 G98 G81 X10.0 Y4.5 R0.1 Z-0.82 F5.0
N14 ...
...
...
N20 G80
...
```

The fixed cycle (G81 in the example) is called in block N13. The last Z axis value *preceding* this block is programmed in block N12 as Z2.0. This is setting of the initial position - *two inches above* Z0 level of the part. The Z level can be selected at a standard general height, if the programs are consistent, or it may be different from one program to another. Safety is the determining issue here.

Once a fixed cycle is applied, the initial Z level cannot be changed, unless the cycle is canceled first with G80. Then, the initial Z level can be changed and the required cycle be called. The initial Z level is programmed as an absolute value, in the G90 mode.

R LEVEL SELECTION

The cutting tool position from which the feedrate begins is also specified along the Z axis. That means a fixed cycle block requires *two positions* relating to the Z axis - one for the start point at which the cutting begins, and another for the end point indicating the hole depth. Basic programming rules do not allow the same axis to be programmed more than once in a single block. Therefore, some adjustment in the control design must be made to accommodate both Z values required for a fixed cycle. The obvious solution is that one of them must be *replaced with a different address*.

Since the Z axis is closely associated with depth, it retains this meaning in all cycles. The replacement address is used for the tool Z position from which the cutting feedrate is applied. This address uses the letter R. A simplified term of reference to this position is the *R level*. Think of the R level in terms of *'Rapid to start point'*, where the emphasis is on the phrase *'Rapid to'* and the letter *'R'*- see *Figure 25-5*.

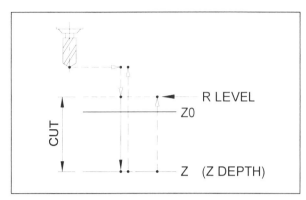

Figure 25-5
R level selection for fixed cycles

Z DEPTH CALCULATIONS

Each fixed cycle must include a depth of cut. This is the depth at which the cutting tool *stops* feeding into the material. Depth is programmed by the Z address in the cycle block. The end point for the depth cut is programmed as a Z value, normally *lower* than the R level and the *initial* level. Again, the G87 cycle is an exception.

To achieve a program of a high quality, always make a special effort to program the calculated Z depth accurately - *i.e., exactly*, without guessing its value or even rounding it off. It may be tempting to round-off the calculated depth of .6979 to .6980 or even to .70 - *avoid it!* It is not a question of triviality or whether one can get away with it. It is a matter of principle and programming consistency. With this approach and attitude, it will be so much easier to retrace the cause of a problem, should one develop later.

Z depth calculation is based on the following criteria:

❑   Dimension of the hole in the drawing (diameter and depth)

❑   Absolute or incremental programming method

❑   Type of cutting tool used + Added tool point length

❑   Material thickness or full diameter depth of the hole

❑   Selected clearances - above and below material (below material clearance for through holes)

On vertical machining centers, the Z0 is typically programmed as the top of finished part face. In this case, the absolute value of Z address will always be programmed as a *negative* value. Recall that the *absence* of a sign in an axis address means a positive value of that address. This method has one strong advantage. In case the programmer forgets to write the minus sign, the depth value will automatically become a positive value. In that case, the cutting tool will move *away* from the part, generally into a safe area. The part program will not be correct, but can be easily corrected, with only a loss of time.

The R level is not only the start point of cutting feedrate, it is also the Z level to which the cutting tool will retract upon cycle completion, if preparatory command G99 was programmed. If G98 was programmed, the retract will be to the initial level. Later, the G87 backboring cycle will be described as an exception, due to its purpose. This cycle does not use G99 retract mode, only G98! However, for all cycles, the R level value must be selected carefully. The most common values are .04-.20 of an inch (1-5 mm) above the part Z0. Part setup has to be considered as well, and adjustments to the setting made, if necessary.

The R level usually increases about three or four times for tapping operations using cycles G74 and G84, to allow the feedrate acceleration to reach its maximum.

➡ Example of R level programming :

```
...
N29 G90 G00 G54 X6.7 Y8.0 S850 M03
N30 G43 Z1.0 H04 M08    (INITIAL LEVEL IS 1.0)
N31 G99 G85 R0.1 Z-1.6 F9.0  ® LEVEL IS 0.1)
N32 ...
...
...
N45 G80
```

The initial level in the example is in block N30, set to Z1.0. The R level is set in block N31 (cycle call block) as .100 inches. In the same block, the G99 command is programmed and never changed during the cycle. That means the tool position will be .100 above part zero at the start *and* end of the cycle. When the tool moves from one hole to the next, it moves along the XY axes only at this *Z height* level of .100 above work.

The R level position is normally *lower* than the initial level position. If these two levels coincide, the start and end points are equivalent to the initial position. The R level is commonly programmed as an absolute value, in the G90 mode, but changed into an incremental mode G91, if application benefits from such a change.

➡ Example of Z depth calculation :

To illustrate a practical example of Z depth calculation, consider the hole detail in *Figure 25-6*. We will use a ∅.75 inch drill to make a hole, with a full depth of 2.25 inches. If a standard twist drill is used, the tool tip has to be taken into consideration. Its design has a typical 118° to 120° point angle and we have to *add* an additional .225 inches to the specified depth:

```
.3 × .75 = .225
2.25 + .225 = 2.475
```

Based on the result, the total Z depth of 2.475 inches can be written in the CNC program:

```
G99 G83  X9.0 Y-4.0 R0.1 Z-2.475 Q1.125 F12.0
```

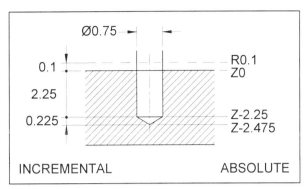

Figure 25-6

Z depth calculation for a drilling fixed cycle

A peck drilling cycle G83 is used in the example for best machining, although the R and Z values would be the same for G81, G82 or G73 cycles. The tool point length calculation is described in detail in *Chapter 26*.

DESCRIPTION OF FIXED CYCLES

In order to understand how each fixed cycle works, it is important to understand the internal structure of each cycle and details of its programming format. In the following descriptions, each fixed cycle will be evaluated in detail. The cycle heading indicates the basic programming format of the cycle, followed by the explanation of the exact operational sequences. Common applications of each cycle will also be described.

All these details are important and should be a help in understanding the nature of each cycle, as well as which cycle to select for the best machining results. As a bonus, the knowledge of the internal cycle structure will help in designing any unique cycles, particularly in the area of custom macro programming.

◆ G81 - Drilling Cycle

G98 (G99) G81 X.. Y.. R.. Z.. F..	
Step	Description of G81 Cycle
1	Rapid motion to XY position
2	Rapid motion to *R level*
3	Feedrate motion to *Z depth*
4	Rapid retract to *initial level* (with G98) *or* Rapid retract to *R level* (with G99)

WHEN TO USE G81 CYCLE - Figure 25-7 :

Mainly for drilling and center drilling, where a dwell at the Z depth is not required. If used for boring, the G81 cycle will produce a scratch mark on the hole cylinder during retract.

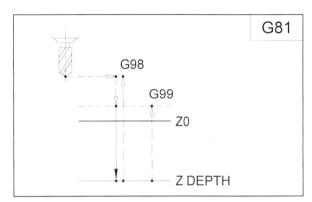

Figure 25-7

G81 fixed cycle - typically used for drilling

◆ G82 - Spot-Drilling Cycle

G98 (G99) G82 X.. Y.. R.. Z.. P.. F..	
Step	Description of G82 cycle
1	Rapid motion to XY position
2	Rapid motion to *R level*
3	Feedrate motion to *Z depth*
4	Dwell at the depth - in milliseconds (P-)
5	Rapid retract to *initial level* (with G98) *or* Rapid retract to *R level* (with G99)

WHEN TO USE G82 CYCLE - Figure 25-8 :

Drilling with a dwell - tool pauses at the hole bottom. Used for center drilling, spot drilling, spotfacing, countersinking, etc. - anytime a smooth finish is required at the bottom of hole. Often used when slow spindle speed needs to be programmed.

If used for boring, the G82 cycle will produce a scratch mark on the hole cylinder during retract.

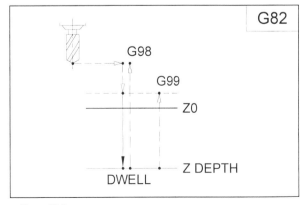

Figure 25-8

G82 fixed cycle - typically used for spot drilling

◆ G83 - Deep Hole Drilling Cycle - Standard

G98 (G99) G83 X.. Y.. R.. Z.. Q.. F..	
Step	Description of G83 cycle
1	Rapid motion to XY position
2	Rapid motion to *R level*
3	Feedrate motion to *Z depth* by the amount of *Q value*
4	Rapid retract to *R level*
5	Rapid motion to the previous depth less a clearance (clearance is set by a system parameter)
6	Items 3, 4, and 5 repeat until the programmed *Z depth* is reached
7	Rapid retract to *initial level* (with G98) *or* Rapid retract to *R level* (with G99)

WHEN TO USE G83 CYCLE - Figure 25-9 :

For deep hole drilling, also known as peck drilling, where the drill has to be retracted above the part (to a clearance position) after drilling to a certain depth. Compare this cycle with the high speed deep hole drilling cycle G73.

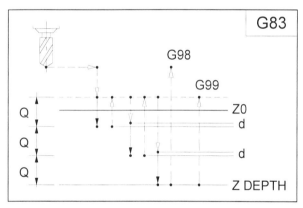

Figure 25-9

G83 fixed cycle - typically used for deep hole drilling (this cycle retracts to R level after each peck)

◆ G73 - Deep Hole Drilling Cycle - High Speed

G98 (G99) G73 X.. Y.. R.. Z.. Q.. F..	
Step	Description of G73 cycle
1	Rapid motion to XY position
2	Rapid motion to *R level*
3	Feedrate motion to *Z depth* by the amount of *Q value*

4	Rapid retract by a clearance value (clearance value is set by a system parameter)
5	Feedrate motion in *Z* axis by the *Q amount* plus clearance
6	Items 4, and 5 repeat until the programmed *Z depth* is reached
7	Rapid retract to *initial level* (with G98) *or* Rapid retract to *R level* (with G99)

WHEN TO USE G73 CYCLE - Figure 25-10 :

For deep hole drilling, also known as peck drilling, where the chip breaking is more important than the full retract of the drill from the hole. The G73 cycle is often used for a long series drills, when a full retract is not very important.

The G73 fixed cycle is slightly faster than the G83 cycle, hence the name 'high speed', because of the time saved by *not* retracting to the R level after each peck. Compare this cycle with the standard deep hole drilling cycle G83.

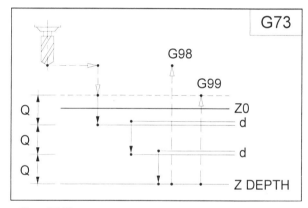

Figure 25-10

G73 fixed cycle - typically used for deep hole drilling (this cycle does not retract to R level after each peck)

Number of pecks calculation

When using cycles G83 and G73 in the program, always have at least a reasonable idea about how many pecks will the tool make in each hole. Unnecessary peck drilling of hundreds or thousands of holes will accumulate the total lost time, which can be can very significant. Try to avoid too many pecks for a single hole. For predictable results, the number of pecks can be *calculated*.

The number of pecks calculation applies equally to both G83 and G73 fixed cycles. Calculation of the number of pecks in cycles G83 and G73 is based on the value of the Q address and the total distance between the R level and Z depth - not from the top of part! Dividing this distance by the programmed Q value will produce a number of pecks the tool will make at each hole location. The number of pecks in a cycle must be an integer and fractional calculations must always be rounded *upwards*:

➲ Example 1 - English data :

```
G90 G98 G83 X.. Y.. R0.1 Z-1.4567 Q0.45 F..
```

In the example, the distance between the R level and Z depth is 1.5567 inches, the Q value is .450, so the number of pecks can be calculated:

```
1.5567 / .45 = 3.4593333
```

The result has too many decimal places and cannot be used as is, because most controls only accept *four* decimal places for English units and *three* decimal places for metric units. The result must be correctly *rounded upwards!*

The nearest higher integer is four, so each hole will require four pecks. The hole depth cannot be changed, so the only other available method to change the number of pecks is to change the R level and/or the depth of each peck. The R level is usually as close to the top face of part as is practical, so there is not much that can be done there. That leaves the Q value, the depth of each peck. By increasing this value, the total number of pecks will be fewer, by decreasing the Q value, the total number of pecks will be higher.

➲ Example 2 - Metric data :

```
G90 G99 G73 X.. Y.. R2.5 Z-42.5 Q15.0 F..
```

In this example, the distance between the *R level* and *Z depth* is exactly 45 mm and the Q value is 15 mm. The number of pecks will be 45 divided by 15, which equals to the exact value of 3. No rounding is necessary and the number of pecks executed per hole will be three.

> In order to *increase* the number of pecks,
> change the current Q value to a *smaller* number.

> In order to *decrease* the number of pecks,
> change the current Q value to a *larger* number.

The Q value setting is more accurate, if it is actually calculated rather than just guessed. To achieve a precise number of pecks, divide the total distance between the R level and Z depth by the required number of pecks. The result will be the Q value programmed for the selected number of pecks. If rounding is necessary, always round off *upwards*, otherwise the number of pecks may increase by one, without receiving any cycle time benefits.

➲ Example 3 - Metric data :

In this example, the distance between the R level and Z depth is 56 mm, and exactly three pecks are required. The calculation of each peck depth is simple:

```
56 / 3 = 18.666667
```

The result of the calculation must be rounded to either 18.667 or 18.666. Although it looks that only *one* micron (0.001 mm) is at stake, it will make a big difference which way the rounding is done. If only three pecks are required, round off *upwards*, to Q18.667:

```
Cut 1    18.667
Cut 2    18.667
Cut 3    18.666
```

Total   56 mm

If the result is rounded downwards, to Q18.666, the number of pecks will be four and practically no cutting will take place during the last peck:

```
Cut 1    18.666
Cut 2    18.666
Cut 3    18.666
Cut 4     0.002
```

Total   56 mm

➲ Example 4 - English data :

In this example, the distance between the R level and Z depth is 2.5 inches and four pecks are required:

```
Q = 2.5 / 4 = .625
```

In this case, no rounding is necessary, and Q0.625 will result in *exactly* four pecks, each of identical depth.

The peck drilling value of Q setting cannot be changed for a single hole - all pecks in a hole will have an equal length, with the possible exception of the last peck. If the last peck amount is greater than the remaining distance to the programmed Z *depth*, only that distance will be drilled.

> No peck depth will ever exceed the Z *depth* coordinate position.

The programmed Q value can be manipulated in any creative way. By changing the Q value skillfully, particular results can be achieved, such as an exact position of the tool tip during material penetration. This method is described in detail in *Chapter 26*.

To determine the 'best' depth of peck, consider the overall operating conditions for the job. The setup rigidity, the part fixturing, the design of cutting tool, the machinability of material and other factors contribute to what the cutting tool can withstand.

The goal in peck drilling is to make an efficient part program under safe conditions. That means programming the deepest Q amount that is reasonable and practical for the particular job and its setup. Always jeep in mind that there are two fixed cycles available, the standard G84 and the often neglected G73 cycle.

◆ G84 - Tapping Cycle - Standard

> G98 (G99) G84 X.. Y.. R.. Z.. F..

The sequence of G84 fixed cycle is based on the *normal* initial spindle rotation - specified by M03.

The tap design must be of the *right hand* design for the G84 cycle with M03 spindle rotation in effect.

Step	Description of G84 cycle
1	Rapid motion to XY position
2	Rapid motion to *R level*
3	Feedrate motion to *Z depth*
4	Spindle rotation stop
5	Spindle reverse rotation (M04) and feedrate back to *R level*
6	Spindle rotation stop
7	Spindle rotation normal (M03) and retract to *initial level* (with G98) *or* remain at the *R level* (with G99)

WHEN TO USE G84 CYCLE - Figure 25-11 :

Only for tapping a right hand thread. At the start of cycle, the normal spindle rotation M03 must be in effect.

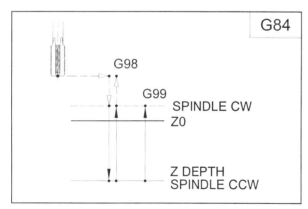

Figure 25-11
G84 fixed cycle - exclusively used for right hand tapping

◆ G74 - Tapping Cycle - Reverse

> G98 (G99) G74 X.. Y.. R.. Z.. F..

The sequence of G74 fixed cycle is based on the *reverse* initial spindle rotation - M04.

The tap design must be of the *left hand* design for the G74 cycle with M04 spindle rotation in effect.

Step	Description of G74 cycle
1	Rapid motion to XY position
2	Rapid motion to *R level*
3	Feedrate motion to *Z depth*
4	Spindle rotation stop
5	Spindle normal rotation (M03) and feedrate back to *R level*
6	Spindle rotation stop
7	Spindle rotation reverse (M04) and retract to *initial level* (with G98) *or* remain at the *R level* (with G99)

WHEN TO USE G74 CYCLE - Figure 25-12 :

Only for tapping a left hand thread. At the start of cycle, the reverse spindle rotation M04 must be in effect.

Chapter 26 describes various techniques of hole machining, including tapping.

The following notes cover only the most important tapping and programming issues and apply equally to both G84 and G74 tapping cycles:

- ❏ R level should be higher in the tapping cycle than in the other cycles to allow for the stabilization of the feedrate, due to acceleration.

- ❏ Feedrate selection for the tap is very important. In tapping, there is a direct relationship between the spindle speed and the lead of the tap - this relationship must be maintained at all times.

- ❏ The override switches on the control panel used for spindle speed and feedrate, are ineffective during G84 or G74 cycle processing.

- ❏ Tapping motion (in or out of the part) will be completed even if the feedhold key is pressed during tapping cycle processing, for safety reasons.

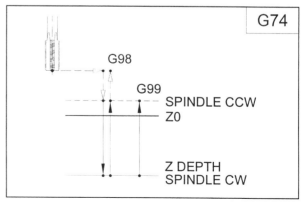

Figure 25-12
G74 fixed cycle - exclusively used for left hand tapping

◆ G85 - Boring Cycle

G98 (G99) G85 X.. Y.. R.. Z.. F..	

Step	Description of G85 cycle
1	Rapid motion to XY position
2	Feedrate motion to *R level*
3	Feedrate motion to *Z depth*
4	Feedrate motion back to *R level*
5	Rapid retract to *initial level* (with G98) *or* remain at *R level* (with G99)

WHEN TO USE G85 CYCLE - Figure 25-13 :

The G85 boring cycle is typically used for boring and reaming operations. This cycle is used in cases where the tool motion *into* and *out of* holes should improve the hole surface finish, its dimensional tolerances and/or its concentricity, roundness, etc. If using G85 cycle for boring, keep in mind that on some parts a tiny amount of stock may be removed while the cutting tool feeds backwards. This physical characteristics is due to the released tool pressure during retract. If the surface finish gets worse rather than improves, try using another boring cycle.

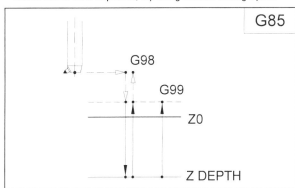

Figure 25-13
G85 fixed cycle - typically used for boring and reaming

◆ G86 - Boring Cycle

G98 (G99) G86 X.. Y.. R.. Z.. F..	

Step	Description of G86 cycle
1	Rapid motion to XY position with spindle on
2	Rapid motion to *R level*
3	Feedrate motion to *Z depth*
4	Spindle rotation stop
5	Rapid retract to *initial level* (with G98) *or* Rapid retract to *R level* (with G99)

WHEN TO USE G86 CYCLE - Figure 25-14 :

For boring rough holes or holes that require additional machining operations. This fixed cycle is very similar to the cycle G81. The difference is the spindle stop at the hole bottom.

NOTE - Although this cycle is somewhat similar to the G81 cycle, it has characteristics of its own. In the standard drilling cycle G81, the tool retracts while the spindle of the machine tool is rotating, but the spindle is stationary in the G86 cycle. Never use the G86 fixed cycle for drilling - for example, to save time - since any deposits of the material on the drill flutes may damage the drilled surface of the part or the drill itself.

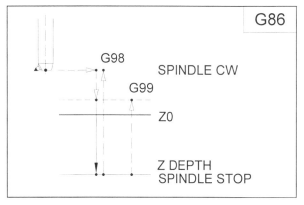

Figure 25-14
G86 fixed cycle - typically used for rough and semifinish boring operations

◆ G87 - Backboring Cycle

There are two programming formats available for the backboring fixed cycle G87 - the first one (using Q) is much more common than the second one (using I and J):

G98 G87 X.. Y.. R.. Z.. Q.. F..	
G98 G87 X.. Y.. R.. Z.. I.. J.. F..	

Step	Description of G87 cycle
1	Rapid motion to XY position
2	Spindle rotation stop
3	Spindle orientation
4	Shift *out* by the Q value *or* shift by the amount and direction of I and J
5	Rapid motion to *R level*
6	Shift *in* by the Q value *or* shift back in the opposite direction of I and J
7	Spindle rotation on (M03)
8	Feedrate motion to *Z depth*

9	Spindle rotation stop
10	Spindle orientation
11	Shift *out* by the Q value *or* shift by the amount and direction of I and J
12	Rapid retract to initial level
13	Shift *in* by the Q value *or* shift back in the opposite direction of I and J
14	Spindle rotation on

WHEN TO USE G87 CYCLE - Figure 25-15 :

This is a special cycle. It can only be used for some (not all) *backboring* operations. Its practical usage is limited, due to the special tooling and setup requirements. Use the G87 cycle only if the total costs can be justified economically. In most cases, reversal of the part in a secondary operation is an option.

NOTE - The boring bar must be set very carefully. It must be preset to match the diameter required for backboring. Its cutting bit must be set in the spindle oriented mode, facing the opposite direction than the shift direction.

G99 is *never* used with the G87 cycle

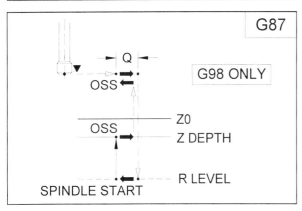

Figure 25-15
G87 fixed cycle - exclusively used for backboring

◆ G88 - Boring Cycle

G98 (G99) G88 X.. Y.. R.. Z.. P.. F..

Step	Description of G88 cycle
1	Rapid motion to XY position
2	Rapid motion to *R level*
3	Feedrate motion to *Z depth*
4	Dwell at the depth - in milliseconds (P..)

5	Spindle rotation stop (feedhold condition is generated and the CNC operator switches to manual operation mode and performs a manual task, then switches back to memory mode). CYCLE START will return to normal cycle
6	Rapid retract to *initial level* (with G98) *or* Rapid retract to *R level* (with G99)
7	Spindle rotation on

WHEN TO USE G88 CYCLE - Figure 25-16 :

The G88 cycle is rare. Its use is limited to boring operations with special tools that require *manual interference* at the *bottom* of a hole. When such a operation is completed, the tool is moved out of the hole for safety reasons. This cycle may be used by some tool manufactures for certain operations.

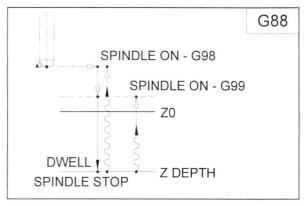

Figure 25-16
G88 fixed cycle - used when manual operation is required

◆ G89 - Boring Cycle

G98 (G99) G89 X.. Y.. R.. Z.. P.. F..

Step	Description of G89 cycle
1	Rapid motion to XY position
2	Rapid motion to *R level*
3	Feedrate motion to *Z depth*
4	Dwell at the depth - in milliseconds (P-)
5	Feedrate motion to *R level*
6	Rapid retract to *initial level* (with G98) *or* remain at *R level* (with G99)

WHEN TO USE G89 CYCLE - Figure 25-17 :

For boring operations, when the feedrate is required for the *in* and the *out* directions of the machined hole, with a specified dwell at the hole bottom. The dwell is the only value that distinguishes the G89 cycle from the G85 cycle.

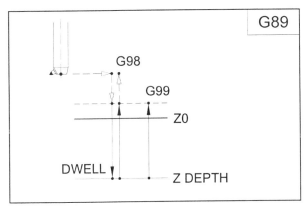

Figure 25-17
G89 fixed cycle - typically used for boring or reaming

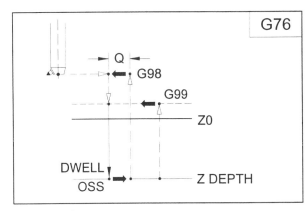

Figure 25-18
G76 fixed cycle - typically used for high quality boring

♦ G76 - Precision Boring Cycle

This is a very useful cycle for high quality holes. There are two programming formats available for the precision boring fixed cycle G76 - the first one (using Q) is much more common than the second one (using I and J):

G98 (G99) G76 X.. Y.. R.. Z.. P.. Q.. F..

G98 (G99) G76 X.. Y.. R.. Z.. P.. I.. J.. F..

Step	Description of G76 cycle
1	Rapid motion to XY position
2	Rapid motion to *R level*
3	Feedrate motion to *Z depth*
4	Dwell at depth - in milliseconds (P-) *(if used)*
5	Spindle rotation stop
6	Spindle orientation
7	Shift *out* by the Q value *or* shift by the amount and direction of I and J
8	Rapid retract to *initial level* (with G98) *or* remain at *R level* (with G99)
9	Shift in by the Q value *or* shift back in the opposite direction of I and J
10	Spindle rotation resumes

WHEN TO USE G76 CYCLE - Figure 25-18 :

Boring operations, usually those for hole finishing, where the quality of the completed hole is *very* important. The quality may be determined by the hole dimensional accuracy, its high surface finish, or both.

The G76 cycle is also used to make holes cylindrical and parallel to their axes.

FIXED CYCLE CANCELLATION

Any fixed cycle that is active can be canceled with the G80 command. The control mode is automatically transferred to a rapid motion mode G00:

```
N34 G80
N35 X5.0 Y-5.75
```

Block N35 does not specify the rapid motion, it only *implies* it. This is a normal programming practice, but specified G00 as well may be a personal choice, although not necessary:

```
N34 G80
N35 G00 X5.0 Y-5.75
```

Both of the examples will produce identical results. The second version of the example may even be a better choice. A combination of the two examples is also a good choice:

```
N34 G80 G00 X5.0 Y-5.75
```

In all three cases, the differences appear rather small, but they are very important to *understanding* the cycles. Although G00 without G80 would also cancel the cycle, it is a poor programming practice that should be avoided.

FIXED CYCLE REPETITION

When a selected fixed cycle is programmed for many holes, this cycle is processed only once at each hole location within a part. This is the normal condition, based on the assumption that most holes require only one cycle per tool. In the CNC program, there is no *self-evident* special command that would indicate how many times to process the fixed cycle. That is true, the command is not evident, but it does exist. In fact, the assumption is that the fixed cycle is to be done just *once - i.e.,* not repeated at all.

Normally, the control system will execute a fixed cycle only once at a given location - it this case, there is no need to program the number of executions, since the system defaults to one automatically. To repeat the fixed cycle several times (more than once), program a special command that *'tells'* the CNC system *how many times* you want the fixed cycle to be executed.

◆ The L or K Address

The command that specifies the number of repetitions (sometimes called *loops*) is programmed with the address L or K for some controls. The L or K address for the fixed cycle repetition is assumed to have a value of one, which is equivalent to a program statement L1 or K1. The L1 or K1 address does not have to be specified in the program.

For example, the fixed cycle call of the following drilling sequence,

```
N33  G90  G99 ...
N34  G81  X17.0  Y20.0  R0.15  Z-2.4  F12.0
N35  X22.0
N36  X27.0
N37  X32.0
N38  G80 ...
```

is equivalent to:

```
N33  G90  G99 ...
N34  G81  X17.0  Y20.0  R0.15  Z-2.4  F12.0  L1  (K1)
N35  X22.0  L1  (K1)
N36  X27.0  L1  (K1)
N37  X32.0  L1  (K1)
N38  G80 ...
```

Both examples will provide the control system with instructions for drilling four holes in a straight row - one at the location of X17.0 Y20.0, the other holes at locations X22.0 Y20.0 and X27.0 Y20.0, and X32.0 Y20.0 respectively - all to the depth of 2.4 inches.

If the L or K value in the second example is increased (or rather added to the first example), for instance, from L1 to L5 (or K1 to K5), the fixed cycle will be repeated five times

at *each* hole location! There is no need for this type of machining. By changing the format only a little, the fixed cycle repetition can be used as a benefit - to make the program more powerful and efficient:

```
N33  G90  G99 ...
N34  G81  X17.0  Y20.0  R0.1  Z-2.4  F12.0
N35  G91  X5.0  L3  (K3)
N36  G90  G80  G00 ...
```

With that change, the advantage of a feature *'hidden'* in the first example is emphasized - the *equal increment* between holes being exactly 5.0 inches. By using the incremental mode, on a temporary basis in block N35 and employing the power of the repetitive count L or K, the CNC program can be shortened dramatically. This method of programming is very efficient for a large number of hole patterns in a single program. A further enhancement is to combine the L or K count with subprograms or macros.

◆ L0 or K0 in a Cycle

In previous discussions, the default for a fixed cycle repetition was specified as L1 or K1, that does not have to be specified in the program. Any L or K value other than L1 or K1 must always be specified, within the allowable range of the L or K address. That range is between L0 and L9999 or K0 and K9999. The lowest L/K word is L0 or K0 - not L1 or K1! Why would we ever program a fixed cycle and then say *'do not do it'*. The address L0 or K0 means exactly that - *'do not execute this cycle'*. The full benefit of the L0/K0 word will be apparent in the examples listed under the section for subprograms, in *Chapter 39*.

By programming the L0 or K0 in a fixed cycle, what we are really saying is not *'do not execute this cycle'*, but *'do not execute the cycle yet, just remember the cycle parameters for future use'*.

For most machining, fixed cycles are quite simple to learn. They do, however, have some complex features, waiting to be discovered and used in an efficient manner - even for a single hole.

MACHINING HOLES

There is a quite good chance that the majority of programs for CNC machining centers include machining of at least one hole, probably more. From a simple spot drill to reaming, tapping and a complex backboring, the field of hole machining is very large. In this chapter, we look at many available programming methods for holes machining, and learn a number of techniques used. Various drilling and boring operations, as well as reaming, tapping and single point boring will be covered.

The most common type of hole machining on CNC machining centers is in the area of drilling, tapping, reaming and single point boring. A typical machining procedure may be to center drill or spot drill a series of holes, then drill them, then tap or bore them. Machining even a single hole will benefit from using the fixed cycles - G81 to G89, G73, G74 and G76, all described in *Chapter 25*.

SINGLE HOLE EVALUATION

Before machining even a single hole on a machine, all required tool paths have to be programmed. Before that, cutting tools have to be selected, speeds and feeds applied, the best setup determined and many other related issues must be resolved. Regardless of the exact approach, always start with a thorough *evaluation of the given hole*.

The first step relates to the drawing data. That will usually define the material to be machined, the hole location and its dimensional values. Holes are often *described*, rather than *dimensioned* and the programmer has to supply the missing details. *Figure 26-1* shows a medium complexity hole that can be machined using a CNC machine.

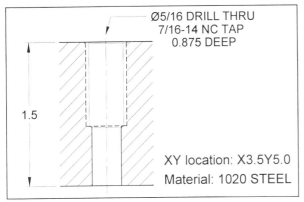

Figure 26-1
Evaluation of a single hole - programming example O2601

All the relevant information is in the drawing, but some searching for details and other requirements is needed. The hole location X3.5Y5.0 was specified in the drawing, as well as the material - mild steel. The Z axis program will be assigned to the top face of part. Drilling and tapping operations are obvious, but is that all there is to know?

How many tools will be needed? What about center drilling to maintain exact location of the hole? Is the spot drill a better choice? What about chamfering the drilled hole for tapping? What about the hole tolerances and surface finish? What about ...?

◆ Tooling Selection and Applications

Based on the drawing information alone, it may seem only two tools will be needed to program this hole. In reality, the implied information must be *interpreted* - it is not the purpose of the drawing to describe *how* to machine the hole - only the hole requirements related to its functionality and purpose. A good CNC machinist will most likely select *four* tools for the best machining results. If four tools are selected, the first tool could be a 90° *spot drill*, followed up by the *tap drill*, then the *through-the-hole drill* and finally, the *tap*. A standard center drill may be used instead of the spot drill, but an additional tool will be required to chamfer the hole diameter at the top. All choices have to be sorted.

For this example, the following four tools are used:

❑ Tool 1 - T01 - 90° spot drill (+ chamfer)
❑ Tool 2 - T02 - Letter U tap drill (∅.368)
❑ Tool 3 - T03 - ∅ 5/16 drill (through the material)
❑ Tool 4 - T04 - 7/16-14 UNC tap

Tool 1 - 90° Spot Drill

The first tool will be a 90° spot drill. Its purpose is dual - it will act as a centering drill and starts up the hole at a highly accurate XY location. A center drill or a spot drill are much more rigid tools than a twist drill and either one will *startup* the hole, so the drill that follows will not deviate from its path (basic hole location and concentricity requirements are guaranteed). The second purpose of the spot drill is its chamfering capabilities. The design of this tool allows a chamfer to be made at the top of the hole, providing the spot drill diameter is *larger* than the chamfer diameter required. In this case, a ∅5/8 spot drill will be used, suitable to chamfer the ∅7/16 hole.

The drawing does not specify a chamfer or its size, but a good machinist will always make a small chamfer, sometimes called a broken corner, unless there is a different requirement. A suitable chamfer will be .015×45°.

Once the spot drill is selected, its cutting depth has to be calculated - yes, *calculated*, not guessed. In order to achieve a .015×45° chamfer for a tap size $\varnothing7/16$ ($\varnothing.4375$), the tap diameter has to be *enlarged* by .015 per side (.03 on diameter), to the .4675 chamfer diameter. *Figure 26-2* shows the relationships of the hole to the tool used (diameters and depth).

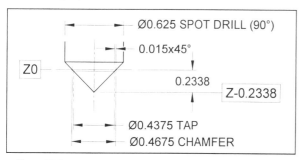

Figure 26-2

Spot drill operation detail - T01 in program O2601

Note, that for a 90° drill, the depth of cut will be *exactly* one half of the chamfer diameter ($\varnothing \times 0.5$):

$$.4675 \ / \ 2$$
or $.4675 \ \times \ .5 \ = \ .23375$
or $Z-0.2338$

Drill point length is discussed later in this chapter.

Tool 2 - Tap Drill

Logically, the second tool will have to be a drill. In the example, *two* drills have to be used for the job - one for the through hole ($\varnothing5/16 = \varnothing.3125$), the other one for the tap (letter U drill = $\varnothing.368$). The question is - which one first? Does it really matter?

It certainly *does* matter which drill is programmed first. The key here is the difference between the two drill diameters. It is a very small difference, only .0555 measured on diameter, in fact. From a machining point of view, it makes sense to use the *larger drill first*, than the smaller drill. The tap drill is larger than the through hole drill, so the T02 will be the tap drill. If the smaller drill is programmed first, the larger drill that follows may produce an inaccurate hole, due to a very small amount of material to remove.

Now comes the question of the first drill size. The *drill* in question is called a *tap drill*. It is the drill that will create a round hole of proper size (*diameter* and *depth*) that can be used for the tap that follows the sequence of operations. Since the machining operation calls for tapping, it makes a

big difference for *what purpose* is the tap used. Not all tapped holes can be done the same way. Some jobs require a loose fit, others a tight fit. The fit for the tap is determined by the size of the tap drill. Most tapping applications fall into the 72-77% full thread depth category. In this case, the T02 (letter U drill) will yield approximately 75% full thread depth. The percentage of the thread depth can be found in catalogues of all tap manufacturers. For example, these are the choices for the 7/16-14 tap:

Drill $\varnothing$	Decimal Value	Full Thread %
T	.3580	86%
23/64	.3594	84%
U	.3680	75%
3/8	.3750	67%
V	.3770	65%

In general terms, for thin material stock, 75 to 80% full thread depth is recommended, for very thin stock even 100%. A thread that has 53% depth will, in most cases, *break* the bolt before it strips it. A full 100% thread is stronger by only 5% than a 75% thread, but the machine *power* required for tapping is three times higher.

The programmed Z depth of the tap drill has to be deep enough to guarantee the required full thread depth of .875. That means the *full* diameter of the drill has to reach a little deeper, for example, to .975 depth. That allows the end chamfer length of the tap to be *below* the full tap depth of .875, specified in the drawing. *Figure 26-3* shows the tap drill values graphically.

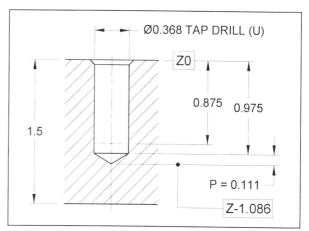

Figure 26-3

Tap drill operation detail - T02 in program O2601

The actual programmed depth for the tap drill will have to take into consideration one more factor - the *drill point length*. The drill or - tool - point length is sometimes abbreviated as TPL or just by the letter P. This chapter contains a table showing various mathematical constants to calculate drill point length - the most common constant uses the drill diameter multiplied by .300, for a 118° drill point angle:

$$P = .368 \times .300 = .1104 = .1110$$

Adding the two calculations (.975+.111), will provide the programmed Z depth of Z-1.086.

Tool 3 - Through Drill

The next tool is a tool that drills the hole through the material. In the example, it is the T03 (tool 3), a ⌀5/16 standard drill.

As for the cutting depth of the through drill, some simple calculations are needed. To do the calculations, the required hole depth has be known, which is 1.5 inches in the example. Then, the calculated drill point length can be added to the required drill depth, usually with an extra clearance.

The calculations for this through drilling operation are illustrated in *Figure 26-4* .

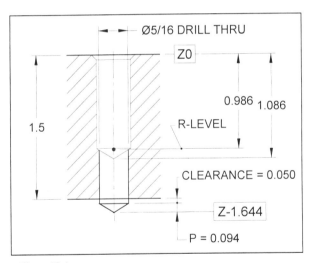

Figure 26-4

Through drill operation detail - T03 in program O2601

First, evaluate the drill point length P. It is calculated from the relationship of two given values - the drill *diameter* and the drill point *angle*. For a standard ⌀5/16 drill (⌀.3125) that has 118° drill point angle, the 0.300 constant is used again, the length of the drill point P is:

```
P = .3125 × .300 = .09375 = .0938 = .094
```

For the through hole in the example, the drawing depth of 1.5 inches *plus* the calculated depth of .094 *seems to be* sufficient to drill the hole using the selected tap drill.

In most through-hole applications, this value will *not* be sufficient - some extra clearance has to be added, applied to the tool penetration (breakthrough), say fifty thousands of an inch (.050). The programmed value for the total drill depth (absolute Z value in the program) is the *sum* of the nominal hole length, plus the tool point angle length, plus the selected clearance. In the program example, amount for the through drill depth will be:

```
Depth = 1.5 + .094 + .05
      = 1.644  or  Z-1.644  in the program
```

One last calculation for this tool still has to be made. Remember that the previous tool had been used to predrill an opening? That means a smaller tool of ⌀.3125 is placed into an existing ⌀.368 hole. The drilling can start from inside of the hole, rather than from a clearance above the part. In the program, the R value is used and selected at R-0.986, which applies .100 clearance above the bottom of the existing hole.

Tool 4 - Tap

There is one more tool left to complete this example. It will be used for tapping the 7/16-14 thread. The thread size as specified in the drawing is 7/16 nominal diameter with 14 threads per inch (1/14 = .0714 = pitch). Anytime a tapping tool is used in the program, watch the programmed depth along the Z axis, particularly in a blind or semi-blind hole. The example shows a semi-blind hole, because the through-hole is smaller than the tapped hole. If there were no through-hole, we would have a blind hole (solid bottom), and if the through-hole were the same size as the tap drill, we would have a 100% through hole.

A through-hole is the most forgiving for the Z depth calculation, closely followed by the semi-through hole. A blind hole has very little latitude, if any, and has to be programmed with a maximum care.

The example drawing for the hole calls for the tap depth of .875 inches. This is the *full* depth of the thread. Full depth of a thread is the actual distance a screw or a nut must travel before stopping (before retract). The programmed depth is, if fact, an *extended* depth, which must be *greater* than the theoretical depth, in order to achieve this goal. To calculate the length of the extended depth, evaluate the tap end chamfer design (its type and length), described in more detail in the tapping section of this chapter.

A reasonable Z depth is Z-0.95 (about one pitch over the depth) and can be optimized after actual machining. This is not really a calculation but an 'intelligent guess' - there is not much else that can be done and extensive experience helps. This completes the section on tooling application for a typical hole and provides enough data to write the actual program. Some of the procedures used in the example will now be explained in more detail.

◆ Program Data

In the example, only one hole is machined. If more holes are needed, they can be added by modifying the following program. For one hole used in the example, the program includes all considerations for the four tools selected earlier. The spindle should be empty at the beginning of program:

```
O2601 (SINGLE HOLE EXAMPLE)
(T01 - 5/8 DIA - 90 DEGREE SPOT DRILL)
N1 G20
N2 G17 G40 G80 T01
N3 M06
N4 G90 G54 G00 X3.5 Y5.0 S900 M03 T02
N5 G43 Z0.1 H01 M08
N6 G99 G82 R0.1 Z-0.2338 P300 F4.0
N7 G80 Z1.0 M09
N8 G28 Z1.0 M05
N9 M01

(T02 - LETTER U DRILL - 0.368 DIA DRILL)
N10 T02
N11 M06
N12 G90 G54 G00 X3.5 Y5.0 S1100 M03 T03
N13 G43 Z0.1 H02 M08
N14 G99 G83 R0.1 Z-1.086 Q0.5 F8.0
N15 G80 Z1.0 M09
N16 G28 Z1.0 M05
N17 M01

(T03 - 5/16 DRILL THROUGH - 0.3125 DIA)
N18 T03
N19 M06
N20 G90 G54 G00 X3.5 Y5.0 S1150 M03
N21 G43 Z0.1 H03 M08
N22 G98 G81 R-0.986 Z-1.644 F8.0
N23 G80 Z1.0 M09
N24 G28 Z1.0 M05
N25 M01

(T04 - 7/16-14 TAP)
N26 T04
N27 M06
N28 G90 G54 G00 X3.5 Y5.0 S750 M03 T01
N29 G43 Z0.4 H04 M08
N30 G99 G84 R0.4 Z-0.9 F53.57   (F = S x LEAD)
N31 G80 G00 Z1.0 M09
N32 G28 Z1.0 M05
N33 G00 X-1.0 Y10.0      (PART CHANGE POSITION)
N34 M30
%
```

This rather detailed example shows that even a simple single hole requires a lot of thought and a great deal of programming and machining skills.

DRILLING OPERATIONS

The example O2601 provides a good illustration of what kind of programming and machining conditions are necessary for a typical hole. Next, let's look at the details of drilling operations in general, as they relate to various tools.

Drilling is one of the oldest operations in a typical machine shop. By definition, drilling is a removal of solid material to form a circular hole of the same diameter as the cutting tool (drill). The material removal is achieved by either rotating the drill (on milling systems) or by rotating the part itself (on turning systems). In either case, a vertical or horizontal machining application is possible. In a rather loose sense of the word, drilling operations also cover the extended areas of reaming, tapping and single point boring. Many programming principles that apply to drilling operations, can be equally applied to all the related operations.

◆ Types of Drilling Operations

The drilling operation is determined by either the *type of hole* or the *type of tool*:

By the type of tool :	By the type of hole :
Center drill	Through hole
Spot drill	Chamfered hole
Twist drill (HSS, cobalt, etc.)	Semi-blind hole
Spade drill	Blind hole
Carbide indexable drill	Premachined hole
Special drill	...

◆ Types of Drills

Drills are categorized by their *design* and by their *size*. The oldest and the most common design is a twist drill, usually made of high speed steel. Twist drill can also be made of cobalt, carbide and other materials. Other drill designs include spade drills, center drills, spot drills and indexable insert drills. The distinction in size is not only between metric and English drills, but also a finer distinction within the category using English units. All metric drills are designated in millimeters. Since the English (imperial) dimensioning is based on inches (which is rather a large dimensional unit), finer distinctions are necessary. The inch dimensions of standard drills in English units are divided into three groups:

❏ FRACTIONAL SIZES :

1/64 minimum, in diameter increments of 1/64

❏ NUMBER SIZES :

Drill size number 80 to drill size number 1

❏ LETTER SIZES :

Drill size letter A to drill size letter Z

Metric sizes do not need any special distinctions. For English sizes, a listing of the standard drills and their decimal equivalents is available from many sources.

◆ Programming Considerations

A standard drill has, regardless of size, two important features - the *diameter* and the *point angle*. The diameter is selected according to the requirements of the drawing, the tool point angle relates to the material hardness. They are both closely connected, since the diameter determines the size of the drilled hole, the tool point angle determines its depth. A smaller consideration is the number of flutes, which is normally two.

◆ Nominal Drill Diameter

The major consideration for a drill is always its diameter. Normally, the drill diameter is selected based on the information in the drawing. If the drawing calls for a hole that needs only drilling and does not need any additional machining, the drill is a standard drill. Its diameter is equivalent to the size specified in the drawing. A drill size of this kind is called a *nominal* or 'off-the-shelf' size.

Most applications involve holes that require other specifications in addition to their diameter - they include tolerances, surface finish, chamfer, concentricity, etc. In those cases, a single regular drill cannot be used alone and still satisfy all requirements. A *nominal* drill alone, even if the size is available, will not guarantee a high quality hole, due to machining conditions. Choosing a *multitool* technique to machine such a hole is a better choice. The normal practice in those cases is to use a drill size a bit smaller than the final hole diameter, then use one or more *additional* tools, which are capable of finishing the hole to the drawing specifications. These tools cover boring bars, reamers, chamfering tools, end mills and others. Using these tools does mean more work is involved, but the quality of the finished part should never be traded for personal conveniences.

◆ Effective Drill Diameter

In many cases, a drill is used to penetrate its *full* diameter through the part. In many other cases, only a *small portion* of the drill end point is used - a portion of the angular drill tip - *Figure 26-5*.

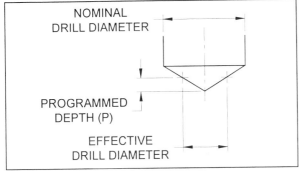

Figure 26-5
Nominal and effective drill diameters (twist drill shown)

During the cut, the drill angular end will be gradually entered into the part, creating an increasingly larger hole diameter, yet still smaller than the drill diameter. At the end, the largest machined diameter will be equivalent to the *effective* diameter of the drill used. The *effective* drill diameter defines the actual hole diameter created *within* the zone of the drill end point. Typical use of this kind of machining is a spot drilling operation for chamfering. The spindle speed and feed must be calculated according to the *effective* drill diameter, not the full diameter. The *r/min* for the effective diameter will be higher and the feedrate lower than the corresponding values for the nominal drill size. For this kind of jobs, selection of a short drill for rigidity is advised.

◆ Drill Point Length

The second important consideration is the *length* of the drill point. This length is very important to establish the cutter depth for the full diameter. With the exception of a flat bottom drill, all twist drills have an angular point whose angle and length must be known in programming. The angles are considerably standard and the length must be calculated rather than estimated, because of its importance to the accurate hole depth - *Figure 26-6*.

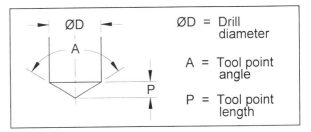

Figure 26-6
Tool point length data for a standard twist drill

On indexable insert drills this length is different, due to the drill construction. The indexable drill is not flat and its drill point length must also be considered in programming. A tooling catalogue shows the dimensions.

The drill point length can be found quite easily, providing the diameter of the drill (nominal or effective) and the drill point angle are known. From the following formula and the table of constants, the required drill point length for standard drills can be calculated. Basic formula is:

$$P = \frac{\tan\left(90 - \dfrac{A}{2}\right)}{2} \times D$$

☞ where ...

P = Length of the drill point
A = Included angle of the drill point
D = Diameter of the drill

The same formula can be simplified and used with a mathematical constant (fixed for each drill point angle):

$$P = D \times K$$

☞ where ...

P = Drill point length
D = Drill diameter
K = Constant (see the following table)

The most common constants are listed in this table:

Tool Point Angle (degrees)	Exact Constant	Practical Constant (K)
60	.866025404	.866
82	.575184204	.575
90	.500000000	.500
118	.300430310	.300
120	.288675135	.289
125	.260283525	.260
130	.233153829	.230
135	.207106781	.207
140	.181985117	.180
145	.157649394	.158
150	.133974596	.134

The constant in the formula is rounded, but its shorter value is sufficient for all programming applications. The value of the constant K for 118° drill angle is .300, the real value is .300430310. The constant value has the advantage of being easy to memorize and there is no formula to solve. For most jobs, only three constants are needed. For 90° (spot drilling and soft materials), 118° (standard materials), and 135° (hard materials). They are easy to memorize:

❏   0.500 ... for a 90° drill angle

❏   0.300 ... for a 118° - 120° drill angle

❏   0.200 ... for a 135° drill angle

◆ Center Drilling

Center drilling is a machining operation that provides a small, concentric opening for a tailstock support or a pilot hole for a larger drill. Chamfering is not recommended with a center drill, because of the 60° angle of the tool.

Never center a hole to be drilled with indexable insert drills !

The most common tool for center drilling is a standard center drill (often called a *combined drill and countersink*), producing a 60° angle. Established North American industrial standards use a numbering system from #00 to #8 (plain type) or #11 to #18 (bell type) for center drills. In the metric system, center drills are defined by the pilot diameter, for example, a 4 mm center drill will have the pilot diameter of 4 mm. In both cases, the higher the number, the larger the center drill diameter. For some pre-drilling operations, such as chamfering, a tool with a 90° point angle, called a *spot drill*, is a better choice.

Many programmers only estimate the depth of a center drill, rather than calculate it. Perhaps a calculation is not necessary for a temporary operation. What is a reasonable compromise between guessing and calculating is a data table, similar to that in *Figure 26-7*.

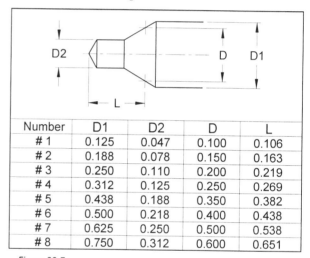

Number	D1	D2	D	L
# 1	0.125	0.047	0.100	0.106
# 2	0.188	0.078	0.150	0.163
# 3	0.250	0.110	0.200	0.219
# 4	0.312	0.125	0.250	0.269
# 5	0.438	0.188	0.350	0.382
# 6	0.500	0.218	0.400	0.438
# 7	0.625	0.250	0.500	0.538
# 8	0.750	0.312	0.600	0.651

Figure 26-7

Standard center drill cutting depth table - #1 to #8 plain type
L is the depth of cut for an arbitrary effective diameter D

In the table, there are all the necessary dimensions for standard English size center drills. The most important of them is the cutting depth L. Its calculation has been based on an *arbitrary selection* of the chamfer diameter D.

For example, #5 center drill has the depth value L that is listed as .382, based on an arbitrarily selected chamfer diameter D of .350 inches. These values can be modified as desired or a different table can be made. A similar table can be developed for metric center drills

◆ Through Hole Drilling

Drilling a hole through the material is a very common operation. It requires the Z depth to include the material thickness, the drill point length and an extra clearance *beyond* the drill penetration point, also known as the *breakthrough* amount.

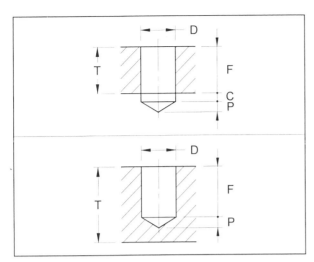

Figure 26-8
Drill depth calculation data
Through hole (top) and Blind hole (bottom)

In *Figure 26-8* is shown that the programmed depth for a through hole is the *sum* of the material thickness T, that is equivalent to the full diameter depth F, plus the breakthrough clearance C, plus the tool point length P.

For example, if the material thickness is one inch and the standard drill diameter D is ∅5/8 (∅.625) of an inch, the programmed depth, including a .050 clearance, will be:

```
1 + .050 + (5/8 × .300) = 1.2375
```

Pay attention to obstructions (machine table, vise, parallels, fixture, machine table, etc.), when programming the tool breakthrough clearance. There is usually a very little space below the bottom face of the part.

◆ Blind Hole Drilling

The major difference between drilling a blind hole and a through hole is that the drill does *not* penetrate the material. Blind hole drilling should not present any more problems than a through hole drilling, but use a peck drilling method for deep holes. Also a choice of a different drill geometry may improve the machining and the hole cleanup may often be necessary as well.

In a typical shop drawing, the depth of a blind hole is given as the *full diameter* depth. The drill point length is *not* normally considered to be part of the depth - it is in *addition* to the specified depth. In *Figure 26-8*, the programmed depth of a blind hole will be the *sum* of the full diameter depth F, plus the tool point length P.

As an example, if a standard ∅3/4 drill (∅.750) is used to drill a full diameter hole depth of 1.25 of an inch, the programmed depth will be:

```
1.25 + (.750 × .300) = 1.4750
```

In the part program, the block will be

```
N93 G01 Z-1.475 F6.0
```

or - in case of a fixed cycle,

```
N93 G99 G85 X5.75 Y8.125 R0.1 Z-1.475 F6.0
```

Metric holes are treated exactly the same way. For example, a ∅16 mm drill is used to machine the full diameter depth of 40 mm. The calculation uses the same constant as for the size in English units:

```
40 + (16 × .300) = 44.8
```

The depth specified in the drawing will have to be *extended* by the calculated drill point length. The programmed block will have the Z axis value equal to the total of the 40 mm specified depth, plus the 4.8 mm calculated point length:

```
N56 G01 Z-44.8 F150.0
```

If the depth appears in a fixed cycle, the same depth value will be used, although in a different format:

```
N56 G99 G81 X215.0 Y175.0 R2.5 Z-44.8 F150.0
```

When machining blind holes, the cutting chips may clog the holes. This may cause a problem, especially if there is a subsequent operation on the hole, for example, reaming or tapping. Make sure you include a program stop code M00 *or* M01 before this operation. The M00 is a better choice, if each hole will have to be cleaned every time the program is executed. Otherwise, the more efficient optional program stop M01 is sufficient.

◆ Flat Bottom Drilling

Flat bottom hole is a blind hole with a bottom at 90° to the drill centerline. There are two common methods of programming such a hole. A good practice is to use a standard drill to start the hole, then use a flat bottom drill of the same diameter and finish the hole to its full depth. Also a good choice is to use a *slot drill* (also known as the *center cutting end mill*), *without* predrilling. This is the best method, but some tool sizes may not be available.

To program a flat bottom hole using a slot drill is quite simple. For example - a ∅10 mm hole should be 25 mm deep (with a flat bottom). Using a ∅10 mm slot drill, the program is quite short (tool in spindle is assumed):

```
O2602 (FLAT BOTTOM - 1)
N1 G21
N2 G17 G40 G80
N3 G90 G54 G00 X.. Y.. S850 M03
N4 G43 Z2.5 H01 M08
```

```
N5 G01 Z-25.0 F200.0
N6 G04 X0.5
N7 G00 Z2.5 M09
N8 G28 Z3.0 M05
N9 M30
%
```

A fixed cycle could be used instead and other improvements added as well, but the program is correct as is.

The next example shows a program for two tools - a ∅1/2 inch standard drill and a ∅1/2 inch flat bottom drill. The required finished depth is Z-0.95 at the flat bottom:

```
O2603  (FLAT BOTTOM - 2)
(T01 - ½ INCH STANDARD DRILL)
N1 G20
N2 G17 G40 G80 T01
N3 M06
N4 G90 G54 G00 X.. Y.. S700 M03 T02
N5 G43 Z0.1 H01 M08
N6 G01 Z-0.94 F9.0
N7 G00 Z0.1 M09
N8 G28 Z0.1
N9 M01


(T02 - ½ INCH FLAT BOTTOM DRILL / END MILL)
N10 T02
N11 M06
N12 G90 G54 G00 X.. Y.. S700 M03 T01
N13 G43 Z0.1 H02 M08
N14 G01 Z-0.74 F15.0
N15 Z-0.95 F7.0
N16 G04 X0.5
N17 G00 Z0.1 M09
N18 G28 Z0.1 M05
N19 M30
%
```

There are three blocks of special interest in program O2603. The first block is N6, indicating the depth of the standard drill. The drill stops short of the full depth by .010 of an inch. Z-0.94 is programmed instead of the expected Z-0.95. A little experiment as to *how* short may be worth it. A reason for *not* drilling to the full depth with the standard drill is to prevent possible dimple mark at the hole center.

The other two blocks appear in the second tool of the program - blocks N14 and N15. In block N14, the flat bottom drill feeds at a heavier feedrate to the depth of only .740 inches. That makes sense, as there is nothing to cut for the flat bottom drill for almost 3/4 of an inch. Follow the calculation of the 0.740 intermediate depth from this procedure:

From the total depth of .94 cut by the standard drill (T01), subtract the length of the tool point P. That is .15 for a 118° drill point angle and ∅.5 drill. The result is .79. From the result, subtract .05 for clearance, and the result is the Z axis value of Z-0.74. In the block N15, the flat bottom drill removes the excessive material left by T01, at a suitable cutting feedrate, usually programmed at a slower rate.

From the machining viewpoint, programming a center drill or a spot drill first to open up the hole may be a better choice. This extra operation will guarantee concentricity for both the standard drill and the flat bottom drill. Another possible improvement would be to use a suitable end mill instead of a flat bottom drill. An end mill is usually more rigid and can do the job much better.

◆ Indexable Insert Drilling

One of the great productivity improvement tools in modern machining is an indexable insert drill. This drill uses carbide inserts, just like many other tools for milling or turning. It is designed to drill holes in a *solid* material. It does *not* require center drilling or spot drilling, it is used with high spindle speeds and relatively slow feedrates and is available in a variety of sizes (English and metric). In most cases, it is used for *through* holes, although blind holes can be drilled as well. This type of a drill can even be used for some light to medium boring or facing.

The design of the indexable insert drill is very precise, assuring constant tool length, as well as elimination of regrinding dull tools. *Figure 26-9* shows the cutting portion of a typical indexable drill.

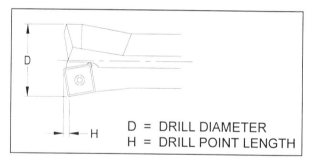

Figure 26-9

Cutting end of a typical indexable insert drill

In the illustration, the diameter D of the drill is the hole size produced by the drill. The tip point length H is defined by the drill manufacturer and its amount is listed in the tooling catalogue. For example, an indexable drill with the D diameter of 1.25, may have the H tip length .055. The indexable drill can be used for rotary and stationary applications, vertically or horizontally, on machining centers or lathes. For best performance, the coolant should be pressure fed *through* the drill, particularly for tough materials, long holes, and horizontal operations. The coolant not only disperses the generated heat, it also helps flush out the chips. When using an indexable insert drill, make sure there is enough power at the machine spindle. The power requirements at the spindle increase proportionally with increased drill diameters.

On a machining center, the indexable drill is mounted in the machine spindle, therefore it becomes a rotating tool. In this setup, the drill should be used in a rigid spindle that runs true - no more than .010 inch (0.25 mm) of T.I.R. *(Total Indicator Reading)*. On spindles that have a quill, try to work with the quill *inside* the spindle, or extend it as little as possible. Coolant provisions may include an internal coolant, and special adapters are available for through the hole cooling, when the drill is used on machining centers.

On a CNC lathe, the indexable drilling tool is always stationary. The correct setup requires that the drill is positioned on the center and be concentric with the spindle centerline. The concentricity should not exceed .005 inch (0.127 mm) of T.I.R.

Always exercise care when the drilling operation starts on a surface that is not flat. For best results, use indexable drills on surfaces that are 90° to the drill axis (flat surfaces). Within limits, the drill can also be used to enter or exit an inclined, uneven, concave, or convex surface quite successfully. The feedrate may need to be reduced for the duration of any interrupted cut. The *Figure 26-10* shows the areas where the feedrate should be slower.

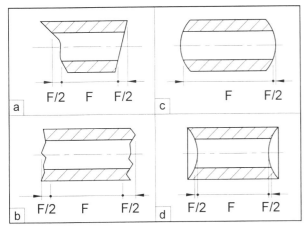

Figure 26-10

Uneven entry or exit surface for indexable drills feedrate: F = normal feedrate, F/2 = reduced feedrate (one half of F)

In the illustration, the letter F identifies the area that is cut with the *normal* feedrate (normal entry/exit), and the F/2 indicates the area that requires a *reduced* feedrate. For the reduced feedrate, programming one half of the normal feedrate is sufficient.

In the illustration, the frame *a* shows a tilted surface entry (inclined surface), the *b* frame shows an uneven surface, and the frames *c* and *d* show convex and concave surfaces respectively.

> An indexable drill should always be used
> in a fully protected machining area.

PECK DRILLING

Peck drilling is also called *interrupted cut drilling*. It is a drilling operation, using the fixed cycles G83 (standard peck drilling cycle) or G73 (high speed peck drilling cycle). The difference between the two cycles is the tool retract method. In G83, the retract after each peck will be to the R level (usually above the hole), in G73, there will only be a small retract (between .02 and .04 inches).

Peck drilling is often used for holes that are too deep to be drilled with a single tool motion. Peck drilling methods also offer several opportunities to improve the standard drilling techniques as well. Here are some possible uses of the peck drilling methods for machining holes:

- ❑   Deep hole drilling
- ❑   Chip breaking - also used for short holes in tough materials
- ❑   Cleanup of chips accumulated on the flutes of the drill
- ❑   Frequent cooling and lubricating of the drill cutting edge
- ❑   Controlling the drill penetration through the material

In all cases, the drilling motions of the G83 or G73 cycle produce an interrupted cut that can be programmed very simply by specifying the Q address value in the cycle. This value specifies the actual depth of each peck. The smaller the Q value, the more pecks will be generated and vice versa. For most deep hole drilling jobs, the *exact* number of pecks is not important, but there are cases when the pecking cycle needs to be controlled.

◆ Typical Peck Drilling Application

For the majority of peck drilling applications, the peck drilling depth Q needs to be only a *reasonable* depth. For example, a deep hole (with the depth at Z-2.125 inches at the tool tip) is drilled with a .250 diameter drill and .600 peck depth. The G83 cycle may be programmed like this:

```
N137 G99 G83 X.. Y.. R0.1 Z-2.125 Q0.6 F8.0
```

These programming values are *reasonable* for the job at hand - and that is all that matters. For most jobs, the number of pecks is usually not too important.

◆ Calculating the Number of Pecks

If the number of pecks the G83/G73 cycle will generate is important, it has to be *calculated*. The knowledge of how many pecks will result with a certain Q value for a given total depth is usually not important. If the program is running efficiently, there is no need for a modification. To find out how many pecks the G83/G73 cycle will generate, it is important to know the *total distance* the drill travels between the R level and the Z depth (as an incremental value). It is equally important to know the peck depth Q value. The Q divided into the travel distance is the number of pecks:

$$P_n = \frac{T_d}{Q}$$

☞ where ...

P_n   =   Number of pecks
T_d   =   Total tool travel distance
Q   =   Programmed peck depth

For example, in the following G83 cycle,

```
N73 G99 G83 X.. Y.. R0.125 Z-1.225 Q0.5 F12.0
```

the total drill travel distance is 1.350, divided by .500, which yields 2.7. Since the number of pecks can only be positive, the nearest *higher* integer will be the actual number of pecks, in this case 3.

◆ Selecting the Number of Pecks

Much more common is the programming of a *desired* number of pecks. If only a certain number of pecks will do the job in the most efficient way, the Q value has to be calculated accordingly. Since the Q value specifies the *depth of each peck* and *not* the number of pecks, some simple math will be needed to select the depth Q, so it corresponds to the desired number of pecks.

For example - we require *3 pecks* in the following cycle - what will the Q depth be?

```
N14 G99 G83 X.. Y.. R0.1 Z-1.238 Q.. F12.0
```

The total drill travel from the R level to the Z depth is 1.338. To calculate the peck depth Q value, the new formula is similar to the previous one:

$$Q = \frac{T_d}{P_n}$$

☞ where ...

Q   =   Programmed peck depth
T_d   =   Total tool travel distance
P_n   =   Number of required pecks

Using the above formula, the result of 1.338/3 is .446. Therefore, G83 block Q depth will be Q0.446:

```
N14 G99 G83 X.. Y.. R0.1 Z-1.238 Q0.446 F12.0
```

No rounding is necessary in this case. Now, let's have a close look at another situation, where the travel distance has changed very slightly:

```
N14 G99 G83 X.. Y.. R0.1 Z-1.239 Q.. F12.0
```

The result of 1.339/3 is .446333333 - a number that has to be rounded to the maximum of four decimal places (English units). Mathematically correct rounding to four decimal places will be .4463. Follow individual peck depths to see what will happen:

Peck 1	.4463	accumulated depth	...	.4463
Peck 2	.4463	accumulated depth	...	.8926
Peck 3	.4463	accumulated depth	...	1.3389
Peck 4	.0001	accumulated depth	...	1.3390

There will be *four pecks* and the last one will only cut .0001 - or practically nothing at all. In those cases, where the last cut is very small and inefficient, always round the calculated Q value *upwards*, in this case to the minimum of .4464 or even to .447:

```
N14 G99 G83 X.. Y.. R0.1 Z-1.239 Q0.447 F12.0
```

Always remember, the cutting tool will *never* go past the programmed Z depth, but it could reach this depth in a very inefficient way that should be corrected.

◆ Controlling Breakthrough Depth

Less frequent programming method, also very powerful, is to use the peck drilling cycle to control the *breakthrough* of the drill through the material, regardless of the drill size or material thickness. Here is some background. In many tough materials, when the drill starts penetrating the bottom of the part (for a through hole), it creates potentially difficult machining conditions. The drill has the tendency to *push* the material out rather than cut it. This is most common when the drill is a little dull, the material is tough, or the feedrate is fairly high. These adverse conditions are also the result of heat generated at the drill edge, the lack of lubrication reaching the drill cutting edge, worn-off flutes and several other factors.

The possible solution to this problem is to relieve the drill pressure when it is *about half way* through the hole, but *not* completely through - *Figure 26-11*.

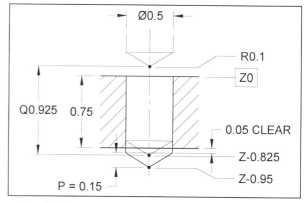

Figure 26-11

Controlled breakthrough of a hole using G83 peck drilling cycle

Peck drilling cycle G83 is great for it, but the Q depth calculation is extremely important. The total number of pecks is not important, only the last two are critical for this purpose. To control the problem associated with the drill penetration, only two peck motions are needed. The illustration shows the two positions for a ⌀1/2 drill drill through a 3/4 thick plate.

For most jobs, such a hole requires no special treatment. Just one cut through (using G81 cycle) and no peck drilling. Let's evaluate the solution to this situation. The ⌀.5 drill has the point length of .300 × .500 = .150. Take one half (.075) of the drill point length as the first penetration amount, which will bring the drill .075 *below* the 3/4 plate thickness, to the Z depth of Z-0.825. *This depth has to be reached with the value of the Q depth.* Keep in mind that the Q depth is an *incremental* value, measured from the R level, in this case R0.1. That specifies the Q depth as Q0.925 (.100 above and .825 below Z0). The programmed Z depth is the final drill depth. If the .05 clearance is added below the plate, the Z depth will be the sum of the plate thickness (.75), the clearance (.05) and the drill point length (.150), for the program value of Z-0.95:

```
G99 G83 X.. Y.. R0.1 Z-0.95 Q0.925 F..
```

This technique does not only solve a particular job related problem, it also shows how *creativity* and *programming* are complementary terms.

REAMING

The reaming operations are very close to the drilling operations, at least as far as the programming method is concerned. While a drill is used to *make* a hole (to open up the hole), a reamer is used to *enlarge* an existing hole.

Reamers are either cylindrical or tapered, usually designed with more than two flutes of different configurations. Reamers made of high speed steel, cobalt, carbide and with brazed carbide tips. Each reamer design has its advantages and disadvantages. Carbide reamer, for example, has a very high resistance to wear, but may be not economically justified for every hole. A high speed steel reamer is economical, but wears out much faster that a carbide reamer. Many jobs do not accept any compromise in the tooling selection and the cutting tool has to be selected correctly for a given job. Sizing and finishing tools, such as a reamer, have to be selected even more carefully.

Reamer is a *sizing* tool and is not designed for removal of heavy stock. During a reaming operation, an existing hole will be *sized* - reamer will *size* an existing hole to close tolerances and add a high quality surface finish. Reaming will *not* guarantee concentricity of a hole. For holes requiring both high concentricity *and* tight tolerances, center drill or spot drill the hole first, then drill it the normal way, then rough bore it and only then finish it with a reamer.

A reaming operation will require a coolant to help make a better quality surface finish and to remove chips during cutting. Standard coolants are quite suitable, since there is not very much heat generated during reaming. The coolant also serves in an additional role, to flush away the chips from the part and to maintain the surface finish quality.

◆ Reamer Design

In terms of design, there are two features of a reamer that have a direct relationship to the CNC machining and programming. The first consideration is the *flute design.*

Most reamers are designed with a left-hand flute orientation. This design is suitable to ream *through* holes. During the cut, the left-hand flute design 'forces' the chips to the bottom of the hole, into an empty space. For blind holes that have to be reamed, the left-hand type of a reamer may not be suitable.

The other factor of the reamer design is the *end chamfer.* In order to enter an existing hole that is still without a chamfer, a lead-in allowance is required. The reamer end provides that allowance. Some reamers also have a short taper at their tip, for the same purpose. The chamfered lead is sometimes called a '*bevel lead*' and its chamfer an '*attack angle*'. Both have to be considered in programming.

◆ Spindle Speeds for Reaming

Just like for standard drilling and other operations, the spindle speed selected for reaming must be closely related to the type of material being machined. Other factors, such as the part setup, its rigidity, its size and surface finish of the completed hole, etc., each contributes to the spindle speed selection.

As a general programming rule, the spindle speed for reaming will be reasonable if you use a modifying factor of .660 (2/3), based on the speed used for drilling of the *same* material. For example, if a speed of 500 r/min produced good drilling conditions, the two thirds (.660) of that speed will be *reasonable* for reaming:

```
500 × .660 = 330 r/min
```

Do not program a reaming motion in the reversed spindle rotation - the cutting edges may break or become dull.

◆ Feedrates for Reaming

The reaming feedrates are programmed higher than those used for drilling. *Double* or *triple* increases are not unusual. The purpose of the high feedrates is to force the reamer *to cut*, rather than to rub the material. If the feedrate is too slow, the reamer wears out rapidly. The slow feedrates cause heavy pressures as the reamer actually tries to enlarge the hole, rather than remove the stock.

◆ Stock Allowance

Stock is the amount of material left for finishing operations. A hole to be reamed must be smaller (*undersize*) than the pre-drilled or pre-bored hole - a logical requirement. Programmer decides *how much* smaller. A stock too small for reaming causes the premature reamer wear. Too much stock for reaming increases the cutting pressures and the reamer may break.

A good general rule is to leave about 3% of the reamer diameter as the stock allowance. This applies to the *hole diameter - not per side.* For example, a 3/8 reamer ($\varnothing$.375), will work well in most conditions if the hole to be reamed has a diameter close to .364 inches:

```
.375 - (.375 × 3 / 100) = .36375 = .364
```

Most often, a drill that can machine the required hole diameter exactly will not be available. That means using a boring bar to *presize* the hole before reaming. It also mean an extra cutting tool, more setup time, longer program and other disadvantages, but the hole quality will be worth the effort. In these cases, for tough materials and some of the 'space age' materials, the stock allowance left in the hole for reaming, is usually decreased.

◆ Other Reaming Considerations

The general approach for reaming is no different than for other operations. When drilling a blind hole, then reaming it, it is inevitable that some chips from the drilling remain in the hole and may prevent a smooth reaming operation. Using the program stop function M00 *before* the reaming operation allows the operator to remove all the chips first, for a clear entry of the reamer.

The reamer size is always important. Reamers are often made to produce either a *press fit* or a *slip fit.* These terms are nothing more than machine shop expressions for certain tolerance ranges applied to the reamed hole.

Programming a reamer requires a fixed cycle. Which cycle will be the most suitable? There is no *reaming cycle* defined directly. Thinking about the traditional machining applications, the most accepted reaming method is the *feed-in and feed-out* method. This method requires a feedrate motion to remove the material from the hole, but it also requires a feedrate motion back to the starting position, to maintain the hole quality - its size and surface finish. It may be tempting to program a rapid motion out of a reamed hole to save cycle time, but often at the cost of quality. For the best machining, the *feed-out* of the reamed hole is necessary. Suitable fixed cycle available for the Fanuc controls is G85, which permits *feed-in* and *feed-out* motions. The cutting feedrate of the cycle will be *the same* for both motions. Any feedrate change will affect both motions - *in* and *out.*

SINGLE POINT BORING

Another sizing operation on holes is called *boring.* Boring, in the sense of machining holes, is a point-to-point operation along the Z axis only, typical to CNC milling machines and machining centers. It is also known as a '*single point boring*', because the most common tool is a boring bar that has only *one* cutting edge. Boring on CNC lathes is considered a *contouring* operation and is not covered in this chapter (see *Chapters 34-35*).

Many jobs requiring precision holes that have previously been done on a special jig boring machine can now be done on a CNC machining center, using a single point boring tool. The modern CNC machine tools are manufactured to very high accuracy, particularly for the positioning and repeatability - a proper boring tool and its application can produce very high quality holes.

◆ Single Point Boring Tool

As for its practical purpose, a single point boring is a *finishing*, or at least a *semifinishing*, operation. Its main job is to enlarge - or *to size* - a hole that has been drilled, punched or otherwise cored. The boring tool works on the diameter of the hole and its purpose is to produce the desired hole diameter, within specified tolerances, often with a quality surface finish as well.

Although there is a variety of designs of boring tools on the market, the single point boring tool is usually designed for the *cartridge* type inserts. These inserts are mounted at the end of the holder (*i.e.,* a boring bar) and usually have a built-in micro adjustment for fine tuning of the effective boring diameter - *Figure 26-12.*

Figure 26-12
Effective diameter of a single point boring tool

The same programming techniques are applied to the boring bars of other designs, for example, a *block tool.* A block tool is a boring bar with two cutting edges, 180° apart. If the adjusting mechanism for the diameter is *not* available on the tool holder, the effective boring diameter must be *preset*, using either a special equipment, or the slow but true and tried *trial-and-error* method. This trial and error setup is not that unusual, considering the setup methods that are available for a single point boring bar.

Just like any other cutting tool, a single point boring bar achieves the best cutting results if it is short, rigid and runs concentric with spindle centerline. One of the main causes of poorly bored holes is the *boring bar deflection*, applying equally to milling and turning. The tool tip (usually a carbide bit), should be properly ground, with suitable cutting geometry and overall clearances. The position of the boring bar in the spindle - or its orientation - is very important for many boring operations on machining centers.

◆ Spindle Orientation

Any round tool, such as a drill or an end mill, can enter or exit a hole along the Z axis, with little programming considerations for the hole quality. Neither of the tools is used for holes that demand high quality surface finish and close tolerances. With boring, the hole surface integrity *is very important*. Many boring operations require that the cutting tool does not damage the hole surface during retract. Since retracting from a hole almost always leaves some marks in the hole, special methods of retract must be used. There is one such method - it uses cycle G76 or G87 with the *spindle orientation* feature of the machine and a *shift* of the boring tool away from the finished surface. This feature was already described in *Chapter 12*, so just a reminder now.

The sole purpose of spindle orientation is to replace the tool holder in exactly the same position after each tool change. Without spindle orientation, the tool tip will stop at a random position of its circumference. Orienting the spindle for boring purposes is only one half of the solution. The other is the setting position of the boring bit. This is usually a responsibility of the operator, since it has to be done during setup at the machine. The boring bar cutting bit must be set in such a way that when the shift takes place in fixed cycle G76 or G87, it will be into the direction *away* from the finished hole wall, ideally by the XY vector relative to the angle of the spindle orientation - *Figure 26-13*.

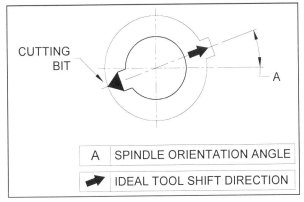

| A | SPINDLE ORIENTATION ANGLE |
| ➤ | IDEAL TOOL SHIFT DIRECTION |

Figure 26-13
Single point boring bar and the spindle orientation angle

Spindle orientation is factory designed and *fixed*. Programmer considers its length and, usually, its direction.

When the machine spindle is oriented, it must be in a stopped mode. The spindle cannot rotate during any machining operation that requires a spindle shift. Review descriptions of the fine boring fixed cycle G76 and the back-boring cycle G87 in *Chapter 25*. Machine operator *must* always know which way the spindle orients *and* into which direction the tool shift actually moves.

Programming a bored hole that will be reamed later requires the boring bar only to assure the concentricity and straightness of the finished hole. The surface finish of the bored hole is not too important. If the boring is the *last* machining operation in the hole, the chances are that the surface finish *will* be very important. It is difficult to retract the boring tool without leaving drag marks on the hole cylindrical surface. In that case, select a suitable fixed cycle, probably the precision boring cycle G76 is the best choice.

◆ Block Tools

When using a single point boring bar for roughing or semifinishing operations, there is an option that is more efficient. This option also uses a boring tool, but one that has *two* cutting edges (180° opposite) instead of one - it is called a *block tool*. Block tools cannot be used for fine finishing operations, because they cannot be *shifted*. The only way of programming a block tool is within the *'in-and-out'* tool motion. Several fixed cycles support this kind of motion. All motions *'in'* are at a specified feedrate. On the way *'out'*, some motions are feedrates, others are rapid, depending on the cycle selection. The cycles that can be used with block tools are G81 and G82 (feed-in-rapid-out), as well as G85 and G89 that feeds in and feeds out while the machine spindle is rotating and another one, G86, when the tool retracts while the spindle is not rotating.

The greatest advantage of a block tool is the increased feedrate that can be programmed for this tool. For example, if the feedrate for a single point tool is .007 per flute, for a block tool it will be at least double, .014 inches per flute or more. Block tools are generally available in diameters from about ∅.750 inch and up.

BORING WITH A TOOL SHIFT

There are two fixed cycles that require the tool shift away from the centerline of current hole. These cycles are boring cycles G76 and G87. G76 is by far the most useful and both are illustrated together in program example O2604.

◆ Precision Boring Cycle G76

The G76 cycle is used for holes requiring a high quality of the size and surface finish. The boring itself is normal, however, the retract from the hole is special. The boring bar stops at the bottom of the hole in an oriented position, shifts away by the Q value in the cycle and retracts back to the starting position, where it shifts back to its normal position.

The G76 cycle has been described in detail in the previous chapter. In this chapter is an actual programming example, shown as a single hole in *Figure 26-14* - ⌀25 mm.

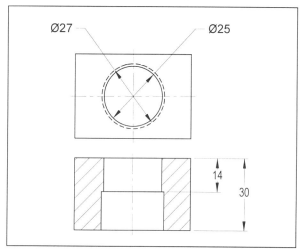

Figure 26-14

Drawing for G76 and G87 programming example - program O2604

From the drawing, only the 25 mm hole is considered, and the program input will be quite simple:

```
N.. G99 G76 X0 Y0 R2.0 Z-31.0 Q0.3 F125.0
```

A hole bored with G76 cycle will have a high quality.

◆ Backboring Cycle G87

Although the backboring cycle has some applications, it is not a common fixed cycle. As the name suggests, it is a boring cycle that works in the reverse direction than other cycles - *from the back of the part*. Typically, the backboring operation starts at the bottom of the hole, which is the 'back of the part', and the boring proceeds from the bottom upwards, in the Z positive direction.

The G87 cycle has been described in the previous chapter. The *Figure 26-14* also shows a diameter of 27 mm, which will be bored during the same setup as the 25 mm hole. This larger diameter is at the 'back side of the part', and it will be backbored, using the G87 cycle.

Figure 26-15 shows the setup of the tool that will bore the 27 mm hole, from the bottom of the hole, upwards. Pay a close attention to the descriptions.

In the illustration, the D1 represents the diameter of the smaller hole, and D2 represents the diameter of the hole to be backbored. D2 is always larger than D1. Always make sure there is enough clearance for the body of the boring bar within the hole and at the hole bottom.

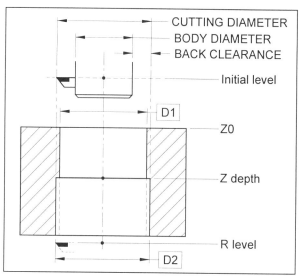

Figure 26-15

Setup considerations for a backboring tool

◆ Programming Example

In order to show a complete program, four tools will be used - *spot drill* (T01), *drill* (T02), *standard boring bar* (T03) and a *back boring bar* (T04). Program is O2604.

```
O2604 (G76 AND G87 BORING)
(T01 - 15 MM DIA SPOT DRILL - 90 DEG)
N1 G21
N2 G17 G40 G80 T01
N3 M06
N4 G90 G54 G00 X0 Y0 S1200 M03 T02
N5 G43 Z10.0 H01 M08
N6 G99 G82 R2.0 Z-5.0 P100 F100.0
N7 G80 Z10.0 M09
N8 G28 Z10.0 M05
N9 M01

(T02 - 24 MM DIA DRILL)
N10 T02
N11 M06
N12 G90 G54 G00 X0 Y0 S650 M03 T03
N13 G43 Z10.0 H02 M08
N14 G99 G81 R2.0 Z-39.2 F200.0
N15 G80 Z10.0 M09
N16 G28 Z10.0 M05
N17 M01

(T03 - 25 MM DIA STANDARD BORING BAR)
N18 T03
N19 M06
N20 G90 G54 G00 X0 Y0 S900 M03 T04
N21 G43 Z10.0 H03 M08
N22 G99 G76 R2.0 Z-31.0 Q0.3 F125.0     (25 DIA)
N23 G80 Z10.0 M09
N24 G28 Z10.0 M05
N25 M01
```

```
(T04 - 27 MM DIA BACK BORING BAR)
N26 T04
N27 M06
N28 G90 G54 G00 X0 Y0 S900 M03 T01
N29 G43 Z10.0 H04 M08
N30 G98 G87 R-32.0 Z-14.0 Q1.3 F125.0 (27 DIA)
N31 G80 Z10.0 M09
N32 G28 Z10.0 M05
N33 G28 X0 Y0
N34 M30
%
```

Make sure to follow all rules and precautions when programming or setting up a job with G76 or G87 fixed cycles in the program. Many of them are safety oriented.

◆ Precautions in Programming and Setup

The precautions for boring with a tool shift relate to a few special considerations that are necessary for successful realization of the two cycles G76 and G87. The following list sums up the most important precautions:

- ❑ The through boring must be done before the backboring

- ❑ The first boring cycle (G76) must be programmed all the way through the hole, never partially

- ❑ For the G76 cycle, only a minimum Q value is required (ex., 0.3 mm or .012 inches)

- ❑ For the G87 cycle, the Q value must be greater than one half of the difference between the two diameters:
  $(D2-D1)/2 = (27-25)/2 = 1$,
  plus the standard minimum Q value (0.3 mm)

- ❑ Always watch for the body of the boring bar, so it does not hit the hole surface during the shift. This can happen with large boring bars, small holes, or a large shift amount.

- ❑ Always watch the body of the boring bar, so it does not hit an obstacle below the part. Remember that the tool length offset is measured to the cutting edge, not to the actual bottom of the boring tool.

- ❑ G87 is always programmed in G98 mode, never in G99 mode !!!

- ❑ Always know the shift direction and set the tool properly

ENLARGING HOLES

An existing hole can also be enlarged from the top. To enlarge an existing hole at the top, we can use one of three methods that will enlarge an existing hole. These methods are common in every machine shop. They are:

- ❑ Countersinking   C'SINK or CSINK on drawings

- ❑ Counterboring   C'BORE or CBORE on drawings

- ❑ Spotfacing   SF, S.F., or S/F on drawings

All three machining methods will enlarge an existing hole, with one common purpose - they will allow the fitting part to be accurately seated in the hole by creating a clean surface. For example, a bolt head that has to be seated on a flat surface will require countersinking or spotfacing operation. All three operations require a perfect alignment with the existing hole (concentricity). Programming technique is basically the same for all three operations, except for the tool used. Speeds and feeds for these tools are usually lower than for drills of equivalent size. Any hole to be enlarged must exist prior to these operations.

◆ Countersinking

Countersinking is an operation that enlarges an existing hole in a conical shape, to a required depth. Countersinking is used for holes that have to accommodate a conical bolt head. From all three similar operations, countersinking requires the most calculations for precision depth. Typical countersinks have three angles:

- ❑ 60 degrees

- ❑ 82 degrees - the most common angle

- ❑ 90 degrees

Other angles are also possible, but less frequent.

To illustrate the programming technique and the required calculations, the cutting tool used must be known first. *Figure 26-16* shows a typical countersinking tool.

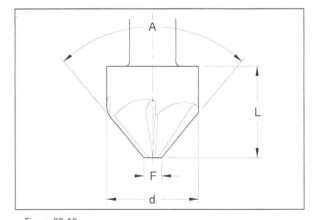

Figure 26-16

Typical nomenclature of a countersinking tool

In the illustration, *d* is the countersink body diameter, A is the countersink angle, F is the diameter of the tool flat (equal to zero for a sharp end), *l* is the body length.

Programming of a countersink requires certain data in the drawing. This information is often provided through a description (leader/text) in the drawing, for example:

```
.78 DIA CSINK - 82 DEG
13/32 DRILL THRU
```

There is one challenge for programming a countersink. The specified countersink diameter must be accurate. That is the ∅.78 in the description. The countersink angle is 82°. The precise diameter can be *created* by carefully calculating the Z depth. That should not be too difficult, because we can use the constant values K for the tool point length (described earlier in this chapter), then calculate the cutting depth, similar to drills. The problem here is that the constant K for a drill point always *assumes a sharp point* at the tool tip. Countersinking tools do not always have a sharp point (except for some small sizes). Instead, they have a diameter of the flat F, normally specified in tooling catalogues.

Figure 26-17 illustrates an example of a countersink requirement, shown in a typical drawing.

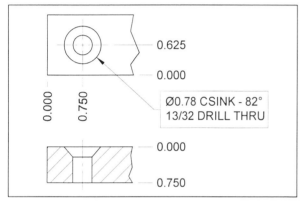

Figure 26-17

Programming example of a countersinking operation

Figure 26-18 shows the known and unknown countersinking dimensions required for depth programming of a countersinking tool.

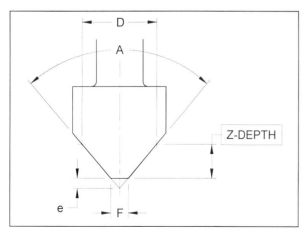

Figure 26-18

Data required for calculating the Z depth of a countersink, for given diameters D and F and the angle A

The process of calculation is simple enough. First, determine the height *e,* for a given flat diameter F. Use the standard constants as applied to a drill point length:

60° = .866
82° = .575
90° = .500

In the illustration, D is the required countersink diameter, A is the countersink angle, F is the flat diameter, *e* is the height of the sharp end, and the Z-DEPTH is the programmed tool depth. In this case, the angle A is 82°, the flat diameter F as per catalogue is 3/16 (.1875). The height of the sharp end *e* can be calculated:

e = .1875 × .575   (K for 82° = .575)
e = .1078

The Z depth for a tool *with a sharp* end will be:

```
Z depth = .78 × .575 = .4485
```

Since that depth *includes* the height of sharp end, all that has to be done to find out the Z depth, is to *subtract* the *e* value from the theoretical Z depth:

```
Z depth = .4485 - .1078 = .3407
```

This is the *programmed* Z depth and the program block for the countersink in the drawing may look something like this:

```
N35 G99 G82 X0.75 Y0.625 R0.1 Z-0.3407 P200 F8.0
```

Incidentally, the R level could be *lowered*, since there is a through hole already machined in the previous operation. Be careful here, the R level will most likely be negative. Always program the G98 command and a small initial level, for example, Z0.1:

```
N34 G43 Z0.1 H03 M08      (0.1 IS INITIAL LEVEL)
N35 G98 G82 X0.75 Y0.625 R-0.2 Z-0.3407 P200 F8.0
```

DO NOT make the R value too deep !

◆ Counterboring

Counterboring is an operation that enlarges an existing hole in a cylindrical shape to the required depth. Counterboring is used for holes that have to accommodate a round bolt head. It is often used on uneven or rough surfaces, or surfaces that are not at 90° to the bolt assembly. As for the proper tool selection, use a counterboring tool specially designed for this type of machining, or a suitable end mill instead. In either case, the program uses G82 fixed cycle. Since the depth of the counterbore is always given, there are no extra calculations required. *Figure 26-19* shows a typical counterboring description.

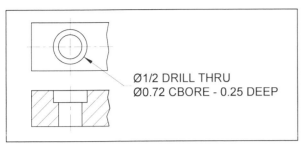

Figure 26-19

Programming example of a counterboring operation

For the example, the ⌀1/2 inch hole had been machined earlier. The program block will be quite simple:

```
N41 G99 G82 X.. Y.. R0.1 Z-0.25 P300 F5.0
```

In counterboring, if a relatively slow spindle speed and fairly heavy feedrates are used, make sure the dwell time P in G82 cycle is sufficient. The rule of thumb is to program the double value or higher of the calculated minimum dwell. Minimum dwell D_m is:

$$D_m = \frac{60}{r/min}$$

For example, if the spindle speed is programmed as 600 r/min, the minimum dwell will be 60/600=0.1, and doubled to 0.2 in the program, as P200. Doubling the minimum dwell value guarantees that even at 50% spindle speed override, there will be *at least* one full revolution of the spindle that cleans up bottom of the counterbored hole. Many programmers choose to use a slightly longer dwell time, for more than one or two revolutions at the bottom.

◆ Spotfacing

Spotfacing is virtually identical to counterboring, except that the depth of cut is very minimal. Often, spotfacing is called *shallow counterboring*. Its purpose is to remove just enough material to provide a flat surface for a head of a bolt, a washer, or a nut. Programming technique is exactly the same as that for counterboring.

MULTILEVEL DRILLING

On many occasions, the same cutting tool will have to be programmed to move up and down between different heights (steps on a part). For example, a drill will cut holes that have the same depth, but start at different heights.

This kind of programming requires two major conditions - the tool *should be* programmed *efficiently* (no time loss) and *must be* programmed *safely* (no collision).

Handling this programming problem is not difficult, once the available options are evaluated. The options are two preparatory commands - G98 and G99, used with fixed cycles exclusively. Recall that the G98 command will cause the cutting tool to return to the *initial level*, the G99 command will cause the cutting tool to return to the *R level*. In practical programming, the G98 command is used only in cases where an obstacle between holes has to be bypassed.

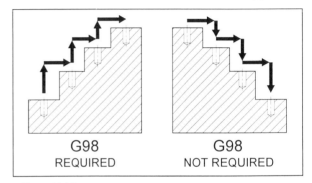

Figure 26-20

Tool motion direction between holes at different heights

Figure 26-20 illustrates two programming possibilities, in a symbolic representation. The front view of a stepped part shows the direction of tool motion between holes. On the left, the motion from one hole to the next could cause a collision with the wall and G98 *is required* for safety. On the right, with no obstacles, G98 *is not required*, and G99 can be used. The setting for the initial level is usually done in the G43 block, where the Z value must represent a clear tool location above *all* obstacles.

A practical example of this technique is illustrated in *Figure 26-21* and program O2605.

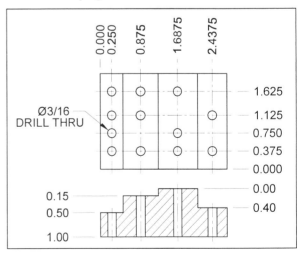

Figure 26-21

Multilevel drilling - programming example O2605

Two tools are used - T01 is a 90° spot drill, cutting to the depth of .108 below *each step face*. T02 is a ∅3/16 drill *through*, programmed to the absolute depth of Z-1.106:

```
O2605 (MULTILEVEL EXAMPLE)
(T01 - 0.375 SPOT DRILL - 90 DEG)
N1 G20
N2 G17 G40 G80 T01
N3 M06
N4 G90 G54 G00 X0.25 Y0.375 S900 M03 T02
N5 G43 Z1.0 H01 M08
N6 G99 G82 R-0.4 Z-0.608 P200 F8.0
N7 Y0.75
N8 Y1.125
N9 G98 Y1.625
N10 G99 X0.875 R-0.05 Z-0.258
N11 Y1.125
N12 G98 Y0.375
N13 G99 X1.6875 R0.1 Z-0.108
N14 Y0.75
N15 Y1.625
N16 X2.4375 Y1.125 R-0.3 Z-0.508
N17 Y0.375
N18 G80 Z1.0 M09
N19 G28 Z1.0 M05
N20 M01

(T02 - 3/16 DRILL THRU)
N21 T02
N22 M06
N23 G90 G54 G00 X2.4375 Y0.375 S1000 M03 T01
N24 G43 Z1.0 H02 M08
N25 G99 G83 R-0.3 Z-1.106 Q0.35 F10.0
N26 G98 Y1.125
N27 G99 X1.6875 Y1.625 R0.1
N28 Y0.75
N29 Y0.375
N30 X0.875 R-0.05
N31 Y1.125
N32 Y1.625
N33 X0.25 R-0.4
N34 Y1.125
N35 Y0.75
N36 Y0.375
N37 G80 Z1.0 M09
N38 G28 Z1.0 M05
N39 G00 X-2.0 Y10.0
N40 M30
%
```

Study the program in detail. Watch the direction of tools - T01 starts at the lower left hole and ends at the lower right hole, in a zigzag motion. T02 starts at the lower right hole and ends at the lower left hole, also in a zigzag motion. Note there are more G98 or G99 changes for the first tool than the second tool. In multilevel hole machining understand three areas of program control, used in O2605:

- ❏ G98 and G99 control
- ❏ R level control
- ❏ Z depth control

WEB DRILLING

Web drilling is a term for a drilling operation taking place between two or more parts, separated by an empty space. The programming challenge is to make such holes *efficiently*. It would be easy to program one motion through all the separate parts as well as the empty spaces. For many holes, this approach would prove to be very inefficient. Evaluate the front view of a web drilling example shown in *Figure 26-22*.

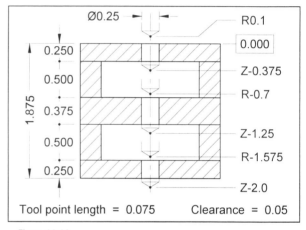

Figure 26-22

Web drilling example (front view) - program O2606

In the program, X1.0Y1.5 is used as the hole position. Drawing will not show the R levels or Z depths, they have to be calculated. In the example, clearances above and below each plate are .05, except the first R level (R0.1). The length of the ∅1/4 drill point is .3 × .25 = .075.

```
O2606 (WEB DRILLING)
(T01 - 90-DEG SPOT DRILL - 0.5 DIA)
N1 G20
N2 G17 G40 G80 T01
N3 M06
N4 G90 G54 G00 X1.0 Y1.5 S900 M03 T02
N5 G43 Z1.0 H01 M08
N6 G99 G82 R0.1 Z-0.14 P250 F7.0
N7 G80 Z1.0 M09
N8 G28 Z1.0 M05
N9 M01

(T02 - 1/4 DIA DRILL)
N10 T02
N11 M06
N12 G90 G54 G00 X1.0 Y1.5 S1100 M03 T01
N13 G43 Z1.0 H02 M08
N14 G99 G81 R0.1 Z-0.375 F6.0      (TOP PLATE)
N15 R-0.7 Z-1.25               (MIDDLE PLATE)
N16 G98 R-1.575 Z-2.0         (BOTTOM PLATE)
N17 G80 Z1.0 M09
N18 G28 Z1.0 M05
N19 M30
%
```

Note that a single hole has required three blocks of the program, rather than the usual one. Each block represents only one plate in the part. Also note the G98 in block N16. Only one hole is done in the example, so the G98 is not really needed. The cycle cancellation command G80 with a return motion in block N17 would take care of the tool retract from the hole. However, if more holes are machined, move the tool to the new XY position *before* the G80 is programmed. In this case, the G98 is needed when the drills penetrates the last plate of the part. This example is not a perfect solution to web drilling cuts, as there is still some wasted motion. The only efficient programming method is to use the optional custom macro technique and develop a unique and efficient web drilling cycle.

TAPPING

Tapping is second only to drilling as the most common hole making operation on CNC machining centers. As it is very common to tap on a CNC mill or a machining center, *two* tapping fixed cycles are available for programming applications on most control systems. They are the G84 cycle for normal tapping (R/H), and the G74 cycle for reverse tapping (L/H):

G84	Normal tapping - for *right hand* threads with M03 spindle rotation
G74	Reverse tapping - for *left hand* threads with M04 spindle rotation

The following example shows that programming a tapping for one hole is similar to other fixed cycles. All the tool motions, including spindle stop and reversal at the hole bottom are included in the fixed cycle:

```
...
N64 G90 G54 G00 X3.5 Y7.125 S600 M03 T06
N65 G43 Z1.0 H05 M08
N66 G99 G84 R0.4 Z-0.84 F30.0
N67 G80 ...
```

Is it possible to tell the tap size used? It should be. In the example, the tap size is a standard, 20 TPI (twenty threads per inch), plug tap. The XY coordinates are missing from the G84 cycle, because the current tool position has been established in block N64. The usual R level is the thread starting position and the Z depth is the absolute depth of thread. The last address in the block is feedrate in inches per minute (*in/min*), programmed with the F address.

Note that the R level of R0.4 has a value that is somewhat *higher* than might be used for drilling, reaming, single point boring and similar operations. Also, the programmed feedrate *appears* to be unreasonably high. There is a good reason for these values - they are both *correct* and selected *intentionally*.

The higher clearance for the R level allows *acceleration* of the feedrate from 0 to 30 inches per minute to take place *in the air*. When the tap contacts the part, cutting feedrate should be at its programmed value, *not less*. A good rule of thumb is to program the tapping clearance about *two to four times* the normal clearance. This clearance will guarantee the feedrate to be fully effective when the actual tapping begins. Try to experiment with a slightly smaller number, to make the program more efficient. Another good method is to *double, triple, or quadruple the pitch of the tap* and use that value as the clearance above the tap. Whichever method is used, its purpose is to eliminate the feedrate problems associated with motion acceleration.

Another question was the feedrate amount. The relatively high value of 30 in/min (F30.0) has also been carefully calculated. Any cutting feedrate for tapping *must* be synchronized with the *spindle speed* - the *r/min* programmed as the S address. Keep in mind that the tap is basically a *form* tool and the thread size and shape are built into it. Later in this chapter, the relationship between the spindle speed and the feedrate is explained in more detail. The cutting feedrate F in the program example was calculated by multiplying the thread *lead* by the spindle *speed* given as *r/min*:

```
F = 1 / 20 TPI × 600 r/min = 30.0 in/min
```

Another way to calculate feedrate is to divide the *spindle speed* (r/min) by the number of *threads per inch* (TPI):

```
F = 600 r/min / 20 TPI = 30.0 in/min
```

The quality of the tapped hole is also important, but it is not influenced solely by the correct selection of speeds and feeds, but by several other factors as well. The material of the tap, its coating, its geometry, the flute clearances, the helix configuration, the type of the start-up chamfer, the material being cut - and the tap holder itself - all have a very profound effect on the final quality of the tapped hole. For best results in tapping, a floating tap holder is mandatory, unless the CNC machine supports *rigid tapping*. The floating tap holder design gives the tap a 'feel', similar to the feel that is needed for manual tapping. A floating tap holder has is often called the *tension-compression* holder and its applications are the same for both milling and turning operations. This type of holder allows the tap to be pulled out of it or pushed into it, within certain range. The only noticeable difference is the mounting method of the tool (tool orientation) in the machine (either vertical or horizontal). High end floating tap holders also have an adjustable *torque*, which can change the feel of the tap and even the range of the tension and compression.

Tapping applications on CNC lathes are similar to those on machining centers. A special *tapping cycle* for a lathe control is not needed, as only one tap size can be used per part. Each tapping motion is programmed with the G32 command and block-by-block method.

CNC lathe tapping is different but not more difficult than tapping for CNC machining centers. Because it does not use fixed cycles, programmers make some common errors. This chapter uses examples for tapping on CNC lathes in a sufficient depth.

◆ Tap Geometry

There are literally dozens of tap designs used in various CNC programming applications. A whole book would easily be filled just on the topic of tapping tools and their applications. For CNC programming, only the *core basics* of tap geometry are important.

There are two considerations in tap design that directly influence the programming and the data input values:

❑ Tap flute geometry

❑ Tap chamfer geometry

Tap Flute Geometry

The *flute* geometry of a tap is described in tooling catalogues in terms such as 'low helix', 'high helix', 'spiral flute', and others. These terms basically describe *how* the tap cutting edges are ground into the body of the tap. When programming a tapping operation, the effectiveness of the flute geometry is tied to the spindle speed. Experimenting with the tapping feedrate is limited by the tap lead (pitch), but there is a greater latitude with the spindle speed selection. The part material *and* the flute geometry of the tap *both* influence the machine spindle speed. Since almost all tooling designs (not limited to taps only) are the results of corporate policies, engineering decisions and philosophies, various trade names and marketing strategies, there is not a one way of saying 'use this tool' or 'use that tool' for a CNC program. The tooling catalogue of a tool supplier is the best source of technical data, but a catalogue from another supplier may provide a better solution to a particular problem. Information gathered from a catalogue is a very good starting base for the data in the CNC program. Keep in mind that the all taps share some common characteristics.

Tap Chamfer Geometry

Tap *chamfer geometry* relates to the end configuration of the tap. For CNC programming, the most important part of the tap end point geometry is the *tap chamfer.*

In order to program a desired hole correctly, the tap must be selected according to the specifications of the hole being tapped. If tapping a blind hole, a different tap is required than for tapping a through hole. There are three types of taps, divided by their end geometry configuration:

❑ Bottoming tap

❑ Plug tap

❑ Taper tap

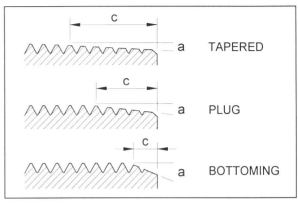

Figure 26-23

Typical tap end - chamfer geometry configuration

The major difference between the taps is the *length* of the tap chamfer. *Figure 26-23* shows how the characteristics of the drilled hole will influence programmed depth of the selected tap.

The tap chamfer length *c* is measured as the number of threads. A typical number of threads for a tapered chamfer is 8 to 10, for a plug tap 3 to 5, and for a bottoming tap 1 to 1.5. The angle of the chamfer *a* also varies for each type; typically 4-5° for the tapered tap, 8-13° for the plug tap and 25-35° for the bottoming tap.

A blind hole will almost always require a bottoming tap, a through hole will require a plug tap in most cases and a taper tap in some rarer cases. Described in different words, the larger the tap chamfer, the greater depth allowance must be added to each drilled hole.

◆ Tapping Speed and Feedrate

The relationship of the machine spindle speed (r/min) and the programmed cutting feedrate is extremely important when programming the cutting motion in feedrate *per time* mode. *Per time* mode is programmed as *in/min* (inches per minute) in programs using English units, and *mm/min* (millimeters per minute) for the metric units programming. This per minute mode is typical to CNC milling machines and machining centers, where virtually all work is done either in *in/min* or *mm/min*. For tapping operations, regardless of the machine tool, always program the cutting feedrate as the *linear distance* the tap must travel during *one spindle revolution*. This distance is always equivalent to the *lead* of the tap, which is the same as the tap *pitch* (for tapping only), because taps are normally used to cut a single start thread only.

When using the *feedrate per revolution* mode, mode that is typical to CNC lathes, the tap lead is always equivalent to the feedrate. For example, the lead of .050 results in .050 in/rev feedrate, or F0.05 in the program.

On CNC machining centers, the typical feedrate mode is always *per time*, measured in *per minute* mode, and the feedrate is calculated by one of the following formulas:

$$F_t = \frac{r/min}{TPI}$$

☞ where ...

F_t	=	Feedrate per time (per minute)
r/min	=	Spindle speed
TPI	=	Number of threads per inch

A similar formula will produce an identical result:

$$F_t = r/min \times F_r$$

☞ where ...

F_t	=	Feedrate per time (per minute)
r/min	=	Spindle speed
F_r	=	Feedrate per revolution

For example, a 20 TPI thread lead for a mill will be:

```
1 / 20 = .0500 inches
```

and the programmed feedrate has to take into consideration the machine spindle speed, for example, 450 r/min:

```
F = 450 × .05 = 22.5 = F22.5 (in/min)
```

A metric tap on a lathe uses the same logic. For example, a tap of 1.5 mm lead (pitch) using 500 r/min is programmed with the feedrate of 750 mm/min:

```
F = 500 × 1.5 = 750.00 = F750.0 (mm/min)
```

The key to successful tapping is to maintain the relationship of the *tap lead* and the *spindle speed*. If the spindle speed is changed, the feedrate per time (*in/min* or *mm/min*) must be changed as well. For many tension-compression type tap holders, adjustment of the feedrate *downwards* (so called underfeed) by about five percent may yield better results. This is because the tension of the tapping holder is more flexible than the compression of the same holder.

If the spindle speed in the above example is changed from S450 to S550 (tap size is *unchanged* at 20 TPI), the spindle speed change must be reflected in a *new* tapping feedrate:

```
F = 550 × .05 = 27.50 = F27.5 (in/min)
```

In the program, the new tapping feedrate will be:

```
F = 27.5 - 5% = 26.125
```

The actual feedrate value would be F26.1 or even F26.0. It is easy to change the spindle speed of the tool in the program, or even directly on the CNC machine, then forget to modify the feedrate for the tapping tool itself. This mistake can happen during program preparation in the office or during program optimization at the machine. If the change is small, there may be no damage, more due to luck than intent. If the change of spindle speed is major, the tap will most likely break in the part.

◆ **Pipe Taps**

Pipe taps are similar in design to standard taps. They belong to two groups:

❑ Taper taps   NPT and API

❑ Straight taps (parallel)   NPS

Their size designation (nominal size), is *not* the size of the tap, but the size of the pipe fitting. *American National Standard* pipe taper (NPT) has a taper ratio of 1 to 16, or 3/4 inch per foot (1.78991061° per side) and the tap chamfer is 2 to 3-1/2 threads.

Programming for pipe taps follows the usual considerations for standard threads. The only common difficulty is how to calculate the Z depth position at least as a reasonable one, if not exactly. The final depth may be a subject of some experimentation with a particular tap holder and typical materials.

A proper tap drill size is very important. It will be different for tap holes that are only drilled and for tap holes that are drilled and reamed (using a 3/4 per foot taper reamer).

The following is a table of taper pipe thread sizes for NPT group and recommended tap drills, data that is useful for CNC programming:

NPT Group		Drilled Only		Taper Reamed	
Pipe Size	TPI	Tap Drill	Dec. Size	Tap Drill	Dec. Size
1/16	27	D	.2460	15/64	.2344
1/8	27	Q	.3320	21/64	.3281
1/4	18	7/16	.4375	27/64	.4219
3/8	18	37/64	.5781	9/16	.5625
½	14	45/64	.7031	11/16	.6875
3/4	14	29/32	.9062	57/64	.8906
1.0	11-1/2	1-9/64	1.1406	1-1/8	1.1250
1-1/4	11-1/2	1-31/64	1.4844	1-15/32	1.4688
1-1/2	11-1/2	1-47/64	1.7344	1-23/32	1.7188
2.0	11-1/2	2-13/64	2.2031	2-3/16	2.1875

For the straight pipe thread sizes (NPS), the following tap drills are recommended:

Pipe Size	TPI	Tap Drill	Decimal Size
1/16	27	1/4	.2500
1/8	27	11/32	.3438
1/4	18	7/16	.4375
3/8	18	37/64	.5781
½	14	23/32	.7188
3/4	14	59/64	.9219
1.0	11-1/2	1-5/32	1.1563
1-1/4	11-1/2	1-1/2	1.5000
1-1/2	11-1/2	1-3/4	1.7500
2.0	11-1/2	2-7/32	2.2188

The tapping feedrate maintains the same relationships for pipe taps as for standard taps.

◆ Tapping Check List

When programming a tapping operation, make sure the program data reflect the true machining conditions. They may vary between setups, but the majority of them are typical to *any* tapping operations on *any* type of CNC machine. Here is a short list of items that relate *directly* to the tapping operations in CNC programming.

- ❏ Tap cutting edges (have to be sharp and properly ground)
- ❏ Tap design (has to match the hole being tapped)
- ❏ Tap alignment (has to be aligned with tapped hole)
- ❏ Tap spindle speed
  (has to be reasonable for the cutting conditions)
- ❏ Tap feedrate (has to be related to the tap lead and the machine spindle speed)
- ❏ Part setup
  (rigidity of the machine setup and the tool is important)
- ❏ Drilled hole must be premachined correctly
  (tap drill size is important)
- ❏ Clearance for the tap start position
  (allow clearance for acceleration)
- ❏ Cutting fluid selection
- ❏ Clearance at the hole bottom
  (the depth of thread must be guaranteed)
- ❏ Tap holder torque adjustment (ease of cutting)
- ❏ Program integrity (no errors)

Many designs of tap holders have their own special requirements, which may or may not have any effect on the programming approach. If in doubt, always check with the tap holder manufacturer for the suggested operation.

With modern CNC machines, the method of *rigid tapping* has become quite popular. There is no need for special tapping holders, such as the *tension compression* type - regular end mill holders or strong collet chucks can be used, saving the cost of tool holders. However, the CNC machine and its control system must support the rigid tapping feature. To program rigid tapping, there is a special M code available - check the machine documentation.

> The rigid tapping mode must be supported by
> the CNC machine before it can be used in a program

HOLE OPERATIONS ON A LATHE

Single point hole operations on a CNC lathe are much more limited than those on a CNC machining center. First, the number of holes that can be drilled or tapped in a single operation on a lathe is only one per part operation (two are rare), while the number of holes for a milling application may be in tens, hundreds and even thousands. Second, the boring (internal turning) on a lathe is a *contouring* operation, unlike boring on a milling machine, which is a *point-to-point* operation.

All the point-to-point machining operations on a CNC lathe are limited to those that can be machined with the cutting tool positioned at the spindle *centerline*. Typically, these operations include center drilling, standard drilling, reaming and tapping. A variety of other cutting tools may also be used, for example, a center cutting end mill (slot drill) to open up a hole or to make a flat bottom hole. An internal burnishing tool may also be used for operations such as precise sizing of a hole, etc. To a lesser degree, other operations, such as counterboring and countersinking may be programmed at the lathe spindle centerline, with a special point-to-point tool - *not a contouring tool*. All operations in this group will have one common denominator - they are all used at the spindle centerline and programmed with the X position as X0 in the program block.

The spindle speed for *all* centerline operations on a CNC lathe must be programmed in actual revolutions per minute (*r/min*), *not* in the constant surface speed mode (*CSS*). For that reason, G97 is used - for example,

`G97 S575 M03`

will assure the required 575 r/min at the normal spindle rotation (at 100% spindle speed override).

What will happen if CSS mode is used with G96 command, rather than the proper G97 command? The CNC system will use the given information, the spindle speed address S, in the program (given in peripheral - or surface - speed per minute, as *ft/min*). The system will then calculate the required spindle speed in *r/min* for the use by the machine tool.

The calculation is based on the standard mathematical formula that relates to the part *diameter*. If the diameter is zero - which is *exactly* what it is at the spindle centerline - the spindle revolutions will always be the *highest r/min* that is available in the currently selected spindle gear range. This calculation is an exception to the standard r/min calculation formula, where the spindle speed at the centerline (diameter zero) would be zero - yes, *0 r/min*!

For example, if the peripheral (surface) speed for a given material is 450 ft/min, the r/min at a $\varnothing$3 inch (X3.0) for the same material will be approximately:

```
S = (450 × 3.82) / 3 = 573 rpm
```

If the same speed of 450 ft/min is applied to the diameter zero (X0 in the program), the formula does not change, but the resulting action does:

```
S = (450 × 3.82) / 0 = 0 r/min          (ERROR)
```

Although the spindle might be expected to stop (because of the mathematical laws), it will do the *exact opposite* (because of the control design). Spindle speed will reach the maximum *r/min* that the current gear range will allow. Be very careful here - make sure that the centerline operations on a CNC lathe are always done in the G97 (r/min) mode and *not* in the G96 mode (CSS) mode.

♦ Tool Approach Motion

A typical geometry offset configuration setup (or the G50 values) on a CNC lathe often have a relatively large X value and relatively small Z value. For example, the geometry offset for a tool may be X-11.8Z-1.0 (or G50X11.8Z1.0). This location indicates a suitable tool change position applicable to a drill. What does it mean to the tool motion for a drilling operation?

It means that the rapid motion will complete the Z axis motion long before completing the X axis motion (with the infamous hockey-stick motion of the rapid command). The result is a tool motion *very close* to the part face:

```
N36 T0200 M42
N37 G97 S700 M03
N38 G00 X0 Z0.1 T0202 M08
N39 ...
```

To avoid a potential collision during the tool approach towards the part, use one of the following methods:

❏ Move the X axis first to the spindle centerline, then the Z axis, directly to the start location for the drilling

❏ Move the Z axis first to a clear position, then the X axis to the spindle centerline, then complete the Z axis motion into the drilling start position

The first method may be practical only in those cases when the tool motion area is absolutely clear and has no obstacles in the way (do not count on such a situation). The second method, and probably the most common in programming, will first move the Z axis close (but not too close) to the part, say .50 inch in the front (Z0.5). The motion that follows is the X axis motion only - directly to the centerline (X0). At this point, the cutting tool (such as a drill) is far from the Z axis face. The last approach motion will be to the Z axis start position, closer to the part face, where the actual drilling cut begins. This method eliminates (or at least minimizes) the possibility of a collision with obstacles along the way. The obstacles are - or at least could be - the tailstock, the parts catcher, the steadyrest, the fixture, the face plate, etc. The example of this programming tool path method is the previous example, modified:

```
N36 T0200 M42
N37 G97 S700 M03
N38 G00 X0 Z0.5 T0202 M08
N39 Z0.1
N40 ...
```

This programming method splits the tool approach along the Z axis into *two* tool positions - one is the safe clearance for approach, the other one is the safe clearance position for the drill start. There is a minor alternative to this motion - the last Z axis approach will be at a cutting feedrate, rather than at a rapid motion rate:

```
N36 T0200 M42
N37 G97 S700 M03
N38 G00 X0 Z0.5 T0202 M08
N39 G01 Z0.1 F0.05
N40 ...
```

For the last approach motion, the Z axis motion has been changed to a linear motion, with a relatively high feedrate of .05 in/rev (1.25 mm/rev). Feedrate override can be used for setup, to control the rate of the feed. During actual production, there will be no significant loss in the cycle time.

♦ Tool Return Motion

The same logical rules of motion in space that apply to the tool approach, apply also to the tool return motion. Remember that the first motion *from* a hole must always be along the Z axis:

```
...
N40 G01 Z-0.8563 F0.007
N41 G00 Z0.1
...
```

In block N40, the actual drill cutting motion takes place. When the cut is completed, block N41 is executed. The drill will rapid out of the hole to the same position it started from (Z0.1). It is not necessary to return to the same position, but it makes the programming style more consistent.

Once the cutting tool is safely out of the hole, it has to return to the tool changing position. There are two methods:

❑ Simultaneous motion of both axes

❑ Single axis at a time

Simultaneous motion of the X and Z axes does not present the same problem as it did on the approach - on the contrary. The Z axis will complete the motion first, moving away from the part face. Also, there is no reason to fear a collision during a return motion if the approach motion was successful and the programming style was consistent:

```
...
N70  G01 Z-0.8563 F0.007
N71  G00 Z0.1
N72  X11.0 Z2.0 T0200 M09
...
```

If in doubt, or if an obstacle is expected to be in the way of a tool motion, for example a tailstock, program a single axis at a time. In most cases, that will move the positive X axis first, as most obstacles would be to the right of the part:

```
...
N70  G01 Z-0.8563 F0.007
N71  G00 Z0.1
N72  X12.0
N73  Z2.0 T0200 M09
...
```

The programming example illustrates the return motion with the X axis programmed first. The fact that the tool is .100 off the front face is irrelevant - after all, the tool started cutting from that distance without a problem.

Other, less traditional, methods for the tool motion towards and away from the part are also possible.

◆ Drilling and Reaming on Lathes

Drilling on a CNC lathe is also quite common operation, mainly as means for a hole opening to be used with other tools, such as boring bars. There are three basic kinds of drilling, typical to a CNC lathe machining:

❑ Center drilling and spotfacing

❑ Drilling with a twist drill

❑ Indexable insert drilling

Each method follows the same programming techniques as those described in the milling section earlier, except that there are no fixed cycles of the milling type used for the lathe work. Keep in mind that on a CNC lathe, the part is rotating, whereby the cutting tool remains stationary. Also keep in mind that most lathe operations take place in a horizontal orientation, causing concerns about coolant direction and chip removal.

◆ Peck Drilling Cycle - G74

On Fanuc and compatible controls, there is a multiple repetitive cycle G74 available, that can be used for two different machining operations:

❑ Simple roughing with chip breaking

❑ Peck drilling (deep hole drilling)

In this section, the peck drilling usage of the G74 cycle is described. The roughing application of the G74 cycle is a contouring operation and is very seldom used.

In peck drilling, just like in any ordinary drilling, select the spindle speeds and cutting feedrates first, then determine the hole starting position and finally, its depth position. In addition, establish (or even calculate) the depth of each peck. The lathe cycle G74 is rather limited in what it can do, but it has its uses. Its format for peck-drilling is:

```
G74 X0 Z.. K..
```

☞ where ...

G74	=	Peck drilling cycle selection
X0	=	Indicates cutting on centerline
Z	=	Specifies the end point for drilling
K	=	Depth of each peck (always positive)

The following program uses illustration in *Figure 26-24*, and shows an example of drilling a 3/16 hole (∅.1875) with a peck drill depth of .300 inches:

```
...
N85  T0400 M42
N86  G97 S1200 M03
N87  G00 X0 Z0.2 T0404 M08
N88  G74 X0 Z-0.8563 K0.3 F0.007
N89  G00 X12.0 Z2.0 T0400 M09
N90  M01
```

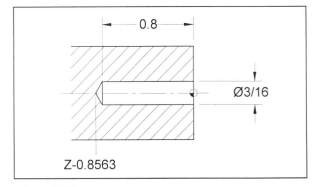

Figure 26-24

Sample hole for the peck drilling lathe example

The peck drilling motion will start from the Z0.2 position in block N87 and continue to the Z-0.8563 position in block N88. That results in a 1.0563 long cut. Calculation of the number of pecks is the same as in milling.

With the .300 length of each peck, there will be the total of three *full* length pecks and one partial length peck, at the following Z axis locations:

```
Z-0.1
Z-0.4
Z-0.7
Z-0.8563
```

Although the first three pecks are .300 deep each, the first one starts at Z0.2 and ends at Z-0.1. That will result in two thirds of the cut being in the air. Programmer has to decide when this approach is an advantage and when another method would be more suitable. At the end of each peck motion using the G74 cycle, the drill will make a small retract by a *fixed* distance. This distance is set by a parameter of the control system and is typically about .020 inches (0.5 mm). A full retraction after each peck out of the hole (similar to the G83 cycle for milling controls) is not supported by the G74 cycle.

Note that there is no programmed motion *out* of the hole when the peck drilling cycle is completed. This return motion is built-in within the G74 cycle. If a tool motion such as G00Z0.2M05 follows block N88, no harm is done. It may give the operator extra confidence when running the job.

♦ Tapping on Lathes

Tapping on CNC lathes is a common operation that follows the same machining principles as tapping on machining centers. The major difference for lathes is the absence of a tapping cycle. There is no real need for a tapping cycle on a lathe, since most of lathe tapping operations machine only *one hole of the same type*. The absence of a tapping cycle may present some unexpected difficulties. Unfortunately, they are more common among programmers with limited experience. Before evaluating these difficulties

closer, it is important to know the tool that holds the tap (the *tap holder*) and the tapping process on lathes in general.

The selected tap should always be mounted in a special tapping holder; the best type is one with tension and compression features, known as the *tap floating holder*. Never use a drill chuck or a similar solid device - it will break the tap quickly and possibly damage the part as well.

Since there is no fixed cycle for tapping on a typical CNC lathe, each tool motion is programmed as a separate block. To do that, and to find out how to tap properly, let's first evaluate the process for a typical right hand tap in general, applied to a lathe operation:

Step 01   Set coordinate position XZ
Step 02   Select tool and gear range
Step 03   Select spindle speed and rotation
Step 04   Rapid to the center line and clearance with offset
Step 05   Feed-in to the desired depth
Step 06   Stop the spindle
Step 07   Reverse the spindle rotation
Step 08   Feed-out to clear of the part depth
Step 09   Stop the spindle
Step 10   Rapid to the starting position
Step 11   Resume normal spindle rotation or end program

Translated into a CNC program carefully, this step by step procedure can be used in everyday programming as a general guide to tapping on CNC lathes.

Figure 26-25 shows the layout of the part and the tool setup, used for programming example O2607. The example program O2607 follows the eleven steps described above literally and is based on a very solid foundation. Technically, the program O2607 is correct - but only technically, not practically. Are there possible problem situations in the example O2607?

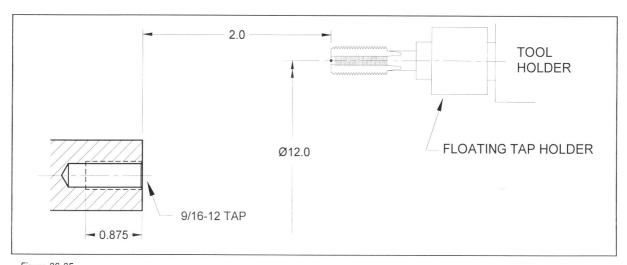

Figure 26-25
Typical setup of a tapping tool on a CNC lathe - program examples O2607 and O2608

```
O2607 (TAPPING ON LATHES)
(ONLY THEORETICALLY CORRECT VERSION)
...
(T02 - TAP DRILL 31/64)
...
...
N42 M01

(T03 - 9/16-12 PLUG TAP)
N43 T0300 M42
N44 G97 S450 M03
N45 G00 X0 Z0.5 M08 T0303
N46 G01 Z-0.875 F0.0833
N47 M05
N48 M04
N49 Z0.5
N50 M05
N51 G00 X12.0 Z2.0 T0300 M09
N52 M30
%
```

A brief look at the program O2607 does not show that anything is wrong. After all, the program covers all the necessary motions and is, therefore, correct. *Yet, this program contains major flaws !*

All earlier tapping steps have been carefully followed. Conducting a more in-depth study of the program will reveal two areas of potential difficulty or even danger. The first problem may arise if the feedrate override setting switch is *not* set to 100%. Remember, the tapping feedrate is always equal to the thread lead (F0.0833 is the feedrate for 12 TPI). If the override switch is set to any other value but 100%, the thread will be stripped at best and the tap broken at worst with related part damage.

The other problem will become evident only in a single block mode run, during setup or machining. Look at blocks N46 and N47. In the N46 block, tap reaches the Z axis end position - while the spindle is still rotating! True, the spindle will be stopped in block N47, but in the single block mode it will be too late. A similar situation will happen during the feed-out motion. The spindle reverses in block N48, but does not move until the N49 block is processed. Therefore, the program O2607 is a *very poor* example of tapping on CNC lathes.

These are some details usually not considered for a fixed cycle application (such as the G84 tapping cycle), when used for milling programs. For milling, all tool motions are built-in, so they are contained *within* the fixed cycle. To eliminate the first potential problem of the feedrate override, programming the M48/M49 functions will temporarily disable the feedrate override switch. Even better way is to replace the feed-in and feed-out tap motion command from the current G01 mode to G32 mode (G33 on some

controls). The G32 command is normally used for single point threading. Two major results will be achieved with the G32 command - the *spindle will be synchronized*, and the *feedrate override will be ineffective* by default (automatically). The second problem will be solved if the spindle M functions are programmed in the *same block* as the tool motion. That means joining the block N46 with N47, and block N48 with N49. This much improved version of the tapping example is in the new program O2608.

```
O2608 (TAPPING ON LATHES)
(PRACTICALLY CORRECT VERSION)
...
(T02 - TAP DRILL 31/64)
...
...
N42 M01

(T03 - 9/16-12 PLUG TAP)
N43 T0300 M42
N44 G97 S450 M03
N45 G00 X0 Z0.5 M08 T0303
N46 G32 Z-0.875 F0.0833 M05
N47 Z0.5 M04
N48 M05
N49 G00 X12.0 Z2.0 T0300 M09
N50 M30
%
```

The block (N48 in the example) containing the spindle stop function M05, is not required if the tap is the *last* tool in the program, although it does no harm in any other program. Compare this program O2608 with program O2607. Program O2608 is a great deal more stable and the possibility of any significant problem is virtually eliminated.

◆ Other Operations

There are many other programming variations relating to machining holes on CNC machining centers and lathes. This chapter has covered some of the most important and the most common possibilities.

Some less common applications, such as machining operations using tools for backboring, or block boring tools, tools with multiple cutting edges and other special tools for machining holes may be quite infrequent in programming. However, programming these *unusual* operations is no more difficult then programming the ordinary everyday tool motions, using ordinary everyday tools.

The real ability of a CNC programmer is measured in terms of applying the past knowledge and experience to a new problem. It requires a thinking process and it required a degree of ingenuity and hard work.

27 PATTERN OF HOLES

In the point-to-point machining operations, consisting of drilling, reaming, tapping, boring, etc., we are often required to machine either a single hole or a series of holes with the same tool, usually followed by other tools. In practice, several holes are much more common than a single hole. Machining several holes with the same tool means machining a *pattern* of holes or a *hole pattern*. An English dictionary defines the word 'pattern' as a *'characteristic or consistent arrangement or design'*. Translated to the hole machining terms, any two or more holes machined with the same tool establish a pattern. The desired hole pattern is laid out in the part drawing either randomly (*characteristic arrangement or design*) or in a certain order (*consistent arrangement or design*). Dimensioning of a hole pattern follows standard dimensioning practices.

This chapter describes some typical hole patterns laid out on a flat part and the various methods of their programming. To make matters simple, all programming examples related to the hole patterns will assume a center drilling operation, using a #2 center drill, with chamfer diameter .150, to the depth of .163 (programmed as Z-0.163). The program zero (program reference point Z0) is the top face of part and the tool is assumed to be already in the spindle. For the purposes of clarity, no hole diameters or material size and thickness are specified in the examples.

From the dictionary definition above, we have to establish what makes a hole pattern *characteristic* or *consistent*. Simply, any series of holes that are machined with the same tool, one hole after another, usually in the order of convenience. That means all holes within a single pattern have the same nominal diameter. It also means that all machining must start at the same R level and end at the same Z depth. Overall, it means that all holes within a pattern are machined the same way for any single tool.

TYPICAL HOLE PATTERNS

Hole patterns can be categorized into several typical groups, each group having the same character. Every hole pattern encountered in CNC programming belongs into one of the following pattern groups:

❑ Random pattern

❑ Straight row pattern

❑ Angular row pattern

❑ Corner pattern

❑ Grid pattern

❑ Arc pattern

❑ Bolt circle pattern

Some groups may be divided further into smaller subgroups. A thorough understanding of each pattern group should help you to program any similar hole pattern.

There are several control systems available that have a built-in hole pattern programming, for example for a *bolt circle pattern*. These programming routines simplify the hole pattern programming quite substantially, but the program structure is usually unique to that particular brand of control and cannot be applied to other controls.

RANDOM HOLE PATTERN

The most common pattern used in programming holes is a *random pattern*. Random pattern of holes is a pattern where all holes share the same machining characteristics, but the X and Y distances between them are inconsistent. In other words, holes within a random pattern share the same tool, the same nominal diameter, usually the same depth, but a variable distance from each other - *Figure 27-1*.

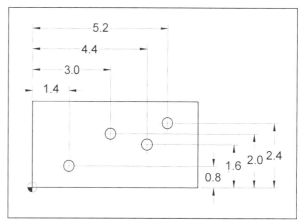

Figure 27-1

Random pattern of holes - program example O2701

There are no special time saving techniques used in programming a random pattern - only a selected fixed cycle used at individual hole locations. All XY coordinates within the hole pattern have to be programmed manually; the control system features will be no help here at all:

```
O2701 (RANDOM HOLE PATTERN)
N1 G20
N2 G17 G40 G80
N3 G90 G54 G00 X1.4 Y0.8 S900 M03
N4 G43 Z1.0 H01 M08
N5 G99 G81 R0.1 Z-0.163 F3.0
N6 X3.0 Y2.0
N7 X4.4 Y1.6
N8 X5.2 Y2.4
N9 G80 M09
N10 G28 Z0.1 M05
N11 G28 X5.2 Y2.4
N12 M30
%
```

STRAIGHT ROW HOLE PATTERN

Hole patterns parallel to the X or Y axis with an equal pitch is a *straight row* pattern. *Figure 27-2* shows a 10 hole pattern along the X axis, with a pitch of .950 inch.

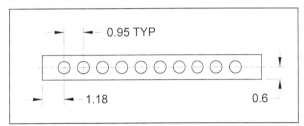

Figure 27-2
Straight row hole pattern - program example O2702

The programming approach takes advantage of a fixed cycle repetition feature, using the L or K address. It would be inefficient to program each hole individually. As always, the tool will be positioned at the first hole in G90 mode, then the cycle will machine the first hole in block N5.

For the remaining holes, G90 mode must be changed to incremental mode G91, which instructs the control to machine the other nine holes incrementally, along the X axis only. The same logic would also apply for a vertical pattern along the Y axis. In that case, the pitch increment would be programmed along the Y axis only. Note that the repetition count is always equal to the *number of spaces*, not the number of holes. The reason? The first hole has already been machined in the cycle call block.

```
O2702 (STRAIGHT ROW HOLE PATTERN)
N1 G20
N2 G17 G40 G80
N3 G90 G54 G00 X1.18 Y0.6 S900 M03
N4 G43 Z1.0 H01 M08
N5 G99 G81 R0.1 Z-0.163 F3.0
N6 G91 X0.95 L9
N7 G80 M09
N8 G28 Z0 M05
N9 G28 X0 Y0
N10 M30
%
```

Two features of program O2702 should be emphasized. In block N6, the dimensioning mode was changed from the absolute G90 to the incremental G91, to take advantage of the equal pitch distance. When all ten holes have been machined, the program has to include return to the machine zero position motion, in the example, along all three axes. However, without a calculation, *we do not know* the absolute position at the tenth hole for the X axis (the Y axis remains unchanged at the position of .60 inches = Y0.6). To solve this 'problem', cancel the cycle with G80, leave the G91 mode in effect and move to the machine zero position in the Z axis first (for safety reasons). Then - still in the incremental mode G91 - return both X and Y axes to the machine zero simultaneously.

Normally, this first tool of the example would be followed by other tools to complete the hole machining. To protect the program and machining from possible problems, make sure that the G90 absolute command is reinstated for every tool that follows.

ANGULAR ROW HOLE PATTERN

Pattern of holes in a row at an angle is a variation of a straight line pattern. The difference between the two is that the incremental pitch applies to both X and Y axes. A hole pattern of this type will be established on the part drawing as one of the two possible dimensioning methods:

❑   X and Y coordinates are given for the first and the last hole

In this method, the pattern angular position is not specified and no pitch between holes is given.

❑   X and Y coordinates are given for the first hole only

In this method, pattern angular position is specified and the pitch between the holes is given.

In either case, all the necessary X and Y dimensions are available to write the program. However, the programming approach will be different for each method of drawing dimensioning.

◆ Pattern Defined by Coordinates

This method of programming is similar to the straight row pattern. Since the pitch between holes is not given, the increment between holes along each of the two axes must be calculated. This axial distance is commonly known as the *delta distance* (delta X is measured along the X axis, delta Y is measured along the Y axis). Such a calculation can be done in two equally accurate ways.

The first calculation method can use a trigonometric method, but it is much easier to use the ratio of sides instead. In the *Figure 27-3*, the pattern length along the X axis is 10.82 and along the Y axis it is 2.0:
(2.625 - .625 = 2.0)

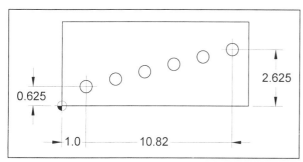

Figure 27-3

Angular hole pattern with two sets of coordinates - program O2703

Pattern of this kind has all the holes spaced by equal distances along X and Y axes. As all holes are equally spaced, the ratio of the sides for individual holes is identical to the ratio of the whole pattern. When expressed mathematically, the increment between holes along the X axis is equal to the overall distance of 10.82 divided by the number of X axis spaces; the increment along the Y axis is equal to the overall distance of 2.0 divided by the number of Y axis spaces. The number of spaces for a six hole pattern is five, so the X axis increment (the delta X) is:

```
10.82 / 5 = 2.1640
```

and the Y axis increment (the delta Y) is:

```
2.0 / 5 = .4
```

The other calculation method uses trigonometric functions, which may also be used as a confirmation of the first method, and vice versa. *Both results must be identical,* or there is a mistake somewhere in the calculation. First, establish some temporary values:

```
A = tan⁻¹(2.0 / 10.82) = 10.47251349°
```
$$A = \tan^{-1}(2.0 / 10.82) = 10.47251349°$$

```
C = 2.0 / sinA = 11.00329063
```

```
C1 = C / 5 = 2.20065813
```

Now, the actual increment along the two axes can be calculated, using C1 dimension as the distance between holes:

```
X increment = C1 × cosA = 2.1640
Y increment = C1 × sinA = .4000
```

The calculated increments match in both methods, calculation is correct, and can now be used to write the program (O2703) - block N6 contains the values:

```
O2703 (ANGULAR ROW 1)
N1 G20
N2 G17 G40 G80
N3 G90 G54 G00 X1.0 Y0.625 S900 M03
N4 G43 Z1.0 H01 M08
N5 G99 G81 R0.1 Z-0.163 F3.0
N6 G91 X2.164 Y0.4 L5 (K5)
```

```
N7 G80 M09
N8 G28 Z0 M05
N9 G28 X0 Y0
N10 M30
%
```

Note that the program structure is identical to the example of the straight row pattern, except the incremental move with L5 (K5) address is along two axes instead of one.

◆ Pattern Defined by Angle

An angular line pattern can also be defined in the drawing by the X and Y coordinates of the first hole, the number of equally spaced holes, the distance between holes and the angle of pattern inclination - *Figure 27-4*.

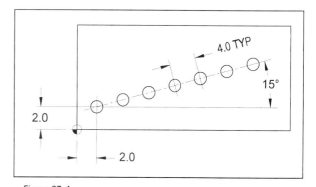

Figure 27-4

Angular hole pattern with coordinates, pitch and angle - O2704

In order to calculate the X and Y coordinate values, use trigonometric functions in this case:

```
X = 4.0 × cos15 = 3.863703305
Y = 4.0 × sin15 = 1.03527618
```

Program can be written after you round off the calculated values - program O2704:

```
O2704 (ANGULAR ROW 2)
N1 G20
N2 G17 G40 G80
N3 G90 G54 G00 X2.0 Y2.0 S900 M03
N4 G43 Z1.0 H01 M08
N5 G99 G81 R0.1 Z-0.163 F3.0
N6 G91 X3.8637 Y1.0353 L6 (K6)
N7 G80 M09
N8 G28 Z0 M05
N9 G28 X0 Y0
N10 M30
%
```

Since the calculated increments are rounded values, a certain accumulative error is inevitable. In most cases, any error will be well contained within the required drawing tolerances. However, for the projects requiring the highest precision, this error may be important and must be taken into consideration.

To make sure all calculations are correct, a simple checking method can be used to compare the calculated values:

➲ Step 1

Find the absolute coordinates XY of the *last* hole:

```
X   =   2.0 + (4.0 × 6 × cos15)
    =   25.18221983 = X25.1822

Y   =   2.0 + (4.0 × 6 × sin15)
    =   8.211657082 = Y8.2117
```

➲ Step 2

Compare these new XY coordinates with the previously calculated increments as they relate to the last hole of the pattern (using rounded values):

```
X   =   2.0 + 3.8637 × 6 = 25.1822
Y   =   2.0 + 1.0353 × 6 =  8.2118
```

Note that both X and Y values are accurate. When rounding, particularly when a large number of holes is involved, the accumulative error may cause the hole pattern out of tolerance. In that case, the only correct way to handle the programming is to calculate the coordinates of each hole as absolute dimensions (that means from a *common* point rather than a *previous* point). The programming process will take a little longer, but it will be much more accurate.

CORNER PATTERN

Pattern of holes can be arranged as a corner - which is nothing more than a pattern combining the straight and/or angular hole patterns - *Figure 27-5.*

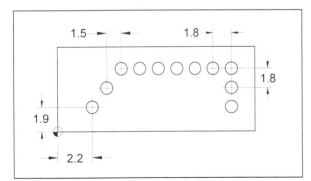

Figure 27-5

Corner pattern of holes - program example O2705

All rules mentioned for the straight and angular hole patterns apply for a corner pattern as well. The most important difference is the corner hole, which is *common* to two rows. A corner pattern can be programmed by calling a fixed cycle for each row. Soon, it will become apparent that each

corner hole will be machined *twice*. Visualize the whole process - the *last* hole of one row pattern is also the *first* hole of the next pattern, duplicated. Creating a special custom macro is worth the time for many corner patterns. The normal solution is to move the tool to the first position, call the required cycle and *remain within* that cycle:

```
O2705 (CORNER PATTERN)
N1 G20
N2 G17 G40 G80
N3 G90 G54 G00 X2.2 Y1.9 S900 M03
N4 G43 Z1.0 H01 M08
N5 G99 G81 R0.1 Z-0.163 F3.0
N6 G91 X1.5 Y1.8 L2 (K2)
N7 X1.8 L6 (K6)
N8 Y-1.8 L2 (K2)
N9 G80 M09
N10 G28 Z0 M05
N11 G28 X0 Y0
N12 M30
%
```

The program offers no special challenges. In block N6, the angular row of holes is machined, starting from the lower left hole, in N7 it is the horizontal row of holes, and in N8 the vertical row of holes is machined. The order is continuous. Just like in the earlier examples, keep in mind that the repetition count L or K is for the number of moves (spaces), not the number of holes.

GRID PATTERN

Basic straight grid pattern can also be defined as a set of equally spaced vertical and horizontal holes, each row having equally spaced holes. If the spacing of all vertical holes is the same as the spacing of all horizontal rows, the final grid pattern will be a square. If the spacing of all vertical holes is not the same as the spacing of all horizontal rows, the resulting grid pattern is a rectangle. A grid pattern is sometimes called a *rectangular hole pattern - Figure 27-6.*

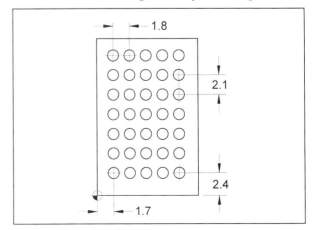

Figure 27-6

Rectangular grid hole pattern - program example O2706

A grid pattern is very similar to a series of corner patterns, using similar programming methods. The major consideration for a grid pattern programming is in its efficiency. Each row can be programmed as a single row pattern, starting, for example, from the left side of each row. Technically, that is correct, although not very efficient due to the loss of time, when the tool has to travel from the last hole of one row, to the first hole of the next row.

More efficient method will look like a *zigzag* motion. To program a zigzag motion, program the first row or column starting at any corner hole. Complete that row (column), then jump to the nearest hole of the next row (column) and repeat the process until all rows and columns are done. The wasted time of the rapid motion is kept to the minimum.

```
O2706 (STRAIGHT GRID PATTERN)
N1 G20
N2 G17 G40 G80
N3 G90 G54 G00 X1.7 Y2.4 S900 M03
N4 G43 Z1.0 H01 M08
N5 G99 G81 R0.1 Z-0.163 F3.0
N6 G91 Y2.1 L6 (K6)
N7 X1.8
N9 Y-2.1 L6 (K6)
N10 X1.8
N11 Y2.1 L6 (K6)
N12 X1.8
N13 Y-2.1 L6 (K6)
N14 X1.8
N15 Y2.1 L6 (K6)
N16 G80 M09
N17 G28 Z0 M05
N18 G28 X0 Y0
N19 M30
%
```

Two features of the program are worth noting - one is the jump from one row of the pattern to another - it has no repetition address L or K, because only *one* hole is being machined at that location. The second feature may not be so obvious right away. To make the program shorter, start along the axis that contains the *larger* number of holes (the Y axis in the program example O2706). This example is a variation on the previous examples and also adheres to all the rules established so far. A special subprogram made for a grid pattern is also a common programming approach and can be used as well.

◆ **Angular Grid Pattern**

Although the straight grid pattern is the most common pattern for square and rectangular hole arrangement, a grid pattern may also be in the shape of a parallelogram, called an *angular grid pattern - Figure 27-7*.

Again, the programming approach remains the same as for the rectangular grid pattern, the only extra work required is the calculation of the angular increments, similar to previous methods:

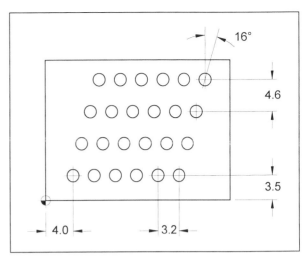

Figure 27-7
Angular grid hole pattern - program example O2707

The unknown increment in the drawing is the distance measured along the X axis, from a hole in one horizontal row to the next hole in following horizontal row:

```
X = 4.6 × tan16 = 1.319028774 (X1.319)
```

The program can be written in a similar way as for the straight row grid, except the extra 'jump' between rows will take place along both axes:

```
O2707 (ANGULAR GRID)
N1 G20
N2 G17 G40 G80
N3 G90 G54 G00 X4.0 Y3.5 S900 M03
N4 G43 Z1.0 H01 M08
N5 G99 G81 R0.1 Z-0.163 F3.0
N6 G91 X3.2 L5 (K5)
N7 X1.319 Y4.6
N8 X-3.2 L5 (K5)
N9 X1.319 Y4.6
N10 X3.2 L5 (K5)
N11 X1.319 Y4.6
N12 X-3.2 L5 (K5)
N13 G80 M09
N14 G28 Z0 M05
N15 G28 X0 Y0
N16 M30
%
```

Many experienced programmers will consider even more efficient way of approaching the programs for grid patterns by using subprograms or even *User Macros*. Subprograms are especially useful for grid patterns consisting of a large number of rows or a large number of columns. The subject of subprograms, including a practical example of a really large grid pattern, is covered in *Chapter 39*. The subject of *user macros* is not covered in this handbook.

ARC HOLE PATTERN

Another quite common hole pattern is a set of equally spaced holes arranged along an arc (not a circle). Such an equally spaced set of holes along any portion of a circle circumference creates an *arc hole pattern*.

The approach to programming an arc hole pattern should be the same as if programming any other hole pattern. Select the first hole as the one that is most convenient. Is it the first hole or the last hole on the arc that is easier to find the coordinates for? Perhaps starting at 0° (3 o'clock or East position) would be better? The illustration in *Figure 27-8* shows a typical layout of an arc hole pattern.

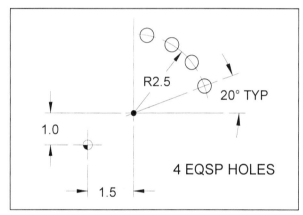

Figure 27-8
Arc hole pattern - program O2708

In the pattern, the arc center locations are known, so is the arc radius, angular spacing between holes and the number of equally spaced holes along the circumference.

A number of calculations is needed to find the X and Y coordinates for each hole center location within the bolt hole pattern. The procedure is similar to that of an angular line in a grid pattern, but with several more calculations. The calculation uses trigonometric functions applied to each hole separately - all necessary data and other information are listed in the drawing.

For any number of holes, exactly the double number of calculations will be required to get the coordinates for *both* axes. In the example, there are four holes, therefore eight calculations will be necessary. Initially, it may seem as a lot of work. In terms of calculations, it is a lot of work, but keep in mind that only two *trigonometric* formulas are involved for any number of holes, so the calculations will become a lot more manageable. Incidentally, this observation can be applied to just about any other similar programming application.

The best way to illustrate the arc pattern programming, is to use the drawing example. First, the programming task will be split into four individual steps:

➲ STEP 1

Start with the calculation of a hole that is nearest to 0° location (3 o'clock position or East direction), then continue for other holes in the counterclockwise direction of the arc.

➲ STEP 2

Use trigonometric functions to calculate the X and Y coordinates of the *first* hole:

Hole #1

```
X = 1.5 + 2.5 × cos20 = 3.849231552 (X3.8492)
Y = 1.0 + 2.5 × sin20 = 1.855050358 (Y1.8551)
```

➲ STEP 3

Use the same trigonometric formulas as in *Step 2* and calculate XY coordinates for the 3 *remaining* holes. For each hole in the pattern, increase the included angle by 20°, so the second hole angle will be 40°, the third 60°, and so on:

Hole #2

```
X = 1.5 + 2.5 × cos40 = 3.415111108 (X3.4151)
Y = 1.0 + 2.5 × sin40 = 2.606969024 (Y2.607)
```

Hole #3

```
X = 1.5 + 2.5 × cos60 = 2.750000000 (X2.75)
Y = 1.0 + 2.5 × sin60 = 3.165063509 (Y3.1651)
```

Hole #4

```
X = 1.5 + 2.5 × cos80 = 1.934120444 (X1.9341)
Y = 1.0 + 2.5 × sin80 = 3.462019383 (Y3.462)
```

➲ STEP 4

If the XY coordinates are calculated in the same order as they will appear in the CNC program, the listing of all hole locations can be used in that order:

```
Hole #1:    X3.8492   Y1.8551
Hole #2:    X3.4151   Y2.6070
Hole #3:    X2.7500   Y3.1651
Hole #4:    X1.9341   Y3.4620
```

Now, the program for the hole arc pattern can be written, using the XY coordinates for each hole location from the established calculations - program O2708:

```
O2708 (ARC PATTERN)
N1 G20
N2 G17 G40 G80
N3 G90 G54 G00 X3.8492 Y1.8551 S900 M03
N4 G43 Z1.0 H01 M08
N5 G99 G81 R0.1 Z-0.163 F3.0
N6 X3.4151 Y2.607
N7 X2.75 Y3.1651
N8 X1.9341 Y3.462
```

```
N9 G80 M09
N10 G28 Z0.1 M05
N11 G28 X1.9341 Y3.462
N12 M30
%
```

There are two other methods (perhaps more efficient) to program an arc hole pattern. The first method will take an advantage of the local coordinate system G52, described in *Chapter 40*. The second method will use the polar coordinate system (optional on most controls), described later in this chapter - in program O2710.

BOLT HOLE CIRCLE PATTERN

A pattern of equally spaced holes along the circumference of a circle is called a *bolt circle pattern* or a *bolt hole pattern*. Since the circle diameter is actually pitch diameter of the pattern, another name for the bolt circle pattern of holes is a *pitch circle pattern*. The programming approach is very similar to any other pattern, particularly to the arc hole pattern and mainly depends on the way the bolt circle pattern is oriented and how the drawing is dimensioned.

A typical bolt circle in a drawing is defined by XY coordinates of the circle center, its radius or diameter, the number of equally spaced holes along the circumference, and the angular orientation of holes, usually in relation to the X axis (that is to the zero degrees).

A bolt circle can be made up of any number of equally spaced holes, although some numbers are much more common than others, for example,

4, 5, 6, 8, 10, 12, 16, 18, 20, 24

In later examples, the 6-hole and the 8-hole patterns (and their multiples) have *two* standard angular relationship to the X axis at zero degrees.

Figure 27-9 is a typical bolt circle drawing. The programming approach for a bolt circle is similar to arc pattern.

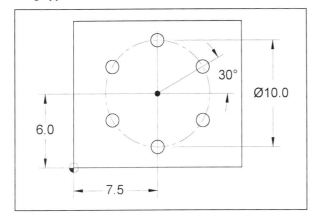

Figure 27-9
Bolt circle hole pattern - program O2709

First, select the machining location to start from, usually at program zero. Then find the absolute XY coordinates for the center of the given circle. In the illustration, the bolt pattern center coordinates are X7.5Y6.0. There will be no machining at this location, but the center of the circle will be the starting point for calculations of all holes on the bolt circle. When the circle center coordinates are known, write them down. Each hole coordinate on the circumference must be adjusted by one of these values. When all calculations for the first hole are done (based on the circle center), continue to calculate the X and Y coordinates for the other holes on the circle circumference, in an orderly manner.

In example O2709 are 6 equally spaced holes on the bolt circle diameter of 10.0 inches. That means there is a 60° increment between holes (360/6=60). The most common starting position for machining is at the boundary between quadrants. That means the most likely start will be at a position that corresponds to the 3, 12, 9 or 6 o'clock on the face of an analog watch. In this example, the start will be at the 3 o'clock position. There is no hole at the selected location, the nearest one will be at 30° in the counterclockwise direction. A good idea is to identify this hole as a hole number 1. Other holes may be identified in a similar way, preferably in the order of machining, relative to the first hole.

Note that each calculation uses exactly the same format. Any other mathematical approach can be used as well, but watch the consistency of all calculations:

Hole #1

```
X = 7.5 + 5.0 × cos30 = 11.830127   (X11.8301)
Y = 6.0 + 5.0 × sin30 =  8.500000   (Y8.5)
```

Hole #2

```
X = 7.5 + 5.0 × cos90 =  7.5000000   (X7.5)
Y = 6.0 + 5.0 × sin90 = 11.0000000   (Y11.0)
```

Hole #3

```
X = 7.5 + 5.0 × cos150 = 3.16987298 (X3.1699)
Y = 6.0 + 5.0 × sin150 = 8.50000000 (Y8.5)
```

Hole #4

```
X = 7.5 + 5.0 × cos210 = 3.16987298 (X3.1699)
Y = 6.0 + 5.0 × sin210 = 3.50000000 (Y3.5)
```

Hole #5

```
X = 7.5 + 5.0 × cos270 = 7.50000000 (X7.5)
Y = 6.0 + 5.0 × sin270 = 1.00000000 (Y1.0)
```

Hole #6

```
X = 7.5 + 5.0 × cos330 = 11.830127  (X11.8301)
Y = 6.0 + 5.0 × sin330 =  3.500000  (Y3.5)
```

Once all coordinates are calculated, the program is written in the same way as for the previous patterns:

```
O2709 (BOLT CIRCLE PATTERN)
N1 G20
N2 G17 G40 G80
N3 G90 G54 G00 X11.8301 Y8.5 S900 M03
N4 G43 Z1.0 H01 M08
N5 G99 G81 R0.1 Z-0.163 F3.0
N6 X7.5 Y11.0
N7 X3.1699 Y8.5
N8 Y3.5
N9 X7.5 Y1.0
N10 X11.8301 Y3.5
N11 G80 M09
N12 G28 Z0.1 M05
N13 G91 G28 X0 Y0
N14 M30
%
```

It would be more logical to select the bolt circle center as program zero, rather than the lower left corner of the part. This method would eliminate modifications of the bolt circle center position for each coordinate value and perhaps reduce a possibility of an error. At the same time, it would make it more difficult to set the work offset G54 on the machine. The best solution is to use G52 *local coordinate offset* method. This method is especially useful for those jobs that require translation of the bolt circle pattern (or any other pattern) to other locations of the same part setup. For details on the G52 command, see *Chapter 40*.

◆ Bolt Circle Formula

In the previous calculations, there are many repetitious data. The methods are the same, only the angle changes. This type of calculation offers an excellent opportunity for creating a common formula that can be used, for example, as the basis of a computer program, calculator data input, etc. *Figure 27-10* shows the basis for such a formula.

Using the following explanation and the formula, coordinates for any hole in any bolt circle pattern can be calculated easily. The formula is similar for both axes:

$$X = \cos((n-1) \times B + A) \times R + X_c$$

$$Y = \sin((n-1) \times B + A) \times R + Y_c$$

☞ where ...

X	=	Hole X coordinate
Y	=	Hole Y coordinate
n	=	Hole number counter - CCW from 0°
H	=	Number of equally spaced holes
B	=	Angle between holes = 360 / H
A	=	First hole angle - from 0°
R	=	Bolt circle radius *or* bolt circle diameter/2
X_c	=	Bolt circle center from the X origin
Y_c	=	Bolt circle center from the Y origin

◆ Pattern Orientation

The bolt circle pattern orientation is specified by the angle of the first hole from the 0° of the bolt circle.

In daily applications, bolt circle patterns will have not only different number of holes, but different orientations as well. The bolt circles most commonly affected are those whose number of equally spaced holes is based on the multiples of six (6, 12, 18, 24, ...) and multiples of eight (4, 8, 16, 24, 32, ...). This relationship is important, since the orientation of the first hole will influence the position of all the other holes in the bolt circle pattern.

Figure 27-11 shows relationship of the first hole position to the 0° location of the bolt circle. 0° location is equivalent to the 3 o'clock position or the East direction.

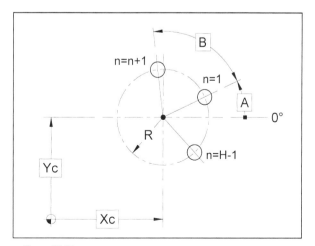

Figure 27-10

Basis for a formula to calculate bolt hole pattern coordinates

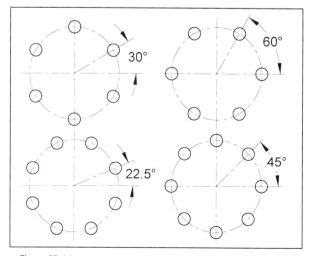

Figure 27-11

Typical orientations of a six and eight hole bolt circles

POLAR COORDINATE SYSTEM

So far, all mathematical calculations relating to the arc or bolt circle pattern of holes have been using lengthy trigonometric formulas to calculate each coordinate. This seems to be a slow practice for a modern CNC system with a very advanced computer. Indeed, there is a special programming method available (usually as a control option) that takes away all the tedious calculations from an arc or bolt circle pattern - it is called the *polar coordinate system*. There are two polar coordinate functions available, always recommended to be written as a separate block:

G15	Polar coordinate system cancel	*OFF*
G16	Polar coordinate system	*ON*

Program input values for bolt hole or arc patterns may be programmed with the polar coordinate system commands. Check first the options of the control before using this method. The programming format is similar to that of programming fixed cycles. The format is, in fact, identical - for example:

```
N.. G9.. G8.. X.. Y.. R.. Z.. F..
```

Two factors distinguish a standard fixed cycle from the same cycle used in the polar coordinate mode.

The first factor is the initial command G that precedes the cycle - no special G code is required for a standard cycle. For any cycle programmed in the polar coordinate system mode, the preparatory command G16 must be issued to activate the polar mode (ON mode). When the polar coordinate mode is completed and no longer required in the program, the command G15 must be used to terminate it (OFF mode). Both commands must be in a separate block:

```
N.. G16                    (POLAR COORDINATES ON)
N.. G9.. G8.. X.. Y.. R.. Z.. F..
N.. ...
N.. ...                    (MACHINING HOLES)
N.. ...
N.. G15                    (POLAR COORDINATES OFF)
```

The second factor is the *meaning* of the X and Y words. In the standard fixed cycle, the XY words define the position of a hole in rectangular coordinates, typically as an absolute location. In the polar mode and G17 in effect (XY plane), both words take on a totally different meaning - specifying a radius and an angle:

❑ The X word becomes radius of the bolt circle

❑ The Y word becomes angle of the hole, measured from 0°

Figure 27-12 illustrates the three basic input requirements for a polar coordinate system.

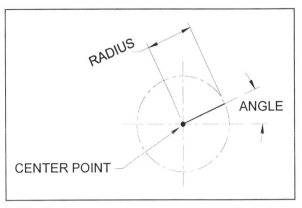

Figure 27-12

Three basic characteristics of polar coordinates

In addition to the X and Y data, polar coordinates also require the *center* of rotation. This is the *last point* programmed before G16 command. Earlier, data in program O2708 and *Figure 27-8* were calculated using trigonometric functions. With the polar coordinates control option, the program can be much simplified - O2710:

```
O2710 (ARC PATTERN - POLAR)
N1 G20
N2 G17 G40 G80
N3 G90 G54 G00 X1.5 Y1.0 S900 M03     (PIVOT POINT)
N4 G43 Z1.0 H01 M08
N5 G16                          (POLAR COORDINATES ON)
N6 G99 G81 X2.5 Y20.0 R0.1 Z-0.163 F3.0
N7 X2.5 Y40.0
N8 X2.5 Y60.0
N9 X2.5 Y80.0
N10 G15                         (POLAR COORDINATES OFF)
N11 G80 M09
N12 G91 G28 Z0 M5
N13 G28 X0 Y0
N14 M30
%
```

In the next program O2711, holes are equally spaced on the bolt circle circumference. Dimensions in *Figure 27-13* are applied to the polar coordinate programming method.

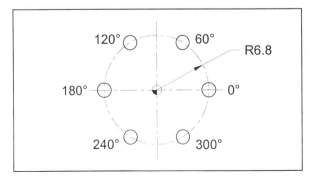

Figure 27-13

Polar coordinate system applied to bolt hole circle - program O2711

```
O2711 (G15-G16 EXAMPLE)
N1 G20
N2 G17 G40 G80
N3 G90 G54 G00 X0 Y0 S900 M03     (PIVOT POINT)
N4 G43 Z1.0 H01 M08
N5 G16                      (POLAR COORDINATES ON)
N6 G99 G81 X6.8 Y0 R0.1 Z-0.163 F3.0
N7 X6.8 Y60.0
N8 X6.8 Y120.0
N9 X6.8 Y180.0
N10 X6.8 Y240.0
N11 X6.8 Y300.0
N12 G15                     (POLAR COORDINATES OFF)
N13 G80 M09
N14 G91 G28 Z0 M05
N15 G28 X0 Y0
N16 M30
%
```

Note that the center of polar coordinates (also called *pivot* point) is defined in block N3 - it is the *last X and Y location* programmed *before* the polar command G16 is called. In the program example O2711, the center is at X0Y0 location (block N3) - compare it with program O2710.

Both, the radius and angle values, may be programmed in either absolute mode G90 or incremental mode G91.

If a particular job requires many arc or bolt hole patterns, polar coordinate system option will be worthy of purchase, even at the cost of adding it later. If the Fanuc *User Macro* option is installed, macro programs can be created *without* having polar coordinates on the control and offer even more programming flexibility.

◆ Plane Selection

Chapter 29, and particularly *Chapter 31*, describe the subject of planes. There are three mathematical planes, used for variety of applications, such as polar coordinates.

G17	XY plane selection
G18	ZX plane selection
G19	YZ plane selection

Selection of a correct plane is *extremely critical* to the proper use of polar coordinates. Always make it a habit to program the necessary plane, even the default G17 plane.

G17 plane is known as the XY plane. If working in another plane, make double sure to adhere to the following rules:

The first axis of the selected plane is programmed with the arc radius value.

The second axis of the selected plane is programmed as the angular position of the hole.

In a table format, all three possibilities are illustrated. Note, that if no plane is selected in the program, the control system defaults to G17 - the XY plane.

G-code	Selected plane	First axis	Second axis
G17	XY	X = radius	Y = angle
G18	ZX	Z = radius	X = angle
G19	YZ	Y = radius	Z = angle

Most polar coordinate applications take place in the default XY plane, programmed with the G17 command.

◆ Order of Machining

The order in which the holes are machined can be controlled by changing the *sign* of the angular value, while the polar coordinate command is in effect. If the angular value is programmed as a positive number, the order of machining will be counterclockwise, based on the 0° position. By changing the value to a negative number, the order of machining will be clockwise.

This feature is quite significant for efficient programming approach, particularly for a large number of various bolt hole patterns. For example, a center drilling or spot drilling operation can be programmed very efficiently with positive angular values (counterclockwise order). The start will be at the first hole and, after the tool change, the drilling can continue in the reverse order, starting with the last hole. All angular values will now be negative, for the clockwise order of a subsequent tool. This approach requires a lot more work in standard programming, when the polar coordinates are not used. The polar coordinate application using the G16 command eliminates all unnecessary rapid motions, therefore shortening the cycle time.

Face milling is a machining operation that controls the height of the machined part. For most applications, face milling is a relatively simple operation, at least in the sense that it usually does not include any difficult contouring motions. The cutting tool used for face milling is typically a multi tooth cutter, called a *face mill,* although end mills may also be used for certain face milling operations, usually within small areas. The top surfaces machined with a face mill are generally perpendicular to the axis of the facing cutter. In CNC programming, the face milling operations are fairly simple, although two important considerations are critical:

❑ Selection of the cutter diameter

❑ Initial starting position of the tool in relation to the part

It helps to have some experience and knowledge of face milling principles, such as the right cutter and insert selection, distribution of cuts, machine power consumption, and several other technical considerations. Some of the basic ones are covered in this chapter, but manufacturers' tooling catalogues and various technical references will be a more in-depth source.

CUTTER SELECTION

Like all milling operations, face milling employs a cutting tool that rotates while the part remains stationary. Face milling requires that a specific amount of material be removed from the top of part, at one or several depth levels, in a single cut or multiple cuts. The programming for face milling is so effortless that, in fact, many programmers do not pay sufficient attention to proper selection of the face milling cutter, proper inserts, do not even consider the machine requirements and capabilities.

A typical face mill is a multi tooth cutter with interchangeable carbide inserts. High speed steel face mills are not recommended for CNC work, although an HSS end mill can be a suitable choice to face mill small areas or areas hard to get to in any other way. Typical to a face milling operation is the fact that not all inserts of the milling cutter are actually working at the same time. Each insert works only within a part of one complete revolution. This observation may be an important consideration when trying to establish an optimum tool life for a face milling cutter. Face milling does require significant power resources from the machine tool. For the insert setup in the cutter body, it is very important to have all inserts properly mounted.

◆ Basic Selection Criteria

Based on the job to be machined, selection of a face mill cutter has to take into account several situations:

❑ Condition of the CNC machine

❑ Material of the part

❑ Setup method and work holding integrity

❑ Method of mounting

❑ Overall construction of the cutter

❑ Face mill diameter

❑ Insert geometry

The last two items, cutter diameter and insert geometry, will influence the actual program development the most, although other items are important as well.

◆ Face Mill Diameter

One of the most important considerations for face milling operations is the selection of cutter size. For a single face cut, the ideal width of the face mill cutter should be about 1.3 to 1.6 times larger than the material width. For example, a single 2.5 inches wide cut will benefit from a ∅4.0 face mill as a suitable size. This 1.3 to 1.6 times ratio will assure a good formation of chips and their clearout from the part. For multiple cuts, always select the largest diameter cutter that can be used for the job, always considering the machine power rating, the cutter and insert geometry, the setup rigidity, depth and width of each cut, and other machining related factors.

The basic purpose of face milling is to machine off the top of a part to the specified height. For this type of machining, select a reasonable face mill diameter size, which often means to use relatively large diameter face mills. Sizes in the range of 2 to 12 inches (50 to 300 mm) are not unusual, depending on the machine and the job.

One important consideration in face milling is the diameter of the tool versus the full width of the cut. Take, for instance, a ∅5.0 inch face mill. All tooling catalogues list the *nominal* size of the face mill (5 inches in the example), although the body diameter can be found in the catalogue as well. The nominal diameter always refers to the *full width of the cut*. There is no way to tell the actual diameter of a tool holder body from the nominal size alone, it has to be looked up in the tooling catalogue. Normally, the actual size of the cutter body is not needed, except in those cases where the face milling takes place close to walls or other obstacles. The size of the cutter body may prevent access to some areas of the part and may interfere elsewhere as well. The *Figure 28-1* shows some typical configurations.

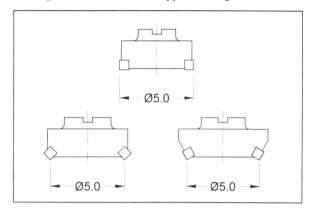

Figure 28-1
Nominal diameter of various face mill cutters

◆ Insert Geometry

Learn and become familiar with the basic terminology of milling cutters in order to understand the terms used in programming. Most of the tooling companies have available catalogues and technical booklets for the cutters and inserts they manufacture that explain the cutter usage as well as all related terms. Keep in mind that cutting tool technology does change quite rapidly and constant improvements are being made. For the programming purposes in this chapter, we look only at the very basic items of insert geometry for face milling cutters.

Insert geometry and insert mounting into the cutter body is determined by a design that controls the position of the insert in the material during a cut. These factors strongly influence quality of the cutting. There are typically *three* general categories, based on the cutting rake angle of the face mill (known as the *rake* angle):

- ❏ Positive geometry   … single or double
- ❏ Negative geometry   … single or double
- ❏ Combination of both   … positive / negative

Any detailed variations are too numerous to list, but a short overview offers at least some basis for further studies.

Positive Geometry

Positive geometry cutters require less machining power than negative cutters, so they may be more suitable on CNC machines with limited power rating, usually small machines. They offer a good chip breaking characteristics and are a good choice for machining steel materials when the cutting load is not too heavy. Positive inserts are generally single sided, therefore less economical.

Negative Geometry

Negative geometry face mills offer very high strength of the insert edge and usually require a heavy duty machine and a robust setup. The side effects are poor formation of the chips for steel but not for some kinds of cast irons, where there is hardly any curling effect during chip formation. Their main benefit is the insert economy, since negative inserts are generally double sided, offering up to eight cutting edges for a single square insert, inserted in pockets of the face mill.

Double Negative Geometry

Double negative geometry can be used only if the machine has sufficient power rating and both the cutting tool and the machined part are firmly mounted within a rigid setup. Cast iron or certain hard materials will usually benefit from using double negative geometry. The chips do have the tendency to concentrate towards the machined part and do not fly away from the part with ease, possibly causing a chip jamming against the insert or wedging themselves in confined areas. Positive/negative inserts should eliminate this clogging problem.

Positive / Negative Geometry

Positive / Negative geometry is the most beneficial to face milling operations where chip clogging could become problematic. This dual geometry design offers strength of the negative insert with the capability of 'curling' the chip into a spiral shape. This design is usually most suitable for full width face milling.

Always consult supplied technical specifications of the cutting tool manufacturers and compare several products before deciding on the most suitable choice for a particular work. Face mills and their inserts come literally in hundreds of varieties and each manufacturer claims superiority over the competition.

CUTTING CONSIDERATIONS

To program a cutting motion for a face mill, it is important to understand how a face mill works best under different conditions. For example, unless a specially designed face milling cutter and proper insert geometry, shape and grade are used, try to avoid face milling a part width that is equal to, or only a slightly larger than, the cutter diameter. Full width face milling cut may cause the insert edge to wear out prematurely and the chip to 'weld' itself to the insert. Not only the insert suffers in form of a wear out, the part surface finish suffers as well. In some more severe cases, the insert may have to be discarded prematurely, increasing the machining cost.

Figure 28-2 shows desirable and undesirable relationship of the cutter diameter to the part width during face milling.

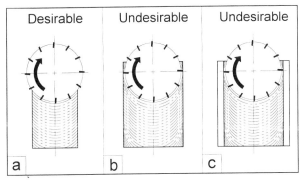

Figure 28-2

Schematic relationship of the cutter diameter and the part width.
Only the cutter size (a) is desirable, although not its position.

The illustration shows only relationship of the cutter dia-meter to the part width - it does *not* suggest the actual method of cutter entry into the material. The most impor-tant consideration for CNC programming of a face mill is the *angle* the milling cutter enters into the material.

◆ Angle of Entry

The face mill entry angle is determined by position of the cutter center line relative to the part edge. If a part can be milled with a single cut, avoid situations where the cutter center line position matches the part center line. This *neu-tral* position causes a chatter and poor finish. Offset the cut-ter *away* from part center line, either for a *negative* cutter entry angle, or a *positive* cutter entry angle. *Figure 28-3* shows both types of entry angles and their effects.

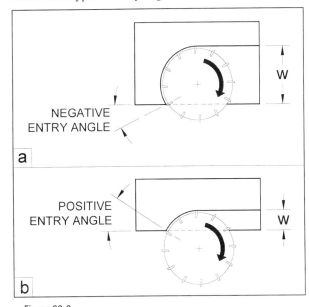

Figure 28-3

Insert entry angle into the part. W = width of cut
(a) at the strongest insert point - negative entry angle
(b) at the weakest insert point - positive entry angle

A neutral angle of entry (not shown) has the cutter center line coincident with the part edge. Needless to say, when the cutting insert enters material, a certain force is required. During positive entry angle, the weak cutting edge has to absorb most of the forces. Since insert edge it is the weak-est part of the insert, a positive entry angle may cause a breakage or at least some insert chipping. Normally, this entry method is *not* recommended.

Negative entry angle of an insert absorbs the entry force at the middle, at the strongest point of the insert. This is the preferred method, as it increases the insert life. It is always a good idea to keep the face mill center *within* the part area, rather than away from it. That way, the insert will always enter at the preferred negative angle.

All these examples assume a solid part material being machined. If the face mill has to travel over some *empty spaces,* the cut will be interrupted. The *entry into* and the *exit from* the part during interrupted cut will cause the cutter entry angle to be variable, not constant. As many other fac-tors have to be considered in face milling, take these rec-ommendations and suggested preferences only as guide-lines. Always consult a tooling representative on the best method of handling a particular face milling job, particu-larly for materials that are difficult to machine.

◆ Milling Mode

In milling, the programmed cutting direction, relative to the table motion direction is always very important. In fact, this factor so important that it is discussed in several sec-tions of this handbook and covers a subject called the *mill-ing mode.*

Traditionally, there are *three* milling mode possibilities available in milling operations:

❑ Neutral milling mode

❑ Conventional milling mode

❑ Climb milling mode

A neutral milling mode is a situation where the cutter fol-lows the center line of a slot or a face, climb milling on one side and conventionally milling on the other side of center line. The *conventional* milling mode is also called the *'up'* mode and the *climb* milling mode is also called the *'down'* mode. These are all correct terms, although the terminol-ogy may be a little confusing. The terms *climb milling* and *conventional milling* are more often used with peripheral milling than with face milling, although exactly the same principles do apply for all milling. For most face milling cuts, the climb milling mode is the best overall choice.

In *Figure 28-4*, example (a) illustrates the neutral *cutting* mode, example (b) shows the so called *down* cutting mode (or *climb* milling mode) and example (c) shows the so called *up* cutting mode (or *conventional* milling mode).

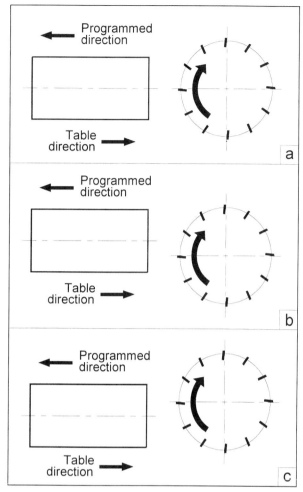

Figure 28-4

Face milling modes:
(a) Neutral milling mode
(b) Climb or 'down' milling mode
(c) Conventional or 'up' milling mode

◆ Number of Cutting Inserts

Depending on the face mill size, the common tool is a multi tooth cutter. A traditional tool called *fly-cutter* has usually only a single cutting insert and is not a normal tool of choice in CNC. The relationship of the number of inserts in the cutter to the effective cutter diameter is often called the *cutter density* or *cutter pitch*.

Typical face mills will belong into one of these three categories, based on the cutter density:

- ❑ Coarse density   … coarse pitch of inserts
- ❑ Medium density   … medium pitch of inserts
- ❑ Fine density   … fine pitch of inserts

As an overall general type, a coarse density cutter is usually a suitable choice. The more cutting inserts are engaged in material simultaneously, the more machining power will be required. Regardless of the insert density, it is important to have sufficient cutting clearances - the chips must not clog the cutter, but fly out freely.

At all times, at least one cutting insert must be in contact with the material, which will prevent heavy interrupted cut, with the possible damage to the cutter and to the machine. This situation may occur if a large face mill diameter is used for a very narrow part width.

PROGRAMMING TECHNIQUES

Although defined earlier as a relatively simple operation, face milling can be programmed much better if some common sense points are observed. Since face milling often covers a large cutting area, it is important to consider carefully the actual tool path from the start position to the end position. Here is a list of some points that should be evaluated for any face milling operation:

- ❑ Always plunge-in to the required depth away from the part (in the air)

- ❑ If surface finish is important, change the cutter direction away from the part (in the air)

- ❑ Keep the cutter center within the part area for better cutting conditions

- ❑ Typically, select a cutter diameter that is about 1.5 times larger than the intended width of cut

Figure 28-5 shows a simple plate used for examples.

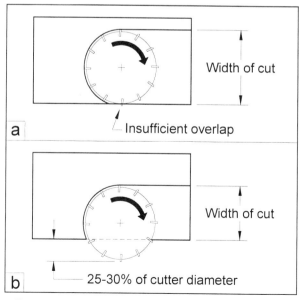

Figure 28-5

Width of cut in face milling - (b) is the recommended method

Figure 28-5a illustrates the *incorrect* and *Figure 28-5b* the *correct* width of a face mill cut. In the example (a), the cutter is engaged in the part with its full diameter, causing friction at the cutting edge and decreasing tool life. The example (b) keeps only about 2/3 of the cutter diameter in the work, which causes a suitable chip thickness, as well as favorable angle of insert entry into the material.

◆ Single Face Mill Cut

For the first face milling programming example, we will use a 5×3 plate (1 inch thick) that has to be face milled along the whole top surface to the final thickness of .800. *Figure 28-6* shows this simple drawing.

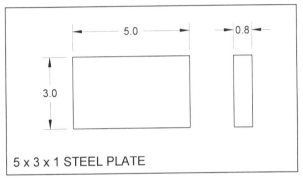

5 x 3 x 1 STEEL PLATE

Figure 28-6

Example of a single face mill cut - program O2801

From the drawing is apparent that the face milling will take place *along* the part, so the X axis horizontal direction will be selected. Before the program can be started, there are two major decisions to be made:

❑ Face mill diameter

❑ Start and end position of the cut

There are other important decisions to make, but these two are the most critical.

The part is only 3 inches wide, so a face mill that is *wider* than 3 inches should be selected. Although a ∅4.0 inch face mill seems like a natural choice, let's see if it conforms to the conditions that have been established earlier. The cutter diameter should be 1.3 to 1.6 larger than the width of cut. In this case, $3 \times 1.3 = 3.90$ and $3 \times 1.6 = 4.80$. With a ∅4.0 face mill, that means only 1.33 times larger. Considering the need for the cutter to overlap both edges of the part, selection of a *five* inch face mill diameter is better.

Once the face mill diameter has been finalized, concentrate on the *start* and *end* positions. For safety reasons, plunging to the depth has to start *away* from the part, in the air. The decision to cut along the X axis (horizontally) has been made, so the question is whether from the left to the right or from the right to the left. It really does not matter, except for the direction of chip flow, so selection from the right to the left is arbitrary.

The part X0Y0 is at the lower left corner. To establish the starting X position, consider the part length of 5.0 inches, the cutter radius (5/2=2.5) and the clearance (.25). The start X axis position will be the sum of these values, X7.75. For the Y axis start position calculation, consider the overhangs on both edges and select climb milling mode at the same time. Actually, the climb milling will be combined with a little of conventional milling, which is quite normal for face milling operations. *Figure 28-7* shows the cutter start position at X7.75Y1.0, and the end position at X-2.75Y1.0, as well as the details of calculations.

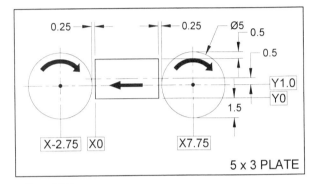

Figure 28-7

Face mill positions for a single face mill cut example

The position Y1.0 was based on the desire to have about one quarter to one third of the cutter diameter overhang at the part edge, for best insert entry angle. As 1.5 inch overhang is 30% of the cutter diameter, the programmed absolute Y position was established at a convenient Y1.0.

Now, part program for the single face milling cut can be written, with the top of part as program zero (Z0). Only one face cut is used - program example O2801.

```
O2801
(SINGLE FACE MILLING CUT)
N1 G20
N2 G17 G40 G80
N3 G90 G54 G00 X7.75 Y1.0 S344 M03
N4 G43 Z1.0 H01
N5 G01 Z-0.2 F50.0 M08
N6 X-2.75 F21.0
N7 G00 Z1.0 M09
N8 G28 X-2.75 Y1.0 Z1.0
N9 M30
%
```

Spindle speed and feedrate are based on 450 ft/min surface speed, .006" per tooth and 8 cutting inserts, used only as reasonable values. Note the Z axis approach in block N4. Although the tool is well above an empty area, the rapid motion is split between blocks N4 and N5, for safety reasons. With increased confidence, rapid to the Z-0.2 directly may be an option, if desired. This example shows the program Z0 at the top of the *unmachined* part, not the more customary finished face.

◆ Multiple Face Mill Cuts

The general principles applying to a single face cut do apply equally to multiple face cuts. Since the face mill diameter is often too small to remove all material in a single pass on a large material area, several passes must be programmed at the same depth.

There are several cutting methods for a large area to be face milled and each may produce good machining conditions under certain circumstances. The most typical methods are *multiple unidirectional* cutting and *multiple bidirectional* cutting (called zigzag) - at the same Z depth.

Multiple unidirectional cuts start from the same position in one axis, but changes the position in the other axis, above the part. This is a common method of face milling, but it lacks efficiency, because of frequent rapid return motions.

Multiple bidirectional cuts, often called zigzag cutting, are also used frequently; they are more efficient then the unidirectional method, but cause the face mill to change the climb milling method to the conventional method and vice versa. This method may work for some jobs, but is not generally recommended.

In the next two illustrations, *Figure 28-8* shows schematically a unidirectional face milling. *Figure 28-9* shows a bidirectional face milling.

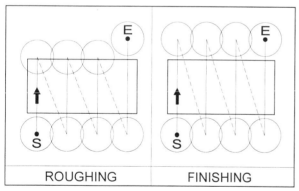

Figure 28-8

Unidirectional approach to a multiple face cut for rough and finish face milling

Compare the XY motions of these two methods. In addition, a tool path difference (cutter position) between roughing and finishing is also shown. The cutting direction may be either along the X or along the Y axis, but the principles of the cutting motion will remain the same.

Note the start position (S) and the end position (E) in the two illustrations. They are indicated by the heavy dot at the center of cutter. Regardless of the cutting method, the face milling cutter is always in a clear position at the start and end of cutting, mainly for safety reasons.

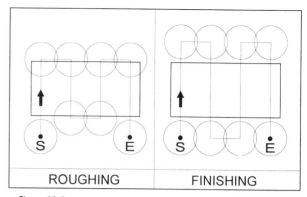

Figure 28-9

Bidirectional approach to a multiple face cut for rough and finish face milling

There is another fairly efficient method that cuts only in one mode, normally in climb milling mode. This method may remind of a circular or a spiral motion (along the XY axes) and is the most recommended method. It combines the two previous methods and is illustrated in *Figure 28-10*.

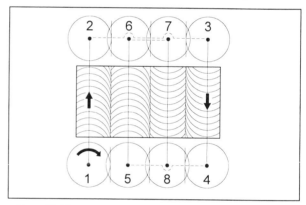

Figure 28-10

Schematic tool path representation for the climb face milling mode, applied to a unidirectional cutting

The illustration shows the *order* and *direction* of all individual tool motions. The idea is to make each cut approximately the same width, with only about 2/3 of the diameter cutting at any time, and always in climb milling mode.

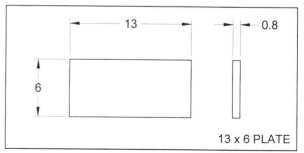

Figure 28-11

Example of a multiple face mill cut - program O2802

The programming example for multiple face milling cuts is based on the drawing shown in *Figure 28-11*. The previously discussed basics are applied and should present no difficulty in understanding the program.

```
O2802
(MULTIPLE FACE MILLING CUTS)
N1 G20
N2 G17 G40 G80
N3 G90 G54 G00 X0.75 Y-2.75 S344 M03       (POS 1)
N4 G43 Z1.0 H01
N5 G01 Z-0.2 F50.0 M08
N6 Y8.75 F21.0                             (POS 2)
N7 G00 X12.25                             (POS 3)
N8 G01 Y-2.75                             (POS 4)
N9 G00 X4.0                               (POS 5)
N10 G01 Y8.75                             (POS 6)
N11 G00 X8.9                  (POS 7 - 0.1 OVERLAP)
N12 G01 Y-2.75                     (POS 8 - END)
N13 G00 Z1.0 M09
N14 G28 X8.75 Y-2.75 Z1.0
N15 M30
%
```

In program O2802, all relevant blocks are identified with tool positions corresponding to the numbers in an earlier *Figure 28-10*.

The 13 inch part width was separated into four equal cutting widths of 3.25 each, which is a little less than 2/3 of a ∅5.0 cutter, its usable width of cut. Clearances of .25 off the part are the same as for the single face cut example. The major deviation from the norm was the motion to position number 7 in *Figure 28-10* and block N11 in the program. The last cutting motion is from position 7 to position 8. In order to make the surface finish better, the expected cut was overlapped at X9.0 by .100 to the programmed value of X8.9. In *Figure 28-12*, the schematics for O2802 program are shown, including block number references.

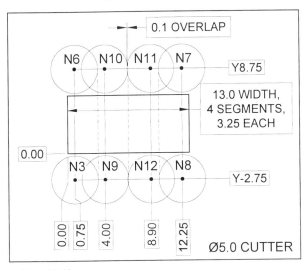

Figure 28-12
Multiple face milling details for program example O2802

Some of the examples could have been done in a shorter way along the X axis, resulting in a smaller program. However, for the purpose of example illustrations, using the Y axis was more convenient.

USING POSITION COMPENSATION

In both previous examples, the starting XY position of the face mill has been calculated, considering its diameter and a suitable clearance. To use O2801 program as an example, the starting position was X7.75 Y1.0. The part was 5.0 inches, plus a clearance of .25, plus the 2.5 inches cutter radius - total of X7.75 absolute value of the cutter center. The big disadvantage of this method is apparent when using a face mill that has a *different diameter* than the one expected by the program. A last minute change of the face mill at the machine may cause problems. Either there will be *too much* clearance (if the new tool is smaller) - or worse - there will be *not enough* clearance (if the tool is larger). There is another way to solve this problem.

As the title of this section suggests, the solution is to use the 'obsolete' *Position Compensation* feature of the control system, already described in *Chapter 17*. It is probably the only practical application of the position compensation on modern CNC machining centers.

Revisit the example O2801 and refer to earlier *Figures 28-6* and *28-7*. Illustrations show that we have to face (with a single cut) a 5×3 plate, using a ∅5 inch face mill. In order to adhere to the safety rules in machining, the face mill has to be positioned in an open area, away from the part. In order to keep the face mill cutting edges away from the part by one quarter of an inch, the clearance of .25 inches has to be incorporated *with the radius of the face mill*, which is 2.5 inches, to achieve the actual tool starting position for the face milling cutter.

In a face milling program, this situation will take on one of the following forms:

❏ The face mill radius is programmed using the actual values

❏ Position compensation method is used

In the first case, the program O2801 may be the result, with the following content:

```
O2801
(SINGLE FACE MILLING CUT - NO COMPENSATION)
N1 G20
N2 G17 G40 G80
N3 G90 G54 G00 X7.75 Y1.0 S344 M03
N4 G43 Z1.0 H01
N5 G01 Z-0.2 F50.0 M08
N6 X-2.75 F21.0
N7 G00 Z1.0 M09
N8 G28 X-2.75 Y1.0 Z1.0
N9 M30
%
```

Block N3 moves the face mill to the actual, calculated start position of the cut. In block N6, the cut is completed - again, at the actual previously calculated position. The program O2803 using position compensation is similar, but it does have some notable differences.

Compare the original program O2801 with the new program O2803, program that uses the position compensation feature - *Figure 28-13*:

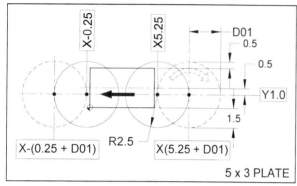

Figure 28-13

Example of the position compensation as applied to face milling - program O2803

```
O2803
(SINGLE FACE MILLING CUT)
(USING POSITION COMPENSATION)
N1 G20
N2 G17 G40 G80
N3 G90 G54 G00 X8.0 Y1.0 S344 M03
N4 G43 Z1.0 H01
N5 G46 X5.25 D01
N6 G01 Z-0.2 F50.0 M08
N7 G47 X-0.25 F21.0
N8 G00 Z1.0 M09
N9 G91 G28 X0 Y0 Z0
N10 M30
%
```

When comparing, note the major differences in block N3 (new X value), in block N5 (compensation G46), and also in block N7 (compensation G47). The situation will benefit from some more detailed evaluation.

The N3 block contains the X position with value of X8.0. That is the initial position. Since the plan is to apply the compensation G46 (single contraction), the tool has to be at a position of a *larger* value than the one expected when the compensation is completed. Therefore, X8.0 is an arbitrary value. Note that if the G45 compensation command were planned, the initial position would have to be a *smaller* value than the one expected when the compensation is completed. This is because the position compensation is always relative to *the programmed direction*.

The N5 block is added to program O2803. It contains the position compensation G46, which is a single contraction in the programmed direction by the compensation amount contained in the register of D01 offset. Note that the programmed coordinate value is X5.25, which is the total of the part length (5.0) and the selected clearance (.250). The face mill radius is totally disregarded in the program. The main benefit of this method is that, within reason, the programmed coordinates will not change, even if the face mill diameter is changed. For example, if a ⌀3.5 inch face mill is used, the job can be done very nicely, but the starting position may have to be changed. In this case, the stored value of the D01 offset will be 1.75, but block N5 will still contain X5.25. The CNC system will do its work.

The last block worth a further look is N7. It contains G47 position compensation command. The X value is equivalent to the selected clearance of X-0.25. The G47 command means a double elongation of the offset value along the programmed direction. This is necessary, because of the need to compensate at the start of cut, as well as at the end of cut. Also note the initial position and the compensated start position *cannot* be the same, otherwise no compensation will take place. With some ingenuity, the face milling can be programmed very creatively, using a rather obsolete programming feature.

CIRCULAR INTERPOLATION

In the majority of CNC programming applications, there are only two types of tool motions related to contouring. One is the *Linear Interpolation*, discussed earlier, the other one is the *Circular Interpolation*, discussed in this chapter. The programming method of controlling a tool path along an arc is similar to the method of programming a tool path along a line. The method of circular contouring is called *circular interpolation*. It is commonly used in profiling on CNC vertical and horizontal machining centers, as well as on lathes and many other CNC machines, such as simple milling machines, routers, burners, water jet and laser profilers, wire EDM, and others.

Circular interpolation is used for programming arcs or complete circles in such applications as outside and inside radii (blend and partial), circular pockets, spherical or conical shapes, radial recesses, grooves, corner breaks, helical cutting, even large counterbores, etc. The CNC unit will interpolate a defined arc with a very high precision, if the necessary information is given in the program.

ELEMENTS OF A CIRCLE

To understand the principles of programming various circular motions, it helps to know something about the basic geometrical entity known as the *circle*. As an entity that is quite common in everyday life, a circle has various properties that are strictly mathematical, only considered in specialized disciplines, such as Computerized Numerical Control, motion control and automation.

The following definition of a circle and several other definitions that are related to a circle are based on some common dictionary definitions - *Figure 29-1*.

> A circle is defined as a closed curve on a plane, where all points have the same distance from an internal point called the circle center point.

There are other similar definitions of a circle that can be found in dictionaries and mathematical books. The general understanding of a circle and its various properties as described in this handbook, provides a sufficient knowledge for general CNC programming. Additional knowledge will be needed for some specialized or complex programming applications. At this time, become at least reasonably familiar with the geometrical and trigonometric relationships for arcs and circles.

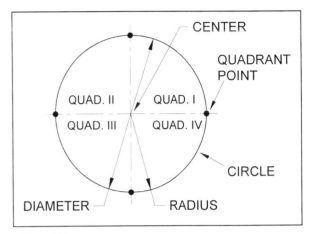

Figure 29-1
Basic elements of a circle

◆ Radius and Diameter

In the simplest mathematical terms, a circle is defined by its center point and its radius. Two of the most important elements of a circle used in part programming are the circle *radius* and the circle *diameter*.

> The radius of a circle is the line segment from the center point to any point on the circle

> The diameter of a circle is the line segment through the center point of the circle and having both end points on the circle

The center point location of the circle is also important for CNC programming. The plural form of the word *radius* is *radii*, although the word *'radiuses'* has been accepted as a colloquial term. In CNC programming, radii and diameters are used all the time, on a daily basis for almost all contouring machines. Drawings used in machine shops use radius and diameter dimensions a lot, with an almost unlimited number of possible applications.

Radii and diameters are also used in relation to the cutting tool insert designation, they are used for measuring and gauging (inspections), as well as in trigonometric calculations and various auxiliary sketches. In programming, the actual application of an arc or circle is not important, only its mathematical characteristics.

◆ Circle Area and Circumference

The *area* of a circle is defined by this formula:

$$A = \pi \times R^2$$

☞ where ...

A = Area of the circle
R = The circle radius
π = Constant (3.1415927)

The *circumference* of a circle is the *length* of a circle if it were a straight line:

$$C = \pi \times D$$

☞ where ...

C = Circumference of the circle
D = The circle diameter
π = Constant (3.1415927)

It is important to note that both the area and circumference of a circle (its actual length) are seldom used in CNC programming, although understanding their concepts presents a rather useful knowledge.

QUADRANTS

A quadrant is a major property of a circle and can be defined mathematically:

> A quadrant is any one of the four parts of the plane formed by the system of rectangular coordinates.

It is to every programmer's benefit to understand the concept of quadrants and their applications for circular motions in milling and turning programs.

A circle is programmed in all four quadrants, due to its nature, while most arcs are programmed within one or two quadrants. When programming the arc vectors I, J and K (described later), the angular difference between the arc start and end points is irrelevant. The only purpose of arc vectors is to define a *unique* arc radius between two points.

For many arc programming projects, the direct radius can be used with the R address, available for majority of control systems. In this case, the angular difference between the start and end points is very important, because the computer will do its own calculations to find the arc center. The arc with the angular difference of *180° or less*, measured between the start and end points, uses an *R positive* value. The arc in which the angular difference is *more than 180°*, uses an *R negative* value. There are two possible choices and the radius value alone cannot define a unique arc.

Also worth mentioning is a *mirrored* tool path and its relationship to the quadrants. Although it is not a subject of the current chapter, mirroring and quadrants must be considered together. What happens to the tool path when it is mirrored is determined by the quadrant where the mirrored tool path is positioned. In the *Chapter 41* are more details about mirror image as a programming subject. For now, it should be adequate to cover a very brief overview only.

For example, if a programmed tool path in *Quadrant I* is mirrored to *Quadrants II* or *IV*, the cutting method will be reversed. That means a *climb* milling will become *conventional* milling and vice versa. The same rule applies to a programmed tool path in *Quadrant II* as it relates to *Quadrants I* and *III*. This is a very important consideration for many materials used in CNC machining, because climb milling in *Quadrant I* will turn into conventional milling in *Quadrants II* and *IV* - a situation that is not always desirable. Similar changes will occur for other quadrants.

◆ Quadrant Points

From the earlier definition should be clear that quadrants consist of two perpendicular lines that converge at the arc center point and an arc that is exactly one quarter of a circle circumference. In order to understand the subject deeper, draw a line from the center of an arc that is parallel to one of the axes and is longer than the arc radius. The line created an *intersection* point between the line and the arc. This point has a special significance in programming. It is often known as the *Quadrant Point* - or the *Cardinal Point* - although the latter term is not used too often, except in mathematical terminology. There are four quadrant points on a given circle, or four intersections of the circle with its axes. The quadrant points locations can be remembered easier by associating them with the dial of a compass or a standard watch with an analog dial:

Degrees	Compass direction	Watch direction	Located between quadrants
0	EAST	3 o'clock	IV and I
90	NORTH	12 o'clock	I and II
180	WEST	9 o'clock	II and III
270	SOUTH	6 o'clock	III and IV

At this point of learning, it may be a good idea to refresh some terms of the angle direction definition. The established industry standard (mathematics, as well as CAD, CAM and CNC) defines an absolute angular value as being *positive* in the *counterclockwise* direction and *always starting from zero degrees*. From the above table, zero degrees correspond to the *East* direction or *three o'clock* position of an analog clock - *Figure 29-2*.

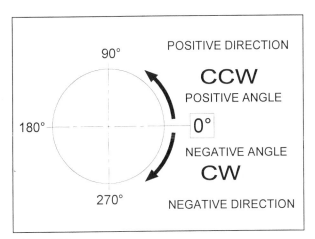

POSITIVE DIRECTION

Figure 29-2
Mathematical definition of the arc direction

There is another reason why the quadrant points are important in CNC programming. In some cases, the quadrant points will be used as the arc *end points*, even if the circular cut covers more than one quadrant. This is particularly true for many older control systems, where crossing the quadrants in a single block is not allowed. The modern controls can generate arc of any length in a single block, with virtually no restrictions.

PROGRAMMING FORMAT

The programming format for a circular interpolation tool path must include several parameters, without which the task of cutting an arc would be impossible. The important parameters are defined as:

❑ Arc cutting direction (CW or CCW)

❑ Arc start and end points

❑ Arc center and radius value

The cutting feedrate must also be in effect, discussed in more detail later in this chapter. Special modal G codes are used for circular motion programming and additional parameters related to the circle radius are also required.

◆ Arc Cutting Direction

A cutting tool may move along an arc in two directions - *clockwise* (CW) or *counterclockwise* (CCW). These two terms are assigned by convention. On most machines, the motion direction is determined by looking perpendicularly at the plane in which the circular motion is programmed. The motion *from* the plane vertical axis *towards* the plane horizontal axis is clockwise, reverse motion is counterclockwise. This convention has *mathematical* origins and does not always match the machine axes orientation. *Chapter 31* describes machining in planes, this chapter will only take a brief look.

In a typical program format, the first statement of a circular cutting motion block is the *cutting direction*. This direction defines the cutting tool motion along the programmed arc, which is either clockwise or counterclockwise. The motion direction along an arc is programmed using preparatory commands in the block.

◆ Circular Interpolation Block

There are two preparatory commands associated with programming an arc direction:

G02	Circular motion clockwise	CW
G03	Circular motion counterclockwise	CCW

Both the G02 and G03 commands are modal, therefore they remain in effect until the end of program or until canceled by another command from the same G code group, usually by another motion command.

The preparatory commands G02 and G03 are the key words used in programming to establish circular interpolation mode. The coordinate words following the G02 or G03 command are always designated within a selected plane. The plane is normally based on the available axes combinations of XY, ZX and YZ for milling or similar applications. Normally, there is no plane selection on a lathe, although some control indicate it as G18, the ZX plane.

The plane selection and the combination of circular motion parameters and the arc cutting direction determine the *arc end point*, and the R value specifies the *arc radius*. Special arc center modifiers (known as vectors) are also available, if the programmer requires them.

When the G02 or G03 command is activated by a CNC program, any currently active tool motion command is automatically canceled. This canceling motion is typically G00, G01 or a cycle command. All circular tool path motions must be programmed with a cutting feedrate in effect, applying the same basic rules as for linear interpolation. That means the feedrate F must be programmed *before* or *within* the cutting motion block. If the feedrate is not specified in the circular motion block, the control system will automatically look for the *last* programmed feedrate. If there is no feedrate in effect at all, many controls usually return an error message (an alarm) to that effect. The feedrate may be specified in one of two ways. Either directly, within the arc cutting block only or indirectly, by assuming the last active feedrate. Circular motion in a rapid mode is not possible. Also not possible is a simultaneous three axes circular motion. For more details on this subject, look up *Chapter 45* describing helical milling.

On the majority of older controls, direct radius address R cannot be specified and the arc center vectors I, J and K must used instead:

```
G02 X.. Y.. I.. J..    Milling program - CW
G02 X.. Z.. I.. K..    Turning program - CW
G03 X.. Y.. I.. J..    Milling program - CCW
G03 X.. Z.. I.. K..    Turning program - CCW
```

Control systems supporting the arc radius designation by address R will *also* accept the IJK modifiers, but the reverse is *not* true. If *both* the arc modifiers IJK and the radius R are programmed in the same block, the radius value takes priority, *regardless of the order:*

```
G02 (G03) X.. Y.. R.. I.. J..
G02 (G03) X.. Y.. I.. J.. R..
```

The controls that accept *only* the modifiers IJK will return an error message in case the circular interpolation block contains the R address (an unknown address).

◆ Arc Start and End Points

The *start point* of an arc is the point where circular interpolation *begins*, as determined by the cutting direction. This point *must* be located on the arc and it can be a tangency point or an intersection, resulting in a blend radius or a partial radius respectively. The instruction contained in the start point block is sometimes called the *departure command - Figure 29-3*.

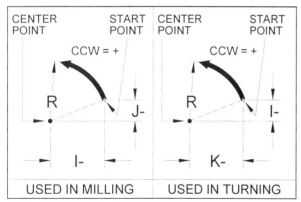

Figure 29-3
Center point and start point of an arc

The arc start point is always relative to the cutting motion direction and is represented in the program by coordinates in the block *preceding* the circular motion. In terms of a definition,

> The start point of an arc is the last position of the cutting
> tool before the circular interpolation command.

Here is an example:

```
N66  G01 X5.75 Y7.5
N67  G03 X4.625 Y8.625 R1.125
N68  G01 X.. Y..
```

In the example, block N66 represents the *end* of a contour, such as a linear motion. It also represents the *start* of the arc that follows next. In the following block N67, the arc is machined, so the coordinates represent the end of arc and start point of the next element. The last block of the example is N68 and represents the end point of the element that started from the arc. The *end point* of the arc is the coordinate point of any two axes, where the circular motion *ends*. This point is sometimes called the *target* position.

◆ Arc Center and Radius

The *radius* of an arc can be designated with the address R or with arc center vectors I, J and K. The R address allows programming the arc radius directly, the IJK arc center vectors are used to actually define the physical (actual) arc center position. Most modern control systems support the R address input, older controls require the arc center vectors only. The basic programming format will vary only slightly between the milling and turning systems, particularly for the R address version:

```
G02 X.. Y.. R..    Milling program - CW
G02 X.. Z.. R..    Turning program - CW
G03 X.. Y.. R..    Milling program - CCW
G03 X.. Z.. R..    Turning program - CCW
```

Why is the arc center location or the arc radius needed at all? It would seem that the end point of an arc programmed in combination with a circular interpolation mode should be sufficient. This is never true. Always keep in mind that *numerical control* means control of the tool path by *numbers*. In this case, there is an *infinite number* of mathematical possibilities and all are corresponding to this incomplete definition. There is virtually an unlimited number of arc radii that will fit between the programmed start and end points and still maintain the cutting direction.

Another important concept to understand is that the cutting direction CW or CCW has nothing to do with the arc center or the radius. The control system needs more information than direction and target point in order to cut the desired arc. This additional information must contain a definition that defines a programmed arc with a *unique* radius.

This unique radius is achieved by programming the R address for the direct radius input, or using the IJK arc center vectors. Address R is the actual radius of the tool path, usually the radius taken from the part drawing.

◆ Arc Center Vectors

Figure 29-4 shows the signs of arc vectors I and J in all possible orientations. In different planes, different pairs of vectors are used, but the logic of their usage remains exactly the same.

Arc vectors I, J and K are used according to the following definitions (only I and J are shown in the illustration):

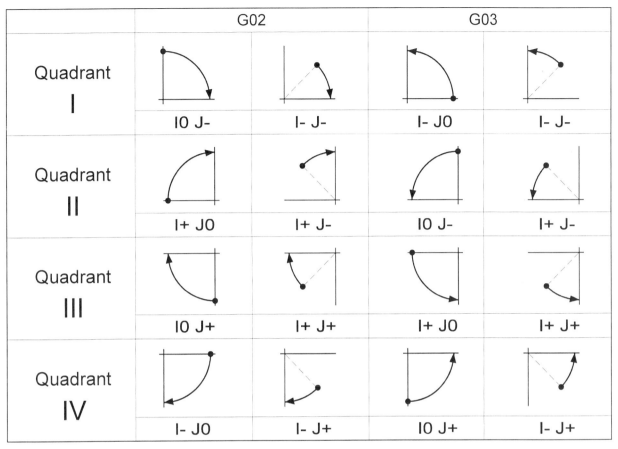

Figure 29-4

Arc vectors I and J (also known as arc modifiers) and their sign designation in different quadrants (XY plane)

> Arc center vector *I* is the distance, with specified direction,
> measured from the start point of the arc,
> to the center of the arc, parallel to the X axis.

> Arc center vector *J* is the distance, with specified direction,
> measured from the start point of the arc,
> to the center of the arc, parallel to the Y axis.

> Arc center vector *K* is the distance, with specified direction,
> measured from the start point of the arc,
> to the center of the arc, parallel to the Z axis.

The distance between the start point of the arc and the center point of the arc (as specified by the IJK vectors) is almost always measured as an *incremental* distance between the two points. Some control systems, for example many Cincinnati designs, use the absolute designation to define an arc center. In those cases, the arc center is programmed as an *absolute* value from the *program zero*, not from the arc center. Always make sure how each of the control systems in the machine shop handles these situations.

The lack of a standard in this respect creates a major difference in the programming format, so be careful to avoid a

possible error. An error can be particularly likely in those cases where both types of controls are installed in the shop. There is *no compatibility* between programs using absolute and incremental designation of the arc center.

The specified *direction* applies only to the incremental designation of arc center. It is the definition of relative position of the arc center from the start point, programmed with a directional sign - absence of the sign always assumes a positive direction, minus sign indicates a negative direction and must always be written. Arcs using absolute center definition follow standard rules of absolute dimensioning.

◆ Arc in Planes

For machining centers, programming an arc in any one of the three geometrical planes is allowed - *Figure 29-5*. The correct arc vectors must be used for each plane:

```
G17 G02 (G03) X.. Y.. R..    (or   I.. J..)
G18 G02 (G03) X.. Z.. R..    (or   I.. K..)
G19 G02 (G03) Y.. Z.. R..    (or   J.. K..)
```

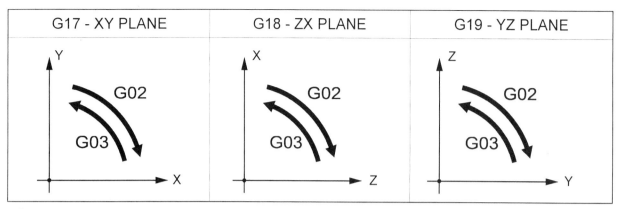

Figure 29-5

Arc cutting direction in three planes - the orientation of the axes is based on mathematical, not machine, planes

If the programmed plane is not aligned with the machine axes, or if the axes used in the program are selected without a plane designation, the circular motion will take place according to the *axis selection in the program*. Always watch when the modal axis motion is omitted. The safest method of avoiding this potentially harmful problem is to follow a simple precaution - *never count on modal values*.

In nonstandard planes, the circular program block should always contain specifications for *both axes*, as well as both arc vectors or the R value. Such a block is *complete* and will always be executed on the basis of axes designation priority. This method is preferable to the selection of a previously defined plane. Even if the plane designation is incorrect, the resulting tool motion *will* always be correct.

RADIUS PROGRAMMING

Programming arc is very common. By definition, an arc is only a *portion* of a circle and there are many ways to program an arc. If the arc is 360°, it must be programmed with the cutting start position being the same as its end position. In this case, a full circle is the result. If only a portion of the circle is programmed, only a radius is programmed. Two kinds of radii are used in CNC programming:

❏ Blend radius

❏ Partial radius

Each radius may be programmed in the CW or CCW direction and each may be external or internal, as well as in any orientation that the cutting tool can handle.

◆ Blend Radius

A point of tangency between an arc and its adjacent element creates a *blend* radius. Blend radius is defined as a radius tangent between a *line and an arc*, an *arc and a line*, or between *two arcs*. A blend arc creates a smooth transition between one contour element and another. The point of tangency is the only contact point between the two elements.

The simplest form of a blend radius is between two perpendicular lines that are parallel to the machine axes. Calculation of the start and end points requires only a few additions or subtractions. More complex calculation is required when even one line is at an angle. In this case, trigonometric functions are used to calculate the start or the end point, or both. Similar calculations are required for blends between other entities as well. A blend arc is also known as a *fillet arc* or a *fillet radius*.

◆ Partial Radius

The opposite of a blend arc is a partial arc - there is no smooth blend between two contour elements, instead, there is an intersection. Mathematically, there are always two possible selections, however, the part drawing should be quite clear as to the shape of any partial radius. Partial radius can also exist between *two lines*, *one line* and *an arc*, or between *two arcs*. Partial radius can be defined as a radius where either the start point or the end point is *not* tangent to the adjacent element, but intersects it in two places. The actual calculation of point coordinates for the arc start or end point is about the same as that for a blend arc, depending which method of dimensioning had been used in the part drawing.

FULL CIRCLE PROGRAMMING

All Fanuc systems and many other controls support a full circle programming. Full circle is an arc machined along 360°. Full circle cutting is possible on the lathes in theory only, since the type of work does not allow it. For the milling work, full circle programming is fairly routine and is required for common operations, such as:

❏ Circular pocket milling

❏ Spotface milling

❏ Helical milling (with linear axis)

❏ Milling a cylinder, sphere or cone

A *full circle cutting* is defined as a circular tool motion that completes 360° between the start and end points, resulting in identical coordinates for the start and end tool positions. This a typical application of one block programming of a full circle - *Figure 29-6*.

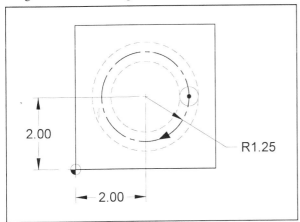

Figure 29-6

Full circle programming using one block of program entry

```
...
G90 G54 G00 X3.25 Y2.0 S800 M03
G01 Z-0.25 F10.0
G02 X3.25 Y2.0 I-1.25 J0 F12.0     (FULL CIRCLE)
G00 Z0.1
...
```

Older controls do not allow a circular interpolation in more than one quadrant per block. In this case, the circular motion has to be divided among four or even five blocks, depending on the starting tool position. Using the previous drawing example, the resulting program will be a little longer, but with the same results - *Figure 29-7*:

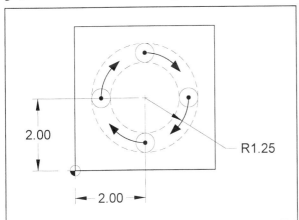

Figure 29-7

Full circle programming using four blocks of program entry

```
...
G90 G54 G00 X3.25 Y2.0 S800 M03
```

```
G01 Z-0.25 F10.0
G02 X2.0 Y0.75 I-1.25 J0 F12.0     (BLOCK 1 OF 4)
G02 X0.75 Y2.0 I0 J1.25            (BLOCK 2 OF 4)
G02 X2.0 Y3.25 I1.25 J0            (BLOCK 3 OF 4)
G02 X3.25 Y2.0 I0 J-1.25           (BLOCK 4 OF 4)
G00 Z0.1
...
```

This is an example of a four block programming that covers a full circle cutting. The arc start and end points are both located at a *quadrant* point of the axis line, which is an important programming consideration. The *quadrant* point in the example is equivalent to 3 o'clock position (0°). Note that the G02 is repeated in each block only for the example emphasis and does not need to be repeated in a production program. The same applies to the occurrences of I0 and J0 - they do not have to be written unless they change.

Try not to make the job more difficult by establishing the starting position of the cut away from any of the four *quadrant* points, which are at 0°, 90°, 180° and 270°. For example, if the cutting starts at 33°, there will be *five* circular blocks, not *four*, and the XY coordinates of the start point of the arc (shown as *xs* and *ys* distances), will have to be calculated using trigonometric functions - *Figure 29-8*:

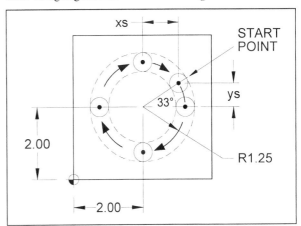

Figure 29-8

Full circle programming using five blocks of program code

```
...
G90 G54 G00 X3.0483 Y2.6808 S800 M03
G01 Z-0.25 F10.0
G02 X3.25 Y2.0 I-1.0483 J-0.6808 (BLOCK 1 OF 5)
G02 X2.0 Y0.75 I-1.25 J0          (BLOCK 2 OF 5)
G02 X0.75 Y2.0 I0 J1.25           (BLOCK 3 OF 5)
G02 X2.0 Y3.25 I1.25 J0           (BLOCK 4 OF 5)
G02 X3.0483 Y2.6808 I0 J-1.25     (BLOCK 5 OF 5)
G00 Z0.1
...
```

Values x_s and y_s were calculated by the following trigonometric functions:

$$x_s = 1.25 \times \cos 33 = 1.0483382$$
$$y_s = 1.25 \times \sin 33 = .6807988$$

From the results, the start point of the cut can be found:

```
X = 2 + x_s = 3.0483382 = X3.0483
Y = 2 + y_s = 2.6807988 = Y2.6808
```

If the control system supports a full circle program input in *one block*, the output program will be shorter, but *will require* the I and J arc center vectors only - the R radius value *cannot* be used in this case. The reason is that I and J vectors are *always* unique in their meaning, the radius R designation can be ambiguous. The following example is correct, using the I and J arc vectors:

```
...
G90 G54 G00 X3.0483 Y2.6808 S800 M03
G01 Z-0.25 F9.0
G02 X3.0483 Y2.6808 I-1.0483 J-0.6808
G00 Z0.1
...
```

The I and J modifiers cannot be arbitrarily replaced with the address R. The next example is *not* correct:

```
...
G90 G54 G00 X3.0483 Y2.6808 S800 M03
G01 Z-0.25 F9.0
G02 X3.0483 Y2.6808 R1.25 F12.0      (* WRONG *)
G00 Z0.1
...
```

The reason? Mathematically, there are *many* options for a full circle programming. If an R value is programmed for a 360° arc, no circular motion will take place and such a block will be ignored by the control. This is a precaution built into the control software, to prevent from cutting an incorrect arc because of the many existing possibilities. In *Figure 29-9*, only a handful of the possible arcs is shown. The circles share the same cutting direction, start point, end point, and radius. *They do not share center points.*

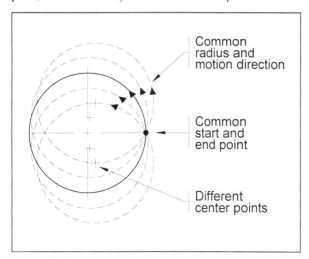

Figure 29-9
Many mathematical possibilities exist for a full circle cutting with R

◆ Boss Milling

As an example of a full circle cutting, a simple boss will be used, as illustrated in *Figure 29-10*.

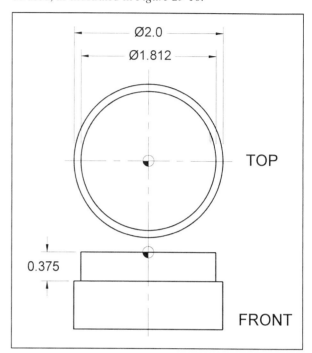

Figure 29-10
Boss milling example - drawing for program O2901

Boss or spigot milling are terms used for *external* milling of a full circle. The opposite is an *internal* milling of a full circle, such as a *circular pocket*. The cutter used will be ∅.75 inch end mill programmed at .375 depth:

```
O2901
(0.75 DIA END MILL)
N1 G20
N2 G17 G40 G80
N3 G90 G54 G00 X-1.0 Y1.5 S750 M03
N4 G43 Z0.1 H01
N5 G01 Z-0.375 F40.0 M08
N6 G41 Y0.906 D01 F20.0
N7 X0 F14.0
N8 G02 J-0.906
N9 G01 X1.0 F20.0 M09
N10 G40 Y1.5 F40.0 M05
N11 G91 G28 X0 Y0 Z2.0
N12 M30
%
```

In program O2901, the tool moves first to the XY position and depth, then the cutter radius offset was started. When reaching the cutting depth, the tool made a straight climb milling motion to the top of boss. Then it swept around the circle to the same point, moved away straight, and by reversing the initial motions, it returned to its Y axis start point - *Figure 29-11* shows the block numbers.

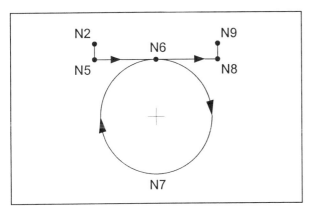

Figure 29-11

Boss milling example - tool motions for program O2901

Alternate applications may include multiple roughing passes, a semifinishing pass, two cutting tools and other selections related to machining.

◆ Internal Circle Cutting - Linear Start

Internal full circle cutting is common and has many applications, such as circular pockets or counterbores. In an example, a ∅1.25 circular cavity is to be machined to the depth of .250 inch, in program O2902. A simple linear motion will be used for the startup, where the entry point blend is not too important. The cutting tool is a center cutting end mill (also known as a slot drill) - *Figure 29-12:*

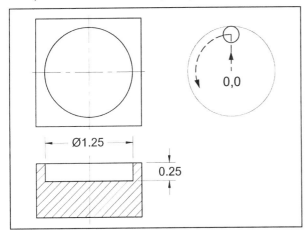

Figure 29-12

Internal circle cutting - linear approach only

```
O2902
(0.5 DIA CENTER END MILL)
N1 G20
N2 G17 G40 G80
N3 G90 G54 G00 X0 Y0 S900 M03
N4 G43 Z0.1 H01
N5 G01 Z-0.25 F10.0 M08
N6 G41 Y0.625 D01 F12.0
N7 G03 J-0.625
```

```
N8 G01 G40 X0 F20.0 M09
N9 G91 G28 X0 Y0 Z2.0 M05
N10 M30
%
```

Program O2902 shows both the arc start point and end point at 90°, programmed at 12 o'clock position. The cutter radius offset started during the motion from the arc center.

> A cutter radius offset cannot start or end in a circular mode.

This is true for almost any circular application, except the very few that use a special cycle.

◆ Internal Circle Cutting - Circular Start

The simple linear approach programming method in the last example will not be practical when smooth blend between the approach and the circular cut is required. To improve the surface finish, the start position of circular motion can be reached on an arc. The usual startup is from the center, first at a 45° linear motion, to apply the cutter radius offset, then on an arc that blends with the full circle. The *Figure 29-13* illustrates the principle and program O2303 shows the complete program.

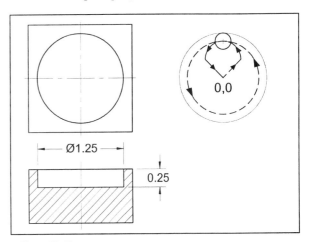

Figure 29-13

Internal circle cutting - linear and circular approach

```
O2903
(0.5 DIA CENTER END MILL)
N1 G20
N2 G17 G40 G80
N3 G90 G54 G00 X0 Y0 S900 M03
N4 G43 Z0.1 H01
N5 G01 Z-0.25 F10.0 M08
N6 G41 X0.3125 Y0.3125 D01 F12.0
N7 G03 X0 Y0.625 R0.3125
N8 J-0.625
N9 X-0.3125 Y0.3125 R0.3125
N10 G01 G40 X0 Y0 F20.0 M09
N11 G91 G28 X0 Y0 Z2.0 M05
N12 M30
%
```

This programming method is slightly longer, but the surface finish quality with a circular approach is much better than with the linear approach.

If a control systems has the *User Macro* option and many circular pockets are required, the O2903 example could also be adapted to a macro. Some controls do have a circular pocket milling cycle built-in.

◆ Circle Cutting Cycle

Several controls, for example some Yasnac or Mitsubishi, but not Fanuc, have a built-in routine (cycle) to cut a full internal circle using special preparatory commands, typically G12 and G13. These cycles are very convenient programming aid and to the surprise of many programmers, Fanuc dropped this feature many years ago.

There is a logical relationship between G02 and G12, as well as between G03 and G13:

G12	Full circle cutting cycle	*CW*
G13	Full circle cutting cycle	*CCW*

A typical programming format for using these two special commands is quite simple:

G12 I.. D.. F..   *Full circle CW*
G13 I.. D.. F..   *Full circle CCW*

In this format description, the I address is the *radius* of finished circle and is programmed as an incremental value with a sign. If the sign is positive (plus sign is assumed), the start point of the cut will be at 0°, which is equivalent to the 3 o'clock position or East direction. If the sign is negative, the start point of the cut will be at 180° position, which is equivalent to the 9 o'clock position or West direction. The command cannot be forced to a start in the Y axis direction.

Programmed D address is the control register number for the cutter radius offset and F is the feedrate address. There are alternate versions of this cycle on some controls, but all very similar in nature.

Other conditions must also be accepted for successful usage of this programming shortcut. The cutting tool must *always* start at the center of a circular pocket, the cutting plane must be set to the XY plane and the arc starting position is usually built-in, generally at 0° or 180° (Y axis start is not possible). There is also a *built-in* cutter radius offset (G12 to the right, G13 to the left). Never program the commands G41 and G42 when using G12 or G13 command. If the cutter radius is in effect, it will be overridden by the selection of G12 or G13. The safest approach is to program these two cycles in G40 mode (cutter radius offset canceled) at all times.

What is not true in any other circular application, is true in this situation. In normal programming of arcs and circles, a cutter radius offset cannot start in an arc tool motion. In G12/G13 programming mode, the start motion from the uncompensated circle center position is *circular* to the compensated start point on the arc circumference. This is all built into the control and there no choice is offered. Consider this situation as a special case, definitely not as a rule.

On some Yasnac CNC models, there is an additional parameter in the G12/G13 format - the radius parameter, or the R parameter. This indicates special rapid motion portion, designed to reduce air cutting time.

As an example of G12/G13 programming, the earlier circular pocket, illustrated in *Figure 29-14,* will be used.

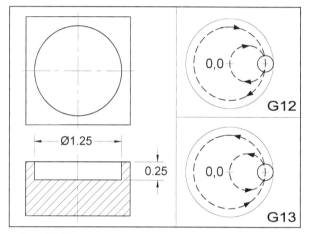

Figure 29-14

Full circle cutting using G12/G13 cycles - program O2904

```
O2904
(0.5 DIA CENTER CUTTING END MILL)
N1 G20
N2 G17 G40 G80
N3 G90 G54 G00 X0 Y0 S900 M03
N4 G43 Z0.1 H01
N5 G01 Z-0.25 F10.0 M08
N6 G13 I0.625 D01 F12.0 M09       (IF AVAILABLE)
N7 G91 G28 X0 Y0 Z2.0 M05
N8 M30
%
```

The program is only two blocks shorter, but it is much simpler to develop. The cutter radius offset is automatic (built-in) and the editing at the machine is much easier. There is also an additional bonus - since the start point on the circle is not a result of a straight line, but a lead-in arc, the surface finish quality will be better than using other types of tool approach. This is a preferable method when the machined surface quality is important. There is also a built-in lead-out arc in the cycle, similar to the lead-in arc, that is effective when the circle cutting is completed.

ARC PROGRAMMING

With a full arc cutting, which means the *complete* 360° motion, the R address cannot be used at all. The arc center vectors I and J have to be applied, even on latest controls.

What if the circle is 359.999°? Well, at first, circle *must* have 360°, therefore the word 'circle' is incorrect. Even a small difference of 0.001° does make a difference between a circle and an arc. Although this difference is much more important mathematically than for practical programming, the distinction is very important. In circular interpolation terms, an *incomplete circle* is nothing more than an *arc*. Look at this arc a little differently. If a 90° arc is made, the R address can be programmed, for example:

```
G01 X2.0 Y5.25 F12.0
G02 X3.75 Y7.0 R1.75
```

If an arc that covers *exactly* 180° is programmed, the program will not be much different:

```
G01 X2.0 Y5.25 F12.0
G02 X5.5 Y5.25 R1.75
```

Note that the Y coordinate is the same for the arc start and end position. The Y value in the circular motion block does not have to be repeated, it is used here only for illustration.

Another example shows programming an arc of 270°, still using the R address. Are the following blocks correct?

```
G01 X10.5 Y8.625 F17.0
G02 X13.125 Y6.0 R2.625
```

The blocks appear to be correct. The calculations, the format, individual words, they all appear to be right. Yet, the program is *wrong!* Its result will be a 90° arc, not 270°.

Study the illustration in *Figure 29-15*. It shows that there is not just one, but *two* mathematical possibilities when the R address is used for arcs. The solid contour is the tool path, the dashes identify the two possible radii.

Programmers do not normally think of these mathematical alternatives, until they program arcs *larger* than 180° (or scrap a part). This is a similar situation to that of a full circle, described earlier. Although the I and J vectors can be used to remedy the problem, a different remedy may be a preferred choice. The R address can still be used in the program, but with a *negative* sign for any arc that is greater than 180°. For arcs smaller than 180°, the usual positive R radius remains in effect. Recall from some earlier explanations that if there is no sign with the R word (or any other word), the word assumes a *positive* value. Compare the two programming examples:

```
G01 X10.5 Y8.625 F17.0
G02 X13.125 Y6.0 R2.625          (90 DEGREES)
```

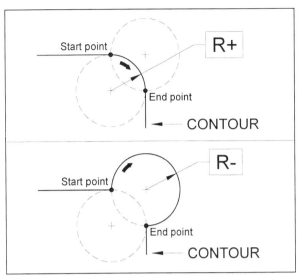

Figure 29-15
Sign of R address for circular cutting - only the center is different

The following example is identical to the previous one, except for the R address sign.

```
G01 X10.5 Y8.625 F17.0
G02 X13.125 Y6.0 R-2.625          (270 DEGREES)
```

If frequently programming arcs that cover more than 180°, establish a particular programming style. If the style is well thought out, it will avoid the costly mistakes associated with the R address sign error.

FEEDRATE FOR CIRCULAR MOTION

In most programs, the feedrate for circular interpolation is determined the same way as feedrate for linear interpolation. The cutting feedrate for arcs is based on established machining conventions. They include the work setup, material machinability, tool diameter and its rigidity, programmer's experience and other factors.

Many programmers do not consider the machined radius when selecting the cutting feedrate for the tool. Yet, if the machined surface finish quality is really important, always consider the size of every radius specified in the part drawing. Perhaps the same feedrate for linear and circular motions programmed so far may have to be adjusted - either upward or downward.

In lathe programming, there is no reason to distinguish between linear and circular tool motions, regardless of the radius size. The tool nose radius is usually small, only averaging .0313 inches (or 0.8 mm) and the equidistant tool path is close to the programmed tool path, taken from a drawing. This is not the case for milling contour programming, where large tool radii are normal and common.

The adjusted arc feedrate is not required in every program. If the cutter center tool path is close to the part drawing contour, no adjustment is needed. On the other hand, when a large diameter cutter is used to contour a small outside radius, a problem that affects the surface finish may occur. In this case, the tool center path generates a much longer arc than one in the drawing. In a similar situation, if a large cutter diameter is used for an inside arc, the equidistant path will be much shorter than the original arc length.

In normal programming, the linear feedrate is used for arcs as well, as determined by the machinability rating for the given material. The formula for linear feedrate is:

$$F_l \;=\; r\,/\,min \times F_t \times n$$

☞ where ...

F_l	=	Linear feedrate (in/min or mm/min)
r/min	=	Spindle speed
F_t	=	Feedrate per tooth
n	=	Number of cutting edges

A linear feedrate for 1000 r/min, .0045 in/tooth load and two cutting edges, the feedrate is 9 in/min. Using a relatively large cutter diameter, such as $\varnothing$.625 (15.875 mm) or larger, the linear feedrate adjustment up or down for circular motion may be necessary to maintain good finish.

The elementary rule of feedrate adjustment for arcs is that the normally programmed linear feedrate is *increased for outside arcs* and *decreased for inside arcs* - Figure 29-16.

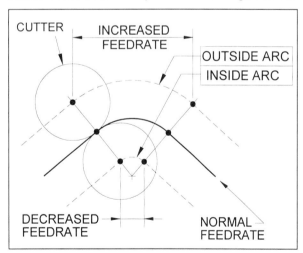

Figure 29-16
Feedrate adjustments for circular tool motion

An outside arc of a tool path is longer than the drawing arc

An inside arc of a tool path is shorter than the drawing arc

Two formulas provide the tools to find the adjusted arc feedrate, mathematically equivalent to the linear feedrate. Both formulas are recommended for external or internal contouring only, *not* for rough machining of solid material.

◆ Feedrate for Outside Arcs

For outside arcs, the adjusted feedrate will be *higher* than the linear feedrate, calculated from this formula:

$$F_o \;=\; \frac{F_l \times (R + r)}{R}$$

☞ where ...

F_o	=	Feedrate for outside arc
F_l	=	Linear feedrate
R	=	Outside radius on the part
r	=	Cutter radius

Based on the linear feedrate of 14 in/min, an outside .375 arc radius requires an *upward* adjustment for a $\varnothing$.50 cutter:

```
Fo  =  14 × (.375 + .25) / .375 = 23.333333
```

The result is a major increase, to F23.3 in the program. Consider the same example with .75 cutter radius ($\varnothing$1.5):

```
Fo  =  14 × (0.375 + 0.75) / 0.375 = 42.0
```

The feedrate changed from 14 in/min to 42 in/min - a 3 times increase. Here, use previous experience to determine whether the adjustment is justified or not.

◆ Feedrate for Inside Arcs

For inside arcs, the adjusted feedrate will be *lower* than the linear feedrate, calculated from this formula:

$$F_i \;=\; \frac{F_l \times (R - r)}{R}$$

☞ where ...

F_i	=	Feedrate for inside arc
F_l	=	Linear feedrate
R	=	Inside radius on the part
r	=	Cutter radius

Based on the linear feedrate of 14 in/min, the feedrate for .8243 inch inside radius with $\varnothing$1.25 cutter, must be adjusted *downward*:

```
Fi  =  14 × (.8243 - .625) / .8243 = 3.384932
```

The result is a feedrate of 3.38 in/min. In the program, this will be the applied feedrate for the F address.

30   *CUTTER RADIUS OFFSET*

The contour of a part - also known as a profile - is normally programmed for milling applications by establishing the depth in the Z axis first, then moving the cutting tool individually along the X axis, Y axis, or both axes simultaneously. For turning applications, either the X axis or the Z axis, or both axes can be used to face, turn or bore a contour. For both types of machining, each contour element contour requires one block of cutting motion. These motions between contour change points can be programmed in inches or millimeters and they can use an absolute value position or an incremental distance. In either case, keep in mind that this type of programming uses the *center line* of the spindle as the X and Y or X and Z tool movements. Although the center line programming is a very convenient method for program development, it is also a method unacceptable for machining. During contact with the material, the *edge* of the cutting tool must touch the programmed part contour, not its *center line*.

The tool path for all contouring operations is always equivalent to the cutting tool motion. Whether used on a CNC machining center or on a CNC lathe, the *cutting tool edge must always be tangent to the contour*, which means the tool motion has to create a path where the center point of the cutter is *always at the same distance* from the contour of the part. This is called the *equidistant* tool path.

The illustration in *Figure 30-1* shows two types of a tool path. One is *not compensated,* the other is *compensated.* Both are applied to a particular contour, with the cutter diameter shown as well, including its positions.

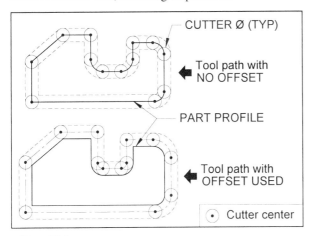

CUTTER Ø (TYP)

Tool path with
NO OFFSET

PART PROFILE

Tool path with
OFFSET USED

⊙ Cutter center

Figure 30-1

Tool path not compensated (above) and compensated (below), by the cutter radius

MANUAL CALCULATIONS

Some realities should become apparent from the *Figure 30-1*. The most noticeable observation is that the machined contour must always take place with the tool path compensated by its radius, which means its center point must be located in positions shown in the lower example. This machining requirement is not matched by the reality of the engineering drawing. In a drawing, all dimensions refer to the part contour, *not* the contour of the tool center. In fact, the drawing is dimensioned to the tool positions illustrated in the upper example of the illustration. The question is - how do the tool center positions get from a drawing to the part contour?

The answer is - they have to be *calculated*. Actually, they do not have to be, if the CNC system is equipped with an advanced built-in feature called the *cutter radius compensation* or *cutter radius offset*. On the CNC turning systems, this feature is called the *tool nose radius compensation* or the *tool nose radius offset*. This advanced and common control feature enables the programmer to apply the offset command, program the part contour *as per drawing* dimensions and let the control do all the necessary calculations and adjustments automatically.

At this point, the current chapter could continue and strictly concentrate on the automatic method of programming, using this exceptional feature. After all, all modern CNC machines do have a cutter radius offset built-in. Once several basic rules are followed, the feature is easy to use.

In order to automate something, we have to first understand how it works. If something is automated already, the knowledge of how it works makes the job so much easier, particularly when encountering a difficulty that has to be resolved very quickly. To really understand cutter radius offset - many programmers and machine operators do not - it is important to understand the principles built in the system, principles that are very much based on basic mathematical calculations, including the often unpopular trigonometry calculations. A very simple drawing is shown in *Figure 30-2* for that purpose.

The program zero will be selected at the lower left corner of the part. Since the cutting will be external, in a climb milling mode, the tool will start along the Y direction first. At the moment, the start and end tool position is not important, only calculations of the individual contour points at intersections and tangency points.

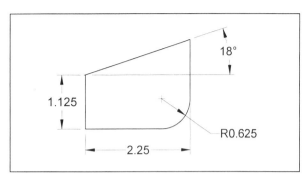

Figure 30-2

Sample drawing for manual calculations (examples)

Note that there are five points on the drawing, one at each contour change. These points are either *intersections* or points of *tangency*. As each point has two coordinates, total of *ten* values will be required.

The drawing always offers some points that need no calculations. It is a good idea to get well organized and mark the points from the drawing first. Then, make a chart in the order of tool path. Study *Figure 30-3* carefully - it shows all five points and all the values that need no calculation, perhaps some addition or subtraction only.

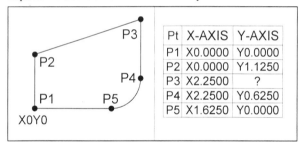

Figure 30-3

Contour change points required by the cutter path

Out of the ten values required, nine of them are given. The missing Y value for P3 is not expected on the drawing. Regardless of whether the cutter radius offset is used or not, some calculations will always be necessary and this is one of them. After all, *manual* programming is done *by hand*. *Figure 30-4* shows the trigonometry method used.

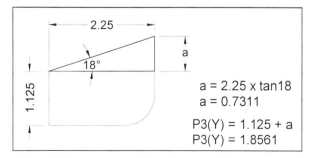

Figure 30-4

Trigonometric calculations to find unknown Y coordinate

All five points can be summed up in a small table:

Point No.	X coordinate	Y coordinate
P1	X0	Y0
P2	X0	Y1.125
P3	X2.25	Y1.8561
P4	X2.25	Y0.625
P5	X1.625	Y0

Once all the coordinates are completed, there is enough data to start the tool path, but only *if* the cutter radius offset feature is used. However, that is not the intention at the moment. *To illustrate, a whole new set of points has to be found - coordinates for the center of the cutter!*

◆ Tool Path Center Points

The cutting tool for milling is always round. An end mill, for example, has a diameter of a certain size. Even tools used for turning and boring have a round end (called the *tool nose radius*), even if it is relatively small. Of course, we all know that any round object has a center. Milling cutter or a lathe tool tip are round objects, so they have a center. This evaluation may sound a bit too elementary and it is, but it is also the basis, *the key element*, the whole concept, of cutter radius offset. Every control system takes it into consideration.

Take, for example, an electric router tool to cut a shape out of wood - how is it used? Using a pencil outline of the desired shape, the router bit is placed into the tool and starts cutting. Where? *It starts cutting **outside** of the outlined shape, otherwise the piece cut will be either too large or too small!* The same procedure is used when cutting a board with a saw - the saw width has to be compensated.

This activity is so simple, it might have been even done automatically, without serious thinking. The *radius* of the router bit (or the width of the saw) was *compensated* for before and during the cut. Just like the outline of the shape in wood is followed, the outline of the machined part, *outline that is offset by the cutter radius is followed as well*.

The tool path generated by the cutting tool center *always* keeps the same distance from the part contour (outline). There is even a special name for this type of tool path - it is called the *equidistant* tool path, which means 'distant by the same amount'. *Figure 30-5* shows the sample drawing with the applied equidistant tool path.

The question now is - what to do about the point coordinates that have just been calculated and stored in the above table. Are they useful? Can they be used in a program? *Yes* to the first question, but *not yet* to the second. A few additional conditions have to be taken into consideration.

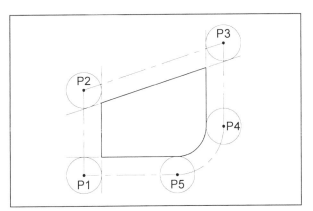

Figure 30-5

Equidistant tool path - cutter center coordinates required

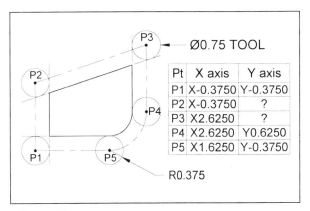

Figure 30-6

Contour change points for the cutter center path

In the program, the *old* set of points will be used to calculate a *new* set of points. Again, try to see which points are easy to calculate and establish them first.

For example, what are the XY coordinates of point P1? It is easy to see that the new P1 has the value of cutter radius in the X minus and also the value of cutter radius in the Y minus direction, from the old P1. The actual value for *any* points cannot be calculated at all, *without knowing the cutter radius first*.

◆ Cutter Radius

The nominal radius of the cutter is always known on new tools, or tools that have been physically measured. For high precision work, the radius of the cutter must be known almost 100%, say within .0001" (0.0025 mm = 2.5 micron). That is not always possible for reground tools, tools previously used, or tools that are undersize or oversize for some reason. All this means that programming the centerline of the cutter requires the exact tool radius to be known at the time of programming, in all cases.

◆ Center Points Calculation

Coordinate points illustrated in *Figure 30-5* above, represent the center of the cutter radius at each contour change point. Now, another requirement can be brought into the picture, the cutter radius size. A new coordinate set of five points can be developed. For the example, a brand new cutter of ⌀.750 will have .375 radius.

Which points can be 'read' from the illustration directly, without any trigonometric calculations? Look at and evaluate *Figure 30-6*. Out of ten values required, only eight have been identified, but also realize that the previous ten calculations had to be done earlier as well, adding to the overall programming effort.

In order to finish the discussion on programming of the cutter center, the two Y values for P2 and P3 have to be calculated. Let's start with point P2.

Figure 30-7 shows the details of point P2 calculation. The trigonometry method itself is a subject programmers have to know how to work with - it is part of mathematics, extended to CNC programming. A similar calculation is required for P3, shown in *Figure 30-8*.

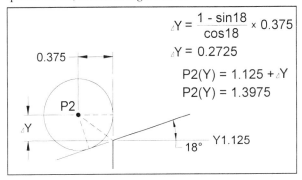

$$\triangle Y = \frac{1 - \sin 18}{\cos 18} \times 0.375$$

$$\triangle Y = 0.2725$$

$$P2(Y) = 1.125 + \triangle Y$$

$$P2(Y) = 1.3975$$

Figure 30-7

Calculation of P2 for the cutter center point

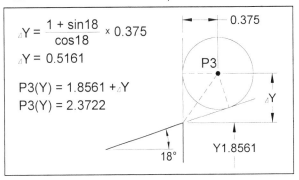

$$\triangle Y = \frac{1 + \sin 18}{\cos 18} \times 0.375$$

$$\triangle Y = 0.5161$$

$$P3(Y) = 1.8561 + \triangle Y$$

$$P3(Y) = 2.3722$$

Figure 30-8

Calculation of P3 for the cutter center point

Now, all XY coordinates are known, for all the center points around the part contour. These points are in the order of machining and they will appear in that same order in the program. Not just the point locations but also various G codes, M codes, feedrates, and other data.

At the moment, it is still too soon to write the program. This section can be closed with the table of the new points, representing the center of ∅.750 cutter but *none other*!

Point No.	X coordinate	Y coordinate
P1	X-0.375	Y-0.375
P2	X-0.375	Y1.3975
P3	X2.625	Y2.3722
P4	X2.625	Y0.625
P5	X1.625	Y-0.375

There is a single digit 1 used in the calculations. It may raise a question where it came into the equation. It represents the value of $sin90°$, which is 1. And that little triangle in front of the Y - it is a symbol for the Greek word 'delta', often used in mathematics to represent an increment, a vector, or a distance.

COMPENSATED CUTTER PATH

The previous examples are typical to the programming methods used on the early numerical controls. These controls (normally of the NC type, not CNC), had no cutter radius offset feature at all. The tool path was developed in such a way that the contour change points had to be calculated *with the cutter radius in effect*. This method of programming added a great amount of time to the part development process, greatly increased the possibility of programming errors and disallowed any flexibility during machining. Even a small difference between the programmed cutter radius and the actual cutter radius required recalculation, a correction of the program and the creation of a new punched tape (there was no CNC memory in those days). With the development of numerical control technology, and the addition of a computer to the control system (the modern CNC systems), the cutter compensation methods have been made not only possible but also greatly simplified for the programmer.

♦ Types of Cutter Radius Offset

As the CNC technology developed, so did the cutter radius offset methods. This development has taken three stages. Today, they are known as the three types of a cutter radius offset - the *Type A*, the *Type B*, and the *Type C*:

❏ Type A offset - oldest - uses special vectors in the program to establish the cutting direction (G39, G40, G41, G42).

❏ Type B offset - old - uses only G40, G41 and G42 in the program, but it does not look ahead.
Overcutting is possible for Type B offset.

❏ Type C offset - current - uses only G40, G41 and G42 in the program, but with the look ahead feature.
Overcutting is prevented for Type C offset.

The *Type C* cutter radius offset - the *look ahead* type (also called the *intersectional* type) - is the one that is used on all modern CNC systems today. There is no need to call it *Type C* anymore, as there are no other types available.

♦ Definition and Applications

Cutter radius offset is a feature of the control system that allows programming a contour without knowing the exact diameter (radius) of the cutter. This very sophisticated feature performs all necessary calculations of contour change points, based on three items:

❏ Points of the drawing contour

❏ Specified direction of the cutter motion

❏ Radius of the cutter stored in the control system

In practical programming - and machining - this feature allows the CNC programmer to develop a program without knowing the exact cutter diameter at the time of programming. It also allows the CNC operator to adjust, to fine tune, the cutter size in the control system (nominal, oversize or undersize), during actual machining. In practical terms, using cutter radius offset (and tool nose radius offset on lathes) should be considered for a number of reasons:

❏ Unknown exact size of the cutter radius

❏ Adjusting for the cutter wear

❏ Adjusting for the cutter deflection

❏ Roughing and finishing operations

❏ Maintaining machining tolerances   ... and many others

Every contouring requires the consideration of a cutter radius.

Some applications may not be too clear at the moment, but with increased knowledge of this topic, it will be easier to understand the subject. The suggestions are only some of the possibilities the automatic cutter radius offset offers. Now let's look at its actual use in programming.

PROGRAMMING TECHNIQUES

In order to program the cutter radius in a compensated mode, the three items mentioned earlier must be known:

❏ Points of the drawing contour

❏ Specified direction of the cutter motion

❏ Radius of the cutter stored in the control system

These items are the actual *data sources*. Computers only work with data and the data has to be provided by the user. For the purposes of this chapter, we assume that all data for the contour change points are based on the drawing - the XY drawing coordinates.

◆ Direction of Cutting Motion

Whenever an external or an internal tool path is programmed, there will always be a choice of *two* directions. For now only, the directions can be called the *clockwise* and the *counterclockwise* direction around the part contour. This general motion direction is compounded by the fact that there is a specific motion of the table (in milling), or the motion of the tool (in turning). These are two very separate groups that need to be clarified - which one to consider - motion of the table or motion of the tool? Keep in mind, that *regardless of the CNC machine type*, it is imperative to follow one basic rule of CNC programming:

> In CNC programming, always consider the cutting tool moving around the part, never the other way around.

This statement is true for CNC lathes, where it is obvious, but *it is also true for CNC machining centers*, where it is not so obvious. It is also true for other types of CNC machines, such as wire EDM, laser cutting machines, waterjet cutters, flame cutters, etc. When it comes to the so called direction *clockwise* versus *counterclockwise*, a closer look is necessary.

◆ Left or Right - not CW or CCW

The first thing to take care of is to eliminate the misleading terms *clockwise* and *counterclockwise*. These terms are reserved exclusively for circular interpolation and have no place in discussion of the cutter radius offset. Instead, the more accurate terms *Left* and *Right* are used for clarity.

Just like in everyday situations, when faced with the directional terms *left* and *right*, we determine the correct position of an object with respect to a certain previously established *viewing direction*. A moving object is said to be to the left or to the right of a stationary object, depending on the *direction of its movement*.

In CNC programming, there is no difference. The compensated cutter tool path is positioned to the *left* or to the *right* of the stationary contour, when looking into the cutter path direction, as illustrated in *Figure 30-9*.

The illustration shows all three options - a cutter without a direction, a cutter with direction specified and positioned to the *left of the contour*, and a cutter with direction specified and positioned to the *right of the contour*.

Out of the two compensation options, which one is better? Compensation *to the left* is preferred on CNC machining centers, because it produces a *climb milling* mode of cutting, assuming that a standard right-hand tool is used with M03 rotation. There might be a case for the compensation to the right, causing so called *conventional milling* mode of cutting. This mode should be used only in special cases, after consultation with a tooling specialist. This only applies to milling systems, not to turning.

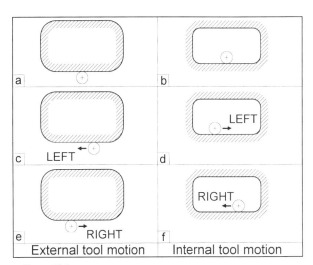

Figure 30-9

Cutter path direction as it relates to a stationary part contour:
(a - b) No motion direction shown - left and right is unknown
(c - d) Cutter positioned to the LEFT of the contour
(e - f) Cutter positioned to the RIGHT of the contour

◆ Offset Commands

In order to program one or the other mode of cutting (cutting direction), there are two preparatory commands available to select the cutter radius offset direction:

G41	Offset (compensation) of the cutter radius to the LEFT of the contouring direction
G42	Offset (compensation) of the cutter radius to the RIGHT of the contouring direction

G41 or G42 mode is canceled by the G40 command:

G40	Cutter radius offset mode CANCEL

Figure 30-10 shows all three radius offset commands:

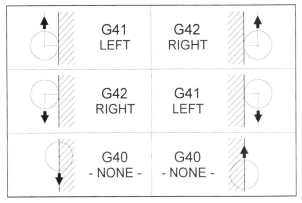

Figure 30-10

Application of G41, G42 and G40, to the cutter path

In terms of the milling method, G41 command is applied to the *climb* milling mode, G42 command is applied to the *conventional* milling mode. This is true *only* if the spindle rotates with M03 function active (spindle CW) and the cutter is right hand. If the cutter is left hand, the spindle must rotate with M04 function active (spindle CCW) and all rules applying to cutter radius offset are the *exact opposite* of those discussed here. There is no cutter radius offset applied when G40 command is in effect.

Figure 30-11 shows the G41 command as a climb milling mode and the G42 command as a conventional milling mode. Climb milling mode is the most common in CNC milling, particularly in contour milling.

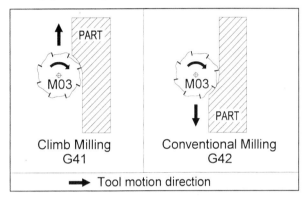

Figure 30-11
Climb milling and conventional milling mode for a right hand cutter and the spindle rotation mode M03

◆ Radius of the Cutter

The benefit of the cutter radius offset that allows to program the tool path as if the part contour were the required cutter path, does not mean the cutter radius should be either forgotten or ignored during programming. The logical question at this stage is - if the actual cutter radius is not specified in the program, *where is it specified?*

First, look at *Figure 30-12* - it illustrates the effect of different cutter radii applied to the same part contour.

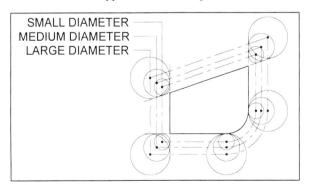

Figure 30-12
Effect of the cutter radius on the actual tool path

The answer to the last question is - *in the control system area called offset settings.* We are already familiar with the offset areas (*offset screens* on the control unit). These terms have been used for the *Position Compensation*, the *Work Offsets* and the *Tool Length Offset* (discussed in earlier *Chapters 17* to *19* respectively). Now is the time to look at the control offsets in more depth and emphasize their relationship to the compensation of the cutter radius. Although this topic may appear to be aimed at the interest of the CNC operator, the programmer *has to* understand the same principles equally well, if not in even more depth.

◆ History of Offset Types

Fanuc controls have developed over the years, and because of their reliability and popularity, many of the older models are still in use by machine shops. To understand the offsets and their application, it is important to know what *type of offset* the Fanuc control has. The rule of thumb is - as expected - the lower level or the older the control is, the lower the flexibility, and vice versa. Notice the word *flexibility* - it is not the *quality* that is lower or higher - just the flexibility. Differences are categorized as *Offset Memory Types*. There are *three* memory types on Fanuc systems:

- ❏ Type A - lowest level of flexibility
- ❏ Type B - medium level of flexibility
- ❏ Type C - highest level of flexibility

Do not confuse these tool offset memory types with the cutter radius offset types! These offset types determine how the *tool length offset* and the *cutter radius offset* will be entered into the control system and nothing else. Work offsets G54 to G59 are not affected.

Tool Offset Memory Type A

The *Type A* tool offset is the lowest level available. Its flexibility is very limited, because this offset type shares the tool length values with the cutter radius values in a *single column*. Because of the data sharing for two different offsets, it is often called the *shared* offset. In practice it means that the tool length offset value is stored in the *same* control registry area as the tool radius value. Addresses H *and/or* D can be used, with details covered later. Controls equipped with this type of tool offset memory are the most economical type in their class.

Tool Offset Memory Type B

While the *Type A* offset has only a single screen column, *Type B* has *two columns*. Now - do *not* assume! The two columns are *not* separate columns for tool length values and tool radius values at all. They are separated for the *Geometry Offset* in one column and the *Wear Offset* in the second column. Apart of this distinction, the *Type B* is still a shared type of offset for both, tool length and tool radius values. Again, the program uses addresses H *and/or* D.

Tool Offset Memory Type C

The *Type C* offset group offers the most flexibility. It is the only offset type available that separates the tool length values from those of the tool radius. It still keeps the distinction of the *Geometry Offset* and the *Wear Offset*, as the *Type B* does. That means the control display will now 2+2 columns - yes, *four* columns in total. In this type, normally addresses H and D will be used for their unique purposes.

It is relatively easy to tell which offset type is available - just look at the control display. *Figure 30-13* shows the typical appearance of each *Offset Memory Type* (all shown with zero values). The actual appearance may be slightly different, depending on the control model.

Offset No.	Offset
01	0.0000
02	0.0000
03	0.0000
...	...

Offset No.	Geometry	Wear
01	0.0000	0.0000
02	0.0000	0.0000
03	0.0000	0.0000
...	...	...

Offset No.	H-offset		D-offset	
	Geometry	Wear	Geometry	Wear
01	0.0000	0.0000	0.0000	0.0000
02	0.0000	0.0000	0.0000	0.0000
03	0.0000	0.0000	0.0000	0.0000
...	...	...	...	...

Figure 30-13

Fanuc tool offset memory types A, B, C from the top down

◆ Programming Format

The minimum information supplied to the control system in the CNC program is the offset command G41 or G42, always combined with the H or D address in effect, usually applied during a single axis motion (multi-axis motion is also allowed, if programmed carefully):

```
G41 X.. D..    or..
G42 X.. D..    or..
G41 Y.. D..    or..
G42 Y.. D..
```

The inclusion or exclusion of the tool motion and how many axes can be used at a time will be discussed in this chapter as well. First, let's resolve the question of which address to use and when. The H address or the D address?

◆ Address H or D ?

With the three types of *Tool Memory Offset,* it is reasonable to expect somewhat different programming methods for each type. Up to a point, this is true.

Both the *Type A* and the *Type B* are shared type offsets, with only a single register, where the tool length offset values are stored along with the cutter radius offset amounts. Normally, the *Type A* and *Type B* are associated with the address H only. That means the H address is used with the G43 command, as well as with the G41 or G42 commands. Many cutting tools do not require the cutter radius offset in the program, but *all* cutting tools require the tool length offset in the program. If a particular cutter requires *both* tool length offset number *and* cutter radius offset number, *two different offset numbers* from the same offset range must be used in the program and stored in the control register. That is the reason these offsets are called *shared* offsets.

For example, programmed tool T05 requires both offsets, which obviously cannot have the same offset number. The solution is to use the tool number as the tool length offset number and increase that number by 20, 30, 40, or so, for the cutter radius offset. The entry for the *Type A* in the offset screen could be similar to the one in *Figure 30-14*:

Offset No.	Offset
...	...
05	-8.6640
...	...
35	0.3750

Figure 30-14

Shared offset register screen for tool offset memory Type A

For the offset *Type B*, there are two columns available, but it is still a shared offset. The entry in the offset screen will be similar to the *Type A*, shown in *Figure 30-15*:

Offset No.	Geometry	Wear
...	...	...
05	-8.6640	0.0000
...	...	...
35	0.3750	0.0000

Figure 30-15

Shared offset register screen for tool offset memory Type B

The *Type C* offset will have two pairs of columns. Since the tool length and the tool radius have each their own columns, the *same offset number* can be used for both - there is no need for the 20, 30, 40 or so, increment. In this case, the H address is reserved for the tool length offset number and the D address is reserved for the cutter radius offset number. *Figure 30-16* shows an input logically corresponding to the *Type A* and the *Type B*:

Offset	H-offset		D-offset	
No.	Geometry	Wear	Geometry	Wear
...	...	...	...	...
...	...	...	...	...
05	-8.6640	0.0000	0.3750	0.0000
...	...	...	...	...

Figure 30-16

Unique offset register screen for tool offset memory Type C

◆ Geometry and Wear Offsets

Similar to the application of geometry and wear offsets for *tool length offset*, described in *Chapter 19*, the identical general rules can be used for the cutter radius offset.

Offsets entered in the *Geometry* offset column should only contain the nominal cutter radius. In the examples, we have used a ∅.750 cutter, with the radius of 0.375. That is the nominal value and that would also be the typical value entered into the *Geometry* offset column. The *Wear* offset column should only be used for adjustments, or fine tuning, relative to the nominal size, as required during setup and/or machining. There is no separate column for adjustment or fine tuning for the *Type A* offset. Adjustments can still be made, the only difference is that the value in the single column will always change with each adjustment, even if it represents the cutter radius.

APPLYING CUTTER RADIUS OFFSET

All programming aids required to apply the cutter radius offset in an actual CNC program are now known. The actual application, the way to use the offset in a CNC program, as well as the methods of proper usage, will be discussed next. There are *four major keys* to a successful use of the cutter radius offset feature:

1. To know how to start the offset

2. To know how to change the offset

3. To know how to end the offset

4. To know what to watch between the start and end

Each item is important and will be discussed in order.

◆ Startup Methods

Starting up the cutter radius offset is much more than using the G41X..D.. in the program (or something similar). Starting up the offset means adherence to two cardinal rules and several important considerations and decisions. The cardinal rule number one is simple - it relates to the start position of the cutter:

> Always select the start position of the cutter
> away from the contour, in the clear area

The cardinal rule number two is also simple and is based on the adherence to the first rule:

> Always apply the cutter radius offset
> together with a tool motion

These two rules are not arbitrary - rules can be broken. The suggestion here is to follow the rules until a better way is found. When selecting a startup tool position, a few questions are worth asking:

❏   What is the intended cutter diameter?

❏   What clearances are required?

❏   Which direction will the tool take?

❏   Is there no danger of collision?

❏   Can other diameter cutter be used if needed?

❏   How much stock is to be removed?

The same drawing used already will be used for this example as well and the cutter radius offset will be applied to the contour. To turn the offset on, to make it effective, the cutter will be *away* from the actual cutting area, in the clear. The intended cutter is ∅.750, the climb milling mode is desired, and .250 clearance is away from the contour. With these numbers, the start position is calculated at X-0.625 Y-0.625. *Figure 30-17* shows the start position that satisfies all rules and answers the questions established earlier.

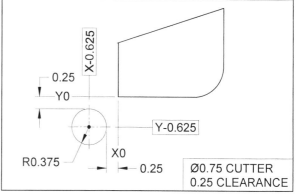

Figure 30-17

Start position of the cutter before radius offset is applied

Of course, the suggested location is not the only one suitable, but it is just as good as other possibilities. Note that the cutter located at the position X-0.625Y-0.625 is *not* compensated, the coordinates are to the *center* of the cutter. Once the start location is established, the first few blocks of the program can be written:

```
O3001 (DRAWING FIGURE 30-2)
N1 G20
N2 G17 G40 G80
N3 G90 G54 G00 X-0.625 Y-0.625 S920 M03
```

```
N4 G43 Z1.0 H01
N5 G01 Z-0.55 F25.0 M08   (FOR 0.5 PLATE THICK)
N6 ...
```

For extra safety, the approach to the depth of Z-0.55 (based on a ½ inch plate thickness) was split into two motions, although the cutter is safely above the clear area. Once the depth has been reached, the first motion can be programmed. The cutting direction is to the left of the part (climb milling) and the G41 command is used. Moving the tool around the part on the left side, means the first target point on the part has to be the X0Y1.125 location. However, this position cannot be reached directly, because the left side of the part has to be machined as well. That means the tool has to reach the X0 position first. Next decision is selecting the Y position, to get the target point. Normally, this is done by programming a so called *lead-in* motion, or a cutter *entry* motion.

Figure 30-18 shows some possibilities, all of them correct and all reaching the X0Y1.125 location eventually. Which one is the best? Here are some possible options:

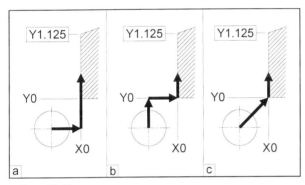

Figure 30-18

Possible lead-in motions to apply the cutter radius offset

The (a) option is simple - the tool moves towards the X0 first and the cutter radius offset is turned on *during* that motion. Then, the tool continues towards the first target point (Y1.125), already in the compensated mode.

These two motions will appear in the program as:

```
N.. G01 G41 X0 D01 F15.0
N.. Y1.125                                   (P2)
N.. ...
```

The option (b) is technically correct, but requires three motions, whereas two motions are quite sufficient. This version will not be selected for the final program, although the program would still be correct:

```
N.. G01 G41 Y0 D01 F15.0
N.. X0
N.. Y1.125                                   (P2)
N.. ...
```

The last possibility - option (c) - is also simple and requires only two motions:

```
N.. G01 G41 X0 Y0 D01 F15.0
N.. Y1.125                                   (P2)
N.. ...
```

In all three versions, the cutter radius offset is started *together with the first motion*, while still away from the actual part contour. Because the option (c) actually ends *on* the part, selecting the option (a) is the preferred programming method of the lead-in. A combination of (a) and (c) is also a good choice, with the Y axis target in the negative area.

Once the offset has been turned on, the contour change points can be programmed along the part and the control computer will do its work by constantly keeping the cutter properly offset at all times. The program O3001 can now be extended up to point P5 in the original illustration:

```
O3001 (DRAWING FIGURE 30-2)
N1 G20
N2 G17 G40 G80
N3 G90 G54 G00 X-0.625 Y-0.625 S920 M03
N4 G43 Z1.0 H01
N5 G01 Z-0.55 F25.0 M08   (FOR 0.5 PLATE THICK)
N6 G41 X0 D01 F15.0            (START OFFSET)
N7 Y1.125                               (P2)
N8 X2.25 Y1.8561                        (P3)
N9 Y0.625                               (P4)
N10 G02 X1.625 Y0 R0.625               (P5)
N11 G01 X..
```

At block N10, the tool has reached the end of the 0.625 radius. The contouring is not yet finished, the bottom side has to be cut, along the X axis. The question is - how far to cut and when to cancel the cutter radius offset?

This is the last cut on the part, so it has to be machined *while the offset is still in effect!* The cutter can end at X0, but that is not a practical position - the tool should move a bit farther, still along the X axis only. How far is further? Why not to the same X-0.625, the original start position? This is not the only clearance position available, but is the safest, most reliable and consistent. The block N11 will be:

```
N11 G01 X-0.625
```

Now the cutter has left the part contour area and the cutter radius offset is not required anymore. It will be canceled shortly, but a little review of the startup may help.

The cutter radius was known for this job, which is not always the case. The programmer needs a suitable tool, because the cutting values depend on it. Within reason, a ∅.750 or ∅.875 cutter are not far apart - *except for clearances*. The clearance of .250 was selected for .375 cutter radius. That means the program is still good for cutters up to and including ∅1.25. CNC operator has this freedom, because the only change is to the D01 offset amount in the control system offset registry. The speeds and feeds may have to be adjusted, if necessary. We will look later at what exactly happens when the cutter radius offset is applied.

The general rule to establish the start position is that it should always be selected with a clearance that is greater than the radius of the *largest* cutter that may be used. This clearance may be increased for a large stock left on the material or for a tool that is above average diameter. In order to complete the program, let's look at methods of canceling the cutter radius offset, when it is no longer needed.

◆ Offset Cancellation

A *lead-in* motion has been used at the startup of the cutter radius offset. To cancel the offset, a *lead-out* motion will be used. The length of the lead-out (just as the length of the lead-in) has to be somewhat greater than or at least equal to the cutter radius. The lead-in and the lead-out motions are also called *ramp-in* and *ramp-out* motions.

The safest place to cancel cutter radius offset, for any machine, is away from the contour just finished. This should always be a clear area position. The start position can also be the end position. *Figure 30-19* shows the offset cancellation in the example. Program O3001 can be now be written.

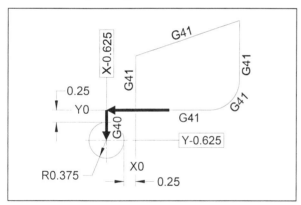

Figure 30-19

Cutter radius offset cancellation - program O3001

```
O3001 (DRAWING FIGURE 30-2)
N1 G20
N2 G17 G40 G80
N3 G90 G54 G00 X-0.625 Y-0.625 S920 M03
N4 G43 Z1.0 H01
N5 G01 Z-0.55 F25.0 M08   (FOR 0.5 PLATE THICK)
N6 G41 X0 D01 F15.0            (START OFFSET)
N7 Y1.125                           (P2)
N8 X2.25 Y1.8561                    (P3)
N9 Y0.625                           (P4)
N10 G02 X1.625 Y0 R0.625            (P5)
N11 G01 X-0.625
N12 G00 G40 Y-0.625           (CANCEL OFFSET)
N13 Z1.0 M09
N14 G28 X-0.625 Y-0.625 Z1.0
N15 M30
%
```

Finally, the program O3001 is completed. There was no need for any change of the tool direction - such an change is rather a rare occurrence, at least for contouring operations using milling controls. Since the directional change may be needed in the future, some comments may be useful.

◆ Cutter Direction Change

During a normal milling cut, there will seldom be a need to change the cutter offset direction from left to right or from right to left. If it does become necessary, the normal practice is to change from one mode to the other *without* canceling the G40 command. This practice is seldom used in milling, because change from G41 to G42 would also be a change from the preferred climb milling to the less preferred conventional milling. However, it is quite common in CNC lathe programming, with examples shown later.

HOW THE RADIUS OFFSET WORKS

Being able to program from given examples is certainly a good way to learn. Learning by a recipe or a sample does help in many cases, but it will not help much in cases where there is no sample, no recipe, and no example. In those cases, it is critical to really *understand* all principles behind the subject, such as principles of the cutter radius offset. The startup method is a good beginning. Next question is - what does happen during the tool motion in block N6?

N6 G41 X0 D01 F15.0

It is not as simple as it looks. We cannot evaluate just one block, such as N6, and know exactly what happens. The programmer has to understand what the control will do. Computers do not think, they only execute the programmed instructions and follow these instructions very diligently. Block N6 is an instruction: *Move to X0, apply the radius value stored in D01 to the left, during a linear motion at 15 in/min*. This is the program instruction to the control system. Where does the tool stop? Look at *Figure 30-20*:

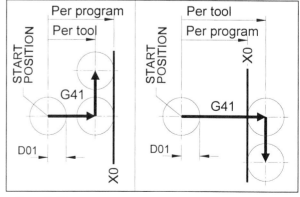

Figure 30-20

Ambiguous startup for a cutter motion in radius offset mode

Yes, there are *two* possibilities and they are both correct! Both versions compensate the cutter to the left of X0 target position. The conditions specified in block N6 have been fully satisfied - the cutting tool moves to X0 as expected, the offset is turned on to the *left* of the part contour, *during* the motion, using the radius value stored in the offset register *D01*. So what is the problem?

The situation is *ambiguous*. There are *two possible outcomes*, while only one is required. Which one? For this job, the one in the left part of the illustration, one where the tool moves along the Y+ direction next, when the radius offset has been applied. *This is the key!* The motion direction that follows G41 or G42 block *must* be known to the control. Look at two different ways the program can be written:

➲ Example 1 - Figure 30-21 left :

The next target position after N6 is Y *positive* direction:

```
N3 G90 G54 G00 X-0.625 Y-0.625 S920 M03
...
N6 G41 X0 D01 F15.0            (START OFFSET)
N7 Y1.125          (POSITIVE Y-MOTION FOLLOWS)
```

➲ Example 2 - Figure 30-21 right :

The next target position after N6 is Y *negative* direction:

```
N3 G90 G54 G00 X-0.625 Y-0.625 S920 M03
...
N6 G41 X0 D01 F15.0            (START OFFSET)
N7 Y-1.125         (NEGATIVE Y-MOTION FOLLOWS)
```

In both cases, the content of block N6 is the same, but the motion that follows the N6 is not - *Figure 30-21*.

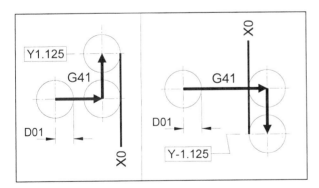

Figure 30-21

Importance of the next tool motion for cutter radius offset. Y+ next direction on the left, Y- next direction on the right

◆ **Look-Ahead Offset Type**

The block N6 alone does not contain sufficient amount of data to successfully apply the radius offset. The *next motion* - in fact, *the direction of the next motion* - must be known to the control system at all times!

How does the control handle such requirement? Controls using the cutter radius offset *Type C* have a built-in feature called the 'look-*ahead*' type of cutter radius offset.

The *look-ahead* feature is based on the principle known as *buffering* or *reading-ahead*. Normally, the control processor executes one block at a time. There will never be a motion caused by any buffered block (next block).

In a short overview, this is the sequence of events:

❑ The control will first read the block containing the startup of the cutter radius offset (that is the block N6)

❑ The control detects an ambiguous situation, and does not process the block as yet

❑ The control advances the processing to the next block (that is N7), to find out into which direction the tool will be programmed next

❑ During the 'next block reading', there is no motion at all - the control will only *register* the direction towards the target point and applies the radius offset on the correct side of the part contour, during the startup block (N6 in the example)

This look-ahead type of the cutter radius offset is very advanced internally in the software, but makes the contour programming so much easier on a daily basis. As maybe expected, there are some situations to be aware of.

◆ **Rules for Look-Ahead Cutter Radius Offset**

Look at the following sample program selection, not related to any previous examples:

➲ Example - single NO MOTION block :

```
N17 G90 G54 G00 X-0.75 Y-0.75 S800 M03
...
N20 G01 X0 D01 F17.0            (START OFFSET)
N21 M08                      (NO MOTION BLOCK)
N22 Y2.5                        (MOTION BLOCK)
...
```

What is the difference in the program structure? Ignore the reason for the coolant ON function in block N21. If it can be justified, there is nothing wrong with it. The fact remains - there is *no axis motion* in block N21, which is the same block the control system will *look ahead to* for the direction of the next tool motion. Look at one more program selection - again, as a new example:

➲ Example - two NO MOTION blocks :

```
N17 G90 G54 G00 X-0.75 Y-0.75 S800 M03
...
N20 G01 X0 D01 F17.0            (START OFFSET)
N21 M08                      (NO MOTION BLOCK)
N22 G04 P1000                (NO MOTION BLOCK)
N23 Y2.5                        (MOTION BLOCK)
...
```

Uncomfortable, perhaps - but not wrong - this time there are *two consecutive blocks* following the cutter radius offset - two consecutive blocks that *do not include any motion.*

Both examples present a program that might be fine if the radius offset were not applied. With an offset in effect, such a program structure can create problems. Controls with the 'look-ahead' feature can look ahead only so many blocks. If the control has the feature, *one block look-ahead is always there.* There are two or more look-ahead blocks available. It all depends on the control features, and not all controls are the same. Here are some basic suggestions:

❑ If the control has a look-ahead type cutter radius offset feature, but the number of blocks that can be processed ahead is not known, assume it is only one block

❑ Make a test program to find out how many blocks the control can read ahead

❑ Once the cutter radius offset is started in the program, try hard not to include any non-motion blocks - restructure the program, if necessary

Keep in mind that the control subjects the program input to the rules embedded in the software. The correct input must be provided first, in the form of an accurate program.

What kind of a response can be expected if the cutter radius offset is programmed wrong? Probably a scrap of the part. If the control system cannot calculate the offset cutter position, it will act as if the offset were not programmed at all. That means, the initial tool motion will be towards the X0 with the *cutter center.* When the necessary information is passed on to the control, the offset will be applied, usually too late, after the cutter has entered the part. Scrap is the most likely result in this case. Such an *incorrect* program is shown in *Figure 30-22:*

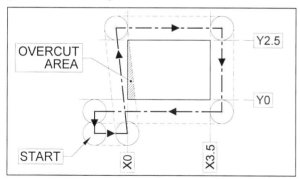

Figure 30-22
Tool path error due to wrong program structure - program O3002

There are two *no-motion* blocks in the program example O3002 that cause this error. They are after the cutter radius offset had been applied - in blocks N7 and N8. If the control system can look ahead only a single block, the program is wrong and the corrupted tool path shown in the above illustration will be the result.

```
O3002 (PROGRAM WITH RADIUS OFFSET ERROR)
N1 G20
N2 G17 G40 G80
N3 G90 G54 G00 X-0.5 Y-0.5 S1100 M03
N4 G43 Z1.0 H01
N5 G01 Z-0.55 F20.0        (FOR 0.5 PLATE THICK)
N6 G41 X0 D01 F12.0              (START OFFSET)
N7 M08                        (NO MOTION BLOCK)
N8 G04 P1000                  (NO MOTION BLOCK)
N9 Y2.5                          (MOTION BLOCK)
N10 X3.5                         (MOTION BLOCK)
N11 Y0                           (MOTION BLOCK)
N12 G01 X-0.5                    (MOTION BLOCK)
N13 G00 G40 Y-0.5             (CANCEL OFFSET)
N14 Z1.0 M09
N15 G28 X-0.5 Y-0.5 Z1.0
N16 M30
%
```

A control that can read only *one* or *two* blocks ahead *will not* process program O3002 correctly - the next motion is in the *third* block when the offset is in effect. In order to avoid an incorrect tool motion, avoid any program structure that contains more than one no-motion block.

◆ Radius of the Cutter

Every milling cutter has a diameter and one half of that diameter is the cutter radius. With new tools, the radius is always known and is sufficiently accurate. The accuracy of the radius depends on the cutter quality as well as on the way it is mounted in the machine spindle. A run-off of .001 or .002 inches may not be a problem for roughing operations, but for a precision finish, much higher accuracy is needed. Also needed is a way to *correct* for a tool wear, or even a slight tool deflection. All this is done through the D offset number, used as a pointer to the actual radius amount stored in the control register.

One simple - but very basic - rule should help to make sure the cutter radius offset will not fail:

The radius of the cutter should be smaller than the programmed length of the tool travel.

For example, in the program O3001, the tool starting position is at X-0.625, the target position is X0. That means the programmed length of the tool travel is .625. The radius selected was .375, which is smaller and adheres to the rule.

There are two other possibilities - one, where the cutter radius is the *same* as the programmed length of the tool travel, and two, where the cutter radius is *larger* than the programmed length of the tool travel.

Figure 30-23 shows a start position of a cutter that has the *same* programmed length of travel as the cutter radius. This is certainly allowed, but definitely not recommended. The reason is it limits the range of adjustments that can be made to the actual cutter radius during machining.

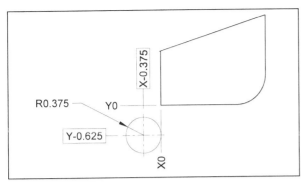

Figure 30-23

Cutter start position is equal to the cutter radius

The following example results in a .375 travel length as programmed along the X axis. If the D01 amount is less than .375, there will be a motion toward X0. If the D01 amount is equal to .375, the difference between the programmed length and the actual length is zero and there will not be any motion along the X axis. In that case, the offset of the radius takes place without a movement and the motion to the target position Y1.125 will continue.

```
N3 G90 G00 G54 X-0.375 Y-0.625 S920 M03
...
N6 G41 X0 D01 F15.0              (START OFFSET)
N7 Y1.125                                (P2)
```

Try to avoid situations like this one - although logically correct, they do not provide any flexibility and can cause serious difficulties at some time in the future.

Figure 30-24 shows a start position where the cutter is partially on the other side of the target position. This is definitely not allowed and the control system will respond with an alarm warning - the infamous *'Cutter radius interference'* alarm or *'CRC interference'* message, *Alarm #041*.

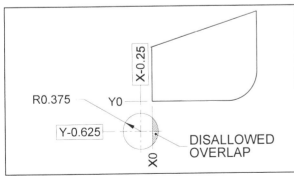

Figure 30-24

Cutter start position is smaller then the cutter radius

The following program sample is very similar to the previous examples, except the X axis start position is too close to the target position, if the cutter is stored in the D01 register in the amount of .3750:

```
N3 G90 G54 G00 X-0.25 Y-0.625 S920 M03
...
N6 G41 X0 D01 F15.0              (START OFFSET)
N7 Y1.125                                (P2)
```

What will happen here? As usually, the control calculates the difference between the programmed travel length of .25 and the cutter radius .375. It will check the direction of the next travel as Y positive and determines that because the cutter is positioned to the left of the intended motion, it has to move .125 in the X minus direction! That does not seem to be a problem, because there is a plenty of free space. *But there is a problem* - the control *does not recognize* the fact that there is a free space! Programmer knows it, but the control does not. The engineers who designed the software could have taken a number of actions; yet, they wisely decided to play it safe. They have decided to let the control system to reject this possibility and issue an alarm. Depending on the control system, the alarm *'Overcutting will occur in cutter radius compensation C'* or *'CRC interference'* or a similar message will appear - the common alarm number for this error is *No. 041* on Fanuc control systems. Many programmers, even with a long experience, have experienced this alarm. If not, they were either very fortunate or have never used cutter radius offset in the program.

Anytime the cutter radius interference alarm occurs, always look at the *surrounding blocks* as well, not just at the one where the control stops processing.

In the next section, we look at the cutter radius interference that occurs during a tool motion, not just at the startup or termination of the cutter radius offset.

◆ Radius Offset Interference

The last example illustrated only one of several possibilities, when the cutter radius offset alarm may occur. Another cause for this alarm is when a cutter radius is trying to enter an area that is *smaller* than the cutter radius, stored as the D offset amount. To illustrate, evaluate the next program O3003, for a simple part shown in *Figure 30-25*.

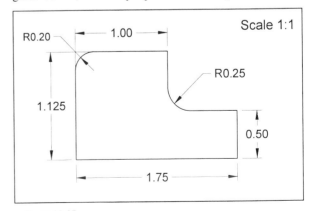

Figure 30-25

Simple drawing for program O3003

```
O3003 (DRAWING FIGURE 30-25)
N1 G20
N2 G17 G40 G80
N3 G90 G54 G00 X-0.625 Y-0.625 S920 M03
N4 G43 Z1.0 H01
N5 G01 Z-0.55 F25.0 M08   (FOR 0.5 PLATE THICK)
N6 G41 X0 D01 F15.0                (START OFFSET)
N7 Y0.925
N8 G02 X0.2 Y1.125 R0.2
N9 G01 X1.0
N10 Y0.75
N11 G03 X1.25 Y0.5 R0.25
N12 G01 X1.75
N13 Y0
N14 X-0.625
N15 G00 G40 Y-0.625              (CANCEL OFFSET)
N16 Z1.0 M09
N17 G28 X-0.625 Y-0.625 Z1.0
N18 M30
%
```

The program is quite simple, it is correct and it follows all rules discussed so far. The key to success is the selection of the cutter diameter and the entry amount of the D address into the control system. Let's see what will happen - the same cutter is used as before, a ∅.750 inch end mill. The amount of D01 stored in the control will be .3750.

The control unit will process the information from the program combined with the offset amounts to determine the tool motion. Then, it executes the blocks as it moves the tool along the part. Suddenly, at block N7 alarm *No. 041* occurs - *cutter radius interference problem*.

What has happened? There is nothing wrong with the program at all. Most CNC operators would look at the program and check it. After careful study, if they find it correct, the cause of the problem must be somewhere else, outside of the program. Try not to blame the computer and don't waste any more time once you are satisfied that the program is OK. Check the offset input in D01. The amount of .375 is stored there. That is also OK for the tool in the spindle. Check the drawing next. That is OK too. So while everything seems OK and there is still a radius offset alarm on the screen, do the next logical step.

Always evaluate the relationships between:

❏ Drawing dimensions   ... *and* ...   Program input

❏ Program input   ... *and* ...   Offset amounts

❏ Offset amounts   ... *and* ...   Drawing dimensions

This circular advice may take a while getting used to. It also requires a fair amount of experience as well. In the program example O3003, the problem is in the relationship of the stored offset amount and the drawing dimension.

Study the drawing carefully - there is an *internal* corner radius of .250 while the offset is set to the cutter radius of .375. This larger stored radius is expected to fit into the .250 part radius. Obviously, it cannot - hence the alarm.

Since the drawing dimension cannot be changed, the size of the cutter diameter must be changed, to a cutter diameter that is *less than* .500 inches. The other drawing radius of .200 is no problem, as external radii can have any size.

The Fanuc controls will *not* allow *gouging* in cutter radius offset *Type C*. This feature is built-in and there is no opportunity to see what would actually happen, if the protection were not there. Nobody wants to see the gouging on the part, but the *Figure 30-26* shows the same effect graphically. In fact, this was a real error in the earlier forms of cutter radius offsets *Type A* and *Type B*.

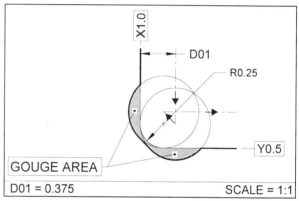

Figure 30-26

Effect of overcutting (gouging) in cutter radius offset mode.
Type C radius offset (look ahead type) does not allow overcutting

◆ Single vs. Multiaxis Startup

There is another possible problem during cutter radius offset startup, particularly if programming the startup motion along two axes, rather than the suggested single axis. We had look at this possibility for an external cutting, with no problems. Now we look at internal cutting.

Evaluate the two approach methods in *Figure 30-27*, using a cutter radius offset startup towards an internal profile, for example, a wall of a pocket or other internal contour.

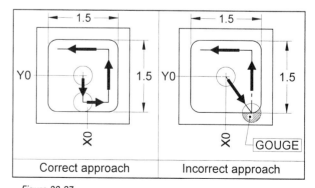

Figure 30-27

Possible problem in cutter radius offset mode during a startup with two axes simultaneously (internal cutting shown)

➔ Correct approach - single axis motion :

The correct programming approach shown on the left side of the illustration contains the following blocks - only the starting program blocks are listed:

```
N1 G20 (CORRECT APPROACH WITH A SINGLE AXIS)
N2 G17 G40 G80
N3 G90 G54 G00 X0 Y0 S1200 M03
N4 G43 Z0.1 H01 M08
N5 G01 Z-0.25 F6.0       (FOR 0.25 POCKET DEPTH)
N6 G41 Y-0.75 D01 F10.0        (START OFFSET)
N7 X0.75
N8 Y0.75
...
```

There is no internal radius in the program to worry about, so the amount stored in the offset register D01 does not have to consider it and will represents the cutter radius as is.

➔ Incorrect approach - multiaxis motion :

The incorrect motion approach shown on the right side of the illustration contains the following initial blocks:

```
N1 G20 (INCORRECT APPROACH WITH TWO AXES)
N2 G17 G40 G80
N3 G90 G54 G00 X0 Y0 S1200 M03
N4 G43 Z0.1 H01 M08
N5 G01 Z-0.25 F6.0     (FOR 0.25 POCKET DEPTH)
N6 G41 X0.75 Y-0.75 D01 F10.0   (START OFFSET)
N7 Y0.75
...
```

There is no way the control system can detect the bottom wall of the pocket at Y-0.75. The startup for the offset is exactly the same as for external cutting, but more damaging.

Compare the two possible startups for the drawing shown in *Figure 30-2*, earlier in the chapter. If the radius offset is started with a single axis motion, the result is shown at the left side illustration in *Figure 30-28*. If the offset is started with a two-axis motion, the result is shown at the right side illustration in *Figure 30-28*.

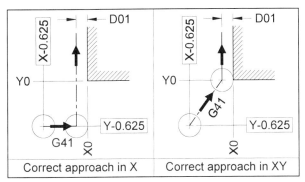

Figure 30-28

Startup of the cutter radius offset for external cutting:
Single axis approach - shown on the left
Two axis approach - shown on the right

Here are the first few correct blocks of each method:

➔ Correct approach - single axis motion :

```
N1 G20 (CORRECT APPROACH WITH ONE AXIS)
N2 G17 G40 G80
N3 G90 G54 G00 X-0.625 Y-0.625 S920 M03
...
N6 G41 X0 D01 F15.0              (START OFFSET)
N7 Y1.125                               (P2)
...
```

➔ Correct approach - multiaxis motion :

```
N1 G20 (CORRECT APPROACH WITH TWO AXES)
N2 G17 G40 G80
N3 G90 G54 G00 X-0.625 Y-0.625 S920 M03
...
N6 G41 X0 Y0 D01 F15.0          (START OFFSET)
N7 Y1.125                               (P2)
...
```

Note that in cases of the cutter radius offset for an external contour, *both* programs listed are correct, because there appears to be *no interference* with any section of the part. In fact, there is *the same* interference as in the internal milling example - the only difference is that this type of 'interference' is of no consequence - it takes place while in the air.

There will always be a problem that cannot be solved in any handbook, regardless of how comprehensive that book may be. The subjects and examples included in this handbook present common basis for a better understanding of the subject. With growing experience, the understanding becomes much deeper. Before going any further, let's review some general rules of the cutter radius offset feature.

OVERVIEW OF GENERAL RULES

Reminders and rules are only important until a particular subject is fully understood. Until then, a general overview and some additional points of interest do come handy. Programming the cutter radius offset is no different. The following items are marked [M] for milling, [T] for turning, and [M-T] for both types of control systems:

❑ [M-T] Never start or cancel the radius offset in an arc cutting mode (with G02 or G03 in effect). Between the startup block and the cancel block, arc commands are allowed and normal, if the job requires them.

❑ [M-T] Make sure the cutter radius is always smaller than the smallest inside radius of the part contour.

❑ [M-T] In the canceled mode G40, move the cutter to a clear area. Always consider the cutter radius, as well as all reasonable clearances.

❑ [M-T] Apply the cutter radius offset with the G41 or G42 command, along with a rapid or a linear motion to the first contour element (G00 or G01 in effect).

❏ [M] Reach the Z axis milling depth in the G40 mode (cutter radius offset cancel mode).

❏ [M-T] Give the preference to a single axis approach from the startup position.

❏ [M] Do not forget the offset number D.. for in the program - it is a small error that can cost you a lot.

❏ [M-T] Make sure to know exactly where the tool command point will be when the radius offset is applied along two axis.

❏ [M-T] In the compensated mode (G41 or G42 in effect), watch for blocks that do not contain an axis motion. Avoid non-motion blocks if possible (missing X, Y and Z).

❏ [M-T] Cancel cutter radius offset with the G40 command, along with a rapid or a linear motion (G00/G01) only, preferably as a single axis motion only.

❏ [M] Retract from the depth (along the Z axis only) after the radius offset has been canceled.

❏ [M] Make sure the cutter radius offset corresponds to the work plane selected (see Chapter 31).

❏ [M-T] G28 or G30 machine zero return commands will not cancel the radius offset (but either one will cancel the tool length offset).

❏ [M-T] G40 command can be input through the MDI to cancel the cutter radius offset (usually as a temporary or an emergency measure).

PRACTICAL EXAMPLE - MILLING

The following in-depth example attempts to present a practical application of the cutter radius offset to both the CNC programmer and the CNC operator. It covers virtually all situations that can happen during the machining process and presents solutions to maintaining the required dimensions of the part. The first subject that has to be well understood is the difference between the *programmed* and the *measured* part size.

◆ Part Tolerances

When the machining is completed on a CNC machine (sometimes even before that), the part has to pass through some inspection process. That means all drawing requirements have to be met for the part to pass the inspection. One of the requirements is to maintain dimensional tolerances, either as specified in the drawing, or as implied in the drawing. Implied tolerances are often company established standard that are based on the number of decimal places used for the dimensions (a method on the decline).

The next example focuses strictly on the effect of cutter radius offset on the part size in the XY plane (top view). For that reason, only a simple application is presented, with the simplest tool path, but not necessarily the best machining method. *Figure 30-29* shows the drawing.

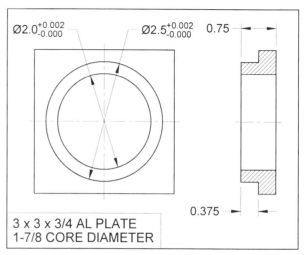

3 x 3 x 3/4 AL PLATE
1-7/8 CORE DIAMETER

Figure 30-29

Drawing to illustrate practical application of a cutter radius offset

The focus will be on the *specified* tolerance in the drawing as being +.002/-.000, for the dimensions of the two diameters - the ∅2.5 external and ∅2.0 internal. Note that the range of all dimensional tolerances is the *same* for both diameters. This statement will be very important later.

◆ Measured Part Size

Every experienced machinist knows that the actual measured size of the part depends on many factors, such as the rigidity of setup, cutting depth, material being used, cutting direction, the selection of tool, its exact size, and so on.

When a part is inspected, the measured size can have only one of the three possible outcomes:

❏ Right on size   ... within specified tolerances

❏ Oversize   ... will be scrap for internal cutting

❏ Undersize   ... will be scrap for external cutting

The first outcome is always ideal, regardless of whether the external or internal cutting takes place. If the measured dimension is *on size* - that is *within* the specified tolerances - there is no need to do anything, the part is good. The second outcome (oversize) and the third outcome (undersize) have to be considered together.

In both cases, the measured dimension is *outside of the specified tolerance* range. This situation requires a look at additional two items that also have to be considered:

❏ External cutting method   ... known as Outside or OD

❏ Internal cutting method   ... known as Inside or ID

Because the cutting tool approaches the machined contour from different directions, the terms *oversize* and *undersize* are always relative to the *type of cutting*. The following table shows the most likely results:

Condition	External	Internal
Oversize	Recut Possible	Scrap Likely
On Size	No Action Required	No Action Required
Undersize	Scrap Likely	Recut Possible

Looking at the table, it is clear that no action is necessary if the measured part size is within tolerances, regardless of whether the external or the internal cutting took place. For the oversize or undersize results, a recut may be possible or a scrap will be the likely result.

A part machined *externally* (∅2.500 inch OD in the example) that is measured as *larger* than the allowed tolerance can likely be recut, but a size that is *smaller* than the allowed tolerance range will result in a scrap.

A part machined *internally* (∅2.000 inch ID in the example) that is measured as *smaller* than the allowed tolerance can likely be recut, but a size that is *larger* then the allowed tolerance range will result in a scrap.

◆ Programmed Offsets

The most attractive feature of the cutter radius offset is that it allows to change the actual tool size right on the machine, *by means of the offset register function D*. In the program example, only one tool is used - *.750 inch diameter* end mill - and one single cut for each contour (external and internal). The program X0Y0Z0 is at the center of the circles and the top of the part:

```
O3004       (T01 - 0.75 DIA END FINISHING MILL)

(**** PART 1 - 2.5 DIA EXTERNAL CUTTING **** )
N1 G20
N2 G17 G40 G80
N3 G90 G54 G00 X0 Y2.5 S600 M03    (START POS.)
N4 G43 Z0.1 H01 M08          (CLEAR+TOOL LG.)
N5 G01 Z-0.375 F20.0         (DEPTH FOR 2.5 DIA)
N6 G41 Y1.25 D01 F10.0       (APPROACH MOTION)
N7 G02 J-1.25                (EXT. CIRCLE CUTTING)
N8 G01 G40 Y2.5              (RETURN MOTION)
N9 G00 Z0.1                  (CLEAR ABOVE)

(**** PART 2 - 2.0 DIA INTERNAL CUTTING **** )
N10 Y0                       (START POS. AT X0Y0)
N11 G01 Z-0.8 F20.0          (DEPTH FOR 2.0 DIA)
N12 G41 Y1.0 D11 F8.0        (APPROACH MOTION)
N13 G03 J-1.0                (INT. CIRCLE CUTTING)
N14 G01 G40 Y0               (RETURN MOTION)
N15 G00 Z0.1 M09             (CLEAR ABOVE)
N16 G28 Z0.1 M05             (Z AXIS MACHINE ZERO)
N17 M01                      (OPTIONAL STOP)
...
```

Figure 30-30 shows the tool path for the first half of the program - the *external* diameter of 2.500 inches. *Figure 30-31* shows the tool path for the second half of the program - the *internal* diameter of 2.000 inches.

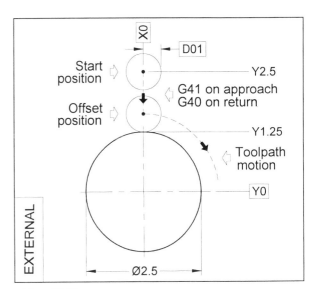

Figure 30-30
Detail for external tool path shown in example O3004

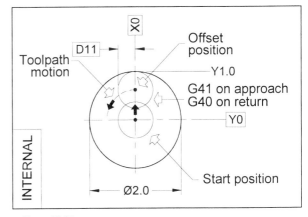

Figure 30-31
Detail for internal tool path shown in example O3004

As is customary in CNC programming and is also used in program O3004, the tool path uses the *drawing* dimensions and the other positions defined by the programmer. This is not only the standard but also the most convenient method to develop a CNC program. Such a program is easy to understand by the machine operator, drawing dimensions are easy to trace (if necessary) and changes can be made, if required. In plain language, the programmer *ignores the cutter radius* and writes the program *as if the cutter were a point* - in effect, a cutting tool with a zero diameter.

◆ D Offset Amount - General Setting

The reality of machining is that a zero diameter cutter is not usually used, except for some engraving work. The majority of cutting tools do have specified diameters and the actual diameters have to be always considered - if not in the program, then on the machine.

One critical fact to be established first is that the CNC system always calculates a specified offset by its cutter *radius, not* by its diameter. It means the programmer provides the *cutter radius offset* in the form of a *D address*. On the machine, the programmed offset D01 will apply to the cutter radius registered in offset 1, D02 to the radius registered in offset 2, etc. What actual amounts are in these registers?

Since no radius of the cutter is included anywhere in the program, the offset register D must normally contain the cutter radius *actual* value. Be careful - some machine parameters may actually be set to accept the cutter *diameter,* although all internal calculations are still set by the radius.

Evaluate program O3004; what will be the stored amount of D01? A ⌀.750 inch end mill is used, so the D01 should be set to .375. This is correct in theory, but factors such as tool pressures, material resistance, tool deflection, actual tool size, tool tolerances and other factors do influence the finished part size. The conclusion is that the D01 registered amount can be .375, but only under *ideal* conditions.

Ideal conditions are rare. The same factors that influence machining will also have a significant effect on part dimensions. It is easy to see that any measured size that is not within tolerances can be only *oversize* or *undersize* and *external* and *internal* cutting method *does* make a difference as to how the offset can be adjusted.

Regardless of the cutting method, there is one major rule applied to the cutter radius offset adjustment in any control system - the rule has two equal parts:

> POSITIVE increment to the cutter radius offset will cause the cutting tool to move AWAY from the machined contour.

> NEGATIVE increment to the cutter radius offset will cause the cutting tool to move CLOSER to the machined contour.

Note the word *'increment'* - it means that the current radius offset amount will be *changed* or *updated* - but *not* replaced - with a new amount. The concept of *'moving away'* and *'moving closer to'* the part refers to the tool motion as the CNC operator will see. The measured size of the part can be controlled by *adjusting* the cutter radius offset value in the control, programmed as the D address, according to these two rules. The most useful rule that applies *equally to the external and internal* adjustments has two alternatives:

> To ADD more material TO the measured size, use LARGER setting amount of the D offset

> To REMOVE material FROM the measured size, use SMALLER setting amount of the D offset

Experienced CNC operators can change offset settings at the machine, providing the program contains the cutter ra-

dius offset commands G41 or G42 as well as the D address offset number - with the appropriate cancellation by G40.

Evaluating what *exactly* happens during the tool motion for each cutting method (*external* or *internal*) offers certain options. In both cases, the cutting tool moves from the *starting* position, within the clear area, to the *target* position of the machining contour. This is the motion where the cutter radius offset is *applied*, so this motion is critical. In fact, this is the motion that *determines the final measured size* of the part. Each method can be considered separately.

◆ Offset Adjustment

Before any special details can be even considered, think about how the offset amount can be changed. In those cases where the size of the part is to be adjusted, the *incremental* change of the offset value is a good choice. Incremental offset change means *adding to* or *subtracting from* the current offset amount (using the +INPUT key on a Fanuc screen) or storing the adjustment in the *Wear* offset screen column. Changes to the program data is *never* the option.

◆ Offset for External Cutting

Evaluate the tolerance range for the *outside* circle ⌀2.5. The tolerance for this diameter is +.002/-0.0, so all sizes between 2.500 and 2.502 are correct. Any size *smaller* than 2.5 is *undersize* and a size *greater* than 2.502 is *oversize*.

There are three possible results of the measured size for *external* cutting. All examples are based on the expected middle size of 2.501 and on D01 holding the amount of .375, which is the radius of a ⌀.750 milling cutter.

➲ External measured dimension - Example 1 :

`2.5010` *with* `D01 = 0.3750`

This is the ideal result - no offset adjustment is necessary. The tool cutting edge touches the intended machining surface exactly. All is working well and the offset setting is accurate. Only standard monitoring is required. This is not such a rare situation as it seems - in fact, it is quite common with a new cutter, rigid setup and common tolerances.

➲ External measured dimension - Example 2 :

`2.5060` *with* `D01 = 0.3750`

The measured diameter is *.005 oversize*. The tool edge has *not reached* the contour and *has to move closer* to it. The radius offset amount has to decrease by one half of the oversize amount, which is on the diameter or width but the offset amount is entered as a radius, per one side. Offset D01 is adjusted incrementally by .0025, to D01=0.3725.

➲ External measured dimension - Example 3:

`2.4930` *with* `D01 = 0.3750`

The measured diameter is *.008 undersize*. The cutting tool edge has *reached beyond* the programmed machining surface and *has to move away* from it. The radius offset amount has to be *increased* by one half of the undersize amount. The undersize is measured on the diameter (or width) of the part, but the offset amount is entered as a radius, per side only. The D01 offset is adjusted incrementally by .004, to D01=0.3790.

◆ Offset for Internal Cutting

Now is the time to look at tolerance range for the *inside* diameter of 2.0 inches. The tolerance range for this diameter is +.002/-0.000, so all part sizes between 2.000 and 2.002 will be correct. A size *smaller* than 2.000 will be *undersize* and a size *greater* than 2.002 will be *oversize*.

There are three possible results of the measured size for *internal* cutting. All examples are based on the expected middle size of 2.001 and on D11 holding the amount of .375, the radius of a ∅.750 cutter.

➲ Internal measured dimension - Example 4 :

`2.2010` *with* `D11 = 0.3750`

This is the ideal result - no offset adjustment is necessary. The tool cutting edge touches the intended machining surface exactly. All is working well and the offset setting is accurate. Only normal monitoring is required.

➲ Internal measured dimension - Example 5 :

`2.0060` *with* `D11 = 0.3750`

The measured diameter is .005 *oversize*. The tool edge has *reached beyond* the intended machining surface and *has to move away* from it. The radius offset value has to be *increased* by one half of the oversize amount. The oversize is on the diameter (or width), but the offset amount is entered as a radius, per side only. The D11 offset must be incremented by .0025, to D11=0.3775.

➲ Internal measured dimension - Example 6 :

`1.9930` *with* `D11 = 0.3750`

The measured diameter is .008 *undersize*. The tool edge has *not reached* the intended machining surface and *has to move closer* to it. The radius offset value has to be *decreased* by one half of the undersize amount. The undersize is on the diameter (or width), but the offset amount is entered as a radius, per side only. The D11 offset must be incremented by .004, to D11=0.3710.

◆ One Offset or Multiple Offsets?

The program O3004 used D01 for the external diameter and D11 for the internal diameter. Only one tool was used and the goal was the middle tolerance of 2.501 for the external diameter and 2.001 for the internal diameter. Are two offsets in the program needed or a will a single offset do?

Keep in mind that the last few examples evaluated only possibilities that were independent from each other, with no common connection. Program O3004 presents a common connection between the two diameters. It is one ∅.750 end mill, used for cutting *both* diameters.

Assume for a moment, that only one offset is used, for example D01, with the stored amount of .375. When measured, the external diameter is 2.001. After continuing cutting the internal diameter of 2.000 inches, when measured again, its size is not 2.001 as expected, but only 1.999. This measurement is .002 undersize then the expected diameter. The reason is that *both* diameters have a +.002/-0.000 tolerance. The results are different - for the external diameter, +.002 means oversize that can be recut, for the internal diameter, +.002 means oversize that is a scrap. Since one offset alone cannot be adjusted to meet the middle tolerance on *both* diameters, *two offsets* have to be used. It follows, that if D01=.3750 and makes a perfect external diameter, D11 should have a smaller setting amount of only .3730.

The CNC operator should always be aware of the offsets used in the program and understand about the stored offset amounts, especially if more than one offset is used for one tool.

In the setup or tooling sheet, the programmer should always list the offsets used in the program and suggest the starting values for each offset as a professional courtesy.

◆ Preventing a Scrap

When it comes to initial offset amounts, some creative techniques can be used here. The goal is to use offset settings in such a way that the part will not likely be a scrap, even with an unproven tool. A good operator can prevent scraps caused by wrong offsets, at least to some degree. The key is to create some *temporary* offset settings. The goal is to force a cut that is *oversize externally* or *undersize internally*, measure it, adjust it, then recut to the right size.

Whether machining an external or internal tool path, even the best setup will not guarantee that the part dimensions will be within tolerances. When machining an *external* contour, the diameter can be cut *intentionally larger* than required - in a controlled way. In this case, the risk that the diameter will be *too small* is present.

In *internal* contour machining, the diameter can be cut *intentionally smaller* than required, in a controlled way. In this case, the risk that the diameter will be *too large* is present. Either case offers benefits but some drawbacks, too.

In the last few examples, the solution is to move the tool *away from* the intended external machined surface by a *positive offset increment*. The increment amount must be greater than the expected error of the tool radius, as well as being suitable for a recut.

In both cases, when the test cut is made, measure the diameter and adjust the offset by one half of the difference between measured and intended diameters. If only one side is cut, the difference is not halved.

◆ Program Data - Nominal or Middle?

Many coordinate locations in the program reflect actual dimensions that are taken from the drawing. The question is - what happens if the drawing dimension specifies a tolerance range? There are two opinions among CNC programmers. One opinion favors the use of the middle value of tolerance range, the other prefers to use the nominal size and ignore the tolerances. Both opinions have some credibility and should not be discarded. In this handbook, the preference is to use the *nominal* dimensional sizes and let the tolerances be handled by proper use of offsets - at the machine. Two reasons prevail. One is that a program using nominal dimensions is easier to read. Two, in case of drawing changes, they will affect the tolerances more often than nominal sizes.

TOOL NOSE RADIUS OFFSET

All the principles and rules described so far also apply to the radius offset for a lathe contouring tool. There are few differences, mainly caused by the shape of the tool.

In milling, the cutting tool is always round. The cutter periphery is the cutting edge and its radius value is the offset. Turning tools have a different design. The most common is a multi-sided carbide insert. An insert may have one or more cutting edges. For strength and longer insert life, the cutting edge has a relatively small corner radius. Typical radii for turning and boring tools are:

```
1/64 = .0156 (English)  or  0.40 mm (metric)
1/32 = .0313 (English)  or  0.80 mm (metric)
3/64 = .0469 (English)  or  1.20 mm (metric)
```

Because the tool cutting edge is often called a *tool nose*, the term *tool nose radius offset* became common.

◆ Tool Nose Tip

The tool nose is usually the corner of the tool, where two cutting edges blend into a nose radius. *Figure 30-32* shows typical corners of a turning tool and a boring tool.

The tool nose reference point in turning is often called the *command point*, the *imaginary point* and, lately, even the *virtual point*. It is the point that is moved along the contour, because it is directly related to X0Z0 of the part.

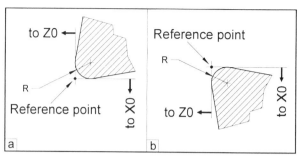

Figure 30-32
Tool reference point for turning and boring - (a) turning, (b) boring

◆ Radius Offset Commands

The same preparatory commands used in milling operations are used for contouring on CNC lathes - *Figure 30-33*:

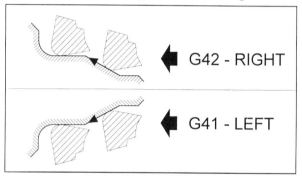

Figure 30-33
Lathe application of the tool nose radius offset

G41	Offset of the tool nose radius to the LEFT of the contouring direction
G42	Offset of the tool nose radius to the RIGHT of the contouring direction
G40	Offset of the tool nose radius CANCEL

For lathes, G codes do not use the D address - offset value is stored in the *Geometry/Wear* offset. *Lathe tools have different cutting edges*, otherwise they are similar to milling.

◆ Tool Tip Orientation

The center of a circle symbolizing an end mill must be equidistant to the contour by its radius. In milling, cutting edges are part of the tool radius, on lathes, they are not. Lathe tools *do* have a radius but separate cutting edges. The nose radius center is also equidistant from the contour, and the edges change their orientation, even for the same insert. Additional definitions are needed in a form of a vector pointing towards the radius center. This vector is called the *tool tip orientation*, numbered arbitrarily. Control uses this number to establish the nose radius center and its orientation. *Figure 30-34* shows two tools and their tip orientation.

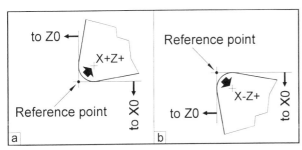

Figure 30-34

Relationship of the tool reference point and the nose radius center

The tip orientation is entered during the setup, according to arbitrary rules. Fanuc controls require a fixed number for each possible tool tip. This number has to be entered into the offset screen at the control, under the T heading. The value of the tool radius R must also be entered. If the tool tip is 0 or 9, the control will compensate to the center. *Figures 30-35* and *30-36* show the standard tool tip numbering for CNC lathes with X+ up and Z+ to the right of origin.

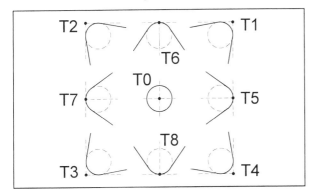

Figure 30-35

Arbitrary tool tip numbers for nose radius offset - rear lathe shown

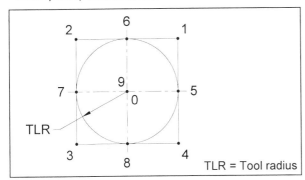

Figure 30-36

Schematic illustration of the tool tip numbering (Fanuc controls)

◆ Effect of Tool Nose Radius Offset

Some programmers do not bother using the tool nose radius offset. *That is wrong!* Theoretically, there is no need for the offset if only a single axis is programmed. However,

single axis motions are part of a contour that also includes radii, chamfers and tapers. In this case, the tool nose radius offset *is* needed, otherwise all radii, chamfers and tapers will not be correct. The illustration in *Figure 30-37* shows what areas of the part would be *undercut* or *overcut*, if the tool nose radius offset were *not* used during machining.

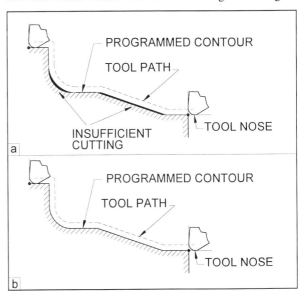

Figure 30-37

Effect of tool nose radius offset - (a) offset not used   (b) offset used

◆ Sample Program

The following program example O3005 shows a simple application of the tool nose radius offset on an external and internal contour, based on the drawing in *Figure 30-38*. Only the finishing cuts are shown - roughing is also necessary, but would most likely use the special G71 multiple repetitive cycle, described in *Chapter 35*.

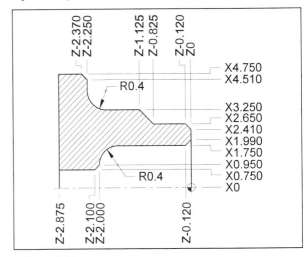

Figure 30-38

Simplified sample drawing for program example O3005

```
O3005
...
N31 T0300                    (EXTERNAL FINISHING)
N32 G96 S450 M03
N33 G00 G42 X2.21 Z0.1 T0303 M08
N34 G01 X2.65 Z-0.12 F0.007
N35 Z-0.825 F0.01
N36 X3.25 Z-1.125
N37 Z-1.85
N38 G02 X4.05 Z-2.25 R0.4
N39 G01 X4.51
N40 X4.8 Z-2.395
N41 U0.2
N42 G00 G40 X8.0 Z5.0 T0300
N43 M01

N44 T0400                    (INTERNAL FINISHING)
N45 G96 S400 M03
N46 G00 G41 X2.19 Z0.1 T0404 M08
N47 G01 X1.75 Z-0.12 F0.006
N48 Z-1.6 F0.009
N49 G03 X0.95 Z-2.0 R0.4
N50 G01 X0.75 Z-2.1
N51 Z-2.925
N52 U-0.2
N53 G00 G40 X8.0 Z2.0 T0400
N54 M01
...
```

Note that the contour start and end positions are in the *clear area* - away from the part. Make sure there is enough clearance. *Cutter radius compensation interference alarm* (alarm #41) is always caused by insufficient clearance.

◆ Minimum Clearance Required

> As a rule, each clearance in program should be large enough to accommodate the double tool nose radius.

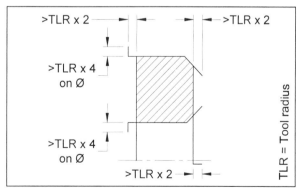

Figure 30-39

Minimum clearance for tool nose radius offset

Figure 30-39 shows minimum clearances when set at the start and end of cut. Make sure the nose radius fits into the clearance two or more times. Symbols *>TLR × 2* and *× 4* mean the clearance should be greater than twice or four times the nose radius. Double radius per side becomes a quadruple radius on a diameter.

In tool nose radius offset, programming the minimum clearance of at least .100 inches per side (2.5 mm), provides a sufficient clearance for all three standard tool nose radii - 1/64, 1/32 and 3/64 (0.40, 0.80 and 1.20 mm respectively).

◆ Change of Motion Direction

On CNC lathes, a change in cutting direction is used much more often than on machining centers. The following example shows a facing cut on a solid face with G41 in effect, changing to a turning cut(-s) with G42 in effect - see *Figure 30-40*. Possible problem is discussed as well.

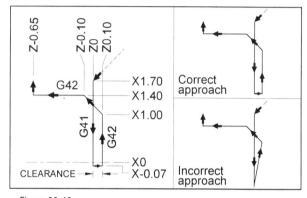

Figure 30-40

Tool nose radius offset change for the same tool

```
...
N21 T0100                    (CORRECT APPROACH)
N22 G96 S400 M03
N23 G00 G41 X1.7 Z0 T0101 M08        (START)
N24 G01 X-0.07 F0.007               (FACE OFF)
N25 G00 Z0.1                     (ONE AXIS ONLY)
N26 G42 X1.0                   (THEN COMPENSATION)
N27 G01 X1.4 Z-0.1 F0.012          (CONTOURING)
N28 Z-0.65
N29 X...
```

Face cutting is a single axis motion and the offset is used for consistency. For solid parts, the face cut must end *below* the center line, X-0.07 in block N24, at a diameter marginally larger than double tool radius. If the cut finishes at X0, the tool leaves a small unfinished tip at the center line and the face will not be flat. Also compare the *correct* and *incorrect* tool motions on the right side of the last illustration. If the above program is modified to the following version,

```
...
N21 T0100                    (INCORRECT VERSION)
N22 G96 S400 M03
N23 G00 G41 X1.7 Z0 T0101 M08        (START)
N24 G01 X-0.07 F0.007               (FACE OFF)
N25 G00 G42 X1.0 Z0.1          (*** WRONG ***)
N26 G01 X1.4 Z-0.1 F0.012          (CONTOURING)
N27 Z-0.65
N28 X..
```

... the face will never be completed! Think about it.

From all available machining operations, *contouring* or *profiling* is the single most common CNC application, perhaps along with hole making. During contouring, the tool motion is programmed in at least three different ways:

❑ Tool motion along a single axis only

❑ Tool motion along two axes simultaneously

❑ Tool motion along three axes simultaneously

There are additional axis motions that can also be applied (the *fourth* and *fifth* axis, for example), but on a CNC machining center, we *always* work with at least *three* axes, although *not always* simultaneously. This reflects the three dimensional reality of our world.

This chapter applies only to CNC milling systems, since turning systems normally use only two axes, and planes are therefore not required or used. Live tooling on CNC lathes does not enter this subject.

Any absolute point in the program is defined by three coordinates, specified along the X, Y and Z axes. A programmed rapid motion G00 or a linear motion G01 can use *any number* of axes simultaneously, as long as the resulting tool motion is safe within the work area. No special considerations are required, no special programming is needed.

That is not the case for the following three programming procedures, where the various considerations change quite significantly:

❑ Circular motion using the G02 or G03 command

❑ Cutter radius offset using the G41 or G42 command

❑ Fixed cycles using the G81 to G89 commands, or G73, G74 and G76 commands

In all three cases - *and only in these three cases* - programmer has to consider a special setting of the control system - it is called a *selection of the machining plane*.

WHAT IS A PLANE?

To look up a definition of a plane, research a standard textbook of mathematics or even a dictionary. From various definitions, plane can be described in one sentence:

> A plane is a surface in which a straight line joining any two of its points will completely lie on that surface.

Planes in the mathematical sense have their own properties. There is no need to know them all, but there are important properties relating to planes that are useful in CNC programming and in various phases of CAD/CAM work:

❑ Any three points that do not lie on a single line define a plane (these points are called non-collinear points)

❑ A plane is defined by two lines that intersect each other

❑ A plane is defined by two lines that are parallel to each other

❑ A plane is defined by a single line and a point that does not lie on that line

❑ A plane can be defined by an arc or a circle

❑ Two intersecting planes define a straight line

❑ A straight line that intersect a plane on which it does not lie, defines a point

These mathematical definitions are only included for reference and as a source of additional information. They are *not* required for everyday CNC programming.

MACHINING IN PLANES

The path of a cutting tool is a combination of straight lines and arcs. A tool motion in one or two axes always takes place in a plane designated by two axes. This type of motion is *two-dimensional*. In contrast, any tool motion that takes place in three axes at the same time is a *three-dimensional* motion.

◆ Mathematical Planes

In CNC machining, the only planes that can be defined and used are planes consisting of a combination of *any two* primary axes XYZ. Therefore, the *circular cutting motion*, *cutter radius offset* and *fixed cycles* can take place only in any one of the three available planes:

XY plane	ZX plane	YZ plane

The actual *order of axis designation* for a plane definition is very important. For example, the XY plane and the YX plane are *physically* the same plane. However, for the purposes of defining a relative tool motion direction (*clockwise* vs. *counterclockwise* or *left* vs. *right*), a clear standard must be established.

This international standard is based on the mathematical rule that specifies the *first* letter of the plane designation always refers to the *horizontal* axis and the *second* letter refers to the *vertical* axis when the plane is viewed. Both axes are always *orthogonal* (horizontal and vertical) and *perpendicular* (at 90°) to each other. In CAD/CAM, this standard defines the difference between the top and bottom, front and back, etc.

A simple way to remember *mathematical* designation of axes for all three planes is to write the alphabetical order of all three axes *twice* and isolate each pair with a space:

> XYZXYZ   ... becomes ...   XY   ZX   YZ

In *mathematical* terms, the planes are defined as:

Plane	Horizontal Axis	Vertical Axis
XY	X	Y
ZX	Z	X
YZ	Y	Z

Note the emphasis on the word *'mathematical'*. The emphasis is intentional, and for a very good reason. As will soon be apparent, there is a great difference between the mathematical planes and the machine planes, as defined by the viewing direction of the machine.

◆ Machine Tool Planes

A typical CNC machining center has three axes. Any two axes form a plane. A machine plane may be defined by looking at the machine from standard operating position. For a vertical machining center, there are three standard views, viewed perpendicularly (straight on):

- ❏ The top view   ... XY plane
- ❏ The front view   ... XZ plane
- ❏ The right side view   ... YZ plane

The illustration in *Figure 31-1* shows the difference between the two definitions, caused by a viewpoints that are not compatible.

It is clear that the XY plane and top view are the same in both definitions, and so is the YZ plane and the right side view. The ZX mathematical plane is *different* from the front plane on the machine, which is XZ, as shown in the middle illustration.

The mathematical plane defined as the ZX plane, where Z is the horizontal axis, is *reversed* on the machine plane for CNC machining centers. On the machine, this plane becomes the XZ plane, where the *X axis* is the horizontal axis - a very important distinction.

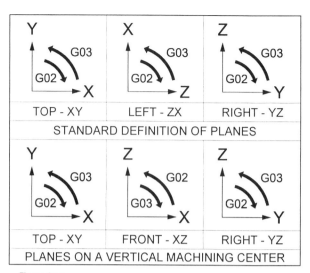

Figure 31-1

Comparison of standard mathematical planes (above), and planes on a CNC machining center (below)

In programming, the selection of planes is extremely important, yet often neglected and even misunderstood by programmers and operators alike. The main reason is that the majority of tool motions (particularly for contouring) are programmed and machined in the standard XY plane. On all CNC machining centers, the spindle is always perpendicular to the XY plane. Vertical and horizontal applications are the same in this respect.

◆ Program Commands for Planes Definition

The selection of a plane for Fanuc and related controls adheres to the *mathematical* designation of planes, *not* the actual CNC machine tool planes. In a part program, each of the three *mathematical* planes can be selected by a special preparatory command - a unique G code:

G17	XY plane selection
G18	ZX plane selection
G19	YZ plane selection

For *all* rapid motions (programmed with G00) *and* all linear motions (programmed with G01), the plane selection command is totally irrelevant and even redundant. That is not the case for other motion modes, where the plane selection in a program is *extremely* important and must be considered carefully.

For machining applications using the circular interpolation mode, with G02 or G03 commands, cutter radius offset mode with G41 or G42 commands and fixed cycles mode with G81 to G89 commands, as well as G73, G74, G76, the plane selection is very critical.

◆ Default Control Status

If the plane is not selected by the program, the control defaults automatically to G17 XY plane in milling and G18 ZX plane in turning. If the plane selection G code is programmed, it should be included at the program beginning. Since the three plane commands only have affect on *circular motions*, *cutter radius offsets* and *fixed cycles*, the plane selection command G17, G18 or G19 can be programmed *before* any of these machining motions take place.

> Always program the appropriate plane selection command.
> Never rely on the control settings !

Any plane selection change is programmed as desired, prior to actual tool path change. Plane can be changed as often as necessary in a program, but only one plane can be active at any time. Selection of one plane cancels any other plane, so the G17/G18/G19 commands cancel each other. Although true in an informative sense, it is most likely that the opportunities to mix all three plane commands in a single program are remote. From all three available motions, only the circular motion is affected by plane selection, but let's have look at the programming of a rapid and linear motions as well, at least for comparison purposes.

STRAIGHT MOTION IN PLANES

Both rapid motions G00 and linear motions G01 are considered straight motions when compared with circular motions. Straight motions can be programmed for a single axis or as a simultaneous motion along two or three axes. The following examples only show typical unrelated blocks:

● Example - Rapid positioning - G00 :

```
G00 X5.0 Y3.0        XY plane - 2D rapid motion
G00 X7.5 Z-1.5       XZ plane - 2D rapid motion
G00 Y10.0 Z-0.25     YZ plane - 2D rapid motion
G00 X2.0 Y4.0 Z-0.75  XYZ - 3D rapid motion
```

● Example - Linear interpolation - G01 :

```
G01 X-1.5 Y4.46 F15.0    XY plane - 2D linear motion
G01 X8.875 Z-0.84 F10.0  ZX plane - 2D linear motion
G01 Y12.34 Z0.1 F12.5    YZ plane - 2D linear motion
G01 X6.0 Y13.0 Z-1.24 F12.0  XYZ - 3D linear motion
```

The examples refer to tool motion along the programmed axes. Plane selection command does not need to be used for any straight motion (along a single axis), unless the cutter radius offset or a fixed cycle is in effect. All tool motions will be interpreted correctly by the control, regardless of any plane in effect. The rules that apply to linear motions are not the same for circular motions.

CIRCULAR INTERPOLATION IN PLANES

In order to complete a circular motion correctly, the control system has to receive sufficient information from the part program. Unlike rapid positioning with G00 in effect or linear interpolation with G01 in effect, the circular interpolation requires a programmed *direction* of motion. G02 is the command for CW direction and G03 is the command for CCW direction. According to general mathematical rules, the *clockwise* direction is always viewed from the vertical axis towards the horizontal axis in any selected plane. *Counterclockwise* direction is always viewed from the horizontal axis towards the vertical axis.

When we compare the mathematical axes designation and the actual orientation of the machine axes (based on a vertical machining center), the XY plane (G17) and the YZ plane (G19) correspond to each other. These two planes normally present no problems to CNC programmers. The ZX plane (G18) may cause a serious problem if not properly understood. Mathematically, the horizontal axis in G18 plane is the Z axis and the X axis is the vertical axis. On a vertical machining center, the order of machine axes orientation is *reversed*. It is important to understand that the clockwise and counterclockwise directions *only appear* to be reversed, but in reality, they are the same. If the mathematical axes orientation is aligned with the machine axes, they *will* indeed match. *Figure 31-2* shows the steps of aligning the mathematical planes with the machine planes:

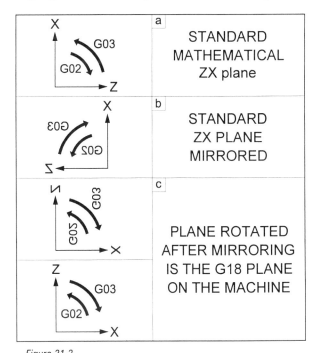

X ↑ G03 G02 → Z	**a** STANDARD MATHEMATICAL ZX plane
X ↑ G03 G02 ← Z	**b** STANDARD ZX PLANE MIRRORED
Z ↑ G03 G02 → X	**c**
Z ↑ G03 G02 → X	PLANE ROTATED AFTER MIRRORING IS THE G18 PLANE ON THE MACHINE

Figure 31-2

Progressive steps in aligning the mathematical ZX plane with the machine XZ plane, using G18 plane selection

Note that the G code direction for arcs *does not* change either within the mathematical plane (a), or the mathematical plane mirrored (b), or even the mirrored plane rotated by negative 90° (c), even if the plane itself is changed. What occurred here is *not* a creation of any new plane (mathematical or otherwise). The view still represents a three dimensional object, viewed from a different directional point (viewpoint).

On horizontal machining centers, the situation is similar. The XY plane (G17) and the ZX plane (G18) match between mathematical designation and the actual axes orientation. It is the G19 plane (YZ) that *appears* to be reversed and may cause some problems before the logical structure of the planes is well understood.

The proper selection of a machining plane will enable programming various contouring operations using circular and helical interpolation, cutter radius offset and fixed cycles. The most common applications of this type of machining are filleted (blend) radii, intersecting radii, circular pockets, profiled counterbores, cylinders, simple spheres and cones, and other similar shapes.

In order to understand the CNC applications of G02 and G03 commands in planes, the illustration in *Figure 31-3* should be helpful.

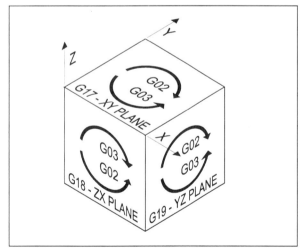

Figure 31-3

Actual circular tool path direction in all three machine planes. Note the apparent inconsistency for the G18 plane

◆ G17-G18-G19 as Modal Commands

The preparatory commands for a plane selection G17, G18 and G19 are all *modal* commands - programming any one of them will activate the selected plane only. The plane selection in the program will be in effect until canceled by another plane selection. The three plane related G codes belong to the G code group number 02 exclusively.

The following format examples show some typical programming applications for circular interpolation:

```
G17 G02 X14.4 Y6.8 R1.4

G18 G03 X11.575 Z-1.22 R1.0

G19 G02 Y4.5 Z0 R0.85
```

Some older control systems do not accept the direct radius designation specified by the R address. Instead, the arc vectors I, J and K must be used. For programming circular motion within a selected plane, correct pair of arc vectors must be selected:

```
G17 G02 (G03) X.. Y.. I.. J..

G18 G02 (G03) X.. Z.. I.. K..

G19 G02 (G03) Y.. Z.. J.. K..
```

From the previous topics, remember that:

❑ XY axes - G17 plane - I and J arc center modifiers
❑ XZ axes - G18 plane - I and K arc center modifiers
❑ YZ axes - G19 plane - J and K arc center modifiers

◆ Absence of Axis Data in a Block

The programming format described here contains *complete data* for the end point of a circular motion. In practice, however, an experienced programmer does not repeat the modal values from one block to another. The major reason for this approach is saving programming time, shortening the program length and increasing the available memory space in the control system.

The portion of the following program example shows a typical application in a program where modal axes values are *not repeated* in subsequent blocks:

```
N..  G20                        English units
...
N40 G17                         XY plane selected
N41 G00 X20.0 Y7.5 Z-3.0        Start position of the tool
N42 G01 X13.0 F10.0             Plane selection irrelevant
N43 G18 G02 X7.0 R3.0           Z axis is assumed as absent
N44 G17 G01 X0                  Plane selection irrelevant
```

Block N43 represents a contour of a 180° arc in the ZX plane. Because of the G18 command in N43, the control will correctly interpret the 'missing' axis as the Z axis, and its value will be equal to the last Z axis value programmed (Z-3.0). Also examine the G17 command in block N44. It is always a good practice to transfer the control status to its original plane selection as soon as the plane changes, although this is not absolutely necessary in the example.

Omitting the G18 command in block N43 will cause a serious program error. If G18 is omitted, the originally selected command G17 will still be in effect and circular interpolation will take place in the XY plane, instead of the intended ZX plane.

In this case, the axis assumed as 'missing' in the G17 plane will be the Y axis and its programmed value of Y7.5. The control system will process such a block as if it were specified in a complete block:

```
N43 G17 G02 X7.0 Y7.5 R3.0
```

An interesting situation will develop if the plane selection command G18 in block N43 is absent, but the circular interpolation block contains *two axes* coordinates for the end point of the circular motion:

```
N43 G02 X7.0 Z-3.0 R3.0      G17 is still in effect
```

Although G17 is still the active plane, the arc will be machined correctly in the G18 plane, even if G18 had not been programmed. This is because of the special control feature called *complete instruction* or *complete data priority*, provided in block N43 of the last example. The inclusion of two axes for the end point of circular motion has a higher priority rating than a plane selection command itself. A complete block is one that includes all necessary addresses without taking on modal values.

> Two axes programmed in a single block
> override the active plane selection command.

◆ Cutter Radius Offset in Planes

The plane selection for rapid or linear motion is irrelevant, providing that no cutter radius offset G41 or G42 is in effect. In theory, it means that regardless of the plane selection, all G00 and G01 motions will be correct. That is true, but seldom practical, since most CNC programs *do* use a contouring motion and they also use the cutter radius offset feature. As an example, evaluate the following blocks:

```
N1 G21
...
N120 G90 G00 X50.0 Y100.0 Z20.0
N121 G01 X90.0 Y140.0 Z0 F180.0
```

When the rapid motion programmed in block N120 is completed, the cutter will be positioned at the absolute location of X50.0 Y100.0 Z20.0. The absolute location of the cutting motion will be X90.0 Y140.0 Z0, after the block N121 is completed.

Adding a cutter radius offset command G41 or G42 to the rapid motion block, the plane selection will become extremely important. The radius offset will be effective only for those two axes selected by a plane selection command.

There will *not* be a 3-axis cutter radius offset taking place! In the next example, compare the absolute tool positions for each plane when the rapid motion is completed and the cutter radius offset is activated in the program. Tool absolute position when the cutting motion is completed depends on the motion following block N121.

The radius offset value of D25=100.000 mm, stored in the control offset registry, is used for the next example:

➲ Example :

```
N120 G90 G00 G41 X50.0 Y100.0 Z20.0 D25
N121 G01 X90.0 Y140.0 Z0 F180.0
```

The compensated tool position when block N120 is completed, will depend on the plane G17, G18 or G19 currently in effect:

❑ If G17 command is programmed with three axes :

   G17 X.. Y.. Z..   XY motion will be compensated

❑ If G18 command is programmed with three axes :

   G18 X.. Y.. Z..   ZX motion will be compensated

❑ If G19 command is programmed with three axes :

   G19 X.. Y.. Z..   YZ motion will be compensated

The following practical programming example illustrates both circular interpolation and cutter radius offset as they are applied in different planes.

PRACTICAL EXAMPLE

The example illustrated in *Figure 31-4* is a simple job that requires cutting the R0.75 arc in the XZ plane. Typically, a ball nose end mill (also known as a spherical end mill) will be used for a job like this.

In the simplified example, only two main tool passes are programmed. One pass is the left-to-right motion - across the left plane, over the cylinder, and over the right plane. The other pass is from right to left - across the right plane, over the cylinder, and across the left plane. A stepover for the tool is also programmed, between the passes. The program of this type for the whole part could be done in the incremental mode and would greatly benefit from the use of subprograms.

Figure 31-5 demonstrates tool motion for the two passes included in the program example. To interpret the program data correctly, note that program zero is at the *bottom left corner* of the part. Both clearances off the part are .100 and the stepover is .050:

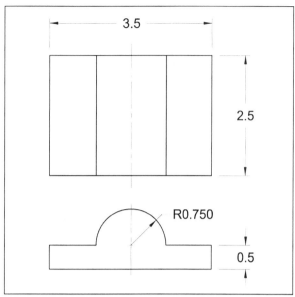

Figure 31-4

Drawing for the programming example O3101

```
O3101
N1 G20
N2 G18                        (ZX PLANE SELECTED)
N3 G90 G54 G00 X-0.1 Y0 S600 M03
N4 G43 Z2.0 H01 M08
N5 G01 G42 Z0.5 D01 F8.0
N6 X1.0
N7 G03 X2.5 I0.75    (= G03 X2.5 Z0.5 I0.75 K0)
N8 G01 X3.6
N9 G91 G41 Y0.05
N10 G90 X2.5
N11 G02 X1.0 I-0.75(= G02 X1.0 Z0.5 I-0.75 K0)
N12 G01 X-0.1
N13 G91 G42 Y0.05
N14 G90 ...
```

When working with this type of CNC program the first time, it may be a good idea to test the tool path in the air, a little above the job. Errors can happen quite easily.

Three axes cutting motion is programmed manually only for parts where calculations are not too time consuming. For parts requiring complex motions calculations, a computer programming software is a better choice.

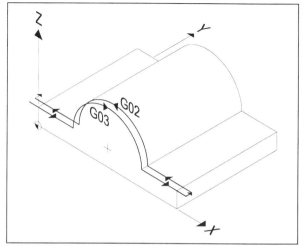

Figure 31-5

Tool path for programming example O3101

FIXED CYCLES IN PLANES

The last programming item relating to plane selection is the application of planes in fixed cycles. For cycles in the G17 plane (XY hole locations), G17 is only important if a switch from one plane to another is contained in the same program. With special machine attachments, such as *right angle heads*, the drill or other tool is positioned *perpendicular* to the normal spindle axis, being in G18 or G19 plane.

Although the right angle heads are not very common, in many industries they are gaining in popularity. When programming these attachments, always consider the tool direction into the work (the *depth* direction). In the common applications of fixed cycles, G17 plane uses XY axes for the hole center location and the Z axis for the depth direction. If the angle head is set to use the Y axis as the depth direction, use G18 plane and the XZ axes will be the hole center positions. If the angle head is set to use the X axis as the depth direction, use G19 plane and the YZ axes will be the hole center positions. In all cases, the R level always applies to the axis that moves along the depth direction.

The difference between the tool tip and the center line of spindle is the actual overhang. This extra overhang length must be known and incorporated into all motions of the affected axis not only for correct depths, but also for safety.

PERIPHERAL MILLING

Even with the ever increasing use of carbide cutters for metal removal, the traditional HSS (high-speed steel) end mills still enjoy a great popularity for a variety of milling operations and even on lathes. These venerable cutters offer several benefits - they are relatively inexpensive, easy to find, and do many jobs quite well. The term *high speed steel* does not suggest much productivity improvement in modern machining, particularly when compared to carbide cutters. It was used long time ago to emphasize the benefit of this tool material to carbon tool steel. The new material of the day was a tool steel enhanced with tungsten and molybdenum (*i.e.,* hardening elements), and could use spindle speeds two to three times faster than carbon steel tools. The term high-speed-steel was coined and the HSS abbreviation has become common to this day.

The relatively low cost of high speed steel tools and their capability to machine a part to very close tolerances make them a primary choice for many milling applications. End mills are probably the single most versatile rotary tool used on a CNC machine.

The solid carbide end mills and end mills with replaceable carbide spiral flutes or inserts are frequently used for many different jobs. Most typical are jobs requiring a high metal removal rates and when machining hard materials. The HSS end mill is still a common cutting tool choice for everyday machining.

Many machining applications call for a harder tooling material than a high speed steel, but not as hard as carbide. As the tooling cost becomes an issue, the frequent solution is to employ an end mill with additional hardeners, for example a cobalt end mill. Such a tool is a little more expensive than a high speed steel tool, but far less expensive than a carbide tool. Cobalt based end mills have longer cutting tool life and can be used the same way as a standard end mill, with a noticeably higher productivity rate.

Solid carbide end mills are also available in machine shops and commonly used as regular small tools. Larger tools made of solid carbide would be too expensive, so special end mills with indexable inserts are the tools of choice. They can be used for both roughing operations and precision finishing work.

This chapter takes a look at some technological considerations when the CNC program calls for an end mill of any type or for a similar tool that is used as a profiling tool for peripheral cutting and contouring. This is an operation when the side of the cutter does most of work.

END MILLS

End mills are the most common tools used for peripheral milling. There is a wide selection of end mills available for just about any conceivable machining application. Traditional end mills come in metric and English sizes, variety of diameters, styles, number of cutting flutes, numerous flute designs, special corner designs, shanks, and tool material compositions.

Here are some of the most common machining operations that can be performed with an end mill - HSS, cobalt, solid carbide or an indexable insert type:

- ❏ Peripheral end milling and contouring
- ❏ Milling of slots and keyways
- ❏ Channel groves, face grooves and recesses
- ❏ Open and closed pockets
- ❏ Facing operations for small areas
- ❏ Facing operations for thin walls
- ❏ Counterboring
- ❏ Spotfacing
- ❏ Chamfering
- ❏ Deburring

End mills can be formed by grinding them into required shapes. The most common shapes are the *flat* bottom end mill (the most common type in machine shops), an end mill with a full radius (often called a *spherical* or a *ball nose* end mill), and an end mill with a corner radius (often called the *bull nose* end mill).

Each type of an end mill is used for a specific type of machining. Standard *flat end mill* is used for all operations that require a flat bottom and a sharp corner between the part wall and bottom. A *ball nose end mill* is used for simultaneous three dimensional (3D) machining on various surfaces. An end mill similar ro a ball nose type is the *bull nose end mill,* used for either some 3D work, or for flat surfaces that require a corner radius between the part wall and bottom. Other shapes are also required for some special machining, for example, a center cutting end mill (called a slot drill), or a taper ball nose end mill.

Figure 32-1 shows the three most common types of end mills used in industry and the relationship of cutter radius to the cutter diameter.

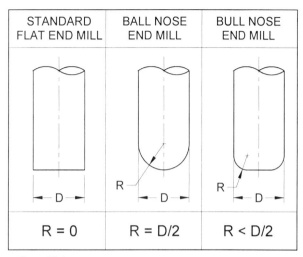

Figure 32-1
Basic configuration of the three most typical end mills

◆ High Speed Steel End Mills

True high speed steel end mills are the 'old-timers' in machine shops. They are manufactured either as a single end or a double end design, with various diameters, lengths and shank configurations. Depending on the cutting tip geometry, they can be used for peripheral motion (XY axes only), plunge motion (Z axis only), or all axes simultaneously (XYZ axes). Either a single end or a double end can be used for CNC machining. When using a double end mill, make sure the unused end is not damaged in the tool holder, when mounted. On a CNC machine, all end mills are normally held in a collet type tool holder, providing the maximum grip and concentricity. Chuck type holders are not recommended for end mills of any kind.

◆ Solid Carbide End Mills

In essence, the solid carbide end mills have the same characteristics as HSS types and vary only in the type of material they are made of. Using a solid carbide end mill requires special machining circumstances. The tool itself is fairly expensive, and from the metallurgical point of view, solid carbide is a brittle material that chips easily, particularly at sharp corners, or when it is dropped or improperly stored. When handled properly, it can remove metal with a great efficiency and produce superior surface finishes.

◆ Indexable Insert End Mills

The indexable insert end mills provide all the benefits of solid carbide end mills, but with the added convenience of replaceable carbide inserts. Many designs are available in this category as well. The holders for these tools match their internal diameter to the tool diameter. The tool has a ground flat area where the holder mounting screw prevents the tool from spinning.

◆ Relief Angles

It is always important to select the proper tool relief angle for different materials being cut. Relief angle is also called the *clearance* angle. For HSS milling cutters, the recommended flute relief angles become larger for softer materials. For example, the primary relief angle for steel is 3° to 5°, whereby the primary angle for aluminum is 10° to 12°. The cutting tool representative will supply additional information for a particular machining application.

◆ End Mill Size

Three very important criteria relating to the size of an end mill have to be considered for CNC machining:

- ❑ End mill diameter
- ❑ End mill length
- ❑ Flute length

For CNC work, the diameter of the end mill must be very accurate. The nominal diameters are those that are listed in the catalogues of various tooling companies. Nonstandard size, such as reground cutters, must be treated differently for CNC work. Even with the benefits of cutter radius offset, it is not advisable to use reground end mills for precision machining, although they may do a good job for emergency situations and for some roughing. That does not mean a reground cutter cannot be used for non-CNC work elsewhere in the shop or for less demanding CNC work.

The length of an end mill projected from the tool holder is also very important. A long projection may cause chatter that contributes to the wear of cutting edges. Another possible side effect for a long tool is deflection. Deflection will negatively influence the size and surface finish quality of the finished part. Flute length is important for determination of the depth of cut.

Regardless of the overall tool length (the tool projection length from the spindle), the length of flutes determines the cutting depth. *Figure 32-2* shows a useful ratio of width vs. depth of a rough side cut in milling:

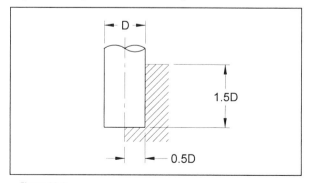

Figure 32-2
Relationship of the end mill diameter to the depth of cut for rough side cuts in milling

◆ Number of Flutes

When selecting an end mill, particularly for a material of average hardness, the number of flutes should be the primary consideration. For profiling, many programmers select (virtually automatically) a four-flute end mill for any required tool size larger than ∅.625 or ∅.750. An end mill that has to *plunge-in* - that is - it has to cut *into* a solid material along the Z axis - has normally only two flutes, regardless of diameter. This 'plunging-type' of end mill is also known under a more technical name as a *center-cutting end mill*, or under a rather old-fashioned name, a *slot drill*. The term slot drill has no relation to the tool called a drill, but to its machining action - just like a drill, a slot drill penetrates into a solid material, parallel to the Z axis.

It is the area of small and medium end mill diameters that requires the most attention. In this size range, the end mills come in two-, three-, and four-flute configurations. So what are the benefits of a two-flute versus a three-flute versus a four-flute design, for example? The *type of material* is the guiding factor here.

In this area material compositions, there is the expected give and take situation or a trade off. On a positive side, the fewer flutes an end mill has, the better conditions exist to avoid a chip buildup between the flutes during heavy cuts. Simply, there is more room. On the negative side, the fewer flutes that work in the material, the slower feedrate has to be programmed as each flute works harder. When cutting soft, nonferrous materials, such as aluminum, magnesium, even copper, preventing a chip buildup is important, so a two-flute end mill type is practically the only choice, even if the feedrate has to be somewhat compromised.

A different scenario is presented for harder materials, because two other factors have to be considered - *tool chatter* and *tool deflection*. There is no doubt, that in ferrous materials, the multi flute end mills will deflect less and chatter less than their two-flute counterparts.

What about the three-flute end mills? They seem to be - and in fact they are - a reasonable compromise between the two-flute and four-flute types. Three-flute end mills have never become a standard choice, even if their machining capabilities are often very good to excellent. Machinists have a difficulty to measure their diameter accurately, particularly with common machine shop tools such as a vernier or a micrometer. However, they do work very well in most materials.

Regardless of the number of flutes, an end mill with a larger diameter will deflect less than a similar end mill with a small diameter. In addition, the effective length of the end mill (measured as its overhang from the holder face) is important. The longer is the tool, the greater is the deflection - and that applies to all tools. Deflection pushes the tool away from its axis (center line). These are all results of common physical laws.

SPEEDS AND FEEDS

In many other sections of the handbook, speeds and feeds are mentioned. Tooling catalogues have very good charts and recommendations on speeds and feeds for particular tools, used with different materials. However, one standard formula (English version) is used for calculating the spindle speed in *r/min* (revolutions per minute):

$$r/min = \frac{12 \times ft/min}{\pi \times D}$$

☞ where ...

r/min	=	Spindle speed (revolutions per minute)
12	=	Constant to convert *feet to inches*
ft/min	=	Surface speed in feet per minute
π	=	Constant for flat to diameter conversion
D	=	Diameter of the tool in inches

For the metric system, the formula is similar:

$$r/min = \frac{1000 \times m/min}{\pi \times D}$$

☞ where ...

r/min	=	Spindle speed (revolutions per minute)
1000	=	Constant to convert *mm to meters*
m/min	=	Surface speed in meters per minute
π	=	Constant for flat to diameter conversion
D	=	Diameter of the tool in millimeters

Sometimes, there may be a benefit from the reverse formula - for example, when cutting at a certain spindle speed *(r/min)* that seems to be the perfect choice for the particular material. Next time a different diameter of the tool for that same material is used, just find out the *ft/min* rating for the material, which is applicable to any cutter size. The next formula will do exactly that (tool diameter is in inches):

$$ft/min = \frac{\pi \times D \times r/min}{12}$$

Metric formula is similar, but the tool diameter is in millimeters (mm):

$$m/min = \frac{\pi \times D \times r/min}{1000}$$

All entries in the formulas are based on previous explanations and should be easy to understand and apply.

To calculate a cutting feedrate for any milling operation, the spindle speed in *r/min* must be known first. Also known has to be the number of flutes and the chip load on each flute (suggested chip load is usually found in tool catalogues). For the English units, the chip load is measured in *inches per tooth* (a tooth is the same as a flute or an insert), with the abbreviation of *in/tooth*. The result is the cutting feedrate that will be in inches per minute - *in/min*.

For a lathe feedrate using standard turning and boring tools, the number of flutes is not applicable, the result is directly specified in inches per revolution *(in/rev)* or millimeters per revolution *mm/rev*.

$$\text{in} / \text{min} \ = \ \text{r} / \text{min} \ \times \ f_t \ \times \ N$$

☞ where ...

in/min	=	Feedrate in inches per minute
r/min	=	Spindle speed in revolutions per minute
f_t	=	Chip load in inches per tooth (per flute)
N	=	Number of teeth (flutes)

For metric system of measurement, the chipload is measured in *millimeters per tooth* (per flute), with the abbreviation of *mm/tooth*. The metric formula is similar to the one listed for English units:

$$\text{mm} / \text{min} \ = \ \text{r} / \text{min} \ \times \ f_t \ \times \ N$$

☞ where ...

mm/min	=	Feedrate in millimeters per minute
r/min	=	Spindle speed in revolutions per minute
f_t	=	Chip load in millimeters per tooth
N	=	Number of teeth (flutes)

As an example of the above formulas, a ⌀.750 four flute end mill may require 100 ft/min in cast iron. For the same cutting tool and part material, .004 per flute is the recommended chip load. Therefore, the two calculations will be:

Spindle speed:

```
r/min = (12 × 100) / (3.14 × .750)
r/min = 509
```

Cutting feedrate:

```
in/min = 509 × .004 × 4
in/min = 8.1
```

For safety reasons, always consider the part and machine setup, their rigidity, depth and/or width of cut and other relevant conditions very carefully.

Feed per tooth f_t (in *inches per tooth*), can be calculated as reversed values from the formula listed above.

The English units version of the formula is:

$$f_t \ = \ \frac{\text{in} / \text{min}}{\text{r} / \text{min} \ \times \ N}$$

Metric units formula is very similar, it calculates the feed per tooth f_t in *mm/tooth*:

$$f_t \ = \ \frac{\text{mm} / \text{min}}{\text{r} / \text{min} \ \times \ N}$$

When using carbide insert end mills for cutting steels, the faster spindle speeds are generally better. At slow speeds, the carbide cutter is in contact with a steel being cold. As the spindle speed increases, so does the steel temperature at the tool cutting edge, producing lower strength of the material. That results in favorable cutting conditions. Carbide insert cutting tools can often be used three times and up to five times faster than standard HSS cutters. The two basic rules relating to the relationship of tool material and spindle speed can be summed up:

> High speed steel (HSS) tools will wear out very quickly, if used at high spindle speeds = high r/min

> Carbide insert cutters will chip or even break, if the spindle speed is too low = low r/min

◆ Coolants and Lubricants

Using a coolant with a high speed steel (HSS) cutter is almost mandatory for cutting all metals. Coolant extends the tool life and its lubricating attributes contributes to the improved surface finish. On the other hand, for carbide insert cutters, coolant *may not* be always necessary, particularly for roughing steel stock.

> Never apply coolant on a cutting edge that is already engaged in the material!

◆ Tool Chatter

There are many reasons why a chatter occurs during peripheral milling. Frequent causes are weak tool setup, excessive tool length (overhang from tool holder), machining thin walls of material with too much depth or too heavy feedrate, etc. Cutter deflection may also contribute to the chatter. Tooling experts agree that well planned experiments with the combination of spindle speeds and cutting feedrates should be the first step. If chatter still persists, look at the machining method used and the setup integrity.

STOCK REMOVAL

Although peripheral milling is mainly a semifinishing and finishing machining operation, end mills are also successfully used for roughing. The flute configuration (flute geometry) and its cutting edge are different for roughing and finishing. A typical roughing end mill will have corrugated edges - a typical example is a *Strasmann end mill*. Strasmann is said to be the original designer and developer of roughing cutters and the trademarked name is now used as a generic description of this type of roughing end mill.

Good machining practice for any stock removal is to use large diameter end mill cutters with a short overhang, in order to eliminate, or at least minimize, the tool chatter and tool deflection during heavy cuts.

For deep internal cavities, such as deep pockets, it is a good practice to pre-drill to the full depth (or at least to the *almost* full depth), then use this new hole for an end mill that is smaller than the drilled hole. Since the end mill penetrates to the depth in an open space, the succeeding cuts will be mainly side milling operations, enlarging the cavity into the required size, shape and depth.

◆ Plunge Infeed

Entering an end mill into the part material along the Z axis alone is called *center-cutting*, *plunging* or *plunge infeed*. It is a typical machining operation and programming procedure to enter into an otherwise inaccessible area, such as a deep pocket, a closed slot, or any other solid material entry. Not every end mill is designed for plunge cutting and the CNC machine operator should always make sure the right end mill is always selected (HSS or carbide or indexable insert type of end mill). Programmer can make it easier by placing appropriate comments in the program.

◆ In and Out Ramping

Ramping is another process where the Z axis is used for penetrating (entering) into a solid part material. This time, however, the X axis *or* the Y axis are programmed simultaneously with the Z axis. Depending on the end mill diameter, the typical ramping angle is about 25° for a 1.000 inch cutter, 8° for a 2.000 inch cutter, and 3° for a 4.000 inch cutter. Ramping approach toward the part can be used for flat type, ball nose type, and bull nose type of end mills. Smaller end mills will use smaller angles (3°-10°). See *Figure 32-3* for an illustration of a typical ramping motion.

Always be very careful from which XYZ tool position the cutting tool will start cutting at the top of part. Considering only the start point and the end point may not produce the best results. It is easy to have a good start and good end tool positions, but somewhere during the cut, an unwanted section of the part may be removed accidentally. A few simple calculations or a CAD system may help here.

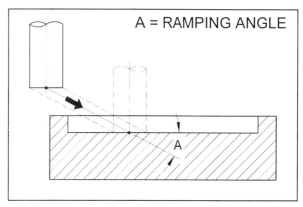

Figure 32-3

Typical entry angle for a ramping infeed into a solid material

◆ Direction of Cut

The direction of a cut for contouring operations is controlled by the programmer. Cutting direction of the end mill for peripheral milling will make a difference for most part materials, mainly in the area of material removal and the quality of surface finish. From the basic concepts of machining, the cutting direction can be in two modes:

❑ Climb milling - also known as the **DOWN** milling

❑ Conventional milling - also known as the **UP** milling

Anytime the G41 command is programmed, cutter radius is offset to the left of part and the tool is *climb* milling. That assumes, of course, that the spindle rotation is normal, programmed with the M03 function, and the cutting tool is right hand. The opposite, G42 offset, to the right of the part, will result in *conventional* milling. In most cases, climb milling mode is the preferred mode for peripheral milling, particularly in finishing operations.

Figure 32-4 illustrates the two cutting directions.

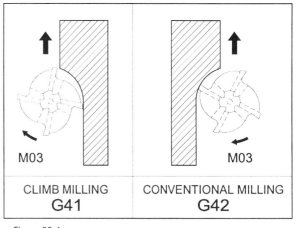

Figure 32-4

Direction of the cut relative to material, with M03 in effect

Climb Milling

Climb milling - sometimes called the *down* milling - uses rotation of the cutter in the feeding direction and has the tendency to push the part *against* the table (or the fixture). Maximum thickness of the chip occurs at the beginning of the cut and upon exit, the chip is very thin. The practical result is that most of the generated heat is absorbed by the chip, and hardening of the part is largely prevented.

> Do not misunderstand the words *climb* and *down* describing the same machining direction.

Both terms are correct, if taken in the proper context.

Conventional Milling

Conventional milling - sometimes called the *up* milling - uses rotation of the cutter against the feeding direction, and has the tendency to pull the part *from* the table (or the fixture). Maximum thickness of the chip occurs at the end of the cut and upon exit, the chip is very thick. The practical result is possible hardening of the part, rubbing the tool into the material, and a poor surface finish.

◆ Width and Depth of Cut

For good machining, the width and depth of cut should correspond to the machining conditions, namely the setup, the type of material being machined and the cutting tool used. Width of cut depends also on the number of flutes of the cutter that are actually engaged in the cut.

Approximately one third of the diameter for the depth of cut is a good rule of thumb for small end mills, a little more for larger end mills.

Peripheral milling requires a solid machining knowledge and certain amount of common sense. If a successful machining operation in one job is documented, it can be adapted to another job with ease.

SLOTS AND POCKETS

In many applications for a CNC machining center, the material has to be removed from the inside of a certain area, bounded by a contour and a flat bottom. This process is generally known as *pocketing*. To have a true pocket, the contour that defines the pocket boundary must be closed. However, there are many other applications, where the material has to be removed from an open area, with only a partial contour defined. An open slot is a good example of this type. This chapter looks at applications of closed pockets, partial pockets, slots and various programming techniques for internal material removal.

OPEN AND CLOSED BOUNDARY

A continuous contour on which the start point and the end point is in a different location, is called an *open contour*. Continuous contour defined in the program that starts and ends at the same point location, is called a *closed contour*. From the machining point of view, the major difference between an open and closed contour is *how the cutting tool reaches the contour depth*.

◆ Open Boundary

An open boundary is not a true pocket, but belongs to a related category. Machining of this kind of a contour is quite flexible, as the tool can reach the required depth in an open space. Any good quality end mill in different varieties can be used to machine an open boundary.

◆ Closed Boundary

The excessive material within a closed boundary can be removed in two ways, depending on the cutting operation. One way is to use an external tool and move it towards the outside of the boundary, another way is to use an internal tool and move it towards the inside of the boundary. In both cases, the actual machining follows. Cutting along the outside of a part is not considered pocketing but peripheral milling (*Chapter 32*). Cutting on the inside of a closed boundary is typical for machining pockets of various regular and irregular shapes. Some typical examples of regular shape pockets are closed slots, rectangular pockets, circular pockets, and so on. Irregular shape pockets can have any machinable shape, but they still use the same machining and programming techniques as regular pockets.

One of the most commonly machined boundary shapes in manufacturing is milling of a simple cavity, usually quite small, called *a slot*.

PROGRAMMING SLOTS

Slots are often considered as special types of grooves. These 'grooves' usually have one or two radial ends. If there are two ends, they are joined by a straight groove. A slot can be either open or closed, with the same size radius on both ends, two different radii, or one radius only. A typical slot that has only one end radius is a keyway.

Slots can be open or closed, straight, angular, circular, with straight walls or shaped walls (using a tapering end mill). Programming slots with accuracy in mind usually requires a roughing operation and a finishing operation. Both operations can be made with the same tool or with two or more tools, depending on the part material, required dimensional tolerances, surface finish, and other conditions.

Certain slots, for example keyways, can be done with special cutters, called slotting cutters, rather than an end mill. To program a slotting cutter is usually a simple process of a linear motion - in and out. More complex - and more accurate - slots are machined with end mills, and the walls of the slot are contoured under program control.

Figure 33-1 shows a drawing of a typical open slot. This drawing will be used to illustrate the programming techniques of an open slot.

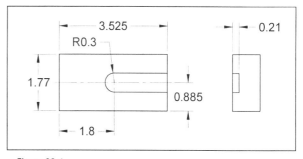

Figure 33-1
An open slot programming example O3301

◆ Open Slot Example

Before programming any tool motion, study the drawing. That way, the machining conditions can be established, as well as setup and other requirements. The program zero can be determined quickly - dimensions are from the lower left corner (XY) and top (Z) of the object. That location will become the program zero.

Next considerations will relate to machining subjects:

- Number of tools
- Tool size
- Speeds and feeds
- Maximum cutting depth
- Method of cutting

Number of Tools

One or two tools can be used to cut the slot. If dimensional tolerances are very critical or the material is hard to cut, use two tools - one tool for roughing, another one for finishing. The tools could have the same diameter or different diameters. For this example, only one tool will be used for both roughing and finishing.

Tool Size

The size of the cutting tool is mainly determined by the width of the slot. In the drawing, the slot has .300 radius, so the width is .600. There is no standard cutter of ∅.600 - but - even if there were - would it be practical? What about a ∅.500 inch cutter for .500 inch wide slot? It is possible, but the resulting cut would not be of the highest quality. Tolerances and surface finish would be hard to control. That means choosing a tool, preferably available off-shelf, that is a little smaller then the slot width. For the slot in the example, a ∅.500 inch end mill is a suitable choice. When selecting the tool size, always calculate how much stock the tool will leave on the slot walls for finishing. Too much stock may require some semifinishing cuts. With the ∅.500 cutter and the slot width of .600, the amount of stock left will be easy to calculate:

$$ S = \frac{(W - D)}{2} $$

☞ where ...

S	=	Stock left on material
W	=	Width of slot (= slot radius times two)
D	=	Cutter diameter

Stock left on the slot wall in the example will be:

S = (.600 - .500) / 2 = .050

This is a suitable stock for finishing with one cut.

Speeds and Feeds

Spindle speeds and cutting feedrates will depend on the exact situation at the CNC machine, so the example only uses a reasonable speed of 950 r/min and cutting feedrate of 8 in/min.

Maximum Cutting Depth

The drawing shows the slot depth as .210. Always check the depth - it may be too deep for a single cut, usually for small cutters or tough materials. Although a single cut can be used for the full depth, some stock at the slot bottom should be left for finishing.

Method of Cutting

Once all the other machining conditions are established, the method of cutting almost presents itself. The tool will be positioned above a clear position and at the slot center line. Then the tool will be fed into the slot depth, leaving some material at the bottom, for finishing. In a linear motion, the tool will rough out the material all the way to the center of the slot radius, then retract above the material. Then it will be moved back to the original starting position and at the full depth for contouring the slot, in climb milling mode. In *Figure 33-2*, the XY tool motions and their program locations are shown.

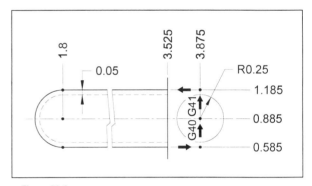

Figure 33-2

Contouring details for the open slot example O3301

To create the program is not difficult at all. The tool is in the spindle and all typical methods explained throughout the handbook are used.

```
O3301 (OPEN SLOT)
N1 G20                                    (INCH MODE)
N2 G17 G40 G80                    (START UP SETTINGS)
N3 G90 G54 G00 X3.875 Y0.885 S950 M03   (START)
N4 G43 Z0.1 H01 M08        (START POSITION ABOVE)
N5 G01 Z-0.2 F50.0         (0.01 LEFT ON BOTTOM)
N6 X1.8 F8.0            (CUT TO SLOT RADIUS CNTR)
N7 G00 Z0.1                (RETRACT ABOVE WORK)
N8 X3.875                      (RETURN TO START)
N9 G01 Z-0.21 F50.0         (FEED TO FULL DEPTH)
N10 G41 Y1.185 D01 F8.0      (APPROACH CONTOUR)
N11 X1.8                       (CUT TOP WALL)
N12 G03 Y0.585 R0.3         (CUT SLOT RADIUS)
N13 G01 X3.875              (CUT BOTTOM WALL)
N14 G00 G40 Y0.885     (RETURN TO START POINT)
N15 Z1.0 M09               (RETRACT ABOVE WORK)
N16 G28 X3.875 Y0.885 Z1.0 M05      (M/C ZERO)
N17 M30                        (END OF PROGRAM)
%
```

The example is quite self evident and the included block comments will offer better understanding of the programming order and procedure. In this example, only one tool has been used. For high precision machining, using two tools will be better, even if it means a longer program.

◆ Closed Slot Example

Closed slot does not differ from an open slot that much. The greatest difference is in the tool entry into the material. There is no outside location - tool has to plunge into the material along the Z axis, unless there is a predrilled hole. One method is to use a *center cutting end mill* (known as *slot drill*). If this type of end mill is not available, or machining conditions are not suitable, tool will have to *ramp* into the material, as a second method. Ramping is a linear cutting motion, usually in the XZ, the YZ, or the XYZ axes.

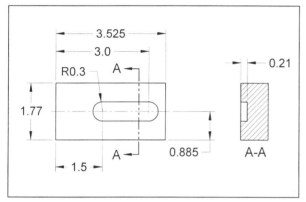

Figure 33-3

A closed slot programming example O3302

The second example is based on the drawing shown in *Figure 33-3*. The drawing is a modification of the open slot drawing. Many considerations already established will apply equally to the closed slot. A ∅.500 inch end mill will be used, this time with a center cutting geometry that allows plunging into a solid material.

Apart from the different tool geometry required for the plunging cut, only the method of cutting will change. For a closed slot (or a pocket), the tool has to move above work, to a certain XY start location. In the example, it will be the center of one of the slot radii. Portion of slot on the right is selected arbitrarily. Then, a plunge at a reduced feedrate will be to the required depth (leaving .010 on the bottom) and, in a linear motion, the slot will be roughed out between the two centers - *Figure 33-4*.

Retracting the tool is not necessary, it can be fed into the final depth at the same tool location. The stock is .050 all around the slot contour. At the final depth, and from the center location of the left part of the slot, the finish contour will start. Contouring will be more complex this time, because the tool is in a rather tight spot.

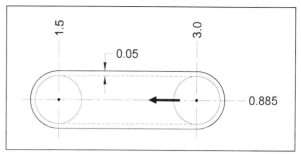

Figure 33-4

Roughing operation detail for a closed slot example O3302

Internal Contour Approach

In the program, the tool is now at the center of the left side of slot, ready to start the finishing cut. Climb milling mode has been selected and the contour approached in such a way that the tool motion continues to its left. One way is the make a straight linear cut from the current tool location at the center, to the 'south' position of the left arc (while applying the cutter radius offset).

This method works, but when approaching an inner contour it is better to use a tangential approach. An internal contour approached at a tangent requires an auxiliary approach arc (so called *lead-in* arc), since the linear approach towards the contour is not a choice.

Although the tangential approach using an arc improves the surface finish of the part, this preference creates another serious problem. *The cutter radius offset cannot be started in a circular interpolation mode!* Therefore, a non-circular motion has to be *added* - there will be *two* motions from the current tool location at the slot radius center to the start point of the contour:

❏ First, a linear motion with cutter radius offset applied

❏ Second, the tangential approach arc motion

This technique is illustrated in *Figure 33-5*.

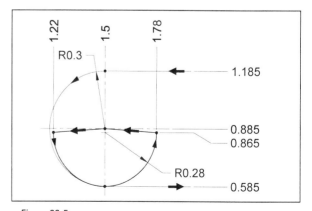

Figure 33-5

Detail of tangential approach towards an inner contour

Now, look closely at how the approach arc was created. The goal is to select the location and radius of the approach arc. Location selection is easy - the arc *must be tangent* to the contour. The radius dimension has to be selected with some logical thinking. When faced with an unknown dimension, always think of its purpose first. The purpose of the approach arc is to lead-in the cutting tool in a smooth curve towards the contour. That means the approach arc radius must be smaller than the cutting tool radius. Finally, there is the slot radius itself, which is defined by the drawing. Relationship of all three radii can be put in perspective:

$$R_t \ < \ R_a \ < \ R_c$$

☞ where ...

R_t = Radius of the cutting tool
R_a = Radius of the approach arc (lead-in arc)
R_c = Radius of the contour (slot radius)

Supply some numeric data and the approach radius can be calculated. The formula shows the relationships of all three radii. The slot contour radius (R_c) is assigned by the drawing. Once the cutting tool size is selected, that radius becomes fixed as well (R_t). That leaves the lead-in approach radius (R_a). That one has to be calculated and calculated *accurately*.

From the formula, it is clear that the approach radius arc must be greater than the cutter radius (.250), while both must be smaller that the contour radius (.300). That means the range (within three decimal places) is .251 to .299. If only increments of .010 are considered, which one is better - .260 or .290? Well, the larger the better. With selection of rather a larger approach radius from the range, the arc tangential approach takes place at a smoother curve than with a smaller radius. The result is an improved surface finish. For program O3302, .280 is selected as the approach radius. This selection meets all the three relationships:

$$.250\,(R_t) \ < \ .280\,(R_a) \ < \ .300\,(R_c)$$

That is all the information needed before writing the program. Note the programming similarities with the open slot listed in program O3301.

```
O3302 (CLOSED SLOT)
N1 G20                                    (INCH MODE)
N2 G17 G40 G80                      (STARTUP SETTINGS)
N3 G90 G54 G00 X3.0 Y0.885 S950 M03    (START)
N4 G43 Z0.1 H01 M08        (START POSITION ABOVE)
N5 G01 Z-0.2 F4.0           (0.01 LEFT ON BOTTOM)
N6 X1.5 F8.0           (CUT TO SLOT RADIUS CENTER)
N7 Z-0.21 F2.0                 (FEED TO FULL DEPTH)
N8 G41 X1.22 Y0.865 D01 F8.0 (LINEAR APPROACH)
N9 G03 X1.5 Y0.585 R0.28    (CIRCULAR APPROACH)
N10 G01 X3.0                     (CUT BOTTOM WALL)
N11 G03 Y1.185 R0.3        (CUT RIGHT SLOT RADIUS)
```

```
N12 G01 X1.5                          (CUT WALL TOP)
N13 G03 Y0.585 R0.3               (CUT RADIUS LEFT)
N14 X1.78 Y0.865 R0.28    (CIRCULAR DEPARTURE)
N15 G01 G40 X1.5 Y0.885      (LINEAR DEPARTURE)
N16 G00 Z1.0 M09            (RETRACT ABOVE WORK)
N17 G28 X1.5 Y0.885 Z1.0 M05         (M/C ZERO)
N18 M30                         (END OF PROGRAM)
%
```

This program example is also a good illustration of how to approach any inside contour for finishing. Slots of other kinds (angular, circular, etc.), use the same principles illustrated in the last two examples.

POCKET MILLING

Pocket milling is also a typical and common operation on CNC machining centers. Milling a pocket means to remove material from an enclosed area, defined by its boundary. This bounded area is further defined by its walls and bottom, although walls and bottom could be tapered, convex, concave, rounded, and have other shapes. Walls of a pocket create the boundary contour. Pockets can have square, rectangular, circular or undefined shape, they can be empty inside or they may have islands.

Programming pockets manually is usually efficient only for simple pockets, pockets of regular shapes, such as rectangular or circular pockets. For pockets with more complex shapes and pockets with islands, the assistance of a computer is usually required.

◆ General Principles

There are two main considerations when programming a pocket for milling:

❑   Method of cutter entry

❑   Method of roughing

To open a space to start milling a pocket (into solid material), the cutter motion has to be programmed to enter along the direction of spindle (Z axis), which means the cutter must be *center cutting* - to be able to *plunge* cut. In cases where the plunge cut is either not practical or not possible, a method called *ramping* can be used very successfully. This method is often used when the center cutting tool is not available. Ramping requires the Z axis to be used together with the X axis, Y axis, or both. This motion will, of course, be a 2 axis or a 3 axis linear motion. All modern CNC machining centers support it.

The method of removing the majority of material from the pocket is called roughing. The roughing method selection can be a little more complex. The location *where* to start the plunge or ramped cut is important, so is the *width* of cut. It may be difficult to do all the roughing in climb milling mode. It may be difficult to leave exactly the same amount of stock for finishing everywhere in the pocket.

Many cuts will be irregular and stock amount will not be even. For that reason, it is quite common to program a semifinishing cut of the pocket contour, before any finishing cut takes place. One or more tools may be used for this situation, depending on exact requirements.

Some typical methods for roughing a pocket are:

❏ `Zigzag

❏ One direction - from the inside of the pocket out

❏ One direction - from the outside of the pocket in

In computer applications, other pocketing options are also possible, such as a true spiral, morph, one way, and others. In many cases, there is a choice of specifying the angle of cut, even a user selected point of entry and finishing left overs. Manually, these more complex methods may be used as well, but it may be a very tedious work.

◆ Pocket Types

The most common pockets are also the easiest to program. They all have a regular shape, without any islands:

❏ Square pocket

❏ Rectangular pocket

❏ Circular pocket

Square pocket and a rectangular pocket are fundamentally the same and apart from their different side lengths, there is no major difference in programming.

RECTANGULAR POCKETS

Rectangular and square pockets are quite easy to program, particularly if they are parallel to the X or Y axes. As an example of a rectangular pocket, the one illustrated in *Figure 33-6* will be used.

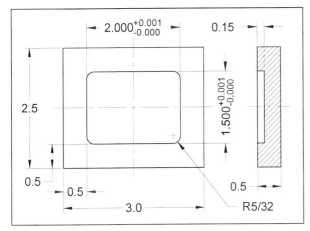

Figure 33-6

Sample drawing of a rectangular pocket - program O3303

To illustrate the complete pocket programming, starting with tooling selection is important. Material is also important and so are other machining decisions. Although rectangular pockets are often drawn with sharp corners, they will always have corners of the tool used, or larger, when machined. The corners in the drawing are 5/32 (.1563), and a ∅5/16 center cutting end mill (∅.3125). For roughing, it may be a good choice, but for finishing, the radius should be a little smaller so the tool can actually *cut* in the corner, not just rub there. Selection of a ∅.250 end mill is reasonable and will be used it in the example.

Since all the material in the enclosed area has to be removed (including the bottom), think about all possible places where the cutting tool can enter into the depth by plunging or ramping. Ramping must always be done in a clear area, but plunging can be done almost anywhere. There are only two practical locations:

❏ Pocket center

❏ Pocket corner

There are some benefits to both selections and the inevitable disadvantages. Starting at the pocket center, the tool can follow a single directional path and, after the initial cut, can cut only in climb milling or conventional milling mode. There are slightly more math calculations involved in this method. The other method, starting at the pocket corner, is quite popular as well, but uses a zigzag motion, so one cut will be in a climb milling mode, the other cut will be in a conventional mode of machining. It is a little easier for calculations, however. In the example, the corner will be used as a start location.

Any corner of the pocket is equally suitable for the start. In the program example O3303, the lower left corner of the pocket will be used.

There are three important factors the programmer has to consider when selecting start location for the cutting tool in an enclosed area:

❏ Cutter diameter (or radius)

❏ Amount of stock left for finishing

❏ Amount of stock left for semifinishing

There are also very important dimensions of the part, as defined in the drawing. They are the *length*, the *width*, and the *corner radius* (or radii) of the pocket - they must always be known, as well as the pocket position and its orientation to other elements of the part.

In the *Figure 33-7*, the starting point is identified as X1 and Y1 distance from the given corner (lower left), and all additional data are shown as well.

The letters identify variable settings that must be done; the programmer chooses their values, depending on the job.

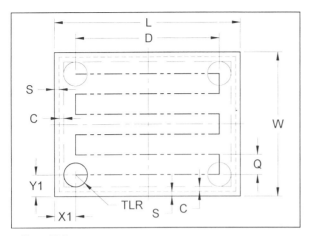

Figure 33-7
Pocket roughing start point in the corner - zigzag method

☞ The meaning of the description letters is :

X_1 = X location of tool at start
Y_1 = Y location of tool at start
TLR = Tool radius (cutter diameter / 2)
L = Pocket length as per drawing
W = Pocket width as per drawing
Q = Calculated stepover between cuts
D = Calculated length of actual cut
S = Stock left for finishing
C = Stock left for semifinishing (clearance)

◆ Stock Amount

There are two stock amounts (values) - one relates to the *finishing* operation, usually done with a separate finishing tool, the other one relates to the *semifinishing* operation, usually done with the roughing tool. The cutter moves back and forth in a zigzag direction, leaving behind so called scallops. In 2D work, the word 'scallops' is used to describe uneven wall surface caused by the tool shape, and is similar in 3D cutting as well. The result of such a zigzag tool path is generally unacceptable for the finish machining, because of the difficulty of maintaining tolerances and surface finish while cutting uneven stock.

To avoid possible cutting problems later, a secondary semifinishing operation is often necessary. It purpose is to eliminate the scallops. Choose semifinishing cut particularly for machining tough materials or when using small size diameter tools. Semifinishing stock allowance, marked as the C value in the illustration, can also be equal to zero. If that is the case, it means no additional stock is required. Typically the stock allowance will have a small value.

Figure 33-8 illustrates the result of a roughing operation of a rectangular pocket, *without* the semifinishing cut. Note the uneven stock (scallops) left for the finishing tool. The high spots create the heaviest obstacles for a subsequent tool, so semifinishing tool path is highly recommended.

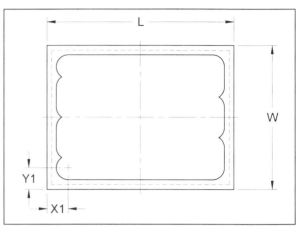

Figure 33-8
Result of a zigzag pocketing, without a semifinish cut

◆ Stepover Amount

The actual shape of the pocket before semifinishing is determined by the amount of *stepover*. A stepover in pocketing is just another name for the *width of cut*. This amount may be selected without actual calculation. A much better way is to *calculate* the stepover amount, based on the number of required cuts. That way, the amount will be equal for *all* cuts. Since it is quite common to think of a width of cut as some *percentage* of the cutter diameter, use this method for reference purposes only, and still *calculate* the cutting width and select one that will be the *closest* to the cutter diameter percentage desired.

In the example, a rather larger than average stepover will be used, based on five required cuts (zigzag type). There is a substantial difference whether the number of cuts is selected as an *even number* or as an *odd number*:

❑ Even number of cuts will terminate the roughing on the opposite side of the pocket relative to the start location

❑ Odd number of cuts will terminate the roughing on the same side of the pocket relative to the start location

Practically, it does not matter which corner is selected to start at or in which direction the first cut begins. What matters is that the stepover is reasonable and, preferably, equal for all cuts. There is a simple way of calculating the stepover, based on a given number of cuts. If the calculated amount is too small or too large, just repeat the calculation with a different number of cuts N.

The calculation can be expressed in a formula:

$$Q = \frac{W - 2 \times TLR - 2 \times S - 2 \times C}{N}$$

In the formula, N is the number of selected stepovers and all other variables have the same meaning as before.

● Example :

In the example, five equal stepovers are needed, based on the pocket width of 1.500 inches, tool diameter 0.250 (TLR is 0.125), finishing stock S as 0.025 and semifinishing stock C as 0.010. The stepover size will be:

```
Q = (1.5 - 2 × 0.125 - 2 × 0.025 - 2 × 0.01) / 5
Q = 0.2360
```

That may be a little too much for a ∅0.250 end mill, but it will make the example a bit shorter. Seven stepovers would result in a more reasonable amount of 0.1686 (rounding to three decimal places to 0.169 does no harm).

The above formula may be modified to use the pocket length, rather than the pocket width. This may be a better choice if the pocket is narrower along the X axis, than it is along the Y axis.

◆ Length of Cut

Before the semifinishing, the length, the incremental distance D of each cut, have to be calculated.

In many respects, the formula to calculate the length of cut is very similar to the stepover calculation:

$$D = L - 2 \times TLR - 2 \times S - 2 \times C$$

In this example, the D value will be:

● Example :

```
D = 2.0 - 2 × 0.125 - 2 × 0.025 - 2 × 0.01
D = 1.6800
```

This is the *incremental* length of cut between the stepovers (no cutter radius offset has been used).

◆ Semifinishing Motions

The only purpose of semifinishing motions is to eliminate uneven stock. Since the semifinishing will be normally done with the same tool as the roughing operation, the place to start the semifinishing cuts is the last tool position of the roughing sequence. In this case, it was the upper left corner of the pocket. *Figure 33-9* shows the motions from the *Start* to *End* (of the semifinishing).

The length L_1 and W_1 are calculated, and the difference between the *Start* position and the *End* position, is the C value, along both axes.

The formula for the length and width of semifinishing cut, its actual cutting distance, is listed next:

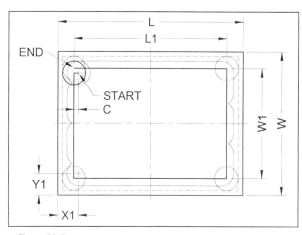

Figure 33-9

Semifinishing tool path begins at the last roughing location, and leaves equal stock for finishing operation

$$L_1 = L - 2 \times TLR - 2 \times S$$

$$W_1 = W - 2 \times TLR - 2 \times S$$

● Example :

```
L1 = 2.0 - 2 × 0.125 - 2 × 0.025
L1 = 1.7000
```

```
W1 = 1.5 - 2 × 0.125 - 2 × 0.025
W1 = 1.2000
```

◆ Finishing Tool Path

Once the pocket is roughed out and semifinished, another tool (or even the same tool in some cases) can be used to finish the pocket to its final size. This programmed tool path will typically provide offsets to maintain machining tolerances and speeds and feeds to maintain required surface finish. Typical starting tool position for a small to medium pocket is at its center, for a large pocket the starting position should be at the middle of the pocket, away from one of the walls, but not too far.

For the finishing cut, the cutter radius offset should be in effect, mainly to gain flexibility in maintaining tolerances during machining. Since the cutter radius offset cannot be started during an arc or a circular motion, linear lead-in and lead-out motions have to be added. In *Figure 33-10* is the illustration of a typical finishing tool path for a rectangular pocket (with the start at the pocket center).

Some conditions do apply in these cases. One is that the leading arc radius must be calculated, using precisely the same method as for slots:

$$R_a > R_t < R_c$$

☞ where ...

R_a = Radius of the approach arc
R_t = Radius of the cutting tool
R_c = Radius of the corner

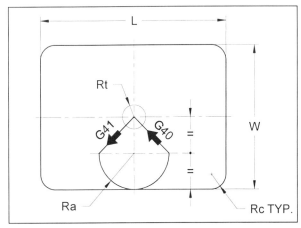

Figure 33-10
Typical finishing tool path for a rectangular pocket

The mode of the milling cut is normally the climb milling mode and the radius offset used will be G41, to the left side of the contour.

Example :

To calculate the approach (lead-in) radius for the example drawing, start with the corner radius. That radius is given as 5/32 (.1563) and the tool radius has been selected as .125, so the condition $R_t < R_c$ is satisfied. In order to also satisfy the condition $R_a > R_t$, choose almost any approach radius larger than the tool radius, as long as it is reasonable. The pocket length and width are also important, as always. If possible, choose the approach radius as *one quarter* of the pocket width W, for a little easier tool motion calculations. In the example,

R_a = W / 4 = 1.5 / 4
R_a = .375

Condition is satisfied, the approach radius is larger than the tool radius, and can be safely used in the program.

◆ Rectangular Pocket Program Example

Once all selections and decisions have been done, the part program can be written for the pocket in example O3303. Two tools will be used, both ∅.250 end mills, the roughing cutter must be able of center cutting. Program zero is the lower left corner of the part. All roughing and semi-finishing steps are documented in the program.

```
O3303 (RECTANGULAR POCKET)
N1 G20
N2 G17 G40 G80 T01   (.250 ROUGHING SLOT DRILL)
N3 M06
N4 G90 G54 G00 X0.66 Y0.66 S1250 M03 T02
N5 G43 Z0.1 H01 M08
N6 G01 Z-0.15 F7.0
(-- ROUGHING START -------------------------)
N7 G91 X1.68 F10.0                    (CUT 1)
N8 Y0.236                         (STEPOVER 1)
N9 X-1.68 F12.0                       (CUT 2)
N10 Y0.236                        (STEPOVER 2)
N11 X1.68                             (CUT 3)
N12 Y0.236                        (STEPOVER 3)
N13 X-1.68                            (CUT 4)
N14 Y0.236                        (STEPOVER 4)
N15 X1.68                             (CUT 5)
N16 Y0.236                        (STEPOVER 5)
N17 X-1.68                            (CUT 6)
(-- SEMIFINISH START -------------------------)
N18 X-0.01               (SEMIFINISH STARTUP X)
N19 Y-0.01               (SEMIFINISH STARTUP Y)
N20 Y-1.19                   (LEFT Y- MOTION)
N21 X1.7                    (RIGHT X+ MOTION)
N22 Y1.2                       (UP Y+ MOTION)
N23 X-1.7                    (LEFT X- MOTION)
N24 G90 G00 Z0.1 M09
N25 G28 Z0.1 M05
N26 M01


N27 T02               (.250 FINISHING END MILL)
N28 M06
N29 G90 G54 G00 X1.5 Y1.25 S1500 M03 T01
N30 G43 Z0.1 H02 M08
N31 G01 Z-0.15 F12.0
(-- FINISHING POCKET -------------------------)
N32 G91 G41 X-0.375 Y-0.375 D02 F15.0
N33 G03 X0.375 Y-0.375 R0.375 F12.0
N34 G01 X0.8437
N35 G03 X0.1563 Y0.1563 R0.1563
N36 G01 Y1.1874
N37 G03 X-0.1563 Y0.1563 R0.1563
N38 G01 X-1.6874
N39 G03 X-0.1563 Y-0.1563
N40 G01 Y-1.1874
N41 G03 X0.1563 Y-0.1563 R0.1563
N42 X0.8437
N43 G03 X0.375 Y0.375 R0.375
N44 G01 G40 X-0.375 Y0.375 F15.0
N45 G90 G00 Z0.1 M09
N46 G28 Z0.1 M05
N47 X-2.0 Y10.0
N48 M30
%
```

Study the program carefully. It follows all the decisions made earlier and offers many details.

In the program, blocks N17 and N18 can be joined together into a single block. The same applies to blocks N19 and N20. They are only separated for the convenience of tracing the tool motions to match the illustrations. There is a slight benefit in using the incremental mode of programming, but the absolute mode would have been just as easy.

CIRCULAR POCKETS

The other common types of pockets are so called *circular* or *round pockets*. Although the word *pocket* somehow implies a closed area with a solid bottom, the programming method relating to circular pockets can also be used for circular openings that may have a hole in the middle, for example, some counterboring operations.

To illustrate a practical programming application for a circular pocket, *Figure 33-11* shows the typical dimensions of such a pocket.

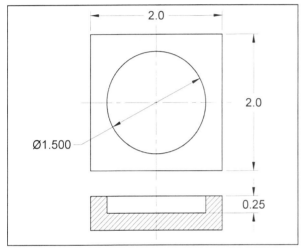

Figure 33-11

Sample drawing of a circular pocket (program examples O3304-06)

In terms of planning, the first thing to be done is the selection of the cutter diameter. Keep in mind, that in order to make the pocket bottom clean, without any residual material (uncut portions), it is important to keep the stepover from one cut to another by a limited distance that should be calculated. For circular pockets, this requirement influences the *minimum* cutter diameter that can be used to cut the circular pocket in a single 360° cut.

◆ Minimum Cutter Diameter

In the following illustration - *Figure 33-12*, the relationship of the cutter diameter to the pocket diameter is shown. There is also a formula that will determine the minimum cutter diameter as *one third* of the pocket diameter. The milling will start at the circular pocket center, with a single 360° tool motion. In practical terms, selecting a cutter slightly larger than the minimum diameter is a much better choice. The major benefit of this calculation is when the pocket has to be done with only one tool motion around. The formula is still valid, even if cutting will be repeated several times around the pocket, by increasing the diameter being cut. In that case, the formula determines the maximum width of the cut.

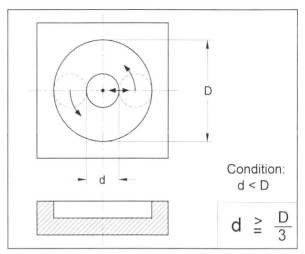

Figure 33-12

Relationship of the cutter diameter to the pocket diameter

For example, the pocket diameter in the sample drawing is 1.5 inches. Using the formula, select a plunging cutter (center cutting end mill), that has the diameter larger than 1.5/3, therefore larger than .500. The nearest nominal size suitable for cutting will be ∅.625 (5/8 slot drill).

◆ Method of Entry

The next step is to determine the method of the tool entry. In a circular pocket, the best place to enter along the Z axis, is at the *center* of the pocket. If the pocket center is also the program zero X0Y0, and the pocket depth is .250, the beginning of the program may be similar to the following example (cutting tool placed in the spindle is assumed):

```
O3304 (CIRCULAR POCKET - VERSION 1)
N1 G20
N2 G17 G40 G80
N3 G90 G54 G00 X0 Y0 S1200 M03
N4 G43 Z0.1 H01 M08
N5 G01 Z-0.25 F8.0
N6 ...
```

In the next block (N6), the cutting tool will move from the pocket center towards the pocket diameter, and apply cutter radius offset along the way. This motion can be done in two ways:

❏ As a simple straight linear motion

❏ As a combined linear motion with a circular approach

◆ Linear Approach

The linear departure from the pocket center can be directed into any direction, but a direction towards a quadrant point is far more practical. In the example, a motion along the Y positive direction is selected, into the 90° position.

Along the way, cutter radius offset for the climb milling mode G41 is programmed, followed by the full 360° arc and another straight motion, back towards the center. During this motion, the cutter radius offset will be canceled. *Figure 33-13* shows the tool path.

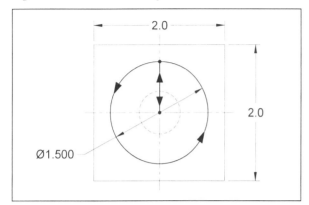

Figure 33-13

Linear approach for a circular pocket milling - program O3304

The graphic representation can be followed by a corresponding program segment - approach a quadrant point, profile the full arc, then return back to the center:

```
N6 G41 Y0.75 D01 F10.0
N7 G03 J-0.75
N8 G01 G40 Y0 F15.0
```

Now, the tool is back at the pocket center and the pocket is completed. The tool must also retract first, then move to machine zero (G28 motion is always in the rapid mode):

```
N9 G28 Z-0.25 M09
N10 G91 G28 X0 Y0 M05
N11 M30
%
```

This method is very simple, but may not always be the best, particularly for very close tolerances or high surface finish requirements. Drawing tolerances may be achieved by roughing operations with one tool and finishing operations with one or more additional tools.

A possible surface tool mark, left at the contact point with the pocket diameter, is a distinct possibility in a straight approach to the pocket diameter. The simple linear approach is quite efficient when the pocket or a counterbore is not too critical. Here is the complete listing for program O3304:

```
O3304 (CIRCULAR POCKET - VERSION 1)
N1 G20
N2 G17 G40 G80
N3 G90 G54 G00 X0 Y0 S1200 M03
N4 G43 Z0.1 H01 M08
N5 G01 Z-0.25 F8.0
N6 G41 Y0.75 D01 F10.0
N7 G03 J-0.75
```

```
N8 G01 G40 Y0 F15.0
N9 G28 Z-0.25 M09
N10 G91 G28 X0 Y0 M05
N11 M30
%
```

Another programming technique for a circular pocket is much more practical - one that makes better surface finishes and also maintains tight tolerances required by many drawings. Instead of a single linear approach directly towards the pocket diameter, the cutting tool can be applied in a combined linear-circular approach.

◆ Linear and Circular Approach

For this method, the cutting motion will be changed. Ideally, a small one half-arc motion could be made between the center and the pocket start point. That is possible only if the cutter radius offset is *not* used. As a matter of fact, some controls use a circular pocket milling cycle G12 or G13, doing exactly that (see an example later in this section). If the Fanuc control has the optional *User Macros*, custom made G12 or G13 circular pocket milling cycle can be developed. Otherwise, a step-by-step method is the only way, one block at a time.

Since the radius offset is needed to maintain tolerances, and the offset cannot start on an arc, a linear approach will be programmed first with the cutter radius offset applied. Then, the circular lead-in approach is programmed. When the pocket is completed, the procedure will be reversed and the radius offset canceled during a linear motion back to the pocket center. The approach radius calculation in this application is exactly the same as described earlier in this chapter, for the slot finishing tool path. *Figure 33-14* shows the suggested tool path.

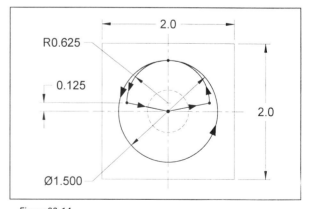

Figure 33-14

Combined linear and circular approach for a circular pocket milling - - program example O3305

This example uses an approach radius of .625. Any radius that is *greater* than the cutter radius (.3125) and *smaller* than the pocket radius (.750) is correct. The final program O3305 complements the above illustration in *Figure 33-14*.

```
O3305 (CIRCULAR POCKET - VERSION 2)
N1 G20
N2 G17 G40 G80
N3 G90 G54 G00 X0 Y0 S1200 M03
N4 G43 Z0.1 H01 M08
N5 G01 Z-0.25 F8.0
N6 G41 X0.625 Y0.125 D01 F10.0
N7 G03 X0 Y0.75 R0.625
N8 J-0.75
N9 X-0.625 Y0.125 R0.625
N10 G01 G40 X0 Y0 F15.0
N11 G28 Z-0.25 M09
N12 G91 G28 X0 Y0 M05
N13 M30
%
```

This programming technique is by far superior to the straight linear approach. It does not present any additional programming difficulty at all, partly because of the symmetry of tool motions. In fact, this method can be - and should be - used for just about any approach towards an internal contour finishing.

◆ Roughing a Circular Pocket

Often a circular pocket is too large for a given tool to guarantee the bottom cleanup in a single cut around. In this case, the pocket has to be enlarged by roughing it first, in order to remove all excessive material, then the finishing tool path can be applied. Some controls have special cycles, for example, a spiral pocketing. On Fanuc controls, custom cycles can be created with the *User Macros* option.

As an example, the same pocket drawing will be used as illustrated earlier in *Figure 33-11,* but machining will be done with a ⌀.375 cutter - *Figure 33-15*.

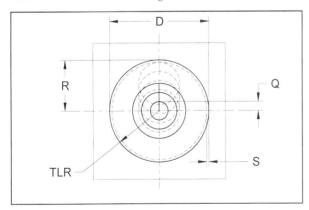

Figure 33-15

Roughing out a circular pocket - program O3306

The ⌀.375 end mill is a small tool that will *not* cleanup the pocket bottom using the earlier method. The method of roughing is shown in *Figure 33-15*, and the value of Q is the equal stepover amount, calculated from the number of steps N, the cutter radius TLR and the stock amount S, left for the finishing tool path.

The calculation is logically similar to the one for the rectangular pocket and the desired amount of the stepover can be achieved by changing the number of steps.

The example for program O3306 uses three stepovers, calculated from the following formula:

$$Q = \frac{R - TLR - S}{N}$$

☞ where ...

Q	=	Calculated stepover between cuts
R	=	Pocket radius (pocket diameter D / 2)
TLR	=	Tool radius (cutter diameter / 2)
S	=	Stock left for finishing
N	=	Number of cutting steps

In our application, the example values are:

➲ Example :

```
R   =  1.5 / 2 = .75          Diameter D = 1.5
TLR =  .375 / 2 = .1875
S   =  .025
N   =  3
```

Using the above formula, the stepover amount Q can be found by calculation:

```
Q = (.75 - .1875 - .025) / 3
Q = .1792
```

Final roughing program is quite simple and there is no cutter radius offset programmed or even needed. Note the benefit of incremental mode G91. It allows the stepover Q to be easily seen in the program, in the G01 linear mode. Every following block contains the arc vector J, cutting the next full circle. Each circle radius (J) is increased by the amount of stepover Q:

```
O3306 (CIRCULAR POCKET ROUGHING)
N1 G20
N2 G17 G40 G80
N3 G90 G54 G00 X0 Y0 S1500 M03
N4 G43 Z0.1 H01 M08
N5 G01 Z-0.25 F7.0
N6 G91 Y0.1792 F10.0                    (STEPOVER 1)
N7 G03 J-0.1792                         (ROUGH CIRCLE 1)
N8 G01 Y0.1792                          (STEPOVER 2)
N9 G03 J-0.3584                         (ROUGH CIRCLE 2)
N10 G01 Y0.1792                         (STEPOVER 3)
N11 G03 J-0.5376                        (ROUGH CIRCLE 3)
N12 G90 G01 X0 F15.0
N13 G28 Z-0.25 M09
N14 G91 X0 Y0 M05
N15 M30
%
```

CIRCULAR POCKET CYCLES

In *Chapter 29*, circular pocketing cycles were described briefly. In this chapter, two more examples will provide additional details. Fanuc does not have the useful G12 and G13 circular pocketing cycle as a standard feature. Controls that do have it, for example Yasnac, have a built-in macro (cycle), ready to be used. Fanuc users can create their own macro (as a special G code cycle), with the optional *User Macro* feature, which can be developed to offer more flexibility than a built-in cycle.

The two G codes are identical in all respects, except the cutting direction. The meaning of the G codes in a circular pocket cycle is:

| G12 | Circular pocket cutting CW |
| G13 | Circular pocket cutting CCW |

Either cycle is *always* programmed with the G40 cutter radius offset cancel mode in effect, and has the following format in the program:

| G12 I.. D.. F..　　(CONVENTIONAL MILLING) |

or

| G13 I.. D.. F..　　(CLIMB MILLING) |

☞ where ...

I　=　Pocket radius
D　=　Cutter radius offset number
F　=　Cutting feedrate

Typically, the cycle is called at the *center* and the *bottom* of a pocket. All cutting motions are arc motions, and there are three of them. There are no linear motions. The arbitrary start point (and end point) on the pocket diameter is at 0° (3 o'clock) - *Figure 33-16*.

Previous example in *Figure 33-11* can be used to illustrate the G12 or G13 cycle. For comparison, here is the program O3305, using a ∅.625 end mill:

```
O3305 (CIRCULAR POCKET - VERSION 2)
N1 G20
```

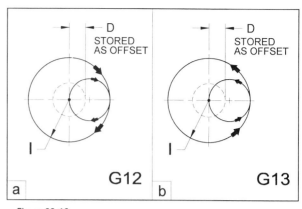

Figure 33-16
Circular pocket cycles G12 and G13

```
N2 G17 G40 G80
N3 G90 G54 G00 X0 Y0 S1200 M03
N4 G43 Z0.1 H01 M08
N5 G01 Z-0.25 F8.0
N6 G41 X0.625 Y0.125 D01 F10.0
N7 G03 X0 Y0.75 R0.625
N8 J-0.75
N9 X-0.625 Y0.125 R0.625
N10 G01 G40 X0 Y0 F15.0
N11 G28 Z-0.25 M09
N12 G91 G28 X0 Y0 M05
N13 M30
%
```

If the G12 or G13 cycle or a similar macro is available, the following program O3306 can be written, using the same tool and climb milling mode:

```
O3306 (CIRCULAR POCKET - G13 EXAMPLE)
N1 G20
N2 G17 G40 G80
N3 G90 G54 G00 X0 Y0 S1200 M03
N4 G43 Z0.1 H01 M08
N5 G01 Z-0.25 F8.0
N6 G13 I0.75 D1 F10.0          (CIRCULAR POCKET)
N7 G28 Z-0.25 M09
N8 G91 G28 X0 Y0 M05
N9 M30
%
```

Macros are very powerful programming tools, but their subject is beyond the limits of this handbook.

34 | *TURNING AND BORING*

There is so much information that can be covered in this section, that a whole book could be written just on the subject of turning and boring. Selected subjects are presented in this chapter, others are covered in chapters dealing with lathe cycles, grooving, part-off, single point threading, etc.

TOOL FUNCTION - TURNING

In terms of distinction, *turning* are *boring* are practically identical operations, except for the area of metal removal where the actual machining takes place. Often, terms *external turning* and *internal turning* are also used, meaning the same as turning and boring respectively. From programming perspective, the rules are virtually the same, and any significant differences will be covered as necessary.

CNC lathes require programming the selected tool by its tool number, using the T address. In comparison with a CNC machining center, the tool function for lathes is more extensive and calls for additional details. One major difference between milling and turning controls is the fact that the T address for CNC lathes *will make the actual tool change*. This is not a case in milling. No M06 function exists on a standard CNC lathe.

◆ T Address

One difference from machining centers is that a tool defined as T01 in the program *must* be mounted in the turret station #1, tool defined as T12 must be mounted in turret station #12, etc. Another difference between milling and turning tools is in the *format* of the T address. The format for turning system is T4, or more accurately, T2+2. The first two digits identify the turret station number and geometry offset, the last two digits identify the wear tool offset number for the selected tool station - *Figure 34-1*.

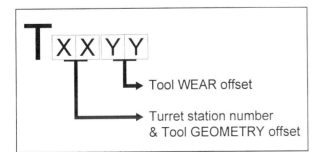

Figure 34-1
Typical tool function address for CNC lathes

Txxyy format represents tool station *xx* and wear offset number *yy*. For example, T0202 will cause the turret to index to the tool station #2 (first two digits) which will become the working station (active tool). At the same time, the associated tool wear offset number (the second pair of digits) will become effective as well.

Selection of the tool number (the first pair of digits), also selects the geometry offset on most modern CNC lathes. In that case, the second pair of digits will select the tool wear offset number. Any tool station selected by the turret station number identification can be associated with any offset number within the available offset range. In most applications, only one tool offset number is active for any selected tool. In such a case, it is wise to program the offset number the same as the tool number. Such an approach makes the operator's job much easier. Consider the following choices:

G00 **T0214**	*Tool station 02, wear offset 14*
G00 **T1105**	*Tool station 11, wear offset 05*
G00 **T0404**	*Tool station 04, wear offset 04*

Although all examples are technically correct, only the last example format is recommended. When many tools are used in a program, the offset numbers for individual tools may be confusing, if they do not correspond to the tool station numbers. There is only one time when the offset number cannot be the same as the tool station number. That happens in the cases when *two or more offsets* are assigned to the same tool, for example T0202 for the first wear offset, T0222 for the second wear offset.

Leading zeros in the tool function can be omitted for the tool number selection, but not for selection of the wear offset number. T0202 has the same meaning when written as T202. Eliminating the leading zero for tool wear offset will result in an incorrect statement:

T22 means *T0022*, which is an illegal format.

In summary, the active side of the turret (tool station) is programmed by the first pair of digits, the wear offset number is programmed by the last pair of digits in the tool function command:

G00 **T0404**

The most useful preference is to disregard the leading zero suppression and use the tool function in its full format, as shown above and in all examples in this handbook.

LATHE OFFSETS

Although the tool offset has been to some extent covered in the previous section describing the tool function, it is a very important feature for turning systems and some review will be beneficial.

The geometry offset is measured for *each tool* as the actual *distance from the tool reference point to the program zero* (Z axis distance will be stored as a negative value and so will be the X diameter) - *Figure 34-2*.

Geometry offset identifies the position of the tool from program zero

Wear offset is used for fine-tuning dimensions

The best way to illustrate the importance of tool wear offset, is to consider a program that does *not* use it. All programmed dimensions are ideal values, based on the drawing. Variable insert tolerances are not considered, neither is the tool wear. Any deviation from programmed dimensions caused by the actual tool size will produce an incorrect dimension when the part is machined, a very important concern for jobs with tight tolerances. The tool wear offset is used to 'fine tune' the actual machined dimensions against intended programmed dimensions.

The purpose of the tool wear offset is to adjust the difference between the programmed dimensions and the actual tool position on the part. If the wear offset is not available on the control, the adjustments are made to the only offset available - that is to the geometry offset.

◆ Offset Entry

The tool offset can be entered into the program in two different ways:

❑ As a command *independent* of the tool motion

❑ As a command applied *simultaneously* with a tool motion statement

◆ Independent Tool Offset

For an independent offset entry in the program, the tool offset is applied *together with the tool indexing:*

```
N34  G00  T0202
```

This command is usually programmed as the first block for each tool (in a clearance position). If the older G50 position register is used, the offset is programmed together with, or immediately following, the coordinate register block. At this point, the tool is still at its indexing position. When the tool offset is activated, it will cause a *physical motion* by the value of the offset, as stored in the offset register. Note the preparatory command G00 before the tool function. This is a very important command, since it will enable the physical offset motion to actually take place. G00 is more important for the first tool, but should be programmed for any tool. In *Chapter 5*, covering the control system, the status of the control when the power is turned on is described. Since the control system usually assumes the G01 command (linear interpolation) at the start up, a feedrate would be required. However, it looks rather absurd to program T0202 F0.025, although it is correct. Rapid motion is far more practical and rather that depending on the current control status, programming the G00 command will *always* get the offset activated.

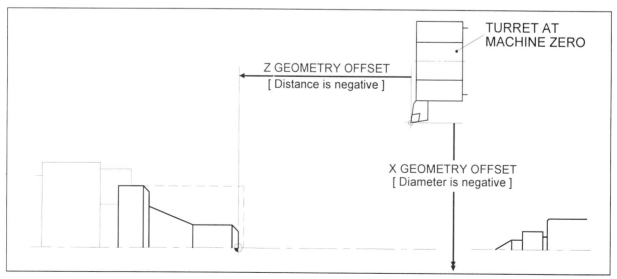

Figure 34-2

Geometry offset is the distance from tool reference point to program zero, measured along an axis from machine zero

◆ Tool Offset with Motion

The second method is to program the wear offset simultaneously with a cutting tool motion, usually during the tool approach towards the part. This is the preferred method. The following two examples illustrate this recommended programming of the T function for turning systems - the offset is activated when the second pair of digits in a tool number call are equal to or larger than 01:

```
N1 G20 T0100
N2 G96 S300 M03
N3 G00 X.. Z.. T0101 M08
...
```

Note the tool change in the first block N1 - it uses no offset number - just the tool number that is also the geometry offset number. The offset is applied two blocks later in N3.

In most cases, it makes no difference, whether the offset is activated with or without a motion command. But some limitations are possible when programming the tool offset entry *without* a motion command. For example, if the wear offset value stored is unusually large and the tool starts from the machine zero position, this type of programming may cause an overtravel condition.

Even in cases of a small offset value, there will always be a 'jump' motion of the turret when the offset is activated. Some programmers do not like this jumpy motion, although it will do no harm to the machine. In these cases, the best approach is to activate the tool wear offset during the first motion, usually as a rapid approach motion towards the part. One consideration is very important when the tool wear offset is activated together with a motion. Earlier in this chapter was a comment that the lathe tool function is also a function causing the tool indexing. Without a doubt, the one situation to avoid is the simultaneous tool indexing and tool motion - it may have dangerous consequences.

The best approach is to start each tool with the tool indexing only, *without* any wear offset:

```
N34 T0200 M42
```

The above example will register the coordinate setting for tool 2, it will also index tool 2 into the working position, but it will *not* activate any offset (T0200 means index for tool 2 without tool wear offset). Gear range function may be added as well, if required. Such a block will normally be followed by the selection of spindle speed, and rapid approach to the first position, close to the part. That is the block where the tool wear offset will be activated - *on the way* towards the first position:

```
N34 T0200 M42
N35 G96 S190 M03
N36 G00 G41 X12.0 Z0 T0202 M08
N37 G01 X1.6 F0.008
...
```

Also note that no G00 is required for a block containing tool indexing with zero wear offset entry. The advantage of programming the tool offset simultaneously with a motion is the elimination of the jumpy motion; at the same time, no overtravel condition will result, even if the wear offset is unusually large. The wear offset value will only *extend* or *shorten* the programmed rapid approach, depending on the actual offset amount stored.

Generally, the tool wear offset register number is entered *before* or *during* the rapid approach motion.

◆ Offset Change

Most lathe programs require one offset for each tool. In some cases, however, the program can benefit if two or even more offsets are assigned to the same tool. Needless to say, only one offset can be active at one time. The current offset can be changed to another offset for the same tool to achieve the extra flexibility. This is useful mainly in cases when individual diameters or shoulder lengths must be machined to exact tolerances. Any new offset must be programmed *without* a cancellation of the previous one. In fact, this is the preferable method for changing from one offset to another. The reason is simple - remember that any offset change serves a purpose only during actual cutting. Offset cancellation could be unsafe if programmed during cutting motion. This is a very important - *and largely unexplored programming technique* - that some detailed examples are justified.

MULTIPLE OFFSETS

Most jobs machined on CNC lathes require very high precision. High precision requires tolerance ranges as specified in the engineering drawing and these ranges may have quite a variety. Since a single offset per tool is often not enough to maintain these tolerances, two or more wear offsets are required for one tool.

The following three examples are designed to present a complete understanding of the advanced subject covering multiple offsets. The same basic drawing will be used for all examples.

The project is very simple - program and machine three diameters as per drawing, and maintain tolerances at the same time. One rule at the beginning - the program *will not use* the middle tolerance of the X or Z value. This is an unfortunate practice that makes changes to the program much more difficult at a later time, if the tolerances are changed by engineers or designers.

In the drawings, the following tolerances can be found:

❑   Tolerances only on the diameter

❑   Tolerances only on the shoulders (faces)

❑   Tolerances on the diameters and shoulders

◆ General Approach

The tolerances in all three examples are for training purposes only and will be much smaller in reality. All chamfer tolerances are ±0.010, and non-specified tolerances are ±0.005. That will allow concentration on the project. Material is a ∅1.5 inch aluminum bar and three tools are used:

T01	For the face and rough contour
T03	For the finishing of the contour to size
T05	0.125 wide part-off tool

The skill of the programmer determines the final result - the correct number of offsets must be entered at the right places within the program and the CNC operator must store the correct values for each offset. In all cases, the main goal will be to aim for the middle tolerance in machining, not in programming.

◆ Diameter Tolerances

The drawing in *Figure 34-3* shows the sample part with variable tolerances only on the *diameters*.

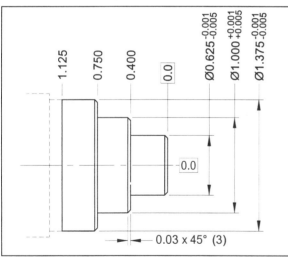

Figure 34-3

Multiple offsets - example for diameters - O3401

The programming solution is to include *two* offsets for finishing, for example, T0313 and T0314. In the control, correct offset amounts have to be set before machining - the ideal amounts for middle tolerance are shown:

```
13    X-0.003    Z0.000
14    X+0.003    Z0.000
```

Note that the Z-offset (which controls shoulders) must be the *same for both wear offsets*.

Here is the complete program - O3401:

```
O3401
(1.5 ALUMINUM BAR - EXTEND 1.5 FROM JAWS)

(T01 - FACE AND ROUGH TURN)
N1  G20
N2  G50 S3000 T0100
N3  G96 S500 M03
N4  G00 G41 X1.7 Z0 T0101 M08
N5  G01 X-0.07 F0.005
N6  Z0.1
N7  G00 G42 X1.55
N8  G71 P9 Q16 U0.04 W0.004 D1000 F0.01
N9  G00 X0.365
N10 G01 X0.625 Z-0.03 F0.003
N11 Z-0.4
N12 X1.0 C-0.03 (K-0.03)
N13 Z-0.75
N14 X1.375 C-0.03 (K-0.03)
N15 Z-1.255
N16 U0.2
N17 G00 G40 X5.0 Z5.0 T0100
N18 M01

(T03 - FINISH TURN)
N19 G50 S3500 T0300
(-- OFFSET 00 AT THE START OF THE TOOL ------)
N20 G96 S750 M03
N21 G00 G42 X1.7 Z0.1 T0313 M08
(-- OFFSET 13 FOR THE 0.625 DIAMETER -------)
N22 X0.365
N23 G01 X0.625 Z-0.03 F0.002
N24 Z-0.4
N25 X1.0 C-0.03 (K-0.03) T0314
(-- OFFSET 14 FOR THE 1.0 DIAMETER ---------)
N26 Z-0.75
N27 X1.375 C-0.03 (K-0.03) T0313
(-- OFFSET 13 FOR THE 1.375 DIAMETER -------)
N28 Z-1.255
N29 U0.2
N30 G00 G40 X5.0 Z5.0 T0300
(-- OFFSET 00 AT THE END OF TOOL ------------)
N31 M01

(T05 - 0.125 WIDE PART-OFF)
N32 T0500
N33 G97 S2000 M03
N34 G00 X1.7 Z-1.255 T0505 M08
N35 G01 X1.2 F0.002
N36 G00 X1.45
N37 Z-1.1825
N38 G01 X1.315 Z-1.25 F0.001
N39 X-0.02 F0.0015
N40 G00 X5.0
N41 Z5.0 T0500 M09
N42 M30
%
```

This is the complete program, using all three tools required. Since T01 and T05 do not change for any forthcoming examples, only T03 will be shown from now on.

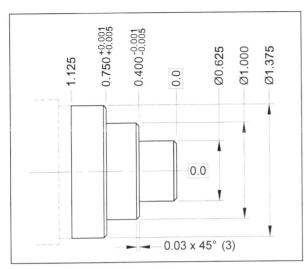

Figure 34-4

Multiple offsets - example for shoulders - O3402

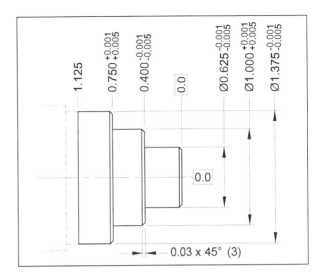

Figure 34-5

Multiple offsets - example for diameters and shoulders - O3403

◆ Shoulder Tolerances

The example drawing shown in *Figure 34-4* illustrates the sample part with variable tolerances specified only on the *shoulders*.

The programming solution is to include *two* offsets for finishing, for example T0313 and T0314. In the control, their amounts have to be set before machining - the ideal amounts for middle tolerance are shown:

```
13    X0.0000    Z+0.0030
14    X0.0000    Z-0.0030
```

Note that in this case, the X offset (which controls size of the diameters) must be the same for both offsets. Here is the T03 for program O3402:

```
O3402
...
(T03 - FINISH TURN)
N19 G50 S3500 T0300
(-- OFFSET 00 AT THE START OF TOOL ----------)
N20 G96 S750 M03
N21 G00 G42 X1.7 Z0.1 T0313 M08
(-- OFFSET 13 FOR THE 0.4 SHOULDER ----------)
N22 X0.365
N23 G01 X0.625 Z-0.03 F0.002
N24 Z-0.4
N25 X1.0 C-0.03 (K-0.03)
N26 Z-0.75 T0314
(-- OFFSET 14 FOR THE 0.75 SHOULDER ---------)
N27 X1.375 C-0.03 (K-0.03)
N28 Z-1.255
N29 U0.2
N30 G00 G40 X5.0 Z5.0 T0300
(-- OFFSET 00 AT THE END OF TOOL -----------)
N31 M01
...
```

◆ Diameter and Shoulder Tolerances

The example drawing shown in *Figure 34-5* illustrates the sample part with variable tolerances specified on both the *diameters* and *shoulders*.

The programming solution is to include *four* offsets for finishing, for example T0313, T0314, T0315 and T0316. In the control, their amounts have to be set before machining - the ideal amounts for middle tolerance are shown:

```
13    X-0.0030    Z+0.0030
14    X+0.0030    Z+0.0030
15    X+0.0030    Z-0.0030
16    X-0.0030    Z-0.0030
```

This is the most intensive version. Not only it is extremely important *where exactly* the offsets appear in the program, but their input amount is also critical.

Note that the four X offsets (which control size of the diameters) tie up with the four Z offsets (which control the length of shoulders). Here is the T03 for program O3403:

```
O3403
...
(T03 - FINISH TURN)
N19 G50 S3500 T0300
(-- OFFSET 00 AT THE START OF TOOL ----------)
N20 G96 S750 M03
N21 G00 G42 X1.7 Z0.1 T0313 M08
(-- OFFSET 13 FROM Z OVER TO Z UNDER ONLY ---)
N22 X0.365
N23 G01 X0.625 Z-0.03 F0.002
N24 Z-0.4
N25 X1.0 C-0.03 (K-0.03) T0314
(-- OFFSET 14 FROM X UNDER TO X OVER ONLY ---)
N26 Z-0.75 T0315
(-- OFFSET 15 FROM Z UNDER TO Z OVER ONLY ---)
```

```
N27 X1.375 C-0.03 (K-0.03) T0316
(-- OFFSET 16 FROM X OVER TO X UNDER ONLY ---)
N28 Z-1.255
N29 U0.2
N30 G00 G40 X5.0 Z5.0 T0300
(-- OFFSET 00 AT THE END OF TOOL -----------)
N31 M01
...
```

The CNC operator must always be aware of the existence of multiple offsets in the program as well as the programmer's reason for using them. Initial settings are always critical, so are all changes during machining. As can be seen in the first two examples, programs O3401 and O3402, one group of offsets must always remain the same (X or Z offsets) . For instance, in the program O3401, offsets 03 and 13 control diameters. That means the Z offset value must be the same - *always!* That also means, if there is a need to shift the shoulders .002 to the left, *all shoulders must be shifted by the same amount:*

```
13    X-0.0030    Z-0.0020
14    X+0.0030    Z-0.0020
```

Failure to do that will result in inaccurate dimensions.

OFFSET SETTING

The OFFSET screen selected by pressing a key on the control panel will initially display the tool geometry and the tool wear offsets. They are identical, except the title at the top of the screen. A typical display will resemble this screen layout (no offsets set):

OFFSET (GEOMETRY)				
NO.	X AXIS	Z AXIS	RADIUS	TIP
01	0.0000	0.0000	0.0000	0
02	0.0000	0.0000	0.0000	0
03	0.0000	0.0000	0.0000	0
...	...	...	...	...

X axis and Z axis are often shown just as X and Z, *Radius* is shown as R, and *Tip* is shown as T.

The NO. is the offset number, either the first pair of the T address - for the *Geometry* offset, or the second pair - for the *Wear* offset. X axis and Z axis are the columns where the offset values are entered for each number, the *Radius* and the *Tip* columns are only used if a tool nose radius offset is programmed. In that case, the *Radius* will be the tool nose radius and the *Tip* will be an arbitrary number, as defined by Fanuc, specifying the tool tip orientation. This subject has been described in *Chapter 30*.

FUNCTIONS FOR GEAR RANGES

A number of CNC lathes are designed to work in several ranges of gear engagement. This feature enables the programmer to coordinate the required spindle speed with specific power requirements of the machine. As a general rule, the higher the requirement for spindle speed, the lower the maximum available power rating will be, and vice versa. The ranges of spindle speed and power ratings for each range are determined by the machine manufacturer, and must never be changed.

Depending on the CNC lathe size, one, two, three, or four gear ranges may be available. Small lathes, or those designed with ultra high spindle speeds, may have no programmable gear range at all, which means only a single default gear range is available. Very large lathes may have all four gear ranges - and the maximum available spindle speed is usually low in comparison. The most common average is two gear ranges.

Miscellaneous functions for gear ranges, are typically M41, M42, M43 and M44, and assume the definition relative to the number of gear ranges available:

Range	Number of available ranges			
	1	2	3	4
Low	-	M41	M41	M41
Medium Low	-	-	-	M42
Medium	-	-	M42	-
Medium High	-	-	-	M43
High	-	M42	M43	M44

Once a certain gear range is selected, the spindle speed range is limited. If the exact range of spindle speed is important, always make an effort to find out the available spindle speeds in each range. Don't be surprised to find out that on most CNC machines, one rpm (1 r/min) is very rare. Typical lowest spindle speed may be around 20 to 30 r/min. Also, don't be surprised to find that there is an overlap, often quite large, for spindle speeds in two ranges. For example, if the *Gear 1* has a range 20 to 1400 r/min, *Gear 2* may have a range of 750 to 2500 r/min. When using spindle speeds available in either range, such as 1000 r/min, selection of gear range is not critical, but low gear range will produce more power.

Here is an actual, although unrelated, example:

Low gear range: 20 - 1075 r/min (M41)
High gear range: 70 - 3600 r/min (M42)

AUTOMATIC CORNER BREAK

In CNC turning and boring, there are occasions where the cut from a shoulder to a diameter (or from a diameter to a shoulder) requires a corner break. Breaking a sharp corner is a common practice when machining between shoulders and diameters. Many engineering drawings specify that all sharp corners are to be broken, often without suggesting their size. It is up to the programmer to decide, usually within the range of 0.005 to 0.020 inches (0.125 to 0.500 mm). The required corner break may be either a *chamfer at a 45° angle*, or a *blend radius* - both usually small. If the size of the corner break is specified, then the programmer must apply it. Corner breaking has three practical reasons:

❑ Functionality
   ... for strength, ease of assembly, and clearances

❑ Safety
   ... sharp corners are dangerous

❑ Appearance
   ... the finished part looks better

In lathe work, many corner breaks apply to cuts between a shoulder and the adjacent diameter (the cut takes a 90° turn in one axis at a time). The start and end points calculation is not difficult but can be time consuming for some jobs, such as shaft turning with many different diameters.

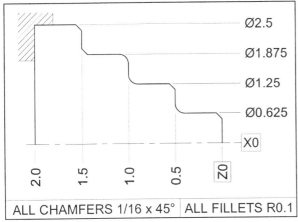

ALL CHAMFERS 1/16 x 45°   ALL FILLETS R0.1

Figure 34-6
Example for an automatic corner break (chamfers and radii)

The drawing in *Figure 34-6* shows a simple external part that contains several corners that will benefit from the automatic corner break programming feature *(not all corners in the drawing qualify)*.

Compare the two methods, to better understand the differences applied in programming. If the programmer *does not* use the automatic corner break feature, each contour change point must be calculated manually and the result will be program O3404:

```
O3404 (MANUALLY CALCULATED CORNER BREAK USED)
...
N51 T0100
N52 G96 S450 M03
N53 G00 G42 X0.3 Z0.1 T0101 M08
N54 G01 X0.625 Z-0.0625 F0.003
N55 Z-0.4
N56 G02 X0.825 Z-0.5 R0.1
N57 G01 X1.125
N58 X1.25 Z-0.5625
N59 Z-0.9
N60 G02 X1.45 Z-1.0 R0.1
N61 G01 X1.675
N62 G03 X1.875 Z-1.1 R0.1
N63 G01 Z-1.4375
N64 X2.0 Z-1.5
N65 X2.375
N66 X2.55 Z-1.5875
N67 U0.2
N68 G00 G40 X10.0 Z5.0 T0100
N69 M01
```

Only the finished contour is programmed (no facing cut), starting at a selected clearance of Z0.1, with the calculated X diameter at X0.3. Each contour change point has to be carefully calculated. At the contour end, the last chamfer has been completed at a clearance of 0.025 above the largest diameter, at X2.55, and calculated Z axis at Z-1.5875.

As always in manual work, the possibility of errors can be significant. For example, one very common error in this type of programming is the target value of X axis. In turning, it is easy to forget to double the chamfer or radius value (or half it for boring). The result is that block N56 may be:

```
N56 G02 X0.725 Z-0.5 R0.1        (ERROR IN X)
```

instead of the *correct* block

```
N56 G02 X0.825 Z-0.5 R0.1        (X IS CORRECT)
```

So what can be done in the program in order to implement the automatic corner break?

Fanuc control system offers two programming methods that relate to automatic corner breaking on lathes:

❑ Chamfering method   ... for a 45° chamfer

❑ Blend radius method   ... for a 90° blend

Both methods work in a very similar manner and certain rules have to be observed in both cases.

◆ Chamfering at 45 Degrees

The automatic corner chamfering will always take place in the G01 mode, and two special vectors I and K are available for this purpose or a C vector on some models.

For the automatic chamfer generation, the vectors I and K specify *the direction and the amount of cut for the required chamfer:*

<table>
<tr><td>The I vector
is used to create a chamfer starting from the X axis,
into the X+Z−, X−Z−, X+Z+, or X−Z+ direction</td></tr>
</table>

<table>
<tr><td>The K vector
is used to create a chamfer starting from the Z axis,
into the Z−X+, Z−X−, Z+X+, or Z+X− direction</td></tr>
</table>

The I and K vector definition is illustrated in *Figure 34-7.*

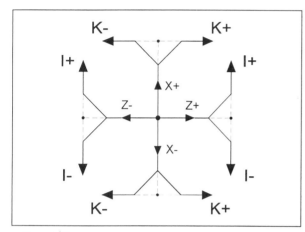

Figure 34-7

Vectors I and K for automatic corner chamfering

When the control system encounters a block containing the chamfering vector I or K, it will automatically *shorten* the active programmed tool path length by the value of the I or K vector, as specified in the program. If not sure whether the I or the K vector should be programmed for automatic chamfering, consult the above illustration, or apply the following rules:

The vector I indicates the *chamfering amount and motion direction* when the tool motion is in the order of *Diameter-Chamfer-Shoulder,* which means cutting along the Z axis before the chamfer. The chamfer deviation can only be from the Z axis towards the X axis, with the I vector programmed:

```
G01 Z-1.75 I0.125      (CUTTING ALONG Z AXIS)
X4.0      (CONTINUING IN X AXIS AFTER CHAMFER)
```

The vector K indicates the *chamfering amount and motion direction* when the tool motion is in the order of *Shoulder-Chamfer-Diameter,* which means cutting along the X axis before the chamfer. The chamfer deviation can only be from the X axis towards the Z axis, when the K vector is programmed:

```
G01 X2.0 K-0.125       (CUTTING ALONG X AXIS)
Z-3.0     (CONTINUING IN Z AXIS AFTER CHAMFER)
```

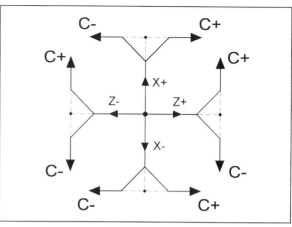

Figure 34-8

Vectors C for automatic corner chamfering

In either case, the sign of I or K vector defines the direction of the chamfer cutting within the coordinate system:

❑   Positive value of I or K vector indicates the chamfering direction into the plus direction of the axis *not* specified in the chamfering block

❑   Negative value of I or K vector indicates the chamfering direction into the minus direction of the axis *not* specified in the chamfering block

The values of I and K commands are always single values (*i.e.*, radius values, not diameter values).

Many latest controls use vectors C+ and C− that replace the I+, I−, K+ and K− vectors - *Figure 34-8.* This is a much simpler programming method and its applications are the same as for the blend radius R, described shortly. There is no distinction between axes vector selection, just the specified direction:

❑   The C vector is used

... to create a chamfer starting from the X axis, into the X+Z−, X−Z−, X+Z+, or X−Z+ direction
- *or* -
... to create a chamfer starting from the Z axis, into the Z−X+, Z−X−, Z+X+, or Z+X− direction

If the unit control allows the C+ or C− vectors, the programming is much easier, as long as the motion direction is watched. The two previous examples will be:

```
G01 Z-1.75 C0.125      (CUTTING ALONG Z AXIS)
X4.0      (CONTINUING IN X AXIS AFTER CHAMFER)
```

```
G01 X2.0 C-0.125       (CUTTING ALONG X AXIS)
Z-3.0     (CONTINUING IN Z AXIS AFTER CHAMFER)
```

As was the case with the I and K vectors, the C vector is also specified as a single value *per side*, not per diameter.

◆ Blend Radius at 90 Degrees

A blend radius between a shoulder and diameter (or vice versa) is programmed in a similar way as the automatic 45° chamfer. *It also takes place exclusively in the G01 mode!* Only one special vector R is used. For automatic blend radius, the vector specifies *the direction and the amount of cut for the radius:*

❑ The R vector is used

   ... to create a blend radius starting from the X axis, into the X+Z–, X–Z–, X+Z+, or X–Z+ direction

   - *or* -

   ... to create a blend radius starting from the Z axis, into the Z–X+, Z–X–, Z+X+, or Z+X– direction

The R vector definition is illustrated in *Figure 34-9.*

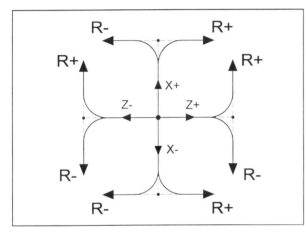

Figure 34-9

Vector R for automatic corner rounding (blend radius)

When the control system encounters the block containing a blend radius vector R, it will automatically *shorten* the active programmed tool path length by the value of the R vector, as specified in the program. If not sure whether the R vector should be programmed for automatic blend radius, consult the above illustration or apply the following rule:

The vector R indicates the *radius amount and motion direction* when the cutting is in the order of *Shoulder-Radius-Diameter,* which means cutting along the X axis before the radius. The same vector is also used when the *radius amount and motion direction* is in the opposite order of *Diameter-Radius-Shoulder,* which means cutting along the Z axis before the radius.

The radius deviation can be from the X axis towards the Z axis, when the R vector is programmed:

```
G01 X2.0 R-0.125        (CUTTING ALONG X AXIS)
Z-3.0        (CONTINUING IN Z AXIS AFTER RADIUS)
```

The radius deviation can also be from the Z axis towards the X axis, when the R vector is programmed:

```
G01 Z-1.75 R0.125        (CUTTING ALONG Z AXIS)
X4.0        (CONTINUING IN X AXIS AFTER RADIUS)
```

In either case, the sign of the R vector defines the direction of the radius cutting within the coordinate system:

❑ Positive value of R vector indicates the radius direction into the *plus* direction of the axis *not* specified in the radius block

❑ Negative value of R vector indicates the radius direction into the *minus* direction of the axis *not* specified in the radius block

◆ Programming Conditions

Breaking corners automatically makes programming for modern CNC lathes a lot easier, as only drawing dimensions are used and no external manual calculations are necessary. Regardless of whether program contains vectors I or K or C for chamfering, or vector R for blend radius corner, the basic conditions and general rules are very similar:

❑ Chamfer or radius must be fully contained in a single quadrant - 90° only

❑ Chamfers must have a 45° angle and radii must have a 90° angle between a shoulder and a diameter or a diameter and a shoulder

❑ The values of chamfering vectors I and K or C, as well as the radius vector R, are always single values - meaning *per side* values, not diameter values

❑ Direction of cut before the corner rounding must be perpendicular to the direction of the cut after rounding, along one axis only

❑ The direction of the cut following the chamfer or radius must continue along a single axis only, and must have the length equivalent to at least the chamfer length or the radius amount - the cutting direction cannot reverse

❑ Both chamfering and blend radius corner breaking takes place in G01 mode (linear interpolation mode)

❑ When writing the CNC program, only the known intersection from the drawing - *the sharp point* - is needed. That is the point between the shoulder and the diameter, without the chamfer or radius being considered

These rules apply equally to turning and boring CNC lathe operations. Study them carefully to avoid problems at the machine later.

◆ Programming Example

The following program O3405 combines the use of the chamfering and blend radius vectors into a complete example. The same drawing is used for this version, as for the traditional method, illustrated earlier in *Figure 34-6.*

In order to fully appreciate the differences between the two programming methods (both are technically correct), compare the following program O3405 with the earlier program O3404. The I and K vectors are used for chamfering, as they are more difficult then the C vectors:

```
O3405 (AUTOMATIC CORNER BREAKS USED)
...
N51 T0100
N52 G96 S450 M03
N53 G00 G42 X0.3 Z0.1 T0101 M08
N54 G01 X0.625 Z-0.0625 F0.003
N55 Z-0.5 R0.1
N56 X1.25 K-0.0625
N57 Z-1.0 R0.1
N58 X1.875 R-0.1
N59 Z-1.5 I0.0625
N60 X2.375
N61 X2.55 Z-1.5875
N62 U0.2
N63 G00 G40 X10.0 Z5.0 T0100
N64 M01
```

Although the program is a little shorter, the five blocks saved in the program offer the least benefit. Where are the G02s and G03s, where are the calculations of each contour change point? Where are the center point calculations?

Except for the contour beginning and end, this type of programming greatly enhances program development and allows for very fast and easy changes during machining, if necessary. If a chamfer or a blend radius is changed in the drawing, only a *single value* has to be changed in the program, without any recalculations. Of course, the rules and conditions mentioned earlier must be always observed. The main benefit of the automatic contouring are the ease of changes and the absence of manual calculations.

ROUGH AND FINISHED SHAPE

The vast majority of material removal on CNC lathe is done by using various cycles, described in detail in the next chapter. These cycles require input of data that is based on machining knowledge, such as a depth of cut, stock allowance, speeds and feeds, etc.

Rough and finished shapes often require manual calculations, using algebra and trigonometry. These calculations should be done on separate sheets of paper, rather than in the drawing itself. That way, the work is better organized. Also, if there is a change later, for example, an engineering design change, it is easier to keep track of what is where.

◆ Rough Operations

A great part of lathe machining amounts to removal of excessive stock to create a part, almost completed. This kind of machining is generally known as *roughing*, rough turning, or rough boring. As a machining operation, rough-ing does not produce a high precision part, that is not the purpose of roughing. Its main purpose is to remove unwanted stock efficiently, which means fast and with maximum tool life, and leave suitable all-around stock for finishing. Cutting tools used for roughing are strong, usually with a relatively large nose radius. These tools have to be able to sustain heavy depths of cut and high cutting feeds. Common diamond shaped tools suitable for roughing are 80° inserts (up to 2+2 cutting corners), and trigon inserts (up to 3+3 cutting corners). 2+2 or 3+3 means on *2 or 3 cutting edges on each side* of the insert. Not all inserts can be used from both sides. *Figure 34-10* shows some typical tools and orientation for rough turning and boring.

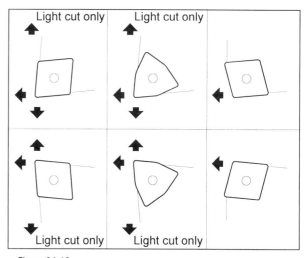

Figure 34-10

Tool orientation and cutting direction for roughing.
Upper row shows external tools, lower row shows internal tools.

Although a number of tools can be programmed in several directions, some directions are not recommended at all, or only for light or medium light cuts.

In practice, always follow one basic rule of machining - this rule is valid for all types of machines:

Always do heavy operations before light operations

This basic rule means that *all* roughing should be done before the first finishing cut is programmed. The reason here is to prevent a possible shift of the material during roughing, after some finishing had already been done.

For example, the requirement is to rough and finish both external and internal diameters. If the above rule is applied to these operations, the roughing out the outside of the part will be first, then roughing out the inside of the part, and only then applying the finishing cuts. It really does not matter whether the roughing is done first externally or internally, as long as it gets done before any finish cuts, which also can be in either order.

Tool wear can be minimized if the depth of cut is sufficient and the cutting radius gets 'under the skin' of the material, usually during the first cut. Coolant is usually a must for most materials and should be applied *before* the tool actually contacts the part.

◆ Finish Operations

Finish operations take place as the final cutting motions, after most of the stock has been removed (roughed out), leaving only a small amount of overall stock for finishing. The cutting tool can have smaller nose radius and, for even a better surface finish, higher spindle speeds and lower cutting feeds are typical.

Many different tools can be used for finishing operations as well, but the most typical finishing tools are two diamond shaped inserts, with a 55° and a 35° insert angle. Their shape, common orientation and cutting directions are shown in *Figure 34-11*.

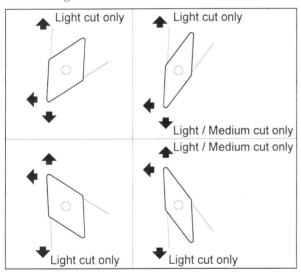

Figure 34-11

Tool orientation and cutting direction for finishing with common lathe tools. Upper row shows external tools, lower row shows internal tools.

Note that some cutting directions are only recommended for light or medium cuts. Why? The answer has a lot to do with the amount of material (stock) the tool removes in the specified direction.

◆ Stock and Stock Allowance

The material machined is often called *stock*. When the tool removes the stock to cut a desired shape, it can only handle a certain amount of it at a time. The insert shape, its orientation toward the part and its cutting direction, the insert size and thickness, all have a profound effect on the allowable stock to be removed. This is especially important in semifinishing and finishing operations. Stock allowance

specifies the amount of material left for these fine operations. If too much material or too little material is left to be cut during finishing, the part accuracy and surface finish quality will suffer. Also, carefully consider not just a stock allowance overall on the part, but *individual stock allowances* for the X and Z axes.

As before, there is a general rule of thumb, that on the X axis, that is for cutting diameters, leave the stock *equivalent to* or *slightly larger* than the radius of the subsequent finishing tool. For example, if a .031 inch tool nose radius (0.80 mm) is used for finishing, leave .030 to 0.040 inch stock (about 1 mm). That is the physical stock, the actual stock amount assigned *per side*, not on diameter!

The amount of stock left on the Z axis (typically for facing shoulders at 90°) is much more critical. If cutting along the positive X axis only (for turning), or the negative X axis only (for boring), with a tool that has a lead angle of 3° to 5°, do not leave more than .003 to .006 inch (0.080 to 0.150 mm) on any straight shoulder. *Figure 34-12* shows the effect of too much stock allowance for certain cutting directions and a method to eliminate it.

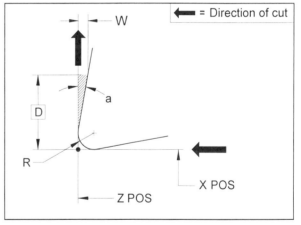

Figure 34-12

Effect of stock allowance W on depth of cut D

In the illustration, the actual depth of cut D at the face Z POS, is determined by the amount of stock W. To calculate the depth D, use this formula:

$$D = \tan\frac{A}{2} \times R + \frac{W}{\tan A} + R$$

☞ where ...

D	=	Actual depth of cut at the face
A	=	Lead angle of the insert
R	=	Radius of the insert
W	=	Stock left on face for finishing
X POS	=	Target position for the X axis
Z POS	=	Target position for the Z axis

The illustration applies equally to the boring, when the X axis direction is opposite the one shown. To understand better the consequences of a heavy stock left on the face, evaluate this example:

➲ Example:

The amount of stock left on face is .030, the tool radius is .031 and the tool lead angle is 3°:

```
W = .030,   R = .031,   A = 3
```

There is enough data available to calculate the unknown depth D, using the above formula:

```
D = tan3/2 × .031 + .030 / tan3 + .031
D = .60425
```

For an insert with a ∅.500 inch inscribed circle (such as DNMG-432, for example), the actual depth of cut at the face will be .60425 - *more than any reasonable amount!*

Since the earlier suggestion was no more than .006, recalculate the example for the largest depth, if the W=.006:

```
D = tan3/2 × .031 + .006/tan3 + .031
D = .14630
```

That is a more reasonable depth of cut at the face, so the Z axis stock allowance of .006 can be used. For facing in the opposite X direction or for not unidirectional faces, leave stock much bigger, usually close to the tool radius.

PROGRAMMING A RECESS

Another very important aspect of programming for CNC lathes is the change of cutting direction. Normally, program a tool motion in such a way that the motion direction from the starting point will be:

❏ Positive X direction for external machining
  … and / or …
  Negative Z direction for external machining

❏ Negative X direction for internal machining
  … and / or …
  Negative Z direction for internal machining

There are also *back turning* or *back boring* operations used in CNC programming, but these are just related and less common variations of the common machining. In the most common machining on CNC lathes, any change of direction in a single axis into the material constitutes an undercut, a cavity, or more commonly known - a *recess*.

A recess is commonly designed by the engineers to relieve - or undercut - a certain portion of the part, for example, to allow a matching part to fit against a shoulder, face, or surface of the machined part.

In CNC lathe programming, a recess can be machined very successfully with any tool that is used with the proper depth of cut, and *a suitable back angle clearance*. It is the second requirement that will be looked at next.

Figure 34-13 shows a simple drawing of a roller. In the middle of the object, there is an undercut (recess) between the ∅1.029 and the ∅.939. The objective is to calculate, not to guess, what is the maximum back angle tool that can be used for cutting the recess in a single operation.

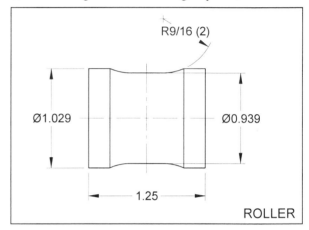

Figure 34-13
Back angle clearance calculation example

The first step is to consider the drawing - that is always the given and unchangeable source of data. The difference between the diameters and the recess radius will be required. *Figure 34-14* illustrates the generic details of the *provided data* (except the angle *b*) from the drawing.

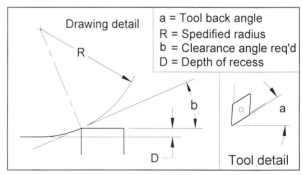

Figure 34-14
Data required to calculate angle 'b'

The formula required to calculate the angle *b* uses simple trigonometric formula. First, calculate the *depth* of the recess D, which is nothing more that one half of the difference between the two given diameters:

$$D = \frac{LARGE\ DIA - SMALL\ DIA}{2}$$

Once the recess depth D is known, the formula to calculate the angle *b* is:

$$b = \cos^{-1} \left(\frac{R - D}{R} \right)$$

For the example, the calculation will be:

$$b = \cos^{-1} \left(\frac{.5625 - .045}{.5625} \right) = 23.07392$$

For actual machining, select a tool with the back angle *a* greater than the calculated angle *b*. For the illustrated drawing (23.07° required clearance), the selected tool could be either a 55° diamond shape (back angle clearance *a* is 30° to 32°), or a 35° diamond shape (back angle clearance *a* is 50° to 52°) - both are greater than the calculated minimum clearance. The actual angles depend on the tool manufacturer, so a tooling catalogue is a good source of data.

This type of calculation is important for any recesses, undercuts and special clearances, whether programmed with the aid of cycles or developed block by block. The example only illustrates one possibility, but can be used for any calculations where the back angle clearance is required.

SPINDLE SPEED IN CSS MODE

From several earlier topics, remember that the abbreviation *CSS* stands for *Constant Surface Speed*. This CNC lathe feature will constantly keep recalculating the actual spindle speed in *revolutions per minute* (r/min), based on the programmed input of surface speed. The surface speed is programmed in *feet per minute* - ft/min (English system) or in *meters per minute* - m/min (metric system).

In the program, the *'per minute'* input uses the preparatory command G96, as opposed to the direct *r/min* input using the command G97.

The *Constant Surface Speed* is a powerful feature of the control system and without it, we would look back many years. There is a rather small problem associated with this feature, often neglected altogether, or at least not considered important enough. This rather 'small problem' will be illustrated in a simple program example.

The program example covers only a few blocks at the beginning, when the cutting tool approaches the part. That is enough data to consider the question that follows.

```
O3406
N1 G20 T0100
N2 G96 S450 M03
N3 G00 G41 X0.7 Z0 T0101 M08
N4 ...
```

The question is this: What is the actual spindle speed (in *r/min*), when the block N2 is executed? Of course, the spindle speed is unknown at the moment. It cannot be known, unless the current *diameter*, the diameter where the tool is located at that moment, is also known. The control system keeps track of the current tool position at all times. So, when block N2 is executed, the actual *r/min* of the spindle will be calculated for the current diameter, as stored in the control, specified in the geometry offset entry. For the example, consider that the current diameter is 23.5 or X23.5.

From the standard *r/min* formula, the spindle speed calculated for 450 ft/min and ⌀23.5 as 73 r/min is rather slow, but correct. At the next block, block N3, the tool position is rather close to the part, at diameter of .700 (X0.7). From the same standard formula, the spindle speed can be calculated for that diameter as 2455 r/min - considerably fast but also correct. The problem? There may not be one for every machine, but if ever there is a problem, the following solution will eliminate it.

The possible problem will be linked to the rapid motion from the ⌀23.5 to the ⌀.700. The actual travel distance (per side of part) is (23.5-.700)/2, which is 11.400. During the rapid travel rate, the cutting tool has to move 11.400 inches and - *at the same time* - change the spindle speed from a slow 73 r/min, to a fast 2455 r/min. Depending on the control system and its handling of such a situation, the tool may actually start cutting *at a slower spindle speed than was originally intended*.

If such a situation does happen and presents a problem, the only step that can be done is to *preprogram* the expected spindle speed in r/min, *before* the cutting tool approach motion, *then* switch to the constant surface speed (CSS) mode and continue.

```
O3407
N1 G20 T0100
N2 G97 S2455 M03                    (R/MIN PRESET)
N3 G00 G41 X0.7 Z0 T0101 M08
N4 G96 S450 M03
N5 ...
```

What had been done requires more evaluation. What had been done is that the spindle was started at the final expected *r/min*, before the tool reaches the part, in block N2. In block N3, the tool moves to the start of cut, while the spindle is already at the peak of the programmed speed. Once the target position along the X axis has been reached (block N3), the corresponding CSS mode can be in effect for all subsequent cuts.

This is an example that does not necessarily reflect everyday programming of CNC lathes. In this situation, some additional calculations have to be done, but if they solve the problem - they are worth the extra effort! Some CAD/CAM system can be set to do exactly that, automatically. If the current X position of the tool is unknown, estimate it.

LATHE PROGRAM FORMAT

In a review of the already presented examples, a certain consistency can be seen in the program output. This may be called a style, a format, a form, a template, as well as several other terms. Each programmer develops his or her own style over a period of time. A consistent style is important for efficient program development, program changes and program interpretation.

◆ Program Format - Templates

Most examples have followed a certain program format. Note that each CNC lathe program begins with the G20 or G21 command and perhaps some cancellation codes. The block that follows is a tool selection, next is spindle speed data, etc. This format will not basically change from one job to another - it follows a certain consistent pattern which forms the basic *template* for writing the program.

◆ General Program Format

To view the format often enough will forge a mental image in the programmer's mind. The details that are not understood yet will become much clearer after acquiring the general understanding of the relationships and details used in various programming methods. Here is a suggested template for a CNC lathe program.

⟳ General Program Pattern - Lathe :

```
O.. (PROGRAM NAME)
N1 G20 G40 G99           (PROGRAM START UP)
N2 T..00 M4..            (TOOL AND GEAR RANGE)
N3 G97 S.. M03           (STABILIZE R/MIN)
N4 G00 [G41/G42] X.. Z.. T.. M08    (APPROACH)
N5 G96 S..               (CUTTING SPEED)
N6 G01 [X../Z..] F..     (FIRST CUTTING MOTION)
N7 ...
...
...                      (MACHINING)
...
N.. G00 [G40] X.. Z.. T..00 (TOOL CHG POSITION)
N.. M01                  (OPTIONAL STOP)
...
N.. M30                  (PROGRAM END)
%
```

This generic structure is good for most lathe programs. Feel free to adjust it as necessary. For example, not every job requires spindle speed stabilization, so block N3 will not be necessary. It also means that M03 rotation has to be moved to block N5. Take the general program pattern as an example only, not as a fixed format.

◆ Approach to the Part

An important part of any lathe program structure is the method of approaching a revolving part. If the part is concentric, the approach can be similar to the A option in *Figure 34-15*. Although a facing cut is illustrated, the approach would be logically the same for a turning or a boring cut. Keep the starting point SP well above the diameter, at least .100 per side and more, if the actual diameter is not known *exactly*. The B option of the tool approach is two single axis at a time. It is a variation of the first example, and the X axis motion can be further split into a rapid and cutting motion, if required. Finally, the C option uses the clearance in the Z axis, far from the front face. Again, the final motion toward the face can be split into a rapid and linear motion.

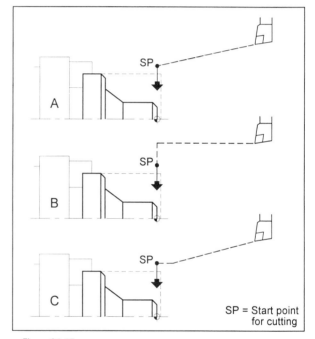

SP = Start point for cutting

Figure 34-15
Safe approach to a part - example for a facing cut shown

There are many variations on these methods, too numerous to list. The main objective of considering the approach to the part in the first place is safety. A collision of a tool with a revolving part can have serious consequences.

Turning and boring is a large subject. Many other examples could have been included in this chapter. Other chapters in this book also cover turning and boring, but in a more specialized way, for example, turning and boring cycles. The examples that were presented in this chapter should be useful to any CNC lathe programming.

In the last chapter, several lathe procedures described programming of a turning and boring tool path. A number of different techniques have been introduced, mainly describing the *finishing* tool path. Virtually no attention has yet been given to the removal of an excessive stock, in such operations as rough turning and rough boring. It is a subject in its own right and this chapter describes various methods of stock removal for roughing and finishing.

STOCK REMOVAL ON LATHES

One of the most time consuming tasks in manual programming for a CNC lathe is the removal of an excessive stock, typically from a cylindrical material, known as rough turning or rough boring - or simply roughing.

To manually program a roughing tool path requires a series of coordinated rough passes, with one block of program for each tool motion. For roughing of a complex contour, such a method is extremely time consuming and very inefficient, as well as prone to errors. Some programmers try to sacrifice programming quality for speed, by leaving an *uneven* stock for finishing, causing the cutting tool to wear out prematurely. The surface roughness of the finished profile often suffers as well.

It is in the area of rough stock removal where the modern lathe controls are very useful and convenient. Almost all CNC lathe systems have a feature that allows the roughing tool path to be processed automatically, using *special cycles*. Roughing is not the only application for these cycles, there are also special cycles available for *threading* and *simple grooving*. The grooving and threading cycles are outside of this chapter, but will be covered in detail in the next three chapters.

◆ Simple Cycles

Fanuc and similar controls support a number of special lathe cycles. There are three rather simple cycles that have been part of Fanuc controls for quite a while. They first appeared with the early CNC units and were limited by the technological progress of the time. Various manuals and textbooks refer to them as the *Fixed Cycles* or *Simple Cycles* or even *Canned Cycles*, similar in nature to their cousins for drilling operations on CNC mills and machining centers. Two of these early cycles are used for turning and boring, the third cycle is a very simple threading cycle. This chapter covers the first two cycles.

◆ Complex Cycles

With the advancement of computer technology, control manufacturers have developed cutting cycles capable of very complex lathe operations and made them an integral part of the lathe control systems. These special cycles are called by Fanuc the *Multiple Repetitive Cycles*. Their major improvement over simple cycles is in their excellent flexibility. Some of these advanced cycles cover turning and boring, others grooving and threading.

Don't get misled by the description *'complex'* - these cycles are only complex in the mathematical sense and even then, only internally. They are complex within the control system only. In fact, these very advanced machining cycles are much easier to program than their simple predecessors. In addition, they can also be very easily changed at the machine control, to optimize them for best performance, right on the job.

PRINCIPLES OF LATHE CYCLES

Similar to drilling operations for CNC machining centers, all cycles for lathes are based on the same technological principles. The programmer only enters the overall data (typically variable cutting parameters), and the CNC system will calculate the details of individual cuts. These calculations are based on the combination of the fixed and variable data. Return tool motions in all these cycles are automatic, and only the values to be changed are specified within the cycle call.

Simple cycles are designed exclusively to cut a straight cut, with no chamfers, tapers or radii and also with no undercuts. The simple cycles can only be used to cut vertically, horizontally, or at an angle, for taper cutting. These original cycles cannot do the same cutting operations as the more modern and advanced multiple repetitive cycles - for example, they cannot rough out a radius or change direction of the cutting. Simply, they cannot contour.

In the category of simple turning cycles, there are two that do allow removal of rough stock from a cylindrical or conical part. Each block of these cycles replaces four *regular* blocks of the part program. In the category of multiple repetitive cycles, there are several cycles designed for complex roughing, one for finishing, as well as cycles for grooving and threading. Multiple repetitive cycles are capable of some very complex contouring.

G90 - STRAIGHT CUTTING CYCLE

Before going further, a reminder. Do not confuse G90 for lathes with G90 for machining centers. In turning, G90 is a lathe cycle, G90 is the absolute mode in milling:

> G90 is absolute mode for milling,
> X and Z axes are absolute mode for turning

> G91 is incremental mode for milling,
> U and W axes are incremental mode for turning

A cycle identified by G90 preparatory command (*Type A* group of G codes) is called the *Straight Cutting Cycle (Box cycle)*. Its purpose is to remove excessive stock between the start position of the cutting tool and the coordinates specified by the X and the Z axes. The resulting cut is a straight turning or boring cut, *normally parallel* to the spindle centerline and the Z axis is the main cutting axis. As the name of the cycle suggests, the G90 cycle is used primarily for removing a stock in a rectangular fashion (box shape). The G90 cycle can also be used for a taper cutting. In *Figure 35-1*, the cycle structure and motions are illustrated.

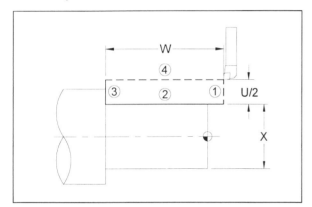

Figure 35-1

G90 simple cycle structure - straight cutting application

♦ **Cycle Format**

The G90 cutting cycle has two predetermined programming formats. The first one is for straight cutting only, along the Z axis, as illustrated in *Figure 35-1*.

❑   Format 1 :

> G90 X(U).. Z(W).. F..

☞ where ...

   X   =   Diameter to be cut
   Z   =   End of cut in Z position
   F   =   Cutting feedrate (usually *in/rev* or *mm/rev*)

The second format adds the parameter I or R to the block and is designed for taper cutting motions, with the dominance of the Z axis - *Figure 35-2*.

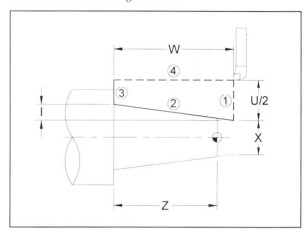

Figure 35-2

G90 cycle structure - taper cutting application

❑   Format 2 (two versions):

> G90 X(U).. Z(W).. I.. F..
> G90 X(U).. Z(W).. R.. F..

☞ where ...

   X   =   Diameter to be cut
   Z   =   End of cut in Z position
   I (R)   =   Distance and the direction of taper
               (I=0 or R=0 for straight cutting)
   F   =   Cutting feedrate (usually *in/rev* or *mm/rev*)

In both examples, the designation of axes as X and Z is used for the absolute programming, indicating the tool position from program zero. The designation of axes as U and W is used for the incremental programming, indicating actual travel distance of the tool from the current position. The F address is the cutting feedrate, normally in *inches per revolution* or *millimeters per revolution*. The I address is used for taper cutting along the horizontal direction. It has an amount equivalent to *one half* of the distance from the diameter at the taper end, to the diameter at the taper beginning. The R address replaces the I address, and is available on *newer controls only*.

To cancel the G90 cycle, all that is necessary to do is to use any motion command - G00, G01, G02 or G03. Commonly, it will be the G00 rapid motion command:

```
G90 X(U).. Z(W).. I.. F..
...
...
G00 ...
```

◆ Straight Turning Example

To illustrate a practical application of G90 cycle, study *Figure 35-3*. It shows rather a simple diameter turning, from a ∅4.125 inch stock down to a final ∅2.22 inch, over the length of 2.56 inches. There are no chamfers, no tapers, and no radii. This fact restricts the practical usefulness of the G90 cycle to a very simple roughing only, but still beats the manual alternative.

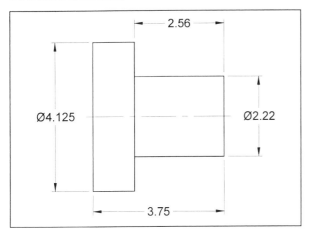

Figure 35-3

Example of G90 cycle in straight cutting - programs O3501 & O3502

Since G90 is a roughing cycle, the *depth* of each cut has to be selected first, then the *stock* amount left for finishing. To decide on the depth of each cut, find out how much stock is actually there to be removed from the diameter. The actual amount of stock is calculated *per side*, as a radius value, along the X axis:

```
(4.125 - 2.22) / 2 = .9525
```

For a .030 stock per side for the finishing cut, the .030 will be subtracted from the total X stock, so the total depth amount to remove will be .9225. Next is the selection of cut segmentation for the total depth. For five even cuts, each depth of cut will be .1845, for six cuts, .1538. Six cuts will be selected and .030 left per side, or 0.06 on the diameter - the first diameter will be X3.8175. Also, .005 stock allowance will be left on the face, so the Z axis end of cut will be at Z-2.555. The actual clearance above the diameter and in front of the part will be the usual .100.

```
O3501
(G90 STRAIGHT TURNING CYCLE - ABSOLUTE)
N1 G20
N2 T0100 M41
N3 G96 S450 M03
N4 G00 X4.325 Z0.1 T0101 M08       (START POINT)
N5 G90 X3.8175 Z-2.555 F0.01             (PASS 1)
N6 X3.51                                 (PASS 2)
N7 X3.2025                               (PASS 3)
N8 X2.895                                (PASS 4)
N9 X2.5875                               (PASS 5)
```

```
N10 X2.28                                (PASS 6)
N11 G00 X10.0 Z2.0 T0100 M09
N12 M01                          (END OF ROUGHING)
```

If preferred, use the incremental programming method. However, it is easier to trace the program progress with the absolute coordinates than the incremental distances. However, here is the incremental version:

```
O3502
(G90 STRAIGHT TURNING CYCLE - INCREMENTAL)
N1 G20
N2 T0100 M41
N3 G96 S450 M03
N4 G00 X4.325 Z0.1 T0101 M08       (START POINT)
N5 G90 U-0.5075 W-2.655 F0.01            (PASS 1)
N6 U-0.3075                              (PASS 2)
N7 U-0.3075                              (PASS 3)
N8 U-0.3075                              (PASS 4)
N9 U-0.3075                              (PASS 5)
N10 U-0.3075                             (PASS 6)
N11 G00 X10.0 Z2.0 T0100 M09
N12 M01                          (END OF ROUGHING)
```

This cycle is quite simple in both versions - all that is needed is to calculate the new diameter for each roughing cut. If the same roughing tool path had been programmed using the block-by-block method (without G90), the final program would be more than three times longer.

◆ Taper Cutting Example

The *Figure 35-4* is a drawing similar to that used for the previous example. In this example, a taper will be cut, also using the G90 simple cycle.

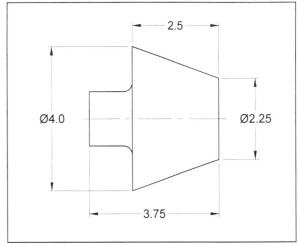

Figure 35-4

Example of G90 cycle in taper cutting - program O3503

In order to distinguish between the straight cutting and the taper cutting methods, using the same G90 cycle, there must be a way to distinguish these two kinds of cut, and there is one indeed.

The difference is the addition of an I parameter to the cycle call, indicating the *taper amount* and its *direction per side*. This value is called *a signed radius value*. It is an I value because of its association with the X axis. For straight cutting, the I value will always be zero and does not have to be written in the program. Its only significance is for taper cutting, in which case it has a non-zero value - *Figure 35-5*.

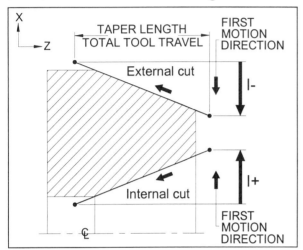

Figure 35-5

The I amount used for G90 turning cycle - external and internal

The illustration shows that the I amount is calculated as a single distance, *i.e.,* as per single side (a radius value), with specified direction, based on the *total* traveled distance and the direction of the first motion from the start position.

There are two simple rules for G90 taper cutting:

❑ If the direction of the first tool motion in X is negative, the I value is negative

❑ If the direction of the first tool motion in X is positive, the I value is positive

On a CNC lathe with the X axis positive direction *above* the spindle center line, the typical I value will be *negative for external* taper cutting (turning) and *positive for internal* taper cutting (boring).

To program the part in *Figure 35-4*, keep in mind that the illustration represents the finished item and does not contain any clearances. Always add all necessary clearances first, then calculate the I amount.

In the example, a clearance of 0.100 will be added at each end of the taper, increasing its length along the axis from 2.5 to 2.7. The I amount calculation requires the *actual length* of tool travel, while maintaining the taper angle at the same time. Either the method of similar triangles or the trigonometric method can be used for such calculation (see *Chapter 52* for details on shop mathematics). *Figure 35-6* and *Figure 35-7* illustrate the details of the known and unknown values for the I amount calculation.

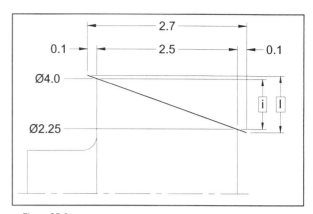

Figure 35-6

Known and unknown values for taper cutting - program O3503 Amount 'i' is known, amount 'I' has to be calculated

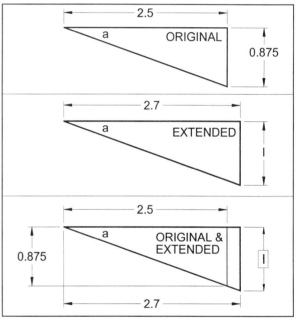

Figure 35-7

The I distance calculation using the similar triangles method

The example shown above almost suggests the simplest method of calculation, a method that is known in mathematics as the *law of similar triangles*. This law has several possible definitions, and the one that applies here is that ...

Two triangles are similar, if the corresponding sides of the two triangles are proportional.

In programming, quite often there is a situation that can be solved by more than one method. Choose the one that suits better a certain programming style, then try the other method, expecting the same result. Both methods will be used here, to confirm the accuracy of the calculation.

➲ Using Similar Triangles Method

First, calculate the difference i between the two known diameters, as per drawing:

```
i = (4 - 2.25) / 2 = 0.875
```

therefore, the ratio of similar triangles will be

```
I / 2.7 = i / 2.5
```

We know i to be 0.875, so the relations can be modified by filling in the known amount:

```
I / 2.7 = 0.875 / 2.5
I = (0.875 × 2.7) / 2.5
I = 0.945        ... is the required amount for programming
```

➲ Using Trigonometric Method

The second method of calculating the I amount requires trigonometry. At this point, it is known that

```
I = 2.7 × tan a
```

and the tangent value has to be calculated first:

```
tan a = i / 2.5
tan a = 0.875 / 2.5
tan a = 0.350
```

The amount of I can be calculated using the result:

```
I = 2.7 × 0.35
I = 0.945        ... is the required amount for programming
```

In both cases, the calculations have the same result, confirming accuracy of the process. The I amount calculation is shown in *Figure 35-6* and detailed in *Figure 35-7*. Program O3503 is the final result - five cuts with 0.03 X-stock left:

```
O3503
(G90 TAPER TURNING EXAMPLE 1 - W/0.03 X-STOCK)
N1 G20
N2 T0100 M41
N3 G96 S450 M03
N4 G00 X4.2 Z0.1 T0101 M08        (START)
N5 G90 X3.752 Z-2.6 I-0.945 F0.01   (1)
N6 X3.374                            (2)
N7 X2.996                            (3)
N8 X2.618                            (4)
N9 X2.24                             (5)
N10 G00 X10.0 Z2.0 T0100 M09     (CLEAR POS.)
N11 M01                          (END OF ROUGHING)
```

◆ Straight and Taper Cutting Example

Another variation of a taper is also common in CNC programming. The *Figure 35-8* shows another simple drawing, this time with a taper *and* a shoulder.

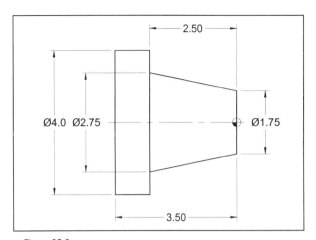

Figure 35-8

Example of G90 cycle used on a taper to a shoulder - O3504

Using the simple cycle G90, the machining requires a tapered cut towards a straight shoulder. A single G90 cycle can be used in this case as well, but could result in some *excessive* or *insufficient* cutting (too much or too little stock). The best approach is to use *two modes* of the cycle - one for the straight roughing, the other for tapered roughing.

Similar to the previous example, the I taper amount has to be calculated, using the same law of similar triangles as before. The height i of the original triangle over the length of 2.5 is calculated as one half of the difference between the $\varnothing2.750$ and the $\varnothing1.750$:

```
i = 2.75 - 1.75 / 2
i = 0.500
```

For the extended taper length, 0.005 stock amount is left at the shoulder for finishing and the taper is extended by 0.100 at the front face, for the total taper length of 2.595:

```
2.5 - 0.005 + 0.100 = 2.595
```

The I amount can now be calculated, based on the original and the extended values:

```
I / 2.595 = 0.500 / 2.5
I = (0.500 × 2.595) / 2.5
I = 0.519              ... negative direction
```

For roughing, a 0.030 stock will be left per side along the X axis, which is 0.060 on diameter.

In roughing operations, it is always important to select a suitable depth of cut, with safety in mind, as well as the cutting conditions. In this example, the depth of cut selection will benefit from one simple programming technique. If the depth of cut is selected arbitrarily, the *last* depth will be whatever is left to cut. A better way is to select a calculated *number of equal* cuts - *Figure 35-9*.

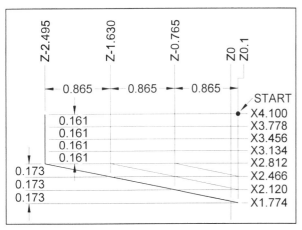

Figure 35-9

Depth of cut calculation for program example O3504

For the calculation, all that is required is to divide the distance per each side by the number of required cuts. The result will be an equal depth of cut for the whole roughing operation. If the cutting depth is too small or too large, just recalculate it with a different number of cuts. Knowing what is a suitable depth of cut is a machining knowledge, expected from CNC programmers.

In *Figure 35-9*, there are four cuts of .161 for the straight roughing and three cuts of .173 for the tapered cutting. All stock allowances are in effect.

The program O3504 will use the calculations:

```
O3504
(G90 TAPER TURNING EXAMPLE - 2)
N1 G20
N2 T0100 M41
N3 G96 S450 M03
N4 G00 X4.1 Z0.1 T0101 M08                (START)
N5 G90 X3.778 Z-2.495 F0.01       (STRAIGHT 1)
N6 X3.456                          (STRAIGHT 2)
N7 X3.134                          (STRAIGHT 3)
N8 X2.812                          (STRAIGHT 4)
N9 G00 X3.0       (CHANGE STRAIGHT TO TAPERED)
N10 G90 X2.812 Z-0.765 I-0.173      (TAPERED 1)
N11 Z-1.63 I-0.346                  (TAPERED 2)
N12 Z-2.495 I-0.519        (TAPERED 3 - FINAL)
N13 G00 X10.0 Z2.0 T0100 M09        (CLEAR POS.)
N14 M01                        (END OF ROUGHING)
```

In a review, to calculate the amount of I or R parameter used in G90 for the taper cutting - *external* or *internal,* use the following formula:

$$I(R) = \frac{SMALLER\ DIA - LARGER\ DIA}{2}$$

The result will also include the *sign of the I amount.*

G94 - FACE CUTTING CYCLE

A cycle that is very similar to G90 is another simple turning cycle, programmed with the G94 command. This cycle is called the *face cutting cycle.* The purpose of G94 cycle is to remove excessive stock between the start position of the cutting tool and the coordinates specified by the X and Z axes. The resulting cut is a straight turning cut, *normally perpendicular* to the spindle center line. In this cycle, it is the X axis that is the main cutting direction. The G94 cycle is used primarily for facing cuts and can be used for simple vertical taper cutting as well, similar to the G90 cycle.

> The G94 cycle is logically identical to the G90 cycle,
> except the emphasis is on the X axis cutting,
> rather than the Z axis cutting.

As the cycle description suggests, the G94 is normally used to perform a rough face-off of the part, towards the spindle center line or to face-off a shoulder.

◆ Cycle Format

Similar to all cycle, the face cutting cycle G94 also has a predetermined programming format. For straight facing, the cycle format is:

> G94 X(U).. Z(W).. F..

For tapered turning, the cycle format is:

> G94 X(U).. Z(W).. K.. F..

The axes X and Z are used for absolute programming, the axes U and W are used for incremental programming, and the F address is the cutting feedrate. The K parameter, if greater than zero, is used for taper cutting along the vertical direction. *Figure 35-10* shows all programming parameters and cutting steps. Apply the same process as for G90 cycle.

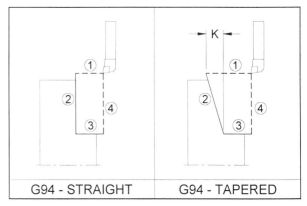

Figure 35-10

G94 turning cycle structure - straight and tapered application

MULTIPLE REPETITIVE CYCLES

Unlike the fixed cycles for various drilling operations on machining centers, or the G90 and G94 simple cycles for turning, the advanced cycles for CNC lathe work are much more sophisticated. The major and most distinctive feature of these cycles is their departure from the repeated order of operations. Lathe work can be very complex and the modern control systems do reflect that need. Not only straight or tapered cuts can be programmed, but also radii, chamfers, grooves, undercuts, etc, simply, several of these cycles are used for contouring. Tool nose radius offset may also be applied, if applicable to the job.

Multiple repetitive cycles, as these cycles are called, require a computer memory in order to be useful, so the old NC machines controlled by a punched tape, *cannot* benefit from them. In tape operation, the control unit reads the tape codes sequentially, in a forward direction only. A CNC control, on the other hand, is far more complex. It can read, evaluate and process information stored in the memory in both directions, forwards and backwards - at all times. It can process mathematical instructions internally in a split of a second, simplifying the programming effort.

◆ General Description

In total, there are seven multiple repetitive cycles available, identified by a preparatory command address G:

Profile cutting cycles - Roughing:

G71	Rough cutting cycle - *Horizontal emphasis*
G72	Rough cutting cycle - *Vertical emphasis*
G73	Pattern repeating roughing cycle

Profile cutting cycles - Finishing:

G70	Finishing cycle for G71, G72 and G73

Chipbreaking cycles:

G74	Peck drilling cycle	in Z axis - horizontal
G75	Peck grooving cycle	in X axis - vertical

Threading cycle:

G76	Threading cycle - straight or tapered

The G76 threading cycle is described separately and in sufficient detail in *Chapter 38*.

◆ Cycle Format Types

Each cycle is governed by very specific rules and has its *do's* and *don'ts*. The following sections describe each of them in detail, except the G76 threading cycle, which will be covered separately in *Chapter 38*.

An important fact to take a note of, is that the format of programming for these cycles, the method of data input, is *different* for the lower level of Fanuc controls, such as the very popular 0T or the 16/18/20/21T series, than for the higher level, such as the 10/11T or the 15T series. These cycles, if they are available for the lower level controls, require their programming format in *two blocks*, not the normal *one block*. Check the parameter settings for each control, to find about compatibility issues. Description of both formats is also included in this chapter.

◆ Cutting Cycles and Part Contour

Probably the most common multiple repetitive cycles in turning and boring are those that are used for *profile cutting* or *contour cutting*. There are three cycles available within the roughing category:

❑ G71, G72 and G73

and one cycle is available for finishing:

❑ G70

The finishing cycle is designed to finish profile generated by *any* one of the three roughing cycles.

In some respects, there is an interesting situation in programming multiple repetitive cycles. So far, the emphasis was to program roughing cuts *before* finishing cuts. This approach makes perfect sense - it is also the only logical way from the *technological* point of view. Don't be surprised if this 'rule' is suddenly broken when computer calculations take over. Yes, the implication here is that when programming the three multiple repetitive roughing cycles, the *finished contour must always be defined first*, then its machining specifications can be applied to the roughing cycle. Sounds strange? At first, perhaps. When working with these cycles longer, it will be easy to see that it is actually quite a clever and ingenious method, although hardly a recent breakthrough.

◆ Chipbreaking Cycles

The two remaining 'chipbreaking' cycles are designed to produce an *interrupted cut*, either along the Z axis (G75), or along the X axis (G74). In practice, the G74 cycle offers more practical applications than G75. The G74 cycle allows to peck-drill on a CNC lathe. Although the need for peck drilling is much lower on a lathe than on a machining center, it is not that rare. However, the G75 cycle, which is really a 'peck-grooving' cycle, is used rarely, as *it does not produce a precision groove*.

CONTOUR CUTTING CYCLES

By far, the profile cutting cycles (contouring cycles), are the most common cycles in CNC lathe programming. They are used for external (turning) and internal (boring) material removal, along almost any machinable contour.

◆ Boundary Definition

The roughing cycles are based on the definition of *two* boundaries, typically called the *material* boundary, which is the outline of the blank, and the *part* boundary, which is the outline of the part contour. This is not a new concept at all, several early programming languages were using this method, such as the Compact II, a very popular language based programming system of the 1970's.

The two defined boundaries create a fully enclosed area that defines the excessive material. From this isolated area, the material is removed in an orderly way, following specified machining parameters in the cycle call block or blocks. Mathematically, the minimum number of points that can define an area is three. These three points must be nonlinear (meaning not on the same line). *Figure 35-11* shows a simple boundary with only three points and a boundary consisting of many points.

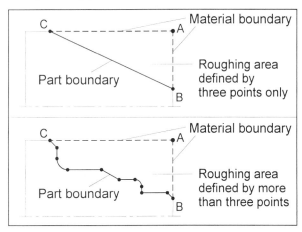

Figure 35-11
Material and part boundaries as applied to turning

In the profile cutting cycles, each point represents a tool position and the points A, B, and C represent the extreme corners of the selected (defined) machining area.

The *material boundary* is not actually defined, it is only implied. It is between points A and B, and points A and C. Material boundary *can not contain* any other points; it must be a straight line, but not always a line parallel to an axis.

The *part boundary* is defined between points B and C, and *may have* any number of points between. For CNC programming, different descriptions will be used rather than the generic ABC points in *Figure 35-11*.

◆ Start Point and the Points P and Q

The point A in the illustration is the *start point* of any profile cutting cycle. It can be defined in simple terms:

> The start point is defined as the last XZ coordinate location of the tool, before the profile cutting cycle is called.

Typically, this start point will be *closest* to the part corner where the rough cutting begins. It is important to select the start point very carefully, because it is more than 'just a start point'. In fact, this special point controls all approach clearances and the actual depth of the *first* roughing cut.

The generic points B and C in the last illustration will become points P and Q in the program, respectively:

> Point P represents the block number of the first XZ coordinate of the finished contour.

> Point Q represents the block number of the last XZ coordinate of the finished contour.

Other in-depth considerations relating to the P and Q boundary points are equally important, and there are quite a few of them:

❑ A number of points may be defined between the P and Q points, representing the XZ coordinates of the finished contour. The contour is programmed using G01, G02, and G03 tool motions, including feedrates.

❑ The material removal defined by the starting point and the P-Q contour must include all necessary clearances.

❑ Tool nose radius offset should not be included between the P and Q points, but programmed before the cycle is called, usually during the motion to the start point.

❑ For roughing, the material to be machined will be divided into a series of machinable cuts. Each roughing cycle accepts a number of user supplied cutting parameters.

❑ For safety reasons, the diameter of the start point should be *above* the stock diameter for external cutting, and *below* the core diameter for internal cutting.

❑ The tool motion between the P point and the Q point must be steadily increasing for external cutting, or steadily decreasing for internal cutting.

❑ Any change in direction between P and Q points is allowed only if Type II cycle is available *and* programmed, and then in one direction only - see the next section for details.

❑ Blocks representing the first XZ coordinate of the contour P, and the last XZ coordinate of the contour Q, must have a sequence number N, not duplicated anywhere else in the program.

TYPE I AND TYPE II CYCLES

In the initial versions of the contour cutting cycles, a change of the contouring direction into the *opposite* direction along one axis was not allowed. That limited these cycles to some extent, because common undercuts or recesses were not possible to use in the program, yet they were common in machine shops.

Presently, this older method is called *Type I* repetitive cycles. The modern controls use many more advanced software features and the change in a *single* direction is now allowed. This newer method is now called *Type II*, allowing more programming flexibility when cutting recesses and cavities (undercuts). *Figure 35-12* compares the two types and shows a disallowed contour change in two directions within a cycle. The example applies to G71 external cutting cycle, but can be modified for any internal cutting.

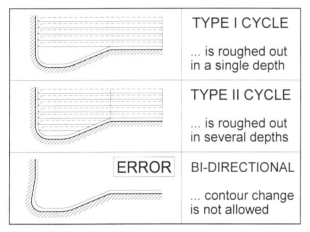

	TYPE I CYCLE ... is roughed out in a single depth
	TYPE II CYCLE ... is roughed out in several depths
ERROR	BI-DIRECTIONAL ... contour change is not allowed

Figure 35-12

Comparison of Type I and Type II cycles -
- bi-directional change along two axes is not allowed

Type I allows a steadily *increasing* profile (for external cutting) or steadily *decreasing* profile (for internal cutting) from the point P to the point Q (typical cutting directions). On older controls, opposite X or Z direction is not allowed. Modern controls do allow an undercut to be machined with *Type I*, but the cutting will be done with a single pass. That may cause some heavy metal removal in certain cases. Make sure to know exactly which type the control system supports. Some experimentation may be necessary.

Type II allows a continually increasing profile or continually decreasing profile from the point P to the point Q. A change into the opposite direction is allowed for a *single axis only*, depending on the active cycle. Rough out process of an undercut will be a multiple tool path. The selection of *Type I* or *Type II* is applicable to the cycle, by programming *both axes* in the block represented by the P point. This is typically the block immediately following the cycle call in the program (after G71, G72, etc.).

◆ Programming Type I and Type II Cycles

If the control system supports the *Type II* metal removal in turning and boring cycles, it also supports the *Type I*, if it needs to be used for some special applications. That means, Fanuc has not replaced one type for another, it has added the *Type II*. Of course, the question is - how to distinguish between the two types in the program? The key to the type selection is in the contents of the block that *immediately follows the cycle call*:

❏   Type I   ... only one axis is specified

❏   Type II   ... two axes are specified

➲ Example - Type I :

```
G71 U.. R..
G71 P10 Q.. U.. W.. F.. S..
N10 G00 X..            (ONE AXIS FOR TYPE I)
...
```

➲ Example - Type II :

```
G71 U.. R..
G71 P10 Q.. U.. W.. F.. S..
N10 G00 X.. Z..        (TWO AXES FOR TYPE II)
...
```

If there is no motion along the Z axis in the first block after the cycle call and the *Type II* cycle is still required, just program W0 as the second axis.

◆ Cycle Formatting

On the next few pages is a description of the six turning cycles, covered in detail. It is important to understand the format of each cycle as it applies to a particular control. Several Fanuc control models are available and for the purposes of programming these multiple repetitive cycles, they can be separated into two groups:

❏   Fanuc system 0T, 16T, 18T, 20T, 21T   ... *lower level*

❏   Fanuc system 10T, 11T, 15T   ... *higher level*

Practically, it only means a change in the way the cycle is programmed, but the subject is also important for solving some incompatibility problems. Note that the tool function T is not specified in any of the examples, although it is also allowed as a parameter in all multiple repetitive cycles. Its only need maybe for a tool offset change.

G71 - STOCK REMOVAL IN TURNING

The most common roughing cycle is G71. Its purpose is to remove stock by *horizontal* cutting, primarily along the Z axis, typically from the right to the left. It is used for roughing out material out of a solid cylinder. Like all cycles, it comes in two formats - a one-block and a double block format, depending on the control system.

◆ G71 Cycle Format - 10T/11T/15T

The one-block format for the G71 cycle is:

> G71 P.. Q.. I.. K.. U.. W.. D.. F.. S..

☞ where ...

P = The first block number of the finishing profile
Q = The last block number of the finishing profile
I = Distance and direction of rough
    semifinishing in the X axis - per side
K = Distance and direction of rough
    semifinishing in the Z axis
U = Stock amount for finishing on the X axis diameter
W = Stock left for finishing on the Z axis
D = The depth of roughing cut
F = Cutting feedrate (in/rev or mm/rev) overrides
    feedrates between the P block and the Q block
S = Spindle speed (ft/min or m/min) overrides spindle
    speeds between the P block and the Q block

The I and K parameters are not available on all machines. They control the amount of cut for semifinishing, the last continuous cut before final roughing motions.

◆ G71 Cycle Format - 0T/16T/18T/20T/21T

If the control requires a double block entry for the G71 cycle, the programming format is:

> G71 U.. R..
> G71 P.. Q.. U.. W.. F.. S..

☞ where ...

First block:

U = The depth of roughing cut
R = Amount of retract from each cut

Second block:

P = The first block number of the finishing profile
Q = The last block number of the finishing profile
U = Stock amount for finishing on the X axis diameter
W = Stock left for finishing on the Z axis
F = Cutting feedrate (in/rev or mm/rev) overrides
    feedrates between the P block and the Q block
S = Spindle speed (ft/min or m/min) overrides spindle
    speeds between the P block and the Q block

Do not confuse the U in the first block, depth of cut per side, and the U in the second block, stock left on diameter. The I and K parameters may be used only on some controls and the retract amount R is set by a system parameter.

The external and internal use of the G71 cycle will use the drawing data in *Figure 35-13*.

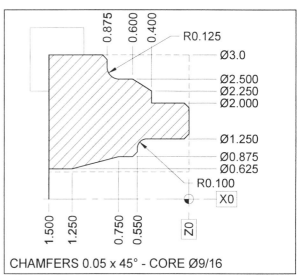

CHAMFERS 0.05 x 45° - CORE Ø9/16

Figure 35-13

Drawing example to illustrate G71 roughing cycle - program O3505

◆ G71 for External Roughing

The stock material in the example has an existing hole of Ø9/16 (.5625). For external cutting of this part, a standard 80° tool will be used for a single cut on the face, as well as for roughing the outer shape.

Program O3505 covers these operations.

```
O3505     (G71 ROUGHING CYCLE - ROUGHING ONLY)
N1 G20
N2 T0100 M41          (OD ROUGHING TOOL + GEAR)
N3 G96 S450 M03       (SPEED FOR ROUGH TURNING)
N4 G00 G41 X3.2 Z0 T0101 M08  (START FOR FACE)
N5 G01 X0.36              (END OF FACE DIA)
N6 G00 Z0.1               (CLEAR OFF FACE)
N7 G42 X3.1          (START POSITION FOR CYCLE)
N8 G71 P9 Q17 U0.06 W0.004 D1250 F0.014
N9 G00 X1.7          (P POINT = START OF CONTOUR)
N10 G01 X2.0 Z-0.05 F0.005
N11 Z-0.4 F0.01
N12 X2.25
N13 X2.5 Z-0.6
N14 Z-0.875 R0.125
N15 X2.9
N16 G01 X3.05 Z-0.95
N17 U0.2 F0.02       (Q POINT = END OF CONTOUR)
N18 G00 G40 X5.0 Z6.0 T0100
N19 M01
```

The external roughing has been completed at this point in the program and the internal roughing can be programmed for the next tool. In all examples that include a tool change between a short tool (such as a turning tool) and a long tool (such as a boring bar), it is important to move the short tool further from the front face. The motion should be far enough to accommodate the incoming long tool. The clearance is 6.0 in the above example (block N18 with Z6.0).

◆ G71 for Internal Roughing

The face has been done with the previous tool and the roughing boring bar can continue the machining:

```
N20 T0300              (ID ROUGHING TOOL)
N21 G96 S400 M03    (SPEED FOR ROUGH BORING)
N22 G00 G41 X0.5 Z0.1 T0303 M08   (START POS.)
N23 G71 P24 Q31 U-0.06 W0.004 D1000 F0.012
N24 G00 X1.55     (P POINT = START OF CONTOUR)
N25 G01 X1.25 Z-0.05 F0.004
N26 Z-0.55 R-0.1 F0.008
N27 X0.875 K-0.05
N28 Z-0.75
N29 X0.625 Z-1.25
N30 Z-1.55
N31 U-0.2 F0.02      (Q POINT = END OF CONTOUR)
N32 G00 G40 X5.0 Z2.0 T0300
N33 M01
```

The part has been completely roughed out, leaving only the required stock on diameters and faces or shoulders. Finishing with the G70 cycle, described later, is possible with the same tool, if tolerances and/or surface finish are not too critical. Otherwise, another tool or tools will be required in the same program, after a tool change.

At this stage, evaluate what has been done and why. Many principles that applied to the example are very common to other operations that also use the multiple repetitive cycles. It is important to learn them well at this point.

◆ Direction of Cutting in G71

The last programming example O3505, shows that G71 can be used for roughing *externally* or *internally*. There are two important differences:

❑ Start point relative to the P point (SP to P versus P to SP)

❑ Sign of the U address for stock allowance on diameter

The control system will process the cycle for *external* cutting, if the X direction from the start point SP to the point P is *negative*. In the example, the X start point is X3.1, the P point is X1.7. The X direction is *negative* or decreasing and an *external* cutting will take place.

The control system will process the cycle for *internal* cutting, if the X direction from start point SP to the point P is *positive*. In the example, the X start point is X0.5, the P point is X1.55. The X direction is *positive* or increasing, and an *internal* cutting will take place.

Figure 35-14 illustrates the concept of G71 cycle, as applied to both, external and internal cutting.

By the way, although the *sign* of the stock U value is very important for the final size of the part, it does *not* determine the mode of cutting. This concludes the section relating to the G71 multiple repetitive cycle. The face roughing cycle G72 is similar, and is described next.

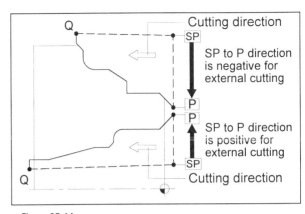

Figure 35-14
External and internal cutting in G71 cycle

G72 - STOCK REMOVAL IN FACING

The G72 cycle is identical in every respect to the G71 cycle, except the stock is removed mainly by *vertical* cutting (facing), typically from the large diameter towards the spindle center line X0. It is used for roughing of a solid cylinder, using a series of *vertical* cuts (face cuts). Like all other cycles in this group, it comes in two formats - a one block and a double block format, depending on the control system. Compare G72 with the G71 structure on examples in this chapter.

◆ G72 Cycle Format - 10T/11T/15T

The one-block programming format for the G72 cycle is:

> G72 P.. Q.. I.. K.. U.. W.. D.. F.. S..

☞ where ...

P	=	The first block number of the finishing profile
Q	=	The last block number of the finishing profile
I	=	Distance and direction of rough semifinishing in the X axis - per side
K	=	Distance and direction of rough semifinishing in the Z axis
U	=	Stock amount for finishing on the X axis diameter
W	=	Stock left for finishing on the Z axis
D	=	The depth of roughing cut
F	=	Cutting feedrate (in/rev or mm/rev) overrides feedrates between the P block and the Q block
S	=	Spindle speed (ft/min or m/min) overrides spindle speeds between the P block and the Q block

The meaning of each address is the same as for the G71 cycle. The I and K parameters are not available on all machines. These parameters control the amount of cut for semifinishing, which is the last continuous cut before final roughing motions are completed.

◆ G72 Cycle Format - 0T/16T/18T/20T/21T

If the control system requires a double block entry for the G72 cycle, the programming format is:

> G72 W.. R..
> G72 P.. Q.. U.. W.. F.. S..

☞ where ...

First block:

W = The depth of roughing cut
R = Amount of retract from each cut

Second block:

P = The first block number of the finishing profile
Q = The last block number of the finishing profile
U = Stock amount for finishing on the X axis diameter
W = Stock left for finishing on the Z axis
F = Cutting feedrate (in/rev or mm/rev) overrides feedrates between the P block and the Q block
S = Spindle speed (ft/min or m/min) overrides spindle speeds between the P block and the Q block

In the G71 cycle for the double block definition, there were two addresses U. In the G72 double block definition cycle, there are two addresses W. Make sure you do not confuse the W in the first block - depth of cut (actually it is a *width* of cut), and the W in the second block - stock left on faces. The I and K parameters may be available, depending on the control.

An example program O3506 for the G72 cycle uses the drawing data in *Figure 35-15*.

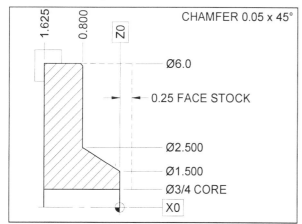

Figure 35-15

Drawing example to illustrate G72 roughing cycle - program O3506

In this facing application, all the main data will be reversed by 90°, because the cut will be segmented along the X axis. Roughing program using the G72 cycle is logically similar to the G71 cycle:

```
O3506 (G72 ROUGHING CYCLE - ROUGHING ONLY)
N1 G20
N2 T0100 M41           (OD FACING TOOL + GEAR)
N3 G96 S450 M03        (SPEED FOR ROUGH FACING)
N4 G00 G41 X6.25 Z0.3 T0101 M08   (START POS.)
N5 G72 P6 Q12 U0.06 W0.03 D1250 F0.014
N6 G00 Z-0.875    (P-POINT = START OF CONTOUR)
N7 G01 X6.05 F0.02
N8 X5.9 Z-0.8 F0.008
N9 X2.5
N10 X1.5 Z0
N11 X0.55
N12 W0.1 F0.02        (Q-POINT = END OF CONTOUR)
N13 G00 G40 X8.0 Z3.0 T0100
N14 M01
```

The concept of G72 cycle is illustrated in *Figure 35-16*. Note the position of the point P as it relates to the start point SP and compare it with the G71 cycle.

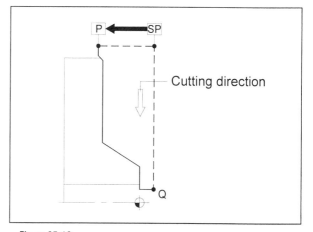

Figure 35-16

Basic concept of G72 multiple repetitive cycle

G73 - PATTERN REPEATING CYCLE

The pattern repeating cycle is also called the *Closed Loop* or a *Profile Copying* cycle. Its purpose is to minimize the cutting time for roughing material of irregular shapes and forms, for example, forgings and castings.

◆ G73 Cycle Format - 10T/11T/15T

The one-block programming format for G73 cycle is similar to the G71 and G72 cycles:

> G73 P.. Q.. I.. K.. U.. W.. D.. F.. S..

☞ where ...

P = The first block number of the finishing profile
Q = The last block number of the finishing profile
I = X axis distance and direction of relief - per side
K = Z axis distance and direction of relief
U = Stock amount for finishing on the X axis diameter

W = Stock left for finishing on the Z axis
D = The number of cutting divisions
F = Cutting feedrate (in/rev or mm/rev) overrides
       feedrates between the P block and the Q block
S = Spindle speed (ft/min or m/min) overrides spindle
       speeds between the P block and the Q block

◆ G73 Cycle Format - 0T/16T/18T/20T/21T

If your control system requires a double block entry for the G73 cycle, the programming format is:

```
G73 U.. W.. R..
G73 P.. Q.. U.. W.. F.. S..
```

☞ where ...

First block:

U = X axis distance and direction of relief - per side
W = Z axis distance and direction of relief
R = The number of cutting divisions

Second block:

P = The first block number of the finishing profile
Q = The last block number of the finishing profile
U = Stock amount for finishing on the X axis diameter
W = Stock left for finishing on the Z axis
F = Cutting feedrate (in/rev or mm/rev) overrides
       feedrates between the P block and the Q block
S = Spindle speed (ft/min or m/min) overrides spindle
       speeds between the P block and the Q block

In the two-block cycle entries, do not mix up addresses in the first block that repeat in the second block (U and W in the G73 example). *They have a different meaning!*

◆ G73 Example of Pattern Repeating

Pattern repeating cycle G73 program example uses the drawing in *Figure 35-17*.

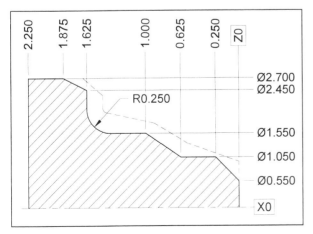

Figure 35-17

Pattern repeating cycle G73 - program example O3507

There are three important input parameters in the G73 cycle - I, K and D. One parameter seems to be missing - *there is no depth of cut specification!* In the G73 cycle, it is not needed. The actual depth of cut is calculated automatically, based on these three parameters:

❑ I ... amount of rough material to remove in the X axis

❑ K ... amount of rough material to remove in the Z axis

❑ D ... number of cutting divisions or number of repeats

Use this cycle with care - its design assumes an *equal amount* of rough stock to be removed along both the X and the Z axes. That is not the typical reality for forgings and castings, where the stock *varies* all over the material - see the illustration in *Figure 37-17*. The cycle can still be used with a reasonable efficiency, but some 'air' cutting may be an unwanted side effect for odd shaped parts.

In the example, the *largest expected* material amount per side will be chosen as .200 (I0.2) and the *largest expected* material amount on the face as .300 (K0.3). The number of divisions could be either two or three, so the program will use D3. Some modification on the control may be necessary during actual setup or machining, depending on the exact condition and sizes of the casting or forging.

This cycle is suitable for roughing contours where the finish contour closely matches the contour of the casting or forging. Even if there is some 'air' cutting, this cycle may be more efficient than the selection of the G71 or G72 cycles. The program O3507 shows roughing and finishing with the same tool (as an example):

```
O3507 (G73 PATTERN REPEATING CYCLE)
N1 G20 M42
N2 T0100
N3 G96 S350 M03
N4 G00 G42 X3.0 Z0.1 T0101 M08
N5 G73 P6 Q13 I0.2 K0.3 U0.06 W0.004 D3 F0.01
N6 G00 X0.35
N7 G01 X1.05 Z-0.25
N8 Z-0.625
N9 X1.55 Z-1.0
N10 Z-1.625 R0.25
N11 X2.45
N12 X2.75 Z-1.95
N13 U0.2 F0.02
N14 G70 P6 Q13 F0.006
N15 G00 G40 X5.0 Z2.0 T0100
N16 M30
%
```

The number of passes D3 may be necessary to accommodate some rotating eccentricity, normally associated with castings and forgings. On the other hand, D2 may be necessary for the heavier cut, to 'bite under the skin' of the material, for better tool life. Schematically, *Figure 35-18* shows three programmed cutting divisions.

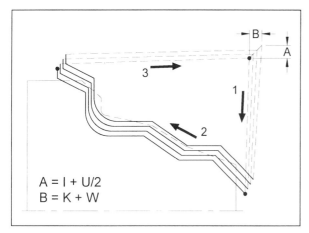

$A = I + U/2$
$B = K + W$

Figure 35-18
Schematic representation of G73 cycle

Note that the pattern repeating cycle does exactly that - it repeats the machining contour (pattern) specified between the P and Q points. Each individual tool path is offset by a calculated amount along the X and Z axes. On the machine, watch the progress with care - particularly for the first tool path. Feedrate override may come useful here.

G70 - CONTOUR FINISHING CYCLE

The last of the contouring cycles is G70. Although it has a smaller G number than any of the three roughing cycles G71, G72 and G73, the finishing cycle G70 is normally used *after* any one of these three roughing cycles. As its description suggests, it is strictly used *for the finishing cut of a previously defined contour.*

♦ G70 Cycle Format - All Controls

For this cycle, there is no difference in the programming format for various controls - it is all the same, and the cycle call is a one-block command.

The programming format for G70 cycle is:

> G70 P.. Q.. F.. S..

☞ where ...

    P = The first block number of the finishing profile
    Q = The last block number of the finishing profile
    F = Cutting feedrate (in/rev or mm/rev)
    S = Spindle speed (ft/min or m/min)

The cycle G70 accepts a previously defined finishing contour from either of the three roughing cycles, already described. This finishing contour is defined by the P and the Q points of the respective cycles, and is normally repeated in the G70 cycle, although it can change.

> For safety, use the same start point for G70
> as for the roughing cycles.

The earlier roughing program O3505, using the G71 repetitive cycle for rough turning and rough boring, can be completed by using another two tools, one for external, one for internal finishing tool path:

```
(O3505 CONTINUED ...)
...
N34 T0500 M42         (OD FINISHING TOOL + GEAR)
N35 G96 S530 M03     (SPEED FOR FINISH TURNING)
N36 G42 X3.1 Z0.1 T0505 M08       (START POS.)
N37 G70 P9 Q17            (FINISHING CYCLE - OD)
N38 G00 G40 X5.0 Z6.0 T0500
N39 M01

N40 T0700                   (ID FINISHING TOOL)
N41 G96 S475 M03       (SPEED FOR ROUGH BORING)
N42 G00 G41 X0.5 Z0.1 T0707 M08    (START POS.)
N43 G70 P24 Q31           (FINISHING CYCLE - ID)
N44 G00 G40 X5.0 Z2.0 T0700
N45 M30                      (END OF PROGRAM)
%
```

Even for the external finishing, the cutting tool is still programmed to start *above* the original stock diameter and off the front face, although all roughing motions have already been completed. A similar approach applies to the internal cut. For safety reasons, this is a recommend practice.

There are no feedrates programmed for the G70 cycle, although the cycle format accepts a feedrate. The defined block segments P to Q for the roughing tool already include feedrates. These programmed feedrates will be ignored in the roughing mode and will become active only for the G70 cycle, during finishing. If the finish contour did not include any feedrates, then program a *common feedrate* for finishing all contours during the G70 cycle processing. For example, program block

```
N37 G70 P9 Q17 F0.007
```

will be a waste of time, since the .007 in/rev feedrate will *never* be used. It will be overridden by the feedrate defined between blocks N9 and N17 of program O3505). On the other hand, if there is no feedrate programmed for the finishing contour at all, then

```
N.. G70 P.. Q.. F0.007
```

will use .007 in/rev exclusively for the finishing tool path.

The same logic described for G71 cycle, applies equally to the G72 cycle. The roughing program O3506, using the G72 cycle for rough turning of the part face, can be completed by using another external tool for finishing cuts using the G70 cycle:

```
(O3506 CONTINUED ...)
...
N15 T0500 M42          (OD FACING TOOL + GEAR)
N16 G96 S500 M03       (SPEED FOR FINISH FACING)
N17 G00 G41 X6.25 Z0.3 T0505 M08   (START POS.)
N18 G70 P6 Q12             (FINISHING CYCLE)
N19 G00 G40 X8.0 Z3.0 T0500
N20 M30
%
```

The rules mentioned earlier also apply for the contour finishing defined by the G72 cycle. Program O3507, using the G73 cycle, can be also be programmed by using another external tool for finishing, applying the same rules.

BASIC RULES FOR G70-G73 CYCLES

In order to make the multiple repetitive stock removal cycles (contouring cycles) work properly and efficiently, observing the rules of their use is very important. Often a small oversight may cause a lengthy delay.

Here are the most important rules and observations:

❑ Always apply tool nose radius offset
   before the stock removal cycle is called

❑ Always cancel tool nose radius offset
   after the stock removal cycle is completed

❑ Return motion to the start point is automatic,
   and must not be programmed

❑ The P block in G71 should not include
   the Z axis value (Z or W) for cycle Type I

❑ Change of direction is allowed only for Type II
   G71 cycle, and along one axis only (W0)

❑ Stock allowance U is programmed on a diameter,
   and its sign shows to which side of the stock it is to
   be applied (sign is the direction in X, to or from the
   spindle centerline)

❑ Feedrate programmed for the finishing contour
   (specified between the P and Q points) will be
   ignored during roughing

❑ D address does not use decimal point, and must be
   programmed for leading zero suppression format:

D0750 or D750 is equivalent to .0750 depth

Only some control systems do allow a decimal point
to be used for the D address (depth of cut)
in G71 and G72 cycles.

G74 - PECK DRILLING CYCLE

The G74 cycle is one of two cycles usually used for non finishing work. Along with G75 cycle, it is used for machining an interrupted cut, such as chips breaking during a long cutting motion. *G74 cycle is used along the Z axis.*

This is the cycle commonly used for an interrupted cut along the Z axis. The name of the cycle is *Peck Drilling Cycle*, similar to the G73 peck drilling cycle, used for machining centers. For the lathe work, G74 cycle application is a little more versatile than for its G73 equivalent on machining centers. Although its main purpose may be applied towards peck drilling, the cycle can be used with equal efficiency for interrupted cuts in turning and boring (for example, in some very hard materials), deep face grooving, difficult part-off machining, and many other applications.

◆ **G74 Cycle Format - 10T/11T/15T**

The one-block programming format for G74 cycle is:

G74 X..(U..) Z..(W..) I.. K.. D.. F.. S..

☞ where ...

X(U)	=	Final groove diameter to be cut
Z(W)	=	Z position of the last peck - depth of hole
I	=	Depth of each cut (no sign)
K	=	Distance of each peck (no sign)
D	=	Relief amount at the end of cut (must be zero for face grooving)
F	=	Groove cutting feedrate (in/rev or mm/rev)
S	=	Spindle speed (ft/min or m/min)

◆ **G74 Cycle Format - 0T/16T/18T/20T/21T**

The two-block programming format for G74 cycle is:

G74 R..
G74 X..(U..) Z..(W..) P.. Q.. R.. F.. S..

☞ where ...

First block:

R   =   Return amount (clearance for each cut)

Second block:

X(U)	=	Final groove diameter to be cut
Z(W)	=	Z position of the last peck (depth of hole)
P	=	Depth of each cut (no sign)
Q	=	Distance of each peck (no sign)
R	=	Relief amount at the end of cut (must be zero for face grooving)
F	=	Groove cutting feedrate (in/rev or mm/rev)
S	=	Spindle speed (ft/min or m/min)

If both the X(U) and I (or P) are omitted in the cycle, the machining is along the Z axis only (peck drilling). In a typical peck drilling operation, only the Z, K and F values are programmed - see *Figure 35-19*.

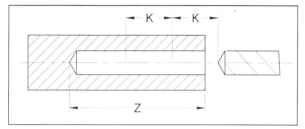

Figure 35-19

Schematic format for G74 cycle example

The following program example illustrates G74 cycle:

```
O3507 (G74 PECK DRILLING)
N1 G20
N2 T0200
N3 G97 S1200 M03              (SPEED IN RPM)
N4 G00 X0 Z0.2 T0202 M08      (START POSITION)
N5 G74 Z-3.0 K0.5 F0.012      (PECK DRILLING)
N6 G00 X6.0 Z2.0 T0200        (CLEAR POSITION)
N7 M30                        (END OF PROGRAM)
%
```

Drilling will take place to a three inch depth, in depth increments of one half of an inch. Note the depth of the first peck is calculated from the start position. Programming of an interrupted groove is very similar in format.

G75 - GROOVE CUTTING CYCLE

The G75 cycle is the other of two lathe cycles available for simple, non precision work. Together with the G74 cycle, it is used for operations requiring an interrupted cut, for example for breaking chips during a long or deep cutting motion. *G75 cycle is used along the X axis.*

This is also a very simple cycle, designed to break chips during a rough cut along the X axis - used mainly for a grooving operation. The G75 cycle is identical to G74, except the X axis is replaced with the Z axis.

◆ G75 Cycle Format - 10T/11T/15T

The one-block programming format for G75 cycle is:

> G75 X..(U..) Z..(W..) I.. K.. D.. F.. S..

☞ where ...

X(U)	=	Final groove diameter to be cut
Z(W)	=	Z position of the last groove
		(for multiple grooves only)

I	=	Depth of each cut (no sign)
K	=	Distance between grooves (no sign)
		(for multiple grooves only)
D	=	Relief amount at the end of cut
		(must be zero or not used for face groove)
F	=	Groove cutting feedrate (in/rev or mm/rev)
S	=	Spindle speed (ft/min or m/min)

◆ G75 Cycle Format - 0T/16T/18T/20T/21T

The two-block programming format for G75 cycle is:

> G75 R..
> G75 X..(U..) Z..(W..) P.. Q.. R.. F.. S..

☞ where ...

First block:

R	=	Return amount (clearance for each cut)

Second block:

X(U)	=	Final groove diameter to be cut
Z(W)	=	Z position of the last groove
P	=	Depth of each cut (no sign)
Q	=	Distance between grooves (no sign)
R	=	Relief amount at the end of cut
		(must be zero for face grooving)
F	=	Groove cutting feedrate
		(usually in/rev or mm/rev)
S	=	Spindle speed (usually ft/min or m/min)

If both the Z(W) and K (or Q) are omitted in the cycle, the machining is along the X axis only (peck grooving).

A practical example of G75 cycle will be presented in the next chapter.

BASIC RULES FOR G74 AND G75 CYCLES

Several notes are common to both cycles:

❏ In both cycles, the X and Z values can be programmed either in the absolute or incremental mode.

❏ Both cycles allow an automatic relief.

❏ The relief amount at the end of cut can be omitted - in that case it will be assumed as zero.

❏ Return amount (clearance for each cut) is only programmable for the two-block method. Otherwise, it is set by an internal parameter of the control system.

❏ If the return amount is programmed (two-block method), and the relief amount is also programmed, the presence of X determines the meaning. If the X value is programmed, the R value means the relief amount.

Groove cutting on CNC lathes is a multi step machining operation. The term *grooving* usually applies to a process of forming a narrow cavity of a certain depth, on a cylinder, cone, or a face of the part. The groove shape, or at least a significant part of it, will be in the shape of the cutting tool. Grooving tools are also used for a variety of special machining operations.

The grooving tool is usually a carbide insert mounted in a special tool holder, similar to any other tool. Designs of grooving inserts vary, from a single tip, to an insert with multiple tips. Inserts are manufactured to nominal sizes. Multi tip insert grooving tools are used to decrease costs and increase productivity.

GROOVING OPERATIONS

The cutting tools for grooving are either *external* or *internal* and use a variety of inserts in different configurations. The most important difference between grooving and turning is the *direction of cut*. Turning tool can be applied for cuts in multiple directions, grooving tool is normally used to cut in a single direction only. A notable exception is an operation known as necking (relief grooving), which takes place at 45°, where the angle of the cutting insert and the angle of infeed *must be identical* (usually at 45°). There is another application of a two axis simultaneous motion in grooving, a corner breaking on the groove. Strictly speaking, this is a turning operation. Although a grooving tool is not designed for turning, it can be used for some light machining, like cutting a small chamfer. During the corner breaking cut on a groove, the amount of material removal is always very small and the applied feedrate is normally low.

◆ Main Grooving Applications

Groove is an essential part of components machined on CNC lathes. There are many kinds of grooves used in industry. Most likely, programming will include many undercuts, clearance and recess grooves, oil grooves, etc. Some of the main purposes of grooving are to allow two components to fit face-to-face (or shoulder-to-shoulder) and, in case of lubrication grooves, to let oil or some other lubricant to flow smoothly between two or more connecting parts. There are also pulley or V-belt grooves that are used for belts to drive a motor. O-ring grooves are specially designed for insertion of metal or rubber rings, that serve as stoppers or sealers. There are many other kinds of grooves. Many industries use grooves unique to their needs, most others use the more general groove types.

◆ Grooving Criteria

For a CNC programmer, grooving usually presents no special difficulties. Some grooves may be easier to program than others, yet there could be several fairly complex grooves found in various industries that may present a programming or machining challenge. In any case, before a groove can be programmed, have a good look at the drawing specifications and do some overall evaluations. Many grooves may appear on the same part at different locations and could benefit from a subprogram development. When planning a program for grooving, evaluate the groove carefully. In good planning, evaluate the selected groove by at least three criteria:

❏ Groove shape

❏ Groove location on a part

❏ Groove dimensions and tolerances

Unfortunately, many grooves are not of the highest quality. Perhaps it is because many grooves do not require high precision and when a high precision groove has to be done, the programmer does not know how to handle it properly. Watch particularly for surface finish and tolerances.

GROOVE SHAPE

The first evaluation before programming grooves is the groove shape. The shape is determined by the part drawing and corresponds to the purpose of the groove. The groove shape is the single most important factor when selecting the grooving insert. A groove with sharp corners parallel to the machine axes requires a square insert, a groove with radius requires an insert having the same or smaller radius. Special purpose grooves, for example an angular groove shape, will need an insert with the angles corresponding to the groove angles as given in the drawing. Formed grooves require inserts shaped into the same form, etc. Some typical shapes of grooving inserts are illustrated in *Figure 36-1*.

Figure 36-1
Typical shapes of common grooving tools

◆ Nominal Insert Size

In many groove cutting operations, the groove width will be greater than the largest available grooving insert of a nominal size (*i.e.,* off the shelf size). Nominal sizes are normally found in various tooling catalogues and typically have widths like 1 mm, 2 mm, 3 mm or 1/32, 3/64, 1/16, 1/8 in inches, and so on, depending on the units selected.

For example, a groove width of .276 inches can be cut with a nearest *lower* nominal insert width of .250 inch. In such cases, the groove program has to include at least two cuts - one or more roughing cuts, in addition to at least one finishing cut. Another grooving tool may be used for finishing, if the tolerances or excessive tool wear make it more practical - *Figure 36-2*.

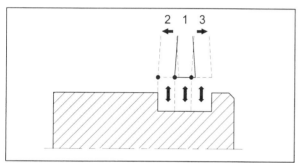

Figure 36-2
Cut distribution for grooves wider than the insert

◆ Insert Modification

Once in a while, programmers encounter a groove that requires a special insert in terms of its size or shape. There are two options to consider. One is to have a custom made insert, if it is possible and practical. For a large number of grooves, it may be a justified solution. The other alternative is to modify an existing insert in-house.

Generally, in CNC programming, off-the-shelf tools and inserts should be used as much as possible. In special cases, however, a standard tool or insert can be modified to suit a particular job. For grooving, it may be a small extension of the insert cutting depth, or a radius modification. Try to modify the groove shape itself only as the last resort. Modification of standard tools slows down the production and can be quite costly.

GROOVE LOCATION

Groove location on a part is determined by the part drawing. The locations can be one of three groups:

❏   Groove cut on a cylinder   … diameter cutting

❏   Groove cut on a cone   … taper cutting

❏   Groove cut on a face   … shoulder cutting

Although some variations are possible, for practical purposes, only these three categories are considered. Each of the three locations may be either *external* or *internal*.

The two most common groove locations are on a cylinder, *i.e.,* on a straight outside - or *external* - diameter, or on a straight inside - or *internal* - diameter. Many other grooves may be located on a face, on a taper (cone), even in a corner. The illustration in *Figure 36-3* shows some typical locations of various grooves.

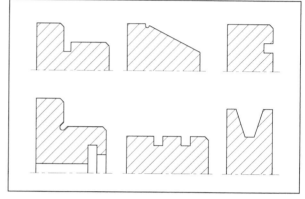

Figure 36-3
Typical groove locations on a part

GROOVE DIMENSIONS

The dimensions of a groove are always important when selecting the proper grooving insert. Grooving dimensions include the *width* and the *depth* of a groove, as well as the corners specifications. It is not possible to cut a groove with an insert that is larger than the groove width. Also, it is not possible to feed into a groove depth that is greater than the depth clearance of the insert or tool holder. However, there is usually no problem in using a narrow grooving insert to make a wide groove with multiple cuts. The same applies for a deep cutting insert used to make a shallow groove. *The dimensions of a groove determine the method of machining.* A groove whose width *equals* the insert width selected for the groove shape, requires only *one* cut. Simple *feed-in* and *rapid-out* tool motion is all that is required. To program a groove correctly, the *width* and *depth* of the groove must be known as well as its *position* relative to a known reference position on the part. This position is the distance to one side - or one wall - of the groove.

Some extra large grooves require a special approach. For example, a groove that is 10 mm wide and 8 mm deep cannot be cut in a single pass. In this case, the rough cuts for the groove will control not only its width, but also its depth. It is not unusual to even use more than one tool for such an operation. Program may also need to be designed in sections. In case of an insert breakage, only the affected program section has to be repeated.

◆ Groove Position

In *Figure 36-4* are shown two most common methods of dimensioning a typical groove. The groove width is given in both cases as dimension W, but the distance L from the front face is different in the example *a* and the example *b*.

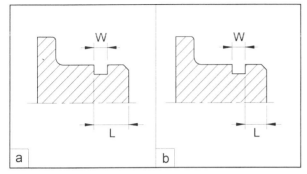

Figure 36-4

Groove position dimensioning - two common methods

In the example *36-4a*, the dimension L is measured to the *left* side of the groove. For programming purposes, this dimension is more convenient, because it will actually appear in the program as specified in the drawing. Normally, the standard tool reference point of a grooving tool is set to the *left side* of the grooving insert.

The example in *Figure 36-4b*, the dimension L is to the *right* side of the groove. The left side dimension can be found easily, by adding the groove width W. The programming considerations will be slightly different, particularly if the dimensional tolerances are specified. Always take the approach that the specified dimension indicates the *more important* dimension. If a tolerance range is specified for any dimension, the tolerance must always be maintained on the finished groove, and it will affect the overall programming method. A groove may also be dimensioned from another reference location, depending on its purpose.

◆ Groove Depth

In *Figure 36-5*, there are two typical methods of dimensioning the groove depth.

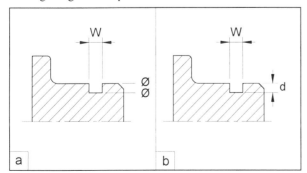

Figure 36-5

Groove depth dimensioning - two common methods

In the *Figure 36-5a*, the top and bottom diameter of the groove are both given. This method has a major benefit that the bottom diameter of the groove will actually appear as an X axis amount in the program. A disadvantage is that the actual depth still has to be calculated and a proper grooving tool selected. The example in *Figure 36-5b* does show the groove depth, but most likely, the bottom diameter will have to be calculated manually. Both dimensioning examples are about equally common in CNC programming.

Deep grooves are usually grooves that have a reasonably normal width but also have a much deeper ratio between the groove top diameter and its bottom diameter.

SIMPLE GROOVE PROGRAMMING

The simplest of all grooves is the one that has the same width and shape as the tool cutting edge - *Figure 36-6*.

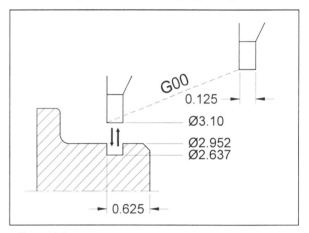

Figure 36-6

Simple groove example - program O3601
Insert width is equal to the groove width

The program such a groove is very straightforward. In rapid mode, move the grooving tool to the starting position, feed-in to the groove depth, then rapid out back to the starting position, and - the groove is finished. There are no corner breaks, no surface finish control, and no special techniques used. Some will say, and no quality either. A dwell at the bottom of the groove may be the only improvement. True, the quality of such a groove will not be the greatest, but a groove it will be. Such a groove is strictly a utility type groove and is hardly ever required in precision manufacturing. At the same time, programming such grooves is a good start to learn more advanced techniques.

The following program O3601, uses a standard .125 inch square grooving insert, for the groove of the same width. The groove depth is the single difference between the two diameters given in the drawing:

```
(2.952 - 2.637) / 2 = .1575
```

The sample program uses the tool T08, as the last tool:

```
O3601 (SIMPLE GROOVE)
(G20)
...
N33 T0800 M42              (TOOL 8 ACTIVE)
N34 G97 S650 M03           (650 RPM SPEED)
N35 G00 X3.1 Z-0.625 T0808 M08   (START POINT)
N36 G01 X2.637 F0.003      (FEED-IN TO DEPTH)
N37 G04 X0.4              (DWELL AT THE BOTTOM)
N38 X3.1 F0.05            (RETRACT FROM GROOVE)
N39 G00 X6.0 Z3.0 T0800 M09    (CLEAR POSITION)
N40 M30                    (END OF PROGRAM)
%
```

The program O3601 does the following. First, the English mode of input is preset from the beginning of the program with G20. Blocks N33 and N34 are startup blocks for the tool T08, with direct *r/min* selected. *Constant Surface Speed* (CSS) in G96 mode can be selected instead. N35 is a block where the tool moves to the position from which the groove will be started (start point). Clearance at this safe location is the clearance above the part diameter, which is .074 inches in the example:

$$(3.1 - 2.952) / 2 = .074$$

The coolant is applied in the same block, during the tool motion. Block N36 contains the actual groove plunging cut, at a slow feedrate of .003 in/rev. Block N37 is a dwell time of 0.4 seconds, followed by the tool return to the starting diameter and completion of the program.

Although this particular grooving example was very simple, let's evaluate the program a little more. It contains several important principles that can be applied to the method of programming any groove, where its precision and surface finish are very critical.

Note the clearance *before* the groove cutting begins. The tool is positioned .074 inches above the part diameter, at ⌀3.100. Always keep this distance to a certain safe minimum. Grooves are usually cut at a slow feedrate and it may take too much time just to cut in the air. Also note the actual cutting feedrate has increased from .003 in/rev in block N36 to a rather high feedrate of .050 in/rev in block N38. Rapid motion command G00 could have been used instead. Feeding out at a heavier feedrate (rather than using a rapid motion), may improve the groove surface finish by eliminating the tool drag on the walls (sides).

The tool width of .125 never appears in the program, directly or indirectly. That means the shape and width of the insert will become the shape and width of the groove. It also means a different groove width, if another insert size is used, although the program structure remains unaffected. The structure will remain unaffected even if the grooving tool *shape* is changed. Combination of the shape *and* the size change will offer endless opportunities, all of them being possible without a single change to the program.

PRECISION GROOVING TECHNIQUES

A simple *in-out* groove will not be good. Its sides may have a rough surface, the outside corners will be sharp and its width is dependent on the insert width and its wear. For most of machining jobs, such a groove is not acceptable.

To program and machine any precision groove requires extra effort, but the result will be a high quality groove. This effort is not always justified, as high quality comes with a price. The next two illustrations show the groove dimensions and program related details. Drawing in *Figure 36-7* shows a high precision groove, although its width is intentionally exaggerated for impact of the example.

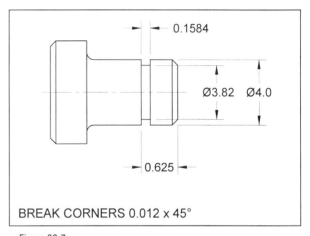

Figure 36-7
Drawing for a precision groove example O3602

What is the best cutting method? One plunge rough cut and two finish cuts, one for each wall, are reasonable; so is the .006 stock added to the bottom diameter. Also, sharp corners will be broken with a .012 chamfer at the ⌀4.0. *Figure 36-8* shows the distribution of the cuts.

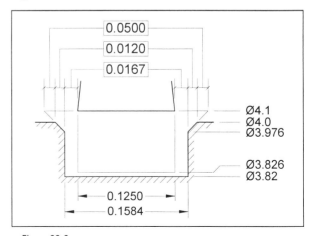

Figure 36-8
Precision groove - distribution of cuts for the example O3602

Before the first block can be programmed, selection of the cutting tool and machining method is a sign of a good planning. These are important decisions because they directly influence the final groove size and its condition.

◆ Groove Width Selection

The grooving tool selected for the example in program O3602 will be an external tool, assigned to the tool station number three - T03. Tool reference point is selected at the *left edge* of the insert, which is a standard selection. The insert width has to be selected as well. Grooving inserts are available in a variety of standard widths, usually with an increment of 1 mm for metric tools, and 1/32 or 1/16 inch for tools in the English system. In this case, the non-standard groove width is .1584 inch. The nearest standard insert width is 5/32 inch (0.15625 inch). The question is - should we select the 5/32 inch insert width? In a short answer, no. In theory, this insert *could* cut the groove, but because the actual difference between the insert width and the groove width is so small (.00215 inch over two walls), there is very little material to cut.

The dimensional difference would allow only slightly more than .001 per each side of the groove, which may cause the insert to rub on the wall rather than cut it. A better choice is to step down to the next lower standard insert width, that is 1/8th of an inch (.1250). There is much more flexibility with 1/8 width than with 5/32 width. Once the grooving tool is selected, the initial values can be assigned - the offset number (03), the spindle speed (400 ft/min), the gear range (M42) - and a note for the setup sheet:

❑   T0303 = .1250 SQUARE GROOVING TOOL

The first few program blocks can now be written:

```
O3602 (PRECISION GROOVE)
(G20)
...
N41 T0300 M42
N42 G96 S400 M03
...
```

◆ Machining Method

Once the grooving tool has been selected and assigned a tool station number (tool turret position), the actual *method* of machining the groove has to be decided. Earlier, the machining method has been described generally, now a more detailed description is necessary.

One simple programming method is *not* an option - the basic *in-out* technique used earlier. That means a *better* method must be selected, a method that will guarantee a high quality groove. The first step towards that goal is the realization of the fact that a grooving insert with the width narrower than the groove width, will have to be plunged into the groove more than once. How many times? It is not difficult to calculate that a groove .1584 wide and ma-

chined with a .1250 wide grooving insert, will need *at least* two grooving cuts. But what about a groove that is much wider than the groove in the example?

There is an easy way to calculate the *minimum number of grooving cuts* (or plunges), using the following formula:

$$C_{min} = \frac{G_w}{T_w}$$

☞ where ...

C_{min} = Minimum number of cuts
G_w = Groove width for machining
T_w = Grooving insert width

Applying the formula to the example, the starting data are the groove width of .1584 of an inch and the grooving insert width of .1250 of an inch. That translates into the minimum of *two* grooving cuts. Always round upwards, to the nearest integer: *.1584/.1250=1.2672=2 cuts*.

A possible decision could be to plunge once to finish the left side of the groove and, with one more plunge, to finish the groove right side. The necessary overlap between the two cuts is guaranteed and the only remaining operation is the chamfering. A groove programmed this way may be acceptable, but will *not* be of a very good quality.

Even if only an *acceptable* quality groove is produced during machining, such a result does not give the programmer much credit. What can be actually done to assure the *highest* groove quality possible?

> In order to write first class programs, make the best efforts to deliver an exceptional quality at the *programming* level, in order to prevent problems at the *machining* level.

How can this suggestion be applied to the example? The key is the knowledge of machining processes. Machining experience confirms that removing an *equal* stock from each wall (side) of the groove will result in better cutting conditions, better surface finish control and better tool life.

If this observation is used in the current example, an important conclusion can be made. If two plunge cuts of uneven width will yield at least *acceptable* results, *three* cuts that are *equally distributed* should yield even better results.

If *at least three* grooving cuts are used to form the groove rather than the minimum two cuts, the CNC programmer will gain control of two always important factors:

❑   Control of the groove POSITION

❑   Control of the groove WIDTH

In precision grooving, these two factors are equally important and should be considered together.

Look carefully at how these factors are implemented in the example. The first factor applied under the program control is the groove position. The groove *position* is given in the drawing as .625 inches from the front face of the part, to the left side of the groove. There is no plus or minus dimensional tolerance specified, so the drawing dimension is used as arbitrary and is programmed directly. The second factor under the program control is the groove *width*. That is .1584 of an inch on the drawing and the selected tool insert width is .1250. The goal is to program the cutting motions in *three steps*, using the technique already selected:

➲ STEP 1

Rough plunge in the middle of the groove, leaving an equal material stock on both groove faces for finishing - also leave small stock on the bottom of the groove

➲ STEP 2

Program the grooving tool operation on the left side of the groove, including the chamfer (corner break)

➲ STEP 3

Program the grooving tool operation on the right side of the groove, including the chamfer (corner break) and sweep the groove bottom towards the left wall.

The last two steps require chamfer cutting or a corner break. The width of the chamfer *plus* the width of the subsequent cut should never be larger than about one half to three quarters of the insert width. In the third step, sweeping of the bottom is desired. That suggests the need to consider stock allowances for finishing.

◆ **Finishing Allowances**

During the first step, the first plunge has to take place at the *exact* center of the groove. To calculate the Z axis position for the start, find first the amount of stock on each wall that is left for finishing. The stock amount will be one half of the groove width minus the insert width - see details in the previous *Figure 36-8*:

(.1584 - .1250) / 2 = .0167

The tool Z position will be .0167 on the positive side of the left wall. If this wall is at Z-0.625, the grooving tool start position will be at Z-0.6083. When the tool completes the first plunge, there will be an equal amount of material left for finishing on both walls of the groove.

Do your best to avoid rounding off the figure .0167, for example, to .0170 inch. It would make no difference for the machining, but it is a sound programming practice to use *only* the calculated values. The benefit of such approach is in eventual program checking, and also with general consistency in programming. Equal stock amounts offer this consistency; .0167 and .0167 is a better choice than .0170 and .0164, although the practical results will be the same.

Next look is at the X axis positions. The first position is where the plunge will start from, the second position is the end diameter for the plunging cut. A good position for the start is about .050 *per side* above the finished diameter, which in this case would be a clearance diameter calculated from the ∅4.0:

4.0 + .05 × 2 = 4.1   (X4.1)

Do not start the cut with a clearance of more than .050 inch (1.27 mm) - with slow feedrates that are typical to grooves, there will be too much air to cut, which is not very efficient. The end diameter is the groove bottom, given on the drawing as 3.82. Dimension of X3.82 could be programmed as the target diameter, but it does help to leave a very small stock, such as .003 per side (.006 on diameter), to make a *sweep finish* of the groove bottom. That will add two times .003 to the 3.82 groove diameter, for the programmed X target as X3.826. Once the plunge is done, the tool returns to the start diameter:

```
N43 G00 X4.1 Z-0.6083 T0303 M08
N44 G01 X3.826 F0.004
N45 G00 X4.1
```

The rapid motion back above the groove (N45) is a good choice in this case, because the sides will be machined later with the finishing cuts, so the surface finish of the walls is not critical at this moment. After roughing the groove, it is time to start the finishing operations.

All the calculated amounts can be added to the previous *Figure 36-8*, and create data for a new *Figure 36-9*:

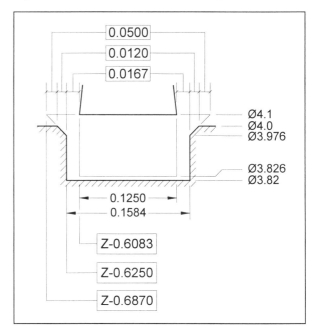

Figure 36-9
Precision groove - groove data used in program O3602

◆ Groove Tolerances

As in any machining, program for grooves must be structured in such a way, that maintaining tolerances at the machine will be possible. There is no specified tolerance in the example, but it is implied as very close by the four-decimal place dimension. A tolerance range, such as 0.0 to +.001, is probably a more common way of specifying a tolerance. Only the dimensional value that falls *within* the specified range can be used in a program. In this example, the aim is the drawing dimension of .1584 (selected intentionally).

A possible problem often encountered during machining and a problem that influences the groove width the most, is a *tool wear*. As the insert works harder and harder, it wears off at its edges and actually becomes *narrower*. Its cutting capabilities are not necessarily impaired, but the resulting groove width may not fall within close tolerances. Another cause for an unacceptable groove width is the *insert width*. Inserts are manufactured within high level of accuracy, but also within certain tolerances. If an insert is changed, the groove width may change slightly, because the new insert may not have exactly the same width as the previous one. To eliminate, or at least minimize, the possible *out of tolerance* problem, use quite a simple technique - program an *additional* offset for finishing operations only.

Earlier, when the precision groove was planned, offset 03 had been assigned to the grooving tool. Why would an *additional* offset be needed at all? Assume for a moment, that all machine settings use just a single offset in the program. Suddenly, during machining, the groove gets narrower due to tool wear. What can be done? Change the insert? Modify the program? Change the offset? If the Z axis offset setting is adjusted, either to the negative or positive direction, that will change the groove *position* relative to the program zero but it will *not change the groove width!* What is needed is a second offset, an offset that controls the groove width only. In the program O3602, the left chamfer and side will be finished with one offset (03), the right chamfer and side will use a second offset. To make the second offset easier to remember, number 13 will be used.

One other step has to be finished first - calculation of the left chamfer start position. Currently, the tool is at Z-0.6083 but has to move by the wall stock of .0167 and the chamfer width .012 as well as clearance of .050 - for a total travel of .0787, to Z-0.687 position. At a slow feedrate, the chamfer is done first and the cut continues to finish the left side, *to the same diameter* as for roughing, which is X3.826:

```
N46 Z-0.687
N47 G01 X3.976 Z-0.625 F0.002
N48 X3.826 F0.003
```

The next step is to return the tool above part diameter. This motion is more important than it seems. In the program, make sure the finished left side is not damaged when the tool retracts from the groove bottom. Also make sure

the grooving tool will not contact the right side wall stock. That means do not retract the tool further then the position of Z-0.6083. It also means do *not* rapid out, because of a possible contact during the 'dogleg' or 'hockey stick' motion, described in Chapter 20 - *Rapid Positioning*. The best approach is to return to the initial start position at a relatively *high but non-cutting* feedrate:

```
N49 X4.1 Z-0.6083 F0.04
```

At this point, the left side wall is finished. To program the motions for the right side wall, the tool has to cut with the *right* side (right edge) of the grooving insert. One method is to change the G50 coordinates in the program, if this older setting is still used, or use a different work coordinate offset. The method used here is probably the simplest and also the safest. All motions relating to the right chamfer and the right side groove wall will be programmed in the *incremental mode*, applied to the Z axis only, using the W address:

```
N50 W0.0787 T0313
N51 X3.976 W-0.062 F0.002
```

In block N50, the tool travels the total distance equivalent to the sum of the right wall stock of .0167, the chamfer of .012 and the clearance of .050. In the same block, the second offset is programmed. This is the only block where offset 13 *should* be applied - one block before, it's too early, and one block, after it's too late.

Block N51 contains the target chamfer position and the absolute mode for the X axis and is combined with the incremental mode for the Z axis.

To complete the groove right side wall, finish the cut at the full bottom diameter, block N52, then continue to remove the stock of .003 from the bottom diameter (block N53) - this is called *sweeping the groove bottom*:

```
N52 X3.82 F0.003
N53 Z-0.6247 T0303
```

Also look at the Z axis end amount - it is a small value that is .0003 short of the .625 drawing dimension! The purpose here is to compensate for a possible tool pressure. *There will not be a step in the groove corner!* Because the sweep will end at the *left* side of the groove, the original offset (03) must be reinstated. Again, the block N53 is the only block where the offset change is correct. Make sure not to change the tool numbers - *the turret will index !*

The intended program O3602 can now be completed. All that remains to be done is the return to the groove starting position, followed by the program termination blocks:

```
N54 X4.1 Z-0.6083 F0.04
N55 G00 X10.0 Z2.0 T0300 M09
N56 M30
%
```

At this point, the *complete* program O3602 can be developed. Note program blocks where the offset has been changed, they are identified in the comment section:

```
O3602 (PRECISION GROOVE)
(G20)
...
N41 T0300 M42                    (NO OFFSET)
N42 G96 S400 M03
N43 G00 X4.1 Z-0.6083 T0303 M08  (OFFSET 03)
N44 G01 X3.826 F0.004
N45 G00 X4.1
N46 Z-0.687
N47 G01 X3.976 Z-0.625 F0.002
N48 X3.826 F0.003
N49 X4.1 Z-0.6083 F0.04
N50 W0.0787 T0313                (OFFSET 13)
N51 X3.976 W-0.062 F0.002
N52 X3.82 F0.003
N53 Z-0.6247 T0303               (OFFSET 03)
N54 X4.1 Z-0.6083 F0.04
N55 G00 X6.0 Z3.0 T0300 M09      (NO OFFSET)
N56 M30
%
```

```
+-----------------------------------------------+
|                  WARNING !                    |
| It is very important to use caution when a double |
| tool offset for a single tool is used during machining |
| ( this warning applies generally - not only for grooving ) |
+-----------------------------------------------+
```

Remember that the *purpose* of the offset in the example is to control the groove *width*, *not* its diameter.

Always follow these precautions, based on the example program O3602:

❑ Start machining with identical initial amounts assigned to both offsets (the same XZ values for offsets 03 and 13).

❑ The X offset amounts of 03 and 13 must always be the same. If the X setting of one offset is changed, the setting of the other offset must be changed to the same value. Adjust both X offsets to control the groove depth tolerance.

❑ If the groove width becomes too narrow and has to be adjusted, only the Z offset amount is changed.

❑ To adjust the groove left side wall position, change the Z offset 03.

❑ To adjust the groove right wall position, change the Z offset 13.

❑ Do not cancel the current offset -
  - change from one to the other offset directly.

❑ Make sure the tool number (the first two digits of the T address) does not change, otherwise, *THERE WILL BE A TOOL CHANGE!*

Other precautions can be added, depending on the exact conditions. Use common sense, and always check the program carefully, before it is released to production.

◆ **Groove Surface Finish**

Programming just about any precision groove should be fairly easy from now on. Only a few last notes on the subject of groove cutting as they relate to the surface finish. Just by following the suggested methods of equal cut distribution, proper spindle speeds and feedrates, good condition of the cutting tool and insert, suitable coolant, and other techniques used in the example, the surface finish will almost take care of itself.

Keep in mind, that the term *'precision groove'* does not only describe the precise groove position and its precise dimensions, it also means a high quality look, a look that often means much more than just a cosmetic feature.

MULTIPLE GROOVES

Multiple grooving is a common term used for cutting the same groove at different positions of the same part. In these cases, the program will most likely benefit from developing a subprogram (subroutine) for multiple grooves, that will be called at various groove locations. Subprograms save valuable programming time, they are easily designed and easily edited. Although subprograms will be discussed in *Chapter 39*, an example of a multiple groove programming using a subprogram is shown at the end of this chapter, at least for reference and basic introduction.

When cutting multiple grooves, more material will be removed. On external diameter grooves, there are no special considerations necessary, gravity will take care of the extra chips. This is not the same situation for internal grooves. The moment several grooves are machined internally, there is a small pile of cutting chips accumulated in the bored hole. These chips can be in the way of a smooth cutting operation and could damage the bored diameter and even the grooving tool itself. To solve this problem, consider machining of only a few grooves, move the tool out and blow out the chips from the internal area. Using the optional stop M01 can be useful in this case. When all chips have been removed, continue with the same tool to cut more grooves.

FACE GROOVES

Face grooving (sometimes incorrectly called *trepanning*) is a horizontal groove cutting process, with the tool moving along the Z axis. The tool is programmed along the same principles as vertical grooving along the X axis. Because of the nature of such a grooving cut, the tool *orientation* presents the most important single consideration in face grooving. The issue is the *radial clearance* of the cutting insert, during a cut. There is no need to worry too much about radial clearance for vertical grooving, because the cutting edge of the insert is on the same plane as the machine center line. However, in horizontal grooving, the insert clearance along the cut radius is of utmost importance.

The next example shows how to program a typical face groove, and is illustrated in *Figure 36-10*.

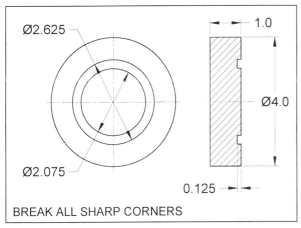

BREAK ALL SHARP CORNERS

Figure 36-10
Face grooving example - program O3603

Although both the external and internal groove diameters are engineering choices in the drawing, the actual groove width is necessary for programming as well. To calculate the groove width, use a simple calculation - find one half of the difference between the two grooving diameters - that is:

```
(2.625 - 2.075) / 2 = .275
```

This is the actual groove width amount, .275 in the given example. Always keep in mind that the program will use a smaller .250 wide face grooving insert. Following the programming examples of a precision groove, listed earlier, three cuts will be made - one rough plunge in the middle of the groove, and two finishing cuts, with a small corner break. But first, let's look at the *radial clearance* of the grooving tool. This is a very important programming consideration; one that is unique to most face grooving operations, yet, it is also one that is easy to be overlooked.

◆ Radial Clearance

Many grooving inserts are high, in order to give them strength. The grooving insert for face grooving operations is mounted at 90° towards the part face (parallel to the spindle center line). A standard grooving insert has virtually no radial clearance and most likely will interfere with the part at its lower end - *Figure 36-11*.

From the illustration is obvious that the grooving insert cannot be used as is and has to be modified. Such a modification is usually done by grinding a suitable radial clearance, as illustrated in *Figure 36-12*.

This is a simple operation, providing the proper grinding tools are available. Make sure that the grinding does not affect the insert width and only minimum necessary material is removed, otherwise the tool loses strength.

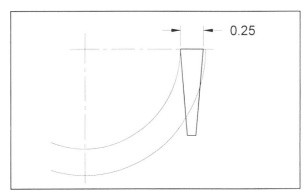

Figure 36-11
Interference of a standard grooving insert on a face groove

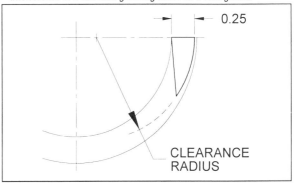

CLEARANCE RADIUS

Figure 36-12
Standard grooving insert modified for face grooving

◆ Face Grooving Program Example

Program O3603 uses modified insert and a .012 corner break, to eliminate sharp corners. Only one offset is used in the program. The tool set point is the lower edge of insert, corresponding to the ∅2.075. All calculations should be easy to retrace, they use exactly the same procedures as those described for vertical grooving:

```
O3603 (FACE GROOVE)
(G20)
...
N21 T0400 M42
N22 G96 S450 M03
N23 G00 X2.1 Z0.05 T0404 M08
N24 G01 Z-0.123 F0.003
N25 Z0.05 F0.04
N26 X1.951
N27 X2.075 Z-0.012 F0.001
N28 Z-0.123 F0.003
N29 X2.1 Z0.05 F0.04
N30 U0.149
N31 U-0.124 Z-0.012 F0.001
N32 Z-0.125 F0.003
N33 X2.0755
N34 X2.1 Z0.05 F0.04 M09
N35 G00 X8.0 Z3.0 T0400
N36 M30
%
```

CORNER GROOVES / NECK GROOVES

Corner grooving is also a grooving operation, one that uses a special grooving insert designed to cut along a 45° angular motion. The groove can be square or with a radius, depending on the tool and insert used and design required. The grooving insert may also be a standard type insert, placed into a 45° angle tool holder. The purpose of this type of a groove is to allow machining of recesses and undercuts, usually in a corner of the part. It assures a shoulder match of two assembled components.

To program a corner groove (neck groove), the radius of the grooving insert must be known, .031 (1/32) of an inch in the example. The cutting depth is established from the drawing data. Normally, the corner groove is specified as a 'minimum undercut'. In this case, the center of the undercut will be at the intersection of the shoulder and the diameter. The cutting motion in and out of the groove must be at 45°, meaning the identical amount of travel in both X and Z axes. *Figure 36-13* illustrates a corner groove with a .031 radius minimum undercut.

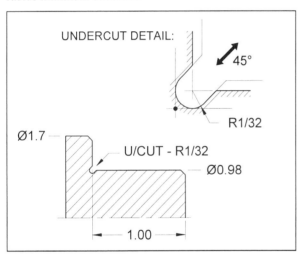

Figure 36-13
Corner groove - undercut program example O3604

The program itself has no hidden surprises and is not difficult to complete or interpret:

```
O3604 (CORNER GROOVE)
(G20)
...
N217 G50 S1000 T0500 M42
N218 G96 S375 M03
N219 X1.08 Z-0.95 T0505 M08
N220 G01 X0.918 Z-1.031 F0.004
N221 G04 X0.1
N222 X1.08 Z-0.95 F0.04
N223 G00 X6.0 Z3.0 T0500 M09
N224 M30
%
```

Block N219 positions the tool in such a way that the center of insert (as well as the setup point) is in on center line of the neck groove (.050 clearance in X and Z axes). Blocks N220 and N222 are the two cutting motions - one into the groove in N220, the other out of the groove in N222. The amount of travel is exactly the same in either direction. The dwell of 0.1 second is added for convenience at the bottom of the neck. The block N220 can also be programmed as an *incremental* motion:

```
N220 G01 U-0.162 W-0.081 F0.004
N221 G04 X0.1
N222 U0.162 W0.081 F0.04
```

GROOVING CYCLES

Fanuc controls for lathes offer two multiple repetitive cycles G74 and G75 that can be used for an interrupted cutting along an axis. The programming formats for both cycles were described in the previous chapter. G74 cycle is used for cutting along the Z axis and is used mostly for peck drilling, the G75 is used for cutting in the X axis, and is used mostly for simple grooving.

◆ G75 Cycle Applications

Although used mainly for grooving, G75 cycle can also be used for an interrupted cut in facing. This cycle is quite simplistic to be of any use for high quality surface finish, but it does have its benefits. Its main purpose is to break chips while cutting along the X axis. This is useful for some grooving and part-off operations, as well as face cutting. Another use is for roughing the core out of deep grooves, so they can be finished later by more precise methods.

In G75, the chip breaking is done by alternating between a cutting motion in one direction and a rapid retract motion in the opposite direction. This means that one cutting motion is always followed by an opposite rapid motion, on the basis of *feed-in-rapid-out* principle and a built-in clearance. *Figure 36-14* illustrates the concept.

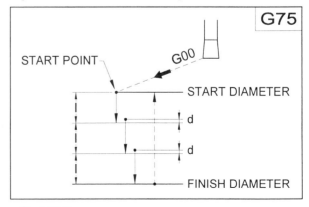

Figure 36-14
Schematic representation of the G75 cycle

The motion retract amount is built within the cycle and is set by an internal parameter of the control system. In *Figure 36-14* it is identified by the value *d*, (usually set to approximately .010 to .020 inches in the control). The next two examples illustrate the practical use of G75 grooving cycle.

◆ Single Groove with G75

A single groove requires the X and Z coordinate of the starting point, the final groove diameter X, and the depth of each cut I. For a single groove, the Z axis position and the K distance cannot be programmed. The Z position is given by the starting point and does not change.

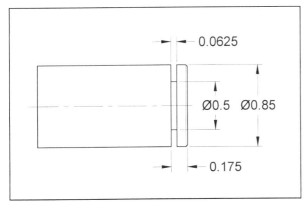

Figure 36-15

Single groove example using the G75 cycle - program O3605

The following program example O3605 cuts a single groove and is based on *Figure 36-15*.

```
O3605
(G75 SINGLE GROOVE)
(G20)
...
N43 G50 S1250 T0300 M42
N44 G96 S375 M03
N45 G00 X1.05 Z-0.175 T0303 M08
N46 G75 X0.5 I0.055 F0.004
N47 G00 X6.0 Z2.0 T0300 M09
N48 M30
%
```

Note that the I value is .055. This is not a value without a meaning. In fact, it is a carefully calculated depth of each groove peck. The tool travel will be from ∅1.050 to ∅.500, or .275 per side (1.05-.50)/2=.275. There will be exactly five grooving pecks of .055 each (.275/5=.055).

◆ Multiple Grooves with G75

It is possible to program multiple grooves very easily, using the G75 cycle. In this case, the groove spacing, the pitch between grooves must always be equal, otherwise the G75 cycle cannot be used. The clearance specification *d* in *Figure 36-14* is normally not programmed.

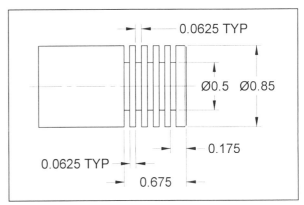

Figure 36-16

Multiple groove example using the G75 cycle - program O3606

The program example O3606 for multiple grooves, using G75 cycle, is based on *Figure 36-16*.

```
O3606
(G75 MULTIPLE GROOVES)
(G20)
...
N82 G50 S1250 T0300 M42
N83 G96 S375 M03
N84 G00 X1.05 Z-0.175 T0303 M08
N85 G75 X0.5 Z-0.675 I0.055 K0.125 F0.004
N86 G00 X6.0 Z2.0 T0300 M09
N87 M30
%
```

The setup and conditions for multiple groove are identical to those for a single groove. The only difference is the additional entries in the G75 cycle call.

This technique may be used not only for multiple grooves separated by solid material, but also for opening up a single groove that is much wider than the grooving insert. The only difference in programming will be the value of K - the distance between grooves. If the K is greater than the insert width, several individual grooves will be cut. If the K is equal to or smaller than the insert width, a single wide groove will be cut. Experiment to find the best amounts.

SPECIAL GROOVES

There are many more types of grooves than can be described in this handbook. They are mainly grooves of special shapes, used by specific industries - grooves that serve a certain purpose. The most typical grooves of this type are round grooves, pulley grooves, O-ring grooves, and several others. Certain grooves, usually those that conform to common industrial standards, can be machined with readily available inserts. A typical example of this kind of grooving is a pulley groove. The programming principles for *'nonstandard'* grooves are no different than those described in this chapter.

GROOVES AND SUBPROGRAMS

Programming multiple grooves with the G75 cycle is usually not the preferred method for precision work. The two main drawbacks are the groove quality and the requirement for an equal spacing between the grooves. There is another method to program multiple grooves, a more efficient method - one that uses subprograms.

Multiple grooves can be programmed very efficiently and with much increased precision by using the technique of subprograms, described in *Chapter 39*. The guiding principle is to program all common groove motions in the subprogram and all motions that vary from groove to groove, in the main program. This way, the same groove can be repeated at fixed intervals or variable intervals, as needed.

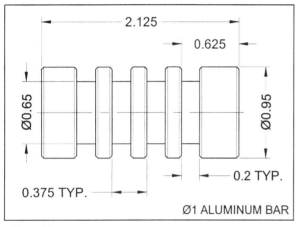

Figure 36-17

*Multiple grooves programming using a subprogram -
O3607 is the main program and O3657 is the subprogram*

In the *Figure 36-17* is a simple example of a multiple groove programming, using a subprogram. Only two cutting tools are used - a turning tool used for facing and turning and a 0.125 wide part-off tool that machines the four grooves, cuts the back chamfer and parts-off the finished part. Part-off operations are described in the next chapter and subprograms are discussed in *Chapter 39*. Note that all the tool motions related to the position of the groove are programmed in the main program O3607, all the tool motions related to the actual groove cutting are programmed in the subprogram O3657. An equal spacing between the grooves is used for the example.

```
O3607 (GRV W/SUB-PROG)
(T01 - 55 DEGREE DIAMOND INSERT)
N1  G20 T0100
N2  G96 S500 M03
N3  G00 X1.2 Z0 T0101 M08
N4  G01 X-0.07 F0.006                  (FACE OFF)
N5  G00 Z0.1
N6  G42 X0.7                   (START OF CHAMFER)
N7  G01 X0.95 Z-0.025 F0.003          (CHAMFER)
N8  Z-2.285                            (TURN OD)
N9  U0.2 F0.03
N10 G00 G40 X4.0 Z4.0 T0100 M09
N11 M01

(T05 - 0.125 PART-OFF TOOL)
N12 G50 S2500 T0500
N13 G96 S500 M03
N14 G00 Z-0.5875 T0505 M08           (POS-GRV1)
N15 X1.0
N16 M98 P3657                        (CUT GRV 1)
N17 G00 W-0.375 M98 P3657            (CUT GRV 2)
N18 G00 W-0.375 M98 P3657            (CUT GRV 3)
N19 G00 W-0.375 M98 P3657            (CUT GRV 4)
N20 G00 Z-2.285           (OPEN UP FOR PART-OFF)
N21 G01 X0.8 F0.006
N22 X1.1
N23 G00 X1.0 Z-2.2        (CHAMFER BACK START)
N24 G01 X0.9 Z-2.25 F0.003            (CHAMFER)
N25 X-0.02 F0.005                    (PART-OFF)
N26 G00 X1.2                           (CLEAR)
N27 G40 X4.0 Z4.0 T0500 M09
N28 M30
%

O3657 (SUB-PROG FOR O3607)
N1  G01 X0.66 F0.004          (FEED TO ROUGH OD)
N2  G00 X1.0                         (CLEAR OUT)
N3  W-0.0875              (SHIFT TO LEFT CHFR)
N4  G01 X0.9 W0.05 F0.002           (LEFT CHFR)
N5  X0.66 F0.004             (FEED TO ROUGH OD)
N6  X1.0 W0.0375 F0.03          (BACK TO START)
N7  W0.0875             (SHIFT TO RIGHT CHFR)
N8  X0.9 W-0.05 F0.002             (RIGHT CHFR)
N9  X0.65 F0.004         (FEED TO FINISH OD)
N10 W-0.075                     (SWEEP BOTTOM)
N11 X1.0 W0.0375 F0.03         (BACK TO START)
N12 M99                     (RETURN TO MAIN)
%
```

This example completes the chapter related to grooving. Although grooving is a relatively simple machining operation, programming grooves can become a significant challenge in certain cases.

PART-OFF

Part-off, sometimes called a *cutoff*, is a machining operation typical to lathe work, usually using a barfeeder attachment. During a part-off, the cutting tool (or part-off tool) separates the completed part from the bar stock. The completed part will fall off the bar, usually into a special bin to protect it from damage.

PART-OFF PROCEDURE

Programming procedure for a part-off tool path is very similar to the grooving procedure. In fact, part-off is an extension of grooving. The purpose of part-off is somewhat different, because the objective is to separate the completed part from the stock material, rather than create a groove of certain width, depth and quality. The material bar stock is usually a long round rod that is 8, 10, 12 or more feet long.

Two most important considerations in part-off are the same as those for standard grooving. One is the chip control, the other is coolant application.

◆ Parting Tool Description

Part-off uses a special cutting tool. Such a tool used for part-off is called a parting tool or a part-off tool. Sometimes the term *cutoff* is used for this kind of a tool, as well as the machining method; it has the same meaning as the term *part-off*. The part-off tool is similar in design to a grooving tool, with one major difference. The *length* of the cutting blade is much longer than that of a grooving tool, making it suitable for deep grooves. A typical example of a part-off tool is illustrated in *Figure 37-1*.

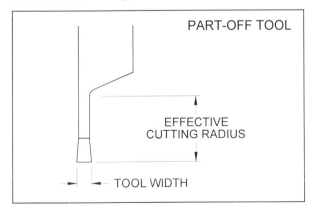

Figure 37-1

Part-off tool - cutting end configuration

At the end of the metal blade is usually a carbide insert, with clearance angles on both sides. The cutting end of the tool is available in several different configurations, always at the end tip of the carbide portion. The most typical tool end configurations are shown in the following illustration - *Figure 37-2*:

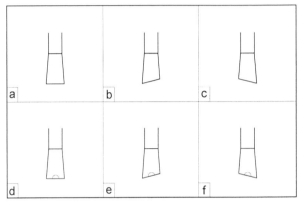

Figure 37-2

Part-off tool - cutting tip configurations

Note the two kinds of each grooving insert design shown - the series without a dimple (items *a, b* and *c*), and the series with a dimple (items *d, e* and *f*). The dimple is an intentional dent pressed in the middle of the cutting edge that deforms the chip and helps in coiling it. The result is a chip that is narrower than the width of the cut. Such a chip does not clog the generated groove and extends the tool life, although it may cost a little more.

Also note a slight angle on *b, c, e* and *f* styles. The angle helps in controlling the size and shape of the stub left on the part when it is separated from the solid bar. It also controls the rim size that is left over on the part when parting-off a tubular bar. Although all designs have their special applications, probably the most versatile choice would be the style *f*, particularly for large cutting diameters. Unlike in the other types of machining, the cutting chips for part-off should *coil*, not break. The cutting insert with a dimple or a similar design is the best suited for that purpose.

It is a common practice amongst programmers to use only one parting tool for all the work. They select the parting tool long enough to accommodate the maximum bar diameter and leave it permanently mounted in the tool holder, even for small diameters. The reasoning for this approach is that it saves a setup time. That is true to some extent but has a downside as well. The long part-off tool usually has a wider insert than a short tool, in order to compensate for strength and rigidity. When the bar size is large, a long part-off tool is necessary, with its relatively wide insert. If such a tool is used for short parts, such as rings or other tubular stock with thin walls, it is the wrong tool selecting that also wastes material. A short part-off tool with a narrower insert will justify the setup change.

A generous supply of coolant should always be made available at the cutting edge, just like for grooving. A water soluble coolant is a good choice, since it offers both the cooling and lubricating qualities. A typical mixture would be one part of soluble oil for 15-20 parts of water or as recommended by the coolant manufacturer. Make sure the coolant is supplied at full pressure, particularly for larger diameters. The pressure helps the coolant to reach the cutting edge and flush off the chips that may accumulate in the opening.

◆ Tool Approach Motion

The first step to program a part-off tool path is to select a part-off tool that has enough capacity to completely separate the part from a solid bar. The next decision is to select the insert width and the location of the tool reference point. A part-off tool that is too short will not reach the spindle centerline safely. A tool too long may not be rigid enough, may cause vibrations, even break during the cut. The width of the insert is also important for good cutting conditions. The width of the tool is directly proportionate to the cutting depth capacity.

Selection of the part-off tool reference point follows the same rules as those for a grooving tool. It is definitely advantageous to have the same side for both, grooving and parting, to maintain setup consistency. The following programs illustrate the difference between the tool reference point on the *left* and on the *right* side of the tool tip - *Figure 37-3* for program O3701 and *Figure 37-4* for program O3702). In both cases, the program zero is the front face of the finished part.

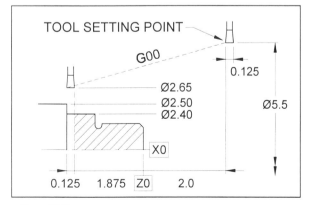

Figure 37-3
Part-off tool approach - left side tool reference - program O3701

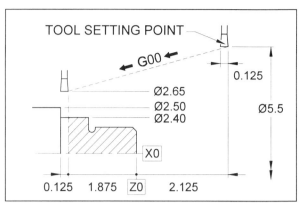

Figure 37-4
Part-off tool approach - right side tool reference - program O3702

In the examples, the tool change position and the final results are identical. Comparison of both programs shows the values of X axis unchanged, but the values for Z axis are different (blocks N122 and N125). This reflects the cutting side of the tool tip.

```
O3701 (PART-OFF / LEFT SIDE TOOL EDGE)
...
N120 G50 S1250 T0800 M42
N121 G96 S350 M03
N122 G00 X2.65 Z-2.0 T0808 M08
N123 G01 X-0.03 F0.004
N124 G00 X2.65 M09
N125 X5.5 Z2.0 T0800
N126 M30
%
```

This example is consistent with the previous suggestions for precision grooving. Setting up the tool reference point on the *left* side of the tool is easier for the CNC operator. If there is a good reason, set the tool reference point on the right side and program according to the *Figure 37-4*.

```
O3702 (PART-OFF / RIGHT SIDE TOOL EDGE)
...
N120 G50 S1250 T0800 M42
N121 G96 S350 M03
N122 G00 X2.65 Z-1.875 T0808 M08
N123 G01 X-0.03 F0.004
N124 G00 X2.65 M09
N125 X5.5 Z2.125 T0800
N126 M30
%
```

The weakness of the left side tool setting is that the insert width has to be always added to the Z position in the program. In the second example, the final length of the part is used directly, but a possible collision with the chuck or collet does exist. Take care selecting the X axis tool approach position and program the tool *above* the stock diameter, even if the previous turning operations removed most of the stock. *Figure 37-5* shows correct and incorrect approaches of a part-off tool.

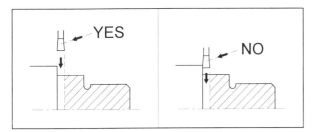

Figure 37-5
Correct and incorrect approach to stock diameter

◆ Stock Allowance

Part-off operation does not always mean all the machining has been completed. Often, part-off may complete only the first operation and additional machining will be necessary on the machined part. In such an event, some extra material (stock) has to be left on the back face, for subsequent finishing. Leave a stock amount of about .010 to .020 inches (0.3 to 0.5 mm). In that case, the block N122 would be changed in both programs - for example, from Z-2.0 to Z-2.01 in the first program example O3701 and from Z-1.875 to Z-1.895 in the second program example O3702.

Another program entry that is important to look at is the X value in block N122 - it is X2.65 in the example. That will leave .125 inches actual clearance above the ⌀2.400. If that seems a little too much, think again. Always consider the *actual* stock diameter, for safety reasons. In the example, the bar stock diameter is 2.500 inches and the actual clearance will be a more reasonable .075 of an inch per side of the stock.

◆ Tool Return Motion

Another safety aspect of programming a part-off tool is the method of returning to the tool change position, when the parting operation is completed. It may be very tempting to replace the two program blocks N124 and N125 with a single block, then return to the tool change position immediately after the part-off:

```
N124 G00 X5.5 Z2.0 (or Z2.125) T0800 M09
```

After all, the part has just been separated, fallen into the bin and one block in the program can be saved. *Don't do this*, it could be a very hazardous procedure. The part *should* have been removed by the tool and it *should* have fallen into the bin - but has all this actually happened? A variety of reasons may cause an *incomplete* part-off. The result is a broken tool, scrapped part, possibly a damage to the machine itself.

> Always return in the X axis first,
> and always above the bar stock diameter.

◆ Part-off with a Chamfer

Not always the machined part will be done during a secondary operation. When the machining has to be completed with a part-off tool, it will require the best quality overall finish possible. One requirement of a good surface finish is broken sharp corners. In the example, the sharp corner is at the intersection of X2.4 and Z-1.875. If the turning tool cannot cut the chamfer during turning operation, part-off tool can be a better choice. Most part-off tools are not designed for cutting sideways (along the Z axis), but chamfering removes only a small amount of material that is within the tool capabilities. Avoid chamfers that are wider than about 75% of the insert width or take several cuts if needed. The chamfer has to be cut *before* the part-off and it should be cut from outside in, *not* from inside out. The correct programming technique for machining a chamfer during part-off is summed up in the following steps:

❑ Position the tool further in the Z axis
    than would be normal for regular part-off

❑ Start the part-off operation and terminate it just
    below the diameter where the chamfer will end

❑ Return to the starting diameter and
    move to the chamfer start position

❑ Cut the chamfer in one block and
    part-off in the subsequent block

To illustrate the programming technique, study the following program example O3703 and illustration shown in *Figure 37-6* - the tool reference point is on the left side, and the required chamfer is .020 inches at 45°:

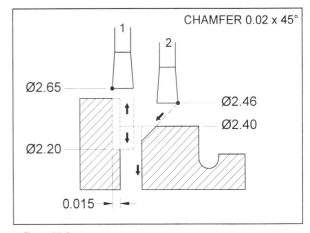

Figure 37-6
Corner breaking with a part-off tool - example O3703

```
O3703 (PART-OFF CHFR)
(G20)
...
N120 G50 S1250 T0800 M42
N121 G96 S350 M03
N122 G00 X2.65 Z-2.015 T0808 M08
```

```
N123 G01 X2.2 F0.004
N124 X2.46 F0.03
N125 Z-1.95                    (LEFT SIDE OF TOOL)
N126 U-0.1 W-0.05 F0.002
N127 X-0.03 F0.004
N128 G00 X2.65
N129 X5.5 Z2.0 T0800 M09
N130 M30
%
```

In block N122, the tool is positioned .015 past the position Z-2.0. Block N123 makes only a temporary groove (to ⌀2.200). The next block N124 is a motion out of the groove, to the starting diameter of the chamfer (⌀2.460). In the following block N125, the tool shifts in the Z axis, to the start position of the chamfer. The value of 1.950 was calculated by additions and subtractions:

```
1.875 - .020 - .030 + .125 = 1.950
```

The value of 1.875 is the back face of the part (as per drawing), the .020 value is the chamfer size; .030 is the clearance, and .125 is the insert width. Note the .125 tool width position adjustment, to maintain the tool reference point on the *left* side of the cutting edge while actually cutting with the right edge. Block N126 is the chamfer cutting, using incremental mode. Using the incremental mode saves a few calculations. If using the absolute mode, block N126 will be:

```
N126 X2.36 Z-2.0 F0.002
```

Also note the decreased feedrate for the chamfer only, to assure a good finish. The feedrate decrease can be quite significant for very small chamfers. The remainder of the program is unchanged.

In some cases, two tools can be justified for part-off operations. The setup of the two tools has to be accurate. A small, rigid grooving tool can do the startup groove and the chamfer, then the part-off tool can do the rest. At the completion of the part-off, the bar stock projecting from the spindle will have a small step. Make sure to program a facing cut for each subsequent part to take this step into consideration.

◆ Preventing Damage to the Part

When the part is separated from the bar, it falls down. On impact, it may suffer a damage severe enough to make a good part a scrap. To prevent the possibility of a damage, the CNC lathe operator may want to place a bucket filled with cutting coolant in the path of the falling part. Another method is to offset the part-off tool away from the centerline, just far enough that it does not separate the part. Then, when the machine is stationary (*i.e.*, not moving), the CNC operator breaks the part manually.

In any case, always follow the safety rules of the company - they exist for protection.

> Never touch the part while the program
> is in operation or the spindle is rotating.

The best solution for part damage prevention is a CNC lathe equipped with a parts catcher, which is often a special machine option, ordered at the time of machine purchase.

For part-off, just like for grooving operations, always make sure there is an adequate supply of inserts on hand. Tools with sharp edges, or with very small radii, are generally weak tools, yet doing some very demanding work. Nobody wants to run out of tools in the middle of a very important rush job.

38 | *SINGLE POINT THREADING*

Threading is a machining process used to produce a helical groove of a particular shape, usually on a cylinder. The major purpose of threads is to connect two parts together without damage during joining and disjoining (assembly and disassembly). The most common applications of threading fall into four major categories:

- ❑ Fastening devices   ... screws and nuts
- ❑ Measuring tools   ... micrometer barrel
- ❑ Motion transmission   ... lead screw, camera lenses
- ❑ Torque increase   ... lifting or supporting jacks

A thread cutting is a very versatile manufacturing process. There are two main groups of thread production - metal cutting and plastic molding. It should not be a surprise that it is the plastic molding method that dominates the manufacturing industry. Given the number of detergent bottles, pop bottles and other plastic products we consume, the number of threaded products employing this method is astronomical.

In the metalworking area of thread production, the group that is the subject of special interest, there are smaller several subgroups:

- ❑ Thread rolling or thread forming
- ❑ Tapping and die work
- ❑ Thread milling
- ❑ Thread grinding
- ❑ Single point thread turning

For a typical CNC programmer, the areas of interest are usually confined to the tapping, milling and single point threading. Methods for tapping operations have been described in *Chapters 25* and *26*, thread milling is described in *Chapter 45*. This chapter covers programming methods described as *single point threading*.

THREADING ON CNC LATHES

CNC lathes can produce a very high quality thread in addition to the variety of turning and boring operations, in a *single setup* of the machined part. This is a very attractive feature for manufacturers and many machine shops have purchased a CNC lathe for that reason alone. Any secondary operation requires additional setup, increasing the cost of production.

Single point thread cutting - typically known as a *single point threading* - uses a tool holder similar to other tool holders, but contains one special threading indexable insert, which may have one, two or three tips. Generally, the shape and size of the threading insert must correspond to the shape and size of the finished thread - *Figure 38-1*.

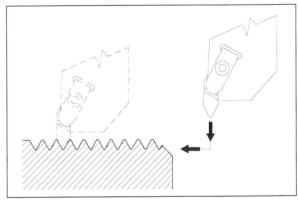

Figure 38-1
Comparison of the thread form and the threading tool shape

By definition, a single point threading is a machining process of cutting a helical groove of a specific shape with a uniform advancement per spindle revolution. The shape or form of the thread is mainly determined by the shape and mounting position of the cutting tool. The uniformity of advancement is controlled by the programmed feedrate.

◆ Form of a Thread

The most common thread form used in CNC programming is the familiar V-thread (in the shape of the letter V) with a 60° included angle. There is a large variety of the V-shape forms in manufacturing, including metric and English threads. Other forms include trapezoidal shapes such as metric trapezoid, ACME and worm threads, square and round threads, buttress threads and many others. In addition to these relatively common forms, there are threads specific to a particular industry, such as automotive, aircraft, military and petroleum industries. To make matters even more complicated, the thread shape can be oriented on a cylindrical surface, a conical surface, it also can be external or internal. The thread can be cut on a face (scroll threads), even on circular surfaces. It can have a single or multiple starts, right or left hand orientation, constant or variable lead, etc.

◆ Threading Operations

This section contains a detailed list of the threading operations that can be programmed for a typical CNC lathe. Several operations require a special type of threading insert and some operations can only be programmed if the control system is equipped with special (optional) features:

- ❑ Constant lead threads
- ❑ Variable lead threads
- ❑ External and internal threading
- ❑ Cylindrical threads (straight treads)
- ❑ Tapered threads (conical threads)
- ❑ Right hand (R/H) and Left hand (L/H) threads
- ❑ Face threads (scroll threads)
- ❑ Single start threads
- ❑ Multi-start threads
- ❑ Circular threads
- ❑ Multi-block threads

In spite of the seemingly endless possibilities and combinations in thread cutting, the programming knowledge and experience gained in one category will be indispensable in other categories. A good threading program is based on a sound knowledge of common threading principles.

TERMINOLOGY OF THREADING

Threading is a relatively large subject, in fact, it is large enough to have a whole book dedicated to it. As subjects of this kind usually are, threading has its own technical terms. These terms appear in books, articles, technical papers, manuals and other sources. To understand them is mandatory for any CNC programmer and operator.

Listed here are some of the most common terms used for threads and thread cutting:

❑ **ANGLE OF THREAD**

... is the included angle between the sides of the thread, measured in an axial plane

❑ **CREST**

... is the top surface of a thread that joins the two sides

❑ **DEPTH OF THREAD**

... generally, the distance between the crest and the root of the thread, measured normal to the axis (in programming, depth is considered as a measurable value per thread side)

❑ **EXTERNAL THREAD**

❑ ... is a thread that is cut on the outside of the machined part, for example as a bolt

❑ **INTERNAL THREAD**

... is a thread that is cut on the inside of the machined part, for example as a nut

❑ **HELIX ANGLE**

... is the angle made by the helix of the thread at the pitch diameter with a plane perpendicular to the axis

❑ **LEAD**

... is the distance the threading tool will advance along an axis during one spindle revolution. The lead always determines the threading feedrate and can have constant or variable form.

❑ **MAJOR DIAMETER**

... is the largest diameter of the thread

❑ **MINOR DIAMETER**

... is the smallest diameter of the thread

❑ **MULTISTART THREAD**

... is a thread with more than one start, shifted by the pitch amount

❑ **PITCH**

... is the distance from a specified point of one thread to the corresponding point of the adjacent thread, when measured parallel to the machine axis

❑ **PITCH DIAMETER**

... on a straight thread, the pitch diameter is an imaginary diameter, "the surface of which would pass through the threads at such points as to make equal the width of the threads and the width of the spaces cut by the surface of the cylinder"

❑ **ROOT**

... is the bottom surface of a thread, joining the sides of two adjacent threads

❑ **SCROLL THREAD**

... is also known as a face thread - it is a thread machined along the X axis, rather than the more common thread machined along the Z axis

❑ **SHIFT**

... in multistart threading, it is the distance by which the cutting tool is displaced to cut another start; this distance is always equal to the pitch of the thread. The number of shifts is always one less than the number of starts

❑ **TAPERED THREAD**

... is a thread on which the pitch diameter is increased or decreased by a constant ratio (such as a pipe thread)

❑ **TPI**

... in English units of measuring, the number of threads counted over the length of one inch (1 / pitch) - metric thread is defined by its pitch - TPI equivalent is not applicable

THREADING PROCESS

Threading is one of the most automated programming tasks in modern machine shop, yet it could be one of the more difficult operations done on a CNC lathe. Initially, it may seem an easy procedure to make a program for a tool path that has the cutting parameters very clearly defined, such as threading. Practical applications, however, could present a big departure from theory. This comment may be arguable, at least until it is time to start searching for solutions to unusual threading problems or even regular threads that just don't seem to be coming out right. An experienced programmer should have the ability to think of yet another solution, when all the other solutions seem to have been used up. This is true for any problem solving process and applies equally to threading problems.

What often makes threading a difficult operation is the cutting tool application. The single point threading tool is unlike any cutting tool. Although the holder is mounted in the turret just like other tools, the cutting insert is unique. Threading tool not only *cuts*, it also *forms* the thread shape. Frequently, the threading insert has the shape of finished thread. The mounting of a threading tool in the turret can be at 90° to, or parallel with, the machine spindle centerline, regardless of thread being cut. The decision which way to mount the tool is determined by the angle of the thread, relative to the spindle center line. It is important that the tool is mounted *square* in the turret. Even a small angular deviation will have an adverse effect on the finished thread.

◆ Steps in Threading

Compare a threading insert with a common 80° diamond tool used for rough turning, and a few oddities will emerge:

Tool radius:

Threading	=	almost sharp edge
Turning	=	typical average is .0313 radius (0.8 mm)

Tool angle:

Threading	=	typically 60° and a weak support
Turning	=	80° and a strong support

Typical feedrates:

Threading	=	up to .25 in/rev (6.5 mm/rev) or more
Turning	=	.015 in/rev to .03 in/rev typical (0.4 mm/rev to 0.6 mm/rev)

Typical depth of cut:

Threading	=	small
Turning	=	medium to large

The comparison shows that even a fine pitch thread cannot be cut with a single threading pass. A single pass would produce a thread of poor quality at best and a unusable thread at worst. The tool life would also be much shorter than expected.

A better approach is to cut the thread in several passes, each pass increasing the thread depth.

For this purpose of multi-pass cutting, the machine spindle rotations must be synchronized for the start of each pass, so each thread depth is at the same position on the threaded cylinder. A quality thread will be completed when the last cutting pass produces the proper thread size, shape, surface finish and tolerances. Since the single point threading consists of several passes to cut a single thread, programmers must understand these passes well.

In programming, the structure of each pass remains the same, only the thread data change from one pass to another. In a most elementary setup, there are at least four motions for each threading pass (as applied to a straight thread):

Motion 1	From the starting position, move the tool to the thread diameter in rapid motion mode
Motion 2	Cut the thread - one axis thread cut (at the feedrate equal to the lead)
Motion 3	Rapid retract from the thread
Motion 4	Rapid return to the starting position

Expending on these brief descriptions, the four step tool motion process will typically include the following considerations that are critical to the CNC program.

Threading Motion 1

Before the first step, the threading tool must move from its indexing position to the position close to the machined part. This is a rapid motion, in the air. Make sure to calculate the XZ coordinates for this position correctly. The coordinates are called the thread *starting position*, because they define where the thread cut will start from and eventually return to. The start position must be defined away from the part, but close to the thread, as the intersection of the X axis clearance and the Z axis clearance.

The first tool motion is directly related to the thread. It is a motion *from* the starting position *to* the cutting diameter of the thread. Since the thread cannot be cut at full depth in a single pass, the total depth must be split into a series of more manageable depths. Each depth will depend on type of tool, the material and the overall rigidity of the setup. This approach motion is programmed in rapid mode.

Threading Motion 2

When the tool reaches the cutting diameter for a given depth, the *second motion* becomes effective. The actual threading pass will be cut during this step, at the specified feedrate and only when the machine spindle is *synchronized* with the threading feedrate. There is no need to take any special steps to maintain the synchronization - in threading mode, the synchronization is automatic. The thread will be cut to the programmed thread end position.

Threading Motion 3

In the *third motion*, when the thread cutting diameter is completed, the tool must retract away from the thread, at the machine rapid rate, to the X axis clearance position. This tool position is normally a diameter programmed *outside* of the threaded area.

Threading Motion 4

The threading process is completed with the *fourth motion*, when the tool returns to the starting position in a rapid mode. All remaining passes are programmed in the same way, just by changing the thread cutting diameter (thread depth control).

Note that only *Threading Motion 2* will be programmed in the threading mode, using a proper G code. Threading motions 1, 3 and 4 will be in G00 (rapid) mode.

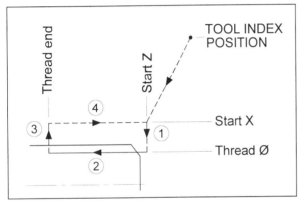

Figure 38-2
Basic steps in single point thread cutting

This typical description illustrated in *Figure 38-2*, is only general in nature and usually not sufficient by itself for high quality thread cutting.

◆ Thread Start Position

The tool starting position is a *clearance position*. For a straight cylindrical thread, the minimum suitable clearance along the X axis is about .100 (2.5 mm) per side, more for coarse threads. For a tapered thread, the clearance is the same, but applied over the larger diameter.

As for the clearance along Z axis, some special considerations are necessary. When the threading tool comes into contact with material, it must be advancing exactly 100% of the programmed feedrate. Since the cutting feedrate for threads is equivalent to the thread *lead*, it will take some time to arrive at the programmed feedrate. Just like a car needs some time to accelerate before reaching its cruising speed, the threading tool has to reach a full feedrate *before* it contacts the material. The effect of acceleration must be considered when deciding the front clearance amount.

When programming coarse threads, the front clearance amount required will generally be much greater than the amount for fine or medium threads. For example, a common thread with 8 TPI requires feedrate of .1250 in/rev! If the Z axis clearance is too small, the machine acceleration process will be incomplete when the tool contacts the material. The result will be an imperfect and unusable thread. To avoid this serious problem, this rule may help:

> Z axis clearance for the starting point should be three to four times the length of the thread lead

This is only a rule of thumb and works well in every day practice. Control manuals may offer a scientific way of calculating the minimum clearance.

In some cases, the Z axis clearance must be reduced because of space shortage, such as when the threading starts very close to a tailstock or machine limits. Since the acceleration time depends directly on the spindle speed, the only remedy for imperfect threads in this case is to *lower* the spindle speed (r/min) - *the feedrate must not be reduced!*

For complex methods of infeed, the starting position is changing for each cut by a calculated amount.

◆ Thread Cutting Diameter and Depth

For cylindrical and conical thread cutting using the block method of programming (no cycles), select the cutting diameter for *each* pass of the threading tool in the program. From the thread starting position, the cutting tool will move *towards* spindle centerline for external threads and *away from* spindle centerline for internal threads. The actual cutting diameter for each pass must be selected not only with respect to the thread diameter, but also with respect to machining conditions.

In threading, the chip load on the insert becomes heavier as the cutting depth increases. A damage to the thread, to the insert, or both, can be averted by maintaining a *consistent chip load* on the insert. One way to achieve the consistency is to *decrease* each subsequent depth of the thread, another way is to apply a suitable *infeed method*. Both threading techniques are often used simultaneously.

To calculate the depth of each pass, complex formulas are not required, just common sense and a bit of experience. All threading cycles have an algorithm (special process) built in the control system that calculates each depth automatically. For manual calculations, the procedure follows a logical approach. The total *depth of the thread* (measured per side) must be known - programmer decides how many *threading passes* will be suitable for the particular thread. Another value to be decided is the *last* cut depth, the cut that actually finishes the thread. These values usually come from experience. The rest is limited to mathematical calculations or available charts.

When the three parameters (values) are established, the total cutting depth must be distributed among the individual threading passes, including the last pass depth. Start with approximation of individual calculations to make each depth smaller than the preceding one and still not exceed the total thread depth. Chances are that each depth will not guarantee absolutely consistent chip load, but the cuts will be well within the margin of acceptability.

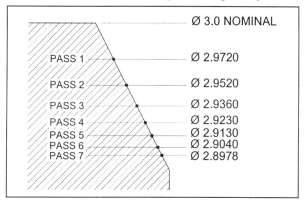

Figure 38-3

Threading diameters distributed for equal chip load

Figure 38-3 shows a typical external thread (diameter is only an example). It is a 12 TPI thread and the programmer has to find single depth of the thread. There is a mathematical way to do it, using a standard formula. The single, full profile external thread depth will be 0.0511 inches, based on the following thread depth formula - for Unified and metric *external* threads only:

$$D = \frac{0.61343}{TPI} = 0.61343 \times P$$

☞ where ...

D	=	Single depth of external thread
TPI	=	Number of threads per inch
P	=	Pitch of the thread (1/TPI)

According to another thread specification standard (UN thread forms), the constant in the formula is 0.64952, which would make the depth 0.0541.

For a full profile internal thread, the formula to calculate the depth will be used for metric and American National threads only - D value is the *internal* depth:

$$D = \frac{0.54127}{TPI} = 0.54127 \times P$$

If seven threading passes are selected, with the last pass of 0.0031 (for programming convenience), the individual depths can be distributed the following way:

Pass #1 depth - 0.0140	Accumulated depth = 0.0140
Pass #2 depth - 0.0100	Accumulated depth = 0.0240
Pass #3 depth - 0.0080	Accumulated depth = 0.0320
Pass #4 depth - 0.0065	Accumulated depth = 0.0385
Pass #5 depth - 0.0050	Accumulated depth = 0.0435
Pass #6 depth - 0.0045	Accumulated depth = 0.0480
Pass #7 depth - 0.0031	Accumulated depth = 0.0511

These calculations will be used when the program is actually written. The thread depth is dependent on the number of threads per inch and whether the thread is external or internal. The thread diameter is irrelevant for the thread depth calculation. Once the diameter is known, the calculated depths can be used and the diameter for each thread pass found. As an example, for an external thread of 3.0-12 size, each threading diameter calculated is based on the nominal thread size of three inches:

Threading diameter #1	3.0 - 2 × 0.0140	= 2.9720
Threading diameter #2	3.0 - 2 × 0.0240	= 2.9520
Threading diameter #3	3.0 - 2 × 0.0320	= 2.9360
Threading diameter #4	3.0 - 2 × 0.0385	= 2.9230
Threading diameter #5	3.0 - 2 × 0.0435	= 2.9130
Threading diameter #6	3.0 - 2 × 0.0480	= 2.9040
Threading diameter #7	3.0 - 2 × 0.0511	= 2.8978

There is nothing wrong with this method of calculating the threading diameters. What this method lacks is a *built-in check for accuracy*. Since each diameter is calculated from the nominal diameter, any error in calculation is *not* accumulative and might be hard to find. A much better method is to calculate each threading diameter based on the *previous* calculation, using single depth of cut, *not* the accumulative depth - compare it with the last method:

Threading diameter #1	3.0000 - 2 × 0.0140 = 2.9720
Threading diameter #2	2.9720 - 2 × 0.0100 = 2.9520
Threading diameter #3	2.9520 - 2 × 0.0080 = 2.9360
Threading diameter #4	2.9360 - 2 × 0.0065 = 2.9230
Threading diameter #5	2.9230 - 2 × 0.0050 = 2.9130
Threading diameter #6	2.9130 - 2 × 0.0045 = 2.9040
Threading diameter #7	2.9040 - 2 × 0.0031 = 2.8978

The advantage of this method is that once the last diameter is found (2.8978 in the example), *add* the double depth to this diameter and the result *must* be equal to the nominal diameter of the thread, or 3.0000 in the example:

$2.8978 + 2 \times 0.0511 = 3.0000$

If the result is *not* the nominal thread diameter, there was an error in the calculations. Using *both* methods and comparing the results is significantly more thorough check

♦ **Thread Cutting Motion**

When the cutting tools reaches the threading pass depth the thread is cut. The cutting starts at the Z axis clearance position and ends at the end of thread with cutting feedrate in effect. Although the threading cut is, in effect, a linea motion, do not use preparatory command G01 for thread-ing. If G01 is used, the start for each pass will *not* be synchronized with the previous thread start. Instead of G01 command, use a G code specifically designated for threading. G32 is the most common code used by Fanuc controls for threading. During a thread cutting motion G32, control system automatically disables the feedrate override. The CNC operator has to be extra careful to set the threading tool exactly, particularly when thread ends close to shoulders of the part. To illustrate the programming process up to this point, here is a typical program section:

```
...
N61 G00 X3.3 Z0.3          (START POINT XZ)
N62 X2.972                 (THREAD DIA START)
N63 G32 Z-1.75 F0.0833     (THREAD TO END)
...
```

◆ Retract from Thread

The moment the thread has reached the end position along Z axis, the tool must leave the material immediately, to avoid making a damage to the thread. This is the third motion in the basic threading process. The retract motion can have two forms - straight away in one axis (normally along the X axis), or a gradual pullout in two axes (simultaneously along XZ axes) - *Figure 38-4.*

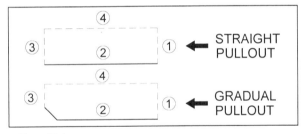

Figure 38-4
Straight and gradual pullout from a thread

Generally, the *straight pullout* should be programmed whenever the tool ends cutting in an *open space*, for example in a relieve or a recess groove. For threads that do not end in an open area, the *gradual pullout* is a better choice. Gradual pullout motion produces better quality threads and prolongs life of the threading insert. To program a straight pullout, the threading mode G32 must be canceled and replaced by a rapid motion mode, using the G00 command:

```
N64 G00 X3.3                      (RAPID OUT)
```

For the gradual pullout, the threading G code and the feedrate *must remain in effect.* When the normal length of thread is completed - but *before* the tool is retracted - the threading tool moves in two axes simultaneously, ending *outside* of the thread. The normal length of the pullout is usually 1 to 1-1/2 times the lead (not the pitch), the suggested angle is 45°. It is also important to pay attention to the clearance diameter.

```
N64 U0.2 W-0.1             (GRADUAL PULLOUT)
N65 G00 X3.3               (RAPID OUT)
```

For *external* threads, the clearance diameter must always be *further* away from spindle center line than the diameter of gradual pullout. For *internal* threads, the clearance diameter must be *closer* to spindle center line than the diameter of gradual pullout. *Figure 38-5* illustrates the concept.

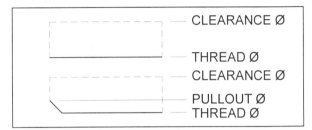

Figure 38-5
Thread pullout and clearance diameter (external example)

◆ Return to Start Position

Regardless of how the tool retraction from the thread is programmed, straight or gradual, the last step in the threading process is always a return to the starting position. This tool motion is entirely in the open space, therefore programmed in the rapid motion mode G00. Normally, the return motion to the starting position is along one axis only, usually the Z axis. This is because in most programs, the tool retraction from the thread has already reached the X axis diameter. Here is a complete program excerpt - gradual pullout is shown:

```
...
N61 G00 X3.3 Z0.3          (START POINT XZ)
N62 X2.972                 (THREAD DIA START)
N63 G32 Z-1.75 F0.0833     (THREAD TO END)
N64 U0.2 W-0.1             (GRADUAL PULLOUT)
N65 G00 X3.3               (RAPID OUT)
N66 Z0.3                   (RETURN TO Z-START)
...
```

THREADING FEED AND SPINDLE SPEED

In threading, the choice of the cutting insert, the spindle speed and feedrate selection are rather restricted. Both, the cutting tool and the feedrate are determined by the finished thread, as specified in the engineering drawing. Threading insert is one of the *weakest* tools used on CNC lathes - yet its applications demand some of the *heaviest* feedrates used in CNC lathe programming for any tool. Other factors that can influence the final thread have to be dealt with as well, such as spindle speed, the depth of each threading pass, the tool edge preparation, setup of the cutting tool and insert, plus similar considerations. Often, a change of only one factor will correct a threading problem. *Figure 38-6* compares feedrates for turning and threading.

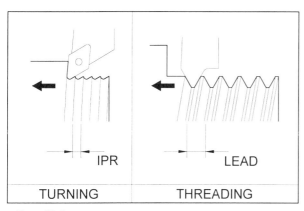

Figure 38-6

Comparison of turning and threading feedrates

◆ Threading Feedrate Selection

The selection of feedrate for general turning or boring is based on such factors as material type, tool nose radius, required surface finish, etc. In this sense, the 'correct' feedrate for turning and boring cover a large range. In threading, this flexibility is limited. The threading feedrate is always determined by the *lead* of the thread - *never* the pitch. In English units drawings, the thread description is given as the number of threads over one inch length, or TPI (TPI = threads per inch), and a nominal diameter. As an example, a thread that is described in the drawing as 3.75-8, means that the thread has *8 threads per inch*, and the nominal diameter (for example, the major diameter) is ∅3.750. All single start *metric* threads have the pitch standardized, depending on the thread diameter. For instance, a thread described as M24×3 is a single start metric thread with the pitch of 3 mm on a 24 mm diameter. A description M24×1 means a single start thread with the pitch of one millimeter.

Regardless of the dimensional unit, the most important terms for selecting the correct feedrate are the *lead of the thread* and the *number of threading starts*.

It may help to review some of the basic relationships of the *thread lead* and the *thread pitch* (see the terminology of threading earlier in this chapter). In a common machine shop conversation (shop talk), the words *lead* and *pitch* are often used incorrectly. The reason is that for a single start thread, the amount of the lead is *identical* to the amount of the pitch. Since most machine shops work with a single start thread on a daily basis, the misuse of the terms is seldom noticed. In addition, virtually all taps have a single start. What may be acceptable in a shop talk language has to be interpreted correctly in CNC programming. Each term has a very specific meaning in threading, so use them in the correct way:

$$\text{LEAD} = \frac{\text{No. of starts}}{\text{TPI}} = F$$

The programmed threading feedrate
is always the lead of the thread, never the pitch!

$$\text{PITCH} = \frac{1}{\text{TPI}}$$

From the last two formulas is easy to deduct that if the number of starts is *one*, both the lead and pitch will always have the same value.

The following formula should be applied for threading feedrate calculation:

$$F = L = P \times n$$

☞ where ...

F	=	Required feedrate (in/rev or mm/rev)
L	=	Lead of the thread (inch or mm)
P	=	Pitch of the thread (inch or mm)
n	=	Number of starts (positive integer)

For example, a thread with a single start and the pitch of three millimeters (3 mm) will require feedrate of

```
3 × 1 = F3.0
```

For threading programs that use English units, the above formula is equally valid, since

$$P = \frac{1}{\text{TPI}}$$

☞ where ...

P	=	Thread pitch
TPI	=	Number of threads per inch

As an example, the thread with one start and 8 TPI will require feedrate of

```
1/8 × 1 = .125 × 1 = F0.125
```

Multistart threads are special in many ways, but the feedrate selection is also the *lead* - not the pitch of the thread.

◆ Spindle Speed Selection

The speed of the spindle for thread cutting is always programmed in direct *r/min*, never as a constant surface speed (CSS). That means the preparatory command G97 must be used with address S, specifying the number of revolutions per minute. For example, G97S500M03, will result in 500 r/min spindle speed. It is true that single point threading takes place over several diameters between the first pass and the root of thread, so G96 selection would seem logical. This is not the case. First, even for fairly deep coarse threads, the difference between the first and last diameter is insignificant. Second - and this reason is even more important - the thread cutting routine requires a perfect spindle and feedrate *synchronization* at the start of each pass. Such synchronization can be more accurately achieved only with constant *r/min* rather than constant surface speed (CSS).

For the majority of threads, the selection of *r/min* requires only consideration of general machining conditions, similar to other turning operations. At the same time, select the spindle speed with some consideration of the feedrate. Because of the heavy feedrates used for threading, there is a distinct possibility that certain threads cannot be cut at *any available spindle speed*. If this is confusing, keep in mind that the feedrate is determined not only by the lead, but also by the *overall capability of the machine*. Every CNC lathe has a programmable feedrate value, specified in either *in/min* or *mm/min*, up to a certain maximum for each axis.

Take a typical maximum programmable feedrate for the X axis may as 250 in/min (6350 mm/min); the maximum for the Z axis may be 450 in/min (11430 mm/min). Recall that there is a direct relationship between the spindle speed and the feedrate per revolution. The result of this relationship is actually *feedrate expressed in terms of time*, not per revolution. The feedrate per time is always the result of the spindle speed in direct *r/min* multiplied by the feedrate per revolution in *in/rev* or *mm/rev*.

◗ English example :

```
700 r/min × .125 in/rev = 87.500 in/min
```

◗ Metric example :

```
700 rpm × 3 mm/rev = 2100 mm/min
```

In CNC lathe programming generally, not only in threading, always make sure that the feedrate per revolution combined with the spindle speed will be *less than or equal to* the maximum *available* feedrate per time for the axis with the lower rating, which is usually the X axis.

Based on this simple rule, the maximum spindle speed for a given lead can be selected according to the following formula:

$$R_{max} = \frac{Ft_{max}}{L}$$

☞ where ...

R_{max} = Maximum allowable r/min
Ft_{max} = Maximum feedrate per time (X axis)
L = Lead of the thread

◗ English example :

If the thread lead L is .125 and the maximum feedrate for the X axis Ft_{max} is 250 *in/min*, then the maximum threading speed R_{max} will be:

```
Rmax = 250 / .125 = 2000 r/min
```

◗ Metric example :

If the thread lead L is 2.5 mm and the maximum feedrate for the X axis Ft_{max} is 6350 mm/min, then the maximum R_{max} threading speed will be:

```
Rmax = 6350 / 2.5 = 2540 rpm
```

The maximum allowable *r/min* only reflects the capabilities of the CNC machine. The feedrate actually used in a program must also take into account the various machining and setup conditions, just like any other tool path operation. In practice, the majority of actual programmed spindle speed (r/min) will be well below the maximum capacity of the CNC machine tool.

◆ Maximum Threading Feedrate

The selection of cutting feedrate in general was discussed earlier, in *Chapter 13*. After studying the section on the maximum *r/min* selection (spindle speed), it should not be surprising that similar limitations apply to the determination of *a maximum threading feedrate* for a given spindle speed (programmed as r/min). Again, the limits of the CNC machine tool are very important, so be aware of them when writing the thread cutting program.

Maximum programmable threading feedrate for a given spindle speed (in *r/min*) can be calculated from the following formula:

$$Fr_{max} = \frac{Ft_{max}}{S}$$

☞ where ...

Fr_{max} = Maximum feedrate for a given spindle speed
Ft_{max} = Maximum feedrate per time (X axis)
S = Programmed spindle speed (r/min)

◗ English example:

If the maximum machine feedrate along X axis is 250 in/min and the spindle speed S is selected as 2000 r/min, then the maximum programmable feedrate will be:

```
250 / 2000 = .125 in/rev
```

Therefore, the maximum thread lead that can be cut at 2000 r/min is .125 inches, which allows 8 threads per inch or finer.

Changing the spindle speed (feedrate remains the same) will allow programming coarser threads on the same CNC lathe. For example, if only 1500 r/min is selected instead of the 2000, the maximum lead will increase to .1670 inches or 6 threads per inch.

● Metric example :

In a similar example, using metric units, the maximum machine feedrate for X axis is 6350 mm/min and the programmed spindle speed S is selected as 1600 rpm. In this case, the maximum programmable threading feedrate will be (in mm/rev):

`6350 / 1600 = 3.969 mm/rev`

That means the maximum lead that can be threaded at 1600 r/min must be less than 4 mm.

The calculated values only indicate the actual capabilities of the control and the machine and do not guarantee a safe job setup or even suitable machining speed.

◆ Lead Error

Normally, the threading feedrate requires the address F, with up to four decimal place accuracy for threads in English units (F2.4 format), and three decimal place accuracy for metric threads (F3.3 format). The majority of threads are short and this accuracy is quite sufficient. There is never problem for metric threads, regardless of the thread length, because the thread is defined by its lead already in the drawing. For threads programmed in the English units, the thread lead must be calculated from the given threads per inch (TPI) in the drawing. For many English threads, the lead is accurately calculated within the four decimals available for the F address. A 10 TPI requires programmed feedrate of F0.1, 16 TPI requires programmed feedrate of F0.0625, etc. These are threads that divide the TPI into one within the four decimal places accurately, such as 8, 10, 16, 20, 40 to name the most common number of threads.

Not all threads fall into this rather convenient group. For many other threads, the calculated value must be properly rounded off.

Take a 14 TPI thread, for example. The exact threading feedrate should be 1/14=.071428571 inches per revolution. The rounded value used in the program should be F0.0714. Over a short thread length there is no noticeable error at all and the thread is well within all tolerances. That is not true if the thread is unusually long or the rounded value has been improperly calculated. An *accumulative error*, known as the *thread lead error*, will result in a possible scrap due to an incorrect thread. By using the rounded value of .0714, the loss is .000028571 inches for *each* thread revolution. Lead error over one inch (or more) can be easily calculated:

$$L_e = (F_a - F_p) \times TPI$$

☞ where ...

L_e = Maximum lead error per inch
F_a = Desired actual feedrate
F_p = Programmed rounded feedrate
TPI = Number of threads per inch

Over one inch, the error in the example will be .0004 of an inch, over fifty inches it will be full .0200 of an inch. Another example, somewhat more critical, is an incorrect rounding value. Ideally, a thread with 11.5 threads per inch should be programmed with the feedrate of .086956522. If this value is rounded to F0.0870, the accumulated error is .0005 per one inch and the error over 50 inches will be .0250 inches. Even if the CNC machine does not allow six decimal places for the threading feedrate, the proper rounding of the calculation is very important.

Compare the following rounded values and the errors they cause (11.5 TPI over 50 inches):

```
.0869  . . .      error of .0325
.0870  . . .      error of .0250
.0871  . . .      error of .0825
```

What a difference for only one ten-thousandths of an inch rounding.

The manufacturers of Fanuc controls recognized this potential problem and introduced the address E for threading feedrate on their earlier CNC controls. The benefit of using the E address for threading is that it allows programming with *six* decimal places instead of the standard *four* for English threads (increased accuracy allowed for metric threads using the E address is seldom used). With proper rounding, the accumulative error is virtually negligible.

Using the same illustration of 14 TPI over 50 inches, the error for the whole length will only be .0003 of an inch, if the F0.0714 is replaced by E0.071429. The second example, using a thread with 11.5 threads per inch, should be programmed with the feedrate of E0.086957. The accumulated error over fifty inches will be only 0.000275 inches, a negligible error.

> The latest CNC systems allow the use of a
> six digit accuracy for the F address as well!

The lead error is always a potential problem when programming long thread leads. Depending on the kind of threading applications in the machine shop, the accumulative error of the thread lead may be critical or it may never be an issue to deal with.

TOOL REFERENCE POINT

A good tool setup is critical to a good machining environment. While a good setup is important to all tools, it is even more important to maintain a good setup of the threading tools, external and internal. The tool cutting edge has to be properly oriented, securely mounted in the insert pocket and it has to be the right type. Its reference point, used for setup, is also very critical.

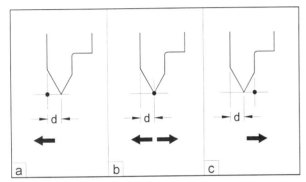

Figure 38-7
Typical reference points for setup of threading tools

The reference point of a threading tool requires more considerations than for turning tools. In the *Figure 38-7*, there are three possibilities, in the order of programming frequency. The third version *(c)* is the rarest and offers virtually no benefit to the programmer except in some cases of left hand threading. For most left hand threads, one of the first two versions is also quite sufficient.

The threading insert setting as in *Figure 38-7a* is the most suitable for general use and for threads that end at a shoulder. Configuration in *Figure 38-7b* is suitable for threads that end on an open diameter. The *Figure 38-7c* shows a possible setting for a left hand threading work.

Selection of the tool reference point (G50 or geometry offset setting) as per illustration in *Figure 38-7a* is the most desirable one, when the intent is to *standardize* tooling setup for *any* type of thread. It is the most convenient setting, regardless of the thread ending. It is also the safest at the same time. In some cases, an allowance must be made for the difference between the programmed edge and the actual edge. The tooling catalogs list this value precisely, or one half of the threading insert width (if applicable) can be used instead.

BLOCK-BY-BLOCK THREADING

The oldest method of single point thread programming is to calculate each and every motion associated with the threading and write it as an individual block of the program. This method is called *block-by-block* threading method, or just a *block threading method*.

Each of the four basic steps occupies one block of program, resulting in the minimum of four blocks per each threading pass. If the gradual pullout from the thread is used for thread cutting, there will be five blocks of program for each threading pass. When cutting coarse threads, threads in hard or exotic materials, even some multi start threads, this method often means quite a long program. The length of the program, difficulty in editing, high possibility of errors, and even small memory capacity of the control system, are the negative sides of using this method.

On the plus side, the programmer has an *absolute programming control over the thread*. Such control placed into capable hands can often be applied to some special threading techniques, for example, cutting a thread shape with a threading tool much smaller than the thread itself or making large knuckle threads with a round grooving tool.

Thread programming method using the block technique for a constant lead thread is available on all CNC lathes that support threading.

G32	Thread cutting command

The preparatory command for this type of threading is G32. Command G33 may exist on some controls, but G32 is the standard G code for Fanuc and compatibles.

In an example, a 3.0-12 TPI external thread will be used. All cuts are distributed in seven passes, for the total depth of .0511:

Pass #1 depth	=	.0140	Total depth	.0140
			Diameter	2.9720
Pass #2 depth	=	.0100	Total depth	.0240
			Diameter	2.9520
Pass #3 depth	=	.0080	Total depth	.0320
			Diameter	2.9360
Pass #4 depth	=	.0065	Total depth	.0385
			Diameter	2.9230
Pass #5 depth	=	.0050	Total depth	.0435
			Diameter	2.9130
Pass #6 depth	=	.0045	Total depth	.0480
			Diameter	2.9040
Pass #7 depth	=	.0031	Total depth	.0511
			Diameter	2.8978

Always make sure all diameters are calculated carefully without errors. Small error can cause big scraps.

The threading operation in program O3801 will use the tool and offset number 5 (T0505), at 450 r/min spindle speed (G97S450):

```
O3801
...
(N45 G50 X12.0 Z4.5)
N46 T0500 M42
N47 G97 S450 M03
N48 G00 X3.2 Z0.25 T0505 M08
```

Now, the thread start point has been reached. The next stage is to implement all four steps, one step per block, for the *first* pass:

```
N49 X2.972                    (PASS 1)
N50 G32 Z-1.6 F0.0833   (or F/E0.083333)
N51 G00 X3.2
N52 Z0.25
```

The remaining six passes can be programmed next, just by changing the diameters. Note that the threading feedrate does *not* repeat - it is modal from block N50 on.

```
N53 X2.952                    (PASS 2)
N54 G32 Z-1.6
N55 G00 X3.2
N56 Z0.25
N57 X2.9360                   (PASS 3)
N58 G32 Z-1.6
N59 G00 X3.2
N60 Z0.25
N61 X2.9230                   (PASS 4)
N62 G32 Z-1.6
N63 G00 X3.2
N64 Z0.25
N65 X2.9130                   (PASS 5)
N66 G32 Z-1.6
N67 G00 X3.2
N68 Z0.25
N69 X2.9040                   (PASS 6)
N70 G32 Z-1.6
N71 G00 X3.2
N72 Z0.25
N73 X2.8978                   (PASS 7)
N74 G32 Z-1.6
N75 G00 X3.2
N76 Z0.25
```

Block N76 terminates the threading routines and the program can be closed as well, if there are no more tools used.

```
N77 X12.0 Z4.5 T0500 M09
N78 M30
%
```

What should strike as odd in the example, is the abundance of repetitions. Observe the three blocks following each new pass diameter - they are always the same. For a thread with many passes these repetitions will be very numerous. This block-by-block method has one main benefit - is it under programmer's full control. Adjustments may be made to the number of threads and depth of each pass. Non-standard infeed method and a gradual pullout from the thread can be added. Actual program editing after it has been completed is much more inconvenient.

BASIC THREADING CYCLE - G92

The control systems can perform many internal calculations and store their results in the control memory for further use. This feature is especially useful for threading, since the repetitiveness of block-by-block tool motion can be avoided and the program shortened significantly.

We will use the same program example that illustrated the G32 command and apply it to a simple threading cycle. This cycle is usually called the G92 threading cycle on Fanuc controls. Incidentally, G92 in the threading context has *nothing* to do with the command of the same name, the traditional and now old-fashioned G92, the *position register* setting command. If the lathe control uses G92 for simple threading cycle, use G50 for the position register command. This applies for older controls only, since modern controls use advanced geometry offsets.

The schematic illustration of a G92 thread cutting cycle is shown in *Figure 38-8*.

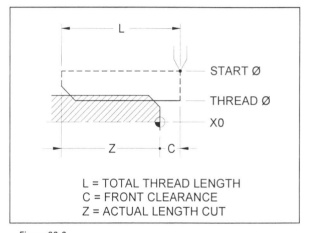

Figure 38-8

G92 - simple thread cutting cycle

For a comparison with the G32 programming method, the same thread will be programmed, 12 threads per inch on a 3.000 inch external diameter. The program will do exactly the same job, except it will have a noticeably different structure.

Using the G92 cycle, the following list shows the calculated diameters for each thread pass, as they will appear in the program (no change at this stage):

Pass #1 depth	=	∅ 2.9720
Pass #2 depth	=	∅ 2.9520
Pass #3 depth	=	∅ 2.9360
Pass #4 depth	=	∅ 2.9230
Pass #5 depth	=	∅ 2.9130
Pass #6 depth	=	∅ 2.9040
Pass #7 depth	=	∅ 2.8978

As before, the threading tool has been assigned a tool number and spindle speed - tool 5 (T0505) and 450 r/min:

```
O3802
...
(N45 G50 X12.0 Z4.5)
N46 T0500 M42
N47 G97 S450 M03
N48 G00 X3.2 Z0.25 T0505 M08    (START POSITION)
```

The first four blocks are identical to the block method of threading. In the next step, the threading tool will be positioned at the first pass diameter, chase the thread, retract from the thread and return to the starting position. The last three blocks are repetitive for each pass. The main benefit of the G92 threading cycle is that it eliminates such repetitive data and makes the program easier to edit.

The format for the G92 straight threading cycle is:

> G92 X.. Z.. F..

☞ where ...

    X = Current diameter of the thread pass
    Z = End position of the thread
    F = Threading feedrate in in/rev

The first threading pass will be programmed in this block - N49. Note the X axis and the Z axis input values as well as the cutting feedrate:

```
N49 G92 X2.972 Z-1.6 F0.0833        (PASS 1)
```

The control system will take the last X value and the last Z value *before the cycle call* as the starting position for the thread. This position is the starting point for the cycle. In the example, the starting position is X3.2 Z0.25 (block N48). The remaining six threading passes can be programmed just by changing the diameters. There is no need to repeat the Z value *or* the feedrate.

```
N50 X2.9520                         (PASS 2)
N51 X2.9360                         (PASS 3)
N52 X2.9230                         (PASS 4)
N53 X2.9130                         (PASS 5)
N54 X2.9040                         (PASS 6)
N55 X2.8978                         (PASS 7)
```

The block N55 will be completed by an automatic return motion to the starting position of the thread. From that position, the program ends the same way as for G32.

```
N56 G00 X12.0 Z4.5 T0500 M09
N57 M30
%
```

One frequent programming mistake that can be made with this cycle is to omit G00 command in block N56. G92 cycle can be canceled only *by another motion command*, in this case by a rapid motion G00. If G00 is missing in the

program, the control system will expect that there are more threads to cut, while they had been actually completed in the previous block.

The simple threading cycle G92 is just that - it is simple, without any frills. It does not have any special infeed methods, in fact, the only feeding method is a straight plunge type. Later in this chapter, the plunge method of infeed will be described as *not* suitable for most threading operations.

An automatic gradual pullout can be programmed with G92 by using M24 function prior to calling the G92 cycle, with examples later in this chapter. If the control system supports the feature (most controls do), always use the much more sophisticated threading cycle - G76, described in the next section.

MULTIPLE REPETITIVE CYCLE - G76

In *Chapter 35*, various lathe cycles were the main subject, normally used for turning and boring. In this section, a similar look will aim at one more of the multiple repetitive cycles, this time used for various threading applications.

In the earlier stages of CNC development, the simple G92 threading cycle was a direct result of the computerized technology of its time. The computer technology has been rapidly advancing and many great new features have been offered to CNC programmers. These new features simplify the program development. One of the major additions is another lathe cycle, used for threading - a multiple repetitive threading cycle G76. This cycle is considered a complex cycle - not because it is difficult to use (on the contrary) but because it has some powerful internal features.

To fully appreciate the impact of G76 threading cycle, compare it with the original G32 threading method, and even the G92 cycle just described. While a program using the G32 method requires four or even five blocks of program for each threading pass, and the G92 cycle requires one block for each threading pass, the G76 cycle will do *any single thread* in *one* block of program code (two blocks are required for some controls). With the G76 cycle, *any number* of threading passes will still occupy only a very small portion of the program, making editing on the machine (if necessary) very easy and fast.

There are two programming formats available, depending on the control model. This is similar to programming of the other lathe cycles.

◆ G76 Cycle Format - 10T/11T/15T

A threading cycle requires initial data input - information provided to the control that defines the thread in machining terms. *Figure 38-9* illustrates the G76 for Fanuc 10/11/15T controls.

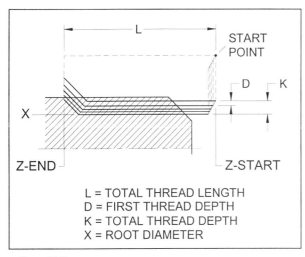

Figure 38-9
G76 - Multiple repetitive thread cutting cycle (10T/11T/15T)

These parameters form the structure of the one-block cycle (for external *or* internal threads):

> G76 X.. Z.. I.. K.. D.. F.. A.. P..

☞ where ...

> X = Diameter of the last threading pass
> Z = Position indicating the thread end
> I = Amount of taper over the total length
> K = Single depth of the thread - positive
> D = Depth of the first threading pass - positive
> A = Included angle of the insert - positive
> P = Infeed method (one of four) - positive

Observe differences in the format structure for the multiple repetitive cycle G76 with the basic G92 cycle. The G76 cycle appears to be simple, but internally, it is very complex - the control system must do a large number of calculations and checks. This is one reason why we use computers - to let them do the hard work. These calculations need data (repetitive information), in the form of input parameters that establish the thread specifications. Yet, in spite of the more input values, the G76 is a very easy cycle to use in CNC programming.

◆ G76 Cycle Format - 0T/16T/18T

On the popular Fanuc controls 0T, 16T and 18T, the G76 cycle is somewhat changed from the 10/11/15T models. Its purpose and function remain the same, the difference is only in the way how program data input is structured. Fanuc 10/11/15T use a single line cycle input, described earlier. Fanuc 0/16/18T control models require a two line input.

If the control system requires a double block entry for a G76 cycle, the two-block programming format is:

> G76 P.. Q.. R..
> G76 X.. Z.. R.. P.. Q.. F..

☞ where ...

First block:

> P = ... is a six-digit data entry in three pairs:
>   Digits 1 and 2 - number of finishing cuts (01-99)
>   Digits 3 and 4 - number of leads for gradual pull-out
>   (0.0-9.9 times lead), no decimal point used (00-99)
>   Digits 5 and 6 - angle of thread
>   (00, 29, 30, 55, 60, 80 degrees only)
> Q = Minimum cutting depth
>   (positive radial value - no decimal point)
> R = Fixed amount for finish allowance
>   (decimal point allowed)

Second block:

> X = (a) Last diameter of the thread (absolute diameter)
>   ... or ...
>   (b) The distance from the start point to
>   the last thread diameter (incremental)
> Z = End of thread along the Z axis
>   (can be an incremental distance W)
> R = Radial difference between start and end
>   positions of the thread at the final pass
>   (R0 used for straight thread can be omitted)
> P = Height of the thread
>   (positive radial value - no decimal point)
> Q = Depth of the first threading pass
>   (positive radial value - no decimal point)
> F = Feedrate of the thread (same as the thread lead)

This format follows the logic of several lathe cycles described earlier in *Chapter 35*. Do not confuse the P/Q/R addresses of the first block with the P/Q/R addresses of the second block. They have their own meaning - within each block only!

➲ Example - English units
(External 1-11/16 thread with 20 TPI) :

```
N10 G76 P011060 Q005 R0.003
N11 G76 X1.6261 Z-1.5 P0307 Q0100 F0.05
```

➲ Example - Metric units
(Internal M76×1.5 thread) :

```
N20 G76 P011060 Q050 R0.05
N21 G76 X76.0 Z-30.0 P812 Q250 F1.5
```

◆ Programming Example

The earlier example of the thread, with 12 TPI on an external diameter of 3.000 inches, can be easily adapted to the G76 programming method. Examples for both types of controls are shown, using only the *minimum* number of program blocks (last tool shown in examples):

```
O3803 (G76 METHOD - ONE BLOCK METHOD)
...
(N45 G50 X12.0 Z4.5)
N46 T0500 M42
N47 G97 S450 M03
N48 G00 X3.2 Z0.25 T0505 M08
N49 G76 X2.8978 Z-1.6 I0 K0.0511 D0140 A60
    P4 F0.0833                    (or F/E0.083333)
N50 G00 X12.0 Z4.5 T0500 M09
N51 M30
%
```

Several points relating to the program may need clarification. The fact that the whole program requires only six or seven blocks is, in itself, significant. Any programming change can be done by a simple modification of a proper parameter in block N49, which is the threading cycle call. For instance, to change the depth of the first threading pass to .0160 from the current .0140, all that has to be modified is the entry of D0140 to D0160.

The comparison of the G76 cycle with G92 cycle is unfair, as each cycle is the product of a different technological era. They coexist in the same control unit even at the present time, mainly to be *downward compatible* with older programs. The two cycles are a good illustration of some significant differences between programming techniques.

For example, in the G92 threading cycle application, input of *each* thread pass diameter is important, in G76 cycle, only the *last* pass diameter input is important.

Internally, the CNC system does all necessary calculations. The supplied information is contained in the program. First, the control registers the thread starting position, the same way as for G92 cycle. In this example (block N48), the position is X3.2Z0.25. The next step the control goes through is the evaluation of all G76 parameters (the programmed data in block N49). The X value is the diameter of the last threading pass, the K value is the single thread depth. That provides enough information for the control to 'know' what is the *theoretical* premachined part diameter (the *actual* premachined diameter cannot be known). This relationship is important for selection of the tool rapid approach *direction*. If the thread start diameter X is *larger* than the last pass diameter, the threading is *external*. If the thread start diameter X is *smaller* than the last pass diameter, the thread is *internal*.

The Z value in the G76 cycle has the same meaning as the Z value in the G32 thread cutting or the G92 threading cycle. It represents the *end* position of the thread and controls the thread length.

Two parameters unique to G76 cycle are the I and the K values. The I value is always a zero if a straight diameter thread is cut. A non-zero value is used for taper threads, where it represents the single difference between the start diameter of the cut and its end diameter (described later in the section dealing with a tapered thread).

In the two block version, the same program will be very similar, applying the same logical thinking.

```
O3804 (G76 METHOD - TWO BLOCK METHOD)
...
(N45 G50 X12.0 Z4.5)
N46 T0500 M42
N47 G97 S450 M03
N48 G00 X3.2 Z0.25 T0505 M08
N49 G76 P011060 Q005 R0.003
N50 G76 X2.8978 Z-1.6 P0511 Q0140 F0.083333
N51 G00 X12.0 Z4.5 T0500 M09
N52 M30
%
```

There are few other parameters to explain, but first look at how the cycle calculates the first thread depth. *The higher level controls using the one-block input will be used for the explanations, unless mentioned otherwise.*

◆ First Thread Calculation

For the G32 block threading, as well as for the G92 simple threading cycle, the thread starting position was always determined as only *reasonable*, applied to both axes for the purposes of supplying a suitable tool clearance. The Z axis clearance in the start position block only takes into consideration the lead of the thread and the spindle speed. Its purpose is to prevent cutting imperfect threads, due to the machine acceleration for the feedrate. The clearance for the X axis is an arbitrary clearance for the tool to move away from the thread. The same principles apply to G76 threading cycle as well and can be used the same way as in the previous threading methods.

There is one *major* difference from programming the G32 and the G92 methods. In the previous threading examples, the starting position for the X axis was X3.2. In the G32 block cutting, as well as in the G92 simple threading cycle, the *first* threading diameter was *always* programmed (in the examples, the value was X2.972). This is not the case in the G76 threading cycle. In this cycle, it is the *last* diameter that is programmed - *not the first* - and that means the first cut diameter must be calculated by the control system internally.

The calculation of the *first* thread diameter is done completely by the control system, providing the following information is supplied:

❏	The root diameter	[X value]
❏	The total thread depth	[K value]
❏	The first thread depth	[D value]

Based on the supplied values, the first diameter T_f of the thread will be calculated as:

$$T_f = X + (K \times 2) - (D \times 2)$$

In the example, X is 2.8978, K is .0511 (or P0511), the first threading depth D is .0140, entered in the program as D0140 or Q0140, depending on the control. Therefore, the first diameter of the thread T_f will be:

$$T_f = 2.8978 + (.0511 \times 2) - (.014 \times 2)$$
$$T_f = 2.9720$$

The result is the same diameter as in the previous simpler examples, but this time it was calculated by the control unit.

THREAD INFEED METHODS

The entry of the threading tool into the material can be programmed in several ways. One of the most important options is the method that controls threading tool approach towards the thread, also known as the *threading infeed*. This is a method detailing the motions of the threading tool, using one of *two* basic methods of infeed, as illustrated in *Figure 38-10*.

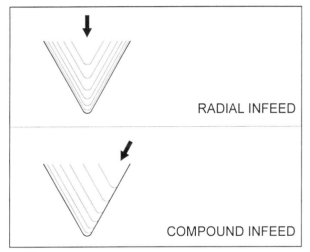

RADIAL INFEED

COMPOUND INFEED

Figure 38-10
Radial and Compound infeed for thread cutting

One common infeed method in thread programming is the plunge method, sometimes called the *radial* method, also known as the *straight* or *perpendicular* infeed; the other method is called an *angular* method, better known as a *compound* infeed or a *flank* infeed.

The need to control the infeed direction in threading is to offer the best cutting conditions for the insert edge. Except for threads with very fine leads and some soft metals, the majority of threading cuts will benefit from a compound infeed (at an angle). Some threaded shapes are excluded for the reason of their geometry - for example, a square thread will always need a plunge infeed (straight radial infeed). The angle of infeed is programmed with the A parameter of the G76 cycle.

Each threading method has its own procedures, using the following features:

- ❏ Constant cutting amount
- ❏ Constant cutting depth
- ❏ One edge cutting
- ❏ Both edges cutting

The P parameter of the G76 cycle selects the feature.

◆ Radial Infeed

Radial infeed method of the threading tool is also one of the most common conventional threading methods. It can be applied to a unidirectional, *straight motion* of the cutting tool, towards the diameter being cut. The threading tool is fed straight for each new pass. This pass diameter is specified as the X data in the program. In G76 threading cycle, A0 parameter is used for a radial infeed. In G32 block programming and G92 simple threading cycle, there is no parameter to program. The Z axis start position is the same for all thread diameters and is easier to program. The radial infeed is suitable for soft materials (brass, some aluminum, etc.), but it could damage threads cut in harder metals.

The result of a radial infeed motion is that *both* insert edges of the threading tool are removing material at the *same time*. Since the edges are opposite to each other, the curling of the chips will also be opposite to each other. In many applications, this will cause high temperature and tool wear problems related to heat. Even decreasing depth for each infeed may not eliminate the problem. If the radial infeed does not produce a high quality thread, a compound infeed approach will generally do a much better job.

◆ Compound Infeed

Compound infeed method, also called a *flank* infeed method, uses an *angular* direction of the tool that moves towards the threading pass diameter. The chip shape produced by compound threading method is similar to the shape of a chip produced by turning. Only one edge of the threading tool does the actual cutting, so the heat dissipates *away* from the tool edge and the chips curl away, extending the tool life. The depth of the chip can be heavier and fewer passes will be required for most threads. In *Figure 38-10* is shown the compound infeed, where one cutting edge is in constant contact with the thread wall. There is no cutting, only undesirable rubbing which may cause a poor surface finish on the thread. To avoid this problem, program the infeed angle a little smaller than the flank angle (one half of the thread included angle). A typical V-thread, with 60° included angle has the flank angle 30° and the infeed angle should be a little less than that, say 29°. *Keep in mind* - the shape or geometry of the thread is not changed - that is built into the shape of cutting insert. What is changing is the way *how* the insert will cut - *Figure 38-11*.

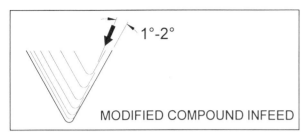

Figure 38-11
Modified compound infeed angle for better thread quality

In the G76 threading cycle, there are very powerful tools in forms of cutting parameters, two of which are related to the infeed method of a threading tool. One is the *address A*, and the other is the *address P*. Only the angle description is available for the two-block method, as the last pair of the P address in the first G76 command - N49 G76 P....60 Q.. R..

◆ Thread Insert Angle - Parameter A

For the compound infeed (all controls), a *non-zero* value is assigned to the parameter that represents the tool angle, a value that is equal to the *included* angle of the threading insert. The tool approach towards the part will be a little less than one half of the included angle A. For example, if the standard A60 is programmed in G76 cycle, the infeed angle will be slightly less 30°, allowing for the extra clearance.

Only the following six A angle settings are allowed in a G76 threading cycle:

A0	A29	A30	A55	A60	A80

A0	Straight or plunge infeed	ISO
A29	ACME type of thread	ANSI
A30	Metric Trapezoidal thread	DIN 103
A55	Whitworth 55° thread	BSW, BSP
A60	Standard 60° V-thread	English or Metric
A80	German PG thread	Panserrohrgewinde

As an additional information and further reference regarding the above listed thread forms, an ACME thread (29° included angle) is very commonly used for transmission of motion, for example a tire changing car jack uses an ACME thread. Some programming notes are described later in this chapter. A *Metric Trapezoid* thread is the metric version of the ACME thread, with 30° included angle. A Whitworth thread has an included angle of 55°, and has its roots in Great Britain; its usage had declined even there, as metric threads become standard worldwide. As far as the A80 PG thread, it is a special German pipe thread *(Panserrohrgewinde)*, with the included angle of 80°, not common in North America.

On the latest Fanuc controls (higher level), there is the P parameter for the G76 cycle, that works very closely with the A parameter and defines the *cutting type*.

◆ Thread Cutting Type - Parameter P

In the G76 threading cycle, the threading infeed can be programmed with the address P, in addition to the address A. The purpose of the threading parameter A is to control the threading infeed method - up to a certain extent - based on the *included angle* of the threading insert. For a more controlled infeed method, a method that controls the threading *depth*, there is also parameter P, programmable in the G76 cycle format and available for the Fanuc controls 10T and higher. It defines the *thread cutting type*, relating to the programmed depth of the thread.

In addition to the radial infeed (straight or plunge), programmed with the A0 parameter and compound infeed (non-zero parameter A), there are two other main cutting types that can be used in programming a thread infeed - a *one side cut* and a *zig-zag cut*. These terms refer to the number of cutting edges employed at one time. The one side cut refers to cutting with *one edge*, the zig-zag cut refers to cutting with *two cutting edges*. Each of them can be used in conjunction with the selected A thread angle parameter and the cutting depth - either as a *constant amount* or a *constant depth*.

Fanuc CNC lathe controls offer four methods of controlling the thread cutting depth infeed *(Figure 38-12)*:

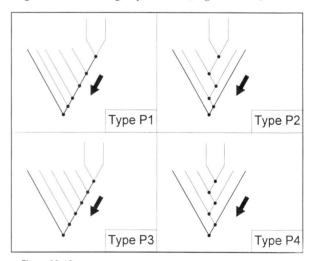

Figure 38-12
Cutting types for G76 threading cycle (parameter P)
Used on Fanuc 10/11/15T models

P1	One edge cutting ...	with constant cutting amount
P2	Two edges cutting ...	with constant cutting amount
P3	One edge cutting ...	with constant cutting depth
P4	Two edges cutting ...	with constant cutting depth

On Fanuc lathe controls manufactured before the model 10T, the P parameter in the G76 cycle was not available. The equivalent of what is now the P1 parameter was the default. On the controls that *do* support the P parameter, if the P number is omitted in G76 cycle call, P1 cutting method is assumed as a default. This is the most common threading application and will be suitable for many jobs. It will apply *one cutting edge* of the threading tool and the *constant cutting amount*. That will result in equal chip volume removal. Feel free to experiment with the other three options as well.

ONE-BLOCK METHOD CALCULATIONS

The compound infeed does not present programming problems when using the advanced G76 threading cycle. If a CNC system has the G76 cycle available, it can be used for about 95% of all work. What about the remaining 5%? What if the G76 cycle cannot be used and a program needs fully controlled compound infeed? How to control other infeed methods available for G76, without having the G76 cycle available or impractical to use?

Unfortunately, there is only one way - take a pocket calculator and calculate each and every tool position and tool motion individually. Is it a lot of work? Yes. Is it worth doing? Absolutely. It has to be a really good job, because even a slight modification at the machine could be very difficult. A top class programming job is always worth the extra time and effort when quality and precision of the final part depends on it. Quality is not instant - programmers (and machine operators) have to invest some work and time into it.

The principles of compound threading as applied to a block-by-block programming are simple, but the programming work may be tedious and editing on the machine may be impractical. Each threading pass has to be calculated in a different Z axis start position. This is called the *shifted* position that must be calculated exactly, otherwise the program will fail. It also had better be right the first time, otherwise the changes could be long and costly. Again, in this example, the same thread will be used as in previous examples (3.0-12 TPI). Program will use the G32 threading command, with a modified compound infeed at 29°.

◆ Initial Considerations

The thread used for the examples in this section is a 3.0-12 TPI external thread. All individual diameters for each threading pass had been calculated earlier, and all single depths for each pass had been established as well at the same time. These values will be used in this example as well. In total, there are seven diameters and seven depths to program the Z value (total depth accumulation is .0511 per side), as illustrated in *Figure 38-13*.

The illustration shows the distribution of each single threading depth for the seven threading passes and matches the table following the illustration:

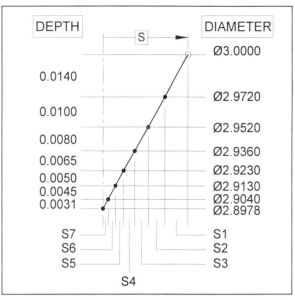

Figure 38-13

Compound infeed calculations for G32 block-by-block threading

Pass #1	depth at	⌀ 2.9720	(single depth .0140)
Pass #2	depth at	⌀ 2.9520	(single depth .0100)
Pass #3	depth at	⌀ 2.9360	(single depth .0080)
Pass #4	depth at	⌀ 2.9230	(single depth .0065)
Pass #5	depth at	⌀ 2.9130	(single depth .0050)
Pass #6	depth at	⌀ 2.9040	(single depth .0045)
Pass #7	depth at	⌀ 2.8978	(single depth .0031)

In addition to the single depths and threading pass diameters, *Figure 38-13* also shows *shifts* as an S and S1-S7 distances. When the Z starting position is shifted, the shift distance must be calculated on the basis of compound angle and the threading pass depth. Any new calculation must be based on the last calculation.

◆ Z Axis Start Position Calculation

The illustrated distance S represents the *total* shift from the nominal Z axis starting position, in the example programmed as Z0.25. The shift is to the Z positive direction.

Theoretically, it makes no difference *which* direction the shift is programmed - towards the thread or away from the thread. Practically, it is always better to program the shift *away* from the thread, if possible - the tailstock may be in the way, so watch its position. This way, the distance for the feedrate acceleration will *increase*, rather than decrease.

Although another approach may also be chosen, the S distance will be calculated first. The total thread depth is .0511 and the selected compound infeed angle is 29°, so using a standard trigonometric formula will provide the S distance value:

```
S = .0511 × tan29 = .028325193
```

The S distance represents the *total* shift of a threading tool. The shift for each threading pass will be its relative share of the S value. Each share is identified as an S_x in the *Figure 38-13*, within the range of S1 to S7.

$$S_x = D \times \tan 29$$

☞ where ...

   S_x = Shift for the current thread pass - incremental
   D = Single depth of the current thread pass

Calculation for each pass uses the same formula, changing the D depth input. Keep in mind that the purpose of this process is to find a new Z start position for *each threading pass - i.e.,* the Z value for a given thread diameter. *Figure 38-14* illustrates the process.

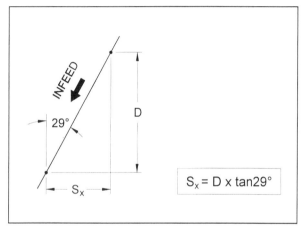

Figure 38-14

Calculation of thread start position - Z axis

Once the modified Z axis start position is known for the first pass depth, it is easy to find start positions for each subsequent pass depth. We know that the modified Z axis starting position for the threading tool will be the already established Z0.25, *plus* the .0283 shift S, rounded from the calculated value of .028325193. The *theoretical* starting position will be Z0.2783, calculated at the ∅3.000, but *never used in the program itself.* This initial value is needed for all the others. For each subsequent calculation, the S_x value has to be *subtracted* from the *current* Z starting position. The following list shows the individual shift values (as rounded numbers in English units):

```
S1 = .0140 × tan29 = .0078
S2 = .0100 × tan29 = .0055
S3 = .0080 × tan29 = .0044
S4 = .0065 × tan29 = .0036
S5 = .0050 × tan29 = .0028
S6 = .0045 × tan29 = .0025
S7 = .0031 × tan29 = .0017
Total .  .  .  .  .  .0283
```

The seven shifted positions for the start Z axis position can be calculated, based on the theoretical starting position of Z0.2783 at the ∅3.000:

```
#1 =      S - S1 = .2783 - .0078 = .2705
#2 = .2705 - S2 = .2705 - .0055 = .2650
#3 = .2650 - S3 = .2650 - .0044 = .2606
#4 = .2606 - S4 = .2606 - .0036 = .2570
#5 = .2570 - S5 = .2570 - .0028 = .2542
#6 = .2542 - S6 = .2542 - .0025 = .2517
#7 = .2517 - S7 = .2517 - .0017 = .2500
```

This example shows the initial position offset away from the thread, then moved one step at a time back to the original Z0.25 position. Using this method offers confidence that the originally set .250 inches minimum clearance will never be smaller. Only the Z axis value will change - other programmed values are not affected by this programming method at all.

The complete program is not short (which is typical with G32 programming), but it does illustrate the compound method of threading when no cycle is available or is practical to use. Only the threading tool is shown in the example.

```
O3805
(COMPOUND INFEED EXAMPLE)
...
(N45 G50 X12.0 Z4.5)
N46 T0500 M42
N47 G97 S450 M03
N48 G00 X3.2 Z0.2705 T0505 M08        (START 1)
N49 X2.972                            (PASS 1)
N50 G32 Z-1.6 F0.0833        (or F/E0.083333)
N51 G00 X3.2
N52 Z0.265                            (START 2)
N53 X2.952                            (PASS 2)
N54 G32 Z-1.6
N55 G00 X3.2
N56 Z0.2606                           (START 3)
N57 X2.9360                           (PASS 3)
N58 G32 Z-1.6
N59 G00 X3.2
N60 Z0.257                            (START 4)
N61 X2.9230                           (PASS 4)
N62 G32 Z-1.6
N63 G00 X3.2
N64 Z0.2542                           (START 5)
N65 X2.9130                           (PASS 5)
N66 G32 Z-1.6
N67 G00 X3.2
N68 Z0.2517                           (START 6)
N69 X2.9040                           (PASS 6)
N70 G32 Z-1.6
N71 G00 X3.2
N72 Z0.25                             (START 7)
N73 X2.8978                           (PASS 7)
N74 G32 Z-1.6
N75 G00 X3.2
N76 Z0.25 M09
N77 X12.0 Z4.5 T0500
N78 M30
%
```

In program O3805, the thread infeed method is equivalent to the P1 parameter in G76 cycle. This cutting type employs only a single edge of the threading insert, with a constant amount per each threading pass. It represents the most common programming method for threads and can be used as a sample for many other thread cutting applications. Block-by-block threads will be longer and will need to be checked for accuracy very carefully.

THREAD RETRACT MOTION

Earlier, a statement had been made that there are only two methods of retracting the tool from the thread - a straight motion along *a single axis*, and a gradual simultaneous motion along *two axes*. Both are used in thread programming. In fact, their frequent applications even justify special miscellaneous functions built into the control system as a standard feature. These thread retract functions are called the *thread chamfering functions* or *thread finishing functions*.

◆ Thread Pullout Functions

When using the threading cycles G92 and G76 for the CNC lathe work, the end of the thread (the Z axis value) will either be in a material that has been previously recessed, or in a solid material. The actual pullout can be programmed along a single axis, or along both axes simultaneously. Typical Fanuc functions designed for this purpose are M23 and M24. They control the pullout of the threading tool at the thread end:

M23	Thread finishing ON	(two axes)
M24	Thread finishing OFF	(one axis)

Other machine controls may have similar functions. The purpose of these functions is to enable or disable the *automatic insertion* of a pullout motion between threading motion sequences 2 and 3, as described earlier in this chapter. *Figure 38-15* illustrates the comparison of the threading motion with and without the pullout.

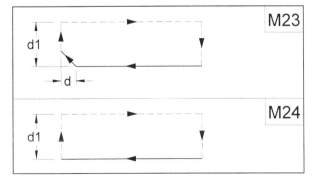

Figure 38-15
Typical miscellaneous functions for gradual thread pullout

◆ Single Axis Pullout

A single axis pullout (thread finishing OFF) is a simple rapid motion programmed at the end of threading pass as the *third motion* of the four basic threading sequences. The pullout direction is always at 90° to the thread. For either threading cycle G92 or G76, this is the default condition, so M24 is not needed, unless M23 function is used as well, usually for another thread in the same program. These two functions cancel each other. If M24 function is used, it must be programmed *before* the threading cycle for which it has been applied. For example, the threading program O3803 using the G76 cycle will be slightly modified in O3806:

```
O3806
...
N45 (G50 X12.0 Z4.5) M24    (THREAD PULLOUT OFF)
N46 T0500 M42
N47 G97 S450 M03
N48 G00 X3.2 Z0.25 T0505 M08
N49 G76 X2.8978 Z-1.6 I0 K0.0511 D0140 A60
    P4 F0.0833            (or F/E0.083333)
N50 G00 X12.0 Z4.5 T0500 M09
N51 M30
%
```

The M24 function appears in block N45, the only block that was available without another M function.

◆ Two-Axis Pullout

Two-axis pullout is a gradual angular tool motion along two axes, away from the thread (thread finishing ON). The example O3807 is similar to the previous example:

```
O3807
...
N45 (G50 X12.0 Z4.5) M23    (THREAD PULLOUT ON)
N46 T0500 M42
N47 G97 S450 M03
N48 G00 X3.2 Z0.25 T0505 M08
N49 G76 X2.8978 Z-1.6 I0 K0.0511 D0140 A60
    P4 F0.0833            (or F/E0.083333)
N50 G00 X12.0 Z4.5 T0500 M09
N51 M24                     (CANCEL M23)
N52 M30
%
```

In this case, M23 was applied in block N45 and an additional block N51 was used to cancel the pullout. The cancellation was not necessary in this program, but it is a good practice to cancel functions used only for specific purposes.

There are some conditions that apply to the M23 function. In *Figure 38-15*, the finishing distance *d* is set by the control parameter, within the range of .100× to 12.700× the thread lead. Normal control setting is equivalent to *one times* the thread lead. The pullout angle from the thread is usually 45°, or a little less because of a delay in the servo system. If the finishing distance *d* is greater than the pullout distance *d1*, the pullout will *not* be done.

HAND OF THREAD

Any thread can be cut in either the *right hand* or the *left hand* orientation. Neither selection has any effect on the profile and/or depth of the thread, but other factors are important. The majority of threading applications use the right hand thread. *Right hand* and *left hand* terms relate to the helix of the thread - *Figure 38-16*.

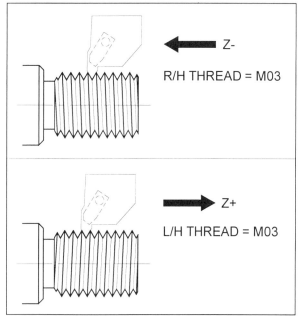

Figure 38-16

Right Hand (top) and Left Hand (bottom) thread cut using a right hand threading holder (reverse mounting)

The hand of thread is determined by two conditions:

❏   Cutting direction of the tool (Z+ or Z-)

❏   Direction of the spindle rotation (M03 or M04)

These conditions are used in combinations to program a particular thread. The factors that influence the programming method for a R/H and L/H thread are:

❏   Threading tool design - right hand or left hand

❏   Spindle rotation direction - M03 or M04

❏   The cutting direction - Z+ or Z-

❏   Tool tip orientation in the turret

Theoretically, either hand of thread can be cut with *any* threading tool, but this approach is not right. A poor choice affects the thread quality, life of the threading insert, additional costs involved, etc. When a thread starts close to a shoulder (in a recess), the clearance for acceleration is limited. The only method to prevent imperfect threads due to acceleration in a small area is to *decrease* the spindle speed.

THREADING TO A SHOULDER

Programming a thread that terminates at a shoulder presents a unique difficulty. The difficulty is the wall - better known as shoulder of the part. It is not enough to program the end point for the thread reasonably - it must be programmed *exactly*. Even then, a collision is possible if the tool setup is not accurate. The three typical problems in this area of thread programming are:

❏   Recess groove is too narrow or non-existent

❏   Threading insert is too wide

❏   Thread is too deep

The first problem of threading towards a shoulder, a narrow width of the recess groove, is easy to correct - just increase the recess width in the program. The majority of recess grooves can be adjusted for the threading tool, without damaging engineering intent behind the design. This may be a justified case of 'overruling' the drawing - but check first anyway!

The second and third problems may not be related, but the solution is usually the same for both. If the threading insert is too wide or the thread is too deep, try to increase the recess width first, if possible. If the recess width cannot be increased, for whatever reason, then there is another choice - to *decrease* the width of the threading insert. The obvious solution is to change the threading tool for a smaller one that can still cut the required thread depth. This may be an insert one size smaller, which usually requires a different tool holder as well.

If a smaller tool cannot be used, program for a *modified existing threading insert*. Modification in this case means grinding off the portion of the insert that is in the way of cutting, without disturbing the portion that actually removes the material. Before deciding on the modification by grinding, consider other options carefully - altering the standard tools designed for CNC work should always be the *last* resort, not the automatic first choice. A coated insert will loose its cutting advantages, if the coating is removed by grinding. Be careful not to grind off coating within the cutting section of the insert. In case the program does use a modified threading insert, a few suggestions may help to do it with more insight.

Always use care with modified tools

◆   Insert Modification

There is a number of standard threading inserts in every tooling catalogue and chances of finding one suitable for the job at hand are good. In case a standard threading insert needs modification, the following example illustrates a few programming considerations - incidentally, it is irrelevant if there *is* or there *is not* a recess groove on the part.

To modify a standard threading insert, look at its normal configuration first. *Figure 38-17* shows a typical threading insert with the known width W and the angular length A, tip radius or flat R, and an unknown angular height H.

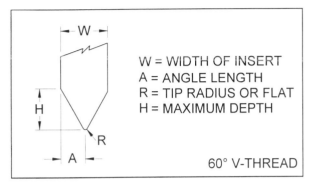

W = WIDTH OF INSERT
A = ANGLE LENGTH
R = TIP RADIUS OR FLAT
H = MAXIMUM DEPTH

60° V-THREAD

Figure 38-17
Essential dimensions of a threading insert

In the example, insert dimension W is .250 and A dimension is .130. The included angle of the threading insert is 60° and the insert flat or tip radius R is .012, not relevant in this case. The dimension H indicates the maximum thread depth and is normally measured to the sharp point of the insert tip. It is calculated using a trigonometric function:

```
H = A / tan30
H = .130 / .577350269
H = .225166605
H = .2252
```

The problem is illustrated in *Figure 38-18*.

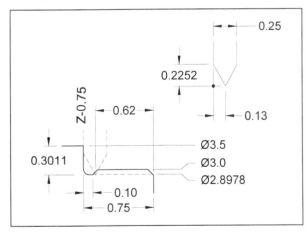

Figure 38-18
Threading insert before modification does not fit in the recess area

The job is to program a thread with a .100 recess groove width, using an insert that has an angular length A of .130. This insert is not suitable for the job, as it cannot finish the minimum full depth thread length - the difference between the shoulder length and the recess width:

```
.750 - .100 = .650
```

The minimum thread length in the illustration is only .620. There are no clearances and the length of the thread is too short. To solve this problem, select a smaller size threading insert if possible. If not, *modification* of a larger insert is the only way.

The modification requires grinding of the insert in the non-critical areas, to allow the tool to complete the minimum .650 thread length. In theory, the minimum amount to be ground off the insert is .030, the difference between the required and the actual thread lengths. This modification does not provide for any *clearance* at the thread shoulder or at the insert tip. Both of these clearances are essential for the best threading results. Even a minor setup error on the machine can cause a serious difficulty.

Always calculate the modification amounts, never guess them

In the example, there are three dimensions that influence the amount of insert modification. The sum of all three will be the amount to be ground off the insert. One, the thread length has to be extended by .030 to achieve the .650 minimum length. Two, the clearance from the shoulder will also be .030, and three, the clearance past the thread end will be .020. The two last clearances are the arbitrary decisions by the programmer. The solution is the total amount of the insert modification being .080. In other words, the amount of .080 must be ground off the original large threading insert. That will shorten the original angular length of .130 to the length of .050. Always make sure the depth of thread can be achieved with the modified insert. The part program will reflect the modification in the thread end position of the Z axis, which will be written as Z-0.8 (setup position of the insert does not change), and is illustrated in *Figure 38-19*.

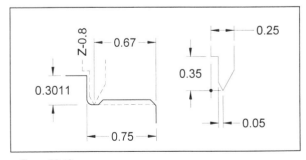

Figure 38-19
Modified threading insert provides enough clearance in the recess

In threading, the thread length is *the actual length of the full depth thread*. The part design often allows a little longer thread, but not shorter. The height of the shoulder is also important. In the example, the shoulder is .3011 high and the insert modification was possible. A large threading insert may not always be modified and the only solution will be to use a smaller insert size.

◆ Program Testing

Whether a threading insert used is based on catalogue dimensions or a modified insert, threading to a shoulder presents a time of anxiety for the CNC operator, when the first part is produced. Since the feedrate override and the feedhold switches are disabled during threading, the program verification on the lathe will become more difficult. Even computer based graphic testing methods may not show the potential collision.

A simple, yet very effective, thread program checking method is always available, right at the CNC lathe. This method requires a skilled CNC lathe operator, who does understand both the program and the threading principles well. Knowledge of the operation panels is also important.

This method employs several features found on the contemporary CNC controls. The purpose of the program test is to find out if the threading tool will collide with the part shoulder *before* actual threading cut takes place.

The following steps are general in nature - adapt them to suit local conditions when testing the threading program:

❑ Use the SINGLE BLOCK mode and step through the program until the thread start position is reached

❑ Switch from the AUTO to the MANUAL mode - spindle stops and the threading tool is in the clearance area

❑ Select the XZ screen display (absolute mode)

❑ Switch to the HANDLE mode for the Z axis

❑ While watching the XZ position display, move the handle in the same direction as the thread, until the tool reaches the programmed Z value, or it cannot move any further, whichever comes first

❑ If the tool reached the programmed Z position first, the tool setup is safe for the threading

❑ If the tool just about touched the part, but has not yet reached the programmed Z end position of the thread, the tool setup needs adjusting by the difference between the programmed position and the actual position, plus some additional clearance

There are other testing methods available, for example, to use temporarily the G01 *linear motion command*, instead of the G32 *threading command*, without a part mounted in the spindle.

> In the non-threading mode, the feed overrides are effective, whereas in the threading mode, they are not !

By reading the current tool position on the screen display and comparing it with the programmed position, it will be possible to know whether the collision will happen or not. During the test, the feedrate can be slowed down or stopped anytime. The purpose of the program test is to establish safe working conditions *before* the threading takes place.

OTHER THREAD FORMS

Although the standard V-shape thread with the 60° included tip angle is the most common thread form, it is by no means the only form. There are many threading forms and shapes programmers encounter in machine shops, too numerous to list.

As an example of a different threading form, look at an ACME thread as a subject for discussion. In metric, there is an equivalent thread, called the *Metric Trapezoidal* thread. From the programming perspective, both threads are almost identical. ACME thread has a 29° included thread angle, the metric trapezoidal thread has a 30° angle and somewhat different geometry definition.

The main application of the trapezoidal type thread is to *transmit a motion*, usually with a disengaging half-nut. Certain types of lead screws for conventional lathes use this type of thread. The programming of a trapezoid threads often requires a steadyrest, since these threads may be quite long. An important consideration is the lead error accumulated over a long distance, discussed earlier.

◆ Thread Depth

Every thread has its formulas and mathematical relationships. There are two basic formulas relating to an ACME thread depth. One is for threads of 10 TPI and *coarser*, the other for threads of 12 TPI and *finer*. For ACME threads 10 TPI and coarser, the thread depth formula is:

$$T_d = .500 \times P + .010$$

For the ACME threads 12 TPI and finer, the thread depth formula is modified only slightly:

$$T_d = .500 \times P + .005$$

☞ where ...

T_d = Thread depth
P = Thread pitch

Other threads in the trapezoidal group are *Stub ACME* or a *60° Stub ACME*. Programming trapezoidal threads is no more difficult than programming any V-shape thread, providing the thread formulas and the geometric details of the thread design are known to the programmer.

There other threads that can be encountered outside of the 60° category - the *Square threads*, *API* threads (used in the petroleum industry), *Buttress* threads, *Aero* threads, *Dardelet* self locking threads, *Round* threads, *Lebus* threads (require special control features), and several others. Thread and threading data can be found in various tooling catalogues and technical publications.

TAPERED THREAD

Programming procedure for a tapered thread is not significantly different than that for a straight thread. Tapered threading motion is along *two axes* simultaneously, rather than a single axis. The four basic motion steps are, therefore, almost identical to those for a straight thread:

Motion 1	Rapid from the start position to the thread diameter
Motion 2	Cut the thread (cutting along two axes)
Motion 3	Retract from the thread
Motion 4	Return to the start position

When compared to a straight thread, the only programming differences for a tapered thread are in the first two motions - *Motion 3* and *Motion 4* remain unchanged. In the *Motion 1*, the starting tool position is determined by the physical orientation of the threading tool - whether it is used for an external or an internal thread.

For *external* thread forms, the starting position of the threading tool must always be *above the largest diameter* of the thread. For *internal* thread forms, the starting position must always be *below the smallest diameter* of the thread. This is the same requirement as for a straight thread, but for a tapered thread it takes on an additional importance. For examples of a tapered thread, evaluate the simplified drawing in *Figure 38-20*.

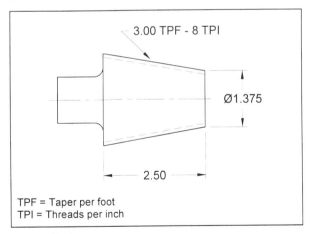

3.00 TPF - 8 TPI

Ø1.375

2.50

TPF = Taper per foot
TPI = Threads per inch

Figure 38-20
Tapered thread example - program O3808

The thread is defined by its overall length (2.500), by the front diameter of the blank part (1.375), by its angle (3.000 inches taper per foot) and by its pitch (8 TPI). It is a single start thread and the program zero will be at the front face of finished part. All premachining operations have been done for the example. The first programming consideration for this type of machining will be the depth of the thread.

◆ Depth and Clearances

From the previously established formula, the external depth D of the thread used in the program will be:

```
D = .61343 / 8 = .0766788 = .0767
```

The thread depth is measured axially and is not related to the thread angle. Once the depth is established, clearances can be set - one in the front and one at the end of the thread. The Z axis clearance amount will depend on the tool acceleration speed. Since the threading feedrate will be programmed as F0.125, four times the lead rule of thumb would require the clearance to be .500 of an inch. Only .400 of an inch is sufficient with a relatively slow spindle speed of 450 r/min. The end clearance can be smaller - there is enough open space at the thread end, and .200 is a reasonable clearance at the thread end, although a smaller value could be used as well.

For a tapered thread, consider the *total* length of the tool travel along each axis, not the actual thread length as per drawing - this is no different than for a single axis thread. The tool travel length in the example will be the combination of the two selected clearances plus the given thread length (along the Z axis):

```
.400 + 2.500 + .200 = 3.100
```

The next step may not be always necessary, depending on the method of programming. If the block threading method is used, *both* the start diameter *and* the end diameter of the thread will be needed for each pass. If a threading cycle G92 or G76 is used - the single distance between the start and end diameter of the tapered thread is needed. This distance will be programmed as parameter I of the threading cycle and is the part of a taper calculation. All previous examples were straight threads and the I value was zero (I0).

◆ Taper Calculation

A thread taper has to be *calculated* to establish its start and end diameters. The calculation method depends on the way the taper is defined and dimensioned in the drawing. Hardly any part drawing will show the dimensions required for programming - they have to be calculated as part of the programming process, using one of two common methods.

One method uses the thread length and angle and can be calculated by applying the standard trigonometric functions. The other method defines taper as the *ratio of its sides*. This method is often confusing to an inexperienced programmer. Typical ratios are defined in the part drawing directly, for example as 1:12, 1:16, etc., or indirectly, for example as the amount of *taper per foot* or, sometimes, as *taper per inch*. Keep one rule in mind:

> Taper is always measured on a diameter

A standard North American pipe thread is a good example of a tapered thread. It is defined by a taper ratio of 1:16, which is equivalent to a *3/4 of an inch per foot* taper, measured on the diameter and perpendicular to an axis. A pipe thread may also be defined with a given angle per side - *one degree, forty seven minutes, twenty three seconds (plus some leftover)*, or *1.789910608 decimal degrees*. For CNC programming, the decimal degrees are preferred to the degrees-minutes-seconds method, and many drawings already reflect this preference. To understand the principles of a taper defined as the ratio of its sides, a definition followed by an example should help.

RATIO indicates the relationship
between two values, expressed as a fraction

Both values in the ratio must be expressed in the same units and should be used in their lowest form of application (1/4 instead of 2/8 or 4/16). For example, the ratio of *3 units* to *4 units* may have these forms:

```
3 : 4   =   3 / 4
```

In terms of a taper definition, it means that for every 3 units change along one axis, there will be 4 units change along the other axis.

TAPER PER FOOT indicates the difference between two
diameters over the length of one foot or 12 inches

The example of a 3 inch taper per foot is equivalent to a 1:4 taper ratio, because

```
3 / 12 = 1 / 4 = 1 : 4
```

In CNC programming, we are only interested in calculating the diameters at the *beginning* and at the *end* of the thread. These calculations can be done either by means of trigonometric functions, or by means of ratio calculations - *Figure 38-21*.

Based on these principles, the required values for the example can be calculated. Note that only the *per side* or *radial* dimensions were used. In many programming applications, only one of the methods described will be needed - either the angle *or* the ratio. There will always be the option to use the other method to verify accuracy of the calculations.

In *Figure 38-22*, the start and end diameters have been calculated using the angle and/or the ratio of sides method. Which results of the calculations will actually be used in the program will depend on the type of selected programming technique, such as using a block-by-block approach versus a cycle method. The details depend on the thread specifications and machine and control features.

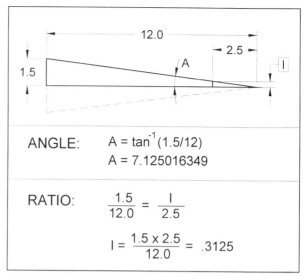

$$\text{ANGLE:} \quad A = \tan^{-1}(1.5/12)$$
$$A = 7.125016349$$

$$\text{RATIO:} \quad \frac{1.5}{12.0} = \frac{I}{2.5}$$

$$I = \frac{1.5 \times 2.5}{12.0} = .3125$$

Figure 38-21

Taper thread calculations - clearances excluded

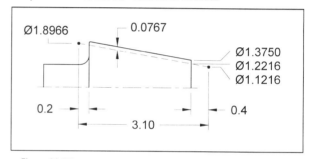

Figure 38-22

Calculated values for the tapered thread program example O3808

◆ Block by Block Taper Thread

In block threading, the taper thread programming is just as simple as programming a straight thread. To simplify the example, a straight infeed and nine threading passes will be used, for the total depth of .0767. The following nine depths must be applied at *both* ends of the thread. The first column lists the depth of thread per pass, the second column lists the front thread diameter, the third column is the end thread diameter. The front diameter is calculated at absolute coordinate of Z0.4, the end diameter at Z-2.7:

Depth	Front⌀	End ⌀
.0165	1.2420	2.0170
.0145	1.2130	1.9880
.0120	1.1890	1.9640
.0100	1.1690	1.9440
.0080	1.1530	1.9280
.0060	1.1410	1.9160
.0040	1.1330	1.9080
.0030	1.1270	1.9020
.0027	1.1216	1.8966

All requirements are available to write program O3808:

```
O3808
...
(G32 - TAPERED THREAD)
(N45 G50 X12.0 Z4.5)
N46 T0500 M42
N47 G97 S450 M03
N48 G00 X2.5 Z0.4 T0505 M08
N49 X1.242                                    (PASS 1)
N50 G32 X2.017 Z-2.7 F0.125
N51 G00 X2.5
N52 Z0.4
N53 X1.213                                    (PASS 2)
N54 G32 X1.988 Z-2.7
N55 G00 X2.5
N56 Z0.4
N57 X1.189                                    (PASS 3)
N58 G32 X1.964 Z-2.7
N59 G00 X2.5
N60 Z0.4
N61 X1.169                                    (PASS 4)
N62 G32 X1.944 Z-2.7
N63 G00 X2.5
N64 Z0.4
N65 X1.153                                    (PASS 5)
N66 G32 X1.928 Z-2.7
N67 G00 X2.5
N68 Z0.4
N69 X1.141                                    (PASS 6)
N70 G32 X1.916 Z-2.7
N71 G00 X2.5
N72 Z0.4
N73 X1.133                                    (PASS 7)
N74 G32 X1.908 Z-2.7
N75 G00 X2.5
N76 Z0.4
N77 X1.127                                    (PASS 8)
N78 G32 X1.902 Z-2.7
N79 G00 X2.5
N80 Z0.4
N81 X1.1216                                   (PASS 9)
N82 G32 X1.8966 Z-2.7
N83 G00 X2.5
N84 Z0.4
N85 G00 X12.0 Z4.5 T0500 M09
N86 M30
%
```

In the example, a straight infeed and pullout is used for clarity. The program will not change very much if a compound infeed is used and/or the angular pullout from the thread. Of course, more calculations will be needed..

◆ Tapered Thread Using a Simple Cycle

In a G92 threading cycle, the thread taper is programmed as the radius I value, with specified direction from the end diameter to the start diameter:

```
G92 X.. Z.. I.. F..
```

The X represents the current thread diameter at the *end of the cut*, Z is the end position of the thread, I is the difference *per side* between the thread diameter at the *end* and the thread diameter at the *start*. The I value must include an algebraic sign (only minus sign must be written), specifying the *direction* of the taper inclination, in this case a negative value. Program O3809 will cut a tapered thread using the G92 threading cycle.

```
O3809
...
(G92 - TAPERED THREAD)
(N45 G50 X12.0 Z4.5)
N46 T0500 M42
N47 G97 S450 M03
N48 G00 X2.5 Z0.4 T0505 M08
N49 G92 X2.017 I-0.3875 Z-2.7 F0.125     (PASS 1)
N50 X1.988                               (PASS 2)
N51 X1.964                               (PASS 3)
N52 X1.944                               (PASS 4)
N53 X1.928                               (PASS 5)
N54 X1.916                               (PASS 6)
N55 X1.908                               (PASS 7)
N56 X1.902                               (PASS 8)
N57 X1.8966                              (PASS 9)
N58 G00 X12.0 Z4.5 T0500 M09
N59 M30
%
```

Note that the I distance of taper inclination is the difference between the *end* diameter of 1.8966 and the *start* diameter of 1.1216, divided by 2. The result is:

```
(1.8966 - 1.1216) / 2 = .3875
```

This I value (.3875) must have a directional sign, to indicate the taper orientation (its direction from the end point). In the example, the I value will be negative because the start diameter of the taper is *below* the end diameter of the taper as viewed on a typical rear lathe. In the program, the entry will be I-0.3875.

◆ Tapered Thread and a Multi Repetitive Cycle

The multiple repetitive threading cycle G76 cycle requires the I value *not to be a zero,* if a tapered thread is cut. The I value in the cycle specifies the difference per side, so called radial distance, as well as the direction between the start and the end diameter of the taper.

Remember that the X diameter is always programmed at the *end* of thread and the I value supplies the taper height and its inclination (taper ratio per side). On CNC lathes with the X+ axis direction upwards from the center line (rear lathes), an increasing taper diameter will require a negative I value, and a decreasing taper will require a positive I value. The I value is always a *single* value, measured on a radius, not a diameter - *Figure 38-23* illustrates the concept for rear lathes.

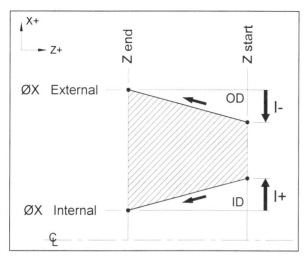

Figure 38-23

Tapered thread inclination direction I used in G76 threading cycles

The basic G76 cycle will be maintained but the I value will be added - a non-zero value must be programmed:

```
O3810
...
(G76 - TAPERED THREAD)
(N45 G50 X12.0 Z4.5)
N46 T0500 M42
N47 G97 S450 M03
N48 G00 X2.5 Z0.4 T0505 M08
N49 G76 X1.8966 Z-2.7 I-0.3875 K0.0767 D0140
    F0.125
N50 G00 X12.0 Z4.5 T0500 M09
N51 M30
%
```

If this method can be used for threading, G76 cycle is the best choice. It offers the fastest program generation as well as the best opportunities for on-machine editing.

MULTISTART THREAD

Most threads have only one start, suitable for most applications. The most common purpose of a multistart thread is to transfer a *precision* motion very rapidly over a relatively long distance. Note the word *precision* - a coarse thread can also be used to transfer a motion rapidly, but with very little precision. An example of precision multistart threads are some internal designs of some camera zoom lenses.

For programmers, there are some unique considerations for a multistart thread. It is important that the start position for each thread is in such a location, that when viewed from the thread end of the screw or the nut, each start on the circumference will be divided in equal angular increments. Also important is to maintain the equal thread profile when viewed from the thread cross section. To achieve these conditions, two programming tools are available.

These tools are the *thread start position* and the *thread feedrate* calculations. *Figure 38-24* shows symbolically the views of the thread cross sections and the end views.

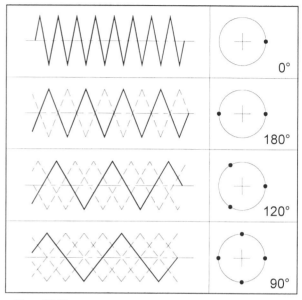

Figure 38-24

Representation of multistart threads (dots indicate thread starts)

In the illustration are four examples of the cross sections (left) and the end views (right) of a single start thread (top), double start (one below), triple start (two below) and a quadruple start (three below).

Although the examples are represented only symbolically, the thread *pitch* is maintained in all examples. Also note the *equal* distribution of each thread start, represented by the heavy dots. Each angle value is the angular spacing of individual starts, when the threaded part is viewed along its center line. The spacing is automatic and only the correct shift value from one thread start to the next has to be programmed, in threading mode.

◆ Threading Feedrate Calculation

The threading feedrate is always the *lead* of the thread, never the *pitch*. For a single start thread, the lead and the pitch have the same value - for a multistart thread, they do not. Take a single start thread of 16 TPI. Here, the lead *and* the pitch are both .0625, so the feedrate is F0.0625. If the drawing specifies the thread as 16 TPI, but indicates a *double start*, (for example 3.0-16 TPI 2 START), that means the *pitch* of the thread will remain unchanged (.0625), but the *lead* of the thread will double to .1250. Therefore, the programmed feedrate for the double start thread with the pitch of .0625 will be F0.125. The multiplication of the pitch will always depend on the *number of thread starts*. That means a triple start thread will have the feedrate three times the pitch, quadruple start thread four times, and so on.

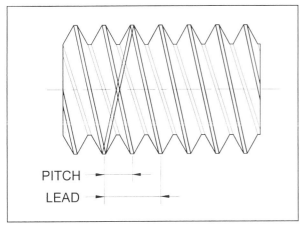

Figure 38-25

Relationship of the pitch and the lead of a double start thread

In *Figure 38-25*, the relationship of pitch and lead of a double start thread is shown. The same logic that applies to a double start thread, also applies to triple, quadruple, etc., threads. The feedrate calculation is identical for all threads:

$$Feedrate = \frac{Number\ of\ starts}{TPI}$$

Figure 38-26 shows the relationships of the pitch and the lead for some common multistart threads - the same pitch-lead relationship is maintained proportionately.

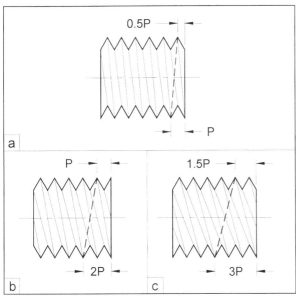

Figure 38-26

Multistart threads - pitch and lead relationships:

(a) Single start thread	*Lead = Pitch = 1P*
(b) Double start thread	*Lead = 2P*
(c) Triple start thread	*Lead = 3P*

◆ Shift Amount

Feedrate is not the only consideration for programming a thread with two or more starts. The other, equally important factor, is the programmed amount of the *tool point shift*. This shift will guarantee that each start will be in the proper relationship to all other starts. When one thread is finished, the starting position of the tool has to be *shifted* (in Z axis only), always by the *pitch* amount. The formula for the tool shift amount will be:

$$Shift\ amount = Pitch$$

The shift has to be programmed *for each* start above the first one. That means the number of shifts in the program is the *one less than the number of starts*:

$$Number\ of\ shifts = Number\ of\ starts - 1$$

Note that the formula is valid even for a single start thread, but there is no shift required (1 - 1 = 0).

A few methods can determine *when* the tool shift is to be programmed. The first method, for a double start thread, is to program one thread to its full depth, then shift out and cut the second thread to its full depth. The second method, for the same thread, is to cut one pass of the first thread, shift *out*, cut the same pass for the second thread, shift *in*, cut the second pass for the first thread, shift *out* again and repeat the process until both threads are completed to the full depth. This approach applies to any number of starts.

The obvious advantage to the first method is the ease of programming. On the negative side, if the tool cutting edge wears out on the first thread, the second thread will not be as accurate. The advantage of the second method is that the tool wear will be equally distributed over both threads, although the programming will require a lot more effort, which presents the negative side. Additional problem is that in many hard materials, the thread edge life may suffer from extensive material removal.

To illustrate a sample multistart thread application, the following general thread specifications will be used:

❑ The number of threads per inch is twelve (12 TPI)

❑ The number of starts is two (double start thread)

❑ The thread is cut as external at 3.000 nominal diameter

❑ The calculated thread depth is .0511 (.61343 / 12)

❑ The number of passes will be seven (for G92 cycle)

Although the block-by-block programming method G32 can be used for special applications, acceptable results can be achieved in many threading applications by using the G92 or G76 cycles, with less programming, as well as the gain of easier editing at the machine.

◆ Application Example

The previous thread with 12 TPI on a ∅3.00 will be used, but as a *double start* thread. The number of passes will be seven, with the same depths as before. The first program O3811 shows the completion of one thread *before* the other (the feedrate is F0.25, *not* F0.125). In the comments, P is the pass number, T is the thread number, first or second:

```
O3811
...
(G92 - DOUBLE START THREAD - 1)
(N45 G50 X12.0 Z4.5)
N46 T0500 M42
N47 G97 S450 M03
N48 G00 X2.5 Z0.4 T0505 M08      (--- THREAD 1)
N49 G92 X2.017 Z-2.7 F0.25       (T1 - P1)
N50 X1.988                       (T1 - P2)
N51 X1.964                       (T1 - P3)
N52 X1.944                       (T1 - P4)
N53 X1.928                       (T1 - P5)
N54 X1.916                       (T1 - P6)
N55 X1.908                       (T1 - P7)
N56 X1.902                       (T1 - P8)
N57 X1.8966                      (T1 - P9)
N58 G00 X2.5 Z0.525              (--- THREAD 2)
N59 G92 X2.017 Z-2.7             (T2 - P1)
N60 X1.988                       (T2 - P2)
N61 X1.964                       (T2 - P3)
N62 X1.944                       (T2 - P4)
N63 X1.928                       (T2 - P5)
N64 X1.916                       (T2 - P6)
N65 X1.908                       (T2 - P7)
N66 X1.902                       (T2 - P8)
N67 X1.8966                      (T2 - P9)
N68 G00 X12.0 Z4.5 T0500 M09
N69 M30
%
```

This version can be modified to alternate between threading cuts of the first thread and the second thread, as shown in program O3812. Applying this technique, the tool wear will be evenly distributed between all threads.

```
O3812
...
(G92 - DOUBLE START THREAD - 2)
(N45 G50 X12.0 Z4.5)
N46 T0500 M42
N47 G97 S450 M03
N48 G00 X2.5 Z0.4 T0505 M08      (--- THREAD 1)
N49 G92 X2.017 Z-2.7 F0.25       (T1 - P1)
N50 G00 Z0.525                   (START 1)
N51 G92 X2.017 Z-2.7             (T2 - P1)
N52 G00 Z0.4                     (START 2)
N53 G92 X1.988 Z-2.7             (T1 - P2)
N53 G00 Z0.525                   (START 1)
N54 G92 X1.988 Z-2.7             (T2 - P2)
N55 G00 Z0.4                     (START 2)
N56 G92 X1.964 Z-2.7             (T1 - P3)
N57 G00 Z0.525                   (START 1)
N58 G92 X1.964 Z-2.7             (T2 - P3)
N59 G00 Z0.4                     (START 2)
```

```
N60 G92 X1.944 Z-2.7             (T1 - P4)
N61 G00 Z0.525                   (START 1)
N62 G92 X1.944 Z-2.7             (T2 - P4)
N63 G00 Z0.4                     (START 2)
N64 G92 X1.928 Z-2.7             (T1 - P5)
N65 G00 Z0.525                   (START 1)
N66 G92 X1.928 Z-2.7             (T2 - P5)
N67 G00 Z0.4                     (START 2)
N68 G92 X1.916 Z-2.7             (T1 - P6)
N69 G00 Z0.525                   (START 1)
N70 G92 X1.916 Z-2.7             (T2 - P6)
N71 G00 Z0.4                     (START 2)
N72 G92 X1.908 Z-2.7             (T1 - P7)
N73 G00 Z0.525                   (START 1)
N74 G92 X1.908 Z-2.7             (T2 - P7)
N75 G00 Z0.4                     (START 2)
N76 G92 X1.902 Z-2.7             (T1 - P8)
N77 G00 Z0.525                   (START 1)
N78 G92 X1.902 Z-2.7             (T2 - P8)
N79 G00 Z0.4                     (START 2)
N80 G92 X1.8966 Z-2.7            (T1 - P9)
N81 G00 Z0.525                   (START 1)
N82 G92 X1.8966 Z-2.7            (T2 - P9)
N83 G00 X12.0 Z4.5 T0500 M09
N84 M30
%
```

The G92 cycle and G00 motion cancel each other. That is the reason for the G code repetitions. The only value that remains in effect is the F0.25 feedrate, which is programmed only once for each example.

THREAD RECUTTING

When threading is done, it should be checked for quality *before* the part is removed from the machine. Once the part is removed, any subsequent reclamping will need a great effort in order to recut the thread. The first threading pass will start at a random place of the cylinder circumference. Each subsequent pass will be automatically synchronized to start at the same position. As long as the threaded part remains clamped, this synchronization is assured.

There are two methods to prevent recutting. First, program a tool wear offset. Second, program M00 function at the end of each threading operation, before any other machining, even for the last tool. If the thread has to be recut after removal, the operator has to follow several steps:

1. Reclamp the threaded part to run concentric w/spindle

2. Set the X axis offset large enough, so the threading tool moves above the thread (external threading) or below the thread (internal threading)

3. Visually align the threading tool tip with the thread already completed (only as accurate as one's eye)

4. Repeat the steps in the air while carefully adjusting the offset so the tool will eventually recut the thread

Thread recutting should be prevented. The difficulty to reset the part precisely is the major quality concern.

39 SUBPROGRAMS

The length of a CNC program is usually measured in the number of characters such program contains. This number is similar to the number of bytes, if the program is stored on a computer disk. The physical length of a program is usually not an issue for most jobs. The program length will vary, depending on the complexity of work, the number of tools used, the method of programming and other factors. Generally, the shorter the program, the less time is needed to write it, and the less space it will occupy in the CNC memory. Short programs also reduce the possibility of a human error, because they are easily checked, modified and optimized. Virtually all CNC systems offer features designed to shorten the length of a program to some extent and make the programming process easier, more efficient and less prone to errors. Typical examples of this type of programming are fixed cycles, multiple repetitive cycles and custom macros. This chapter describes the structure, development and applications of another method of efficient program preparation - the use of *subprograms*.

MAIN PROGRAM AND SUBPROGRAMS

A CNC program is a series of instructions, assigned to different tools and operations. If such a program includes two or more *repetitive* instructions, its structure should be changed from a single long program to two or more separate programs. Each repetitive instruction is written only once and called when required. This is the main concept of subprograms. *Figure 39-1* shows a typical part layout repeated at different locations.

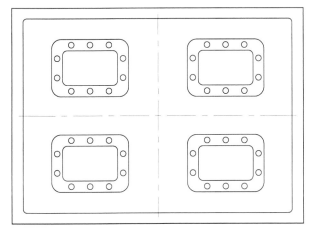

Figure 39-1
Example of a part requirement suitable to be used as a subprogram

Each program must have its own program number and is stored in the control memory. The programmer uses special M code function to call one program from another. The *first* program that calls another program is called the *main program*, all other programs are called *subprograms*. The main program is never called by a subprogram - it becomes the top level of all programs. Subprograms can also be called from other subprograms, up to a certain number of nesting levels. When a program containing subprograms is used, always select the main program, never the subprogram. The only time a subprogram is selected on the control is for editing purposes. In some reference materials, subprograms are also called *subroutines* or *macros*, but the term *subprogram* is used most often and the word *macro* could have a different meaning altogether.

◆ Subprogram Benefits

Any frequently programmed order of instructions or unchanging block sequences, can benefit from becoming a subprogram. Typical applications for subprogram applications in CNC programming are:

- Repetitive machining motions
- Functions relating to tool change
- Hole patterns
- Grooves and threads
- Machine warm-up routines
- Pallet changing
- Special functions   ... and others

Structurally, subprograms are similar to standard programs. They use the same syntax rules and look and feel the same. Often, it may not be easy to see the difference between a regular program and a subprogram at a casual glance. A subprogram can use the absolute or incremental data input, as necessary. Subprograms are loaded into the CNC system memory just like other programs. When properly implemented, they offer several benefits:

- Program length reduction
- Program error reduction
- Programming time and effort reduction
- Quick and easy modifications

Not every subprogram will provide all the benefits, but even one benefit should be a reason to use subprograms.

◆ Identification of Subprograms

The first step towards a successful application of subprograms is the *identification and isolation of repetitive* programming sequences. For example, the next six program blocks represent a machine zero return for a typical horizontal machining center, at the start of program:

```
N1 G20
N2 G17 G40 G80              (STATUS BLOCK)
N3 G91 G28 Z0             (Z AXIS RETURN)
N4 G28 X0 Y0        (X AND Y AXES RETURN)
N5 G28 B0                 (B AXIS RETURN)
N6 G90                    (ABSOLUTE MODE)
N7 ...
```

These blocks represent a typical sequence of commands that will be repeated *every time* a new program for that machine is written. Such a program may be written many times a week, each time repeating the same sequence of instructions. To eliminate the possibility of an error, the frequently used order of blocks can be stored as a separate program and identified by a unique program number. Then, it can be called up at the top of any main program. This stored programming sequence will become a *subprogram* - a branch or an extension of the main program.

SUBPROGRAM FUNCTIONS

A subprogram must be recognized by the control system as a *unique* type of program, *not as a main program*. This distinction is accomplished with two miscellaneous functions, normally applicable to subprograms only:

M98	Subprogram call function
M99	Subprogram end function

The subprogram *call function* M98 must always be followed by the subprogram number P--. The subprogram *end function* M99 terminates the subprogram and transfers the processing back to program it originated from (a main program or a subprogram). Although M99 is mostly used to end a subprogram, it may also be rarely used in the main program, replacing the M30 function. In this case, the program will run 'forever', or until the *Reset* key is pressed.

◆ Subprogram Call Function

The function M98 calls up a previously stored subprogram from another program. If used only by itself in a block, it will result in an error. M98 is an incomplete function - it requires two *additional* parameters to become complete, therefore effective:

❑ The address P identifies the selected subprogram number

❑ The address L or K identifies the number of subprogram repetitions (L1 or K1 is the default)

For example, a typical subprogram call block includes the M98 function and the subprogram number:

```
N167 M98 P3951
```

In block N167, the subprogram O3951 is called from the CNC memory, to be repeated *once* - L1 (K1) counter is the default, depending on the control. The subprogram must be stored in the control before being called by another program.

The M98 blocks that call subprograms may also include additional instructions, such as rapid tool motions, spindle speed, feedrate, cutter radius offset number, etc. On most controls, if included in the same block as the subprogram call, the additional data will be passed to the contents of the subprogram. The following subprogram call block also contains a tool motion in two axes:

```
N460 G00 X28.373 Y13.4193 M98 P3951
```

The block executes the rapid motion first, *then* it calls the subprogram. The order of words in a block makes no difference to the block execution:

```
N460 M98 P3951 G00 X28.373 Y13.4193
```

results in the same machining order as if the tool motion preceded the subprogram call, but looks illogical.

◆ Subprogram End Function

When the main program and the subprogram coexist in the control, they must differ by their program numbers. During processing, they will be treated as one continuous program, so a distinction must be made for the program end function as well. The end of program function is M30 or, less frequently, M02. The subprogram must be terminated by a different function. Fanuc uses M99 for that purpose:

```
O3951 (SUB-1)        Subprogram start
...
...
M99                  Subprogram end
%
```

When a subprogram terminates, the control returns the processing to the program of origin - it will *not* terminate the main program - that is the exclusive function of M30. Additional parameters may also be added to the M99 subprogram end, for example a block skip code, a block number to return to upon exit, etc. Note that the stop code symbol (the % sign) is used in the same manner for a subprogram, as for a main program. The subprogram termination is important and must always be done right. It sends two very important instructions to the control system:

❑ To terminate the subprogram

❑ To return to the block following the subprogram call

Never use the program end function M30 (M02) to terminate a subprogram - it will *immediately cancel all program processing* and reset the control. The program end function does not allow program execution of any blocks beyond the block that contains it.

Normally, the subprogram end M99 returns the processing to the block *immediately following* the subprogram call M98. This concept is illustrated in *Figure 39-2* (without block numbers) and described next.

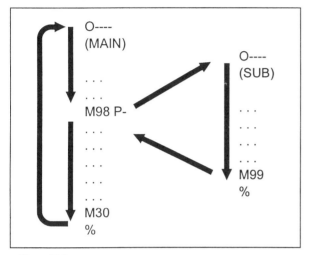

Figure 39-2

Flow of a program processing with a single subprogram

◆ Block Number to Return to

In most programs, the M99 function is programmed as a stand alone entry and as the *last* instruction in the subprogram. Usually, there are no other commands included with it in the block. The M99 function causes the subprogram to terminate and transfers its execution to the next block of the program it originated from. For example,

```
N67 M98 P3952              (SUBPROGRAM CALL)
N68 ...         (BLOCK TO RETURN TO FROM O3952)
N69 ...
N70 ...
```

executes block N67 by calling subprogram O3952. When the subprogram O3952 is processed, the control returns to the original program and continues processing instructions from the block N68, which is the block to return to.

Special Applications

For some special applications, it may be necessary to specify a *different* block number to return to, rather than using the *next block* default. If the programmer finds this option useful for certain jobs and uses this technique, the P address must be included in the M99 block:

```
M99 P..
```

In this format, the P address represents the block number to return to - *from* the completed subprogram. The block number must be present in the program of origin. For example, if the main program contains these blocks,

```
(MAIN - PROGRAM)
...
N67 M98 P3952
N68 ...
N69 ...
N70 ...
```

and the subprogram O3952 is terminated by

```
O3952 (SUB)
...
...
M99 P70
%
```

the calling program processing will continue from the N70 block (the main program in the example), *bypassing* blocks N68 and N69. This kind of application is not very common - it requires suitable type of work, in addition to the thorough understanding of subprogramming principles.

> The address P has a different meaning
> when used with M98 and M99 functions

Daily applications of this powerful programming method are not common, but the feature is an item to be explored by inquisitive programmers. The associated applications will include other programming tools, such as a combination with a block skip function, using the slash code /.

◆ Number of Subprogram Repetitions

A very important subprogram call feature is the address L or K, depending on the control model. This address specifies the number of subprogram repetitions - *how many times* the subprogram has to be repeated before the processing resumes in the original program. In most programs, the subprogram will be called only once, then the original program will continue.

Programs that require a *multiple* subprogram repetition before proceeding with the rest of the original program are common. To compare, a single use of the subprogram O3952 could be called up from the program of origin as:

```
N167 M98 P3952 L1 (K1)
```

This is a correct program block, but the L1/K1 counter does not have to be programmed at all. It can be safely ignored - the control unit defaults to only *one* repetition.

> If no address L/K is specified, the default value is always L1/K1

N167 M98 P3952 L1 (K1) *is identical to* N167 M98 P3952

Note - In the following examples, substitute K for every L listed, if required by the control system.

Number of repetitions for some control models range between L0 and L9999 and the L address other than L1 must always be programmed. Some programmers write the full block, even for a single repetition, rather than counting on the default conditions of the control system. The choice is a personal preference.

Repetition Count Variation

Some Fanuc controls *do not* accept the L/K address as the number of repetitions and use a different format. On these controls, a single subprogram call is the same:

N342 M98 P3952

This block calls the subprogram only once, as no special request has been used. In order to repeat the subprogram call four times, instead of programming

N342 M98 P3952 L4 (K4)

program the requested number of repeats *directly after* the P address, in a single statement:

N342 M98 P43952   *is the same as*   **N342 P00043952**

The result is identical to the other version - the subprogram will be repeated four times. *The first four digits are reserved for the number of repeats, the last four digits define the subprogram number.* For example,

M98 P3950   *is the same as*   **M98 00013950**

assumes a single repetitiomn of subprogram O3950. In order to repeat O0050 subprogram 39 times, program

M98 P390050   *or*   **M98 P00390050**

The maximum number of repetitions does not change for the 0/16/18/20/21 controls - it is represented by the first four digits, to the maximum of 9999.

M98 P99993952

repeats the execution of subprogram O3952, nine thousand, nine hundred and ninety nine times, the maximum number of repetitions available (some old models may have the maximum of only 999 times).

◆ L0/K0 in a Subprogram Call

There is no mystery in using the L/K counter greater than one to repeat a subprogram. This is a common application. Fanuc also offers a *zero* number of repetitions, in the form of L0/K0. When can the L0/K0 be programmed? Would anybody want to repeat a subprogram *zero times?*

There are some good reasons. Observe *Figure 39-3*. The five hole pattern has to be spot drilled, drilled and tapped.

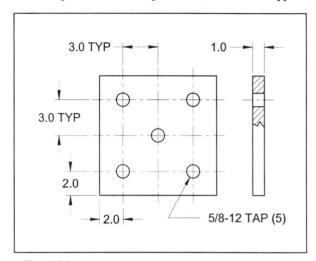

Figure 39-3

*Sample drawing used for a subprogram development
Used in programs O3901, O3902 and O3953*

For the spot drill (∅0.750), G82 cycle is used with the 0.2 seconds dwell to Z-0.3275 depth. For the tap drill, G81 cycle is used, and for tapping 5/8-12 tap, G84 cycle is used. The spot drill prepares the hole for drilling and makes a 0.015 chamfer. The tap drill will be 35/64 drill (∅0.5469), used to open up the hole for 5/8-12 tap:

```
O3901
(TOOL 1 - 90-DEG SPOT DRILL - 3/4 DIA)
N1 G20
N2 G17 G40 G80 T01
N3 M06
N4 G90 G00 G54 X2.0 Y2.0 S900 M03 T02
N5 G43 H01 Z1.0 M08
N6 G99 G82 R0.1 Z-0.3275 P200 F3.0      (LL HOLE)
N7 X8.0                                  (LR HOLE)
N8 Y8.0                                  (UR HOLE)
N9 X2.0                                  (UL HOLE)
N10 X5.0 Y5.0                        (MIDDLE HOLE)
N11 G80 Z1.0 M09
N12 G28 Z1.0 M05
N13 M01

(TOOL 2 - 35/64 DRILL)
N14 T02
N15 M06
N16 G90 G00 G54 X2.0 Y2.0 S840 M03 T03
N17 G43 H02 Z1.0 M08
N18 G99 G81 R0.1 Z-1.214 F11.0
N19 X8.0
N20 Y8.0
N21 X2.0
N22 X5.0 Y5.0
N23 G80 Z1.0 M09
N24 G28 Z1.0 M05
N25 M01
```

```
(TOOL 3 - 5/8-12 TAP)
N26 T03
N27 M06
N28 G90 G00 G54 X2.0 Y2.0 S500 M03 T01
N29 G43 H03 Z1.0 M08
N30 G99 G84 R0.4 Z-1.4 F41.0
N31 X8.0
N32 Y8.0
N33 X2.0
N34 X5.0 Y5.0
N35 G80 Z1.0 M09
N36 G28 Z1.0 M05
N37 G28 X5.0 Y5.0
N38 M30
%
```

This type of program uses repeating XY coordinates for each tool (spot drilling, drilling, tapping). In order to make the program more effective, all *repeating* blocks of the program will be collected into a subprogram and used much more efficiently. Here is the pattern of holes separated from the long program that also includes the G80Z1.0M09, as the standard end of *any* active fixed cycle:

```
X2.0  Y2.0
X8.0
Y8.0
X2.0
X5.0  Y5.0
G80  Z1.0 M09
```

Only a small effort is needed to reformat the existing program and separate it into a main program and a subprogram that stores the repeating machining pattern. Isolated XY coordinates of all five holes in the pattern are included:

```
O3953 (SUBPROGRAM)
(FIVE HOLE PATTERN)
N1 X2.0 Y2.0
N2 X8.0
N3 Y8.0
N4 X2.0
N5 X5.0 Y5.0
N6 G80 Z1.0 M09
N7 M99
%
```

This subprogram can be called from the main program, in this example, from a new program O3902. The L0 prevents double cutting of the first hole:

```
O3902 (MAIN PROGRAM)
(TOOL 1 - 90-DEG SPOT DRILL - 3/4 DIA)
N1 G20
N2 G17 G40 G80 T01
N3 M06
N4 G90 G00 G54 X2.0 Y2.0 S900 M03 T02
N5 G43 H01 Z1.0 M08
N6 G99 G82 R0.1 Z-0.3275 P200 F3.0 L0
N7 M98 P3953
N8 G28 Z1.0 M05
N9 M01
```

```
(TOOL 2 - 35/64 DRILL)
N10 M06
N11 T02
N12 G90 G00 G54 X2.0 Y2.0 S840 M03 T03
N13 G43 H02 Z1.0 M08
N14 G99 G81 R0.1 Z-1.214 F11.0 L0
N15 M98 P3953
N16 G28 Z1.0 M05
N17 M01
```

```
(TOOL 3 - 5/8-12 TAP)
N18 M06
N19 T03
N20 G90 G00 G54 X2.0 Y2.0 S500 M03 T01
N21 G43 H03 Z1.0 M08
N22 G99 G84 R0.4 Z-1.4 F41.0 L0
N23 M98 P3953
N24 G28 Z1.0 M05
N25 G28 X5.0 Y5.0
N26 M30
%
```

In the program, the initial XY tool motion for each cutting tool will position the cutter at the *first* hole of the machining pattern. All fixed cycles used in the program start at the first hole of the pattern. Since the first hole definition is included in the subprogram, *as well as in the main program*, program L0 in the fixed cycle call is mandatory, else the first hole of the pattern will be machined twice. This is a classic application of the L0 relating to fixed cycles, but not subprograms. Also included in subprogram O3953 can be the standard machine zero return block G28Z1.0M05, as it repeats after each M98 call in the main program O3902. This practice correct but not recommended, as it lacks in a clearly structured program.

SUBPROGRAM NUMBERING

To keep track of subprograms is much more important than keeping track of regular programs. Always make sure to know *exactly* what subprograms are available and how they are used, what is their purpose. A single subprogram may be used in many other programs and proper subprogram identification technique is extremely important.

Control unit directory of programs does not distinguish between program numbers and subprogram numbers. The control system recognizes a subprogram call only by its programmed format, the miscellaneous function M98, followed by the P.. subprogram number statement.

All this means that the subprogram number is assigned at the *programming* level, not at the *machine operation* level. It is the programmer's responsibility, *not* the CNC operator's, to assign subprogram numbers. Programmer has a great flexibility in organizing the subprograms and their identification - in fact, any programmer can design and set up certain basic rules and related standards. Many of the rules governing the format of main programs also apply to subprograms. Remember these four main points:

- ❏ If used in a program, the program number is commonly specified by the letter O, followed by four or five digits, depending on the control system

- ❏ If used in a program, the program number can be specified by the colon symbol, commonly : for the ISO format, followed by up to four or five digits, depending on the control system setting

- ❏ The main program number - O or : - cannot be negative or equal to zero

- ❏ The subprogram number cannot be negative or equal to zero

Within the allowed range, *any* number can be assigned to *any* main program or a subprogram. Some programmers do not use program numbers at all. This approach is acceptable for some controls, but only if the application does not require subprograms. In most cases, the main program numbers can be assigned by the machine operator. On the other hand, to maintain control of subprograms, program numbers become very important. The first step is to get organized. This is even more important if the subprograms are designed to be called up by *many other programs at different times*. There is no one *best* method, but some proven suggestions offer an idea how to approach the subject of program numbering and develop a personal approach.

For example, in this handbook, all main programs are numbered consecutively, with the first two digits corresponding to the chapter number. In this chapter, the method also applies to subprograms, but the last two digits are arbitrarily increased by fifty, for example O3953 will be the third subprogram example in the chapter. Feel free to adapt this method to any reasonable format.

◆ Organized Approach

The suggested programming approach is based on the understanding that the CNC memory is *not* used as a storage media for all part programs made. The control system memory capacity is *always* limited. At one point, this limit will be reached and there will be no more space left to accommodate more programs. A good program organization is one that uses the CNC system memory only for the *current program*, perhaps a few more that are to be used soon.

If the unique program number is assigned by the *machine tool operator* during setup, the situation needs some control as well. On some controls, the main program number on the written copy will not always load automatically, so it is not really needed. That means, if an arrangement is made with the shop supervisor that the CNC operator stores the main programs using only *three* digits for the regular program numbers 1-999; then there will be the four digit numbers 1000-9999 available for subprograms. This available range is more than enough for most manufacturing applications. Such an approach presents a good control over those subprograms whose numbers selected. All four-digit subprogram numbers can be documented, logged and subse-

quently called from any program, main or another subprogram, without a fear of duplication or a part program number mismatch.

Subprograms should always be documented in some log book, complete with detailed descriptions, independently from all programs of origins. This way, the subprograms can be used when needed, often at a short notice, regardless of the program for which they have been originally written. Such a method allows to organize all the subprograms by their series number (*i.e.,* 1000, 2000, 3000, etc., or 1100, 1200, 1300, etc.), for either the type of CNC machine, the type of subprogram, or the type of machining operation.

Individual subprograms have to have assigned program numbers that are unique. The program number assigned to a subprogram is called together with the M98 function and the P address. Such a combination of the two words, M98 P.., is the *minimum requirement* for a subprogram call from another program.

Using the example (early in this chapter) of the machine zero return sequence for a four axis vertical machining center, a subprogram can be created (with an assigned number O3954) for the blocks representing all needed commands - units selection G20 or G21 is *not* included:

```
O3954 (MACHINE ZERO RETURN)
N101 G17 G40 G49 G80
N102 G91 G28 Z0
N103 G28 X0 Y0
N104 G28 B0
N105 G90
N106 M99
%
```

The units selection should be used in the main program, for flexibility. Once the machine zero return subprogram has been designed and stored into the memory, every main program can start by calling the subprogram O3954:

```
O3903 (MAIN PROGRAM)
(PART ABC-123)
N1 G20                          Units used for this program
N2 M98 P3954                    Subprogram O3954 call
N3 G90 G54 G00 X.. Y..          Normal program start
N4 ...
...
< ... Machining ... >
...
N45 M30                         Main program end
%
```

To visualize the execution of the two programs by the CNC system, follow all operational steps in the order of program execution. During the program O3903 execution, the control system will follow the following order of operations (instructions):

1. Set program number O3903 as the current program number

2. Display comment on the display screen

3. Set the units of measurement (inches in the example)

4. Branch out to the top of subprogram O3954

5. Execute all blocks in the subprogram O3954

6. When M99 is processed, the subprogram ends and returns to the main program

7. The main program is processed, beginning with the block N3

8. When M30 is processed, the main program ends and returns to the beginning

9. When the CYCLE START switch is activated, steps 1 to 8 are repeated

As the example shows, the main program uses increments of 1, the subprogram also uses increments of 1, but starting with N101 block number. There are two reasons for it. The first reason is that a properly designed subprogram will not likely be a subject to any major changes - there should be no need to add any extra blocks into the subprogram once it has been debugged. The second reason is even more important. The lack of duplicated sequence numbers will be visible on the control display screen. The display of active block numbers will quickly inform the CNC operator whether the main program or a subprogram is being processed. Fanuc controls are very forgiving about the block numbers and allow identification of block sequences freely, within a specified range.

To illustrate the described concept, here is an example. In a simple application, where a main program calls a single subprogram, there should be no problem in block numbering. Even if the sequence numbers are duplicated in both the main program and the subprogram, it is not likely there will be any confusion. On the other hand, when several subprograms are called from the same main program, the duplicated block numbers appear during the main program processing, as well as when subprograms are processed. Such a situation may confuse the CNC operator to the extent of losing track of what is really happening in the control system at any given time.

To avoid this problem, consider assigning *unique block numbers* to each subprogram, thus preventing a duplication. One method is to identify the subprogram numbers in the high thousands series, for example O6100, O6200, O6300, etc. Then, the block numbering in a subprogram can be based on the subprogram number. For example:

```
O6100 (SUB 1)
N6101 ...
N6102 ...
N6103 ...        ... and so on
```

```
O6200 (SUB 2)
N6201 ...
N6202 ...
N6203 ...        ... and so on
```

This method works only with the maximum of one hundred blocks, suitable for many subprograms. The operator finds it easy to monitor a program with several subprograms. This is not a foolproof method for all programs, but the idea will work for most jobs.

◆ Protected Subprograms

Subprograms are special programs designed to be used frequently. Special subprograms may be even stored in the system memory permanently, to be called by all or many other programs. Any interference with these subprograms, accidental or intentional, can prove to be disastrous. If only a single subprogram is lost from the memory, it may halt literally hundreds of programs that depend on the use of this ill fated subprogram.

Fanuc controls address this potential problem by allowing an assignment of a certain specified series of program numbers that can be locked up by a system parameter setting. As a typical example, a program number series 9000 (within the range of O9000 to O9999), will not display on the control screen, when locked by the system parameter. Also, programs in this series cannot be edited or printed out, etc. If the locking parameter is not set, the programs of the 9000 series behave normally, like any other program. In order to take advantage of this feature to protect some important programs from unauthorized editing or even viewing, consult the Fanuc documentation for further details.

SUBPROGRAM DEVELOPMENT

Before a subprogram can be developed, it must be well thought out and planned. Since the most common application for subprograms is repetitive pattern of machining, the programmer should have the ability to *recognize* the machining pattern to be used in a subprogram.

◆ Repeating Pattern Recognition

This ability to recognize a repeating pattern is a matter of experience. The first indications come when writing a conventional program block by block. Visually scan the written copy first. If there are *repeating* clusters of consecutive blocks containing the *same* data, it is a very good reason to evaluate the program more carefully and possibly develop a subprogram.

An experienced programmer will not write a program the long way first. That is a waste of time. The programming experience enhances the ability to recognize a potential for subprograms at the early stages of the program planning. However, for a programmer with limited experience, there is no damage done by developing the long program first. It takes more time and it is not efficient. However, this is how a professional experience is gained. With limited experience, be willing to re-write a program from a single long form to a main program and one or more subprograms. Programmer should be able to identify those sections of a long program that can qualify as subprograms. Once such a series of repetitive data is identified in the conventional program, it is only a matter of small adjustments to separate these repetitive clusters and define them as subprograms.

◆ Tool Motion and Subprograms

One of the most common subprogramming applications is a tool path machined at different locations of the part. For example, a ten hole rectangular pattern needs to be programmed - *Figure 39-4*.

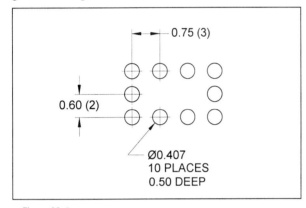

Figure 39-4
Detail of the hole pattern used in program O3904

This hole pattern is repeated at four specified locations of the part, as illustrated in *Figure 39-5*.

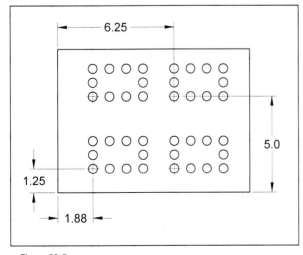

Figure 39-5
Hole pattern layout for program examples O3904 and O3905 (both using subprogram O3955)

Subprogram O3955 contains this pattern and uses the L address to establish the number of fixed cycle repeats. In the first main program O3904, the tool motion precedes the subprogram block. To start the program development, concentrate on the hole pattern. First, select the G91 incremental mode for the pattern. Then program the X and Y incremental values, starting from any hole, such as the lower left hand corner and continue in one direction - *Figure 39-6*.

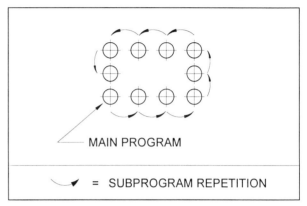

Figure 39-6
Subprogram O3955 processing flow

```
O3955 (SUBPROGRAM)
(FOUR-CORNER LOCATIONS)
N551 G91 X0.75 L3
N552 Y0.6 L2
N553 X-0.75 L3
N554 Y-0.6
N555 M99
%
```

The subprogram is designed to machine *nine* holes in a rectangular pattern. The tenth hole - actually it is the first hole - is machined in a block with the cycle call or the rapid motion. The four pattern locations are not included in the subprogram - they must be included in the main program. Since the main program is using absolute mode G90, the individual locations can be established:

```
O3904 (MAIN PROGRAM)
(FOUR-CORNER PATTERN)
N1 G20
N2 G17 G40 G80
N3 G90 G00 G54 X1.88 Y1.25
N4 G43 Z1.0 S350 M03 H01
N5 G99 G81 R0.1 Z-0.269 F3.5      (LL  HOLE 1)
N6 M98 P3955                      (LL  PATTERN)
N7 G90 X6.25 Y1.88                (LR  HOLE 1)
N8 M98 P3955                      (LR  PATTERN)
N9 G90 X6.25 Y5.0                 (UR  HOLE 1)
N10 M98 P3955                     (UR  PATTERN)
N11 G90 X1.88 Y5.0                (UL  HOLE 1)
N12 M98 P3955                     (UL  PATTERN)
N13 G80 G90 G28 Z1.0 M05
N14 G91 G28 X0 Y0
. . .
```

Only one cutting tool was used for this example, other tools will follow the same programming procedure. This method of the last example is more common - in the absolute mode from the main program, the tool is positioned at the lower left hand corner of the pattern and the first hole of the pattern is drilled at that location. Then the subprogram is called and the remaining nine holes are drilled, using an incremental positioning commands and the number of cycle repetitions. The number of repetitions in the subprogram is the number of spaces, *not* the number of holes.

A simpler way, particularly useful for a great number of pattern locations, is to *combine* the rapid motion to the pattern starting location with the subprogram call. This is acceptable for most control systems:

```
O3905 (MAIN PROGRAM)
(FOUR-CORNER PATTERN)
N1 G20
N2 G17 G40 G80
N3 G90 G00 G54 X1.88 Y1.25
N4 G43 Z1.0 S350 M03 H01
N5 G99 G81 R0.1 Z-0.269 F3.5 M98 P3955
N6 G90 X6.25 Y1.88 M98 P3955
N7 G90 X6.25 Y5.0 M98 P3955
N8 G90 X1.88 Y5.0 M98 P3955
N9 G80 G90 G28 Z1.0 M05
N10 G91 G28 X0 Y0
...
```

The major advantage of O3905 is shortening the length of program O3904 - either method produces the same results and the selection is a matter of personal preference. Note the seemingly unnecessary repetitions of the modal G90 and X and Y axes. Modal values have to be followed extra carefully for subprograms.

◆ Modal Values and Subprograms

> All modal values in effect when the subprogram is called will remain in effect for that subprogram, unless changed within.

In the examples O3904 and O3905, note repetitions of G90, X6.25 and Y5.0. They are *very important*. The subprogram O3955 changes the control status to the incremental mode G91 and the last hole of the ten hole pattern is *not* the same as the first one. The first hole of the pattern is machined when the rapid motion to that hole is completed in the absolute mode. That happens in the main program, *not* within the subprogram.

Here is another common problem. A finish contour subprogram uses cutter radius offset G41 or G42 with the D address. If the same subprogram is to be used for semifinishing and leave some stock, for example, it will not work. The reason is that the D address is fixed and is stored in the control as the full cutter radius. The solution? Use two D offsets and take the D address out of the subprogram, then call it together with M98, for example:

```
M98 P.. D..
```

This way, the offset number D can be changed anytime the subprogram is called, without change to the subprogram itself. This method is useful if the programmed contour requires *two* or *more different* offset values, but it does not work on all controls. Here is the content of a simple contouring subprogram, with embedded D offset. D51 setting value is equal to the cutter radius:

```
O3956 (CONTOUR SUBPROGRAM - A)
N561 G41 G01 X0 D51 F10.0          (D.. INCLUDED)
N562 Y1.75
N563 G02 X0.25 Y2.0 R0.25
N564 G01 X1.875
N565 Y0
N566 X-0.75
N567 G00 G40 Y-0.75
N568 M99
%
```

For contour finishing, the subprogram will be called by normal means, from the main program:

```
M98 P3956
```

The same subprogram can be used for finishing *as well as* for semifinishing, leaving some material stock, but two D offsets have to be used, such as D51 and D52. In this case, offset D51 stores the amount of the cutter radius and contain the stock allowance (D51 = cutter radius + stock), D52 stores the finishing radius only (D52 = cutter radius). For a ∅.500 end mill, the set values could be:

```
D51 = .250 radius + .007 stock = .257
D52 = .250 radius + .000 stock = .250
```

Next, the D.. has to be *removed* from the subprogram:

```
O3957 (CONTOUR SUBPROGRAM - B)
N561 G41 G01 X0 F10.0          (D.. NOT INCLUDED)
N562 Y1.75
N563 G02 X0.25 Y2.0 R0.25
N564 G01 X1.875
N565 Y0
N566 X-0.75
N567 G00 G40 Y-0.75
N568 M99
%
```

The control *does* require the D offset but not necessarily in the same block as G41/G42. As long as the D is specified before G41/G42, it can be passed on to the subprogram from the main program, depending on the operation:

```
M98 P3957 D51      ... for semifinishing
M98 P3957 D52      ... for finishing
```

This is a very powerful method of using subprograms for more than one operation, if the control supports it.

Return from a Subprogram

The current modal values should be clear in the main program when a subprogram is completed. Values that may have changed in the subprogram are absolute or incremental mode, motion command, coolant and others. Subprogram is always a *branch* of another program - it is a *continuous extension* of the program of origin and its integral part. All modal values set anywhere in the program are valid until changed or canceled by a command of the same group. The M99 subprogram end function will not cancel any modal values that are currently active.

As the O3904 and O3905 examples show, a fixed cycle is called from the main program only once. All the modal cycle data are carried forward to the subprograms. The main program clearly shows current modal values.

MULTI LEVEL NESTING

The last example has shown the main program that calls only one subprogram and the subprogram does not call another subprogram. This is called *one level* nesting, or nesting at one level deep. Modern controls allow nesting up to four levels deep. That means, if the main program calls a subprogram number one, this subprogram can call a subprogram number two, that can call a subprogram number three, and that can call a subprogram number four. This is called a *four level* nesting. All four levels are rarely needed for any practical application, but these are the programming tools available, just in case. The following examples show program processing flow of each nesting level.

◆ One Level Nesting

One level nesting means that a main program calls only one subprogram and nothing more. Subprogram that is nested one level deep is the most common in CNC programming. The program processing starts at the top of the

main program. When a subprogram is called from the main program by M98 P.. block, the control forces a branch to the beginning of the called subprogram, processes its contents, then it returns to the main program to process the remaining blocks of the main program - *Figure 39-7*.

◆ Two Level Nesting

The processing of a subprogram that is nested two levels deep also starts at top of the main program. When the control encounters a subprogram call for the first level, it will branch from the main program and starts processing the blocks in the first subprogram, starting from its top. During processing of the first level subprogram, CNC system encounters a call for a second level subprogram.

At this point, processing of the first level is *temporarily* suspended and CNC system branches to the second level. Since there is no subprogram call from the second level, all blocks in the subprogram will be processed. Anytime the block containing M99 function is encountered, the CNC system will automatically return to the program *it branched out of*. It will *resume* processing of that program, temporarily suspended before.

The return to the program of origin will normally be to the block immediately following the subprogram call block in that program. All remaining blocks in the first subprogram will be executed until another M99 function is encountered. When that happens, the control system will return to the program it branched out of (program of origin), in this case to the main program.

Since there are still some blocks left in the main program, they will be processed until the M30 function is encountered. M30 terminates the execution of the main program. *Figure 39-8* illustrates schematically the concept of a two level subprogram nesting.

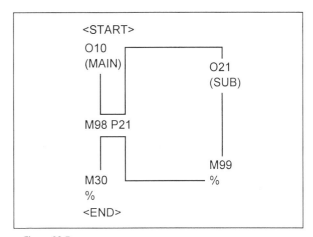

Figure 39-7
One level subprogram nesting

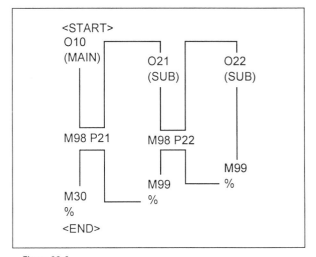

Figure 39-8
Two level subprogram nesting

◆ Three Level Nesting

The nesting up to three levels deep is the next logical extension of the two level nesting. As before, starting at top of the main program (program O10 in the example illustrated in *Figure 39-9*), the first branch will be to the first level (O21), another branch follows (O22) and there is an additional branch to O23. Each subprogram is processed up to the next subprogram call, or the end of subprogram. The program processing will always return to the block following the subprogram call, ending in the main program.

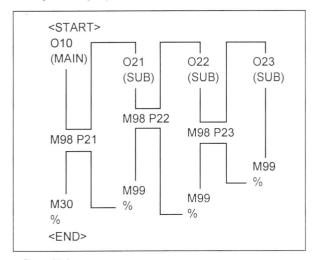

Figure 39-9
Three level subprogram nesting

◆ Four Level Nesting

The logic of multi level subprogram nesting should be pretty clear by now. Four level nesting is just a multiple extension of a single nesting and is logically identical to all the previous examples.

Unnecessary addition of more branches for a multi depth subprogram nesting makes any programming application that much more complex and more difficult to master.

Programming the subprogram nesting into the four level depth (or even the three level depth) will require a full understanding of the program processing order - and having a suitable application for it. In typical machine shop programming, there is seldom the need to use level three and level four nesting. If a good example of a four level nesting application is found, the typical program flow will conform to the format illustrated in *Figure 39-10*.

◆ Nesting Applications

Considering the reality that each subprogram can be repeated up to 9999 times in any program that calls it, shows the enormous programming power available to use and explore. Always be aware of potential difficulties, even dangers, when developing subprograms with several multi

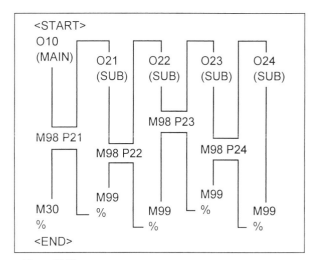

Figure 39-10
Four level subprogram nesting

nested subprograms. Such a programming approach may result in a short program, but at the cost of a long development time. The program preparation time, its development and debugging often take more time than writing conventional programs. Not only the logical development is complex and more time consuming, a significant portion of programming time must be spent on careful and thorough documentation of the process flow of all programs, setting up the initial conditions, checking the validity of data, etc.

There are many fairly experienced CNC programmers in the machining trades field, who try to use a multi level nesting at all costs, and the more levels, the better programmers they feel they are. These programmers, more often then not, use such complex programming technique as the means of expressing their so called 'professional skill', usually measured against other programmers. Often, this is nothing more than a unnecessary contest, a frustration perhaps, and definitely an expression of a little ego trip.

When a programmer becomes obsessed with making the program as short as possible, *at any and all costs,* he or she is taking the wrong trek. Such programs, even if they are technically flawless and logically correct, are not always very easy to use by a CNC operator. A CNC machine operator with limited or no programming knowledge will find these programs extremely intimidating - even skilled and experienced operators will find them hard to read, hard to interpret and most likely, they will be unable to make any substantial changes to them, in order to modify or optimize the programs for a better performance.

A simple general rule for multi level nesting technique - use it only in those cases, when the frequency of their future deployment justifies the extra time spent for their development. Like anything else, many nesting levels offer advantages and the inevitable disadvantages.

CONTOURING WITH A SUBPROGRAM

So far, a number of programming examples have been using a subprogram. They all related to machining holes and, hopefully, offered enough material to understand the concept of subprogramming (there will be one more - a rather special one - at the end of this chapter, so look for it). There are other examples found throughout the handbook that make generous use of subprograms.

Here is one more example relating to this chapter, this time applying a simple XY contouring work to a multiple Z depth - evaluate *Figure 39-11*.

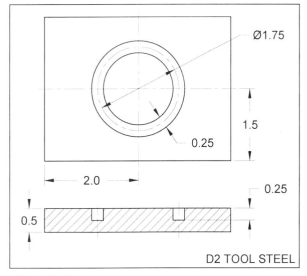

Figure 39-11
Main program O3906 using subprogram O3958

The job requires a groove with a ∅1.750 pitch to be machined to the depth of .250. It is a utility or rough groove, so there is no need for precision tolerances, or even the high quality of the surface finish. All needed is a ∅.250 center cutting end mill (slot drill), plunge to the depth, program a 360° circular tool path, and job is done. Well, almost.

Even in a material that cuts well, for example brass, splitting a single depth cut of .250 into two depth cuts of .125 may prove beneficial. The material is *D2 tool steel,* rather a tough material. The tool will run at only 630 r/min and only plunge into the material .010 at a time, repeating the groove

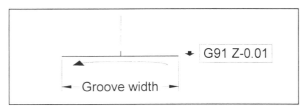

Figure 39-12
Detail of the subprogram O3958 - front view shown

profile 25 times, for 25 × .010 = .250 total required depth. Preference for a subprogram in such a case is without a question. Symbolic detail of the depth cut for a single increment is illustrated in *Figure 39-12*.

The subprogram O3958 will contain only the *tool motions common to all the groove cuts.* That means the .010 incremental plunge cut and the 360° circular cut. All other motions will be in the main program O3906. Note the word *incremental* for the plunge depth. The .010 must be programmed incrementally, otherwise it will cut at the absolute depth of Z-0.01 - all twenty five times! Here is the complete main program O3906, followed by a single related subprogram O3958 (tool T01 is assumed to be in the spindle):

```
O3906 (MAIN FOR SIMPLE DEEP GROOVE)
(T01 - 0.250 DIA CENTER CUTTING END MILL)
N1 G20
N2 G17 G40 G80
N3 G90 G54 G00 X2.875 Y1.5 S630 M03
N4 G43 Z0.1 H01 M08
N5 G01 Z0 F10.0      (START Z POSITION AT Z0 !)
N6 M98 P3958 L25     (CALL SUBPROGRAM 25 TIMES)
N7 G90 G00 Z1.0 M09
N8 G28 Z1.0 M05
N9 M30
%

O3958 (SUB FOR O3906)
N581 G91 G01 Z-0.01 F0.5   (INCREMENT BY -0.01)
N582 G03 I-0.875 F2.0   (FULL CIRCLE CONTOUR)
N583 M99
%
```

Intentionally, the presented program is simple. It does show, however, two important considerations that have to be maintained in any subprogram development. These considerations relate to maintenance of a *continuous relationship between the main program* and *the subprogram.* They can be described as special requirements:

❏   … to maintain a transfer from the main program to a subprogram (before subprogram is called)

❏   … to maintain a transfer from the subprogram, back to the main program (after a subprogram is completed)

The first requirement is met in block N5. The Z axis position *must be at Z0*, nowhere else! Being at Z0, it will enable the tool to increment 25 times the distance of .010, resulting in .250 groove depth. Described differently, the tool start position before a subprogram is called must be at a position that results in a correct tool path.

The second requirement is met in block N7. It is the G90 command that makes this block special. Why? Because the subprogram uses G91 incremental mode. When the subprogram processing returns back to the main program, it no longer benefits from the incremental mode, and the G90 changes the incremental mode back to absolute mode.

TOOL CHANGE SUBPROGRAM

The programming sequence for a typical automatic tool change (ATC) is usually short and simple. For a CNC milling system, the M06 function will normally do the job and for the CNC lathes, it is the T function that does the same thing. The tool change cannot be programmed without establishing certain conditions. Program functions relating to machine zero return, coolant cancellation, spindle stop and others, are all an integral part of the tool change routine. It may take three, four, five or more program blocks to establish the right conditions - *every* time the automatic tool change is programmed, which can be quite often. Even more significant is the fact that the blocks always have the same contents, regardless of the program being used.

As an example of this concept, consider the following sequence of operations, they are quite typical, required to program a tool change for several tools in a single program.

The example is based on a typical vertical CNC machining center, and uses automatic tool change function (ATC):

1. Turn off the coolant
2. Cancel a fixed cycle mode
3. Cancel a cutter radius offset mode
4. Turn off the spindle
5. Return to Z axis machine reference position
6. Cancel offset values
7. Make the actual tool change

The seven individual operations will occur in *every program* that requires this particular tool change and they will occur for *every tool* in *each program*. That is a lot of programming for a simple tool change sequence. To make the programming easier, develop a subprogram that includes all seven operations, then call it in the main program whenever a tool change is required:

```
O3959 (TOOL CHANGE VERTICAL MACHINING CENTER)
N1 M09
N2 G80 G40 M05
N3 G91 G28 Z0
N4 G49 D00 H00
N5 G90 M06
N6 M99
%
```

This example can be easily modified for a different machine design or for a CNC horizontal machine. It may even include special requirements, such as certain manufacturer's options. The tool change may even be programmed at a certain machine table position. The only modification would be the addition of a G53X..Y.. block before the tool change block. Another example is a special code for tool change *and* the coolant ON function. Some machine manufacturers create a special M function, combining the two standard functions, for example, M16 - which is the combination of M06 and M08 standard functions.

Also note the various cancellation functions - there are quite a few of them in subprogram O3959. When designing such a subprogram, the programmer has absolutely no idea whether the coolant will be ON or OFF; no idea if a fixed cycle or the cutter radius offset is active or not. Also, the programmer has no idea as to what the current status of G90 or G91 modes is.

Their *actual* status is really not that important. These cancellations are included in the subprogram, taking advantage of the fact that a cancellation of a function that is already canceled will be ignored by the control system. As the example shows, even a 'simple' tool change sequence requires some serious thinking.

100 000 000 HOLE GRID

In the last section of this chapter, perhaps a little deviation from the handbook seriousness will be tolerated. This section will look at subprograms from a different angle, but with a *real* example. The following exercise takes the subprogramming power to the very extreme. Although it is presented primarily on a light note, it does serve a very practical purpose - it shows the power of subprograms and, hopefully, makes a strong case for their use.

The example illustrates how one hundred million holes, (yes, *one hundred million* holes), can be spot drilled *and* drilled using a program of only 29 blocks for the two cutting tools. These 29 blocks even include the program numbers and stop codes (% signs). *Figure 39-13* shows a simple grid pattern of 10000 rows (X) and 10000 columns (Y).

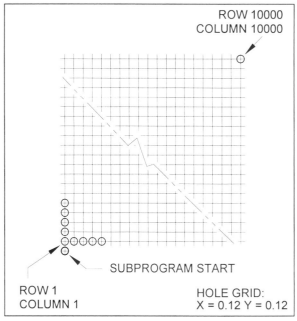

Figure 39-13

100 000 000 holes - rectangular grid pattern

To make the example reasonable, simple, and interesting at the same time, the holes are very small, only ∅5/64 (.0781), with a pitch of .120 along each axis, resulting in a square grid pattern of holes very close to each other.

Only two tools are used, a *spot drill* with a 90° tool point angle to startup the hole for drilling and a ∅5/64 drill. Both cutting tools start machining from R0.06 cycle position above the plate to their respective depths: Z-0.04 for the spot drill and Z-0.215 for the drill.

From the programming point of view, the program design is not difficult at all - it will use a main program and one subprogram. The programming procedure is the same for 100 000 000 holes, as if the grid were only 100 holes. The main program contains the standard settings and also calls the subprogram. The subprogram will repeat the active fixed cycle 9999 times, for two rows, one in each direction.

The start position for the first tool motion is at an arbitrary location X1.0Y1.0 (shifted by .120 along the minus Y axis). A fixed cycle drills the first hole, repeats itself 9999 times, shifts in the positive Y axis once, drills a hole and repeats along the negative X axis 9999 times again. This subprogram pattern repeats 5000 times in the body of the main program:

```
O3960 (SUBPROGRAM)
N601 G91 Y0.12
N602 X0.12 L9999
N603 Y0.12
N604 X-0.12 L9999
N605 M99
%

O3907 (MAIN PROGRAM)
N1 G20
N2 G17 G40 G80 T01               (SPOT DRILL)
N3 M06
N4 G90 G00 G54 X1.0 Y1.0 S3000 M03 T02
N5 G43 Z1.0 H01 M08
N6 G99 G82 R0.06 Z-0.04 P30 F5.0 L0
N7 M98 P3960 L5000
N8 G90 G80 Z1.0
N9 G28 Z1.0
N10 M01

N11 T02                          (5/64 DRILL)
N12 M06
N13 G90 G00 G54 X1.0 Y1.0 S3000 M03 T01
N14 G43 Z1.0 H02 M08
N15 G99 G81 R0.06 Z-0.215 F4.0 L0
N16 M98 P3960 L5000
N17 G90 G80 Z1.0
N18 G28 Z1.0
N19 G91 G28 X0 Y0
N20 M30
%
```

The program design takes an advantage of the subprogram nesting and the maximum number of repetitions.

What makes the program even more interesting is the estimate of *machining time*. This may go a little too far, but let's finish the fun. Before reading the whole page, make a guess - *how long will it take to machine all holes with the two tools?* The speeds and feeds are reasonable for most materials, so are the clearances and the dwell time for spot drilling. A rapid traverse of 475 in/min is assumed in all axes, a reasonable speed. It is worth the few calculations? Motions between the machine zero and the first location are disregarded in both directions for convenience.

The first calculation finds the time it takes to make a rapid motion between all holes. One hundred million spaces (less one space) multiplied by .120 divided by 475 in/min is 25,263.1576 minutes. These motions will be multiplied by two, for two tools, therefore 50,526.3153 minutes.

The spot drill will move .060 from the clearance to the top of part and .040 depth of cut, for the total length of .100, multiplied by one hundred million holes at the rate of 5.0 in/min, therefore cutting time for spot drilling will be 2,000,000 minutes. The spot drill will rapid out of the hole one hundred million times the distance of .100 at the rate of 475 in/min, totaling 21,052.6316 minutes; the dwell time at each location is 0.030 seconds, translated into minutes will take another 50,000 minutes.

The actual drilling will take place to the depth of .215 from .060 clearance level, for the total travel of .275 at the rate of 4.0 in/min - which is another 6,875,000 minutes. The drill will rapid out of one hundred million times by the distance of .275, at the rate of 475 in/min, adding another time of 57,894.7368 minutes.

The grand total of all results is 9,054,473.6837 minutes, which is 150,907.8947 hours, which is 6,287.829 days, which is 17.2269 *years*. Believe it or not, it will take more than seventeen years of *uninterrupted* machining, to spot drill and drill one hundred million holes - and all that can be done with the main program and a subprogram totaling just over two dozen blocks of input.

Going into related details, size of the plate without margins would have to be 100 × 100 feet, so the actual machine travel would have to be greater than 100 feet along the X axis *as well as* the Y axis. Hardly any CNC machine on the market can handle this monstrous task. How would the plate be mounted, for example? That is another question.

To make the example even more fun for the last time, consider the time spent on programming, doing it *without* a subprogram and *without* the repetition count (address L). Assuming that each block will take 6 seconds to write and 55 blocks will fit on a standard paper (hard copy), it would take about 19 years (yes, *nineteen years !*) just to write the program for the two tools (no interruptions, of course). As far as the paper is concerned, it would end up with 'only' 1,818,182 sheets, or a stack of approximately 705 feet (215 meters) thick. Enough of that - *subprograms do work*.

The majority of CNC programs will be programs for a single job - a job that is relative to a specific machine available in the shop. Such a particular job will have its unique characteristics, its special requirements as well as its own tool path. The tool path is the most important of all the features of a CNC program.

It is the CNC programmer's main responsibility to develop a functional tool path for any given job, without errors and in the most efficient way. The tool path development is very important, because it represents a machining pattern unique to the job at hand. In most programming jobs, this machining pattern is executed for the given job only and is irrelevant to any other CNC program. Often, programmers encounter opportunities, where an existing machining pattern can be used for many new jobs. This discovery will encourage development of the programs more efficiently and produce CNC programs for many additional applications and without errors.

The programming technique that addresses this issue is known as the *Translation of a Machining Pattern* or, more commonly, a *Datum Shift*. The most typical example of this technique is a temporary change of the program reference point (program zero) from the original position to a new position, so called *work shift*. Other programming techniques include *Mirror Image*, described in the next chapter, *Coordinate Rotation* and *Scaling Function*, described in the chapters that follow.

This chapter describes in detail the advanced subject of *Datum Shift*, also known as the *Machining Pattern Translation*. This is a basic feature of all CNC systems that can be applied in a variety of ways.

DATUM SHIFT WITH G92 OR G50

In essence, a datum shift is a temporary or permanent relocation of the part zero (program reference point) inside of the program. When this programming technique is used, it relocates an existing machining pattern (tool path) in the program at different locations within the CNC machine work area.

In an earlier section *(Chapter 16)*, explanation of G92 (milling) and G50 (turning) commands was covered. Review these commands now, before continuing further. In particular, recall that these commands *do not* cause any direct tool motion, but they do influence any tool motion that *follows* it. Also keep in mind that the position register command G92 and G50 registers the *absolute* coordinates of the current tool position and have no influence whatsoever on the incremental dimensions, when using the G91 command for milling or the U/W axes for turning. Its normal purpose is to 'tell' the control system the *current tool position*. This step is necessary at least once at the beginning of each tool to establish the relationship between the fixed program zero (part origin) and the actual position of the cutting tool. For example,

```
G92 X10.0 Y6.5
```

is 'telling' the control system that the cutting tool is set at positive 10.0 units away from the program zero in the X axis and positive 6.5 units away in the Y axis.

What happens if a wrong position is registered? What if the values in the G92 or G50 statement do not accurately reflect the *true*, the physical position of a cutting tool? As may be expected, the tool path will occur at the wrong place and the result is quite likely a scrap of the machined part, tool breakage, even a damage to the machine itself. Certainly not a desirable situation.

A imaginative CNC programmer always tries to find ways and special methods that take advantage of the available programming tools. G92 and G50 commands are only two of many tools that offer a tremendous power to a creative CNC programmer.

For simple jobs, there is no need for special or creative manipulations. It is not very economical to invest precious time on adding features to the program that will never provide real advantages. If such a need is well justified, the program can be optimized later.

◆ Program Zero Shift

If the G92 command is used on machining centers or the G50 command for lathes at all, rather than the more current and very efficient G54 to G59 work offsets, only *one* G92 (G50) position register command is needed *for a single tool* - assuming that work offsets are not used.

Any occurrence of more than a single position register command per each tool in one program is called *a program zero shift*.

To illustrate the concept of the program zero shift, a simple but relevant drawing will be used. The drawing is illustrated in *Figure 40-1*.

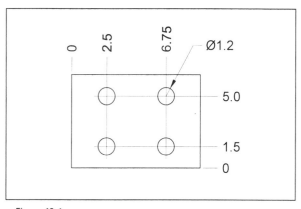

Figure 40-1

A sample drawing for zero shift illustration - program O4001

```
O4001
(G92 USED FOR TWO TABLE LOCATIONS)
N1 G20 G90
N2 G92 X22.7 Y19.5 Z12.5      (TOOL AT M/C ZERO)
N3 S1200 M03
N4 M08
N5 G99 G82 X2.5 Y1.5 R0.1 Z-0.2 P200 F8.0
N6 X6.75
N7 Y5.0
N8 X2.5              (TOOL AT LAST HOLE OF PART A)
N9 G80 Z1.0
N10 G92 X-8.7 Y-4.7     (SET AT LAST HOLE OF A)
N11 G99 G81 X2.5 Y1.5 R0.1 Z-0.2 P200
N12 X6.75
N13 Y5.0
N14 X2.5             (TOOL AT LAST HOLE OF PART B)
N15 G80 Z1.0
N16 G92 X-9.0 Y-4.8        (TOOL FROM M/C ZERO)
N17 G00 Z12.5 M09
N18 X0 Y0                  (TOOL AT M/C ZERO)
N19 M30
%
```

Based on this drawing, the four holes will be machined at *two independent* locations of the machine table setup, as illustrated in *Figure 40-2*.

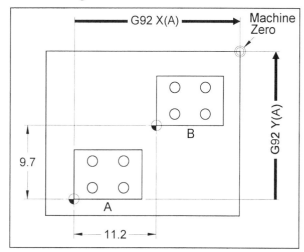

Figure 40-2

Program zero shift using G92 command for two parts - O4001

The G92 X(A) indicates the X distance from the part zero of *Part A* to the machine zero, G92 Y(A) indicates the Y distance from the part zero of *Part A* to the machine zero. Note that the distances are from program zero to machine zero. They could terminate anywhere else if necessary, but *must* start from part zero. In order to use G92, the distances between both parts *must be known*. Rounded values are used to simplify the example:

Part A: **G92 X22.7 Y19.5 Z12.5**

Part B: **X-11.2 Y-9.7 Z0**   *from Part A*

Also note that the Z value is the *same* for both *Part A* and *Part B*, because the same tool is used for both parts. To spot drill the four holes at two locations, the program may be written this way - program O4001:

Several blocks require clarification, namely blocks N2, N8, N10, N14, N16 and N18. Each of them relates to the current tool position in some way. *Be very careful here.* Not understanding the principles behind G92 calculations have caused programmers many troubles.

The cutting tool starts from the machine zero position for each program execution. It is also mounted in the spindle before machining. In block N2, the part zero (reference point) for *Part A* is established. The cutting tool at this point is 22.7 inches from program zero along the X axis, and 19.5 inches along the Y axis. The coordinate setting in block N2 reflects this fact. In blocks N7 and N8, the tool has completed the last hole of *Part A* (at X2.5Y5.0 of the current G92 setting).

The next critical block is N10. At this point in the program, the *Part A* is completed, but the *Part B* has not yet been started. Think a little now and see where exactly the tool is after executing block N9. It is at the position of X2.5Y5.0 of *Part A*. If the tool has to move to the first hole of *Part B*, which is also the position of X2.5Y1.5, the program has to 'tell' the control where the tool is at that exact moment - but in relation to *Part B*! That is done by a simple arithmetic calculation:

```
G92 (X) = 11.5 + 2.5 - 22.7 = -8.7
G92 (Y) =  9.8 + 5.0 - 19.5 = -4.7
```

Evaluate *Figure 40-3* to visualize the calculation. The direction of arrows in the illustration is important for determining the axis sign in the G92 block.

Blocks N13 and N14 contain the coordinates for the last tool location of *Part B*. From the illustration, it should be easy to understand the meaning of the coordinate values in block N16. In order to complete the program, the cutting tool has to return to the home position (machine zero). This

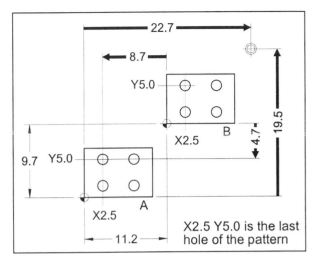

Figure 40-3

Calculations of G92 coordinates (XY) for program example O4001

return will take place from X2.5Y5.0 of the *Part B*, which is 9.000 inches from the machine zero along the X axis and 4.800 inches along the Y axis:

```
G92 (X) = 11.2 + 2.5 - 22.7 = -9.0
G92 (Y) =  9.7 + 5.0 - 19.5 = -4.8
```

Both programmed coordinates X and Y will be negative.

Once the current tool position is set at the last hole of *Part B*, a return to the machine zero can be made. This return is necessary, because it is the location of the first tool. The target position for machine zero is X0Y0 not because it is a machine zero, but because the G92 coordinates were measured from there! The actual X and Y motion to machine zero is programmed in block N18.

LOCAL COORDINATE SYSTEM

The G92 command for position register is as old as absolute programming itself. In time, it has been supplemented by additional commands that control the system of coordinates. The work coordinate system (G54 to G59 work offsets) has been discussed and a suggestion made that G92 should not be used when any work offset is in effect. Such a situation prevents changing the program zero on the fly, when needed only temporarily. Fortunately, there is a solution in the form of a programmable *subset* of the work coordinate system (work offsets) called the *local coordinate system* or the *child coordinate system*.

There are many cases, when a drawing is dimensioned in such a way that the work offsets G54 to G59 become somewhat impractical. A good example is a bolt hole pattern. If the overall machined component is round, chances are that the program zero will be selected at the *center* of the bolt hole pattern, which offers a certain benefit in calculations.

However, if the bolt pattern is within a rectangular area, the part zero maybe at the *edge* corner of the work.

Normally, absolute locations of the bolt holes will have to be calculated from program zero, unless either a *shift* of the program zero is used (using G92 described earlier), or a special coordinate system is selected.

When working with work offsets, three programming methods are available to make the job a lot more convenient and perhaps even less prone to miscalculations:

❑ Use the center of the bolt circle as program zero. This will be convenient for the CNC programmer only, as it causes more work during setup

❑ Use two different work offsets in the program, for example, G54 for the reference to the part edge and G55 for the reference to the center of the bolt circle pattern

❑ Use a local coordinate system, within the current work coordinate system (work offset) selected at the beginning of program

In all cases, one significant advantage has been gained - the programmer uses calculations relating to the bolt circle center coordinates, *directly in the CNC program,* without the need of extra additions and subtractions. This method may even simplify setup on the machine. Which method is better to select and when is addressed next.

The first method, programming to the bolt circle center, is a common method and no comments are necessary.

The second method, using the changes from one work offset to another, is also quite common. Its usage is not difficult. The limitation of this method is the reality that only *six* work offsets are available as a standard feature on typical Fanuc control - G54 to G59. If all six offsets are needed for some work, none is left as a 'spare', to use for situations such as a bolt circle pattern. (There are additional work offsets available as an optional feature of the control system).

The third method, using the *local coordinate system* method, has the main advantage that it allows the use of a *dependent* - also called a *child* - coordinate system *within* the current work offset - also called the *parent* work offset. Any number of local coordinate systems can be defined within any parent work offset. Needless to say, work is always done in one coordinate system at a time. *Note:*

> The local coordinate system is not a replacement for, *but an addition to*, the work coordinate system.

Local coordinate system is a supplement, or a subset, or a 'child' of the current work offset. It must be programmed only when a standard or additional work offset has been selected. There are many applications that can take advantage of this powerful control feature.

◆ G52 Command

What exactly is the *local coordinate system*, and how does it work? Formally, it can be defined as a system of co-ordinates *associated* with the active work offset. It is pro-grammed by the preparatory command G52.

G52	Local coordinate system

The G52 command is always complemented by the ac-tual known work coordinates that set a new - that is *tempo-rary* - program zero as illustrated in *Figure 40-4*.

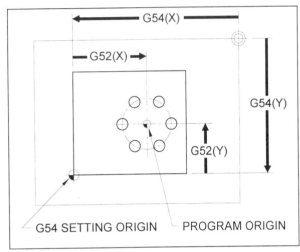

| G54 SETTING ORIGIN | PROGRAM ORIGIN |

Figure 40-4

Local coordinate system definition using the G52 command

The illustration shows a bolt circle of six holes located in a rectangular plate. The typical program zero is at the lower edge of plate and the bolt circle center is located X8.0 and Y3.0 inches from that edge, which will become the G52 shift amount. The bolt circle is ⌀4.500 inches and the first hole is at the 0° position of the bolt circle. Subsequent holes are machined in the CCW direction as holes 2, 3, 4, 5 and 6.

What the program will do is to temporarily transfer the part zero from the lower edge of plate to the bolt circle cen-ter, *in the program*. Using the illustration as a guide, follow the programming blocks, as they relate to the bolt circle and in the logical order they would appear in a program:

```
G90 G54 G00 X8.0 Y3.0      (BOLT CIRCLE CENTER)
(-- WORK COORDINATE SYSTEM POSITION --------)

G52 X8.0 Y3.0
(-- NEW PROGRAM ZERO ESTABLISHED -----------)

(G81) X2.25 Y0 (HOLE 1 LOCATION FROM NEW ZERO)
(-- COORDINATES FROM NEW ZERO --------------)

G52 X0 Y0
(-- CANCEL LOCAL OFFSET AND RETURN TO G54 ---)
```

The modal G52 command is active until it is canceled in the program. To cancel a local coordinate system and to re-turn to the previously active work offset mode, all that has to be done is to program zero values with G52:

```
G52 X0 Y0        ... last example
```

All tool motions that follow the cancellation will be rela-tive to the original work offset, which was specified by the G54 selection earlier in the example.

The bolt circle program uses the techniques described. Think about the benefit of this type of programming, as op-posed to letting the lower left corner be the only part zero.

First, a possible error by the CNC operator during setup has been minimized. True, the operator still has to set the G54 reading at the lower left corner of the plate, but does not have to do any adjustments for the bolt circle center. Programming is also easier, because the coordinate values of the bolt circle originate from the center of the bolt circle, not from the plate edge.

```
O4002 (G54 AND G52 EXAMPLE)
N1 G20
N2 G17 G40 G80 T01
N3 M06
N4 G90 G54 G00 X8.0 Y3.0 S1200 M03 T02   (CNTR)
N5 G43 Z1.0 H01 M08
N6 G52 X8.0 Y3.0     (TEMP PRG ZERO AT BC CNTR)
N7 G99 G82 R0.1 Z-0.2 P100 F10.0 L0   (NO HOLE)
N8 X2.25 Y0                           (HOLE 1)
N9 X1.125 Y1.9486                     (HOLE 2)
N10 X-1.125                           (HOLE 3)
N11 X-2.25 Y0                         (HOLE 4)
N12 X-1.125 Y-1.9486                  (HOLE 5)
N13 X1.125                            (HOLE 6)
N14 G80 Z1.0 M09
N15 G52 X0 Y0                (RETURN TO G54 SYSTEM)
N16 G28 Z1.0 M05
N17 M01

N18 T02
N19 M06
N20 (... Machining continues ...)
```

MACHINE COORDINATE SYSTEM

So far, the *work coordinate system* (G54 to G59 work off-sets) have been discussed, as well as the *local coordinate system* G52. They are both very powerful and extremely useful programming tools. Fanuc control system offers yet another coordinate system, not commonly used. It may be called the *third coordinate system*.

Selection of this coordinate system is exclusively with the machine coordinates and preparatory command G53.

G53	Machine coordinate system

Machine coordinate system uses the coordinates measured from the machine zero as an input - *always!*

At first, benefits in using this unique coordinate system may not be apparent. Before jumping to conclusions, evaluate the rules for machine coordinates system, perhaps some applications will become clear:

❏ Command G53 is effective only in the block where it is specified

❏ Programmed coordinates are always relative to machine zero position

❏ It is only used in the absolute mode (G90)

❏ Current work coordinate system (work offset) is not canceled by G53 command

❏ Cutter radius offset should always be canceled prior to G53 command

At least one possible usage emerges from these rules. The machine coordinate system can be used to guarantee tool changes at the *same machine table location* every time the tool change is programmed, regardless of which work is on the table and which work offset is active. This can be applied to a single program, or as a standard for all programs for a particular machine tool. Remember, the tool change position will always be determined by the actual tool distance from the machine zero position, *not* the program zero and *not* from any other position. On many machines, or during complex setups, it is advisable to establish a fixed tool change position, regardless of the part position. A good example is the machining with a rotary table or any other permanent or semi-permanent fixture located on the machine table.

The following program illustrates the use of the G53 command. It makes the tool change at a fixed position of the machine table, position that is *not* directly related to the program or the job - see *Figure 40-5*.

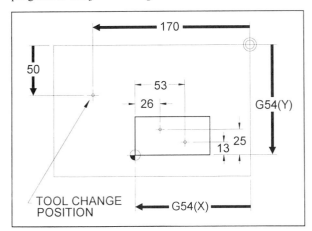

Figure 40-5

Machine coordinate system G53 - program example O4003

```
O4003 (G53 COMMAND USAGE)
N1 G20
N2 G17 G40 G80 T01
N3 G91 G28 Z0
N4 G90 G53 G00 X-170.0 Y-50.0   (TOOL CHG POS)
N5 M06                          (ACTUAL TOOL CHANGE)
N6 G54 G00 X26.0 Y25.0 S1000 M03 T02
N7 G43 Z1.0 H01 M08
N8 G99 G82 R0.1 Z-0.2 P100 F8.0
N9 X53.0 Y13.0
N10 G80 G28 Z1.0 M05
N11 G53 G00 X-170.0 Y-50.0.    (TOOL CHANGE POS)
N12 M01

N13 T02
N14 M06                         (ACTUAL TOOL CHANGE)
M15 G90 G54 G00 X53.0 Y13.0 S780 M03 T03
N16 G43 Z1.0 H02 M08
N17 G99 G81 R0.1 Z-0.836 F12.0
N18 X26.0 Y25.0
N19 G80 G28 Z1.0 M05
N20 G53 G00 X-170.0 Y-50.0     (TOOL CHANGE POS)
N21 M01

N22 T03
N23 M06                         (ACTUAL TOOL CHANGE)
...
%
```

The fourth item of the rules mentioned earlier states that the current work offset is *not* canceled by the machine coordinate system command. Since the programming example O4003 does not illustrate this situation, the following sequence of tool motions (not related to program O4003) shows the independence of G53 from G54:

```
N1 G21                                     (METRIC)
...
N250 G90 G54 G00 X17.7 Y35.3
N251 G01 Z-5.0 F200.0
N252 G00 Z500.0
N253 G53 X-400.0 Y-100.0        (FIXED POSITION)
N254 M00                      (MANUAL TOOL CHANGE)

N255 S1200 M03
N256 X50.0 Y35.0       (IN ORIGINAL WORK OFFSET)
```

N257 *(... Machining continues ...)*

The machining sequence in the program is quite simple. The cutting tool moves to the XY position of the part in N250, performs the required machining operation, such as drilling to depth in N251, rapids to a clear Z position in N252, *then* moves to the fixed tool change position in N253. In the next block, the CNC operator changes the tool manually, in block N254, then the spindle speed and rotation are re-established in N255. In the block N256 only the X and Y coordinate positions are specified. All other values are default values, including the G54 work offset command. The previous block N256 has the same meaning as:

```
N256 G90 G54 G00 X50.0 Y35.0
```

As a good programming practice, always program the complete block that contains all setting information, and do it for each new tool called. There are other practical uses for the machine coordinate system, waiting to be discovered.

DATA SETTING

In a small or medium machine shop, job shop, or any other environment where stand alone CNC machines are used, the machine operator typically sets all offset values that have to be input into the CNC system during the job setup. This common method is very useful when CNC programmer does not know the setting values - the actual values - of various offsets at the time of program development.

In a tightly manufacturing environment, for example agile manufacturing or very large volume production, this method is very costly and inefficient. An agile or large volume production uses modern advanced technology, such as CAD/CAM systems for design and tool path development, concept of cells, robots, preset tools, automatic tool changing and tool life management, pallets, programmable auxiliary equipment, machine automation, and so on. In such an environment, there cannot be any unknown elements - relationships of all reference positions are always known and the need for offsets to be found and set at each individual machine is eliminated. All offset values must be always known to the programmer, before the actual machine and tool setup takes place.

There is an advantage in such information being known - the offset data can be included in the program and be channeled into appropriate registers through the program flow. There is no operator's interference and machining is fully automated, including the maintenance of tools and related offsets. All offsets are under constant program control, *including their updates* required for position changes and changes in tool length or radius.

All this high tech automation is possible with an optional control feature called *Data Setting*. Many controls have this special feature available, a feature that should never be underestimated. Even a small shop with only one stand alone CNC machine can benefit from Data Setting feature, provided it is supported by the control system.

◆ Data Setting Command

To select the data setting option and to set offset data through the program, Fanuc offers a basic G command:

G10	Data setting

In its basic form, the preparatory command G10 is a *non modal* command, valid only for the block in which it is programmed. If it is needed in any subsequent blocks, it has to be repeated in that block.

G10 command has a simple format that is different for machining centers and lathes. Be prepared to encounter minor differences in format for various Fanuc controls, although the programming methods are logically the same. Formats also vary for the different types of offsets, for example, work offsets as opposed to tool length offsets.

Examples in this section are for typical Fanuc controls, and have been tested on Fanuc 16 Model B, a common milling and turning control.

◆ Coordinate Mode

Selection of the absolute or incremental programming mode has a great impact on the offset values input throughout the CNC program. Regardless of which type of offset is entered with the G10 command, the programmed offset amount will *replace* the current offset amount stored in the control, if the program is in absolute mode (G90 for milling controls and XZ for turning controls).

In G91 incremental mode for milling controls and UW axes for turning, the programmed offset amount does *not replace* but *update* the offset amount stored in the control:

G90 with G10 = offset amount will be REPLACED

G91 with G10 = offset amount will be UPDATED

G90 or G91 can be set anywhere in the program, as long as the block containing the selected command is assigned *before* the G10 data setting command is called.

All types of available offsets can be set through the program, using G10 command:

❏	Work offsets	G54 to G59 and G54.1 P..
❏	Tool length offsets	G43 and G44
❏	Cutter radius offsets	G41 and G42

This group includes all associated offsets, if available.

WORK OFFSETS

Before studying this section, review *Chapter 18* that describes the concept of the work offsets in detail.

◆ Standard Work Offset Input

The standard six work offsets G54 to G59 are available for both the milling and turning controls. Due to the machining requirements, they are typically associated with milling controls. The programming format is the same:

```
G10 L2 P.. X.. Y.. Z..        Machining centers   = mills
G10 L2 P.. X.. Z..            Turning centers     = lathes
```

The L2 is a *fixed offset group number* that identifies the input as the work offset setting. The P address in this case can have a value from 1 to 6, assigned to the G54 to G59 selection respectively:

```
P1=G54   P2=G55   P3=G56
P4=G57   P5=G58   P6=G59
```

for example,

```
G90 G10 L2 P1 X-450.0 Y-375.0 Z0
```

inputs X-450.0Y-375.0Z0 coordinates into the G54 work offset register (all examples for this section are metric).

```
G90 G10 L2 P3 X-630.0 Y-408.0
```

inputs X-630.0Y-408.0 coordinates into the G56 work coordinate offset register. Since the Z amount has not been programmed, the current amount of the Z offset is retained.

```
L2 = Standard work offsets
```

◆ Additional Work Offset Input

In addition to the standard six work offsets for milling controls, Fanuc offers an optional set of additional offsets, G54.1P1 to G54.1P48. G10 command can also be used to input offset values to any one of the 48 additional work offsets and the command is very similar to the previous one:

```
G10 L20 P.. X.. Y.. Z..
```

Only the fixed offset group number has changed to L20, which only selects the additional work offsets.

```
L20 = Additional work offsets
```

◆ External Work Offset Input

Another offset that belongs to the work coordinate system is called either *External* or *Common*. This offset cannot be programmed with any standard G code and is used to update all work offsets globally, affecting all work offsets.

To input offset settings into the external offset, G10 uses the L2 offset group and P0 as the offset selection:

```
G90 G10 L2 P0 X-10.0
```

will place X-10.0 into the external work offset, while retaining all other settings (the Y axis, the Z axis and any additional axis as well). In practice, when using the shown setting, *each* work offset used in a particular program will be shifted by 10 mm into the X negative direction.

TOOL LENGTH OFFSETS

Tool length offset value for milling controls can be programmed with the G10 command combined with the L offset group. Depending on the *type* of control memory, the L offset group will have different meanings.

There are three types of memory on Fanuc controls for the tool length and tool radius offsets:

Memory A - only one column for tool length offset	
Input:	Combined Geometry + Wear offset
Values:	Value set by G10 L11 P.. R.. block

Memory B - two columns for tool offset	
Input 1:	Separate Geometry offset value
Values 1:	Values set by G10 L10 P.. R.. block
Input 2:	Separate Wear offset value
Values 2:	Values set by G10 L11 P.. R.. block

Memory C - two columns for tool offset and two columns for radius offset	
Input 1:	Separate Geometry offset value
Used for:	H offset code
Values 1:	Values set by G10 L10 P.. R.. block
Input 2:	Separate Geometry offset value
Used for:	D offset code
Values 2:	Values set by G10 L12 P.. R.. block
Input 3:	Separate Wear offset value
Used for:	H offset code
Values 3:	Values set by G10 L11 P.. R.. block
Input 4:	Separate Wear offset value
Used for:	D offset code
Values 4:	Values set by G10 L13 P.. R.. block

In all cases, the L number is an arbitrarily assigned offset group number by Fanuc and the P address is the offset register number in the CNC system. The R value is the amount of the offset to be set into the selected offset number. Absolute and incremental modes have the same effect on tool length programmed input as for work offsets.

As an example for a CNC machining center, the following block will input the amount of negative 468 mm into the tool length offset register number 5 (five):

```
G90 G10 L10 P5 R-468.0
```

If the offset has to be adjusted in order to make the cut 0.5 mm less deep for the tool length offset 5, change to the incremental mode G91 and program:

```
G91 G10 L10 P5 R0.5
```

Note the G91 incremental mode. If the last two examples are used in the order listed, the final amount of offset number 5 will be -467.5 mm.

> Older Fanuc controls were using the address L1 instead of the newer L11. These controls did not have a wear offset as a separate entry. For a compatibility with the older controls, L1 is accepted on all modern controls in lieu of L11.

◆ Valid Input Range

On most CNC machining centers, the range of tool length offset values is limited:

± 999.999 mm	Metric Geometry offset input
± 99.9999 inches	English Geometry offset input
± 99.999 mm	Metric Wear offset input
± 9.9999 inches	English Wear offset input

The number of available offsets is also limited, depending on the control model. There is a minimum 32 offset numbers available. Optionally, the CNC system can have 64, 99, 200 or 400 offsets available (even more), most of them as a special option.

CUTTER RADIUS OFFSETS

For the offset memory type C, the amount of the cutter radius offset (D) may be input through the program, using G10 command with L12 and L13 offset groups:

```
G90 G10 L12 P7 R5.0
```

will input 5.000 radius value into the cutter radius geometry offset register number 7.

```
G90 G10 L13 P7 R-0.03
```

will input -0.030 radius amount into the cutter radius wear offset register number 7.

If the existing offset amount needs to be only *adjusted*, use the incremental programming mode. The last example of a wear offset will be updated by adding 0.010 mm:

```
G91 G10 L13 P7 R0.01   (NEW SETTING IS 0.02 MM)
```

Be careful with the G90 and G91 mode - remember to restore the mode for subsequent sections of the program.

LATHE OFFSETS

Tool length offset does not apply to the lathe controls, because of a different offset structure. G10 command can be used to set offset data for a lathe control, using this format:

```
G10 P.. X(U).. Z(W).. R(C).. Q..
```

The P address is either the *geometry offset* number or the *wear offset* number to be set. The addresses X, Z and R are absolute values, the addresses U, W and C are their respective incremental equivalents. No G90 or G91 mode is available, using the standard G codes of the *A Group*.

To tell apart the geometry offset and the wear offset, the geometry offset number must be increased by an arbitrary value of 10000:

```
P10001    will be geometry offset number 1
P10012    will be geometry offset 12             ... and so on
```

If the value of 10000 is not added, the P number will then become the number of the wear offset.

Here are some typical examples of offset data setting for a CNC lathe, along with expected results. All examples are consecutive, based on the order of input:

```
G10 P10001 X0 Z0 R0 Q0
```

. . . clears all geometry offset for *G 01* settings
(Geometry offset register 1)

```
G10 P1 X0 Z0 R0 Q0
```

. . . clears all wear offset for *W 01* settings
(Wear offset register 1)

Note - *Q0 also cancels value of tool tip number in G 01*

```
G10 P10001 X-200.0 Z-150.0 R0.8 Q3
```

. . . sets the contents of *G 01* geometry offset to:
        X-200.0   Z-150.0   R0.8   T3

. . . also sets T3 in the wear offset - *automatically !!!*

```
G10 P1 R0.8          Current T setting assumed
```

. . . sets R0.8 value in *W 01* wear offset

Note, that it may be safer to program:

`G10 P1 R0.8 Q3`   *Current setting not assumed*

`G10 P1 X-0.12`

. . . wear offset *W 01* is set to X-0.12,
regardless of its previous setting

`G10 P1 U0.05`

. . . updates X-0.12 by +0.05,
to the new value of X-0.07

Note that the tool tip number (programmed in the G10 application as the Q entry) will always change the geometry offset and the wear offset simultaneously, whatever the amount or the offset type is. The reason is a control built-in safety that attempts to eliminate data entry error.

MDI DATA SETTING

Programming various offset values through the program setting requires full understanding of the input format for a particular control system. It is too late when an incorrect setting causes a damage to the machine or the part.

One method that can be used to make sure the offset data setting is correct, is a simple test. Test the G10 entries in the MDI mode on the CNC unit first, and check the results:

❑   Set the Program mode

❑   Set the MDI mode

❑   Insert the test data

For example, enter:

`G90 G10 L10 P12 R-106.475`

❑   Press INSERT

❑   Press CYCLE START

To verify, check the tool length offset H12 - it should have the stored value of -106.475.

While still in the MDI mode, insert another test data, for example:

`G91 G10 L10 P12 R-1.0`

❑   Press INSERT

❑   Press CYCLE START

Again, to verify, check the setting of tool length offset H12 - it should have the new value of -107.475.

Develop other similar tests to follow the same routine. It is always better to start a program with confidence.

PROGRAMMABLE PARAMETER ENTRY

This section covers yet another aspect of programming the G10 command - this time as a *modal* command. It is used to change a system parameter, through the program. This command is sometimes called the *'Write to parameter function'*, and is definitely not very common in daily programming. Timid programmers should skip this section altogether. It is very important to understand the concept of control system parameters, otherwise this section will not help much. Authorization to change parameters for the machine tool, regardless of other professional qualifications, is equally important to apply this section.

```
                    W A R N I N G !
          Incorrect setting of CNC system parameters
        may cause irreparable damage to the CNC machine!
```

Typical uses of this command are common to changes of machining condition, for example, spindle and feedrate time constants, pitch error compensation data, and others. This command usually appears in the so called *User Macros* (applied by the G65 command) and its purpose is to control certain machine operations. The concept and explanation of *User Macros* is not covered by this handbook.

◆ Modal G10 Command

When the G10 command was used for the offset data setting earlier, it had to be repeated in each block. G10 for the offset entry can only be used as a *non-modal* command. Modern Fanuc controls also allow to do another type of change through the program - the change of *CNC system parameters* through a modal G10 command.

Many entries used in programs are automatically converted to a system parameter by the control. For example, programming G54, the set value is seen on the work offset screen. Yet, the actual storage of G54 value takes place in a system parameter, identified by a certain parameter number. The G54 setting can be changed either through the offset data or through a parameter change, and the parameter number must be known. Some system parameters cannot be changed as easily (and some cannot be changed at all), so the modal G10 command can be very useful. In fact, two related commands are required - G10 to start the setting and G11 to cancel the setting:

`G10 L50`
(... data setting ...)
`G11`

The data setting block has three entries:

```
G10 L50
.. P.. R..
G11
```

In case of a modal G10 and G11 combination, the commands have this meaning:

G10	Data setting mode
L50	Programmable parameter entry mode fixed
.. P.. R..	Data entry specification
G11	Data setting mode cancel

Between the block G10L50 and the block G11 is the list of system parameters that are to be set, one parameter per block. The parameter number uses the N address and the data use P and R addresses. There are several types of parameter input:

Type of parameter input	Allowed input range
Bit type	0 or 1
Bit axis type	0 or 1
Byte type	0 to ±127
Byte axis type	0 to 255
Word type	0 to ±32767
Word axis type	0 to ±32767
Two word type	0 to ±99999999
Two word axis type	0 to ±99999999

Watch the bit types parameters - a single data number is always assigned 8 bits. Each bit has a different meaning, so exercise care when changing one bit but not another.

Word type is also called an *integer type* and the two-word type is also called a *long integer type*.

◆ Parameters Notation

Numbering of bit type and bit axis type parameters is standard from 0 to 7 *(computers start counting from zero not from one)*, from right to left:

Number	#7	#6	#5	#4	#3	#2	#1	#0

where *Number* is the four digit parameter number and the #7 to #0 are individual bit positions - *note the order of numbering and the counting method*. Other, non bit type parameters are input as a byte, word, or two word entries (there are axis and non axis versions).

P Address

The P address is used only for parameters relating to axes (bit axis, byte axis, word axis and two word axis). If the parameter does not relate to an axis, the P address is redundant and does not have to be programmed.

If more than one axis is required to be set at the same time, use multiple .. P.. R.. entries between G10 and G11 - see examples further in this section.

R Address

The address R is the new value to be registered into the select parameter number and must always be entered. The valid range listed above must be observed. Note the lack of decimal points in the entries.

◆ Program Portability

Programs containing even a single programmable parameter entry should be used only with the machine and control for which they were designed.

> Use extreme care when a program that modifies system parameters is used on several machines.

Parameter numbers and their meaning on different control models are not necessarily the same. The exact control model and its parameter numbers must be known during programming. For example, on Fanuc control Model 15, the parameter controlling the meaning of an address without a decimal point is number 2400 (Bit #0). The parameter controlling the same setting on Fanuc control Model 16 will use number 3401 (Bit #0).

The following examples illustrate various programmable parameter entries and have been tested on a Fanuc 16 Model B CNC control - lathe and mill version. The selected parameters are used for illustration only, not necessarily as typical applications. *Testing these parameters on the machine is not recommended!*

The first example changes the baud rate setting of an Input/Output device with RS-232 interface, if the *I/O Channel* is set to *0*):

```
G10 L50
N0103 R10
G11
```

Parameter that controls the baud rate setting for the selected device has a number #103. From a table supplied by Fanuc, the R value can be input:

Setting R-value	Description
1	50 baud
2	100 baud
3	110 baud
4	150 baud
5	200 baud

Setting R-value	Description
6	300 baud
7	600 baud
8	1200 baud
9	2400 baud
10	4800 baud
11	9600 baud
12	19200 baud

In the previous example,

```
G10 L50
N0103 R10
G11
```

4800 characters per second baud rate has been selected.

In another example, the parameter #5130 controls the chamfering distance for thread cutting cycles G92 and G76 (gradual pullout distance applicable to lathe controls only). The data type is a non axis byte, unit of the data is 0.1 of a pitch and the range is from 0 to 127:

```
G10 L50
N5130 R1
G11
```

This program segment will change parameter #5130 to the value of 1. The chamfering amount will be equivalent to one pitch of the thread. Do not confuse byte with a bit - byte is a value 0 to 127 or 0 to 255 for the byte axis type, bit is a state only (0 or 1, OFF or ON, DISABLED or ENABLED), offering selection of only one of two options available. The word BIT is actually an abbreviation of two words:

Bit = *Bi*nary dig*it*   ('binary' means *based on two*)

Another example is for the entry of a two word parameter type. It will change the work offset G54 to X-250.000:

```
G90
G10 L50
N1221 P1 R-250000
G11
```

Parameter #1221 controls G54, #1222 controls G55, and so on. P1 refers to the X axis, P2 refers to the Y axis, and so on, up to 8 axes. Because the valid range of a long integer (two word type) is required, a decimal point cannot be used. Since the setting is in metric system and one micron (0.001 mm) is the least increment, the value of -250.000 will be entered as -250000. The following example is *NOT* correct and will result in an error:

```
G90
G10 L50
N1221 P1 R-250.0     (DECIMAL POINT NOT ALLOWED)
G11
```

Proper input is *without* the decimal point. An error condition (alarm or fault) will also be generated if the P address is not specified at all. For example,

```
G90
G10 L50
N1221 R-250000
G11
```

will generate an error condition. The next example is changed for two axes input:

```
G90
G10 L50
N1221 P1 R-250000
N1221 P2 R-175000
G11
```

If this example is used on a lathe control, P1 is the X axis, P2 is the Z axis. On a machining center, the P1 is the X axis, P2 is the Y axis and P3 will be the Z axis, if required. In either case, the first two axes of the G54 work offset setting will be -250.000 and -175.000 respectively.

Sometimes it is necessary to set all axes to zero. This may be done with a standard offset setting:

```
G90 G10 L2 P1 X0 Y0 Z0        (MILLING CONTROL)
```

or write to a parameter, also for a milling control:

```
G90
G10 L50
N1221 P1 R0        (SET G54 X COORDINATE TO 0)
N1221 P2 R0        (SET G54 Y COORDINATE TO 0)
N1221 P3 R0        (SET G54 Z COORDINATE TO 0)
G11
```

◆ Bit Type Parameter

The next example is quite harmless and may be used as a test, but be careful with any other parameters. Its only purpose is to set automatic block sequencing ON while entering a CNC program at the control. It also serves as an illustration of a bit type parameter and is a good example of some general thoughts and considerations that go into program preparation using programmable parameter mode.

On Fanuc 16 Model B (and most of the other models as well) is a feature that allows automatic entry of sequence numbers, if the program is entered from the keyboard. This feature is intended as a time saving device for manual entry of program data. In order to enable this feature, select the parameter that controls the ON and OFF status of the feature. On Fanuc 16 it is a parameter number 0000 (same as 0). This is a bit-type parameter, which means it contains eight bits. Each bit has its own meaning. Bit #5 (SEQ) controls the state of the automatic sequence numbering (ON or OFF is the same as 1 or 0, but only a number can be input). An individual bit cannot be programmed, only the single data number of all eight bits. That means all the other bits must be known in order to change one. In this example, the current setting of parameter 0 is as follows:

0000			SEQ			INI	ISO	TVC
	#7	#6	#5	#4	#3	#2	#1	#0
	0	0	0	0	1	0	1	0

Specific meaning of the other parameters is irrelevant for this example. The bit #5 is set to 0, which means the automatic block numbering is disabled.

The following program segment will turn on the bit #5, without changing the other bits:

```
G10 L50
N0 R00101010
G11
```

The resulting entry in the parameter screen will reflect that change:

0000			SEQ			INI	ISO	TVC
	#7	#6	#5	#4	#3	#2	#1	#0
	0	0	1	0	1	0	1	0

Note that all bits had to be written. The job is not done yet, however. Fanuc offers an additional feature - the *increment* for the numbering can be selected as well, for example, selection of 10 will use N10, N20, N30 entries, selection of 1 will use N1, N2, N3, and so on. The example will select increments of five, for N5, N10, N15, etc. The increment has to be set - yes - by another parameter number. On Fanuc 16, the parameter number that contains the automatic numbering value is #3216. This is a word type parameter and the valid range is 0 to 9999. This parameter can only be activated by setting the bit #5 in parameter 0000 to 1. Program segment will look like this:

```
G10 L50
N3216 R5
G11
```

Once these settings are completed, there will be no need to enter block numbers in any program entered via the con-

trol panel keyboard. Anytime the End-Of-Block key EOB is pressed, the N number will appear automatically on the screen, in the increments of five, saving keyboarding time during manual program input.

The idea behind the G10 being modal in the programmable parameter entry mode is that more than one parameter can be set as a group. Since the two parameters are logically connected, a single program segment can be created, with the same final results as the two smaller segments earlier. The modal G10 command comes handy here:

```
G10 L50
N0000 R00101010
N3216 R5
G11
```

As neither parameter is the axis type, the address P was omitted. The N0000 is the same as N0, and was used only for better legibility.

Note to Fanuc 15 users (Fanuc 15 system is *higher* than Fanuc 16) - the parameter number that selects whether the automatic sequencing will enabled is 0010, bit #1 (SQN). There is more flexibility on Fanuc 15 - the starting sequence number can be controlled with parameter #0031, and the parameter number that stores the increment amount is #0032, with the same program entry styles as shown. Also, on Fanuc 15, the allowable range of sequence numbers is up to 99999. This is a typical example of a difference between two control models, even whey were produced by the same manufacturer.

◆ Effect of Block Numbers

Many programs include block numbers. It would be perfectly natural to assign block numbers to the last example:

```
...
N121 G10 L50
N122 N0000 R00101010
N123 N3216 R5
N124 G11
...
```

Will the program work? There are now *two* different N addresses in blocks N122 and N123. How does the control handle this situation? Rest easy - there will be no conflict. In case of two N addresses in a single block *within the G10 to G11 segment,* the first N address is the block number, the second one in the same block will be interpreted as the parameter number.

MIRROR IMAGE

The main purpose of a CNC program development is to create a cutter tool path in a specific location of the part or machine. If the tool path requires both the right and left hand orientation, the programming time can be shortened by using a feature called the *Mirror Image*.

Any sequence of machining operations can be repeated symmetrically by using the *mirror image* feature of the control system. There is no need for new calculations, so this technique of programming reduces the programming time as well as the possibility of errors. Mirror image is sometimes called the *Axis Inversion* function. This description is accurate up to a point. Although it is true that in mirror image mode the machine axes will be inverted, but several other changes will also take place. This makes the *Mirror Image* description more accurate. Those who are familiar with a CAD system will find that the mirror image function in CNC is based on the same principles.

Mirror image is based on the principle of symmetrical parts, sometimes known as the *Right Hand* (R/H) and the *Left Hand* (L/H) parts *(Figure 41-1)*.

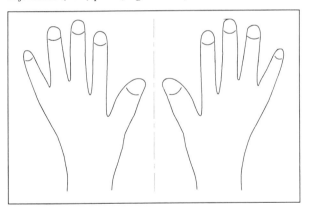

Figure 41-1

Right hand vs. Left hand as the principle of mirror image

Programming mirror image requires understanding of the basic rectangular coordinate system, particularly how it applies to quadrants. It also requires good grasp of circular interpolation and applications of cutter radius offset.

Earlier discussions established that there are four quadrants on a plane. The upper right area creates *Quadrant I*, the upper left area is *Quadrant II*, the lower left area is *Quadrant III*, and the lower right area is *Quadrant IV*. If the program zero is at the lower left corner of the part, you are programming in the first quadrant.

BASIC RULES OF MIRROR IMAGE

The basic rule of a mirror image is based on the fact that machining a given tool path in one quadrant is not much different than machining the same tool path in another quadrant. The main difference is the *reversal* of certain motion directions. That means a given part machined in one quadrant can be repeated in another quadrant *using the same program with the mirror image function in effect*.

The principle of the *Right Hand* vs. *Left Hand* orientation can be applied to a machined part orientation - *Figure 41-2*.

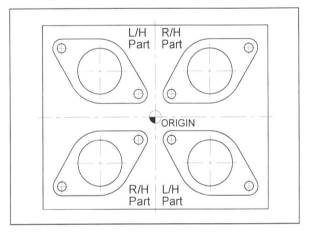

Figure 41-2

The right hand / left hand principle applied a machined part

It was also established earlier that each quadrant requires different sign of axes. Mirror image function allows the reversal of axes and other directional changes automatically.

◆ Tool Path Direction

Depending on the quadrant selected for the mirror image, the tool path directional change may affect *some* or *all* of these activities:

- ❑ Arithmetic sign of axis (plus or minus)
- ❑ Milling direction (climb or conventional)
- ❑ Arc motion direction (CW or CCW)

One or more machine axes may be affected. Normally, these axes are only the X and Y. The Z axis is generally *not* used for mirroring applications.

Not all activities are affected at the same time. If there is no circular interpolation in the program, there is no arc direction to consider. *Figure 41-3* shows the effect of the mirror image on the tool path, in all four quadrants.

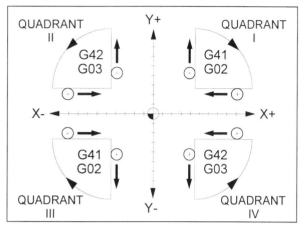

Figure 41-3
Effect of mirror image on tool path in different quadrants

◆ Original Tool Path

The original tool path program may be developed in any quadrant. If there is no mirror image applied (the default condition), the tool path is machined in the defined quadrant only. This is how the majority of all applications is programmed. Once mirroring is started, it always mirrors the *original* machining pattern - the original tool path - regardless in which quadrant it has been defined.

Mirroring will always transfer the machining pattern (the tool path) to another quadrant or quadrants. That is the purpose of the mirror image function. Programming mirror image requires that certain conditions are met. One of the conditions is definition of the mirror axis.

◆ Mirror Axis

Since there are *four* quadrants, they provide in fact *four* available machining areas. These areas are divided by *two* machine axes. Mirroring axis is the machine axis about which all programmed motions will *'flip'* over. *Figure 41-4* shows the mirror axes and their effect on part orientation in quadrants. The mirror axis can be defined in two ways:

❑   At the machine   ... by the CNC operator

❑   Through the program   ... by the CNC programmer

The typical person who is responsible for the *'flip'* is also listed. Either method allows one selection of the following possibilities:

1.   Normal machining - no mirror image set
2.   Mirrored machining about the X axis
3.   Mirrored machining about the Y axis
4.   Mirrored machining about the X and Y axes

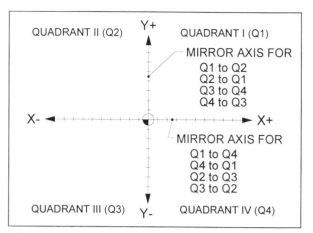

Figure 41-4
Mirror axis and its effect on part orientation

> Programmable mirror image must be
> supported by the control system

Normal machining follows the program as is. For example, if the programmed path takes place in the second quadrant (using absolute mode G90), the normal X values will be negative and the normal Y values will be positive. The sign of coordinate points is always normal within the original quadrant programmed, when no mirror image is used. Once the machining takes place in a mirrored quadrant, one or both signs will change.

◆ Sign of Coordinates

The 'normal' sign depends on the quadrant of the coordinate system used in programming. If programming in the *Quadrant I*, both the X and Y axes have positive absolute values. Here is the complete list for absolute values in all four quadrants:

Quadrant I	X+ Y+
Quadrant II	X- Y+
Quadrant III	X- Y-
Quadrant IV	X+ Y-

Mirroring the programmed tool path, the control system will *temporarily* change one or both signs, depending on the mirroring axis. For example, if the tool motion is programmed in Quadrant I (X+Y+), and is mirrored about the X axis, it will assume the signs of Quadrant IV (X+Y-). Only the X axis is the mirroring axis in this case. In another example, also based on the original program in Quadrant I, the mirroring axis is the Y axis. In this case, the temporary signs will be those of Quadrant II (X-Y+). If mirroring the program defined in Quadrant I along both axes, the program will be executed in Quadrant III (X-Y-).

◆ Milling Direction

Peripheral milling can be programmed in either *conventional milling* or *climb milling* mode. When looking at the original tool motion defined in climb milling mode within *Quadrant I*, the mirrored machining in the remaining quadrants will be as follows:

❏ Mirrored in Quadrant II   ... Conventional mode

❏ Mirrored in Quadrant III   ... Climb mode

❏ Mirrored in Quadrant IV   ... Conventional mode

It is important to understand the machining mode when using mirror image. A conventional machining mode may not yield good results. In may negatively affect the surface finish and the dimensional tolerances.

◆ Arc Motion Direction

Another change to the tool path that will happen only when a single axis is mirrored, is the rotation direction of an arc. Any clockwise arc programmed will become counterclockwise arc when mirrored along one axis, and vice versa. Here is the result of the arc motion direction, again, based on *Quadrant I*:

❏ Quadrant I - Original arc is CW:

    Quadrant II - cutting CCW
    Quadrant III - cutting CW
    Quadrant IV - cutting CCW

❏ Quadrant I - Original arc is CCW:

    Quadrant II - cutting CW
    Quadrant III - cutting CCW
    Quadrant IV - cutting CW

The control system will automatically perform G02 as G03 and G03 as G02 when required. For the majority of machining applications, the arc motion direction change should not affect the machining quality. For both the milling direction and the arc direction, refer again to the earlier *Figure 41-3*.

◆ Program Start and End

When a part is programmed with the intent to use the mirror image, make sure to use a carefully thought out programming method, that uses a slightly different technique than when programming in a single quadrant (without the mirror image). During the mirror image, all motions in the program, with the exception of machine zero return, will be mirrored, when the mirror image is turned on. That means the following considerations *do* matter:

1.   HOW the program is started
2.   WHERE the mirror image will be applied
3.   WHEN the mirror image will be canceled

Start and end of the program that is to be mirrored is usually at the same location, typically at the part X0Y0.

MIRROR IMAGE BY SETTING

A mirror image can be set at the control unit. No special codes are required. Program is relatively short, since it contains tool motion for one quadrant only. Not every program can be mirrored without a good plan first - it must be structured with mirror image in mind.

◆ Control Setting

Most controls have a screen setting or switches dedicated to mirror image set at the control. Both designs allow the operator to set certain parameters in a friendly way, without the danger of overwriting other parameters by error. In case of a screen setting, a display similar to this will appear:

```
MIRROR IMAGE X-AXIS = 0   (0:OFF 1:ON)
MIRROR IMAGE Y-AXIS = 0   (0:OFF 1:ON)
```

This is the default display, where mirroring for both axes is turned off (cancel mode). To apply X axis mirroring only, make sure the display shows

```
MIRROR IMAGE X-AXIS = 1   (0:OFF 1:ON)
MIRROR IMAGE Y-AXIS = 0   (0:OFF 1:ON)
```

To apply only the Y axis mirror, the display must show

```
MIRROR IMAGE X-AXIS = 0   (0:OFF 1:ON)
MIRROR IMAGE Y-AXIS = 1   (0:OFF 1:ON)
```

And finally, in order to mirror about both axes simultaneously, the setting will be ON for both axes:

```
MIRROR IMAGE X-AXIS = 1   (0:OFF 1:ON)
MIRROR IMAGE Y-AXIS = 1   (0:OFF 1:ON)
```

To cancel the mirror image and to return to the normal program mode, the setting for both X and Y axes is zero:

```
MIRROR IMAGE X-AXIS = 0   (0:OFF 1:ON)
MIRROR IMAGE Y-AXIS = 0   (0:OFF 1:ON)
```

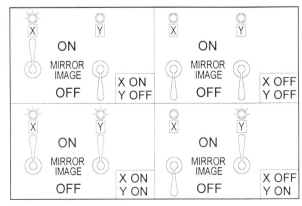

Figure 41-5

Toggle switches for manual setting of mirror image

Figure 41-5 shows mirror image settings using toggle switches ON/OFF mode. Most machines have a confirmation light that is turned ON for the currently mirrored axis.

◆ Programming - Manual Mirror Setting

Figure 41-6 is a drawing with 3 holes to be machined in all four quadrants. It will be used to illustrate the process of setting and programming of the mirror image.

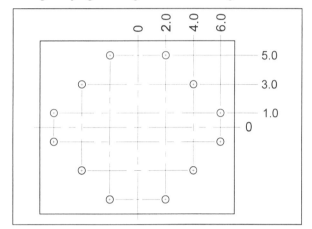

Figure 41-6

Drawing to illustrate manual mirror image programming

For a manual mirror image, the tool motion will be in one quadrant only - *Figure 41-7*, then mirrored into the other quadrants - *Figure 41-8* and example O4101:

```
O4101 (CENTER DRILL THREE HOLES)
N1 G20
N2 G17 G40 G80
N3 G90 G54 G00 X0 Y0 S900 M03              (X0Y0)
N4 G43 Z1.0 H01 M08
N5 G99 G82 X6.0 Y1.0 R0.1 Z-0.269 P300 F7.0
N6 X4.0 Y3.0
N7 X2.0 Y5.0
N8 G80 Z1.0 M09
N9 G28 Z1.0 M05
N10 G00 X0 Y0              (MUST RETURN TO X0Y0)
N11 M30
%
```

Look at the first tool motion in N3. It locates the cutting tool at X0Y0, *where there is no hole!* This is the most important block in the program for a mirror image, because it is this location that is *common to all four quadrants*!

PROGRAMMABLE MIRROR IMAGE

Most controls have mirror image that can be set but not *programmed*. Mirror image activated by the control setting is done on CNC machine, not in the program. On the other hand, programmable mirror image uses the M functions (or sometimes G codes) and almost always uses subprograms.

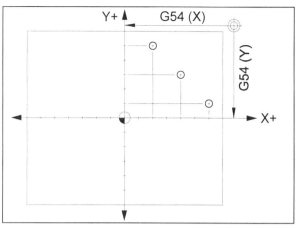

Figure 41-7

Programmed tool motion for the three holes located in Quadrant I

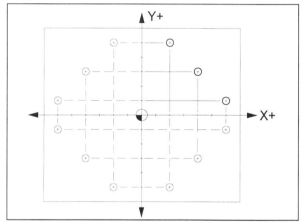

Figure 41-8

Resulting tool motion in all four quadrants using mirror image

Control settings are automatic by the program. The actual program codes for mirror image vary between machines, but the application principles are the same.

◆ Mirror Image Functions

In the examples, these functions will be used:

M21	Mirror image along the X axis
M22	Mirror image along the Y axis
M23	Mirror image cancel (OFF for either axis)

Mirror image is set for each axis by an M function. If one function is in effect when another function is programmed, they will *both* be effective. To make only one axis effective, the mirror function must be canceled first.

Cancel mirror image mode when the tool motion is completed

◆ Simple Mirror Image Example

Program O4102 for the 3 holes in *Figure 41-6,* can be changed to the programmable mirror image. Holes absolute locations are stored in subprogram O4151:

```
O4151
N1 X6.0 Y1.0
N2 X4.0 Y3.0
N3 X2.0 Y5.0
N4 M99
%
```

The main program O4102 calls the subprogram O4151 in different quadrants, using the mirror image functions. Note the X0Y0 location is common to all four quadrants.

```
O4102                              (MAIN PROGRAM)
N1 G20
```

```
N2 G17 G40 G80
N3 M23                                  (MIRROR OFF)
N4 G90 G54 G00 X0 Y0 S900 M03               (X0Y0)
N5 G43 Z1.0 H01 M08
N6 G99 G82 R0.1 Z-0.269 P300 F7.0 L0
N7 M98 P4151                            (QUADRANT I)
N8 M21                               (X-MIRROR ON)
N9 M98 P4151                           (QUADRANT II)
N10 M22                              (Y-MIRROR ON)
N11 M98 P4151                          (QUADRANT III)
N12 M23                                 (MIRROR OFF)
N13 M22                              (Y-MIRROR ON)
N14 M98 P4151                          (QUADRANT IV)
N15 G80 Z1.0 M09                     (CYCLE CANCEL)
N16 M23                                 (MIRROR OFF)
N17 G28 Z1.0 M05                   (Z MACHINE ZERO)
N18 G00 X4.0 Y6.0              (CLEAR ATC LOCATION)
N19 M30                                (PROGRAM END)
%
```

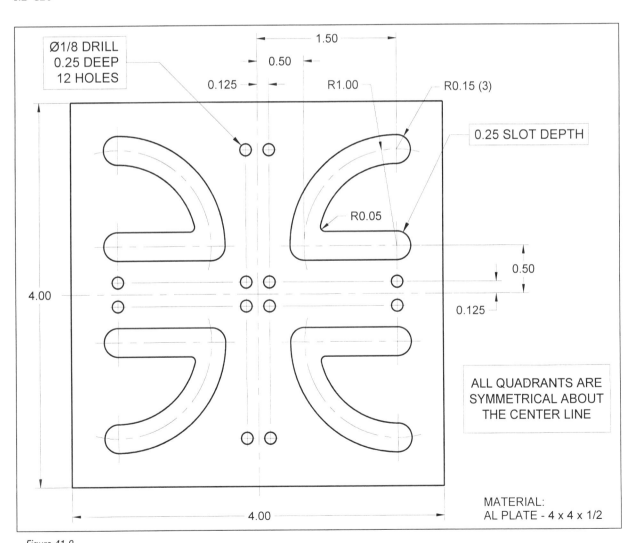

Figure 41-9

Comprehensive example of programmable mirror image. Uses main program O4103 and subprograms O4152 and O4153

◆ Complete Mirror Image Example

A complete example of a mirror image application with more involved tool motions will use two cutting tools to develop the program as per drawing in *Figure 41-9*. Program will also use coordinate shift G52, automatic tool change, a fixed cycle, interpolation motions and cutter radius offset. Two subprograms are needed - one for drilling the three holes in O4152, one for the slot milling in O4153.

```
O4152 (SUBPROGRAM - DRILLING)
N1 X0.125 Y0.125                    (MIDDLE HOLE)
N2 X1.5                             (HOLE IN X)
N3 X0.125 Y1.5                      (HOLE IN Y)
N4 X0 Y0 L0          (NO HOLE AT PLATE CENTER)
N5 M99               (SUBPROGRAM O4152 END)
%
```

Subprogram O4152 contains only the three hole locations in *Quadrant I*. The cycle call is not included in the subprogram and the return to the center of the plate (N4) is still in a cycle mode but with the L0 modifier.

```
O4153 (SUBPROGRAM - MILLING)
N1 G00 X1.5 Y1.5                    (CENTER OF SLOT)
N2 G01 Z-0.25 F3.0
N3 G03 X0.5 Y0.5 I0 J-1.0 F5.0
N4 G01 X1.5
N5 G41 D01 X1.365 Y0.485           (SLOT START)
N6 G03 X1.5 Y0.35 I0.135 J0
N7 X1.65 Y0.5 I0 J0.15
N8 X1.5 Y0.65 I-0.15 J0
N9 G01 X0.7254
N10 G02 X0.6754 Y0.7 I0 J0.05
N11 X0.677 Y0.7125 I0.05 J0
N12 X1.5 Y1.35 I0.823 J-0.2125
N13 G03 X1.65 Y1.5 I0 J0.15
N14 X1.5 Y1.65 I-0.15 J0
N15 X0.35 Y0.5 I0 J-1.15
N16 X0.5 Y0.35 I0.15 J0
N17 G01 X1.5
N18 G03 X1.635 Y0.485 I0 J0.135
N19 G01 G40 X1.5 Y0.5              (SLOT END)
N20 G00 Z0.1
N21 X0 Y0             (MOTION TO PLATE CENTER)
N22 M99              (SUBPROGRAM O4153 END)
%
```

Quadrant I is also used in subprogram O4153 for one slot. The machining starts with the cutter at the slot centerline, roughing the radius and the walls. Then, cutter radius offset is used and slot is finished to size. The subprogram ends at the plate center in N21, the same as in drilling. The program O4103 uses the two subprograms. If more tools are used, the programming technique will not change.

```
O4103 (MAIN PROGRAM)
(USES SUBPROGRAMS O4152 AND O4153)
(X0 Y0 LOWER LEFT CORNER - Z0 WORK TOP)
(M21 = X-MIRROR ON -------------------------)
(M22 = Y-MIRROR ON -------------------------)
(M23 = MIRROR OFF -------------------------)
```

```
(T01 - 1/8 DIA SHORT DRILL)
N1 G17 G20 G40 G80 G49             (STARTUP BLOCK)
N2 T01 M06                         (TOOL CHANGE)
N3 G52 X2.0 Y2.0 M23               (MIRROR OFF)
N4 G90 G54 G00 X0 Y0 S1800 M03 T02
N5 G43 Z1.0 H01 M08
N6 G99 G81 R0.1 Z-0.269 F4.0 L0
N7 M98 P4152                       (QUADRANT I)
N8 M21                             (X-MIRROR ON)
N9 M98 P4152                       (QUADRANT II)
N10 M22                            (Y-MIRROR ON)
N11 M98 P4152                      (QUADRANT III)
N12 M23                            (MIRROR OFF)
N13 M22                            (MIRROR ON)
N14 M98 P4152                      (QUADRANT IV)
N15 G80 M09                        (CYCLE CANCEL)
N16 M23                            (MIRROR OFF)
N17 G52 X0 Y0
N18 G28 Z0.1 M05
N19 G00 X4.0 Y6.0          (CLEAR ATC LOCATION)
N20 M01                            (OPTIONAL STOP)

(T02 - 1/4 DIA CENTER CUTTING END MILL)
N21 T02 M06               (TOOL T02 TO SPINDLE)
N22 G52 X2.0 Y2.0 M23              (MIRROR OFF)
N23 G90 G54 G00 X0 Y0 S2500 M03 T01
N24 G43 Z0.1 H02 M08
N25 M98 P4153                      (QUADRANT I)
N26 M21                            (X-MIRROR ON)
N27 M98 P4153                      (QUADRANT II)
N28 M22                            (Y-MIRROR ON)
N29 M98 P4153                      (QUADRANT III)
N30 M23                            (MIRROR OFF)
N31 M22                            (Y-MIRROR ON)
N32 M98 P4153                      (QUADRANT IV)
N33 M23                            (MIRROR OFF)
N34 G52 X0 Y0 M09
N35 G28 Z0.1 M05
N36 G00 X4.0 Y6.0          (CLEAR ATC LOCATION)
N37 M30                            (PROGRAM END)
%
```

Note how the G52 is used. In order to use the mirror image correctly, the program zero must be defined on the mirror line (mirror axis). Since two lines (axes) are required for this project, the plate center plate must be the program zero. There is no need to return to the X and Y machine zero, either at the end of the tool or at the end of the program. Location in a clear area for the tool change is all that is needed.

MIRROR IMAGE ON CNC LATHES

Mirror Image function has its main application on a CNC machining center. On lathes, this application is limited to a lathe with two turrets, one on each side of the spindle center line. The actual mirroring will use the X axis (the spindle center line) as the mirror axis and, in effect, allows the same programming method for both turrets.

Machining with mirror image can be used alone or combined with other time saving features, such as *Coordinate Rotation* and *Scaling Function*.

A programmed tool motion creates a pattern, contour or a pocket that can be rotated about a defined point by specified angle. With this control feature, there are many opportunities to make the programming process much more flexible and equally efficient. This very powerful programming feature, usually a special control option, is called the *Coordinate System Rotation*, or just *Coordinate Rotation*.

One of the most important applications of coordinate rotation is a program that is defined in an orthographic orientation but machined at an angle (as required by the drawing specifications). Orthographic mode defines only horizontal and vertical orientation, which means that the tool motion takes place parallel to the machine axes. To program orthographic mode is much simpler than calculating tool positions for many contour change points in an angular orientation. Compare the two rectangles shown in *Figure 42-1*.

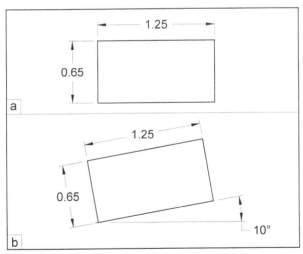

Figure 42-1
Original orthogonal object (a) and a rotated object (b)

The above figure (a) shows an orthogonal orientation of a rectangle, the figure below (b) shows the same rectangle, rotated by 10° in the counterclockwise direction. Manually, it is much easier to program the tool path for figure (a) and let the control system change it to a tool path represented in figure (b). The coordinate rotation feature is a special option and must be the part of the control system.

Mathematically, the coordinate rotation is a feature that requires only three items to define a rotated part - the center of rotation, the angle of rotation, and the tool path to rotate.

ROTATION COMMANDS

The coordinate rotation uses two preparatory commands to turn this feature ON or OFF. The two G commands controlling the coordinate rotation are:

G68	Coordinate system rotation ON
G69	Coordinate system rotation OFF

The G68 command will *activate* the coordinate system rotation, based on the *center of rotation* (also known as the *pivot point*) and the *degrees of rotation*:

G68 X.. Y.. R..

☞ where ...

> X = Absolute X coordinate of the center of rotation
> Y = Absolute Y coordinate of the center of rotation
> R = The angle of rotation

◆ Center of Rotation

The XY coordinates are normally the center of rotation (pivot point). This is a special point about which the rotation takes place - a point that can be defined by two different axes, depending on the selected working plane. X and Y is the absolute rotation center selection for the G17 active plane. G18 will use XZ as the rotation point coordinates and G19 will use YZ as the rotation point coordinates. The plane selection command G17, G18 or G19 must be entered into the program anytime *before* the rotation command G68 is issued.

If the X and Y coordinate locations are not specified with the G68 command as the center of rotation (in the G17 plane), the *current tool position* will be used as the *default* center of rotation. This method is neither a practical nor recommended approach in any circumstances.

◆ Radius of Rotation

The G68 angle representation is specified by the amount of R. The units are *degrees*, measured from the *defined center*. The number of decimal places of the R amount will become the amount of the angle. Positive R defines a CCW rotation, negative R defines a CW rotation - *Figure 42-2*.

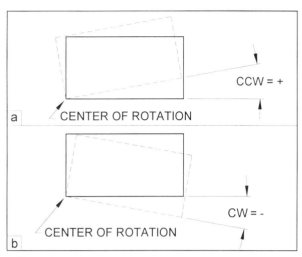

Figure 42-2

Direction of coordinate rotation, based on the center of rotation:
(a) Counterclockwise direction has a positive angle R
(b) Clockwise direction has a negative angle R

For a basic programming example, we use a simple part shape that is easy to visualize, such as a rectangular shape with a fillet corner radius - *Figure 42-3*.

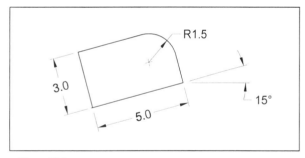

Figure 42-3

Part oriented as per engineering drawing specification

The actual tool path, including the approach towards the part and the departure from the part, is not normally included in the engineering drawing. *Be careful here* - if the approach and/or departure motions are included in the rotation, the *program zero may also be rotated*. In the *Figure 42-4*, the orientation of the part is 15° counterclockwise, based on the lower left corner.

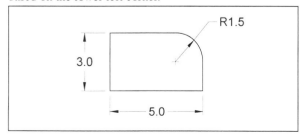

Figure 42-4

Part oriented as per program, using the G68 command

For a moment, ignore the rotation angle and program the part as if it were oriented in an orthogonal position, that is perpendicular to the axes, as shown earlier in *Figure 42-4*.

For actual cutting, decide whether the approach tool motions will be included in the rotation or not. This is a very important decision. In *Figure 42-5* are the two possibilities and the effect of coordinate rotation on program zero. In both cases, the approach tool path starts and ends at the same location of X-1.0 and Y-1.0 (clearance location).

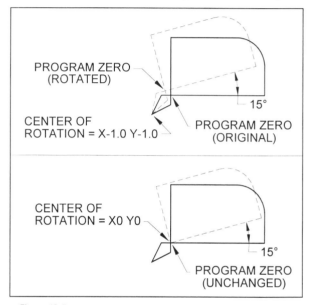

Figure 42-5

Comparison of the programmed tool path (solid line)
and the rotated tool path (dashed line):
(a) Program zero included in the rotation
(b) Program zero not included in the rotation

The following program O4201 illustrates the above example (a) in *Figure 42-5*, which *does* include the program zero rotation. If the program zero is not to be rotated, include only the part profile tool path between the G68 and G69 commands, and exclude the tool approach or departure motions. Also note the G69 in block N2 - the cancellation is included there for added safety.

```
O4201
N1 G20
N2 G69              (ROTATION CANCELED IF NEEDED)
N3 G17 G80 G40
N4 G90 G54 G00 X-1.0 Y-1.0 S800 M03
N5 G43 Z0.1 H01 M08
N6 G01 Z-0.375 F10.0
N7 G68 X-1.0 Y-1.0 R15.0
N8 G41 X-0.5 Y-0.5 D01 F20.0
N9 Y3.0
N10 X3.5
N11 G02 X5.0 Y1.5 R1.5
N12 G01 Y0.5
N13 X-0.5
```

```
N14 G40 X-1.0 Y-1.0 M09
N15 G69                         (ROTATION CANCELED)
N16 G28 X-1.0 Y-1.0 Z1.0 M05
N17 M30
%
```

The program is developed for an orthogonal orientation of the part (= 0° rotation), but machined at 15°, using the coordinate system rotation option.

In the example, block N8 contains cutter radius offset G41. Any tool offset or compensation programmed will be included when the coordinate rotation takes place.

◆ Coordinate Rotation Cancel

Command G69 *cancels* the coordinate rotation function and returns the control system to its normal orthogonal condition. Always specify the G69 command in a separate block, as in the O4201 example.

◆ Common Applications

As mentioned already, the majority of CNC machines do not have the coordinate rotation function available at all or they may have it available as an optional feature. This function can be very useful in two particular areas of machining applications.

❑ If the nature of the work includes orthogonal parts machined at an angle (as per drawing requirement). The earlier example belongs to this category.

❑ If there is a short X and/or Y travel on the machining center and the part is positioned on table at a known angle, because of the limited machine travel.

The second application is very useful example of the coordinate system rotation, provided that two major conditions are satisfied:

❑ Rotated part must fit within the work area

❑ The angle of the setup must be known

In the *Figure 42-6*, a part cannot fit within the work area orthogonally, but it can fit when rotated.

This method is quite interesting but it is not always possible to be implemented. A hundred inch long part cannot be placed within the work area length of only 20 inches. However, there are cases when this programming technique can be very useful, even if it is not too common. The illustration only shows the general principles of the application. If the positioning angle is not known, use an indicator at two locations of the mounted part and calculate it trigonometrically. In some cases, a special fixture may be required for such a setup.

> Do not confuse the term work area with the term table size

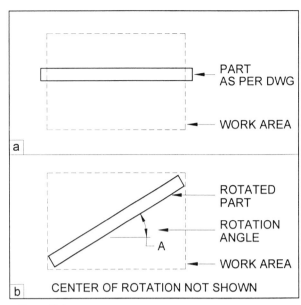

Figure 42-6

Coordinate rotation applied to fit a long part within the work area

Table size is typically larger than the actual work area, to allow for setup and additional space. Work area is used for programming and often the setup as well, and is always defined by the limits of tool motions. Work area must be able to accommodate all programmed tool motions and clearances, including the ones with cutter radius offset in effect.

PRACTICAL APPLICATION

In many cases, subprograms can be used very efficiently together with coordinate rotation. Applications such as milling polygonal shapes or machining at bolt circle locations are only typical possibilities. The following detailed example in *Figure 42-7* shows a part drawing that looks deceptively simple but involves quite a bit of programming.

The requirements and conditions for the program development must be evaluated. The core of the program will machine all 7 pockets with a ∅.250 end mill (center cutting type). To make the program more realistic, rather than plunging to the full depth of .235, we select .050 as the maximum depth of cut. The program will also leave some stock for finishing of the pocket walls (.0075 per side). In addition, all sharp edges must be broken with a minimum chamfer. In all, only three tools will be used:

∅ 3.0 FACE MILL
∅ 1/4 CENTER CUTTING END MILL
∅ 3/8 CHAMFERING TOOL

This is definitely a very advanced programming application. Not understanding the program at first is expected. With growing experience, it will be easier to interpret the program. Hopefully, the enclosed notes will help.

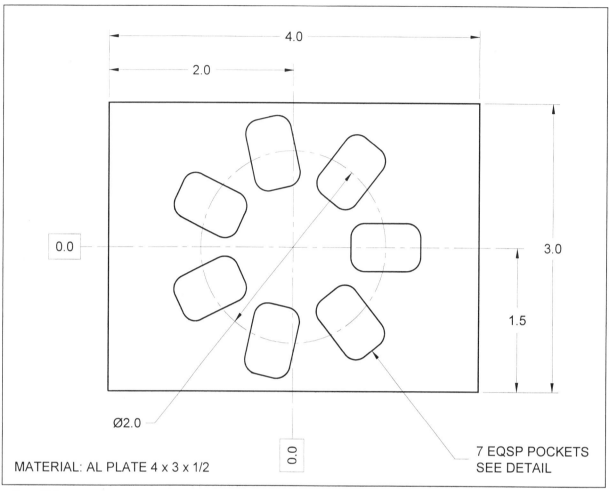

Figure 42-7
Comprehensive example of coordinate system rotation - program O4202

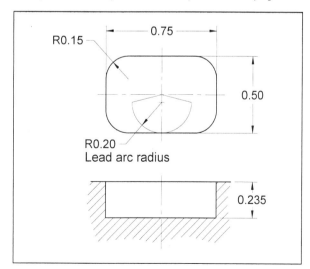

Figure 42-8
Top and front view of the pocket detail for program O4202

The main program O4202 will be developed with the aid of four subprograms. Although some parts may be a little difficult to understand, one key element is absolutely critical. In two subprograms will be the following block:

```
G91 G68 X0 Y0 R51.429
```

Its purpose is to shift to the next pocket, at an angle. The X0Y0 remain the same - they will *always* be absolute, only the angle will increment, because of the G91.

This example is not only a good illustration of the coordinate system rotation, but also shows more advanced techniques of using subprograms and several additional features. Without the advanced programming techniques, the program could be done as well, but it would take much longer and it would be virtually impossible to optimize it at the machine. The complete program that follows (O4202) is heavily documented and should present no problem to follow its progress and structure.

```
O4202 (COORDINATE SYSTEM ROTATION)

(7 POCKETS - PETER SMID - VERIFIED ON FANUC 15M CNC SYSTEM)
(PARAMETER #6400 BIT #0 - RIN - MUST BE SET TO 1 TO ALLOW G90 AND G91)
(MATERIAL 4 X 3 X 1/2 ALUMINUM PLATE - HORIZONTAL LAYOUT)
(X0Y0 IS CENTER OF 2.0 DIA CIRCLE - Z0 AT THE FINISHED TOP OF THE PLATE)


(T01 ....... 3.0 DIA FACE MILL - SKIM CUT TO CLEAN TOP FACE)
(T02 ....... 1/4 DIA CENTER CUTTING END MILL - MAX DEPTH OF CUT 0.05)
(T03 ....... 3/8 DIA CHAMFERING TOOL - 90 DEGREES - MINIMUM CHAMFER)
(T02 / D51 - OFFSET FOR ROUGHING POCKET WALLS .... 0.140 SUGGESTED - 0.0075 PER SIDE)
(T02 / D52 - OFFSET FOR FINISHING POCKET WALLS ... 0.125 SUGGESTED)
(T03 / D53 - OFFSET FOR CHAMFERING .............. 0.110 SUGGESTED - TO BE ADJUSTED)
(INCREMENT OF ROTATION ........................ 360/7 = 51.429 DEGREES)


(T01 - 3.0 DIA FACE MILL - SKIM CUT TO CLEAN TOP FACE)
N1 G20                                   (ENGLISH UNITS)
N2 G69                                   (CANCEL COORDINATE ROTATION IF ACTIVE)
N3 G17 G40 G80 T01                       (SEARCH FOR T01 IF NOT READY)
N4 M06                                   (T01 TO THE SPINDLE)
N5 G90 G54 G00 X-1.375 Y-3.25 S3500 M03 T02 (XY START POSITION FOR FACE MILLING)
N6 G43 Z1.0 H01 M08                      (Z CLEARANCE FOR SETUP - COOLANT ON)
N7 G01 Z0 F30.0                          (TOP OF FINISHED PART FOR FACE MILLING)
N8 Y3.125 F15.0                          (FACE MILL LEFT SIDE)
N9 G00 X1.375                            (MOVE TO THE RIGHT SIDE)
N10 G01 Y-3.25                           (FACE MILL RIGHT SIDE)
N11 G00 Z1.0 M09                         (Z AXIS RETRACT - COOLANT OFF)
N12 G28 Z1.0 M05                         (Z AXIS HOME FOR TOOL CHANGE)
N13 M01                                  (OPTIONAL STOP)


(T02 - 1/4 DIA CENTER CUTTING END MILL - MAX DEPTH OF CUT 0.05)
N14 T02                                  (SEARCH FOR T02 IF NOT READY)
N15 M06                                  (T02 TO THE SPINDLE)
N16 G69                                  (CANCEL COORDINATE ROTATION IF ACTIVE)
N17 G90 G54 G00 X1.0 Y0 S2000 M03 T03    (XY START POSITION FOR THE CENTER OF POCKET 1)
N18 G43 Z1.0 H02 M08                     (Z CLEARANCE FOR SETUP - COOLANT ON)
N19 G01 Z0.02 F30.0                      (CONTROLS 0.005 LEFT ON THE POCKET BOTTOM)
N20 M98 P4252 L7                         (ROUGH AND FINISH MILLING OF SEVEN POCKETS)
N21 G69                                  (CANCEL COORDINATE ROTATION IF ACTIVE)
N22 G90 G00 Z1.0 M09                     (Z AXIS RETRACT - COOLANT OFF)
N23 G28 Z1.0 M05                         (Z AXIS HOME FOR TOOL CHANGE)
N24 M01                                  (OPTIONAL STOP)


(T03 - 3/8 DIA CHAMFERING TOOL - 90 DEGREES)
N25 T03                                  (SEARCH FOR T03 IF NOT READY)
N26 M06                                  (T03 TO THE SPINDLE)
N27 G69                                  (CANCEL COORDINATE ROTATION IF ACTIVE)
N28 G90 G54 G00 X-2.5 Y-2.0 S4000 M03 T01 (XY START POSITION FOR PERIPHERAL CHAMFERING)
N29 G43 Z1.0 H03 M08                     (Z CLEARANCE FOR SETUP - COOLANT ON)
N30 G01 Z-0.075 F50.0                    (ABSOLUTE DEPTH FOR CHAMFERING Z-0.075)
N31 G41 X-2.0 D53 F12.0                  (APPROACH MOTION AND RADIUS OFFSET)
N32 Y1.5                                 (CHAMFER LEFT EDGE)
N33 X2.0                                 (CHAMFER TOP EDGE)
N34 Y-1.5                                (CHAMFER RIGHT EDGE)
N35 X-2.5                                (CHAMFER BOTTOM EDGE)
N36 G00 G40 Y-2.0                        (RETURN TO START POINT AND CANCEL OFFSET)
N37 Z0.1                                 (CLEAR ABOVE PART)
N38 X1.0 Y0                              (MOTION TO THE CENTER OF POCKET 1)
N39 M98 P4254 L7                         (CHAMFER SEVEN POCKETS)
N40 G69                                  (CANCEL COORDINATE ROTATION IF ACTIVE)
N41 G90 G00 Z1.0 M09                     (Z AXIS RETRACT - COOLANT OFF)
N42 G28 Z1.0 M05                         (Z AXIS HOME FOR TOOL CHANGE)
N43 X-2.0 Y8.0                           (PART CHANGE POSITION)
N44 M30                                  (END OF MAIN PROGRAM O4202)
%
```

```
O4251                                   (POCKET TOOL PATH AT ZERO DEGREES - POCKET 1)
N101 G91 Z-0.05                         (START AT POCKET CENTER - FEED-IN BY 0.05)
N102 M98 P4253                          (POCKET CONTOUR - O4253 USED FOR ROUGHING)
N103 M99                                (END OF SUBPROGRAM O4251)
%


O4252                                   (SUBPROGRAM FOR MILLING POCKETS)
N201 M98 P4251 D51 F5.0 L5              (ROUGH TO ABS. DEPTH Z-0.230 IN FIVE STEPS)
N202 Z-0.005                            (FINISH TO FINAL ABSOLUTE DEPTH Z-0.235)
N203 M98 P4253 D52 F4.0                 (POCKET CONTOUR - O4253 USED AT FULL DEPTH)
N204 G90 G00 Z0.02                      (RETURN TO ABS. MODE AND Z AXIS CLEAR POS.)
N205 G91 G68 X0 Y0 R51.429              (NEXT POCKET ANGLE INCREMENT)
N206 G90 X1.0 Y0                        (MOVE TO NEXT ROTATED XY AXES START POSITION)
N207 M99                                (END OF SUBPROGRAM O4252)
%


O4253                                   (POCKET TOOL PATH AT ZERO DEGREES - POCKET 1)
N301 G41 X-0.2 Y-0.05                   (LEAD-IN LINEAR MOTION)
N302 G03 X0.2 Y-0.2 I0.2 J0            (LEAD-IN CIRCULAR MOTION)
N303 G01 X0.225 Y0                      (CONTOUR BOTTOM WALL ON THE RIGHT)
N304 G03 X0.15 Y0.15 I0 J0.15          (CONTOUR LR CORNER RADIUS)
N305 G01 X0 Y0.2                        (CONTOUR RIGHT SIDE WALL)
N306 G03 X-0.15 Y0.15 I-0.15 J0        (CONTOUR UR CORNER RADIUS)
N307 G01 X-0.45 Y0                      (CONTOUR TOP SIDE WALL)
N308 G03 X-0.15 Y-0.15 I0 J-0.15       (CONTOUR UL CORNER RADIUS)
N309 G01 X0 Y-0.2                       (CONTOUR LEFT SIDE WALL)
N310 G03 X0.15 Y-0.15 I0.15 J0         (CONTOUR LL CORNER RADIUS)
N311 G01 X0.225 Y0                      (CONTOUR BOTTOM WALL ON THE LEFT)
N312 G03 X0.2 Y0.2 I0 J0.2             (LEAD-OUT CIRCULAR MOTION)
N313 G01 G40 X-0.2 Y0.05               (LEAD-OUT LINEAR MOTION)
N314 M99                                (END OF SUBPROGRAM O4253)
%


O4254                                   (SUBPROGRAM FOR CHAMFERING POCKETS)
N401 G91 G01 Z-0.175 F50.0             (CHAMFERING DEPTH FOR POCKET AT ABS. Z-0.075)
N402 M98 P4253 D53 F8.0                 (POCKET CONTOUR - O4253 USED FOR CHAMFERING)
N403 G90 G00 Z0.1                       (RETURN TO ABS. MODE AND Z AXIS CLEAR POS.)
N404 G91 G68 X0 Y0 R51.429              (NEXT POCKET ANGLE INCREMENT)
N405 G90 X1.0 Y0                        (MOVE TO NEXT ROTATED XY AXES START POSITION)
N406 M99                                (END OF SUBPROGRAM O4254)
%
```

Normally, a programmed tool motion for a CNC machining center represents the dimensions of the drawing, perhaps with cutter radius offset in effect. Occasionally, there may be times when the machining tool path that has already been programmed once must be repeated, but machined as *smaller* or *larger* than the original, yet still keep it proportional at the same time. To achieve this goal, a control feature called the *Scaling Function* is used. Note the following two important items:

❑ Scaling function is an option on many controls and may not be available on every machine

❑ Some system parameters may be used for this function as well

For even greater flexibility in programming, the scaling function can be used together with other programming functions, namely with *Datum Shift*, *Mirror Image* and *Coordinate System Rotation* - subjects described in the last three chapters.

DESCRIPTION

The control system applies a specified scaling factor to all programmed motions, which means the programmed value of all axes will change. Scaling process is nothing more than multiplying the programmed axis value by the scaling factor, based on a scaling center point. The programmer must supply both the *scaling center* and the *scaling factor*. Through a control system parameter, scaling can be made effective or ineffective for each of the three main axes, but not for any additional axes. The majority of scaling is applied to the X and Y axes only.

It is important to realize that certain values and preset amounts are not affected by the scaling function, namely various offsets. The following offset functions are *not* changed if the scaling function is active:

❑ Cutter radius offset amount   … G41-G42 / D

❑ Tool length offset amount   … G43-G44 / H

❑ Tool position offset amount   … G45-G48 / H

In fixed cycles, there are two additional situations also *not* affected by the scaling function:

❑ X and Y shift amounts in G76 and G87 cycles

❑ Peck drill depth Q in G83 and G73 cycles

❑ Stored relief amount for G83 and G73 cycles

◆ Scaling Function Usage

In industry, there are many applications for scaling the existing tool path. The result is many hours of extra work saved. Here are some of the typical possibilities when a scaling function can be beneficial:

❑ Similar parts in terms of their geometry

❑ Machining with built-in shrinkage factor

❑ Mold work

❑ English to metric and metric to English conversion

❑ Changing size of engraved characters

Regardless of application, scaling is used to make a new tool path larger or smaller than the original one. Scaling is therefore used for *magnification* (increasing size) or *reduction* (decreasing size) of an existing tool path - *Figure 43-1*.

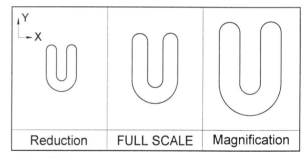

| Reduction | FULL SCALE | Magnification |

Figure 43-1

Comparison of a part reduction (left) and magnification (right) with a part in full scale (middle)

PROGRAMMING FORMAT

To supply the control unit the required information, programmer must provide the following data of information:

❑ Scaling center   … Pivot point

❑ Scaling factor   … Reduction or Magnification

The most common preparatory command for the scaling function is G51, canceled by the command G50:

G50	Scaling mode cancel	Scaling OFF
G51	Scaling mode active	Scaling ON

Scaling function uses the following program format:

> G51 I.. J.. K.. P..

☞ where ...

I	=	X coordinate of the scaling center (absolute)
J	=	Y coordinate of the scaling center (absolute)
K	=	Z coordinate of the scaling center (absolute)
P	=	Scaling factor (0.001 or 0.00001 increment)

The G51 command should always be programmed in a separate block. Commands related to the machine zero return, namely G27, G28, G29 and G30 should always be programmed in scaling OFF mode. If the G92 is used for position register, make sure it is also programmed in scaling OFF mode. Cutter radius offset G41/G42 should be canceled by G40 before scaling function is activated. Other commands and functions can be active, including the work offsets commands G54 through G59.

◆ Scaling Center

> Scaling center determines the location of the scaled tool path

Fanuc 15M uses I/J/K to specify the center point of scaling in X/Y/Z axes respectively. These values are always programmed as absolute values. As the center point controls the location of the scaled tool path, it is important to know one major principle:

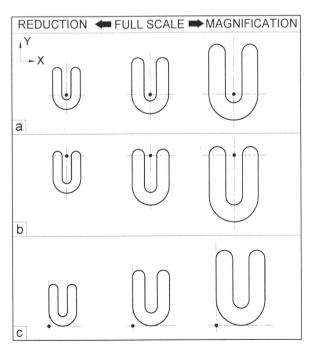

Figure 43-2
Comparison of scaled part location based on the scaling center

The scaled part will always *expand away from* and *reduce towards* the scaling point equally along the axes, as illustrated in *Figure 43-2*.

In order to understand a contour shape that is somewhat more complex, compare the original and the scaled contours in an overlay in *Figure 43-3*. It shows two machine tool paths (A and B) and the scaling center C. Depending on the scaling factor value, the result will be either path A1 to A8 or path B1 to B8.

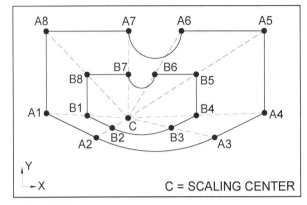

Figure 43-3
Effect of scaling point on the scaled part

Points A1 to A8 and points B1 to B8 in the illustration represent contour change points of the tool path.

➲ If the tool path A1 to A8 is the original path, then tool path B1 to B8 is the scaled tool path about center C, with a scaling factor LESS than 1.

➲ If the tool path B1 to B8 is the original path, then tool path A1 to A8 is the scaled tool path about center C, with a scaling factor GREATER than 1.

The dashed lines connecting individual points are used for easier visualization of the scaling function. Starting from the scaling center C, the line always connects to the contour change point. The B point is always a midpoint between the center point C and the corresponding A point. In practice, it means that the distance between C and B5 and B5 and A5 is exactly the same.

◆ Scaling Factor

> Scaling factor determines the size of the scaled tool path

The maximum *scaling factor* is related to the smallest scaling factor. The more advanced CNC systems can be set internally - through a system parameter - to preset the smallest scaling factor to either 0.001 or 0.00001. Some older models can only be set to 0.001 as the smallest scaling factor. Scaling factor is independent of the units used in the program - G20 or G21.

When the smallest scaling factor is set to 0.001, the largest scale that can be programmed is 999.999. When the smallest scaling factor is set to 0.00001, the largest programmable scale is only 9.99999. Given the choice, the programmer has to decide between large scales at the cost of precision and precision at the cost of large scales. For the majority of scaling applications, the 0.001 scaling factor being the smallest, is quite sufficient. Common terms for scaling factors are:

- ❏ Scaling factor > 1 ... Magnification
- ❏ Scaling factor = 1 ... No change
- ❏ Scaling factor < 1 ... Reduction

If the P address is not provided within the G51 block, the system parameter setting will become effective by default.

◆ Rounding Errors in Scaling

Any conversion process should be expected to result in at least some inaccuracies, mainly due to rounding of calculated values. For example, the inch-to-metric conversion uses the standard multiplying factor of 25.4, which is an *exact* conversion factor. In order to convert a programmed value of 1.5 inches to its equivalent in millimeters, the value in inches must be multiplied by the constant of 25.4:

```
mm = 1.5 inches × 25.4 = 38.1 mm
```

The conversion in this case is 100 percent accurate. Now try to convert the value of 1.5625 inches:

```
mm = 1.5625 inches × 25.4 = 39.6875 mm
```

So far, there is no problem. The resulting metric value as shown is also 100 percent accurate within the four decimal places for normal programming in English units.

Scaling from millimeters to inches is much different. The scaling factor for millimeters to inches (within a nine place accuracy) is 0.039370079. However, scaling factor may only be programmed with a three or five decimal place accuracy. That means *rounding* the scaling factor will result in an inaccurate conversion. In many cases, the rounded result will be quite acceptable, but it is very important to consider the possibility of an error, in case it does matter.

Compare the error amount with different rounded scaling factors for 12.7 mm, which equals exactly to 0.500 inch:

➲ Using 0.001 minimum scaling factor:

```
mm > Inch = 12.7 mm × 0.039      ... preferred
          = 0.4953 inches        ... error of 0.0047
mm > Inch = 12.7 mm × 0.038
          = 0.4826 inches        ... error of 0.0174
mm > Inch = 12.7 mm × 0.040
          = 0.5080 inches        ... error of 0.0080
```

➲ Using 0.00001 minimum scaling factor:

```
mm > Inch = 12.7 mm × 0.03937    ... preferred
          = 0.499999 mm          ... error of 0.000001
mm > Inch = 12.7 mm × 0.03938
          = 0.500126 mm          ... error of 0.000126
mm > Inch = 12.7 mm × 0.03936
          = 0.499872 mm          ... error of 0.000128
```

These examples are rather extreme applications. If a 5% shrinkage factor is to be applied, for example, the scaling factor of 1.05 (magnification) or 0.95 (reduction) is well within the expected accuracy of the final part precision.

PROGRAM EXAMPLES

The first example is very simple - *Figure 43-4*.

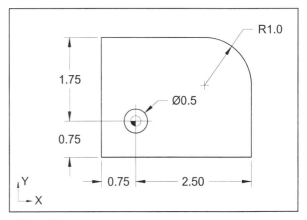

Figure 43-4
Drawing to illustrate scaling function - programs O4301 and O4302

Program O4301 is a basic contouring program, using a single cutting tool and only one cut around the part periphery. It is programmed normally, without any scaling.

```
O4301 (BASIC PROGRAM USING G54 - NOT SCALED)
N1 G20
N2 G17 G40 G80
N3 G90 G00 G54 X-1.25 Y-1.25 S800 M03
N4 G43 Z1.0 H01 M08
N5 G01 Z-0.7 F50.0
N6 G41 X-0.75 D01 F25.0
N7 Y1.75 F15.0
N8 X1.5
N9 G02 X2.5 Y0.75 I0 J-1.0
N10 G01 Y-0.75
N11 X-1.25
N12 G40 Y-1.25 M09
N13 G00 Z1.0
N14 G28 Z1.0
N15 G28 X-1.25 Y-1.25
N16 M30
%
```

Program O4302 is a modified version of O4301. It includes a scaling factor value of 1.05 - or 5% magnification - and scaling center at X0Y0Z0. K0 can be omitted in G51.

```
O4302
(PROGRAM O4301 SCALED BY 1.05 FACTOR)
N1 G20
N2 G17 G40 G80
N3 G50                         (SCALING OFF)
N4 G90 G00 G54 X-1.25 Y-1.25 S800 M03
N5 G43 Z1.0 H01 M08
N6 G51 I0 J0 K0 P1.050         (FROM X0Y0Z0)
N7 G01 Z-0.7 F50.0
N8 G41 X-0.75 D01 F25.0
N9 Y1.75 F15.0
N10 X1.5
N11 G02 X2.5 Y0.75 I0 J-1.0
N12 G01 Y-0.75
N13 X-1.25
N14 G40 Y-1.25 M09
N15 G50                        (SCALING OFF)
N16 G00 Z1.0
N17 G28 Z1.0
N18 G28 X-1.25 Y-1.25
N19 M30
%
```

Program O4303 is more complex. *Figure 43-5* is the original contour. *Figure 43-6* shows contour details with new scales and depth. Program starts with the smallest scale and works down. Note the very important blocks N712 and N713. Each contour must start from the original start point!

```
O4303 (MAIN PROGRAM)
(SCALING FUNCTION - VERIFIED ON YASNAC I80)
(T01 = 1.0 DIA END MILL)
N1 G20
N2 G50                         (SCALING OFF)
N3 G17 G40 G80 T01
N4 M06
N5 G90 G54 G00 X-1.0 Y-1.0 S2500 M03
N6 G43 Z0.5 H01 M08
N7 G01 Z-0.125 F12.0              (SET DEPTH)
N8 G51 I2.0 J1.5 P0.5       (0.5X AT Z-0.125)
N9 M98 P7001            (RUN NORMAL CONTOUR)
N10 G01 Z-0.25                   (SET DEPTH)
N11 G51 I2.0 J1.5 P0.75    (0.75X AT Z-0.250)
N12 M98 P7001          (RUN NORMAL CONTOUR)
N13 G01 Z-0.35                   (SET DEPTH)
N14 G51 I2.0 J1.5 P0.875  (0.875X AT Z-0.350)
N15 M98 P7001          (RUN NORMAL CONTOUR)
N16 M09
N17 G28 Z0.5 M05
N18 G00 X-2.0 Y10.0
N19 M30
%

O7001 (SUBPROGRAM FOR G51 SCALE)
(D51 = CUTTER RADIUS)
N701 G01 G41 X0 D51
N702 Y2.5 F10.0
N703 G02 X0.5 Y3.0 R0.5
N704 G01 X3.5
```

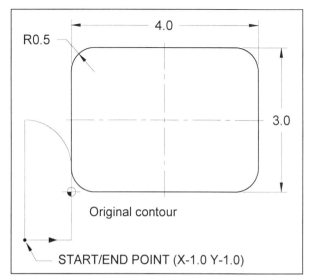

Figure 43-5

Original contour in full scale

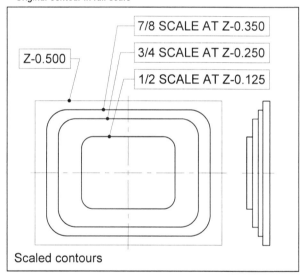

Figure 43-6

Scaled contours at three depths

```
N705 G02 X4.0 Y2.5 R0.5
N706 G01 Y0.5
N707 G02 X3.5 Y0 R0.5
N708 G01 X0.5
N709 G02 X0 Y0.5 R0.5
N710 G03 X-1.0 Y1.5 R1.0
N711 G01 G40 Y-1.0 F15.0
N712 G50                        (SCALING OFF)
N713 X-1.0 Y-1.0    (RETURN TO ORIGINAL START)
N714 M99
%
```

The scaling function offers many possibilities. Check the related control parameters and make sure the program reflects the control settings. There are significant differences between various control models.

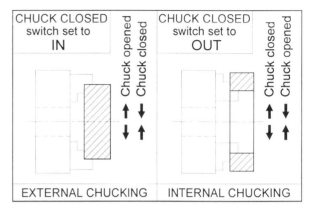

CNC LATHE ACCESSORIES

Any CNC machine can be equipped with additional accessories, to make it more functional or functional in a particular way. In fact, most CNC machines have at least some additional accessories, either as a standard equipment or as an option. Machining centers have indexing and rotary tables, pallets, right angle heads, etc. All these are complex accessories and require a certain amount of time to understand them well. Many CNC lathes are also equipped with a number of additional accessories that are usually quite simple to program. Some of the most noteworthy and typical programmable additions (or features) of this kind are:

❑ Chuck control

❑ Tailstock quill

❑ Bi-directional turret indexing

❑ Barfeeder

Several other features may also be available as programmable options:

❑ Parts catcher (unloader)

❑ Pull-out finger

❑ Tailstock body and quill

❑ Steady rest / follower rest

❑ Part stopper

❑ ... others as per machine design

Some of these accessories are fairly common, so it is worth looking at them in some detail and with a few examples of their programming applications.

CHUCK CONTROL

In manual operations, a chuck, a collet or a special fixture mounted on the headstock of a lathe normally opens and closes when the CNC operator presses a foot pedal. For safety reasons, a chuck that is rotating cannot be opened, because it is protected by an special safety interlock. Another important feature of chucks is that the terms *open* and *close* depend on the method of chucking - external or internal. A key switch is available to select the type of chucking. *Figure 44-1* shows the difference.

Note that the terms *opened* and *closed* are relative to the setting of a toggle switch or a key switch, found on the machine itself, usually marked CHUCK CLOSED - that has two settings - IN and OUT.

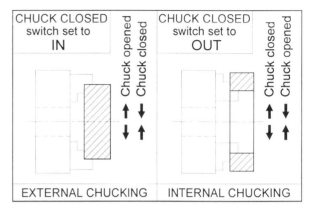

Figure 44-1

Part chucking - external and internal applications.
Note the setting of the CHUCK CLOSED switch

In some applications, such as barfeeding, it is necessary to open and close the chuck under program control. Two M functions that control the chuck or collet opening and closing are normally available.

◆ Chuck Functions

Although the assigned numbers (normally miscellaneous functions) may vary for different machines, the programming application is exactly the same. One function cancels the other. Typical M functions related to chuck control are:

M10	Open chuck
M11	Close chuck

➲ Example :

Typical programming procedure would include spindle stop and dwell:

```
M05          (STOP SPINDLE)
M10          (OPEN CHUCK)
G04 U0.1     (DWELL 1 SECOND)
M11          (CLOSE CHUCK)
M03          (RESTART SPINDLE)
```

This is a very simplified sequence, in which the dwell is the time required for the bar (for example) to go through to the stop position. Some barfeeders do not require the spindle to be stopped to feed the bar through and others have a special programming routine of their own.

M10 and M11 functions can also be used on the machine, during setup, using the MDI setting mode in manual mode. Later in this chapter, M10 and M11 functions will be used for applications associated with barfeeding.

◆ Chucking Pressure

The amount of force required to clamp a part in the chuck is called the *chucking pressure*. On most CNC lathes the pressure is controlled by an adjustable valve, usually in the tailstock area. Once the chuck pressure has been set, it is not changed very often. However, there are jobs that require the chucking pressure to be *increased* (tighter grip) or *decreased* (looser grip) frequently, usually within the same operation. Such special jobs will benefit from a programmable chuck pressure control.

A very few CNC lathe manufacturers offer a programmable chucking pressure. If they do, it is in the form of two *non-standard* miscellaneous function, for example:

M15	Low chucking pressure
M16	High chucking pressure

Typically, the part has to be reclamped in the chuck before either function can replace the other, which may disturb its position in the holding device. If the chucking pressure feature is present on the lathe, read the documentation supplied by the lathe manufacturer.

> When changing chucking pressure
> - manually or through the program -
> always make sure the part is safely clamped.

◆ Chuck Jaws

This section is not directly related to programming, but does covers tips useful to the programmer. Most chucks have three jaws, spaced 120° apart - see *Figure 44-2*.

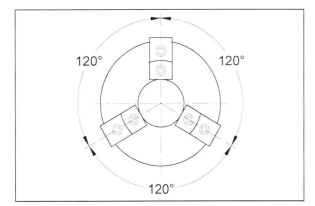

Figure 44-2
Typical three-jaw chuck for a CNC lathe

The jaws may be *hard* (usually serrated for better grip) or *soft* (normally bored by the CNC operator to suit the work diameter. Only soft jaws can be modified.

There is not much that can be done with hard jaws, except purchasing a suitable type for external or internal grip. Soft jaws are designed to be *bored* and the ability to do that is one of the basic skills a CNC operator must have. There are various techniques to bore soft jaws, all beyond the scope of this handbook. What is important is the understanding of what happens if the jaws are not bored correctly.

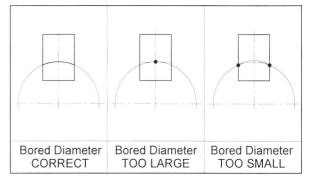

Bored Diameter CORRECT	Bored Diameter TOO LARGE	Bored Diameter TOO SMALL

Figure 44-3
Soft jaws diameter bored correctly (left) and incorrectly

Figure 44-3 shows three versions - one correctly bored jaws and two incorrectly bored jaws. In both incorrect versions, the grip, the concentricity, or both, may suffer.

TAILSTOCK AND QUILL

Tailstock is a very common accessory on a CNC lathe. Its main purpose is to support a part that is too long, too large, or needs to be pressed extra firmly against the jaws, for example, in some rough turning operations. A tailstock may also be used to support a finishing operation of a thin tubular stock, or to support a part that has a shallow grip in the jaws, to prevent it from flying out. On the negative side, tailstock is usually in the way of tool motions, so make sure to avoid a collision. A typical tailstock has three main parts:

❑   Tailstock body

❑   Quill

❑   Center

All parts are important in programming and setup.

◆ Tailstock Body

Tailstock body is the heaviest part of the lathe tailstock. It is mounted to the bed of the lathe, either manually during a setup, or through a programmable option, hydraulically. Programmable tailstock is normally available only as a factory installed option and must be ordered at the time of machine purchase.

◆ Quill

Quill is the shiny cylinder that moves in and out of the tailstock body. It has a fixed range of travel, for example, a 3 inch travel may be found on medium size lathes. When the tailstock body is mounted to the lathe bed in a fixed position, the quill is moved *out* to support the part, or *in*, to allow a part change. The part itself is supported by a center, mounted in the quill.

◆ Center

Center is a device that is placed into the quill with a tapered end, held by a matching internal taper and is physically in contact with the part. Depending on the design, if the tailstock has an internal bearing, a dead center can be used. If the tailstock has no internal bearing, a live center must be used instead. Machined part has to be pre-centered (on the CNC lathe or before), using the same angle of the tool as the tailstock center (normally 60°). A typical tailstock is illustrated in *Figure 44-4*.

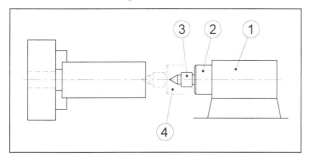

Figure 44-4

Typical tailstock for a CNC lathe:
 (1) Tailstock body   (2) Quill - OUT (retracted for work change)
 (3) Center   (4) Quill - IN (in work support position)

◆ Quill Functions

Programming the tailstock quill motion is just about the same for the majority of CNC lathes. There are two miscellaneous functions that work the same way for a programmable and non-programmable tailstock body. The two typical functions are:

M12	Tailstock quill IN or ON = active
M13	Tailstock quill OUT or OFF = inactive

If the quill is supporting the part, it is *in*, using the M12 function. If the quill is not supporting the part, it is *out*, using the M13 function. For the setup, the M12 and M13 functions may be used, and on many lathes, a toggle switch on the control is provided to operate the quill.

Spindle should be ON when the quill fully supports the part

◆ Programmable Tailstock

Tailstock body is normally not programmable (only the quill is), but this feature is available for many CNC lathes as a *factory installed option*. That means it has to be ordered it when making the initial purchase; the dealer cannot adapt the option to the machine at a later date. Many different types of programmable tailstocks are available, for example, a slide-type that moves left and right only, or a swing out type, that is out of the way when not needed.

A typical tailstock defined as *programmable* can be programmed using two *non-standard* M functions (check these functions). For the example, a CNC lathe will use these two M functions:

M21	Body of tailstock forward
M22	Body of tailstock backward

On some CNC lathes, there may also be two additional M functions available, one of them for *clamping* the tailstock, the other for *unclamping* it. In many cases, the two tailstock functions have the clamp/unclamp functions built-in.

Here is a typical programming procedure to move a tailstock towards the part, do some machining and move it back. Rather than presenting an actual programming example, let this procedure serve as a guide - fill-in the M functions required for a particular CNC lathe:

1. Unclamp the tailstock body
2. Move tailstock body forward
3. Clamp the tailstock body
4. Move quill forward into the part

5. ... *do the required machining operations ...*

6. Move quill backward from the part
7. Unclamp the tailstock body
8. Move tailstock backward
9. Clamp the tailstock body

Some procedures take certain amount of time to complete, even if the time is measured in seconds. It is generally recommended to program a dwell function to guarantee the completion of one step, before the next step starts. A review of *Chapter 24* may help.

◆ Safety Concerns

When programming a job that uses the tailstock, safety is at least as important as for other operations. The tool motion towards the part at the tool path beginning and its return to the tool change position is critical. The safest is an approach from the tool change position towards the part along the Z axis first, *then* the X motion. On return from a clear position close to the work, reverse the order - first retract the X axis above the part, *then* move the Z axis (both axes usually move to a safe tool change position).

BI-DIRECTIONAL TURRET INDEXING

Another efficiency feature is a *bi-directional turret indexing*. Many CNC lathes have a so called *bi-directional indexing* built-in, that means an automatic method of the turret indexing (the control decides the direction). However, there is a certain benefit in having a *programmable* indexing direction. If that feature is available on the CNC lathe, there will be two miscellaneous functions available to program turret indexing. Both functions are non-standard, so check the machine tool manual.

Typical M functions for turret indexing are:

M17	Indexing forward:	T01-T02-T03 ...
M18	Indexing backward:	... T03-T02-T01

Figure 44-5 shows an example of M17 and M18 functions for an 8-sided turret.

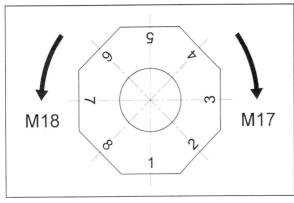

Figure 44-5

Programmable bi-directional turret indexing

In an example, a programmer is working with a lathe that has an eight station turret. Tool T01 will be used first, then tool T08 and then back to tool T01 again. There is no problem to index from T01 to T08 *or* from T08 to T01, using the automatic turret indexing direction. It makes sense, that a bi-directional turret indexing should be used for efficiency. After all, T01 and T08 may be far apart in numbers but they are next to each other on a polygonal turret with eight stations. The control system will *always choose the shortest method*, in this case, from T01 to T08 in backward direction, then from T08 to T01 in forward direction.

If the automatic bi-directional indexing is not built in the machine, it has to be programmed, assuming the control allows that. Otherwise, in normal programming, when going from T08 to T01, the indexing motion will *pass all other six stations*, which is rather an inefficient method. The next example shows how and where to place the M functions.

◆ Programming Example

This example is a complete program incorporating the bi-directional indexing and also shows hoe to use a fully programmable tailstock. All tool motions are realistic but not important for the example. The *order of numbering the tools on the turret may not be consistent* from one machine to another! The terms *forward* and *backward* are related to such order. M functions described earlier are used here:

```
O4401
(BI-DIRECTIONAL INDEXING AND TAILSTOCK)
N1 G20 G99 M18              (SET INDEX BACKWARD)
N2 G50 S1200               (LIMIT MAX RPM)
N3 T0100      (SHORT FROM T02 TO T01 WITH M18)
N4 G96 S500 M03
N5 G00 G41 X3.85 Z0.2 T0101 M08
N6 G01 Z0 F0.03
N7 X-0.07 F0.007
N8 G00 Z0.2
N9 G40 X10.0 Z5.0 T0100
N10 M01


N11 T0800     (SHORT FROM T01 TO T08 WITH M18)
N12 G97 S850 M03
N13 G00 X0 Z0.25 T0808 M08
N14 G01 Z-0.35 F0.005
N15 G04 U0.3
N16 G00 Z0.25
N17 X15.0 Z3.0 T0800
N18 M05           (SPINDLE STOP FOR TAILSTOCK)
N19 M01                     (OPTIONAL STOP)


N20 M21                   (TAILSTOCK FORWARD)
N21 G04 U2.0                  (2 SEC. DWELL)
N22 M12                        (QUILL IN)
N23 G04 U1.0                  (1 SEC. DWELL)


N24 G50 M17   (NO MAX RPM - SET INDEX FORWARD)
N25 T0100     (SHORT FROM T08 TO T01 WITH M17)
N26 G96 S500 M03
N27 G00 G42 X3.385 Z0.1 T0101 M08
N28 G01 X3.685 Z-0.05 F0.008
N29 Z-2.5 F0.012
N30 U0.2
N31 G00 G40 X10.0 Z5.0 T0100
N32 M01                     (OPTIONAL STOP)


N33 T0200     (SHORT FROM T01 TO T02 WITH M17)
N34 G96 S600 M03
N35 G00 G42 X3.325 Z0.1 T0202 M08
N36 G01 X3.625 Z-0.05 F0.004
N37 Z-2.5 F0.006
N38 U0.2 F0.015
N39 G00 G40 X15.0 Z5.0 T0200
N40 M05           (SPINDLE STOP FOR TAILSTOCK)
N41 M01                     (OPTIONAL STOP)


N42 M13                        (QUILL OUT)
N43 G04 U1.0                  (1 SEC. DWELL)
N44 M22                  (TAILSTOCK BACKWARD)
N45 G04 U2.0                  (2 SEC. DWELL)
N46 M30                    (END OF PROGRAM)
%
```

This example first uses T01 to face stock to the spindle center line. Then T08 comes in, the center drill, and makes a center hole. When the center drill moves in a clear position, tailstock body moves forward and locks, then the quill moves into the work. T01 comes back to rough out the chamfer and diameter, after which T02 comes to finish the chamfer and diameter. When the finishing is completed, spindle stops, quill moves out, then the tailstock body moves backward. The operator sets the tailstock position.

At the end of the job, T02 is in the active position. That means M18 has to be programmed at the program *beginning,* to get a short indexing from T02 to T01.

Watch how the M17 or M18 functions are programmed - their location in a particular block is very important. Either function by itself will not cause the turret to index - it only sets the direction! *Txx00 will make the actual indexing.*

All this leads to one question - how do we find out if the available CNC lathe has a built-in automatic indexing direction (shortest direction) or a programmable direction? There is a good chance that on CNC lathes where only the *forward* direction takes place (automatic indexing is not available), there is a feature called the *programmable direction,* available in the form of M17 and M18 miscellaneous - or similar - functions.

Although the tendency on modern CNC lathes is to incorporate the automatic turret indexing direction into the control system (which means that the *control system makes the decision*), there are some benefits in having the programmable method available for special machining occasions. As an example, think of an oversize tool mounted on the turret. The tool is perfectly safe, as long as it *does not* index the full swing of the turret. Automatic indexing has *no provision* for such a situation!

With a programmable indexing, the programmer has a complete control. Programming such a setup in a way that will never cause the turret to index full 360° at any time is possible. This may not be a typical situation - it will take a few seconds extra time, but it can happen quite often.

BARFEEDER ATTACHMENT

Barfeeder is an external attachment to a CNC lathe that allows small and medium cylindrical parts to be machined without interruption, up to the number that can be machined from a single bar of several feet long. There are many advantages of using barfeeders, particularly those of the modern hydrodynamic design type, rather than the old mechanical design. For example, sawing operations are eliminated (replaced with a much more precise part-off tool), no soft jaws to bore, unattended operation is possible (at least for an extended period of time), stock material economy and high spindle speeds can be achieved on many models with many other advantages.

Bars of material are stored in a special tube that guides the bar (by pushing it or pulling it) from the tube to the area where machining takes place. The only limitations are the bar length and the bar diameter. They are specified by the barfeeder manufacturer and the spindle bore diameter of the CNC lathe.

Many ingenious designs of barfeeders do exist nowadays and the programming method is heavily influenced by the design of the particular barfeeder.

The functions controlling the chuck opening and closing, the block skip function, the M99 function and several special functions, are typical aids and tools available for programming barfeeders. Many of these functions had been discussed earlier.

◆ Bar Stopper

Although the bar movement from the guide tube is controlled by the *chuck open* and *chuck close* functions (M10 and M11), the target position for the bar still has to be provided, in terms of how far it has to move out of the guide tube. This position should be lower than the bar diameter and on the positive side of the Z axis (.025 shown). This is the amount to be faced off (Z0 at the front face assumed). *Figure 44-6* shows the example.

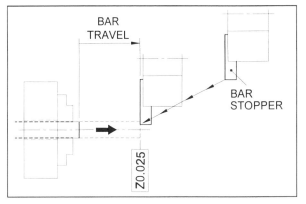

Figure 44-6
Bar stopper position for bar travel

The program is quite simple. It will use the M10 and M11 functions, but also another two functions that *may or may not be required* for a particular barfeeder. These non-standard miscellaneous functions are (in the example):

M71	Barfeeder ON - start
M72	Barfeeder OFF - stop

These functions are only examples and may be different for a certain barfeeding mechanism or unnecessary altogether. Here is the sample program:

```
O4402
N1 G20 T0100 M05              (T01 IS BAR STOPPER)
N2 G00 X0.125 Z0.025 T0101      (STOP POSITION)
N3 M10                             (CHUCK OPEN)
N4 G04 U1.0                       (1 SEC. DWELL)
N5 M71                            (BARFEEDER ON)
N6 G04 U2.0                       (2 SEC. DWELL)
N7 M11                            (CHUCK CLOSE)
N8 G04 U1.0                       (1 SEC. DWELL)
N9 M72                           (BARFEEDER OFF)
N10 G00 X10.0 Z5.0 T0100        (CLEAR POSITION)
N11 M01                         (OPTIONAL STOP)
```

A few important notes relating to a bar stopper may be helpful to develop a better program:

❏ Tool station 1 (T01) holds the bar stopper (N1)

❏ Initially, the chucking of the bar (for each first piece from the bar) is done manually

❏ Spindle rotation must be stopped prior to the chuck opening

❏ All miscellaneous functions related to the barfeeding should be programmed as separate blocks

❏ Dwell should be sufficient for the task but not excessive

These are some general considerations for programming a bar stopper, but always check the recommended procedure for the barfeeder design.

ADDITIONAL OPTIONS

There are many other options (non-standard features) on a CNC lathe that will qualify as programmable accessories. Some maybe be rather unusual, such as a programmable *chip conveyer*, or programmable *tailstock pressure*, others may not be that rare, for example, a programmable *follower rest* (a moving version of a steady rest), used for ahead-of-tool support for long parts. Steady-rest and a follower-rest help prevent chatter or deflection on a relatively long part or a part with thin walls.

Another two other accessories that are also often related to each other - *and to barfeeding as well* - are:

❏ Part Catcher also known as Part Unloader

❏ Pull-Out Finger

Both are commonly used together with barfeeding operations and use two miscellaneous functions.

◆ Part Catcher or Part Unloader

A very common accessory for a continuous machining, using a barfeeder, is *part catcher* or *part unloader*, as it is sometimes called. Its purpose is to catch the completed part after it had been parted-off. Instead of letting the completed part fall into the machine area and possibly causing dam-

age, this attachment will safely intercept the part and move it into a receiving box. The receiving box is often in the area of CNC lathe where the operator can reach without danger, and without having coolant in the way. There are two non-standard miscellaneous functions for a parts catcher:

M73	Part catcher advance	... in or forward
M74	Part catcher retract	... out or backward

The following program example illustrates how each function is programmed for a part-off tool.

```
O4403
N1 G20                            (TOP OF PROGRAM)
...
N81 T0700                    (PART-OFF TOOL ACTIVE)
N82 G50 S1500               (LIMIT MAXIMUM RPM)
N83 G96 S350 M03              (SPINDLE SPEED)
N84 G00 X2.2 Z-2.625 T0707 M08   (START POS.)
N85 M73                     (PART CATCHER ADVANCE)
N86 G01 X-0.01 F0.004         (PART-OFF MOTION)
N87 G00 X2.2 M09         (MOVE ABOVE STOCK DIA)
N88 X10.0 Z5.0 T0700        (SAFE XZ POSITION)
N89 M74                     (PART CATCHER RETRACT)
N90 M01                         (OPTIONAL STOP)
/ N91 M30               (CONTROLLED END OF PROGRAM)
N92 M99        (RESTART FROM THE TOP OF PROGRAM)
%
```

The T07 in the program is a .125 wide part-off tool, parting off a ∅2.0 stock diameter to 2.5 length, a standard process. In the program, there is a special programming technique used, relating to continuous operation. Concentrate on the last three blocks, N90, N91 and N92.

◆ Continuous Operation

Block N90 is an optional stop, typically used for setup and random checking. Block N91 contains M30 - the end of program function. Note the slash symbol in front of the block. This is a *block skip* function, described earlier in *Chapter 23*. When the block skip switch on the control panel is set to the ON position, the control system will *not* process the instructions in block N91. That means the program will not end there and the processing will continue to the block N92, where M99 is programmed.

Although the M99 function is mainly defined as the end of subprogram, it can also be used in the main program (as in this example). In that case, it causes a *continuous* processing loop. The M99 function will make the program to return to the top, and - without interruption - repeat the processing again. Since the first tool will normally have a bar stopper programmed, the barfeeder moves the stock out of the tube and the whole program repeats indefinitely - well, until the block skip switch on the control panel is set to the OFF position. Then the M30 takes over and M99 in the subsequent block will not be processed.

◆ Parts Counter

This kind of unattended lathe machining often uses another feature of the control system - *parts counter*. Parts may be counted via a program (usually a user macro), or by setting the number of required parts on the control system. They may also be programmed by non-standard miscellaneous functions, for example:

M88	Count up	... ascending order
M89	Count down	... descending order

The preset number for the count is usually the bar capacity or the required number of parts from a single bar. The programming example at the end of this chapter will illustrate the counter function and other features.

◆ Pull-Out Finger

As the name suggests, a pull-out finger is a device (CNC lathe accessory) that *grabs and pulls* the bar out of the barfeeding guide tube (while the chuck is open). This is a typical method for barfeeders of the *'pull-type'*. Normally, the pull-out finger is mounted in the turret, either as a separate 'tool', or as an add-on to an existing tool, in order to preserve the number of available tool stations. Since these activities cannot be used with the spindle rotating, yet they often need a feedrate, they are programmed in the G98 mode - feedrate per time (in/min or mm/min).

Regardless of the exact pull-out finger model available, the programming procedure is just about the same - no bar projects from the spindle longer than its face after part-off:

01. At a safe start position, index to the tool station where the pull-out finger is mounted. Spindle must be stopped at this time with M05!

02. At a rapid rate, move to the spindle centerline (X0), and a Z axis position about half-way of the overall bar projection.

03. In 'feed-per-time' mode, feed-in towards the bar as projected after part-off.

04. Dwell for about 0.5 second for the finger to catch the bar stock.

05. Open the chuck with M10.

06. Pull out the bar stock from the guide tube.

07. Dwell for about 0.5 second for the finger to complete the pull-out.

08. Close the chuck with M11.

09. Dwell for about 1 second to complete chuck closing.

10. Move the pull-out finger away from the bar stock.

11. Return the pull-out finger to the safe start position.

12. Reinstate the 'feed-per-revolution' mode.

In programming terms, the general structure will be similar to this format (item numbers correspond the list):

```
O4404
...
N.. ...
N.. Txx.. M05                         (ITEM 01)
N.. G00 X0 Z..                        (ITEM 02)
N.. G98 G01 Z.. F..                   (ITEM 03)
N.. G04 U0.5                          (ITEM 04)
N.. M10                               (ITEM 05)
N.. G01 Z.. F..                       (ITEM 06)
N.. G04 U0.5                          (ITEM 07)
N.. M11                               (ITEM 08)
N.. G04 U1.0                          (ITEM 09)
N.. G00 Z..                           (ITEM 10)
N.. X.. Z.. Txx00                     (ITEM 11)
N.. G99                               (ITEM 12)
N.. ...
```

Feel free to modify the program structure to suit the requirements of any unique setup in the machine shop.

PROGRAMMING EXAMPLE

The following programming example illustrates a complete program for an unattended barfeeding operation, until the number of parts have been machined. The lathe operator sets the required number of parts when starting a new bar stock. This program requires a careful study. It does contain some very practical and advanced features, all of them already discussed, mostly in this chapter:

```
O4405
(N1 TO N18 FOR NEW BAR ONLY - 1.5 CUT-OFF)
N1 M18                         (INDEX T03 TO T01)
N2 G20 T0100 M05              (T01 - BAR STOPPER)
N3 G00 X0.1 Z1.5 T0101         (NEW BAR OUT 1.5)
N4 M10                               (CHUCK OPEN)
N5 G04 U1.0                        (1 SEC. DWELL)
N6 M71                             (BARFEEDER ON)
N7 G04 U1.0                        (1 SEC. DWELL)
N8 M11                              (CHUCK CLOSE)
N9 X5.0 Z2.0 T0100              (CLEAR POSITION)
N10 M01                            (OPTIONAL STOP)

N11 M17                        (INDEX T01 TO T03)
N12 T0300      (T03 - 0.125 WIDE PART-OFF TOOL)
N13 G97 S1400 M03                    (CUTTING RPM)
N14 G00 X1.25 Z0 T0303 M08      (START POSITION)
N15 G01 X-0.02 F0.004          (PART-OFF BAR END)
N16 G00 X1.25 M09               (MOVE ABOVE BAR)
N17 X5.0 Z2.0 T0300            (CLEAR POSITION)
N18 M01                            (OPTIONAL STOP)

N19 M18                        (INDEX T03 TO T01)
N20 T0100 M05                 (T01 - BAR STOPPER)
N21 G00 X0.1 Z0.05 T0101   (0.05 STOCK ON FACE)
N22 M10                              (CHUCK OPEN)
N23 G04 U1.0                       (1 SEC. DWELL)
N24 M71                            (BARFEEDER ON)
N25 G04 U1.0                       (1 SEC. DWELL)
```

```
N26 M11                          (CHUCK CLOSE)
N27 X5.0 Z2.0 T0100          (CLEAR POSITION)
N28 M01                       (OPTIONAL STOP)
N29 M17                    (INDEX T01 TO T02)
N30 T0200       (T02 - FACE-CHAMFER-TURN OD)
N31 G96 S400 M03              (CUTTING SPEED)
N32 G00 G41 X1.25 Z0 T0202 M08   (START FACE)
N33 G01 X-0.07 F0.007       (FACE-OFF FRONT)
N34 G00 Z0.1                      (CLEAR Z+)
N35 G42 X0.67                (CHAMFER START)
N36 G01 X0.92 Z-0.025 F0.003   (CUT CHAMFER)
N37 Z-1.26 F0.01              (CUT DIAMETER)
N38 U0.2 F0.02               (CLEAR ABOVE BAR)
N39 G00 G40 X5.0 Z2.0 T0200  (CLEAR POSITION)
N40 M01                       (OPTIONAL STOP)

N41 T0300     (T03 - 0.125 WIDE PART-OFF TOOL)
N42 G97 S1400 M03               (CUTTING RPM)
```

```
N43 G00 X1.25 Z-1.125 T0303 M08   (START POS.)
N44 G01 X-0.02 F0.004      (PART-OFF TO LENGTH)
N45 G00 X1.25                  (MOVE ABOVE BAR)
N46 X5.0 Z2.0 T0300          (CLEAR POSITION)
N47 M01                       (OPTIONAL STOP)
N48 M89           (INCREASE PART COUNTER BY 1)
/ N49 M30        (CONTROLLED END OF PROGRAM)
N50 M99 P19          (RESTART FROM BLOCK N19)
%
```

As it usually goes with accessories and options, the machine tool manufacturers use a number of M functions to activate and deactivate a particular accessory. It is not possible to cover any specific procedures into a general reference material. Hopefully, the ideas presented in this chapter will help to adapt any manufacturer's recommendations and understand them better.

Helical milling uses an optional control system feature called *helical interpolation*. In its simplest definition, helical interpolation is an operation where a circular interpolation uses *three axes simultaneously*. This could be a misleading statement because it implies a three dimensional arc or a circle. Such an arc or circle does not exist anywhere in the field of mathematics. Yet, it is true that the G02 or G03 circular interpolation command does use *all three* axes - for example,

```
G03 X.. Y.. Z.. .. F..
```

This type of operation is only available for CNC machining centers as an optional feature. Let's look at the subject of helical milling a little closer.

HELICAL MILLING OPERATION

What exactly is helical milling? Essentially, it is a form of a circular interpolation - it is a programming technique to machine arcs and circles *combined* with a linear interpolation in the same block, during the same motion .

Previous topics that were related to circular interpolation presented one major feature of that subject. In circular interpolation, there are *two* primary axes used within the selected plane, with the intent to program an arc motion or a circular motion.

For example, in the G17 XY plane (the plane that is most common), a typical format of the circular interpolation will be in two forms:

➲ Using arc centers IJK for CW and CCW motion :

```
G02 X.. Y.. I.. J.. F..
G03 X.. Y.. I.. J.. F..
```

➲ Using radius R for CW and CCW motion :

```
G02 X.. Y.. R.. F..
G03 X.. Y.. R.. F..
```

Note that there is no Z axis programmed. As a matter of fact, if the Z axis were included in the same block as a circular milling, it will not work - normally. That means it will not work, *unless* the control system has a special feature called *the helical interpolation* option.

◆ Helical Interpolation

Helical interpolation is usually a special control system option that is designed to be used for cutting a circle or an arc with a third dimension. The third dimension is always determined by the active plane:

❏ In G17 XY plane - the third dimension is the Z axis

❏ In G18 ZX plane - the third dimension is the Y axis

❏ In G19 YZ plane - the third dimension is the X axis

In the active plane G17 (XY), the third dimension is the Z axis. In the active plane G18 (ZX), the third dimension is the Y axis and in the active plane G19 (YZ), the third dimension is the X axis.

In all cases, the third dimension - *the third axis motion* - will *always* be a linear motion that is perpendicular to the active plane.

A more formal definition of a helical interpolation can be made, based on the previous statement:

> Helical interpolation is a simultaneous
> two-axis circular motion in the working plane,
> with the linear motion along the remaining axis.

The three axis motion is always synchronized by the control and all axes reach the target location at the same time.

◆ Programming Format

The *general* formats for helical interpolation in a program are similar to the formats available for a circular interpolation - plane selection is *very* important:

➲ Using arc centers IJK for CW and CCW motion :

> G02 X.. Y.. Z.. I.. J.. K.. F..
> G03 X.. Y.. Z.. I.. J.. K.. F..

➲ Using radius R for CW and CCW motion :

> G02 X.. Y.. Z.. R.. F..
> G03 X.. Y.. Z.. R.. F..

The plane selection programmed before the helical interpolation block determines which axes will be active in the program and what their function will be.

◆ Arc Modifiers for Helical Interpolation

The arc modifier functions are programmed using the same principles as in circular interpolation but will be *different for each plane*. Here is a summary in a table:

Active Plane	Circular motion	Linear motion	Arc vectors
G17	X and Y	Z	I and J
G18	X and Z	Y	I and K
G19	Y and Z	X	J and K

Note that the arc vectors apply to the two axes that form the *circular motion* - the linear motion has no influence whatsoever. If the control system supports the direct radius entry R (instead of the traditional IJK vectors), the physical center of the arc motion is calculated automatically, within the current plane.

◆ Applications and Usage

Although the helical interpolation option is not the most frequently used programming method, it may be the only method used for a number of rather special machining applications:

❑ Thread milling

❑ Helical profiling

❑ Helical ramping

From the three groups, the *thread milling* is by far the most common method of helical interpolation applied in industry and is described next. The last two applications are quite similar, although used less frequently and will be described later in this chapter as well.

THREAD MILLING

There are two familiar methods of producing a thread on a CNC machine. On machining centers, the predominant method of thread generating is tapping, normally using fixed cycle G84 or G74. On CNC lathes, a tap is also used frequently (without the use of a cycle), but the majority of threads are machined by the single point threading method, using the block method of G32, the simple cycle G92 and the multiple repetitive cycle G76.

◆ Applying Thread Milling

There are many cases in manufacturing, where either the tapping or the single point threading method is impractical, difficult, or impossible in a given situation. Many of these difficulties can often be overcome by choosing the *thread milling method* instead. Thread milling is probably the most common industrial application of helical interpolation feature of the control.

Thread milling can be used in programming to achieve special benefits. These benefits are quite numerous:

❑ A large thread diameter - virtually any diameter can be thread milled (with high concentricity)

❑ Smoother and more accurate thread generation (only thread grinding can be more accurate)

❑ Combination of thread milling within a single setup eliminates secondary operations

❑ Full depth thread can be cut

❑ Tap is not available

❑ Tapping is impractical

❑ Tapping is difficult and causes problems

❑ Tapping is impossible in hard materials

❑ Blind hole tapping causes problems

❑ Part cannot be rotated on a CNC lathe

❑ Left hand and right hand threading has to be done with one tool

❑ External and internal threading has to be done with one tool

❑ Thread deburring minimized or eliminated

❑ Gain of high surface finish quality, particularly in softer materials

❑ Extended life of the threading tool

❑ Elimination of expensive tapping heads

❑ Elimination of expensive large taps

❑ No need for spindle reversal (as in tapping)

❑ Better power rating of the tool versus the cut (about 1/5th is not unusual)

❑ One tool holder can accept inserts for different thread pitch size

❑ Reduction of overall threading costs

Thread milling enhances other threading operations, it does not replace them. It uses special threading cutters, called *thread hobs*, or special multi tooth *thread milling cutters*. In both cases, there is one common feature for both types of cutters - *the pitch of thread is built into the cutter.*

◆ Conditions for Thread Milling

For successful thread milling, three conditions must exist before writing a program:

❑ Control system must support the operation

❑ Diameter to be threaded must be premachined

❑ Suitable thread milling tool must be selected

All three conditions must exist simultaneously.

◆ Thread Milling Tool

The thread milling cutters are available in at least two varieties - some are made of a solid carbide, some use carbide interchangeable inserts. In either design, the threading tool pitch must match the pitch of a thread required by the drawing. The tool has to be small enough to fit into the available internal space and large enough to guarantee suitable rigidity while cutting externally. For internal thread milling, cutters are available for thread milling in holes as small as .250 inch (6.35 mm).

Unlike a tap, thread milling tools do not have the helix angle built in, only the pitch. The helix angle is required for threading and is controlled during helical interpolation motion by the linear movement. Typical thread milling tools are illustrated in *Figure 45-1*.

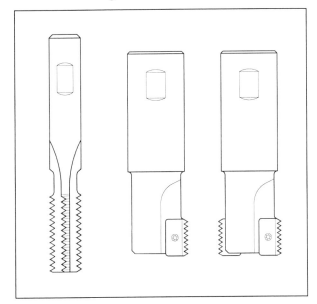

Figure 45-1
Typical thread milling cutters.
Solid carbide (left), single insert (middle) and a double insert (right)

◆ Premachining Requirements

A hole for a tap cannot have the same diameter as the tap itself. It has to be smaller to accommodate the *depth of the thread*. The same rule applies to helical milling:

❑ If the thread is milled on the inside diameter of the part (internally), the premachined diameter must be smaller that the nominal thread size

❑ If the thread is milled on the outside diameter of the part (externally), the premachined diameter must be equal to the nominal thread size

Either diameter (internal or external) may be slightly larger or slightly smaller than the 'normal' size, but this deviation is decided by the required 'fit' of the thread.

◆ Clearance Radius

Clearance radius protects the thread from damage by the cutting tool. Each cutting edge on the threading tool (hob) or indexable insert is ground with a *decreasing angle* in the direction of the cut - this is called the *clearance angle*. This clearance angle guarantees smooth cutting conditions during thread milling.

◆ Productivity of Thread Milling

One of the reasons programmers choose the thread milling operation could be the desire to improve machining productivity. There are many sizes of thread cutting tools available, with just about all pitch variations. In order to achieve the highest level of efficiency in thread milling, use a threading tool that is large enough to cut the required thread in a single revolution (in a 360° sweep). At the same time, the tool must have all necessary clearances.

A great deal of influence on thread milling productivity will be the total length of travel and the selection of cutting feedrates. A large diameter cutter can cut more efficiently (heavier feedrates), but cannot fit into confined areas. Small diameter cutter has the opposite effect - it can be used in a tight areas, but at lower feedrates. A smaller cutter may also be used with higher spindle speeds and the corresponding feedrate - the combined effect may shorten the cutting cycle time.

THE HELIX

The words *helical* and *helix* are quite common in CNC programming and appear in this and other publications quite frequently. Perhaps it is time to look at the terms relating to thread milling in more detail.

The main word that is used in this context is the word *helix*. The word *helix* is based on the original Greek word for *spiral*. A dictionary definition gives us some clue as to its meaning - it suggests that a helix is anything in the shape of the thread of a screw. Helix is defined in the *"Machinery's Handbook"* by Industrial Press, Inc., New York, NY, USA, this way:

"A helix is a curve generated by a point moving about a cylindrical surface (real or imaginary) at a constant rate in the direction of the cylinder's axis."

This quite detailed definition means that the helix is *a curve created by a circular motion of a point on a cylinder or a cone, combined with a simultaneous linear advance*. A curvature of a common screw thread is a typical example of a straight helix.

A cutting tool motion based on the mathematical definition (using three axes), results in a helical motion, also known as helical interpolation.

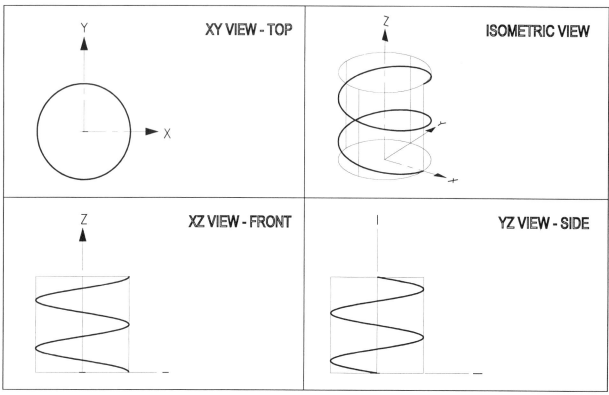

Figure 45-2
A typical helix shown in four standard views - two revolutions are shown between the top and the bottom of the helix

A helix is a machine motion that has *four* varieties:

❑ Clockwise circular cut with positive linear motion

❑ Clockwise circular cut with negative linear motion

❑ Counterclockwise circular cut with positive linear motion

❑ Counterclockwise circular cut with negative linear motion

The typical helix illustration in *Figure 45-2*, shows a common helix (which is a three-dimensional object) in four standard views. The helix is shown in these views:

❑ The top view (XY) shows only a circle.

❑ The front view (XZ) shows the helix from the front.

❑ The side view (YZ) shows the helix from the standard right side view.

❑ The isometric view (XYZ) shows a three-dimensional appearance of a two-turn helix.

Another view of a helix that is often very useful, is the *flat view* (also called the *flat layout*). This view is commonly used to illustrate a helix as a flat object that can wrap around a cylinder. *Figure 45-3* shows a flat layout of a right hand helix (one revolution).

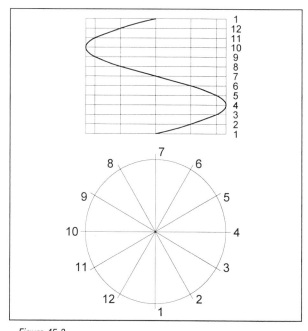

Figure 45-3
*Flat view representation of a right-hand helix.
One revolution of 360° is illustrated*

THREAD MILLING EXAMPLE

A thread milling operation on CNC machining centers can be programmed very efficiently by using the helical interpolation feature of the control system. The easiest way to describe and explain the straight thread milling, is to show an illustrated example - *Figure 45-4*.

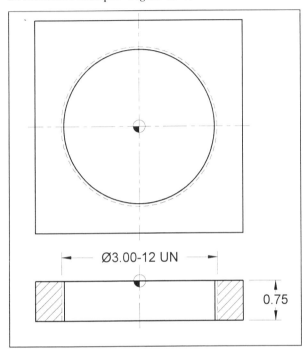

Figure 45-4
Internal thread milling example - program O4501

◆ Straight Thread

The following information is the collected initial data, based on the given drawing and the available tools:

1. Internal thread is 3.00 diameter - through the plate
2. Plate thickness is 0.75
3. 12 TPI = 12 threads per inch
4. 1.500 diameter thread hob
5. Tool T03 and offsets H03 and D03
6. Bored diameter is 2.9000 inches

This summary sets the stage for programming.

◆ Initial Calculations

In the example are six items to consider. Items 1, 2 and 3 were supplied by the drawing, but items 4, 5 and 6 were selected or calculated as part of the programming process. We look at the selected or calculated items individually.

Item 4 is the threading cutter size. Two main characteristics of a threading cutter must be considered - its diameter and the *pitch* of cutting edges (teeth). Selecting a cutter diameter must be done carefully - it must be smaller than the *bored* diameter. Another challenge is to choose a thread milling cutter that has the correct number of teeth per inch (pitch). The thread mill diameter is more important for internal threads but the pitch of threading cutter must be maintained, regardless of whether the thread cutting is internal or external.

Item 5 deals with the tool number and its related offset numbers. In this case, the tool number is 3, programmed as T03. The tool length offset number is H03 and the cutter radius offset number is D03. The D03 offset setting will contain the *radius* of threading cutter, in this case, the nominal value will be .6250. The offset numbers are arbitrary numbers for this example, others may be different. Just keep in mind that the diameter machined for an internal thread cutting *must* be smaller than the thread nominal size - just like predrilling a hole for tapping. That introduces the last item - *Item 6*.

Item 6 lists the bored diameter as 2.9000 inches. Why this number and not other? Remember that the internal thread depth is established by a common formula. A generic formula to calculate the depth D of an internal thread multiplies the pitch by a constant:

$$D = PITCH \times .54127$$

If the formula is applied to a ⌀3.000, with 12 TPI thread (1/12 = .0833333 pitch), the single depth of the thread is:

```
.0833333 × .54127 = .0451058
```

When the formula is applied to the calculation of a pre-bored diameter, this amount has to be *subtracted twice* for the required nominal diameter:

```
3.0000 - 2 × .0451058 = 2.9097884
```

Therefore, the bored diameter for the thread should be ⌀2.9098 inches.

At this point, another consideration must be made. The threading cutter itself. Essentially, the threading cutter is a forming tool. Its crest and its root will be formed on the finished thread. This feature presents a certain advantage. By programming the internal diameter a little smaller, the final size will be formed and result in a smooth surface finish. Leaving about .003 to .006 stock per side will do the trick. For the example, the .003 to .006 range is used and the calculated ⌀2.9097884 can be rounded to an even ⌀2.9, leaving only .0097884 stock on diameter, or .0048942 per side, for finishing. No doubt, the difference is reasonable, but it did take advantage of a rounding to a 'nice' number, such as 2.9000.

◆ Starting Position

After all required data have been collected and properly calculated, another step can be made, this time to calculate the thread starting position.

That is easy for the X and Y axes - the center of the thread diameter is as good start as any - better, in fact. In this example, and for simplicity, this XY position is also equivalent to the X0Y0 position.

Starting position of the threading cutter measured along the Z axis is much more important in helical milling than in any other type of milling. The Z axis start position must always be *synchronized* with the pitch of thread, as the cutting will proceed in three axes simultaneously. The Z axis zero (Z0) will be at the top of part.

The start position of the Z axis is determined by several factors - the *size of the thread mill* (in this case a tool with an indexable insert), the *pitch of the thread* (in this case .0833333), the *direction of the Z axis* motion (up or down) and the *method of the infeed* along the XY axes.

When a thread is cut using the helical interpolation feature, all three axes used must be considered equally. Just like defining the approach arc for circular interpolation, the approach arc for a helical interpolation can be defined the same way - the procedure is exactly the same.

◆ Motion Rotation and Direction

In helical interpolation it is *extremely important* to coordinate, to synchronize, the following three program items:

❑ Spindle rotation

❑ Circular cutting direction

❑ Z axis motion direction

Why are the three items so important? Why do they have to be coordinated at all? Evaluate them, one by one.

Spindle Rotation

Spindle rotation can be either M03 (clockwise) or M04 (counterclockwise).

Circular Cutting Direction

Circular direction follows the rules of circular interpolation - G02 is the clockwise direction, G03 is the counterclockwise direction.

Z Axis Motion Direction

For vertical machining, the direction of cutting along Z axis may be along two directions:

❑ Up or positive

❑ Down or negative

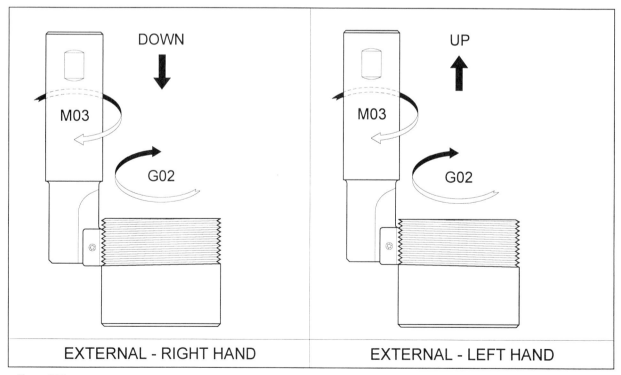

Figure 45-5

EXTERNAL thread milling using the climb milling mode - right and left hand threads, spindle rotation and cutter motions shown

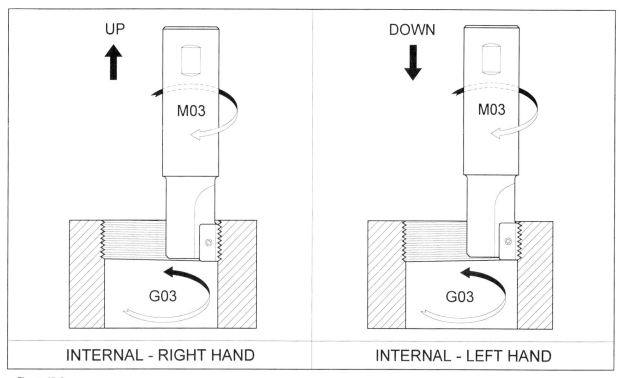

Figure 45-6

INTERNAL thread milling using the climb milling mode - right and left hand threads, spindle rotation and cutter motions shown

Each motion item by itself is important, but it is the *coordination of all motions* that makes the thread to match engineering purposes. These motions together determine the *hand of thread* (left hand vs. right hand), and whether applied externally or internally. *Figures 45-5* and *45-6* show the possibilities for the most common method of threading - in the climb milling mode.

◆ Lead-In Motions

In the example, the thread to be milled is a *right hand* thread *and* an *internal* thread. The spindle rotation is normal, using M03. The last figure indicates that the thread has to be milled from the bottom upwards, using the counterclockwise (G03) tool motion.

There is one last consideration, the thread milling insert - mainly its height. The insert height determines how many revolutions are required to cut the thread at full depth. A single insert cutter will be used, and by consulting the tooling catalogue, determined that two revolutions will be sufficient to mill the required thread.

To start the thread milling example, the cutter has to be positioned at X0Y0 part origin and at a clear Z depth. Since a multi tooth insert cutter is used and there is space available, start will be a little *below* the bottom of the part, say .200, at Z-0.95 (the plate thickness is .750, as per drawing). This extra clearance provides an even entry into the thread. The program start includes all current considerations:

```
O4501 (INTERNAL RIGHT HAND THREAD MILLING)
N1 G20
N2 G17 G40 G80
N3 G90 G54 G00 X0 Y0 S900 M03
N4 G43 Z0.1 H01 M08
N5 G01 Z-0.95 F50.0
...
```

Similar to a program using circular interpolation, the next step to be done is determination of the linear approach to the lead-in arc (in climb milling mode). This is also the motion that applies the cutter radius offset.

> In helical milling, the cutter radius offset applies
> only to the two axes of the selected plane.

In the example, the radius offset is entered in block N6:

```
N6 G01 G41 X0.75 Y-0.75 D01 F10.0
```

The next block is the lead-in arc, with .750 approach radius. Only the motions along X and Y axes will be needed:

```
N7 G03 X1.5 Y0 R0.75            (or I0 J0.75)
```

Since only two axes are used, the motion is planar (on a plane). *Figure 45-7* illustration shows the tool motion. Note that the motion *appears* to be the same as for a circular interpolation (circular pocket application). This can be misleading - there is also the Z *axis* involved.

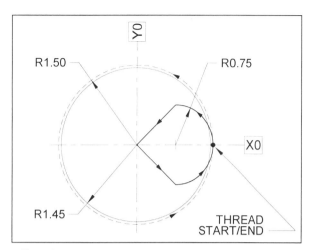

Figure 45-7

Lead-in and lead-out motions for thread milling example O4501 (top view is shown)

However, this programming approach would bring the threading cutter into the material straight! Since the cutter has the threading teeth parallel, it would cut a series of grooves, not threads. This, of course, is unacceptable.

To make a better cut, start *with a helical motion* for the lead-in arc. That means adding the Z axis to the circular motion, in the upwards direction. The amount of the Z target position must be *calculated*, not guessed. Helical approach has to consider the thread *pitch* and the *degrees* of travel on the circumference of the lead-in arc.

The thread pitch in the example is

```
1 / 12 = .0833333
```

and the degrees traveled on the circumference total 90°, from X0.75Y-0.75 to X1.5Y0.

Considering that the thread mill has to advance .0833333 for every 360°, it has to advance one quarter of that distance for each 90°.

The calculation of the linear travel can be made from the following formula:

$$L_t \;=\; \frac{A \times P}{360}$$

☞ where ...

L_t = Linear travel in helical interpolation
A = Amount of degrees interpolated (angle)
P = Thread pitch (1 / TPI)

The advance for 90° in the example will be:

```
Lt = 90 × .0833333 / 360
Lt = .0208333                                    (.0208)
```

The cutting motion takes place along the positive Z direction (up), so the target position absolute value will be *above* the start position, and a corrected block N7 can be written:

```
N7 G03 X1.5 Y0 Z-0.9292 R0.75      (or I0 J0.75)
```

At this point, the tool is in a position when the complete 360° helical motion can begin. Always try to start lead-in and lead-out arcs at quadrant positions (0°, 90°, 180° and 270°). These calculations are much easier to work with.

◆ Thread Rise Calculation

Some technical brochures or product catalogues may base their calculations on the helix angle of the threading insert, but one fact still remains unchanged. The thread milling cutter must advance by the distance that is equivalent to the pitch amount in one revolution (360°). If a lead-in arc is used, only a portion of the pitch is programmed. The amount of linear travel has to be calculated as a ratio per degrees traveled (see previous example). Following formula is another version of the earlier one. It also calculates the amount of linear travel, this time based on the number of threads per inch (TPI):

$$L_t \;=\; \frac{A}{360 \times TPI}$$

☞ where ...

L_t = Linear travel in helical interpolation
A = Amount of degrees interpolated (angle)
TPI = Threads per inch

◆ Milling the Thread

Because of the cutter size and the thread size, *two full revolutions* have been selected to complete the specified thread. For each revolution, that is for each 360°, the linear position of the cutter must be changed by the pitch amount. That is the .0833333 value in the example. Thread motion is a helical milling and either absolute or incremental programming method can be used.

First, the absolute method will be selected, then the incremental method:

```
N8 G90 G03 X1.5 Y0 Z-0.8459 I-1.5      (TURN 1)
N9 G03 X1.5 Y0 Z-0.7626 I-1.5          (TURN 2)
```

The repetitious data will not appear in the final program. For comparison, try to program the two motions in incremental mode:

```
N8 G91 G03 X0 Y0 Z0.0833 I-1.5        (TURN 1)
N9 G03 X0 Y0 Z0.0833 I-1.5            (TURN 2)
```

When the two helical motions are completed, the cutter had traveled .1666 along the positive direction of the Z axis and the total of 720° (two revolutions). The last part of the program will be the ending of the cut.

◆ Lead-Out Motions

For the same reason why the tool approached the thread using helical interpolation over a 90° arc, the exit from the thread will be treated the same way. This departure from the completed thread (lead-out motion) will move the threading cutter *away* from the finished thread, again using a quarter turn motion that is still in the helical mode. The calculation is the same as before and so is the amount:

$$L_t = 90 \times .0833333 / 360$$
$$L_t = .0208333$$

This incremental value will bring the tool up and away from the thread (programmed in absolute mode):

```
N10 G03 X0.75 Y0.75 Z-0.7418 R0.75
```
$$\textit{or} \quad (\text{I-0.75 J0})$$

At this point, the cutter is in a position that is clear of the thread, so the linear motion can resume and cancel the cutter radius offset, then move back to the center of the bore, retract the tool above part, move to machine zero and terminate the program:

```
N11 G40 G01 X0 Y0
N12 G00 Z1.0 M09
N13 G28 X0 Y0 Z1.0 M05
N14 M30
%
```

The thread cutting job is done and the complete program can be written.

◆ Complete Program

The complete program that follows, combines the individual calculations and includes all motions for the threading cutter:

```
O4501 (INTERNAL RIGHT HAND THREAD MILLING)
N1 G20
N2 G17 G40 G80
N3 G90 G54 G00 X0 Y0 S900 M03
N4 G43 Z0.1 H01 M08
N5 G01 Z-0.95 F50.0
N6 G41 X0.75 Y-0.75 D01 F10.0
N7 G03 X1.5 Y0 Z-0.9292 R0.75
N8 Z-0.8459 I-1.5                     (TURN 1)
N9 Z-0.7626 I-1.5                     (TURN 2)
N10 X0.75 Y0.75 Z-0.7418 R0.75
N11 G40 G01 X0 Y0
N12 G00 Z1.0 M09
```

```
N13 G28 X0 Y0 Z1.0 M05
N14 M30
%
```

This program is only a small sample of one thread milling method. The calculations are logical and program code is clear. Reading various technical specifications for a thread milling cutter presents a wealth of information (including programming tips), suggested by the tool manufacturer. These recommendations always take on a more important role than any other method.

Figure 45-8 illustrates isometric view of the sample thread milling program O4501.

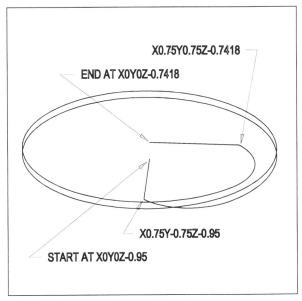

Figure 45-8
Isometric view of tool motions for the thread milling example

◆ External Thread Milling

The external thread milling is often used for large threads with a carbide indexable threading insert. The lead-in and lead-out motions are very important in this situation as well. Their calculations and those for the thread follow the same rules as for an internal thread. A straight linear lead-in and lead-out may be used, similar to the ones described in *Chapter 29 (Circular Interpolation)*. Otherwise, follow the motions shown in *Figure 45-9*.

◆ Tapered Thread Milling

It is possible, but much more difficult, to manually program a tapered thread (such as NPT or NPTF) using a thread milling cutter. For threads with a small pitch, soft material and very narrow taper angle, a tapered cutter may be used and programmed as if it were a straight cutter, in a single revolution. For larger threads, the only method is a *simulation* of the helical milling (software is required in

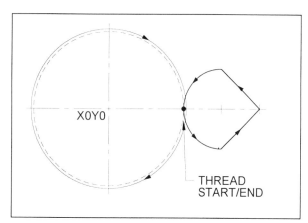

Figure 45-9

Lead-in and lead-out motions for an external thread milling

this case), using very small increments in *linear interpola-tion* mode only. Holders and inserts should be selected by the nominal size of the thread.

Tapered threads are sometimes called *conical* threads and will require different tool holders for right-hand threads and left-hand threads. This is a special application of heli-cal interpolation that does not really belong in the manual programming area.

◆ Further Considerations

Two additional considerations are necessary to cover the subject of general thread milling in a reasonable depth. One is the application of the *cutter radius offset* and the other one is the selection of the *cutting feedrate*.

Cutter radius offset will only be active for the two axes selected by the active plane (for example, in G17, it will be the X and Y axes). Always select the climb milling method, it is the preferred method for the majority of thread milling applications.

Feedrate selection is similar to the feedrate for outside and inside arcs described in the *Circular Interpolation* chapter (*Chapter 29*). Since a precision thread is the goal, the cutting feedrate will be 10 to 30 percent slower. A good start is at about .001 per tooth and up by experimenting.

THREAD MILLING SIMULATION METHOD

There is an interesting way to mill a thread *without* the benefit of helical interpolation option available on the con-trol. This may be a case for many CNC machines, or in such cases where the machine shop needs to mill a thread only once in a while and the helical interpolation is not worth the cost of a control update.

To mill a thread (external or internal) under these condi-tions, a helical milling *simulation* will be used. Simulation

of the helical motions requires a simultaneous three-axis *linear* cutting motion, within the acceptable tolerance of the thread. That means each motion will be a very small three-axis *linear* motion (using the X, Y and Z axes). The more accurate thread needed, the longer program will be generated. This method is practically impossible to do manually, as the development time could hardly be justified in any case. What is needed is a program software that will do the calculations in a matter of seconds. Many manufac-turers of thread milling cutters provide such a software free or for only a small cost.

To illustrate this topic, the same thread will be used as in program O4701. Needless to say, a simulated program may be extremely long - at least a few hundreds blocks. Here is an example of such a program - it shows only a few blocks of the beginning and a few blocks where the tool completes the lead-in arc. It only relates to the straight line and the part of the lead-in arc. Practically, the program is incorrect, be-cause the tool radius is not compensated. The radius com-pensation would be done in the software, not with G41 or G42 in the program - this is a linear interpolation in three axes and cutter radius offset may not be used. The complete program had been done by using a CAD/CAM software, and was 463 blocks long, comparing to just 14 blocks for the complete program using helical interpolation.

```
G20
G17 G40 G80
G90 G54 G00 X0 Y0 S900 M03
G43 Z0.1 H01 M08
G01 Z-0.95 F10.0
X0.75 Y-0.75
X0.7846 Y-0.7492 Z-0.9494
X0.8191 Y-0.7468 Z-0.9488
X0.8536 Y-0.7428 Z-0.9482
X0.8878 Y-0.7373 Z-0.9476
X0.9216 Y-0.7301 Z-0.9470
X0.9552 Y-0.7214 Z-0.9464
X0.9883 Y-0.7112 Z-0.9457
...
...
...
X1.4967 Y-0.0697 Z-0.9304
X1.4992 Y-0.0350 Z-0.9298
X1.5000 Y0.0000 Z-0.9292
...
```

What the program output shows is a series of very small line segments, in a very precise order and increment. Fol-low at least a few blocks and visualize the actual motion. By the way, it took about three seconds to generate the 463 blocks of code in CAD/CAM. Knowing a high level lan-guage (such as Visual Basic®, Visual C++® and similar lan-guages), writing similar utility software can be done very efficiently. Typically, when the utility is executed, the user inputs the number of revolutions, the radius, thread lead and resolution. The length of the program can be shortened but the threading quality may not be acceptable.

Regardless of the method used to generate the tool path for thread milling, this is a machining and programming area that deserves a lot more attention than it normally gets in many machine shops.

HELICAL RAMPING

Although the thread milling is probably the most common application of helical interpolation, it is not the only one. One very useful application of this control feature is called *helical ramping*.

Helical ramping is used primarily as a replacement for a plunge cut into solid materials. Recall that a roughing operation in an enclosed area (for example a pocket), requires the cutting tool to reach a certain Z depth, *before* the actual material removal. This Z axis motion can be in an open space, if the material had been *predrilled*, for instance. The Z axis motion can also be cutting *into a solid* material, if the cutting tool is of the *center cutting* type (using the so called slot drill). Well, there is another possibility - *helical ramping* - that allows using any flat cutter and reach the required Z depth as a series of relatively small helical cutting motions. The cutter can be *flat* and *non-center cutting*, because all the cutting action is done by the cutter sides, not its bottom. Once the required Z depth has been reached, a full circular interpolation is often used to clean up after the last helical cut. A high level CAD/CAM software can do this very efficiently.

● Example :

To illustrate the programming technique for this type of milling application, a standard, flat bottom, ∅.500 inch end mill will be used (there is no need for a center cutting type) and open the start hole to the ∅.750. The pocket depth is .250 and in each helical motion the tool will be moved by .050. The pocket center is X0Y0 and the start Z position (clearance) is .050 above the top of part (which is the Z axis program zero). The total number of helical motions (revolutions) is six (one above the top of work, plus another five below the top of work).

Any increment value can be chosen for the depth, depending on cutting conditions. The smaller the increment, the more helical passes will be necessary and the longer cutting time will be required.

The program can be in either absolute or incremental mode and, in this case, the incremental mode is a little easier to program. The cutting will be done in the climb milling mode - program O4502.

```
O4502 (HELICAL RAMPING)
N1 G20
N2 G17 G40 G80
N3 G90 G54 G00 X0 Y0 S700 M03
N4 G43 Z1.0 H01 M08
N5 G01 Z0.05 F50.0          (APPROACH TO Z-START)
N6 G41 X0.375 D01 F15.0    (START COMPENSATION)
N7 G91 G03 I-0.375 Z-0.05      (CUT ABOVE WORK)
N8 I-0.375 Z-0.05         (CUT 1 BELOW TOP FACE)
N9 I-0.375 Z-0.05         (CUT 2 BELOW TOP FACE)
```

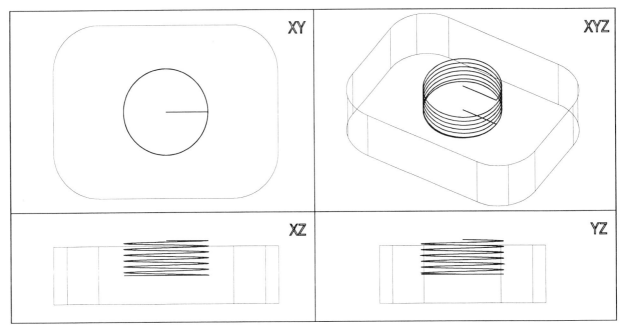

Figure 45-10

Schematic illustration of a helical motion used for ramping - program O4502

```
N10 I-0.375 Z-0.05          (CUT 3 BELOW TOP FACE)
N11 I-0.375 Z-0.05          (CUT 4 BELOW TOP FACE)
N12 I-0.375 Z-0.05          (CUT 5 BELOW TOP FACE)
N13 I-0.375               (CIRCULAR BOTTOM CLEANUP)
N14 G90 G01 G40 X0            (RETURN TO XY START)
N15 G00 Z1.0 M09
N16 G28 Z1.0 M05
N17 M30
%
```

Two items are worth a note here. One, because the incremental mode is used, the Z axis start is *extremely* important (block N4). The cutter radius offset is applied during a simple straight motion from the center to the start of the first

helical motion. *Figure 45-10* shows the schematics of the program in four different views.

Helical interpolation can be a very powerful programming tool, often irreplaceable by any other methods. Although it is a control option, its main benefit is the short program output and the possibility of quick changes may justify its extra cost.

Throughout the handbook, there have been dozens of programming examples. They all shared one common feature - they were aimed at the vertical machining centers. There was a reason for this approach. First, there are more vertical machining centers in machines shops overall, and mixing two different types of machines would make all reference material more complex. Second, almost every subject covered so far for the vertical models is equally applicable to the horizontal models. So what are the differences?

The horizontal machining center mainly differs from a vertical machining center in its general functionality. While a vertical machine is mostly used for only one face type of work, a horizontal machine is used for work on many faces of the part during a single setup. This feature alone makes a horizontal machining center a much more versatile machine - and also more expensive. *Figure 46-1* shows the comparison of the axis orientation.

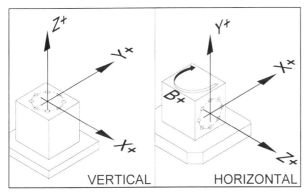

Figure 46-1

Axis orientation differences between vertical and horizontal machines

From the illustration is clear that all the XY plane is used for the primary plane of work and the Z axis is used to control cutting depth. There is no difference whatsoever between the two machine types in this respect.

Between programming and setup, there are three major differences on a horizontal machining center:

❏ Presence of a fourth axis, typically an indexing B axis

❏ Presence of a pallet changer

❏ Richer variety of setup and offset settings

First, a brief look at the fourth axis of a typical CNC horizontal machining center.

INDEXING AND ROTARY AXES

All programming concepts that ve been discussed so far, apply equally to CNC horizontal machines. The XY axes are used mostly for drilling and contouring operations, the Z axis controls the cutting depth.

Horizontal machining centers differ from the vertical machining centers not only in the axes orientation and the type of work that can be machined. One of the major differences is an additional axis.

This is an indexing or a rotary axis, usually designated as the *B axis*. Although the two terms are often used interchangeably, there is a difference between them.

❏ An *indexing* table will rotate the part that is mounted on it, but it cannot be used simultaneously with any kind of cutting motion. This type supports a positioning motion.

❏ A *rotary* table will also rotate the part that is mounted on it, but a simultaneous cutting action is possible. This type supports a contouring motion.

The most common fourth axis on a horizontal machining center is the indexing type, called the B axis.

INDEXING TABLE (B AXIS)

Indexing axis, as the name suggests, is used to index a table, if the machine is equipped with this feature. The horizontal machining centers and boring mills have an indexing table as a standard feature. A full rotary table is an option on a both types of machining centers.

◆ Units of Increment

The indexing axis is programmed in the number of degrees that is required by the job. For example, to index a table to a 45° position, program:

```
G90 G00 B45.0
```

The minimum increment depends on the machine design. For indexing, a typical minimum unit of increment could be 1 degree or even 5 degrees. However, for more flexibility - and for rotary machining - much finer increment is required. Most machine manufacturers offer 0.1, 0.01 and 0.001 of a degree as the minimum indexing increment. In all cases, the programming of the indexing motions can be done in two directions.

◆ Direction of Indexing

The B axis can be programmed to index either clockwise or counterclockwise, looking from top down at the table, which is the XZ plane - *Figure 46-2*.

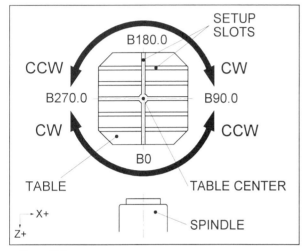

Figure 46-2
B axis direction and general descriptions

The table size including the size of corners is important to determine the clearances before indexing.

◆ Table Clamp and Unclamp Functions

In order to maintain a rigid setup, the indexing table must be clamped to the main body of the machine during a cut. For indexing motions, the table must be unclamped. This is true of most machining centers. For this purpose, manufacturers offer special miscellaneous functions - two functions will be used in the examples:

❏ Table Clamp   … for example M78
❏ Table Unclamp   … for example M79

The function numbers may greatly with different machine designs, so check the manual for proper coding.

Normally, the unclamp function is programmed before the indexing, followed by the B axis motion and another block containing the clamp function:

M79   *Unclamp table*
G00 B90.0   *Index table*
M78   *Clamp table*

Some designs require other M codes, for example to control the clamping pin or a table ready confirmation.

The B axis is programmed logically the same way as the linear axes, including the mode of dimensioning. Either the absolute or the incremental mode can be used for indexing, using standard G90 and G91 commands respectively.

◆ Indexing in Absolute and Incremental Mode

Just like any other axis, the B axis can be programmed in the absolute mode or incremental mode, with the same behavior as the linear axes.

The following example is in the absolute mode, showing two table columns. The first column is the programmed indexing motion in G90 mode, the second column shows the actual resulting indexing motion (Distance-To-Go) and its direction. All rotational directions are based on the perpendicular view to the XZ plane.

➲ Absolute Mode - consecutive indexes :

Programmed motion in G90	Actual indexing motion
G90 G28 B0	Machine B zero position
G00 B90.0	CW 90 degrees
B180.0	CW 90 degrees
B90.0	CCW -90 degrees
B270.0	CW 180 degrees
B247.356	CCW -22.644 degrees
B0	CCW -247.356 degrees
B-37.0	CCW -37 degrees
B42.0	CW 79 degrees
B42.0	No motion (0 degrees)
B-63.871	CCW -105.871 degrees

The next table is similar. The first column is the programmed indexing motion in G91 mode, the second column shows the motion directions and the actual resulting absolute position. All rotational directions are based on the perpendicular view to the XZ plane.

➲ Incremental Mode - consecutive indexes :

Programmed motion in G91	Actual absolute position
G90 G28 B0	Machine B zero position
G91 G28 B0	Machine zero - no motion
G00 B90.0	CW 90.000
B180.0	CW 270.000
B90.0	CW 360.000
B270.0	CW 630.000
B0	No motion
B125.31	CW 755.310
B-180.0	CCW 575.310
B-75.31	CCW 500.000
B-75.31	CCW 424.690
B-424.69	CCW 0.000

Study both tables block by block, in the listing order. The results are always important for understanding. Note the B-37.0 in the first table - exactly the same result could be achieved if the block read B323.0 as a positive value.

In the second table, the first block is in the absolute mode to guarantee a start at B0. One occurrence that is interesting - when the rotation in the same direction reaches 360° (a full circle), it continues to increase. It does not become a zero degrees again. That is something to watch. If indexing (in the incremental mode) takes place twice around, the absolute table position will be 720.000°. Indexing twice will also be necessary in the opposite way in order to reach absolute zero. A small example is illustrated in *Figure 46-3*.

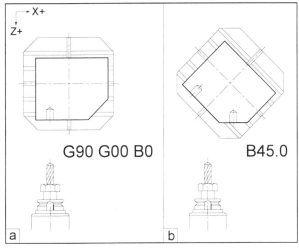

Figure 46-3

B axis direction from B0 to B45.0 in the absolute mode - O4601

To program the two positions shown, the typical block sequence would be, for example O4601:

```
O4601
G90 G54 G00 X.. Y.. Z..
M79
B0
M78
...
< DRILL HOLE AT B0 >
...
G90 G55 G00 X.. Y.. Z..
M79
B45.0
M78
...
< DRILL HOLE AT B45.0 >
...
```

The dimensions relating to the drilling are not important for the example.

> Always observe safe clearances when indexing the B axis

B AXIS AND OFFSETS

One of the most important differences between vertical and horizontal machining centers is the way we program and particularly set the two major offsets:

❑ Work offset

❑ Tool length offsets

Cutter radius offset is *not* affected by the B axis and is programmed the same way as in vertical machining.

The relationship of offsets to the machined face of a part is very important and is also more complex than for the vertical approach.

◆ Work Offset and B Axis

The work offset is measured the same as before - from the machine zero to the program zero. What is different now is the reality of several faces used for machining rather than just one. That means the tool path for each face has to have its own program zero, therefore its own work offset. *Figure 46-4* shows a typical setting, looking at the part from the direction of the spindle.

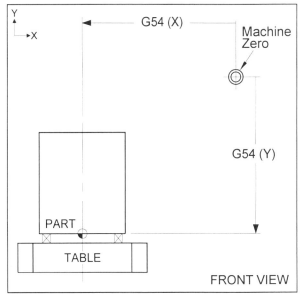

Figure 46-4

Work offset for a horizontal application - front view shown

Although the illustration shows the part zero at the center of the indexing table, the part zero may as well be at the top of each part face or even elsewhere. There are benefits in either approach and there is no 'best' method. Often, it is the specified requirement of the job, the fixture design, nature of the work and - of course - the programmer's personal preferences.

When changing from one face to another, remember to change the work offset. For example, if there are four faces to machine, each face will have its own work offset, such as G54, G55, G56 and G57. The B axis is usually not dependent on work offset, so the best block to program a new offset is during the first rapid motion. The previous short example illustrates the method. The next section describes the work offset setting for the Z axis and tool length offset.

◆ Tool Length Offset and B Axis

It should easy to understand the concept of multiple work offsets for multiple faces. Setting the tool length can be quite complicated, depending on many factors that influence the decision. The first factor is the method of setting tool length. There are at least two methods to set the tool length offset. Both have already been covered in *Chapter 19*, but now they take on a new significance.

Touch-Off Method

One method is to *touch-off* the Z0 of the machined face and register the distance from the tool tip as a negative length offset. This was the preferred method for vertical machines. The touch-off method may be acceptable for a small number of tools and indexes. Although it is possible to select the center of indexing table as Z0, it is not a practical solution. *Figure 46-5* shows the principle of touch-off setup in general terms, and *Figure 46-6* shows a practical example. Note that the setup is exactly the same as for vertical machining. A program block

`G43 Z2.0 H01`

will move the tool Z-298.0, if H01 is set to -300.0.

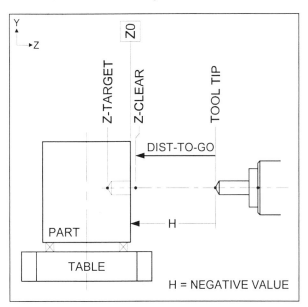

Figure 46-5
Touch-off tool length offset method - layout with H as negative

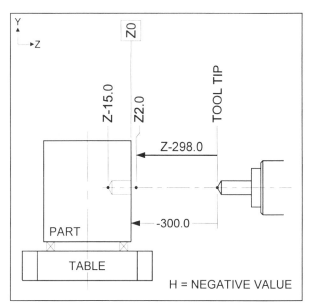

Figure 46-6
Touch-off tool length offset method - example with H as negative

Preset Method

Tool length set on vertical machining centers is often a touch-off method but it could also be the preset method. The *preset* method uses a special tool length presetter device and is done off machine. There is a good reason why the preset method is much more practical for horizontal machining than for vertical machining.

Recall that one tool normally requires one tool length offset. Now, consider a very typical situation for a horizontal machining - a single tool has to machine six faces, followed by other four tools that also do machining on the same six faces. Each of the five tools requires a unique tool length for each face - for the total of 30 different length offsets! This is not an isolated example, but there are several solutions to such a situation.

All solutions use the *preset tool* length measurement and *one additional setting*. The tool mounted into the holder is placed in the presetting device. Through a computerized optical reader, the presetter is calibrated to match the machine gauge line. Then, the tool length is accurately measured. It is a positive value representing the actual tool length from its tool tip to the machine gauge line. This is the amount that will be input into the corresponding tool length offset register. There is only one problem - *where is the relationship of this measured amount to the part position?* In the touch-off method, the tool touches the part and the relationship is direct. The preset method has no contact - *one additional setting* mentioned earlier has to be made.

This setting is an entry of the distance between machine gauge line and the Z0 of the current work offset Z address - *Figures 46-7* and *46-8*.

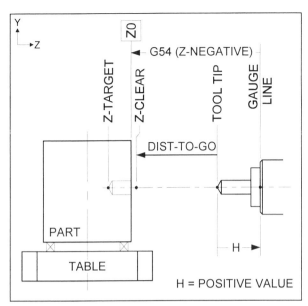

Figure 46-7

Preset tool length offset to Z0=face - layout with H as positive

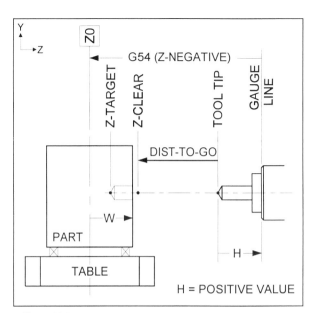

Figure 46-9

Preset tool length offset to Z0=center - layout with H as positive

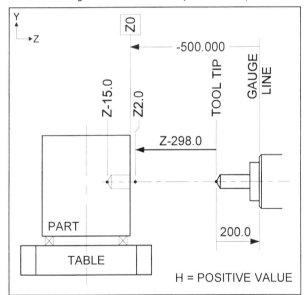

Figure 46-8

Preset tool length offset to Z0=face - example with H as positive

The illustration shows the offset amount entered into the Z register of G54 work offset as -500.0. This is the distance from gauge line to part zero. To prove the method works, use H01=200.0 and the Z clear position as G43Z2.0H01. The *distance-to-go* is calculated the same way as always:

```
G54(Z) + Z clear + H01 =
= -500.0 + 2.0 + 200.0
= -298.0
```

The tool then continues normally to the Z-15.0 depth.

The last example was measured to the Z0 position at the part front face. Another option exists if Z0 is set as the center of indexing table. In fact, it is only the *perception* of a change, the example is the same in reality. *Figures 46-9* and *46-10* show the apparent change from the last two figures.

The setup has changed only because of the additional dimension W, specifying the distance from the program zero to the part face. The Z axis values in the program will also change, as all dimensions are taken from Z0 at the table center, not the face of part.

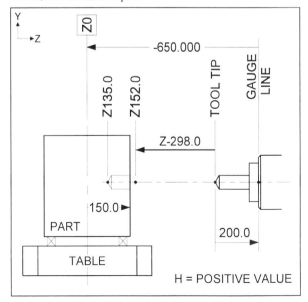

Figure 46-10

Preset tool length offset to Z0=center - example with H as positive

In the program is block that moves the tool to the Z clear position - G43Z152.0H01. To calculate the *distance-to-go* in this case, include the W distance that must always be known during setup (fixture drawing or actual measurement). That makes the W=150.0, no change for the length offset H01=200.0, but an important change to the G54 - now it is measured from the table center (Z0). The Z clear position includes the W length and the physical clearance of 2 mm, same as in the previous case. In this example, the amount of Z-650.0 for G54 is used:

```
G54(Z) + Z clear + H01=
= -650.0 + 152.0 + 200.0
= -298.0
```

The tool then continues normally to the Z-135.0 depth. Overall, this setup application is exactly the same as the previous one. The operator must know where is Z0 located for every job. This information always originates from the CNC programmer in the form of a program comments or - even better - through a setup sheet.

RETURN TO MACHINE ZERO

In vertical machining, the return to machine zero has been programmed after every tool in majority of cases. The return was along the Z axis only. The reason was simple - on a vertical machining center, Z axis machine zero is synchronized with the automatic tool changer. This is *not* the case on a horizontal machining center.

Due to its design, the required machine zero return motion before each tool change is along the *Y axis*. In all other respects, programming of the machine zero commands is exactly the same.

Here is a comparison of a typical ending before a tool change for the two machine types:

Vertical: `G91 G28 Z0`

Horizontal: `G91 G28 Y0 Z0`

The question is what is the Z axis return doing in the block when only the Y return is required. The answer is a one word - *safety*. Although only the Y axis is required to make a successful automatic tool change, the tool has to be *away* from the part at the same time. The return along the Z axis makes it easier. Of course, programming only a sufficient clearance in the Z axis would also achieve the same goal. That may prove more difficult than it appears. With the table in an index position other than zero, long and short tools, different part faces, fixture in the way, etc., it may be difficult to always know exactly how far to retract the Z axis. That is why a simple rule is worth remembering:

> Return in Y axis because it is necessary and in Z axis for safety.

INDEXING AND A SUBPROGRAM

To describe all combinations of various setup methods and their influence on the program format is virtually impossible. The subject of horizontal machining, particularly the setup portion, can be quite complex and requires some experience. The layout presented in this chapter should offer at least the general understanding of the subject. A suitable programming example may help.

To illustrate how indexing can be used in an efficient way, the program example in this section will spot drill and drill 612 holes on a cylinder - *Figure 46-11*. The spot drill will also break a chamfer of .400 × 45°, measured from the high spot of the cylinder. All the depth calculations are real.

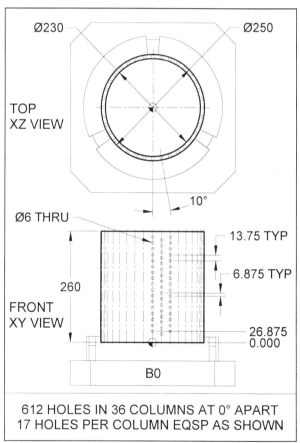

612 HOLES IN 36 COLUMNS AT 0° APART
17 HOLES PER COLUMN EQSP AS SHOWN

Figure 46-11

Practical example for indexing using subprograms - example O4602

Don't get discouraged by the large number of holes required. Using a subprogramming approach will minimize the program length. The program does not use any clamp and unclamp sequences, which is typical to the rotary type B axis. If the machine requires unclamping the table before indexing and clamping it after indexing, use suitable M functions for clamp and unclamp the table.

Before getting into the program itself, the tools and their use need to be selected. Only two tools will be required, a 10 mm spot drill and a 6 mm drill. *Figure 46-12* shows the critical positions of the two tool tips.

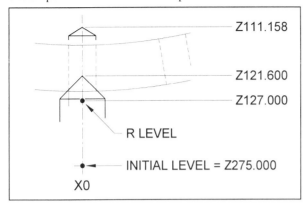

Z111.158

Z121.600

Z127.000

R LEVEL

INITIAL LEVEL = Z275.000

X0

Figure 46-12
Detail of tool data used in program O4602

The R level is the same for both tools and the depth for the spot drill also includes a small chamfer to deburr the holes. Drilling depth guarantees a full drill penetration. Actual calculations are not important here, but they do follow the same rules established in the earlier chapters.

Development of the subprogram needs some work. Two subprograms will be used. They are virtually the same, except for the fixed cycle selection. Several other methods could have been also used, but this chapter concentrates on the indexing table only. The two subprograms will start at the bottom of the pattern, at the B0 location (0°). This hole will be used as the start position only but will not be drilled until all other holes have been done. The hole is not drilled yet, but the 10° indexing has to be included in the subprogram. That is the reason for starting one column away. Two columns are part of each subprogram with a 10° index between them. Comments in the subprograms explain the process. Note the area marked in *Figure 46-13*, indicating the subprogram contents.

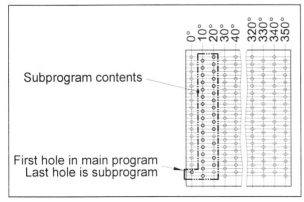

Subprogram contents

First hole in main program
Last hole is subprogram

Figure 46-13
Flat cylinder layout - both ends shown for subprogram development

```
O4602 (MAIN PROGRAM)
(START FROM MACHINE ZERO - T01 IN THE SPINDLE)
(X0Y0 = FIXTURE CENTER / Z0 = BOTTOM OF PART)
(T01 - 10 MM DIA SPOT DRILL)
(T02 - 6 MM DIA DRILL THRU)

N1 G21
N2 G17 G40 G80
/N3 G91 G28 Z0
/N4 G28 X0 Y0
/N5 G28 B0
N6 G90 G54 G00 X0 Y26.875 S1000 M03 T02
N7 G43 Z275.0 H01 M08
N8 M98 P4651 L18
N9 G28 Y0 Z0
N10 G28 B0
N11 M01

N12 T02
N13 M06
N14 G90 G54 G00 X0 Y26.875 S1250 M03 T01
N15 G43 Z275.0 H02 M08
N16 M98 P4652 L18
N17 G28 X0 Y0 Z0
N18 G28 B0
N19 M06
N20 M30
%

O4651 (SUBPROGRAM FOR SPOT DRILL)
N101 G91 G80 Y-6.875        (MOVE DOWN BY PITCH)
N102 G90 Z275.0                    (CLEAR Z)
N103 G91 B10.0             (ROTATE BY 10 DEGREES)
N104 G99 G82 R-148.0 Z-5.4 P200 F120.0    (DRL)
N105 Y13.75 L16     (16 MORE HOLES IN Y PLUS)
N106 G80 G00 Y6.875          (MOVE UP BY PITCH)
N107 G90 Z275.0                    (CLEAR Z)
N108 G91 B10.0             (ROTATE BY 10 DEGREES)
N109 G99 G82 R-148.0 Z-5.4 P200         (1 HOLE)
N110 Y-13.75 L16    (16 MORE HOLES IN Y MINUS)
N111 M99            (END OF SUBPROGRAM O4651)
%

O4652 (SUBPROGRAM FOR 6MM DRILL)
N201 G91 G80 Y-6.875        (MOVE DOWN BY PITCH)
N202 G90 Z275.0                    (CLEAR Z)
N203 G91 B10.0             (ROTATE BY 10 DEGREES)
N204 G99 G83 R-148.0 Z-15.84 Q7.0 F200.0 (DRL)
N205 Y13.75 L16     (16 MORE HOLES IN Y PLUS)
N206 G80 G00 Y6.875          (MOVE UP BY PITCH)
N207 G90 Z275.0                    (CLEAR Z)
N208 G91 B10.0             (ROTATE BY 10 DEGREES)
N209 G99 G83 R-148.0 Z-15.84 Q7.0      (1 HOLE)
N210 Y-13.75 L16    (16 MORE HOLES IN Y MINUS)
N211 M99            (END OF SUBPROGRAM O4652)
%
```

The initial level of Z275.0, used in all three programs, is reasonable for safe indexing. To select a suitable Z axis clearance is very important and knowing the indexing table size and the size of its corners is imperative. For the record, the table for this job will be 400 × 400 mm square with 50 × 50 mm corners. The part setup is concentric with the indexing rotation and there are no interfering elements.

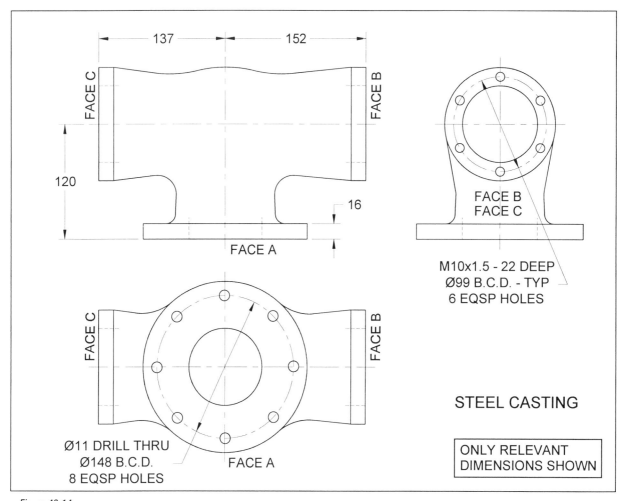

Figure 46-14

A typical multi sided part suitable for a horizontal machining operation - program O4603 (subprograms O4653 and O4654)

COMPLETE PROGRAM EXAMPLE

A typical part for a horizontal machining center requires material removal from several sides in the same setup. Such a part, a type of a housing, is shown in *Figure 46-14*.

For the example, only the holes will be machined at three different faces. For the tools, a spot drill, two drills and a tap will be used. The first step is to decide where to locate program zero. For the ease of programming and setup, the center of each bolt circle and the front of each face (Z) is a good selection. Each face will have its own work offset - G54 for face A, G55 for face B and G56 for face C. The second step is to develop two subprograms for the for the hole locations. All dimensions have been calculated accurately but no details are necessary. First tool is in the spindle at startup. The part is located in a fixture mounted on the indexing table. Pallet changing has been omitted from the example, but is explained in the section that follows. The subprograms contain bolt pattern coordinates.

```
O4653 (SUBPROGRAM FOR 8 HOLES AT 148 MM BCD)
N101 X74.0 Y0
N102 X52.326 Y52.326
N103 X0 Y74.0
N104 X-52.326 Y52.326
N105 X-74.0 Y0
N106 X-52.326 Y-52.326
N107 X-0 Y-74.0
N108 X52.326 Y-52.326
N109 M99
%

O4654 (SUBPROGRAM FOR 6 HOLES AT 99 MM BCD)
N201 X49.5 Y0
N202 X24.75 Y42.868
N203 X-24.75 Y42.868
N204 X-49.5 Y0
N205 X-24.75 Y-42.868
N206 X24.75 Y-42.868
N207 M99
%
```

```
O4603 (MAIN PROGRAM)
(FACE A = G54 = B0 = 8 HOLES)
(FACE B = G55 = B90.0 = 6 HOLES)
(FACE C = G56 = B270.0 = 6 HOLES)

(T01 - 15 MM DIA SPOT DRILL)
(T02 - 8.4 MM TAP DRILL)
(T03 - M10 X 1.5 TAP)
(T04 - 11 MM DIA DRILL)

(T01 - 15 MM DIA SPOT DRILL - ALL HOLES)
N1 G21
N2 G17 G40 G80
/N3 G91 G28 Z0
/N4 G28 X0 Y0
/N5 M79
/N6 G28 B0
/N7 M78
N8 G90 G54 G00 X74.0 Y0 S868 M03 T02
N9 G43 Z10.0 H01 M08
N0 G99 G82 R2.0 Z-5.8 P200 F150.0 L0
N11 M98 P4653           (SPOT DRILL FACE A)
N12 G80 Z300.0
N13 M79
N14 B90.0
N15 M78
N16 G55 X49.5 Y0 Z10.0
N17 G99 G82 R2.0 Z-5.3 P200 L0
N18 M98 P4654           (SPOT DRILL FACE B)
N19 G80 Z300.0
N20 M79
N21 B270.0
N22 M78
N23 G56 X49.5 Y0 Z10.0
N24 G99 G82 R2.0 Z-5.3 P200 L0
N25 M98 P4654           (SPOT DRILL FACE C)
N26 G80 Z300.0 M09
N27 G91 G28 Y0 Z0 M05
N28 M01

(T02 - 8.4 MM TAP DRILL)
N29 T02
N30 M06
N31 G90 G56 G00 X49.5 Y0 S1137 M03 T03
N32 G43 Z10.0 H02 M08
N33 G99 G83 R2.0 Z-24.8 Q6.0 F200.0 L0
N34 M98 P4654           (TAP DRILL FACE C)
N35 G80 Z300.0
N36 M79
N37 B90.0
N38 M78
N39 G55 X49.5 Y0 Z10.0
N40 G99 G83 R2.0 Z-24.8 Q6.0 L0
N41 M98 P4654           (TAP DRILL FACE B)
N42 G80 Z300.0 M09
N43 G91 G28 Y0 Z0 M05
N44 M01

(T03 - M10 X 1.5 TAP)
N45 T03
N46 M06
N47 G90 G55 G00 X49.5 Y0 S550 M03 T04
N48 G43 Z10.0 H03 M08
N49 G99 G84 R5.0 Z-23.0 F825.0 L0
N50 M98 P4654           (TAP FACE B)
```

```
N51 G80 Z300.0
N52 M79
N53 B270.0
N54 M78
N55 G56 X49.5 Y0 Z10.0
N56 G99 G84 R5.0 Z-23.0 L0
N57 M98 P4654           (TAP FACE C)
N58 G80 Z300.0 M09
N59 G91 G28 Y0 Z0 M05
N60 M01

(T04 - 11 MM DIA DRILL)
N61 T04
N62 M06
N63 M79
N64 B0
N65 M78
N66 G90 G54 G00 X74.0 Y0 S800 M03 T01
N67 G43 Z10.0 H04 M08
N68 G99 G81 R2.0 Z-20.3 P200 F225.0 L0
N69 M98 P4653           (DRILL FACE A)
N70 G80 Z300.0 M09
N71 G91 G28 X0 Y0 Z0 M05
N72 M30
%
```

Only a few comments to the example. Both the main program and the two subprograms are quite plain. Compared to vertical machining applications, the Z axis safety clearance may seem a little too high with Z300.0 programmed before each indexing. Large clearances are for safety - they allow the part and the indexing table to index within a safe area, without any obstacles in the way. It is not convenient to actually calculate the minimum Z clearance, but it is important that it is far enough for *all* faces. A CAD software can help here quite a bit. Other features and programming techniques are the same as used elsewhere in the handbook.

AUTOMATIC PALLET CHANGER - APC

One of the greatest concerns in CNC machining is the unproductive time required for the initial part setup and remounting the part when running a batch job. Many features incorporated in the control system or the machine design itself can shorten the unproductive time to a great degree. They include tool length offset, work offsets, cutter radius offset, etc. However, none of them solves the problem of the time used up when mounting individual parts on the table. Probably the major breakthrough was an introduction of a pallet table to the CNC machine. Pallets are not a new idea in machining. For horizontal machining centers, interfaced pallets have become very practical feature to minimize the setup time.

Traditionally, one machine has one work table. Such a design of a machine tool has one major flaw - while the machine is working (and the CNC operator is virtually idle), no other work can be performed. That means a setup for the next part is done at the expense of the machine being idle, resulting in an unproductive time.

By definition, an automatic pallet is a work table that can be moved into and out of the machining position by a program command. If a purpose of such a design is to improve a nonproductive setup time, it is necessary to have at least two independent pallets available - while the part on one pallet is being machined, the other pallet is available for changing the setup for the next job or for unloading and loading individual parts. In this way, the machining and the setup can be done simultaneously, shortening or even totally eliminating the unproductive time.

Although a two pallet system is the most customary for horizontal machining centers, designs with up to twelve pallets are not uncommon.

◆ Working Environment

For a typical dual pallet changer, two major areas should be distinguished:

❏ Machining area   ... within the machine

❏ Setup area   ... outside of the machine

One pallet is normally located in the machining area, the other in the setup area. When a program starts, it normally starts with Pallet #1 (with the part) located in the machining area and Pallet #2 (with no part) in the setup area. There are many designs of pallets, but they all share three major parts:

❏ Pallet

❏ Machine locator

❏ Transfer System

Pallet is the portable work table with a ground surface to which we mount the fixtures and parts. The table can have T slots, tapped holes or both.

Machine locator (also known as a receiver) is a special device located inside of the machine. Its purpose is to accept and firmly hold the pallet loaded with a part ready for machining. Its design must be very robust and accurate at the same time.

Transfer system (also known as a pallet loader) is the system that transfers pallets between the load area and the machine work area.

Often the terms *load* and *unload* are used. *Load* means to move the pallet into the machining area, *unload* means to move the pallet into the setup area. The transfer system determines the type of the pallet.

◆ Types of Pallets

There are two general types of pallets, based on their transfer system:

❏ Rotary type

❏ Shuttle type

The popular *rotary* type works on the principle of a turntable, where one pallet is outside of the machine, the other pallet is in inside of the machine. The pallet change command rotates the pallets 180° and its programming is very simple. *Figure 46-15* illustrates the rotary type..

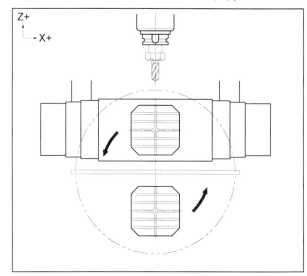

Figure 46-15
Typical rotary type of a pallet changer

Also popular is the *shuttle* type. This design incorporates double rails between the load area and the receiver inside the machine - *Figure 46-16*. Its programming is still simple but more involved than for the rotary type.

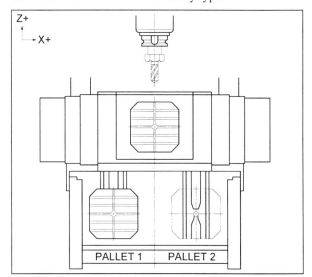

Figure 46-16
Typical shuttle type of a pallet changer

Both pallet types are loaded from the machine front area. Other pallet types are also available for some special machining applications.

◆ Programming Commands

The fairly standard miscellaneous function for automatic pallet changing is M60.

M60	Automatic Pallet Changer (APC)

This command works properly only when the pallet position is at one of two machine reference points:

G28	Machine return to the primary reference point
G30	Machine return to the secondary reference point

G28 command is a familiar command. G30 command is used exactly the same, except it moves the selected axes to the secondary machine reference position.

◆ Pallet Changing Program Structure

The following program emphasizes the pallet changing on a typical shuttle pallet system. It can be easily adapted to a rotary pallet system. In both cases, one pallet is in the machining area, the other is in the setup area.

```
O4604
G91 G28 X0 Y0 Z0
G28 B0
M60                             (LOAD PALLET 1)
<... machining on Pallet 1 ...>
G91 G28 X0 Y0 Z0
G28 B0
M60                             (UNLOAD PALLET 1)
G30 X0
M60                             (LOAD PALLET 2)
<... machining on Pallet 2 ...>
G30 X0
M60                             (UNLOAD PALLET 2)
M30
%
```

HORIZONTAL BORING MILL

The chapter on horizontal machining would not be complete without at least some comments relating to the machine called a *horizontal boring mill*. A CNC boring mill is similar to a CNC horizontal machining center, usually a little larger in size. It may or may not have an automatic tool changer, and usually has the spindle motion split into two axes - Z and W. The following is a typical setup of a 4-axis horizontal boring mill with an indexing B axis and a Fanuc or similar control with:

❑ Six work offsets   ... G54 to G59

❑ Two machine reference points   ... G28 and G30

Although there are five axis designated, a horizontal boring mill is still only a four axis machine. The axes are:

❑ X axis   ... table longitudinal

❑ Y axis   ... column

❑ Z axis   ... spindle quill

❑ W axis   ... table traverse

❑ B axis   ... indexing or rotary table

Settings are similar to a machining center, plus the W axis. During setup, typical work offset values will be set as:

❑ G54 X = Negative
       Y = Negative
       Z = Zero
       W = Negative
       B = Zero

As many horizontal boring mills do not have an automatic tool changer, the G30 position should be set conveniently for the operator to perform a tool change manually (X,Y,W axes). This position is set by a system parameter. Z axis value is the length of quill travel out of the spindle.

Programming format is based on the principle that all motions into the depth are done in the W axis, rather than the Z axis. The quill that is controlled by the Z axis, is pulled out only for clearance purposes; its extension from the spindle must be long enough to guarantee enough clearance for the shortest tool used in the program.

Typical programming format is followed by a more detailed explanation. [nn] lines are for reference only and match the comments that follow:

```
[01] O4605 (PROGRAM NAME)
[02] (MESSAGE OR COMMENT)
[03] N10 G21
[04] N20 G91 G30 W0
[05] N30 G90 S.. M03
[06] N40 G54 G00 X.. Y..
[07] N50 G30 Z0
[08] N60 G43 W.. H..
[09] N70 G01 W.. F..
[10] . . .
[11] . . .
[12] . . . < machining >
[13] . . .
[14] . . .
[15] N350 G00 W..
[16] N360 M05
[17] N380 Z0
[18] N390 G30 G49 W0
[19] N400 G91 G30 X0 Y0
[20] N410 M06
[21] . . .
[22] . . .
[23] . . .
[24] N600 M30
[25] %
```

The following comments reference the line identification numbers in the example:

[01] Program number (name up to 16 characters)
[02] Message to the operator - only between parenthesis
[03] Metric or English units selection
[04] W axis moves to a tool change position
(incremental motion for safety)
[05] Selection of absolute mode and spindle functions
[06] Rapid motion to the starting position in XY within
the G54 work coordinate system
[07] Quill pulls out by the value of parameter 1241 (Z)
[08] Tool length offset (set from the tool tip to the
program zero) and motion to the clearance level
[09] Feedrate motion to the required depth
[10] . . .
[11] . . .
[12] . . . (machining the part) . . .
[13] . . .
[14] . . .
[15] Rapid motion back to the clearance level (see [08])
[16] Spindle stop
[17] Rapid motion of the quill back to the spindle
[18] Rapid motion to the tool change position along the
W axis and cancellation of tool length offset
[19] Rapid motion to the tool change position along the X
and Y axes - in incremental mode for safety
[20] Manual tool change
[21] . . .
[22] . . . (additional machining, following the above format . .)
[23] . . .
[24] End of program
[25] End of record (stop code)

47 WRITING A CNC PROGRAM

Writing a CNC program is the final result of manual programming. This last step requires a sheet of paper, or many sheets of paper, that contain the program. The program is composed of individual instructions related to machining and arranged in a series of sequential blocks. Writing does not mean using only a pen or pencil. Modern writing methods employ a computer and a text editor, but the result is still a written copy of a manually generated part program.

Manual program development is the result of a lot of hard work. A short program with a few lines of code may be as easily entered into the control directly as to be written down on paper. However, the written copy will often be required for documentation and other reference purposes.

The need to program by hand seems somewhat backwards in the age of computers, printers and other hi-tech wonders, but it is a method that will not disappear any time soon. Writing a part program manually requires time and is always subject to errors. Manual work means work by hands, so it seems that a need for special computer skills is not required. Is that a correct assessment?

In the traditional way, a program can be written with a pencil and a paper (and a five pound eraser, as an old cartoon claimed). Its final form is transferred to the control unit, a short program may be keyed into the system directly, by pressing various keyboard keys. For long programs, this approach is a waste of time. The modern alternative to a pencil is the keyboard of a computer, using a simple text editor to make a plain ASCII text file, with no formatting. The computer creates a CNC program as a file stored on the hard drive. This file can be printed or send directly to the CNC machine. The only difference is that the computer keyboard has replaced the pencil and the editing features of the text editor have replaced the eraser. Even today, a great amount of manual programming work in is still done in writing, using a devices such as pens, pencils, calculators and erasers.

Regardless of the media used, learn how the computer - the control system - interprets the written program, what syntax to use, what to avoid and what format is correct. Even if not programming manually at all, it is important to know the principles of program writing techniques, in order to make changes in any program that was developed by a CAD/CAM system, if necessary.

> CNC program should be written in such a way
> that it can be interpreted without a difficulty

PROGRAM WRITING

Writing all collected data into a final version of the CNC part program is one of the last items inside of the programming process. To get to this stage requires hard work through all other stages - when all thoughts have been collected, all decisions have been made and a certain level of comfort has settled in. In the previous chapters, the emphasis was on the program development as a logical process. Now, the focus will shift at the actual method of writing the CNC program, following this logical process.

Writing the program is based on two initial factors:

- ❑ The corporate standards ... company decides
- ❑ The personal style ... you decide

Both factors can be adapted simultaneously in a single program - they are fully compatible. It is unreasonable to expect any *industry or world-wide standards* relating to the various techniques of developing a program. It may be even less reasonable to set any *company based standards,* unless there is a general set of rules and recommendations already in existence.

The final result is that the first guiding factor - company standards - is replaced by the second factor - personal style. From an objective point of view, there is nothing wrong at all with a personal style of programming. If the program works, who cares how it was done. From a revised point of view, it needs to be acknowledged that a CNC programmer can never succeed in isolation. Programming involves at least one user of the final program - the CNC operator - and that makes it, in effect, a team work.

The most common problem with uncontrolled personal style is *inconsistency.* Any CNC machine shop that employs - or plans to employ - more than one programmer, should establish certain minimum standards for the preparation of a part program. Adherence to these standards allows any team member to pick up where another member has left. Often, the personal style of the first programmer in the company will carry on and on and eventually becomes the company standard, for better or worse. Such a situation may well be very positive, but in most cases it needs revaluation or at least a bit of modernizing.

To define a company standard, first evaluate some suggestions and practical observations that may be helpful to prepare the program efficiently for any style that may be suitable to follow and useful in the future.

◆ Legibility of Handwriting

Writing a program without any assistance of a computer and a text editor, means writing a CNC program in pencil, by hand. A hand written program (preferably by pencil) is easier to correct without a mess and it should be double or even triple spaced when written on a sheet of paper. Individual words in a program block should be separated by a space, to further enhance legibility. This way, any additions or future changes (if they become necessary) can be made quite easily, yet still keeping the overall appearance of the paper copy clean and neat. Problems with legibility of a manually generated program are much less of a factor, if the program is typed directly into a computer text file. Even in those cases, the printed copy may be illegible for technical reasons, such as a low printer toner, for example.

◆ Programming Forms

In the early years of numerical control, special programming forms were popular with pre-printed columns for each address in the block. Those were the days when only the numerical values were entered into the appropriate column and the column position itself determined the address meaning. These programming forms were often issued by control and machine manufacturers, as an aid to program writing and a little promotion on a side. Today, a ruled sheet of a standard size paper is fully sufficient. No special columns are needed and if a column or two is justified, it can be drawn easily enough. Modern programs use alphanumeric representation, writing the whole word - alpha characters as well as numeric characters and special symbols. This process is much more economical, hardly any machine manufacturers print programming forms any more.

CNC programmers in some companies do not necessarily do the final program version themselves. Many managers consider such work a secretarial responsibility. That means somebody else (a clerk or an assistant) will read the hand written copy and has to be able to read it correctly, the way it was intended. Such a person may have absolutely no knowledge of CNC programming and may not be able to detect even simple syntax errors.

◆ Confusing Characters

The legibility of programmer's handwriting is very important. Make a special effort when writing certain characters (alphabetical or numeric) that can be interpreted in more than one way. Depending on personal handwriting, some characters can be confusing to the reader. For example, a handwritten letter O and digit 0 can look the same. Digit 2 and the letter Z can be also confusing. The letter I and the digit 1 as well as a low case letter *l* are other examples. These are only some of the most obvious examples, but many other characters can also be confusing, depending on each person's handwriting. Try to develop a consistent writing technique to distinguish potentially confusing characters (confusing is a relative term, of course).

For instance, all computers and printers (even the old tape preparation systems) use a special method to distinguish individual characters on the screen and in print. In this handbook there is an obvious difference between a wide letter O (as in O1111) and a narrow digit 0 (as in 00001).

The same technique should be applied to personal handwriting. Take advantage of the fact that there is no letter O used on most controls anywhere, except as a program number and in a comment section, where a misprint will not create a problem anyway. If preferred, find a special designation only for the letter O and the rest is all digits 0 by default - unless hundreds of zeros are identified specifically and in a unique way in every part program.

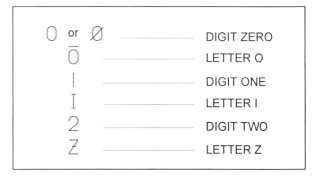

Figure 47-1
Special form of characters written to prevent ambiguity

The illustration in *Figure 47-1* shows some *suggested* methods of common character distinction in handwriting. Find a personal way to write any characters that may improve handwriting legibility. Regardless of which method is selected as personally preferable, adhere to it - *be consistent*. There is nothing worse than adopting different 'standards' for every new program. The programmer or the person who prepares the program final version, will be more than confused and eventually may make a serious error.

Handwritten method can be bypassed entirely by keying in the program data via control keyboard. The program can then be optimized, the part machined and verified program sent out when the job is finished. This procedure may tie up the machine for a while and is not recommended as an everyday method of inputting programs into the control system. The best and fastest method is to prepare the program in a text editor on a computer and send it directly to the CNC machine, through a cable connection.

The majority of CNC users today do not use a punched tape anymore, and if they still do, it is usually for old machines only. More modern methods are available, such as disk storage of a desktop or laptop computer. Through an interface between the computer and the machine, data can be transferred either way very reliably, thus eliminating the punched tape and other methods altogether. Whichever method is selected as preferable, the program still has to be properly formatted.

PROGRAM OUTPUT FORMATTING

Those who followed this handbook from the beginning, chapter by chapter, should be well familiar with programming by now. This section deals with the actual program format - not *its contents*, but how it *appears* on the printed paper or screen of the computer. It will evaluate four versions of the *same* program. Identical in every respect, except the outward appearance. Feel free to be the judge as to which of the four format versions is the most suitable and why. A rather long program is presented - intentionally - this is an actual program. It is not important what it does, only how it looks when printed or displayed. Each new version will be improved over the previous version.

➲ Program Version 1 :

```
G20
G17G40G80G49
T01M06
G90G54G00X-32500Y0S900M03T02
G43Z10000H01M08
G99G82X-32500Y0R1000Z-3900P0500F80
X32500Y32500
X0
X-32500
Y0
Y-32500
X0
X32500
G80G00Z10000M09
G28Z10000M05
M01
T02M06
G90G54G00X-32500Y0S750M03T03
G43Z10000H02M08
G99G81X-32500Y0R1000Z-22563F120
X32500Y32500
X0
X-32500
Y0
Y-32500
X0
X32500
G80G00Z10000M09
G28Z10000M05
M01
T03M06
G90G54G00X-32500Y0S600M03T01
G43Z10000H03M08
G99G84X-32500Y0R5000Z-13000F375
X32500Y32500
X0
X-32500
Y0
Y-32500
X0
X32500
G80G00Z10000M09
G28X32500Y-32500Z10000M05
M30
%
```

This is the most primitive way of writing a program. Although it may offer some doubtful benefits, it is the least friendly version. Good program with a very poor appearance. It is extremely difficult for the CNC operator to read this mess.

➲ Program Version 2 :

```
N1G20
N2G17G40G80G49
N3T01M06
N4G90G54G00X-3.25Y0S900M03T02
N5G43Z1.0H01M08
N6G99G82X-3.25Y0R0.1Z-0.39P0500F8.0
N7X3.25Y3.25
N8X0
N9X-3.25
N10Y0
N11Y-3.25
N12X0
N13X3.25
N14G80G00Z1.0M09
N15G28Z1.0M05
N16M01
N17T02M06
N18G90G54G00X-3.25Y0S750M03T03
N19G43Z1.0H02M08
N20G99G81X-3.25Y0R0.1Z-2.2563F12.0
N21X3.25Y3.25
N22X0
N23X-3.25
N24Y0
N25Y-3.25
N26X0
N27X3.25
N28G80G00Z1.0M09
N29G28Z1.0M05
N30M01
N31T03M06
N32G90G54G00X-3.25Y0S600M03T01
N33G43Z1.0H03M08
N34G99G84X-3.25Y0R0.5Z-1.3F37.5
N35X3.25Y3.25
N36X0
N37X-3.25
N38Y0
N39Y-3.25
N40X0
N41X3.25
N42G80G00Z1.0M09
N43G28X3.25Y-3.25Z1.0M05
N44M30
%
```

This is definitely a much improved version of the program. Look at what a simple block numbering and inclusion of the decimal points do to the program legibility. This program version is far from being final, but it does offer some tangible improvement. Decimal points in programs are of course standard, except for old controls.

The next program version applies all improvements done so far and addresses some additional issues.

◉ Program Version 3 :

```
N1 G20
N2 G17 G40 G80 G49
N3 T01 M06
N4 G90 G54 G00 X-3.25 Y0 S900 M03 T02
N5 G43 Z1.0 H01 M08
N6 G99 G82 X-3.25 Y0 R0.1 Z-0.39
N7 X3.25 Y3.25
N8 X0
N9 X-3.25
N10 Y0
N11 Y-3.25
N12 X0
N13 X3.25
N14 G80 G00 Z1.0 M09
N15 G28 Z1.0 M05
N16 M01
N17 T02 M06
N18 G90 G54 G00 X-3.25 Y0 S750 M03 T03
N19 G43 Z1.0 H02 M08
N20 G99 G81 X-3.25 Y0 R0.1 Z-2.2563 F12.0
N21 X3.25 Y3.25
N22 X0
N23 X-3.25
N24 Y0
N25 Y-3.25
N26 X0
N27 X3.25
N28 G80 G00 Z1.0 M09
N29 G28 Z1.0 M05
N30 M01
N31 T03 M06
N32 G90 G54 G00 X-3.25 Y0 S600 M03 T01
N33 G43 Z1.0 H03 M08
N34 G99 G84 X-3.25 Y0 R0.5 Z-1.3 F37.5
N35 X3.25 Y3.25
N36 X0
N37 X-3.25
N38 Y0
N39 Y-3.25
N40 X0
N41 X3.25
N42 G80 G00 Z1.0 M09
N43 G28 X3.25 Y-3.25 Z1.0 M05
N44 M30
%
```

This version is much improved. It uses all improvements of the previous version, yet adds a significant improvement - *spaces between words*. Still, it is difficult to visually identify the start of a tool. The next version will add a blank line between tools. The spaces do not impose an extra drain on the CNC memory, yet the program is much easier to read.

◉ Program Version 4 :

```
(DRILL-04.NC)
(PETER SMID - 07-DEC-01 - 19:43)

(T01 - 1.0 DIA - 90DEG SPOT DRILL)
(T02 - 11/16 TAP DRILL - THROUGH)
(T03 - 3/4-16 TPI PLUG TAP)
```

```
(T01 - 1.0 DIA - 90DEG SPOT DRILL)
N1 G20
N2 G17 G40 G80 G49
N3 T01 M06
N4 G90 G54 G00 X-3.25 Y0 S900 M03 T02
N5 G43 Z1.0 H01 M08            (INITIAL LEVEL)
N6 G99 G82 X-3.25 Y0 R0.1 Z-0.39 P0500 F8.0
                                        (HOLE 1)
N7 X3.25 Y3.25                          (HOLE 2)
N8 X0                                   (HOLE 3)
N9 X-3.25                               (HOLE 4)
N10 Y0                                  (HOLE 5)
N11 Y-3.25                              (HOLE 6)
N12 X0                                  (HOLE 7)
N13 X3.25                               (HOLE 8)
N14 G80 G00 Z1.0 M09
N15 G28 Z1.0 M05
N16 M01

(T02 - 11/16 TAP DRILL - THROUGH)
N17 T02 M06
N18 G90 G54 G00 X-3.25 Y0 S750 M03 T03
N19 G43 Z1.0 H02 M08
N20 G99 G81 X-3.25 Y0 R0.1 Z-2.2563 F12.0
                                        (HOLE 1)
N21 X3.25 Y3.25                         (HOLE 2)
N22 X0                                  (HOLE 3)
N23 X-3.25                              (HOLE 4)
N24 Y0                                  (HOLE 5)
N25 Y-3.25                              (HOLE 6)
N26 X0                                  (HOLE 7)
N27 X3.25                               (HOLE 8)
N28 G80 G00 Z1.0 M09
N29 G28 Z1.0 M05
N30 M01

(T03 - 3/4-16 PLUG TAP)
N31 T03 M06
N32 G90 G54 G00 X-3.25 Y0 S600 M03 T01
N33 G43 Z1.0 H03 M08
N34 G99 G84 X-3.25 Y0 R0.5 Z-1.3 F37.5(HOLE 1)
N35 X3.25 Y3.25                         (HOLE 2)
N36 X0                                  (HOLE 3)
N37 X-3.25                              (HOLE 4)
N38 Y0                                  (HOLE 5)
N39 Y-3.25                              (HOLE 6)
N40 X0                                  (HOLE 7)
N41 X3.25                               (HOLE 8)
N42 G80 G00 Z1.0 M09
N43 G28 X3.25 Y-3.25 Z1.0 M05
N44 M30
%
```

The final version *(Version 4)* may be a luxury for some users, but it is the most elegant of all four. It adds initial descriptions and messages to the operator. It includes programmer's name and the date of the last update. It also includes the description of all tools at the program beginning. It also uses the same tool descriptions for individual tools, at the beginning of each tool section, where it matters most.

Some lower level controls do not accept comments in the program. If there are comments in the program, such a control system will strip them automatically during loading.

LONG PROGRAMS

Those who ever worked with a punched tape, used to run the program directly in a reel-to-reel operation. The maximum program length was the maximum length of tape that fitted on the reel, about 900 feet - or 275 meters - or 108000 characters. With today's modern equipment, there is no need for a tape anymore, most part programs will run from memory of the CNC system. Unfortunately, that memory capacity is finite as well, often well below what the tape capacity used to be. It all means that a situation may arise, when a particularly long program will not fit into the memory. In addition to a good directory cleanup, there are two other possibilities to eliminate this problem.

◆ Program Length Reduction

A simple way to reduce the program length is to eliminate all *unnecessary* characters from the program. Since the problem is related to a *long program*, the reduction in length will be much greater than can be illustrated here. There is a number of areas that should be considered before taking the red pen and starting the changes:

- ❑ Eliminate all unnecessary leading or trailing zeros (G00 = G0, X0.0100 = X.01, ...)
- ❑ Eliminate all zeros programmed for convenience (ex.: X2.0 = X2.)
- ❑ Eliminate all or most of the block numbers
- ❑ If using block numbers, increments by one will make a shorter program
- ❑ Join several single tool motions into a multiaxis tool motion, if safety allows
- ❑ Use default control settings, but check them first
- ❑ Do not include program comments and messages to the CNC operator
- ❑ Use comments and various descriptions on a separate piece of paper

Organizing the programming process will definitely help - for example, include as many instructions in a single block as possible, rather than dividing them into many individual blocks. Use subprograms if possible, use fewer tool changes, even fewer tools, if that is possible, etc. At the same time, watch for undesirable side effects when eliminating or deviating from an established program format.

There is no doubt that many of these measures will result in some compromise between convenience and necessity. When thinking well ahead and organizing the work properly, the results will be worth the effort.

> These methods are shortcuts and should be used for emergency situations only, not as standard programming procedures.

To illustrate some of the shortcut methods, compare the two following examples - both will have the same results - well, *almost* the same results:

```
O4701 (TYPICAL PROGRAM)
N10 G21 G17 G40 G80 G90
N20 G54 G00 X120.0 Y35.0
N30 G43 Z25.0 H01
N40 S500 M03
N50 M08
N60 G99 G81 X120.0 Y35.0 R3.0 Z-10.0 F100.0
N70 X150.0
N80 Y55.0
N90 G80 G00 Z25.0
N100 M09
N110 G28 X150.0 Y55.0 Z25.0
N120 M30
%
```

A grand total of 194 characters have been programmed. The condensed version of the program needs only 89 characters, with a minor compromise. Program in this form is more memory efficient but much harder to read - remember this is only a short sample, not a real long program in its entirety, where the difference would be more impressive:

```
O4702
G90 G0 X120. Y35.
G43 Z25. H1 S500 M3
M8
G99 G81 R3. Z-10. F100.
X150.
Y55.
G80 Z25. M9
G91 G28 X0 Y0 Z0
M30
%
```

A total of 54.12% of the program length have been saved in a rather very short program. Shortening the program length may become useful in some cases, so here are several methods that have been used in the above example:

- ❑ Program description has been eliminated
- ❑ Block numbers have been eliminated
- ❑ G21, G17 and G54 have been eliminated (correct settings assumed on the control - be careful !)
- ❑ Zeros following a decimal point in a full number have been canceled
- ❑ Some blocks were joined together
- ❑ G80 G00 has been replaced by G80 only (G00 is redundant, although commonly used)
- ❑ Leading zeros in G00, M08, M09, H01, and M03 have been removed
- ❑ Machine zero return has been changed from absolute mode to incremental mode
- ❑ ... Keep in mind, this is a no-frills program

Although both program examples will result in a part machined according to the drawing specifications, some programming instructions will be processed differently. A very important change can be achieved in the tool approach towards the part. In the first example (standard version), the motion command positions the X and Y axes first, with the Z axis motion following in a separate block. In the shorter example, the order of motions has been preserved for safety reasons. If machining conditions allow, these two motions can be combined into one. The G43 and G54 commands can work together in the same block, without a problem:

```
G90 G0 G43 G54 X120. Y35. Z25. H1 S500 M3
```

Always be careful to consider the setup first and guarantee a safe approach towards or away from the part. If any obstacles come in the way because of the shortcut, the condensed example would be a wrong programming method.

Program preparation and its actual writing will become a routine very soon after establishing a personal programming style. If using a computer, learn how to write the program directly at the keyboard - it is a waste of time to write it by hand first. It may take a little time getting used to, but it is well worth it.

◆ Memory Mode and Tape Mode

Most CNC system have a special *Mode Switch* selector to choose from at least two options - the MEMORY mode and the TAPE mode. The Memory mode is used the most frequently - program is loaded into the CNC memory, it is edited from the memory, and is run from the memory. The Tape mode is, of course, to run a program from a tape and many users ignore the possibilities this mode offers. Even if not using punched tapes in the machine shop anymore, (most companies do not), the Tape mode can be used to *em-*

ulate a tape with many added benefits. Tape mode is not to be taken literally. Think of the Tape mode as an *external* mode, not in its actual old fashioned sense.

In order to use this external mode requires a little extra hardware and software. On the hardware side, only a reliable micro computer is needed, with a fair size hard disk storage capacity. A properly configured cable that will connect the computer with the CNC is also required. The computer may be a 'abandoned' slower computer from the office - all that is needed is a minimum configuration, nothing fancy. On the software side, an inexpensive communications software is necessary, to send the program from the computer to the CNC and back.

Once everything is configured to work together, store the CNC program or programs on the hard disk of a personal computer, load the software and work with the CNC system as usually! The major difference is in editing. Since the program actually resides on the hard disk of the PC, use the computer and a text editor to edit the CNC program, not the control system. The capacity of current hard drives is *x-times* more than will ever be needed. Aircraft companies, mold shops, tool and die shops and other industries that require extremely long programs have embraced this technology a while ago, and very successfully, too.

Also consider this method for the *High Speed Machining* programs. This relatively new technology uses very high spindle speeds and feedrates but very small depths of cut. This combination means extremely long programs, many that will not fit into any memory configuration of any CNC system. So before investing into rather expensive memory updates, investigate this method of running a program from a personal computer, if the transfer speed is fast enough.

48 PROGRAM DOCUMENTS

During the program preparation, quite a number of various pieces of documentation will accumulate. All sketches, calculations, setup sheets, tooling sheets, job descriptions, instructions to the operator and related notes contain valuable information. This information should be stored as part of the program documentation folder. Any changes to the finished program at a later date, for whatever reason, can be done much easier if the documentation is complete, well organized and stored in one place. A good documentation makes a review of the program at a later date much easier. If somebody else has to review a program, the documentation will save much of the valuable time. The way programmers document programs reflects not only their personal programming style, it also becomes a reliable indication of their sense of discipline and organizational capabilities.

A simple definition relating to program documentation can be presented:

> Program documentation is a set of all records
> necessary to retrace the program development

Many CNC programmers, even machine shop supervisors, underestimate the importance of good program documentation. Their main arguments are that the paperwork is not worth the time, that it takes too long to collect all documents and prepare the documentation, that it is essentially a nonproductive effort, etc. These arguments are true, to a point - in order to make a good documentation, yes, some time will be required. Not an excessive amount of time, but enough time to do a good job. If there are prepared blank forms available, they just need to be filled. It does not take any more time than writing the same information on any other piece of paper - it can actually take a lot less time. If a CAD system is available, use it to develop a customized tooling library and setup sheet. A variety of blank forms can be predefined, then filled quickly whenever they are needed. CAD system will save time, it makes the program documentation neat, and every sketch drawn in scale can be easily retraced. Using a word processing or a spreadsheet software is another way to save time for documentation.

In essence, the purpose of a program documentation is to communicate programmer's ideas to somebody else or to review them at a later date. Creating documentation is not a directly productive work, but does not have to take too much extra time. Documentation may me a good investment in time management, it can save a lot of time one day in the future.

DATA FILES

A complete part program is not only a hard copy of the program or the program data (usually stored on a disk). The examples of the important documents mentioned here are all a vital part of the program. They create a set of all files used for programming, called the *data files*.

All of these files are useful to the programmer, but only some are important to the CNC machine operator or the setup person. A number of files are only for reference, and are not normally sent to the machine shop. Two basic rules for data files can be established:

❑ Programmer keeps all the files

❑ Machine operator gets copies of relevant files only

These two rules guarantee that the ultimate responsibility for the program remains with the CNC programmer. There is no need to duplicate every piece of documentation for the machine shop - only those items that relate to the actual machining have a place in the shop. Unnecessary duplication is counterproductive and should be avoided. The only items of documentation needed in the machine shop are:

❑ Part drawing

❑ Program printout

❑ Setup sheet

❑ Tooling sheet

The *part drawing* serves as a reference for comparison of the intended shape, dimensions, tolerances, etc., with the actual product. Only the drawing version that was actually used for programming should be considered. The *program printed copy* is the program listing made available to the machine operator. Normally, it is the printed output of the program. The remaining two items, the *setup sheet* and the *tooling sheet* describe the programmer's decisions relating to the part setup and the selection of cutting tools. Each also is a complement to the program itself. In some cases, other documentation has to be included as well, not mentioned here. Any piece of paper that is considered important should be included.

Those who have written programs in a high level language (*C++*, *Visual Basic*, etc) or in older languages such as *Basic*, *Pascal* or even *AutoLISP* (the original programming language for AutoCAD), know that they can add comments within the body of the program.

These comments are usually terse, just long enough to remind the user of what is happening in the program. If more information about the program is necessary, most likely there would be additional instructions, even a user's manual. This kind of external and internal program documentation applied in software development, is also adaptable to a CNC program.

PROGRAM DOCUMENTATION

The difference between an external and internal program documentation deserves some explanation. Is one better than the other? Which one should be used?

The best documentation is the one that combines both types for maximum effect. To distinguish between the two types, let's evaluate them individually.

◆ External Documentation

External documentation of a CNC program generally consists of several items and always of their latest version - this last statement is very important.

The following items are typical to any program documentation. They can be used as desired:

❑ Program copy printout

❑ Methods sheet, if available

❑ Part drawing

❑ Working sketches and calculations

❑ Coordinate sheet

❑ Setup sheet

❑ Tooling sheet

❑ Program data (disk or other media)

❑ Special instructions

The program copy printout is the final version of the programming process. It should be the *exact* contents of the program stored on a disk or other media. In those machine shops that use routing or methods sheets, the programmer should make it a policy to include a copy of the methods sheet in the documentation as well. Part drawing (or its copy) is extremely important to be kept together with the program. It is the ultimate reference source in the future. All sketches and calculations, together with a sheet of coordinates, are also useful at a later date, particularly if the program has to be revised for some reason. The setup sheet and the tooling sheet will be discussed shortly.

That only leaves the program data source (usually stored on a disk or similar media) to be included in the documentation folder and any special instructions that may be required by the programmer, the CNC machine operator or somebody else.

◆ Internal Documentation

Internal documentation is contained *within the body of a program*. When writing a program, make an effort to strategically place comments into the program. Such messages are integral part of the program and are categorized as *internal program documentation*. These messages are either separate blocks of a program or additions to individual blocks (delimited by parentheses) and can be actually seen on the display screen during program execution (on most controls). They are also printed in the copy of the program. The biggest advantage of internal documentation is the convenience offered to the machine operator. The only disadvantage is that when loaded into the CNC memory, the comments do occupy memory space. If the available memory is scarce, be modest with program comments and place more emphasis on the external instructions. All program comments, messages, directions and instructions must be enclosed in parentheses:

```
(THIS IS A COMMENT, MESSAGE OR INSTRUCTION)
```

This is the required program formatting. Either comment, message or instruction can be an individual block in the program or it can be part of a program block. The control system will ignore all characters between the parenthesis. To avoid long descriptions internally, use pointers to external documentation instead - for example:

```
N344 ...
N345 M00 (SEE ITEM 4)
N346 ...
```

The ITEM 4 in the program comment section will be a detailed description that relates to block N345, somewhere else in the program documentation, such as in a setup sheet. This kind of reference is useful when the message or comment would be too long to be stored in the program body.

For example, the CNC operator may find the referenced ITEM 4 in the setup sheet, under the heading of *Special Instructions*:

```
. . .
ITEM 4. - Remove part, clean the jaws, reverse and
clamp on the 120 mm diameter
. . .
```

Properly prepared internal documentation should always briefly describe each cutting tool used:

```
N250 T03
N251 M06
(T03 = 1 INCH DIA 4-FLT E/M)
N252 ...
```

Note that T03 is the *current* tool. This designation will vary greatly, depending on the tool changing systems of the particular machine tool builder. Also note the use of abbreviations in the program comment - 4-FLT E/M is a short form for a *4-flute end mill*.

Every time the *Program Stop* command M00 is used in the program, document the reason why it is used:

```
N104 G00 Z1.0
N105 M00 (CHECK DEPTH = .157 INCHES)

N106 ...
```

To gain an extra space in the comment block, make it a separate block:

```
N104 G00 Z1.0
N105 M00
(DEPTH TO SHOULDER MUST BE .157 INCHES)

N106 ...
```

Comments can be in the same block as program data:

```
N12 G00 X3.6 Z1.0
N13 M05 (CLEAN CHIPS FROM THE HOLE)
N14 ...
```

If the comment is written as a separate line, the block number for the comment block is usually not used. The instructions that are part of the documentation should be clearly understood. Enigmatic or cryptic messages will not do. Distinct messages translate into a time saving by the CNC operator or the setup person and they contribute to a quicker turnaround between individual jobs.

◆ Program Description

On many Fanuc systems, program *description* can also be documented. This is a special kind of a comment, also included in parentheses. There are some conditions that make the program description special.

❏ The description must be included in the same block as the program number

❏ The description must have no more than 15 characters

❏ Low case characters will not be accepted

A typical example of program description may include a drawing name and/or number in the comment section:

```
O4801 (FLANGE-DWG.42541)
```

Once these conditions are followed, the program number can be viewed along with its description right on the directory screen (program listing) of the control system.

If an additional description that does not fit the 15 characters is needed, enter more comments in subsequent blocks. They will not be seen on the directory screen, but can still be handy for internal program documentation. They will be displayed during the program processing on all controls that accept the comments. The length of these comments is not usually limited to 15 characters:

```
O4802 (RING-OP.1)
(DWG. A-8462 REVISION D)
(PETER SMID - 07-DEC-01)
N1 ...
```

The main purpose of program documentation is to transfer all important decisions and ideas from the developer to the user. In the CNC programming environment, the documentation transfers the ideas from the programmer's desk to the machine shop. It serves as an important link within the communication process.

SETUP AND TOOLING SHEETS

Apart from the program printed copy and the part drawing, the *setup sheet* and *tooling sheet* are the other two most important pieces of good program documentation. The major difference between the setup sheet and tooling sheet is the subject emphasized. The setup sheet is a sketch or a drawing that shows the layout and orientation of the part on the table or in a fixture, possibly even the description of individual operations. The tooling sheet usually lists only the cutting tools and their mounting positions, with spindle speeds, feedrates and offsets for each tool. Examples of both types are shown in this chapter.

The ongoing question many programmers always have reminds one of a *Catch 22* situation:

'Do I make the setup sheet and the tooling sheet before or after writing the CNC program?'

As is usual in many programming applications, adherents and foes can be found on both sides of the issue. The favorite reasoning for making the documentation *before* writing a single line of a program is quite simple. The setup sheet and tooling sheet are the guiding forces for writing the program, the forces of being well organized. Adhering to this method implies a well organized programmer or a team, it implies that everything is under control. It also suggests that all fixtures and tools and holders and inserts and other tools are already available in the machine shop, ready and waiting to be used. No doubt, if possible, always make both, the setup sheet and the tooling sheet, *before* starting the program. The logic behind the reasons for this method is very strong indeed.

Logic, however, does not take into consideration the machine shop realities, even if they are essentially not the best or even dead wrong. A small conflict between different departments, a delay in material delivery, a tool on a back order and similar problems, all contribute to the frustrations of a CNC programmer in many companies. Being under pressure from all sides, the programmer has no choice but to improvise, even in times of crisis. Programmer has to *compromise* reflects the reality a little more accurately. If there is no choice, always try to find a *reasonable* compromise, but never as an excuse for being sloppy.

The freedom in programming is considerable but it is not unlimited. A normal part program cannot be written without knowing the machine setup and the tooling to be used. In many cases, the nature of the job offers many solutions. Even if the *exact setup*, or the *exact tool* to be used are not known, think of some ideas, have some opinions - but have ideas and opinions based on experience. The compromise does not rest with the 'now or later' situation, it rests in the selection of the most *likely* possibility. If something has to be changed, make sure the changes will be minimal. In any case, it is quite possible that the setup sheet and/or the tooling sheet will have to be modified *after* the program has been proven and optimized.

◆ Setup Sheet

In many shops, setup sheets are a luxury. It is a simple statement of fact, but many setup sheets are quite poorly prepared if they are prepared at all. Often, they do not reflect the latest program changes and adjustments, they are not consistent between individual machines and even programmers. Although the time spent on preparing a setup sheet is considered nonproductive from the cost angle, it is a time far from being wasted. The setup process can be organized, certain rules can be set and adhered to and they can be applied to the preparation of a good setup sheet.

The golden rule of a good setup sheet is *to make it in scale*. Setup sheet using an outline of the material, fixtures layout, finished shape, tool path, etc., should always be done in scale. Scale, even an approximate scale, is very important for visual comparison. Clamps and other mounting devices should be drawn in positions corresponding to the actual setup. Tool change location should be marked accurately, different views shown, if necessary. Critical positions should be dimensioned, indicating the maximum or minimum distances.

If a cutter radius offset is used, the speeds and feeds reflect a certain *nominal* cutter radius. At the discretion of the operator, the cutter radius may be changed within a reasonable range. This range should appear in the setup sheet, including a note on the adjustment of speeds and feeds.

In many cases when the cutter exceeds a certain length, it may interfere with the part or other tools. In these cases, the setup sheet should include the maximum cutter length allowed within that setup. For a chuck work on a lathe, the maximum grip of the material should be specified in the setup sheet as well.

The main purpose of a setup sheet is to document all details of how the part is mounted on the machine. That means it has to cover the part holding method and reference point relationships (part, machine, and the cutting tool). It has to describe the positions of auxiliary devices used, for example, a tailstock, a barfeeder, a vise, a face plate, hard and soft jaws, and many others. A master form for a setup

sheet may have to be done for every machine or at least for every machine type. A very simple setup sheet is shown in *Figure 48-1*. Feel free to improve it as necessary.

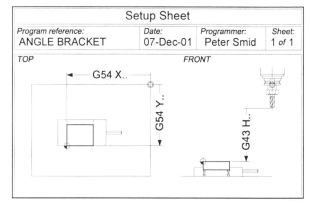

Figure 48-1

Simple setup sheet form - only basic data shown

A well designed setup sheet should also include information about the material used for machining, material the program is based on. Not only the type of material, also its rough dimensions, amount of stock for machining, its condition, and other features that are important to include in program documentation. This information is very valuable at its conception and will be even more valuable in the future, mainly for repeated jobs. Many times, a program is made when the blank material is not yet available. If the programmer finds out later that there is too much deviation from the estimated conditions, the necessary changes are easier to make with good program documentation.

Although not a strict requirement, some programmers include the cutting time for each machining operation on the setup sheet. When the job is run for the first time, the actual cutting time is unknown. As the program is used and optimized on the machine, it becomes proven and eventually finalized, the cutting time becomes known with more precision. Knowing the cutting time may help in planning the load work on the CNC machine. The most useful cutting time for an individual part is the *chip-to-chip* time that includes all the supplementary times (for example the tool change time, part replacement time, etc.), not only the cutting time itself.

◆ Tooling Sheet

Although the tooling is really part of the setup, it requires a separate set of data, that may or may not fit on the setup sheet. If the setups and tools used are constantly simple, it may be more convenient to have only one sheet, describing them both. However, for large or complex setups, making a separate tooling sheet is more practical. Both, the setup sheet and tooling sheet, are part of the same documentation and complement - *not replace* - each other.

Machine unit and the CNC system influence the contents of a tooling sheet. A tooling sheet for a lathe will be different than a tooling sheet for a machining center. The data gathered for either machine will have some similarities and some unique items. A contents of a typical tooling sheet will include description of the following items:

- ❏ Machine and program identification
- ❏ Type of the cutting tool
- ❏ Tool coordinate data
- ❏ Tool diameter
- ❏ Insert radius and the tip number
- ❏ Offsets associated with the tool
- ❏ Tool length
- ❏ Tool projections from the holder
- ❏ Block number of the tool being indexed
- ❏ Brief description of the tool operation
- ❏ Basic speed and feed of the tool
- ❏ Tool holder description
- ❏ Tool number and/or tool station number
- ❏ Special instructions

In addition to the most common items, also include any unique information in the tooling sheet, for example, to inform the operator about non-standard tools, tools that require modification, premachined condition of the material, etc. An example of a simple tooling sheet is in *Figure 48-2*.

Tooling Sheet							
Program reference: ANGLE BRACKET		Date: 07-Dec-01		Programmer: Peter Smid			Sheet: 1 of 1
T#	Tool Description		Tool Ø	RPM	Feed	H..	D..
T01	12 mm Spot Drill		8.7	1650	100.0	H01	---
T02	7.5 mm Drill		7.0	1800	220.0	H02	---
T03	10 mm End Mill - 4 flute		10.0	1210	300.0	H03	D03

Figure 48-2
Simple tooling sheet form - only basic data shown

◆ Coordinate Sheet

The idea of a coordinate sheet is not new. It has been used in programming from the beginning and it was mentioned in this handbook many times already. A simple printed form containing the X, Y and Z axes can be used for both machining centers and lathes. *Figure 48-3* shows an example of a simple coordinate sheet.

Coordinate Sheet			
Program reference: ANGLE BRACKET	Date: 07-Dec-01	Programmer: Peter Smid	Sheet: 1 of 1
P#	X coordinate	Y coordinate	Z coordinate
P1	X0.000	Y0.000	

Figure 48-3
Simple coordinate sheet form - only basic data shown

The Z axis column will be usually blank for machining centers and Y axis column will be blank for lathe programs. Modify the sheet to add additional axes or make separate sheets for each machine type.

DOCUMENTATION FILE FOLDER

All records that have been collected during program preparation are quite likely important enough to be kept for future reference. They may be stored all over the place, sometimes very hard to find. So, now is the time to put them all together and organize them. It is time to make a file folder, identify it, fill it up and store it properly.

◆ Identification Methods

Before some better methods of identifying program documentation can be suggested, think about a very popular, yet quite an impractical method. Some programmers use the program number as a reference for all related material. The basic thinking behind this idea is that the available program number range between 1 and 9999 will take forever to use up, therefore becomes very useful for other purposes. This is a shortsighted thinking, usually by not a very busy programmer who has only one machine to take care of.

Look at possible problems with this thought. True, to make almost ten thousand programs for one machine, it would take 'almost forever'. Even if more machines are available, at a rate of 25 programs a week, numbers will run out in a little more than 7 years. Is that the time to scrap the machine and buy a new one? And if 25 programs a week seems a bit steep, remember that *each* program will have to have a number. That may be three or more separate operations for a single job, there may be dozens and dozens of subprograms that also need their own program number. So the figures are not so unreasonable after all, and some better method should be sought from the beginning. It could be a manually generated method, or a comprehensive computerized database.

The point of this evaluation is that all program number assignment (with the exception of subprograms), should be left to the *CNC operator,* if possible. That means, another way has to be found to identify the documentation containing all records.

One of the first decisions is the program name selection. Regardless of the number of machining operations or included subprograms, there should be only one folder for one job and only one name for one folder. The name of one folder should share the common denominators with any other folder. Try to make such a name *meaningful.*

With an access to a personal computer, the chances are that all files relating to each program are stored in the computer. In that case, the only limiting factor is the software format structure to name the files. For example, the old DOS software files accepted up to eight alpha numeric characters for the file name and another three alpha numeric characters for the file extension. Since Windows 95, long file names are allowed, up to 255 characters plus extension - try to take advantage of this feature. Regardless of the CNC system used, establish a file naming convention conforming to any possible restrictions. There are several methods for this approach.

One is an independent, sequential order. In this simplest form, all documentation related to the first program would be - for example - P0000001, the next program would be P0000002, etc. If the zeros are removed from this format, the files will *not* be listed in a correct alphabetical order on a computer display. No practical limit is imposed. The second approach would be to use the drawing number as the basis of documentation identification. This may a good method for the many companies that are not jobbing shops. Dealing with many different customers also means dealing with many different types of drawing numbers. The variety may be so great that it is almost impossible to find some common ground for standardization. Another variation on the same theme is a *job number,* rather than a drawing number. In many jobbing shops, each job gets a number the moment the order is received. This *Job Number* is always unique, therefore a good candidate to be used as the number identifying the program documentation.

Hopefully, the presented ideas will stimulate many additional ideas that will suit a particular work environment. There are no given rules on the methods of identifying an individual program; there are no rules governing the standard of part program documentation. The old reliable rule is always use the old fashioned *common sense* that is often not so common. Common sense and foresight help any standardization effort. The quality of any standard is measured by its usefulness in the future. For the longer period of time a particular standard can be useful, the better quality of thought has gone into its development.

◆ Operator's Suggestions

When the CNC machine operator runs the job, he or she may have comments, ideas, corrections and variety of other suggestions. It may be a good idea to consider establishing a log book, a card system, a computer database, or a similar method of communicating the operator's ideas back to the CNC programmer. Whatever system may be selected, it should be available at the machine, so the operator has a primary access to it. The main benefit of such a system is that all communication goes into one source and is easier to keep under control.

Apart from the nature of the particular comment or idea, the log should have the operator's name, current date, perhaps even current time, the machine and job description, as well as any other details that may be relevant and useful anytime in the future.

◆ Filing and Storage

Program file folders can be quite bulky, particularly when they contain computer media, such as disks, large size paper drawings, long program printouts, etc. The storage of file folders is usually confined to standard office steel filing cabinets, which should be accessible to *every work shift,* although only qualified and authorized persons should be given the actual access.

If using any kind of computer media for storing the part programs, make sure they are safely stored in a separate container, rather than the file folder itself. Magnetic devices are particularly sensitive to adverse conditions and should be stored away from any heat source and magnetic field (including a telephone, for example). They should be kept in a dry and dust free environment. Keeping duplicates (or even triplicates) in a separate place is also a good and safe procedure. A very practical - and much less bulky - is storage of proven programs on a CD (Compact Disk) or a DVD (Digital Versatile Disk), using a special recording hardware and software. Although these disks still require to be stored away from all heat sources, they are not a subject of magnetic fields.

Individual sheets or pages of the part program documentation should be either numbered consecutively, or have a reference number on each page. Drawers of the filing cabinets should be identified as to their contents. These are common enough requirements, but very often ignored all together, usually *because there is no time.* The main philosophy behind an orderly filing system is the speedy access to a required program that provides instant and accurate information.

PROGRAM VERIFICATION

When the program is completed, there will be a written copy or a file copy stored somewhere on a computer. Now, the program development is completed. It may be a perfect program with no errors. Of course, that was the intent from the beginning - to make an error free program. What happens if - in spite of the best efforts - there is an error? Even a small typing error can cause a severe problem when the program runs on the machine. Could an error be prevented? And if so, how?

Every program should be checked against all errors *before* it reaches the machine. Checking can be quite simple, such as a visual comparison of the written copy and the printed copy. The main purpose of a program check is to detect obvious mistakes - mistakes that can be seen by *concentrating* and *looking for them*. The kind of errors detected first are mostly syntax errors. Of course, there is no guarantee that the program *is* error free, but when it does leave the programmer's desk, the effort should be to *make it* error free. All programs arriving at the machine should gain the confidence of CNC operator. The operator should be free to concentrate all efforts on proving the program sequences and run the first part. The operator has no time to check for program errors that could - and should - have been detected in the office. To do all program proving on the machine is very nonproductive and should be avoided.

DETECTION OF ERRORS

Before an error of any kind can be corrected, it must be discovered first. In CNC programming, errors can be found either *before* or *after* the program copy leaves the office. The intentions of any good programmer are undoubtedly to discover any errors *before* they are detected during the program execution on the CNC machine. This is a *preventive* effort. If the error has to be corrected at the machine, during the run of the program, the CNC operator has to do something that should not normally be part of the operator's duties. Whatever action it is the operator must take, it is a *corrective* action. Therefore, measures that can be taken to help eliminate errors in a CNC program are of two types:

❑ Preventive measures ... proactive measures

❑ Corrective measures ... reactive measures

Preventive measures are those that all parties involved should participate in with suggestions and constructive criticism. On the other hand, corrective measures require certain skills, knowledge, and even authority.

◆ Preventive Measures

All errors should be detected and corrected by the programmer, who has taken a certain amount of preventive measures. The first preventive measure is to *get organized*. Set up *procedures,* set up *standards,* set up *rules*. Then, follow them diligently. Errors that can be found before the program is used on the machine are numerous. Yet, it takes some techniques to become successful in their detection.

The first method any programmer should use is simple - *check your own work*. Read the program and *evaluate* it. If the rules of consistency had been followed, the error check is easy. Programmers know the appearance of the program, the established standards, the order of commands at the beginning and end of each tool. Checking the program should not take much time at all.

The second method can be used when working with other programmers. Ask a co-worker to look through any new or changed program. Do not be surprised to see mistakes such a check can reveal. A fresh, detached, and impartial look is often very productive. Sometimes, even taking a little break or fresh air first will rejuvenate the brain cells.

With a computer and a special simulation software, there is a third option - make a graphic representation of the tool path, either on the screen or on paper.

A major part of preventive measures is finding *syntax errors*. A syntax error is one that can be detected by the control unit. For example, if a dollar sign appears in the program, the control will reject it as illegal. The control returns an error message or an *'alarm'*. If the digit 2 is entered in the program instead of the intended digit 7, that is *not* a syntax error. That is a *logical error,* since both are legitimate characters the control can accept.

◆ Corrective Measures

If an error is discovered at the control, during the program processing, it was missed and the preventive check did not uncover it. An error that is found at the machine slows down production. It forces the machine operator to take corrective measures and eliminate the error. The operator can take one of two actions. One will be to return the program to the programmer, the second action will be to correct the error at the machine. Which choice is better depends on the seriousness of the error. An error can be *soft* or *hard*. A soft error is one that does not require stopping the program from being processed by the CNC system.

For example, a missing *coolant* function M08 in the program can be switched on manually at the machine, without interrupting the program processing. That is an example of a soft error - it is still an error, but classified as a *minor* error.

A hard error occurs when the program processing must be stopped by the operator, as the only available choice, and without doing a damage to the machine, cutting tool, part, or all of them. A common example of a hard error is a programmed tool motion that cuts in the wrong direction. The program itself is wrong and must be corrected. This is an example of a hard error, classified as a *major* error.

Most CNC operators do not like delays, especially delays caused by somebody else. A dedicated machine operator will do anything possible to correct a problem without any assistance. For program errors, the operator will try to take corrective measures to eliminate the problem. Not every operator is *qualified* to do even a simple change to the program. On the other hand, some qualified operators may not be *authorized* to do program changes as a matter of policy.

Every company benefits greatly, if the CNC operator has *at least* a basic training in CNC programming. The purpose of such a training is not to make the machine operator a fully qualified CNC programmer. Its purpose is to highlight how a part program influences CNC machining, the setup, tooling and all the other relationships between programming and machining. Its purpose is to offer the operator tools that can be used for minor program changes, etc. Such a training, if it is designed and delivered in a professional manner, is always a worthwhile investment. It may be a relatively short training that will pay for itself very quickly. Time delays on CNC machines are costly and the sooner the program is made functional, the less damage to the production control has been done.

Whenever a program has been changed at the machine, the program documentation must reflect these changes, particularly if they are permanent. Even a small permanent change should be always be documented in *all* copies of program documentation.

GRAPHIC VERIFICATION

Programming errors can be costly, even if their cause is a minor human error. Omitted minus sign, a misplaced decimal point, an illegal character - all are minor oversights that cause major errors. Although a visually checked program should be error free, that may not always happen. The human eye is weaker when it evaluates nongraphic elements.

One of the most reliable methods of part program verification is a graphic display of the tool path as it appears in the program. Almost all errors relating to the tool path can be detected early, by one of three available graphic verification methods.

One method of graphic verification of a CNC program is a screen plot. This optional control feature will show all programmed tool motions on the screen. The motion will be represented as lines and arcs. The feedrate motions will appear as a solid line of the selected color, the rapid motion will appear as a dashed line. The display of the tool path will appear on the screen of the control.

Many controls offer a graphic simulation option, where the tool path is simulated on the screen. Each cutting tool can be shows by a different color or density, making the visualization easier. Some graphic simulation uses actual tool shape and the part for a realistic display. The negative part of any graphic verification is that it can only be used when the program is loaded into the control.

The second verification method is much older than the first. It is a hard copy plotted representation of the cutting tool motions. Hard copy plotting has been available in computer programming for a long time. To get the benefits of hard copy plotting, a pen plotter and a suitable software will make it work. The plotter is seldom a problem in companies using CAD software but may not be available to small machine shops. The required software is also part of a large computer based programming system and can be quite expensive. A simple version of a pen plotted tool path is a screen dump, usually to a printer.

There is a third method of graphic verification and can be done in the office. It uses a computer and software specially designed to read a manually generated program, then displays the tool path on the screen. Some software even uses a solid model like features, so the actual surface of the part after machining can be seen as well. This is very useful for 2-1/2D and 3D tool path verification.

AVOIDING ERRORS

The goal of every programmer is to write error free programs. That is almost impossible, since any human activity is subject to errors. Programmers with all levels of experience make mistakes, at least once a while.

Since the prevention of errors should be the main goal of any programmer, this section looks at the subject in more depth. The most common mistakes will be evaluated, along with suggestions to prevent, or at least to minimize, their happening. First, what exactly is a program error?

> Program error is the occurrence of data in a program that will cause the CNC machine to work contrary to the intended plan or not to work at all.

All errors can be classified into two groups:

❏ Syntax errors

❏ Logical errors

Although the average distribution of programming errors could be generally split at 50/50 between the syntax and logical errors, certain conditions may swing the balance. A programmer with limited experience will make all kinds of errors. An experienced programmer makes more syntax errors. Let's look at each error group.

◆ Syntax Errors

Errors in this group are usually easy to deal with, once they are identified. Syntax error is simply one or more characters in the program that are either misplaced or do not belong there. This error covers program entries that do not conform to the programming format (known as *syntax*) of the control system. For example, a lathe control systems do not accept the character Y. If the control encounters the letter Y in a lathe program, it will reject it as a syntax error and the program won't run. The same result will happen when the letter U is programmed for most milling controls. Some other letters cannot be used with either system - for instance the letter V - it is an illegal character for most milling and turning controls. Yet, it is very legal character in a four axis wire EDM control.

Syntax errors also occur if valid characters are used in combination with an option not supported by the control. A good example is a custom macro, an option on most Fanuc controls. Custom macro uses a number of standard letters, digits and symbols, but also a number of special symbols, for example a sharp sign #, brackets [ ], asterisk *, etc.

Macros also use special words, such as COS, SIN, GOTO and WHILE, words not allowed in a standard program. Macro symbols or words in a non macro program result in a syntax error. The error will also occur when the custom macro option is available, but some of the symbols are used incorrectly or the special words are misspelled.

The control system handles syntax errors arbitrarily - *it simply rejects them*. The rejection is displayed as an error message on the screen. Program processing will stop. Syntax errors are irritating and embarrassing, but almost harmless. Scrap as a result of a syntax error is possible but rare. The second group - the *logical* errors - is much different.

◆ Logical Errors

Logical errors are more serious than syntax errors. A logical error is defined as an *error causing the machine tool to act in a way contrary to the programmer's intentions*. If a motion is programmed to the absolute coordinate of X1.0, but program states X10.0, the control will go ahead but the tool position will be wrong. The same error will happen when Z10.0 is programmed, although the intent was X10.0. The control *does not* and *cannot* have any built-in protection against logical errors. Programmer has the responsibility to exercise all necessary care and caution. Logical errors can be serious - they may not only result in a scrap, they can damage the machine and even harm the operator.

Logical errors cover an unlimited number of possibilities. For example, the following lathe program *is wrong*:

```
O4901
(EXAMPLE WITH ERRORS)
N1 G20 G40 G99
N2 G50 S2500 T0400 M42
N3 G96 S530 M03
N4 G00 G41 X12.0 Z0.1 M08
/ N5 G01 X-0.06 F0.012
/ N6 G00 Z0.2
/ N7 X12.0
N8 Z0
N9 G01 X-0.06
N10 G00 Z0.1 M09
N11 X20.0 Z5.0 T0400 M01
...
```

There are three errors in the O4901 example. Try to identify them before studying any further.

The first error should be easy - a *tool offset* is missing. In the block N2, tool T0400 is selected without an offset. This block is correct. Block N11 is the return to the indexing position and the tool offset cancellation, which was never programmed. The error is in block N4 - it should be:

```
N4 G00 G41 X12.0 Z0.1 T0404 M08
```

The second error is rather hidden and requires a trained eye to spot it. Note that a block skip symbol was used in blocks N5, N6 and N7. When running the program with the setting of the block skip function ON, the cutting feedrate is missing in block N8. In this case, the control would issue an error message, but only for the *first time* the program is processed. The correct block N8 should be:

```
N8 Z0 F0.012
```

The third error is the missing cutter radius offset cancel in block N11. This block should correctly be written as:

```
N11 G40 X20.0 Z5.0 T0400 M01
```

A error of this kind may have a serious implications for the next tool. Even worse, this error may not be discovered during the first part run. The correct program is O4902:

```
O4902
(EXAMPLE WITHOUT ERRORS)
N1 G20 G40 G99
N2 G50 S2500 T0400 M42
N3 G96 S530 M03
N4 G00 G41 X12.0 Z0.1 T0404 M08
/ N5 G01 X-0.06 F0.012
/ N6 G00 Z0.2
/ N7 X12.0
N8 Z0 F0.012
N9 G01 X-0.06
N10 G00 Z0.1 M09
N11 G40 X20.0 Z5.0 T0400 M01
...
```

After evaluating the three errors, what chances are there that the control will return an error message? *Nil, zero, zilch.* All errors in the example are good illustrations of logical errors. They may not always be easy to find, but they can create a lot of additional problems if not found early.

COMMON PROGRAMMING ERRORS

Strictly speaking, there are no *'common'* programming errors. Every programmer makes some unique mistakes. It is difficult to list any errors as being more common than others. It is also true, that some mistakes are made more frequently than others and in that sense they are more common. Focusing on this group should be beneficial.

Both syntax and logical errors share the same cause - *the person who writes the program.* The most important step towards eliminating errors is the identification of a problem - ask yourself *'what mistake do I do repeatedly ?'* Everybody makes some 'favorite' mistakes, the solution lies in the correct answer to this simple question.

Most errors are a result of insufficient program planning and a lack of precise programming style. Planning offers a sense of direction, style offers tools and organization.

The simplest - and the most frequent - error is an omission of some fundamental instruction. It may be a coolant function, program stop, a missing minus sign and others. Even the whole block may get lost, mainly when preparing the program from poor sources. Many errors are caused by the programmer's inability to *visualize* what will exactly happen when the program is processed. To this category belong all errors relating to setup, tooling and machining conditions - cuts that are too heavy or too light, insufficient clearances and depths, incorrect spindle speeds and cutting feedrates, even the selection of wrong tools for a given job.

◆ Program Input Errors

Most programs are hand written or typed and have to be transferred to the control system or a computer file. Many errors are caused by the *incorrect input of intended data.* Keep in mind that if somebody else is using the program, its legibility and syntax is very important.

Input errors also include errors caused by forgetting to input significant characters in the program. These strings can be almost anything and can cause a serious problem. A missed coolant function is not likely to cause a big problem; a missed decimal point or a wrong tool retraction will. Other errors are insufficient tool clearances, a depth that is too shallow or too deep, errors relating to cutter radius offset (this is always a big group). Be also careful when canceling or changing modal program values. One common error is to cancel one kind of motion by replacing it with another type of motion in one block, then forgetting to reinstate the previous motion later.

◆ Calculation Errors

Using math functions and formulas is a part of developing CNC programs manually. The type of calculation errors include a wrong numeric input, even when a pocket calculator is used. Keying a wrong formula, wrong arithmetic sign or placing parentheses in a wrong position, all represent a serious error.

Rounding Error

A special type of an error is caused by *incorrect rounding.* This error is an accumulative error that results from too many dependent calculations. A rounded value used in other calculations may lead to an error. In many cases the error will be too small to cause any problems, but never count on it. It may become a very bad habit.

Calculations check

To prevent math errors when using formulas for calculations, it is a good idea to check the calculated result once more, using a *different* formula. Math is a generous science and more than one calculation method is usually possible.

◆ Hardware Errors

The last type of program errors is by the *malfunction of a hardware element* of the control system or machine. In CNC, even a bug in the software is possible. Their occurrence is rare, as modern controls are very reliable. When encountering an error, don't blame the control or the machine *as the first and only possible cause.* It shows ignorance and unwillingness to address the problem responsibly. Before calling for a service, make sure to exhaust all other possibilities of error detection first.

◆ Miscellaneous Errors

Some errors can be traced to the part drawing. An error in the drawing is possible, but first make sure to interpret the drawing correctly. Drawing errors include too many or too few dimensions, poor tolerances, etc. Also make sure to work with the latest drawing version only.

Other errors may be caused by the wrong setup, tooling or material. These are not programming errors, but they have to be considered as possibilities. With some common sense and suitable precautions, many programming problems can be eliminated. For example, to prevent an unproven program to be processed as a proven program, just mark it as unproven. Mark it at the *beginning* of the program and leave it there until the program is checked.

A complete elimination of errors is not realistic. Mistakes do not happen - *but mistakes are always caused.* Inexperience, negligence, lack of concentration, poor attitudes, are just some causes. Always program with the attitude to eliminate programming errors altogether. That will be the first step to making fewer errors.

CNC MACHINING

When a part program is completed and sent to the machine shop, the programming process is over. All the calculations have been done, program has been written, documented, and the production file is on the way to the CNC machine. Is the programmer's job really finished? Is there some reason that could bring the program back, perhaps with operator's comments, suggestions, or even criticism?

If the delivered part program is *perfect*, programmer will not hear a word from any direction. No doubt, programmer will hear negative comments from *all* directions. The question is - when is the programmer's responsibility *really* over? At what point in the process of manufacturing can the programming results be evaluated? When can the program qualify as a *good* program?

Probably the fairest and the most reasonable answer would be *whenever the part has been machined under the most optimized working conditions*. This means that the programming responsibility *does not* end with the program and documentation delivery to the shop. The program at this stage is still very much in the development process. It still has to be loaded to the CNC system, the machine has to be set up, cutting tools mounted and measured and a variety of small jobs done before the first part can be started. True, all these tasks are the responsibility of CNC operator, so there is no need for the programmer to care what happens during machining, right?

Wrong! Every CNC programmer should make an effort to be in constant touch with the actual production. In the field of business software development, it is quite normal to have a team of people to work on a certain large programming project. After all, most programming ideas come from talking to colleagues and the actual users of the program or particular software. The same is true for CNC programs used in machine shops. The users of programs are typically CNC machine operators - they can be a gold mine of constructive ideas, improvements and suggestions. Talk to them, ask questions, make suggestions, and - most important - *listen* to what they have to say. Programmers who never put their foot in the machine shop or go there reluctantly, programmers who may go there with their eyes closed and ears plugged, programmers who take the attitude that they are always right, are all on the wrong track. Exchanging ideas with machine operators, asking questions and seeking answers is the only way to be fully informed about what is actually going on in the machine shop. It is in programmer's interest to know how the CNC operators *feel* about the program, the programming style and the approach to programming overall. *Do* exchange ideas and *do* communicate with each other - that is the best advice for becoming a better CNC programmer. Machine shop offers tremendous resources, take advantage of them.

CNC technology is an instrument to improve productivity with a minimal human involvement, measured at least by the physical level. As any other technology, it must be managed intelligently and by qualified people with experience. Without a firm grip and good control, without good management, the technology will not yield the expected results - in fact - in will become counterproductive.

The function and responsibilities of a CNC programmer has been covered. Now, let's look at what happens when the completed program and related material actually reach the machine shop.

MACHINING A NEW PART

The most expensive part done on a CNC machine is always the *first one* of the batch. After the machine setup is completed, the CNC operator is ready to test the program and the machining conditions. Setup time is always non productive and testing a program is non productive as well. It takes quite a bit of time and effort, even if a good part comes out of the first run, as it should. These activities are necessary and must be done, but doing too many 'first' parts for one batch is not productive either.

Generally, there are two groups of CNC programs, each having a different effect on program proving. The first group covers all programs that have *never* been used on the CNC machine. These programs must be proved for accuracy, as well as optimized for best performance. The second group covers the *repetitive* jobs - programs that have been used at least once before and have been proved to be correct in all respects. Programs in this group have most likely been optimized for the best performance under the given conditions. In both cases, the CNC operator must take a good care when running the first part of the batch. However, there are differences between a *new job run* versus a *repetitive job run*.

In either case, two qualities relating to the part program have to be established first:

❏ Setup integrity

❏ Program integrity

These two considerations are equally important - if only one of them is weak, the final result is not satisfactory. Always aim at the highest level in either category. Also keep in mind that the setup integrity has to be established again with each run in the future. The program integrity has to be established correctly only once.

◆ Setup Integrity

The machine setup is only a general description of the type of work actually done to get the CNC production going. The whole process covers the setup of the cutting tools, as well as the part setup and many related tasks. No single check list can ever cover all points that have to be considered during a CNC machine setup. The major look here is at the most important considerations, in a form of a brief check list. Adjust the individual points according to the machines and CNC systems in the shop. Adjust the list to reflect personal working methods and/or programming style. The main purpose of this check list, or any other for that matter, is to cover as many details as possible and not to omit an important item, operation, procedure, etc. Even a small omission may cause an accident and part damage or even a scrap due to a faulty machine tool setup.

Cutting Tools Check

❏	Are the tools properly mounted in holders
❏	Are the proper inserts used (radius, grade, chipbreaker, coating)
❏	Are all the tools the right size
❏	Are the tools placed in the proper magazine station
❏	Are the offsets set correctly (set zero to unused offset values)
❏	Is there an interference between individual tools
❏	Is the boring bar properly oriented (milling)
❏	Are all the tools sharp

Part Setup Check

❏	Is the part mounted safely
❏	Is the part properly oriented on the table (milling)
❏	Is the projection of the part from the chuck safe (turning)
❏	Is the part lined up for squareness (milling)
❏	Are the clearances sufficient
❏	Are all the clamps away from the cutting path
❏	Is the machine at its start (home) position before you press Cycle Start
❏	Does the tool change take place in a clear area

Control Settings Check

❏	Is the coordinate setting registered (for G54 to G59)
❏	Are all the offsets entered correctly
❏	Is coolant necessary
❏	What is the status of the BLOCK SKIP switch
❏	Is the optional program stop M01 active (ON)
❏	Is the DRY RUN off if the part is mounted
❏	Do you start with a SINGLE BLOCK mode set to ON
❏	Do you start with spindle speed and feedrate overrides set to LOW
❏	What is the status of MANUAL ABSOLUTE switch (if applicable)
❏	Has the position read-out on the screen been set from zero (origin preset)

Machine Tool Check

❏	Is the slide lubrication container filled with the proper type of oil (lubricant)
❏	Is the coolant tank filled
❏	Is the chuck and tailstock pressure set correctly (turning)
❏	Has the machine been zeroed before running a job - is the read-out set to zeros
❏	Is there enough pressure for the air attachments (air hose, etc.)

◆ Program Integrity

Any new and unproved program is a potential source of problems. In manual CNC programming, mistakes are a lot more common than in a CAD/CAM program. A good way to look at a new program is through the machine operator's eyes. Experienced CNC operators take a direct approach when running a new program - *they take no chances*. That does not mean the CNC programmer is not to be trusted - it simply reflects the fact that the machine operator is ultimately responsible for the expected quality of the work and is aware of it. He or she has a sense of great responsibility. Whether the damage to the part or even the scrap is caused by the program or for some other reason is a little consolation when the work is rejected.

What does the CNC operator look for in a new part program? Most machine operators would agree that the first and the most important thing is the *consistency* in programming approach. For example, are all tool approach clearances the same way as always? If not - is there a reason? Is the basic programming format maintained from one program to another program and from one machine to another? A good operator scans the written program *twice* - once on the paper copy, the second time when the program is loaded into the control system. It is surprising what can be seen on the screen that was not seen during the paper copy check. The reverse is also true. The common mistakes such as a missing minus sign or an address, a misplaced decimal point or a programmed amount extra large or extra small can usually be detected on the screen easier than on the paper. If using a computer for manual programming, print out the program and check it visually. Using a double check, many costly mistakes can be prevented. There is software available to graphically check the program on a computer, using simulation and file comparisons.

The consistency in a programming style is very important and cannot be overemphasized. Consistency is the major method to gain confidence of the CNC operator in the program integrity, time after time.

RUNNING THE FIRST PART

The CNC machine operator usually starts a new job by studying the documentation included with the program, mainly the drawing, setup sheet and tooling sheet. The next few steps describe the standard setup procedures that will vary occasionally, but generally, they will remain the same for most jobs.

❑   *STEP 1* - Set the cutting tools

This first step uses the tooling sheet or the tooling information from the part program. The CNC operator sets the cutting tools into their holders and respective tool stations and registers all tool numbers into the control memory. Make sure the tools are sharp and mounted properly in the holders.

❑   *STEP 2* - Set the fixture

The fixture that holds or supports the part is mounted on the machine, squared and adjusted, if necessary, but the part is not mounted at this point. Setup sheet serves as the documentation, particularly for complex setups. A fixture drawing may often be required as well.

❑   *STEP 3* - Set the part

Place the part into the fixture and make sure it is safely mounted. Check for possible interferences and obstacles in the setup. This step represents the end of the most initial steps of CNC machine operation.

❑   *STEP 4* - Set the tool offsets

Depending on the type of machine, this step takes care of setting the tool geometry and wear offsets, tool length offset and cutter radius offset, if applicable. One of the most important parts of this step is the setting of the work coordinate system (work offsets G54 to G59) or the tool position registers (G92 or G50), but *not* both. Work offset setting is by far the best and most convenient selection of modern CNC machine tool setup.

❑   *STEP 5* - Check the program

This step is the first evaluation of the part program. The part may be removed from the fixture temporarily. Since all offsets are already set in the control, the program is checked accurately, with all considerations. Program override switches on the control panel may be used, if required. Watch for tool motions in general and be sure to watch for tool indexes specifically. Repeat this step, if not absolutely sure with any aspect of the programmed tool path.

❑   *STEP 6* - Reset the part

If the part was removed in the previous step, now is the time to mount the part in the fixture again. The successful completion of all previous steps allows continuation with proving the first part. At this point, check the tooling once more, also check the oil and air pressure, clamps, offsets, switch settings, chucks, etc., just to be sure.

❑   *STEP 7* - Make a trial cut

An actual trial cut may be required in order to establish whether the programmed speeds and feeds are reasonable or not and if the various offsets are set properly. Trial cut is a temporary or an occasional cut that is designed to identify minor deviations in the offset settings and allows their change. Make sure the trial cut leaves enough material for actual machining. Trial cut also helps to establish tool offsets to keep dimensional tolerances within limits.

❑   *STEP 8* - Adjust the setup

At this point, any necessary adjustments are finalized in order to fine tune the program before production begins. This step includes final offset adjustment (usually a wear offset). It is also a good time to adjust spindle speeds and feedrates, if necessary.

❑   *STEP 9* - Start the production batch

A full batch production can start now. Again - a quick second double check may prove to be worth the time.

The ideal way to run a new program is to run it first through the control graphic display, if available. It is fast and accurate, and offers a lot of confidence before actual machining. This test can be done with a variety of override modes in effect, for example, *Machine Lock* or *Single Block*. Do not underestimate features such as *Zero Axis Neglect* and *Dry Run*, when testing a new program.

If using graphic options of the control system, most likely, there will be two kinds of graphic representation of the tool path:

❑   Tool path simulation

❑   Tool path animation

They have been described in the last chapter.

The first type of graphic representation, the *tool path simulation*, shows the outline of a finished part and the tool motions. The part outline is identified by a single color, the tool motions are identified by a dashed line (rapid motion) and a solid line (cutting motion). During program processing, the order of machining is shown on the display screen as either dashed lines or solid lines, depending on the motion type. The solid area of the part is not shown, neither are the tools, the chuck or the tailstock for lathe applications. With a color screen, the colors can be preset for each tool to further enhance the flexibility of the graphics display.

The more descriptive method of verifying the tool path prior to machining is *tool path animation*. In many respects similar to the tool path simulation, the tool path animation offers a few additional benefits. The part can be seen as a shaded form, rather than an outline only. The tool shape can be preset and seen on the screen display; the shape and size of the chuck can also be preset, as well as the outline of the tailstock, fixture, etc., all in a shaded form. The result is a very accurate representation of the actual setup conditions. As an additional benefit, the display is also proportional in scale. During the actual cut, the material can be seen as being 'machined', right on the screen. The tool path animation is a significant improvement over tool path simulation.

Do not expect 100% accurate display of the tool path. No graphic display can show every single detail and no simulation will show flying chips. What it does show is quite impressive, however. For CNC machining centers, the control with graphics can be set to one of several selectable views. More than one view can be set at the same time on the display screen, using a split screen method, also called *windows* or *viewports*. Many CNC operators run the graphic display twice, especially for milling systems - once in the XY view, the second time in the ZX or YZ view. Make sure the rapid motion display is turned on. The display can be run in a single step mode, areas that are either too small or especially critical can be enlarged (or reduced for large parts) with zooming features. Cutter radius offset, tool length offset and other functions can be turned on or off for the graphic display. Make sure the simulated conditions are as close to real conditions as possible. Also, do not forget to have all tools and offsets set *before* the program is tested. Unfortunately, this graphic option also adds to the overall cost of the control system and many companies choose not to purchase it.

Many programming instructions cannot be tested by using the graphics only. On most controls, there will be no clamps, no spindle speed or feedrates. Many other important activities cannot be seen, but what does show will make the actual cutting so much easier. Since all motions have been tested in the graphic mode of the control, all that has to be done during the actual run, is to concentrate on those details that could not be seen on the display. The tasks to be checked have been narrowed down and the program is easier to follow.

PROGRAM CHANGES

Even if a part program is proven, tested and the first part made and inspected, a good CNC operator looks at ways of improvement. Some improvements may be done immediately on the machine, before the whole job is completed. Some improvements may require a different setup, tooling or fixturing. Often, it would not be practical or even possible to implement those changes on the current job, but they should be applied the next time the job is done. Some changes to the part program are result of a design modification and have nothing to do with the program optimization. Others are strictly steps taken for the best productivity rating. Regardless of the reason, virtually any change required by the machine shop involves the CNC programmer who has to apply any new changes to the new program.

All changes to a program should be for the better, they should *improve* the program. Often a major change will require a complete program rewrite, but more likely, a program can be modified to a reasonable extent. When a program is changed for the better, it is said to be *optimized*, it is upgraded. That can be compared to another type of program change - *program update*.

◆ Program Upgrading

Upgrading a CNC program means to *strengthen* it, to *enrich* it, to make it *better* than it was before. It means to change it in a way that the cost of the part production is decreased. The cost reduction must be achieved with no compromise in quality of the part or machining safety.

The most common form of program upgrading (optimization) is minor changes to spindle speeds and feedrates. The process is called the *cycle time optimization*. Milling operations may require a different approach then turning operations. Jobs that are repeated frequently, as well as large size lots, should be scrutinized with even more care. Keep in mind that only *one second* saved on a cycle time will save one hour for each batch of 3600 pieces, half an hour for each 1800 pieces, and so on.

In the following check list are some major points to consider when optimizing a CNC program. The list is far from complete, but it should serve as a guide to what areas can be looked into and be explored. Some items in the list apply only to the milling operations, others only to turning. There are also some items that apply to both systems. Several of them require a special option of the control system or the machine tool to be available.

❑	Fine-tune the spindle speed and/or feedrate
❑	Choose the heaviest depth of cut possible
❑	Choose the largest tool radius possible
❑	Experiment with new cutting materials

❏	Rearrange tool order for faster tool changes
❏	Program bi-directional turret rotation
❏	Let one tool do as much work as possible
❏	Use the M01 rather than the M00 whenever possible
❏	Avoid excessive dwell times
❏	Eliminate 'air cutting' situations
❏	Shorten rapid motions where applicable
❏	Use multiaxis motion whenever safe
❏	Apply fewer passes for threading
❏	Look for block skip applications
❏	Avoid spindle direction change
❏	Shorten the tailstock travel distance
❏	Do not return to machine zero after each piece
❏	Program tool changes close to part
❏	Reassess the setup and/or design a new one
❏	Re-evaluate your knowledge and skills
❏	Consider upgrading the CNC system

This check list is a typical sample only, although developed from experience. Many more items can be added to this list and many item can be modified in their description. Even programs that are to be used only once should be carefully audited. There may be an improvement that can be applied to a different job, sometimes in the future.

◆ Program Updating

In contrast to program upgrading (optimization), the reason for program updating has nothing to do with decreasing the part cost. In the end, the part *may* cost less, due to a change in the engineering design or similar interventions, but not because of a program change. A program needs to be updated *after any change is made in the drawing that affects the CNC machining*. Even programs that have been previously upgraded may still have to be updated.

Engineering changes in part design are more common in companies that manufacture their own product line. In a job shop, the design changes are typically initiated by the customer, but have the same overall effect. The only difference is in the source and origin of the change.

A specific change that will affect the upgrade of a CNC program may be as small as a change in a single dimensional tolerance or as large as a complete part redesign. Personal experience may be somewhere between the two. An upgraded CNC program will reflect the magnitude of the change - whether it is a minor correction or a complete program rewrite.

◆ Documentation Change

The documentation that is associated with a particular CNC program is not much useful if it does not reflect the program changes done during part machining. Just like a well documented engineering drawing or other important data source, all revisions, updates, upgrades and many other changes should be recorded. Changes in mathematical calculations should be especially well documented and supplemented with formulas and sketches if possible. If there are several existing copies of the documentation, they too, should be replaced to make them current and up to date. The programmer's name, the nature of the change, the date, even the time of the day, should be used to indicate *when* such a change took place. Keeping the old version for reference (at least for a while) may also be a good idea. Sometimes one or two experiments may be necessary before deciding on the best documentation, on the final documentation suitable to particular needs.

ALTERNATE MACHINE SELECTION

Even with the best planning, things can go wrong, at least occasionally. What happens in a machine shop when the only CNC machine is suddenly out of commission? Of course, this never happens, except when a rush job is just about to be set up on that very machine. It usually happens when it is expected the least.

Every production manager has to have an alternate plan of action. One of the most common actions is to do the job on another CNC machine. Of course, such a machine has to be available, but there is more to consider.

Usually a job is programmed for a specific machine and a CNC system. If two or more such machines have been installed in the shop, the program can be executed on any one of them. Comparably, if two or more machines and/or controls are totally incompatible, programs are not transferable and a new program must be developed. The best opportunity for compromise exists if two machines are different in size, but with the same control type. The existing program may be usable as is, or with only very minor modifications.

The major considerations for alternate machine selection involve tooling and setup. First, the cutting tools and holders, as well as fixtures, must be available. Tools must all be the same size, even if holders are different. The part position on the table, clamps locations, data holes, clearance areas, etc., must also be the same. In addition to these general considerations, specific conditions such as spindle speed, feedrates, the power rating of the machine tool and other factors must also be carefully examined. The accuracy and rigidity of the alternate machine is also very important.

Those machine shops where many part programs have to be portable have adapted various standards for both programming and setup operations.

MACHINE WARM UP PROGRAM

Any precision equipment is guaranteed by its manufacturer to work accurately not only if it is handled properly, but also if it operates within a certain environment. Computers - CNC systems included - are particularly sensitive to rapid changes in temperature, humidity, dust level, external vibrations, etc. All potential hazards are clearly specified in the manufacturers' literature. Every CNC operator knows from experience that the part precision depends a great deal on the spindle temperature. Some ultra high precision machines even have an internal cooling system to keep the spindle temperature constant. In cold climates, on a cold morning in the winter, when the machine was sitting all night in an unheated shop, the experienced CNC operator turns the spindle on for a few minutes, to let it warm up. At the same time, in order to make the slide lubricant freely moving along the guide ways, the operator makes a few free motions in both directions of all axes. If this process is repeated every day in the winter months, it may be worth to automate it. A short program will do the job.

To write such a program is simple, but there are several important points to consider. First, make sure that the machine motions will always be in the area where there is no possibility of a collision. This program will be used with many jobs and modifying it every time a new job is set up is not an option. Another point to consider is the spindle speed in r/min. Avoid programming an excessively high r/min - a tool mounted in the spindle for the warm up could have a small or large diameter. To make the program to repeat itself indefinitely, use M99 function at the end. Program also function M30 for the program end, but with a block skip symbol [/]. When the warm up is to terminate, simply turn the block skip switch off. All machine motions will be completed and the program will end naturally.

The example O5001 is a typical warm up program for a milling system and uses English units. The program can be easily adapted to any other machine:

```
O5001 (WARM-UP FOR A MILL)
N1 G20
N2 G40
N3 G91 G28 Z0
N4 G28 X0 Y0
N5 S300 M03
N6 G00 X-10.0 Y-8.0
N7 Z-5.0
N8 S600
N9 G04 P2000
N10 X10.0 Z5.0
N11 Y8.0
N12 S750
N13 G01 X-5.0 Y-3.0 Z-2.5 F15.0
N14 X-2.0 Y-2.0
N15 Z-2.0 S800
N16 G04 P5000
N17 G28 Z0 M05
```

```
N18 G28 X0 Y0
/ N19 M30
N20 G04 P1000
N21 M99 P5                          (REPEAT FROM BLOCK 5)
%
```

The example is simple in structure, yet well thought out. There are several intentional programming techniques contained in the sample program:

❏   The whole program is in the incremental mode

❏   The first motions are to the machine zero

❏   The Z axis motion is the first motion

❏   Spindle speed is increased gradually

❏   Dwell is used to lengthen the current action

❏   The end tool motion is to the machine zero

❏   The end of program M30 is 'hidden' by
     a block skip function

❏   Each repetition of the program starts at block N5

Several program versions can be developed, depending on the machine and the type of work expected on that machine. For example, if developing a warm up program for a CNC lathe, incorporate functions that are typical to a CNC lathe - for example, changing the gear range, moving the tailstock in and out, opening and closing the chuck jaws, doing a tool change, etc. On a horizontal machining center, include the indexing table motion; on a boring mill, the in and out spindle quill motion may also be programmed. Modify the program to suit any particular purpose, but keep in mind its goal - to warm up a machine that had been idle for a relatively long period of time in a cold temperature. Also keep in mind the safety of operations - the goal is *a generic program for a specific machine type*, a program that can be used with all jobs, without modifications.

CNC MACHINING AND SAFETY

Machine shop safety is everybody's responsibility. Some basic safety issues have already been introduced in the first chapter of this handbook. The programmer has to apply safety at the programming level, the operator at the CNC machine, and so on. Many companies have established numerous safety rules and procedures that work well. Follow them and try to improve them.

Generally, the safety concerns of a machine operator are almost the same as those operators running conventional equipment. Safety starts with a clean work place and organized approach to programming, setup and machining. Many do's and don'ts can be itemized, but no list will satisfy all the safety concerns. Here is an attempt at a typical list of safety concerns in CNC shop. There are several general groups in the incomplete list. Many suggestions can be in different groups.

Personal Safety

- Wear suitable clothing (tucked-in shirt, buttoned-up sleeves)
- Remove watches, rings, bracelets, and similar jewelry before machine operation
- Keep long hair under a net or tied up
- Protect your feet by wearing approved safety shoes
- Protect your eyes - wear approved safety glasses with protective side shields at all times
- Wear an approved safety helmet if that is the company policy
- Always protect your hands - never reach towards the part while the spindle is rotating
- In some cases, protection may also be needed for head and ears, perhaps even nose
- Never remove cutting chips by hand, with or without gloves on
- Do not use rags or gloves around moving or rotating objects
- When lifting heavy objects, ask for help, use a crane or do not lift

Machine Environment Safety

- Make sure the floor is swept, free from oil, water, chips, and other hazards
- Check the walkways, so they are not blocked from any direction
- See whether all the material is safely stored and finished parts are in proper containers

Machine Tool Safety

- Do not remove guards and protective devices
- Read and follow operating manuals
- Check fixtures and tools before they are used
- On the machine, make sure all the tools are tight in the holders, that the tools are sharp and selected properly for the job on hand
- Stop all machine motions when measuring or inspecting finished work
- Do not leave objects on top of machines
- Use only a suitable coolant mixture, and keep the coolant tank clean at all times
- Never use a file for breaking corners or a sand paper for surface polishing during the program execution
- Deburr sharp edges before handling a part
- Stop all machine power for maintenance
- Do not operate a faulty machine

- Do not alter design or functionality of the machines or controls
- Electrical or control maintenance should be done by authorized personnel
- Do not use a grinding machine near the CNC machine slides
- Do not use a welding equipment on CNC machine under power
- Behave responsibly - do not engage in pranks and horseplay around machinery

These are only some common sense suggestions, not a comprehensive list for CNC machining safety.

> Always observe the company safety policies, as well as safety laws of a particular jurisdiction

SHUTTING DOWN A CNC MACHINE

When the CNC machine is not used for an extended period of time, it should be shut down. Many users assume that shutting down a CNC machine means just to turn the power off. There is more than that to shutting down a machine tool with a power switch.

◆ Emergency Stop Switch

The purpose of the emergency switch is to *stop all machine motions immediately*, regardless of the current operational mode. When pressed, it will lock in place and must be rotated manually in the opposite direction to release. It should be used sparingly and only in real emergencies, such as when:

- An imminent situation that is unsafe to the human being is about to occur
- An imminent collision of the machine tool elements is about to occur

In certain situations, it is possible to cause damage to the machine and tooling when pressing the *Emergency Switch*. Depending on the machine design, there may be several emergency stop switches available, located at convenient places. The CNC operator should always know the locations of each emergency stop switch. Emergency switch is also called the *E-switch*.

> **WARNING!**
>
> Although the emergency stop switch disconnects all power to the machine axes, the electrical power is still supplied to the CNC machine.

For a complete safety shut-down, always follow proper procedures as enacted by company policies.

When the *Emergency Stop* switch is released or unlocked, the machine does not restart automatically. The machine setup conditions and other conditions have to present before the automatic start can be selected. This condition is usually achieved by pressing the *Power On* switch.

◆ Parking Machine Slides

Several chapters have mentioned a comment that a CNC program cannot be executed unless the machine had been zeroed first. Recall that zeroing the CNC machine while the machine slides are at - or almost at - the machine zero, is impractical and may result in an overtravel. The machine zero return needs about *one inch* minimum (or 25 mm), to be *away* from the machine zero position in each axis. This position is often easier to reach at the *end* of work than at its beginning. A practical CNC machine operator knows that to shut off the machine when the slides are at the machine zero position causes the subsequent start up to take a little more time.

To avoid any potential problems in the future, some programmers make a small program to bring the machine slides into a safe position at the end of work, before the power is turned off. Although the idea is good, the solution to one problem may cause another problem. If the machine slides are 'parked' repeatedly at the *same* position for a lengthy period of time, various dirt deposits will collect *under* the slides, possibly causing staining or even rusting in and around the 'parking' area. A better way is to let the CNC operator do the positioning of the slides manually. It does not take any more time and the slides will never be too long at any one position. All that is needed is a motion of one axis at a time, to a different position every time. Since it is done manually, there is a better chance that the machine position will be always different.

◆ Setting the Control System

Control panel of the CNC unit has many switches set to a certain state at the time of a shut down. Again, variations exist as to what is the proper procedure, but a good CNC operator will leave the control system in such a state that it does minimize a potentially dangerous situation, when used by the next person. Here are only some possibilities to apply before leaving the control system for a break, or a complete shut down:

❑ Turn down the feedrate override switch to the lowest setting

❑ Turn down the rapid override switch to the lowest setting

❑ Set mode to JOG or HANDLE

❑ Set the handle increment to X1

❑ Set the Single Block switch ON

❑ Set the Optional Block switch ON

❑ Set operation mode to MDI

❑ If available, remove the Edit key from the lock

Several other precautions could be also be used, but the ones listed are the most typical and should ensure reasonable safety precautions.

◆ Turning the Power Off

Procedures vary from one machine to another, so always consult the machine manual first. However, there are some procedures pretty common to all machines. General rule is to *reverse* the procedure of turning the power on. For example, if the procedure to turn the power on is

1. Main switch on
2. Machine switch on
3. Control switch on

then the power off procedure will be

1. Control switch off
2. Machine switch off
3. Main switch off

Note that in either case, there is no one switch to do all work. This is for the safety of the sensitive electronic system of the CNC unit. Also check the exact function of the emergency switch (described earlier), as it relates to the machine shut down procedure.

EQUIPMENT MAINTENANCE

To maintain a CNC equipment is a professional discipline of its own. In general, it is better to leave any kind of maintenance to qualified technicians. The CNC machine operator should only be concerned with the basic preventive maintenance, just by taking care of the machine in general. Modern control systems require very little maintenance, usually consisting of the air filter change and similar simple tasks.

The manufacturer of the CNC unit and the machine manufacturer supply reference manuals, including special ones for maintenance, with their products. These publications should be a compulsory reading for any person involved with maintaining machine tools in working order, electrical, electronic, or mechanical. Many machine manufacturers, and even dealers, also offer training courses in maintenance and general troubleshooting.

51   *INTERFACING TO DEVICES*

A completed CNC program, debugged and optimized for the best performance, should be stored for *future use* or *reference*. Before such a program can be stored, it must be first loaded into the CNC memory, tested and optimized. There are many ways of loading the completed part program into the CNC memory. The most basic, and also the most time consuming, method is to simply key in the program at the machine directly, using the control panel and the keyboard. Without a doubt, this is also the least efficient method, prone to errors. It is true, that Fanuc controls offer a feature called *Background Edit*. This is a standard feature on the majority of controls that allows the CNC operator to key in (and/or edit) one part program, while the control runs the machining operations for another part program. In practice, however, the majority of operators simply don't take advantage of this feature for various reasons.

In order to load a part program into the CNC memory or unload a program from the CNC memory, a hardware connection called a *data interface* is needed. An interface is usually an electronic device that is designed to communicate with the computer of the CNC unit.

Typical interfaces and storage media are:

❏   Tape reader and tape puncher

❏   Data cassettes

❏   Data cards

❏   Bubble cassettes

❏   Floppy disks

❏   Hard (fixed) disks

❏   Removable devices

❏   ROM (read-only-memory) devices

❏   ... and others

Many of these devices are proprietary, many require not only a special cabling, but also a software drivers that can run these devices. The focus of this chapter will be on the connections that can be easily assembled and those that use standard configurations. There is one industrial standard most of these devices have in common - a standard called an *RS-232C* interface. Well - almost a standard. There is a number of variations that follow the standard in principle, but deviate from it to some extent. This handbook is not an in-depth discussion of CNC communications, it only does an overview of the standard as a guideline, not as a solution to all CNC communications.

RS-232C INTERFACE

Data transfer between two electronic devices (computers and controls) requires a number of settings that use the same rules for each device. Since each device may be manufactured by a different company, there must be a certain independent standard that all manufacturers adhere to. The RS-232C is such a standard - the letters *RS* stand for *'Recommended Standard'*. Almost every CNC system, a computer, a tape puncher and tape reader, has a connector (known as a *port*) that is marked RS-232C or similar. This port exists in two forms, one with a 25 *pin* configuration, the other with a 25 *socket* configuration. The one with the pins is known as the DB-25P connector, the one with the socket as DB-25S connector (male/female respectively). *Figure 51-1* illustrates the layout.

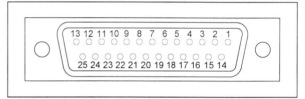

Figure 51-1

Typical 25-pin RS-232C port - DB type

The RS-232C port on the CNC unit is usually a standard feature and uses the DB-25S type (the letter S means it is a socket type). An external computer, usually a desktop computer or a laptop, together with a suitable cable and a communications software is also needed to transfer CNC programs. The external devices use mainly the DB-25P type connector (the letter P means it is a pin type). The price tag for such a setup (hardware and software) is well below the cost of any suitable alternative. It is also a very convenient method. The CNC program is sent to the system memory and is stored there as long as needed to run the job. The CNC operator usually makes some changes and when the job is completed, all changes that are to remain permanent are sent back to the desktop computer or a laptop computer and stored on the hard disk. This method works well with a single CNC machine as well as several machines.

Although terms such as *Transmit* (or *Send*) and *Receive* are more common in software, even the latest CNC systems use the terms *Punch* (which is equivalent to *Send*) and *Read* (which is equivalent to *Receive*). These terms go back to the days of punched tape.

To make this very popular method of communications work, only a suitable cable has to be installed between the computer port and the CNC system port. Loading and configuring a communications software that runs the complete operation also has to be done first. In addition, both devices must be set in a way they can 'talk' to each other.

Later in this chapter will be a few notes relating to the basic principles of using a personal computer as an interface with the CNC system. First, a short look at the original interface device - the venerable punched tape - as a media used for many years but rarely used anymore.

PUNCHED TAPE

Since the beginning of the numerical control technology, a punched tape has been the primary media for sending the part program instructions to the control system. In the late 1980's, the punched tape has lost almost all its splendor and has been replaced by desktop and laptop computers loaded with inexpensive software.

A punched tape is fragile and often bulky. It can get dirty easily, but it had been very popular. It is economical to use and is still available (although the price per roll could be high). The majority of new CNC machines do not have tape reader any more. Used older machines may have it. Many of these old controls accept tape only as an input device, not to run the job from the tape. The tape only loads the CNC memory. Changes to the program can be done through the CNC and a corrected tape may be punched out later.

◆ Tape Reader and Puncher

One of the original facilities for data transfer was a tape reader built into the old NC and CNC machines. Its function on a CNC machine is quite different than on the early non CNC equipment. Rather than using the tape reader as the source for running the program, the tape reader on a CNC machine is used to *load* the program stored on a paper tape into the system memory. Once loaded, the program is executed from the memory, in the *Memory mode* setting, and the paper tape is no longer needed. There is one great weakness with this method. Working on a CNC machine often means some inevitable changes to the program *after* it had been loaded. Since these changes cannot be readily reflected on the tape, there can be confusion at a later date, possibly when the job is repeated. This is an organizational problem and can be resolved relatively easily.

One option is to make all the necessary changes and corrections on the CNC unit, then punch out a new tape, using the RS-232C port. The difficulty of this approach is that while a built-in tape reader was common, a built-in tape puncher was virtually non-existent. A significant amount of money had to be spent on an external portable tape puncher, that usually incorporates the tape reader anyway, and causing duplication.

Modern machine shops do not use tapes, tape punchers and tape readers of any kind. These once powerful tools have been replaced by the inexpensive microcomputer technology and inexpensive communication software.

Even if the punched tape technology is obsolete by any modern standards, it may justify a short sideline for those who still use it and also for those who are interested in the 'historical' aspects of numerical control.

Tape Media

Punched tape is the oldest media for storing programs. The tape is made of good quality, enforced paper. The punched tape is 1.0000 inch wide (25.4 mm) and about 900 feet long (about 274 meters) in a single roll, manufactured to exact standards. The most useful descriptions and dimensions are illustrated in *Figure 51-2*.

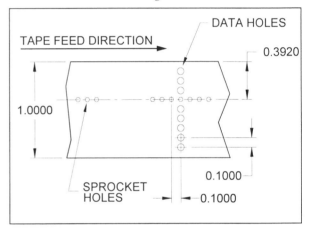

Figure 51-2

Punched tape detail - basic dimensional standards

Paper tape is generally available in black color, for its 100% opacity was required by most tape readers. Tape materials other than paper are also used, such as Mylar®, which is a paper tape sandwiched between two layers of plastic. The plastic makes the tape stronger, which is an important factor when the tape is to be used constantly, for example, in a reel-to-reel operation. An aluminum - or metal - tapes are also available. Both types of tape, Mylar® and metal, are relatively expensive and are used only for critical jobs and long programs. This was typical to the aircraft, defense, nuclear and mold industries. Many companies also used the durable tape material for storing programs that have been proven on the machine.

The punched tape is generally available in a roll form, although a folded strips version may still be available. It has two main purposes:

❑ To store the program data for use at a later date

❑ To serve as a media for transferring the program data into the control system via a tape reader

Tape Coding

A punched tape consists of a series of holes, laid across the tape width, where each row represents one character of the program - *a character is the smallest unit of input*. The punched characters are transferred through the tape reader to the control system in a form of electric signals. Each character can be composed of up to eight signals, represented by a unique combination of holes punched across the width of the tape in .1000 (2.54 mm) increments. A character can be any capital letter of the English alphabet, any digit, plus some symbols, such as a decimal point, minus sign, slash, and others.

ISO and EIA Tape Format

When preparing the tape, try to understand two methods of standard tape coding - one, which employs the *even* number of punched holes, and the other, that uses the *odd* number of punched holes. The technical terms for these two systems are *Even Parity*, when a character is composed of 2, 4, 6 or 8 punched holes, and *Odd Parity*, when the character is composed of 1, 3, 5 or 7 punched holes. There is also coding that is a mixture of the two, called *No Parity*, that has no application for the machine tools. For illustration of a partial tape coding, see *Figure 51-3*.

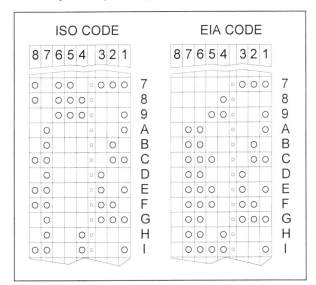

Figure 51-3

Tape coding standards
Even parity (ISO) on the left, odd parity (EIA) on the right

Even parity of the punched tape corresponds to the International Standards Organization coding, called ISO in a common abbreviation, formerly known as the ASCII code (*American Standard Code for Information Interchange*). Odd parity is the standard of the Electronic Industries Association, EIA in short, that is slowly on the decline, mostly due to the limited number of available characters.

The even parity format ISO is also known as the standard DIN 66024 (ISO) or RS-358 (EIA) or ISO code R-840. The odd EIA format is the standard number RS-244-A.

Most modern numerical controls, providing they have a tape reader interface, will accept either tape coding automatically, based on the parity of the *first end-of-block* character punched on the tape.

Parity Check

While punching a tape, make sure that the process is consistent for the whole length of the program tape. Mixing ISO and EIA codes on any one tape will result in a rejection by the control tape reader. Such a fault is normally called a *parity error*. The system check for correct parity is automatically performed by the control unit, when the punched tape is loaded into the CNC memory or processed in a reel-to-reel operation. The control will check for the occurrence of *odd* characters in an ISO tape and the occurrence of *even* characters in an EIA tape. The purpose of such a check is to detect malfunction of the punching or reading equipment, which can be very costly if it causes a character of one coding to become a character of the other coding.

Control In and Out

On ISO tapes (even format), a pair of punched codes representing *parenthesis* identifies a section that is *not* to be processed by the control system. Whatever information is contained between the parenthesis will be ignored by the control. This is a section that may include program comments; they will appear in the hard copy printout, but will not be processed when the tape is read.

Blank Tape

Blank tape is the tape purchased and is completely free of any holes. Often, it may be overprinted with directional arrows, to indicate the feeding direction or the top of tape. The new blank tape is sometimes called a *virgin tape*.

Blank tape can also be one that has only sprocket holes punched but no holes representing individual program characters. The sprocket holes are small size holes, located between the third and the fourth channel of the tape. Blank section of a tape is used at the beginning (leader) and at the end (trailer) of a punched tape, to make it easier to handle. The blank section also provides protection to the coded section when the tape is stored rolled up.

Significant Section

The section of punched tape that contains the program data is often called the *significant data section*. Another term used in conjunction with the significant data section is a *label skip function*. It means that everything up to the first EOB (end-of-block) character, that is punched on the tape will be ignored. That means the significant data section of a tape is the section following the first EOB character.

The first occurrence of a carriage return (caused by the *Enter* key on a computer keyboard) is the first occurrence of the end-of-block character. This signal identifies the beginning of the *significant data section* - section where the actual program is stored. The significant data section is terminated by a stop code, identified usually by a percent sign, acting as the *end-of-file* character. When the stop code is read by the reader, tape reading is completed. That is why no information is ever placed past the percent sign.

◆ Leader and Trailer

The *blank* section of a punched tape is used as a leader and a trailer. The blank section preceding the coded program data (significant data section) is called a *leader*, the section following the data is called a *trailer*. The suitable length of the leader or the trailer is usually about 10 inches (250 mm) for memory operation (without reels), but should be about 60 inches (1500 mm) when the tape is on reels. For smaller diameter reels, the leader and trailer section can be shorter than for large reels. Sometimes the length of the leader section must be extended to allow space for tape identification. Stickers or bright pencils can be used to supply information about the tape in its leader section.

◆ Tape Identification

Each punched tape should be identified as to its contents. Hand written data, adhesive labels or readable characters can be used within the leader section of the punched tape. Adhesive labels may not be a good choice because of their tendency to peel and fall off. Hand written notes may present difficulty when writing on a black background. The identification usually contains the program or tape number, drawing number and the part name - other information may also be included.

So called readable characters - *Figure 51-4* - seem to be the best solution, since they can be generated on the majority of tape preparation equipment.

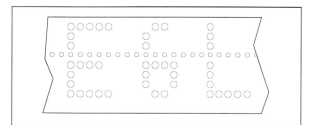

Figure 51-4

Example of readable characters on a punched tape

These special characters are actual punched holes representing real characters, namely letters, digits and symbols, rather than tape codes. An end-of-block character or the stop code may not be used in the readable section, if that section will go through the tape reader.

◆ Non-printable Characters

Most program characters stored on a punched tape will print normally. They are called the *printable* characters and include all capitals A to Z, the numerals 0 to 9, and most symbols. Although alpha numerical characters are printable, these symbols cannot be printed:

❑ Stop code in EIA format

❑ Delete character

❑ Carriage return (or Enter key)

❑ Line feed

❑ Tab codes

One character appears on the display screen as a semicolon (;). This is a symbol for the end-of-block character and is never written. It is a control system *representation* of the carriage return in the part program.

◆ Storage and Handling

Paper tape is punched in a tape puncher. Punchers come with only the basic features, some have advanced features such as keyboard, printer, tape reader, setting switches, Input/Output ports, etc. Additional equipment, such as a tape winder, splicer, digital tape viewer, etc., is also available.

Storage of tapes requires a fair amount of space which increases with more tapes. Tapes are normally stored in plastic boxes, small enough to fit in specially designed metal cabinets with dividers. Tapes can be transferred into computer files to save space and expensive cabinets.

If still using paper tapes, handle them carefully by the edges only. Insist on the same treatment by the operator and others. Take a special care for paper tapes, particularly when they are manipulated by winding or unwinding. In order to prevent curling, the tape should never be wound into a small tight roll, which is very tempting for saving storage space. Heat and direct sunlight are also enemies of the tape, as is water. A reasonable amount of moisture keeps the tape from becoming too dry.

Tapes can be damaged if placed into the tape reader incorrectly. Long tapes require more care than short tapes. Grease and dust are the worst enemies of paper tapes and should be guarded against. Any tape that is to be used many times over, should be duplicated or even triplicated.

DISTRIBUTED NUMERICAL CONTROL

The Input/Output (I/O) port RS-232C on a CNC machine is used to send and receive data. The external sources are usually a hard disk or a paper tape. In many shops, programs are transferred through the means of DNC, which means *Distributed Numerical Control*. The control has features available to make data transfer possible.

To communicate between one CNC machine and one computer using the RS-232C port, all equipment required is a cable between the two devices and a software. To communicate with two or more machines, using the same single RS-232C port, each machine must be connected to a split box with a cable. The split box is available with two or more outlets, selectable by a switch. This is the simplest form of DNC. It requires well organized procedures to make it work efficiently. DNC is not a part of the control unit and is not covered here. Commercial DNC packages are available at various levels of sophistication and cost.

Some DNC software also allows a useful feature called 'drip-feeding', which is a method used when the program is too large to fit into the CNC memory.

TERMINOLOGY OF COMMUNICATIONS

Communications have their own terminology. There are many terms, but five terms are commonly used in CNC:

❏ Baud Rate

❏ Parity

❏ Data Bits

❏ Start Bit

❏ Stop Bit

◆ Baud Rate

Baud rate is the data transmission speed. It is measured as the amount of data bits per second, written as *bps*. Baud rates are only available in fixed values. Typical rates for older Fanuc controls are 50, 100, 110, 200, 300, 600, 1200, 2400, 4800 and 9600 bps. Modern controls can have the baud rate set to 2400, 4800, 9600, 19200, 38400, 57600 and 76800 bps. In terms of time, the higher the rate, the faster the transmission. Single data bit transfer rate will be the result of one divided by the baud rate:

$$S_b = \frac{1}{B}$$

☞ where ...

S_b = Time required to transfer a single bit in seconds
B = Baud rate in seconds

A single bit transferred at 300 bps will take 0.03333 of a second, but a single bit transferred at 2400 bps will take only 0.00042 of a second. In practice, it takes about 10 bits to transfer one character (see *Stop Bits* section below), so at 2400 bps setting, the transmission will be at a rate of about 240 cps (characters per second). 4800 bps is a good setting once everything is working well. Higher settings are necessary for 'drip-feed' methods.

◆ Parity

Parity is a method of checking that all transmitted data were sent correctly. Just imagine what would happen if some characters or digits of a CNC program were not transferred correctly or not transferred at all. Parity can be *even, odd,* or *none,* and *even* is the most common selection for CNC communications.

◆ Data Bits

A *bit* is an acronym for **B**inary dig**it**, and is the smallest unit that can store information in a computer. Each binary digit can have a value of either one (1) or zero (0). One and zero represent the ON and OFF status respectively, so a *bit* is something like a toggle switch that can be turned on and off as needed. In the computer, every letter, digit, and symbol used in the CNC program is represented by a series of bits, eight bits to be precise, that create a unit called a *byte*.

◆ Start and Stop Bits

To prevent loss of data during communication, each byte is preceded by a special bit called the *start bit*, which is low in voltage level signal. This signal is sent to the data receiving device and informs it that a byte of data is coming next.

A bit similar to the start bit, but at the *end* of the byte, has exactly the opposite meaning. It sends a signal to the receiving device that the byte has ended or stopped being transmitted. This bit at the end of a byte is called the *stop bit*. Because the start and stop bits go together, they are often teamed up together as the stop bits and set the devices to *two stop bits*.

Many terms exist in communications. With growing interest, this is a very rich field to study.

DATA SETTING

The data used for communications must be set properly before the data transfer can begin. The setting at one end (computer or the CNC system) must match the setting at the other end. For baud rate, consult the machine manual - a good start is at 2400 bps. Newer models have a higher default. Typical software setting is done through the configuration at the computer end and through the CNC system parameters at the CNC end. Settings at both ends must match. Typical Fanuc settings are:

❏ 4800 bps baud rate

❏ Even parity

❏ 7 data bits (seven data bits)

❏ 2 stop bits (two stop bits)

Proper connection depends mainly on the configuration of the connecting data cables.

CONNECTING CABLES

The most common cable for communication between a CNC machine and a computer is a shielded and grounded cable, containing several small wires (at least eight), each one enclosed in a colored plastic sleeve. The purpose of making a communication cable is to connect the CNC port (usually 25 sockets) with the computer port (usually 25 pins), using a properly configured cable. Always use a cable of high quality. Shielded cables can reach farther distances and are generally better choice to withstand interferences during data transmission. Wires are identified by their gauge value, for example a 22-gauge or a 24-gauge wire is a good choice for communications.

The 25-pin port has each pin or socket numbered (see the first page of this chapter) and the individual wires of the cable have to be connected to proper numbers at each end. It is quite common to 'cross' the wires between each end. Typical crossing would be between the pin number 2 and the socket number 3, and a pin 3 and socket 2. Some numbered positions have to be connected *at the same end* of the cable. This is called 'jumping'.

◆ Null Modem

A very common cable wiring that is used in general communications is called a *null modem*. The connection of the two ends follows a certain standard, shown in *Figure 51-5*. Each number represents the pin or the socket on the DB-25 connector. Note the jumps between connections 6 and 8 at both ends. *Figure 51-6* shows the same null modem configuration in a graphic way. This is a very popular method showing cable configurations.

PIN DB-25P	SOCKET DB-25S
1	1
2	3
3	2
4	5
5	4
7	7
6 and 8	20
20	6 and 8

Figure 51-5

Null modem pin connections

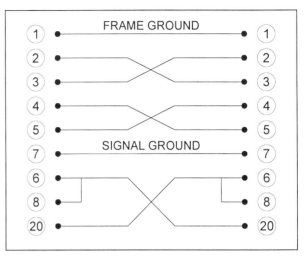

Figure 51-6

Graphic representation of null modem connections

◆ Cabling for Fanuc and PC

As the most common communication will be between a Fanuc control and a desktop computer or a laptop, *Figure 51-7* illustrates a typical cable configuration. Note the similarity to the null modem configuration.

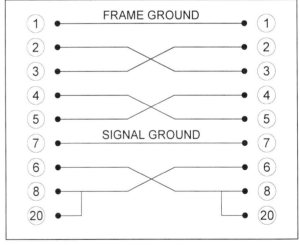

Figure 51-7

Typical cable configuration for Fanuc controls

Regardless of what cable configuration will be used, a good communication software that will run the whole operation is also needed. Some companies use a software specially designed for CNC work, others purchase very inexpensive communications software, sold by the majority of computer stores.

Math in programming - the single word 'math' often appears to be so powerful that it strikes a weak chord in many programmers. It is surprising how many new programmers, manual programmers in particular, are afraid of the often numerous calculations associated with CNC programming. This fear is really not substantiated. Let's look very briefly at what kind of mathematical knowledge is really necessary to handle typical programming calculations for manual program preparation.

First, the basic arithmetic functions - *addition, subtraction, multiplication* and *division* - are at the core of any mathematical activity. Going a bit further, the knowledge of common algebraic functions is definitely useful, mainly *square roots* and *powers of a number*.

Second, since CNC programming is based on the relationship of points within a system of rectangular or polar coordinates, a good knowledge of *basic geometry* is also imperative. The scope of this knowledge should cover understanding many principles of *angles*, the concept of *degrees* and their subsets, tapers, polygons, properties of an arc and circle, the *pi* constant (π), and other associated topics. Knowledge of *planes* and *axial orientations* is important in many cases as well.

Without a doubt, the most important part of geometry, one that absolutely *must* be mastered, is the solution of *right angle triangles*, using *trigonometric functions*. Very seldom there will be a problem or calculation that will require a solution using oblique triangles, although these problems may arise.

> The knowledge of trigonometry is essential
> to any serious CNC programming.

Most difficulties in solving trigonometric problems are not as much in the ability to use a specific formula and solve the triangle - but in the *inability to see* the triangle to be solved in the first place. Often, programming involves a drawing that is very complex in terms of geometrical definitions of the part. Such a drawing will have so many elements, that overlooking the obvious is possible, even likely.

Any specific knowledge of analytic and spacial geometry is not really required for a 2 and 2-1/2 axis work, but it is essential for a work in all three axes, particularly for complex surfaces, 3D tool path and multi surface machining or surface manipulation. However, this kind of programming is not done without a computer and CAD/CAM software.

There are several specific mathematical subjects to learn and to know in depth. All of them have been selected only for their importance in CNC programming and are described here in the necessary detail.

BASIC ELEMENTS

◆ Arithmetic and Algebra

The subject of *arithmetic* deals with handling numbers involving the four basic operations:

- ❑ Addition
- ❑ Subtraction
- ❑ Multiplication
- ❑ Division

Algebra is an extension of arithmetic and deals with handling numbers in terms of equations and formulas. Typical usage will involve:

- ❑ Square roots
- ❑ Powers of a number
- ❑ Trigonometric functions
- ❑ Solving formulas and equations
- ❑ Variable data

In algebra, typical work involves several known values and one or two unknown values. Using various formulas and equations, unknown values can be solved (calculated) to achieve the desired result.

◆ Order of Calculations

In the field of mathematics, there is a precisely defined order in which the calculations are performed. Every electronic calculator is based on these centuries old rules. In a combination of various algebraic operations, the order of calculations will follow these rules:

- ❑ Multiplications and divisions are always calculated first
- ❑ Additions and subtractions follow, order is not important
- ❑ Any roots, powers to a number, and operations within parentheses are always calculated before multiplications and divisions.

The following calculation will have the same result with or without parentheses:

```
3 + 8 × 2 = 3 + (8 × 2) = 19
```

The multiplication is always performed first, regardless of whether it is enclosed in parentheses or not. If addition must be done first, it *must* be enclosed within parentheses:

```
(3 + 8) × 2 = 11 × 2 = 22
```

These two examples show that an innocently looking small omission may have significant consequences.

GEOMETRY

For all practical purposes, there are only three entities in the engineering drawing:

❏ Points

❏ Lines

❏ Circles and Arcs

Points have no parts and are represented by the XY coordinates in a 2D plane or by XYZ coordinates in 3D space. Points are also created by an intersection of two lines, two circles or arcs, and a line and a circle or arc.

Point is also created by a line tangent to a circle, line tangent to an arc, a circle or an arc tangent to another circle or an arc.

Lines are straight connections between two points creating the shortest distance between the points.

Circles and *Arcs* are curved elements that have at least a center and a radius.

Other elements such as *splines* and *surfaces* are too complex for manual programming, although they are also based on the same fundamental elements.

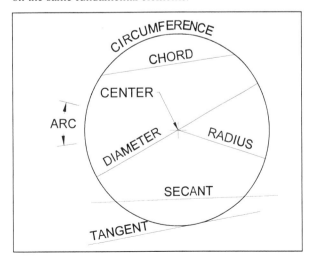

Figure 52-1

Basic elements of a circle

◆ Circle

Circle is mathematical curve, where every point on the curve has the same distance from a fixed point. This fixed point is called a *center*.

Several terms are directly related to a circle - *Figure 52-1*:

❏ CENTER - is a point from which a circle or an arc is drawn with a given radius.

❏ RADIUS (radii in plural) - is a line from the center to any point on the circumference of the circle.

❏ DIAMETER - is a line through the center between two points on the circumference of the circle.

❏ CHORD - is a straight line joining any two points on the circumference of the circle.

❏ ARC - is any part of the circle between two points on the circumference of the circle.

❏ CIRCUMFERENCE - is the length of the circle (length of the line that bounds a circle)

❏ TANGENT - is a point where a line, an arc or another circle touches the circumference of the circle but does not cross it. This point is known as the point of tangency.

❏ SECANT - is a straight line that passes through a circle and divides it into two sections.

Two area sections of a circle have their own names. They are called the *sector* and the *segment* of a circle, and are shown in *Figure 52-2*:

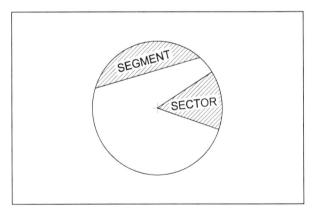

Figure 52-2

Segment and sector of a circle

❏ SECTOR - is an area within a circle formed by two radii and the arc they intercept

❏ SEGMENT - is an area within a circle formed by the chord and its arc

Neither the sector nor the segment of a circle play any significant role in CNC programming.

◆ PI Constant

PI is a Greek letter used in mathematics to represent the ratio of the circle circumference to the circle diameter. Its symbol is π, it is pronounced 'pie', and has the value of *3.141592654....*, and regardless of how many decimal places will be used, it will always represent only an approximate value. For programming purposes, use the value returned by a calculator or computer, usually with six to nine decimal places. In both cases, the internal value is a lot more accurate than the displayed value. In many cases, the rounded value of 3.14 is sufficient for most results.

◆ Circumference of a Circle

The length of a circle - or its *circumference* - is seldom needed for programming and is included here only to enrich the general theory. It can be calculated from the following formula using the *pi* constant:

$$C = 2 \times \pi \times r$$

or

$$C = \pi \times D$$

☞ where ...

C = Circle circumference
π = Constant 3.141592654...
r = Circle radius
D = Circle diameter

◆ Length of Arc

The length of an arc is also a rare requirement and can be calculated from the following formula:

$$C = \frac{2 \times \pi \times r \times A}{360}$$

☞ where ...

C = Circle circumference
π = Constant 3.141592654...
r = Circle radius
A = Arc angle

There are two other very important calculations relating to a circle. They are used in programming very often and should be understood well. One is based on the *chord* of a circle, the other on the *tangency* of a circle. As both calculations require the knowledge of trigonometry, they will be described later in the chapter.

◆ Quadrants

Quadrant - is the part of a circle formed by the system of rectangular coordinates, described in *Chapter 4*, where the axes pass through the center of the circle. There are four equal quadrants in a circle, identified by Roman numerals I, II, III and IV, starting at the upper right quadrant along the counterclockwise direction - *Figure 52-3*.

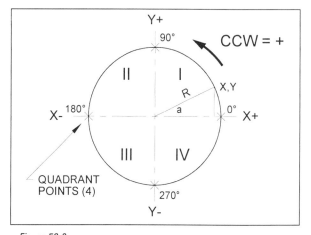

Figure 52-3
Quadrants of a circle and the mathematical definition of angular direction

Each quadrant is exactly 90°, crossing at circle quadrant points. Therefore, a circle has the sum of all four angles equal to 360°. Angles are counted counterclockwise as positive, starting from zero degrees (0°).

Individual quadrant points (also known as the cardinal points) are often compared to a hand direction on the face of an analogue clock or as a direction of a compass pointer. 0° is arbitrarily located at the equivalent position of 3 o'clock or East direction, 90° at 12 o'clock or North direction, 180° at 9 o'clock or West direction, and 270° at 6 o'clock or South direction - *Figure 54-4*.

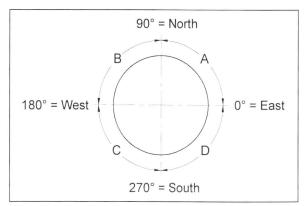

Figure 52-4
Angles and quadrants - 0° is East direction or 3 o'clock direction on the face of a standard analogue clock

POLYGONS

Polygon is a common geometric element defined by a number of straight line segments that are joined at the end points. These line segments are the *sides* or *edges* of the polygon - *Figure 52-5*.

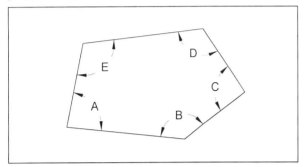

Figure 52-5
Sum of angles in a polygon

The sum of all angles in a polygon can be calculated from the following formula:

$$S = (N - 2) \times 180$$

☞ where ...

    S = Sum of the angles
    N = Number of sides in the polygon

For example, a five sided polygon shown in the illustration has the total sum of angles:

```
S = (5 - 2) × 180
S = 540°
```

There are several different polygons used in geometry, but only one special kind is of interest to CNC programming. This polygon is called a *regular polygon*, all others are irregular polygons. Regular polygon is a polygon where all sides are of equal length, called *equilateral sides*, and where all angles are also equal, called *equilateral angles* - *Figure 52-6*.

A single angle in a regular polygon can be calculated from this formula:

$$A = \frac{(N - 2) \times 180}{N}$$

☞ where ...

    A = Single angle in degrees
    N = Number of sides in the polygon

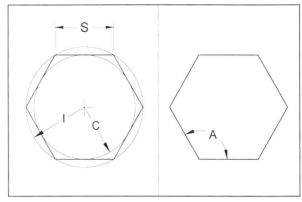

Figure 52-6
Regular polygon
Inscribed and circumscribed circles and a single angle

For example, a six sided polygon (commonly known as the *hexagon*) has a single angle of 120°:

```
A = (6 - 2) × 180 / 6
A = 120°
```

A regular polygon is quite often defined by the number of its sides and its center, located within an *inscribed* or *circumscribed* circle. *Figure 52-6* above illustrates the concept of inscribed and circumscribed polygon, as it applies to a hexagon.

Although regular polygons may have virtually unlimited number of sides, some polygons are so common that they have a special descriptive mathematical name:

Number of sides	Common name
3	Triangle
4	Square (Quadrilateral)
5	Pentagon
6	Hexagon
7	Heptagon
8	Octagon
9	Nonagon
10	Decagon
12	Dodecagon
n	*n*-gon

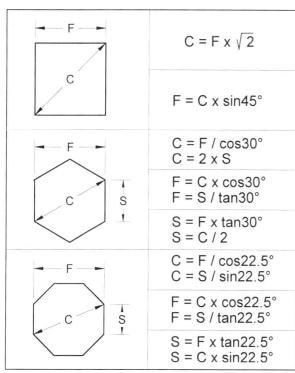

	$C = F \times \sqrt{2}$
	$F = C \times \sin 45°$
	$C = F / \cos 30°$ $C = 2 \times S$
	$F = C \times \cos 30°$ $F = S / \tan 30°$
	$S = F \times \tan 30°$ $S = C / 2$
	$C = F / \cos 22.5°$ $C = S / \sin 22.5°$
	$F = C \times \cos 22.5°$ $F = S / \tan 22.5°$
	$S = F \times \tan 22.5°$ $S = C \times \sin 22.5°$

Figure 52-7

The most common regular polygons - square, hexagon and octagon

In *Figure 52-7* are three most common regular polygons - a *square*, a *hexagon* and an *octagon*. Calculations of the distance between opposite corners C, the distance between flats F and the length of each side S are given. Note that a hexagon may have two different orientations (two horizontal sides or two vertical sides), which have no effect on the calculations. Hexagon orientation can be compared in *Figure 52-6* with the hexagon orientation in *Figure 52-7*.

TAPERS

All taper calculations are virtually confined to the lathe machining exclusively. Infrequently, tapers also appear in milling applications. All tapers in this section relate to the lathe applications (so called *circular tapers*), but can be modified to milling. The main purpose of tapers is to provide a match between assembled parts. By definition,

> A taper is a uniformly created conical surface on a pin or in a hole.

Many tapers are industry standards and are used for small tool holders (shanks), such as a Morse taper or a Brown and Sharpe taper. In addition, there standard tapered pins, machine spindle tapers, tool holder tapers, etc. In most cases, the taper is normally defined by the large end diameter, its length and a special note describing the taper.

The description varies between English and the metric standards. For example, AMER NATL STD TAPER NO. 2 (American National Standard Taper number 2) is a specific taper description. Another common description in English units is a *taper per foot*. Metric system is much simpler, using only a ratio. Ratio is used in English drawings as well. In both measuring systems, there is one common rule:

> Taper on diameter is the difference
> in diameter per unit of length

◆ Taper Definition

Most drawings define a taper in two common ways:

❑ One diameter and length with taper description or note

❑ Diameter at both ends and the length with taper description or note

If a single diameter is defined, it is often the larger one.

The description of the taper is a note with an arrow pointing to the taper. In English measurements, the note may identify a standard taper or a taper per foot (TPF). In metric, the taper is always a ratio. *Figures 52-8* and *52-9* show the differences between the two units, which is only within the taper identification.

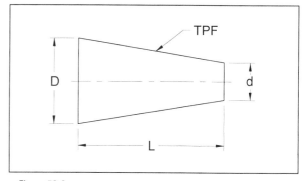

Figure 52-8

Circular taper - English description

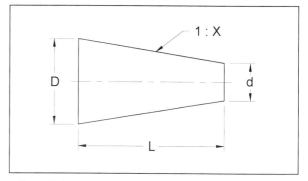

Figure 52-9

Circular taper - Metric description

In the *Figure 52-8,* showing English units method, the letters have the following meaning:

☞ Dimensions ...

D	=	Diameter at the large end in inches
d	=	Diameter at the small end in inches
L	=	Length of taper in inches
TPF	=	Taper per foot in inches
X	=	Ratio value 1 : X (not shown)

In the *Figure 52-9,* showing metric units method, the letters have the following meaning:

☞ Dimensions ...

D	=	Diameter at the large end in millimeters
d	=	Diameter at the small end in millimeters
L	=	Length of taper in millimeters
X	=	Ratio value 1 : X

All formulas in this section use these designations.

◆ Taper Per Foot

Taper per foot is defined as:

> Taper per foot is the difference in diameter in inches over one foot of length.

For example, a taper defined as 3.000 inches per foot, abbreviated as 3.0 TPF or 3 TPF in the drawing, is a taper that will change the conical diameter by 3 inches for every 1 foot of length.

◆ Taper Ratio

Metric definition of a taper is similar:

> Taper is defined as the ratio of difference between the large diameter and the small diameter over the given length of the cone.

The metric specification of a taper is the ratio:

$$\frac{1}{X} = \frac{D - d}{L}$$

The ratio 1 : X means that over the length of X mm, the diameter of the cone will change (either as an increase or as a decrease) by 1 mm.

For example, a taper specified as 1 : 5 will increase 1 mm on diameter, every 5 mm of length.

For milling, the taper is defined as the difference in width over a given length (per side).

◆ Taper Calculations - English Units

Missing drawing dimensions in *Figure 52-8* may be calculated from the given data. If the taper ratio is not specified (the normal case), but we want to know what the ratio is, the following formula will help. To calculate the taper ratio amount X, when D, *d* and L are known:

$$X = \frac{L}{D - d}$$

To calculate the small diameter *d,* with D, L and TPF:

$$d = D - \frac{L \times TPF}{12}$$

To calculate the large diameter D, with *d,* L and TPF :

$$D = \frac{L \times TPF}{12} + d$$

To calculate the length L, if D, *d,* and TPF are known:

$$L = (D - d) \times \frac{12}{TPF}$$

◆ Taper Calculations - Metric Units

Missing drawing dimensions in *Figure 52-9* may be calculated from the given data. In metric system, the taper ratio is normally known, other dimensions can be calculated.

To calculate the small diameter *d,* with D, L and X:

$$d = D - \frac{L}{X}$$

To calculate the large diameter D, with *d,* L and X:

$$D = d + \frac{L}{X}$$

To calculate the length L, if D, *d,* and X are known:

$$L = (D - d) \times X$$

To calculate the ratio X (if unknown), with *d,* D and L:

$$X = \frac{L}{D - d}$$

CALCULATIONS OF TRIANGLES

The most common geometrical entity in programming is a triangle. All triangles are polygons, but not all triangles are regular polygons. All triangles have three sides, although not always of the same length. There is a number of different triangles in geometry, but only a handful are used in everyday CNC programming.

◆ Types of Angles and Triangles

The main groups of triangles can be grouped together by their angles - *Figure 52-10*.

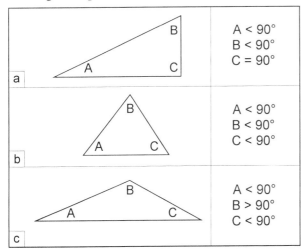

Figure 52-10

Typical triangles
(a) Right triangle   (b) Acute triangle   (c) Obtuse triangle

Some more detailed definitions may be useful:

❑   RIGHT angle means that the given angle is equal to 90°

❑   ACUTE angle means that the given angle is greater than 0° and smaller than 90°

❑   OBTUSE angle means that the given angle is greater than 90° and smaller than 180°

❑   A right triangle is also called a right angle triangle. It defines a triangle that has one right angle (90°)

❑   An acute triangle is also called an acute angle triangle. It defines a triangle that has three acute angles.

❑   An obtuse triangle is also called an obtuse angle triangle. It defines a triangle that has one obtuse angle.

In addition, there is also an *oblique* angle, which is not a new type of an angle, just a new definition:

❑   OBLIQUE angle can be either an acute or an obtuse angle, which means it cannot be 90° or 180°

All triangles share a single feature - the *sum* of all angles in a given triangle is always equal to 180° - *Figure 52-11*.

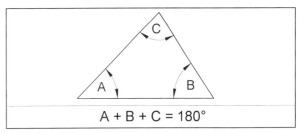

Figure 52-11

Sum of all angles in a triangle is always 180 degrees

The *oblique* triangle - and its close cousin the *isosceles* triangle - are types of triangles seldom ever needed in programming. However unlikely, it is always possible. These triangles can be solved only if at least three dimensions are known, and one of them must always be a side:

❑   One side and two angles must be known

❑   Two sides and the angle opposite one of them

❑   Two sides and the included angle

❑   Three sides

Isosceles triangle has two sides of equal length. Each side - or leg - is joined by a line called the base. The two angles at the base are always equal - *Figure 52-12*.

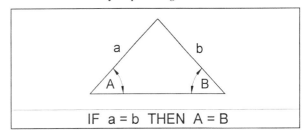

Figure 52-12

Isosceles triangle

A triangle that has all sides of equal length is called an *equilateral* triangle. An equilateral triangle is also always an *equiangular* triangle, because all internal angles are the same - each angle is 60° - *Figure 52-13*.

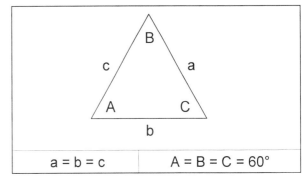

Figure 52-13

Equilateral triangle

◆ Right Triangles

A *right triangle* - or a *right angle triangle* - is triangle that has one angle equal to 90° (a triangle with two or more right angles is impossible). As there are 180° in any triangle (sum of all angles), that means the sum of the two remaining angles must also be 90°. There is a number of mathematical relationships that form the base of all calculations. Here is a look at those that are important in CNC programming. Learn these relationships well enough to be able to apply them to daily situations. Keep in mind that 99.9% of all triangles to be solved are right triangles.

The side of a right triangle that is opposite the right angle is called the *hypotenuse* and is also the longest side of the triangle. The other two sides are called *legs*. The illustration in *Figure 52-14* shows a right triangle, where C angle is the right angle (90°) and the side *c* is the hypotenuse. The sides opposite to angles have a low case identification corresponding to the angles described in capital letters.

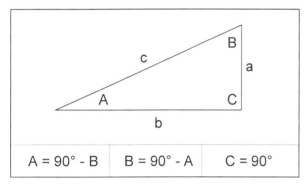

| A = 90° - B | B = 90° - A | C = 90° |

Figure 52-14
Right angle triangle and the relationship of angles

A circle drawn inside of a right triangle that is tangent to all three sides *a, b, c* - *Figure 52-15* - has a diameter D calculated from this formula:

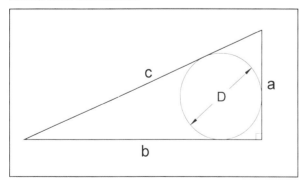

Figure 52-15
Circle inscribed in a right triangle

$$D = a + b - c$$

An inscribed angle in a semicircle is always 90°, as shown in *Figure 52-16*. Line AB is the circle diameter.

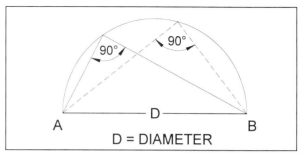

D = DIAMETER

Figure 52-16
Inscribed angle in a semi-circle

In *Figure 52-17* is a line from point A to the center of circle B. A line from point A to the tangency of the circle will create either a point C or point D. The angle *a* is created between lines AC and AD, where the line AB is a *bisector* of the angle *a*, creating two equal angles. The two angles *a1* and *a2* as well as triangles ABC and ABD are identical.

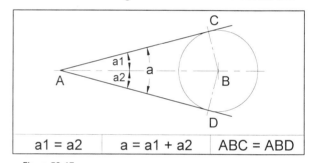

| a1 = a2 | a = a1 + a2 | ABC = ABD |

Figure 52-17
Bisector creates two equal angles

◆ Similar Triangles

Triangles are considered similar if they have their corresponding angles *equal* and their corresponding sides *proportional*. Two triangles are similar, if:

❑ Two angles of one triangle are the same as two angles of the other triangle

❑ An angle of one triangle is the same as the angle of the other triangle and the including sides are proportional

❑ Both triangles are similar to another triangle

❑ The corresponding sides of the two triangles are proportional

In CNC programming, mathematical relationship of triangles are used quite often, for example, when machining tapers or similar angular items. A taper specified in the drawing must frequently be extended at one or both ends, to allow for the necessary tool clearances.

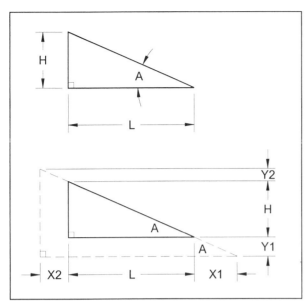

Figure 52-18
Similar triangles - 1

The illustration in *Figure 52-18* shows the relationship between two *similar* triangles. The same illustration also shows several important dimensions:

☞ where ...

L	=	Original length
H	=	Original height
A	=	Common (shared) angle
X1	=	Front clearance in the X axis
X2	=	Back clearance in the X axis
Y1	=	Front clearance in the Y axis
Y1	=	Back clearance in the Y axis

Figure 52-19 shows the same two triangles in a simplified way. In the upper part of the illustration, the values X and Y are sums of the extensions (clearances) from the previous example:

```
X = X1 + X2
Y = Y1 + Y2
```

The bottom part of the figure shows the relationship of the opposite sides H and U to the adjacent sides L and W. The formula of the relationship is:

$$\frac{H}{U} = \frac{L}{W}$$

If three of the values are known rather than two, the unknown value can be calculated using a new formula. For example, the values H, L and W are known, and the value U has to be calculated. H is 0.500, L is 1.750 and W is 2.250. To calculate the side U, the above formula is reversed:

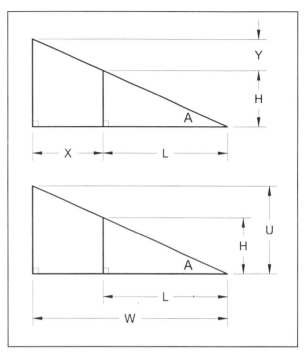

Figure 52-19
Similar triangles - 2

$$U = \frac{W \times H}{L}$$

With known values entered, the U side can be calculated. If the U is isolated on the left and the known values on the right of the equation, the calculation is simple:

```
U = (2.250 × 0.500) / 1.750
U = 0.6428571
```

◆ Sine - Cosine - Tangent

Figure 52-20 shows the most important relationships of sides and angles of a right triangle.

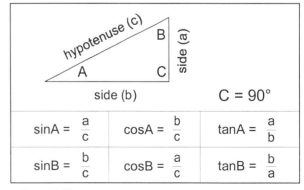

$\sin A = \dfrac{a}{c}$	$\cos A = \dfrac{b}{c}$	$\tan A = \dfrac{a}{b}$
$\sin B = \dfrac{b}{c}$	$\cos B = \dfrac{a}{c}$	$\tan B = \dfrac{b}{a}$

Figure 52-20

Trigonometric functions - sine, cosine, and tangent

This relationship has its own terminology and is defined as a ***ratio of sides,*** using the *sine, cosine* and *tangent* functions of the given angle. Other available functions, namely *cotangent, secant* and *cosecant* are normally not used in CNC programming.

❑   Sine of an angle - abbreviated as *sin* - is a ratio of side opposite the angle to hypotenuse of the triangle

❑   Cosine of an angle - abbreviated as *cos* - is a ratio of side adjacent to the angle to hypotenuse of the triangle

❑   Tangent of an angle - abbreviated as *tan* - is a ratio of side opposite the acute angle to the side adjacent

◆ Inverse Trigonometric Functions

From the definitions, the value of sine, cosine and tangent is expressed as a ratio of two sides. The angle that depends on this value is the result of an *inverse trigonometric function.* An inverse function is sometimes symbolized with the word *arc,* preceding the normal function. For example *arcsin* of an angle A is the *angle whose value* is the ratio of the side *a* to the hypotenuse *c*.

Most pocket calculators indicate the inverse function as *sin, cos* and *tan* raised to the *power of minus 1* as the secondary key function. Just enter the ratio of the function:

If ...   `sinA = a / c`
Then ...   `A = arcsin(a / c)`
Or...   `A = sin⁻¹(a / c)`

If ...   `cosA = b / c`
Then ...   `A = arccos(b / c)`
Or...   `A = cos⁻¹(b / c)`

If ...   `tanA = a / b`
Then ...   `A = arctan(a / b)`
Or...   `A = tan⁻¹(a / b)`

While there is only a single result for each trigonometric function, there could be several results for the inverse function. For example, absolute value of 0.707106781 is the sine of 45°, as well as the sine of 135°.

◆ Degrees and Decimal Degrees

Another type of calculation used in programming is conversion of angles. It relates to a drawing using *minutes* and *seconds* to describe the precision of angular degrees required. There are two methods of dimensioning angles in a drawing. The older and method is the angle designation in DMS or D-M-S, which means *degrees-minutes-seconds.* The modern methods are associated with CAD drawings and use DD or D-D, which means *decimal degrees.* Decimal degrees are needed for calculations of coordinate points, so DMS must always be converted to DD.

The following formula converts degrees-minutes-seconds designation to decimal degrees:

$$DD = D + \frac{M}{60} + \frac{S}{3600}$$

☞ where...

DD = Decimal degrees
D = Degrees
M = Minutes
S = Seconds

Therefore,

`64°48'27"`   *... is equivalent to:*
`64 + (48 / 60) + (27 / 3600) = 64.8075°`

The abbreviations DMS/D-M-S and DD/D-D are commonly used on scientific calculators. Much less useful conversion is to change decimal degrees to DMS. It is not needed in CNC programming, except perhaps to perform a double check, to verify that the original converted result is correct. The calculation of DD to DMS is nothing more than isolating the fractional part of the number in three steps. For example, in order to convert 29.545021° to degrees-minutes-seconds format, three steps are necessary.

The first step is to isolate the whole degrees amount from the decimal degrees:

`29.545021 - 0.545021 = 29°`

The seconds step is to take the decimal portion and multiply it by sixty, to get the minutes:

`0.545021 × 60 = 31.701126 = 32'`

The third and final step is to take the decimal portion of the last result, and multiply it by sixty to get seconds:

`0.701126 × 60 = 42"`

The final DMS value of the example will be 29°32'42", with a slight rounding error.

◆ Pythagorean Theorem

The well known work of the ancient Greek mathematician Pythagoras (6th century B.C.), known today as the *Pythagorean Theorem,* is taught and emphasized fairly early in a high school mathematics classes. This mathematical theory relates to the solution of right triangles and states:

> In any right triangle, the square of the hypotenuse
> is equal to the sum of squares of the other two sides

Pythagorean Theorem is used in programming to find the length of any side in a right triangle, if two other sides are known. *Figure 52-21* shows the calculation of side *a* or side *b* or the hypotenuse *c* in a right triangle.

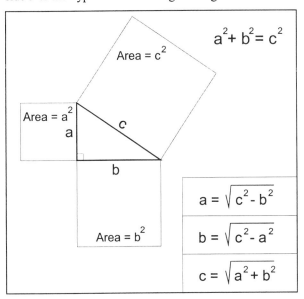

$$a^2 + b^2 = c^2$$

Area = c^2

Area = a^2

Area = b^2

$$a = \sqrt{c^2 - b^2}$$

$$b = \sqrt{c^2 - a^2}$$

$$c = \sqrt{a^2 + b^2}$$

Figure 52-21
Pythagorean Theorem

⬤ *Example ...*
If the length of hypotenuse *c* is 3 units and the side *b* is 2.75 units, the side *a* can be calculated - the *c* squared is 9.0, *b* squared is 7.5625, so the side *a* is:

```
a = √(3 × 3 - 2.75 × 2.75)
a = √(9 - 7.5625) = √1.4375
a = 1.1989579
```

The symbol $\sqrt{\ }$ represent the square root.

◆ Solving Right Triangles

The solutions of right triangles using the Pythagorean Theorem or any other method are equally important. These common methods use the *sin, cos* and *tan* trigonometric functions. As always, start with the known data. In trigonometry, any triangle can be solved, providing one of the two data sources is known:

❑ Two sides of a right triangle

❑ One side and one angle of a right triangle

The 90° angle is always given and never used in calculations. *Figure 52-22* covers all right triangle solutions. If more than a single solution is available, use both methods to double check the result.

			Trigonometry Relationships
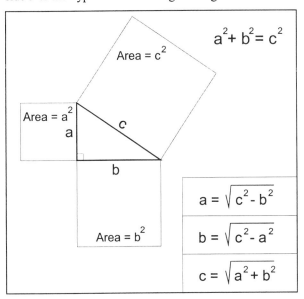	a / c = sinA = cosB		**SOH** Sine = $\dfrac{\text{Opposite}}{\text{Hypotenuse}}$
	b / c = cosA = sinB		**CAH** Cosine = $\dfrac{\text{Adjacent}}{\text{Hypotenuse}}$
	a / b = tanA = cotB		
	b / a = cotA = tanB		**TOA** Tangent = $\dfrac{\text{Opposite}}{\text{Adjacent}}$
a = c x sinA	b = c x cosA	c = a / sinA	
a = c x cosB	b = c x sinB	c = a / cosB	
a = b x tanA	b = a x tanB	c = b / sinB	
a = b / tanB	b = a / tanA	c = b / cosA	
$a = \sqrt{c^2 - b^2}$	$b = \sqrt{c^2 - a^2}$	$c = \sqrt{a^2 + b^2}$	Sin = $\dfrac{p}{h}$ ➡ Peter / Has
A = 90° - B	B = 90° - A	C = 90°	Cos = $\dfrac{b}{h}$ ➡ Broken / His Tan = $\dfrac{p}{b}$ ➡ Pop / Bottle

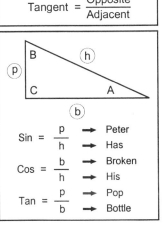

Figure 52-22
Trigonometric functions - formulas for solving right angle triangles

ADVANCED CALCULATIONS

The last two charts show formulas for calculations of the chord C or the tangent T of a circle. Trigonometric formulas can be used as well, but the formulas can make the same calculations faster. With only one exception, there are two solutions, dependent on the available data. The formulas can also calculate the radius R, angle A and the deviation *d*. Calculations relative to the chord of a circle are shown in *Figure 52-23*. Calculations relative to the tangent of a circle are shown in Figure *52-24*.

CONCLUSION

In this chapter, only the most important and commonly used mathematical subjects have been presented. Many more solutions and shortcuts are used by programmers and operators every day, showing their ingenuity in solving math problems. Author will appreciate any formula, shortcut or a solution to any programming problem, and will be considered for the next edition of this handbook.

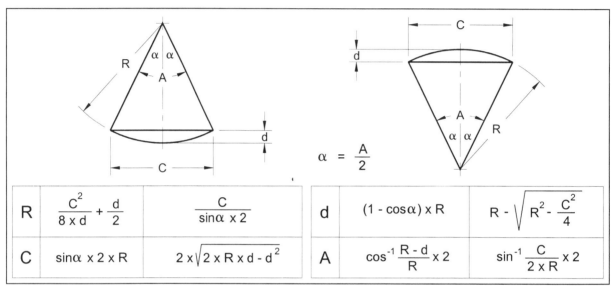

Figure 52-23
CHORD of a circle - calculations of chord, radius and deviation

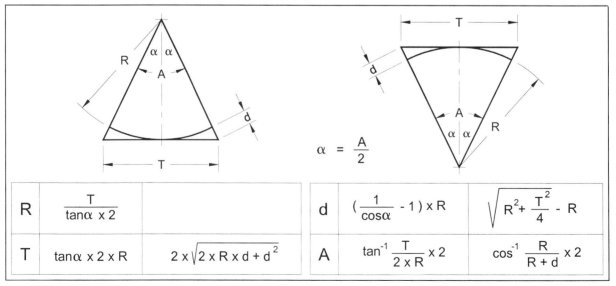

Figure 52-24
TANGENT of a circle - calculations of tangent, radius, angle and deviations

Up to this point, all topics related to *manual* programming of CNC machines - all fifty-two chapters. In the last chapter, we look briefly at an area where manual programming is replaced by a computer, a suitable software and some *additional* skills. Note the word *additional*. Studying the handbook has certainly not been a waste of time. On the contrary - the handbook covers subjects that every CNC programmer should know, *regardless of the programming method used.* Programming with a computer is always desirable but to know the basic skills is the most important prerequisite. The basic skills are in understanding the manual process. All subjects and methods learned do not have to be applied by a pencil and paper. They could be applied by a CAD/CAM - or just CAM - programming. A simple statement may summarize it all:

> Top class programming using CAM software requires
> solid knowledge of manual programming methods.

PROGRAMMING MANUALLY ?

In the area of CNC programming application techniques, computers at all levels, from a personal computer to workstations are capable to produce most CNC machine programs in a time much shorter than any manual programming method. So, why is the high importance of manual programming methods so emphasized? Is the manual programming still alive, and if so, how healthy is it?

There are at least two important reasons why manual programming for CNC machines it is not dead yet and will not disappear anytime soon.

The first reason is that in manual programming, the programmer is able to do what computers cannot - and never will be - *programmers can think.* Manual programming teaches the invaluable lessons of discipline - a very important quality of a professional CNC programmer. Discipline means to concentrate, to constantly evaluate, to make decisions - to think all the time. In manual programming, there is a total, absolute and unequivocal control over the final product - the part program. Only a programmer can evaluate a given situation, analyze the problem and adapt to unforeseen circumstances. Only a programmer can feel that something may not be right. Only people use instruments known as thinking process, intelligence, instinct, gut feel, common sense and experience. Those are instruments inherent to humans, not computers. CNC programming is like the work of an artist - it can never be fully automated.

◆ CAM Software

Current CNC software, commonly known as *CAM software*, has many features that translate into a CNC program, corresponding to individual ideas of how the part program should be written. It can produce a program closely matching a particular direction of thinking, closely matching a particular programming style. But *closely* does not always mean close enough. Here comes the second reason.

The second reason is that when programming manually, the programmer *understands* the programming process and the resulting output. A program generated by a computer has to be in the format compatible with the CNC machine and its control system. If all goes well, there is no need to look at the program at all - it's there, in the files, ready to be loaded into the CNC machine. On the other hand, what if there is a problem - what then? Going back to the computer and reprogram the part may solve the problem on hand. The question is at what price. Ability to read the CNC program code, to really understand it, also means the ability to change it. Spending a valuable computer time just to add a forgotten coolant function seems excessive. Would it not be better just to edit the program by adding M08 function in the right place? Although the example is oversimplified, it also shows that real *understanding* of the programming process is very important. The best way to understand the process is to bypass the computer and get the same results. That can be achieved with manual programming.

It would be unfair to compare or promote manual programming against computer programming and vice versa. What is necessary to promote is the knowledge and understanding of manual programming principles. Without such knowledge, one can not become a good CNC programmer.

Most of the CNC programming can be done quite well on personal computers. The existing technology is progressing very rapidly and many 2D and 3D programming applications are available for a fraction of the cost when compared to just a few years ago. This trend will continue well into the future.

◆ Desktop Computer Programming

The complete computer system - that means the hardware, software and peripherals - suitable for CNC programming is changing at such a rapid pace that any in-depth discussion of the hardware would be obsolete in a matter of weeks. Almost the same speed of obsolescence applies to software as well. New features, new capabilities, new tools are constantly arriving on the market and are offered to users in both areas. Because hardware and software has to be considered together, the first question is what to select first - the hardware or the software?

Such a decision must be based on the required application. What will the computer be used for? What kind of work needs to be computerized, automated? What results are expected? These are the *primary* considerations - *not* the kind of monitor or printer or the capacity of hard disk. They are also very important - but only *after* establishing the application needs.

Certain programming applications are typical to all machine shops. Others are unique to a particular type of manufacturing and the kind of work or the product manufactured. The following short list itemizes the major groups that a typical computer based CNC programming system should have:

- ❏ Tool path geometry creation environment
- ❏ Tool path generation
- ❏ Complete programming environment
- ❏ Post processing
- ❏ Training and technical support

It is important to understand *why* these features are important. Before investing into a technology that is all or partially new to the user, it helps to know *what* tools the software offers and *how* they can be used in everyday work.

TOOL PATH GEOMETRY DEVELOPMENT

Most CNC programming systems require a tool path geometry creation before the actual path of a cutting tool can be generated. The key words here are *tool path geometry*. A common misconception among programmers is that they have to re-create everything in the original drawing. That is a wrong approach.

When it comes to tool path geometry, two scenarios must be faced. One will be work form a paper drawing, the other from a CAD drawing stored in the computer. Although there are differences in approach, the fact remains that either a new geometry is created or an existing geometry is modified.

Modern CAM systems allow drawing in a CAD like manner - using editing features such as trim, fillet, break, copy, move, rotate, offset, mirror, scale, and so on.

Typically, programmer will define what is normally not on the drawing, at least not on a two dimensional representation of the part. Adding depth, separating entities by color, by levels (layers), adding clearances or a special lead-in and lead-out tool motion, and so on.

TOOL PATH GENERATION

The key requirement of a CNC software is to produce a program of an accurate tool path for a specific CNC machine. The tool path creation, with all its calculations, is the most time consuming task in manual programming. It makes sense to make it the most important item to consider when planning to automate the CNC programming process. Only high level CNC software supports a large variety of tool paths. For example, helical milling or a full 3D machining are not always standard features in the software.

One mistake in software selection is to consider only the *existing* CNC machines and existing machining methods and practices. This rather narrowly focused approach is not always successful. Consider future plans in both strategies and capital investment. What about the product? How will the product change in five years? Knowing the philosophy and focus of the company, its policies and management strategies and yes - even its politics - will help to make a more accurate estimate of future needs.

The computer technology has grown a lot, yet it is so new that it is in the state of constant development. Nobody can predict with absolute accuracy what the future will offer in terms of CNC machining and CNC programming. If the current and the future needs are well established *before* purchasing a programming system, there is a good chance to beat obsolescence for a long time. CNC software developers offer periodical updates to their product, with more features added as computing power increases. The updates (new versions of the software), usually reflect developments of the technology, both on the hardware and software sides. It does not mean purchasing every new update offered, but it is important to select a CNC software developed by a solid and well established company that has the best chance to be still in existence when the need to update a system comes up. The computer industry is very active, and mergers, acquisitions and takeovers are as common as bankruptcies and failures.

COMPLETE ENVIRONMENT

A typical high quality CNC programming software allows all programming and relating tasks to be done from a structured menu, using a mouse or similar pointing device. The important thing is that once the software is loaded, it can complete all tasks without returning to the operating system level. Some programming systems are based on modules and files that are not accessible from a menu, or they do not cover all the steps in programming.

The following list is meant only as a very brief guide to some of the main features that apply to CNC programming on personal computers. These are the expected features from any CAM software:

- ❏ Multi machine support (machining centers, lathes, EDM)
- ❏ Associative operations for flexible editing
- ❏ Job setup and material blank definition
- ❏ Tooling list and job comments (setup sheets)
- ❏ Connection between computers (communications feature)
- ❏ Program text editor (with CNC oriented features)
- ❏ Printing capabilities (text and graphics)
- ❏ Pen plotting (plotters)
- ❏ Interface with CAD software (DXF, IGES, CADL, STL, ...)
- ❏ Support for solid modeling
- ❏ Software specifications and features (including customizable post processing)
- ❏ Support for generally available hardware
- ❏ Utilities and special features, open architecture

Each described item will point out its significance. Although all items are useful in a programming system, it does not mean that all items are always necessary. Some features require an additional hardware equipment, such as a printer, plotter, cabling, small peripherals, etc.

◆ Multi Machine Support

When it comes to support of different machine types, CNC software can be divided into two groups:

- ❏ Dedicated software
- ❏ Integrated software

The *dedicated* software supports only one kind of machines. For example, a software that is designed specifically to produce programs for CNC fabrication equipment, cannot be used for lathes, machining centers or EDM.

Dedicated software is often developed for a rather narrow and very specialized field of applications or when it applies to a particular machine only. CNC punching, forming and press brake equipment are good examples of such software.

The *integrated* software allows the programmer selection of several types of machine tools. Such a selection usually offers milling, turning and wire EDM. It is also common to use the software for machines such as burners, routers, laser cutters, waterjets, and profilers. For metal cutting, this is the preferred type of software.

Another reason also speaks clearly in favor of integrated software, and that is its *interface*. It is much easier to get used to one display for a lathe work and have the same display for a milling work or an EDM work. The software menus look the same, the navigational operations share common menu items, the customization of the software (including post processors) is much simplified.

◆ Associative Operations

When a tool path is developed, it is attached to the previously defined tool path geometry. For many reasons, it is not unusual to change the tool path geometry later. The traditional method has been (and for many software vendors still is) to recreate the geometry, then recreate the tool path.

Associative operation avoids the creation of a new tool path, it updates it automatically. It is fast and accurate. It works the other way as well - many tooling parameters can be changed quickly, on demand.

◆ Job Setup

Job setup is a feature that describes the material blank of the part - its shape, dimensions, zero origin, and many other related items. Tools and related speeds and feeds can be often selected from the job setup, as well as various program parameters. Libraries that store common data for tools, materials and operations are also powerful software features.

◆ Tooling List and Job Comments

CNC programming is a process covering several steps. Whether programming manually or with a computer, the selection of cutting tools is a manual task. Once selected, each tool is assigned its identifications, speed and feed values. Several tools can be grouped into a tooling library file and stored. Then, the order of their usage within the program is selected. Some parts require more than one machining operation. Complex setups require special instructions to the machine operator (setup sheet), describing the programmer's intents. All these programming decisions must be recorded and the documentation sent out to the machine shop. It is only reasonable to expect that any CNC programming software will support a tooling list, perhaps in a form of a tool library file and the process list. Material library file is also very useful, as it can store surface speeds for many materials and the programming software will calculate the exact spindle speed and feedrate, based on the tool selected. This is a good example of interaction between the tool library and material library.

◆ Connection Between Computers

A programming system should also include a connection (communications option) between the personal computer and the CNC machine. This feature allows the program data exchange via a cable. Programs can be sent from the computer to the memory of the CNC machine and back.

An important point is that not all CNC machines have the port (outlet) and the capability to take advantage of direct connection. Even if all machines in the shop have this capability, it requires additional hardware and organizational discipline to make all elements work in harmony. The existence of a direct connection in a programming software is a must, even if it is not used immediately after the purchase.

◆ Program Text Editor

A CNC program generated by the software should be 100% complete and ready for use by the machine. The implication is that such a program is so perfect that it needs no further editing. This is the ideal way, the way it should happen. If a change in the program is needed, it should be done *within* the design of the part shape and that means through the CNC software - *not outside of it*. The reason is that any manual change to the generated program does not correspond to the program data as generated by the computer. In the environment where the data is shared by many users, such a practice will cause a lot of problems.

That brings up a question - why does a CNC software have a built-in text editor? There are two reasons. One, the editor can be used for creating or modifying various text files such as setup sheets, tooling sheets, operation data, post processor templates, configuration files, special instructions, procedures, etc. These files can be updated and otherwise modified as required, *without* a damage to the program database. The second reason is that in some *special* circumstances, a CNC program can be edited outside of the computer model, providing the change does not modify significant data. For example, to add a missing *coolant* function M08 to the part program is much faster done in the text editor, than repeating the program generating process with the computer. Purists are right, it is not the right way of using the text editor, but at least the significant data (tool locations) are not tampered with and the database is otherwise completely accurate.

Many programmers use various external text editors or even word processors in text mode. These types of editors are not oriented towards the CNC programming, since they lack some features typical to the CNC program development. Only a CNC oriented text editors can handle automatic block number sequencing, removing the block numbers, adding cosmetic spaces in the program and other functions. The editor should be accessible from the main menu or from within the software.

◆ Printing Capabilities

Any text saved into a file, CNC programs included, can be printed using a standard printer. The paper copy is often necessary as a reference for the CNC operator, for stored documentation, or just for convenience. The printer does not need to be top of the line, just one with a standard paper width. Some programming software supports an option that is known as a printer plot or a hard copy. Hard copy is a graphic image of the screen transferred to the printer. The image quality is usually more than adequate. This hard copy is an excellent aid during program development stage. Better quality printer provides better quality print plot. The printer support is provided by the Windows environment, as most PC based CAM software is developed for the Windows operating system.

◆ Pen Plotting

Pen plot will usually produce image quality superior to the printer plot but for a CAM programming it is an unnecessary luxury. The only time when a pen plotter can be beneficial is for plotting to paper size that is not supported by standard printers. Other reasons will be the need for a color output, a special requirement by customers, or special documentation development. Before the graphics software appeared on the market, plotters were widely used to verify the tool path. Now, the tool path is verified directly on the computer display screen, during interactive programming process, including different views and zooms.

Most plotters are HPGL compatible. HPGL is an acronym for *Hewlett-Packard Graphics Language*, and is currently the most supported plot file exchange format.

◆ CAD Software Access

If an engineering drawing is generated by a CAD software, all drawing information is stored in a computer database. This database can be accessed by several programming software packages, through a file format translation utility (more on the subject later). Once the CNC software accepted and processed the database from the CAD system, the CNC programmer can concentrate on generation of the tool path itself, rather than defining the tool path geometry from scratch. Some modifications are usually necessary, so expect them. The most significant advantage of a quality CAD/CAM system is the avoidance of duplication. Without CAD system, the CNC programmer has a lot of extra work to do, much of it is duplicated.

A high quality CNC software also allows the existing program file to be translated the other way, to a file that a CAD system can accept. This option is called *reversed processing*, and can be a benefit to companies that want to translate existing programs generated manually to an electronic form. Usually some additional work is required in these cases.

High level CNC software is a *stand alone* type. Stand alone software means that it does not need an access to a CAD system - the tool path geometry and the tool path itself can be developed from within the CAM software, independently of other software.

◆ Support for Solids

Solid modeling for 3D applications had been for a long time the domain of large computer systems. With the advance of powerful microcomputers, solid modeling is now part of high level CNC software.

With solid models, the machining process of complex surfaces is much more streamlined. In addition, solid models offer the benefits of supplying engineering data, easier manipulation of objects, and many other features.

◆ Software Specifications

Another benefit of a high level CNC software is that it comes well supplied with a variety of useful features. What makes each system unique, is usually the method of how the programming process is executed. In the early years of development, programming was done by using special programming languages, such as APT™ or Compact II™. Some languages are still available but heavily on the decline. Modern interactive graphics programming has virtually eliminated the need for languages in just about all manufacturing fields. The more popular kind of programming is based on *interactive graphics*. The programmer defines geometry, typically as the tool path geometry, followed by the tool path itself. Any error in the process is immediately displayed on the graphic screen and can be corrected before too much other work is done.

◆ Hardware Specifications

Specification of the software will determine the hardware selection. Hardware is a common term for the computer, monitor, keyboard, printer, modem, plotter, mouse, scanner, disk drive, storage media, CD writer, and many others. The hardware referred to in this chapter is based on the *Windows™* operating systems. Modern operating systems are based on a *graphical user interface (GUI)*. Some software can run under a different operating system, for example Unix (used mainly by workstations) or different Windows versions. It is always to the advantage of the user that the latest version of the operating system and the CAM software is installed on the computer.

When thinking of purchasing a computer hardware, consider carefully at least three major criteria:

❏ Performance   ... computer speed

❏ Data storage   ... type and size

❏ Input / Output   ... ports

Computer Speed

Performance of the computer system is typically measured by the relative speed of the main processor. The higher the number, the faster the computer can process data. To make the comparison easier, the original IBM PC, model year 1983, had a 4.77Mhz processor speed. Later model AT had 6mhz processor speed, improved further to 8 and 10Mhz. Later, computers used the so called 386 microchip (generally Intel 80386 or 80486) and reached 25Mhz, 33Mhz and more. Pentium processors followed, and the process is ongoing. Chips in thousand plus MHZ speed are a reality. For serious CAD/CAM work, the latest fully featured processors should be used. Newest processors offer much higher processing speed, and the more processing speed is available, the better performance of the CNC programming system.

RAM and Data Storage

Data is stored in the computer in two forms - memory storage and disk storage (file). When an application such as CNC programming is started, the CAM software is loaded into the computer memory. The more powerful the application software, the more memory it requires. This memory is known as *Random Access Memory*, usually called *RAM*. Every software specification identifies the minimum available RAM required. RAM of today high level computers around the gigabyte range is not uncommon. Any extra memory will speed up processing quite significantly. The data in the RAM is volatile, which means the data is lost when the application is ended or the computer power is interrupted. To save important data from RAM into disk files, a hard disk or similar media can be used. For a micro computer CAD/CAM work, the absolute minimum requirement is high density removable drive and one large size hard drive. Floppy drives of any kind are not suitable.

The hard drive should have a fast access time and a high storage capacity. Another option is a tape drive, CD-R and CD-RW disks or recordable DVD disks for backup.

Input and Output

Input and *Output* (I/O) computer features, cover hardware items such as monitor, graphic card, keyboard, digitizer, scanner, printer and plotter. Monitor suitable for CAD/CAM work should be a large size color monitor providing very high resolution. The monitor and the graphic card do relate to each other. The card must be able to generate the image, the monitor must be able to display the image. Speed of the video output is also very important.

A keyboard is a standard feature of a computer and serves as a basic input device. Mouse (or a digitizer on larger systems) are also input devices, but much faster than keyboard input. In CAD/CAM, where a lot of work is done in graphic mode under a menu system, the item from the menu is user selected. In most cases it can be selected with a pointing device. The user points at the menu item desired, presses a button on the device and the menu item is executed. The pointing device most suitable for CAM work in the Windows environment is a mouse.

Both the printer and plotter are theoretically optional, but generally worth some consideration. For CNC work alone, a printer is more important than a pen plotter. If the setup is a true CAD/CAM, both peripheral devices may be needed.

All peripherals are interfaced with the computer using specially configured cables connected to the *Input/Output (I/O)* outlets called *ports*. The modem is normally not required for CNC programming, except for data exchange with a remote computer or Internet access. The laser or ink jet printers generally use a parallel interface known as the *Centronics* standard, but many other devices use a serial interface. There are also other I/O options, such as the USB (Universal Serial Bus) interface.

◆ Typical Hardware / Software Requirements

Currently, the most popular hardware for CNC programming is the Windows based computer system. It is not possible to make a simple 'shopping list' for all hardware requirements that every CNC machine shop can use. Here are some rules applicable to any system and are not subject to becoming outdated very quickly. A typical list of minimum hardware requirements and options may be compiled:

❏ Hardware compatibility with IBM (Windows based)
- Apple computers havevery limited CAD/CAM applications

❏ The latest version of the Windows operating system
(must be supported by the CAM software)

❏ High central processor speed - higher = better
(measured in MegaHertz units - MHz)

❏ Fast memory cache

❏ The requirement of a numeric (math) co-processor
(normally part of the higher end processors)

❏ Random Access Memory (RAM) - as much as possible

❏ Enough of hard disk space for program and data storage
(measured in gigabytes or higher - with a fast access time)

❏ Backup system for data protection
(tape cartridge, removable drive, CD, DVD, ...)

❏ High resolution graphics adapter (graphics card)
(should have a rapid refreshing for the video output)

❏ Large high resolution color monitor - non-interlaced
(measured in pixels - the more pixels per screen size,
the finer the display, and the smaller the pixel size,
the better the display)

❏ Pointing device - normally a mouse - is a current standard

❏ Pen plotter is required only in special circumstances
(not needed for CNC work) - B size maximum is usually
enough, if needed

❏ Working real time calendar clock (stamps all created
files with the current date and time - standard feature)

❏ A good quality printer with a parallel or USB port
(for hard copy documentation)

❏ CD or DVD drive & various multimedia features
(sound card necessary)

❏ Access to additional global information
(Internet, E-mail, user groups, newsgroups, ...)

❏ Two or more serial and USB ports

❏ Text editor - usually part of the software (or optional)

It is smart to keep abreast of the micro computer technology. It develops rapidly and even a few weeks may change some fundamental approaches and decisions. Following the development of computer technology creates awareness of the latest improvements, therefore a more educated user and/or buyer.

◆ Utilities and Special Features

Even the most updated version of the operating system is never as powerful and flexible as many users would like it to be. For that reason, many software developers came up with literally thousands of programs and utilities that supplement the readily available features. Many of these utilities are available as shareware or freeware from the Internet and other sources. Access to the Internet and the World Wide Web provides a great source of CNC and machine shop related topics and general information. These utilities are not necessary to use a CAM software, but they are a great time saver for many tasks associated with using a computer.

POST PROCESSORS

CNC software must be able to output a program in a format unique to each control unit. The most important part of a tool path generation is the data integrity. The computer generated program must be accurate and ready for the CNC machine. That means the completed program should require no editing, no optimization, no merging with other programs or similar manual activities. Such a goal can be achieved only by a well developed programming style - and a properly configured post processor for each different CNC machine.

A top quality post processor is probably the most important customized feature of a CNC software. When entering data into the software, values describing the part shape, cutting values, spindle speeds, and many other data are stored for further processing. The software analyzes this data, sorts it and creates a database. The database represents the part geometry, the toolpath geometry and other functions. The CNC system cannot *understand* the data, regardless of its accuracy. To complicate things even more, every CNC system is different. Some program codes are unique to a single machine, some are quite common to many machines. The purpose of a post processor is to process the generic data and convert them to the machine code for individual control systems.

◆ Customizing Post Processor

Typically, a supplied post processor is more or less generic and has to be customized, at least to some extent. To develop a post processor in-house, usually means to customize the generic post processor supplied with the CAM software. The typical process depends on the type of post processor and its format. Small changes may take minutes, large changes days. Post processors can be very expensive.

The CNC programmer must know the machine and control features extremely well. A deep and thorough knowledge of manual programming methods is a must - how else can a useful programming format be developed? Also important is the knowledge of machining methods. Finally, knowing any high level language can make the post processor development much more efficient and powerful.

IMPORTANT FEATURES

There are several important features to look into when investing into a CNC programming software. They do have an impact on the final functionality of the program, at the machine level. All these features are important and should be considered carefully.

◆ Input from User

One of the important features of a CAM programming software is its ability to handle input from the user. This input can be a special sequence of commands that cannot be handled by the post processor at all, or would require too much effort. These commands are usually small in size and can be called and used in the graphics mode whenever required. Examples of such applications are a barfeeder sequence on a lathe or a pallet changing routine on a horizontal machining center. If the software supports some type of variable type of user commands, it adds an extra flexibility and power to the system.

◆ Machining Cycles

Another very important feature of a CAM software is its ability to generate a variety of fixed and repetitive cycles, that modern controls support. These cycles make a manual programming simpler and faster. The modern CNC systems take advantage of such cycles are available with a limited memory capacity. For that reason, support for such cycles is very important in a CNC software, as it provides easy editing at the machine.

◆ User Interface

Customizing the display is also a useful feature. It is not as critical as others, but a facility to customize fonts, colors, toolbars, even menus adds the extra power to the software. Colors are very important in CAD/CAM work. The color settings should be changeable to provide better distinction. The screen appearance may be changed by a different combination of colors for the foreground, background and the text. The result is the visual emphasis on what is important.

The last user interface feature is the selection of verification options in the software. When the tool path simulation is shown on the screen, a circle represents the tool diameter for milling applications, and the shape of a turning tool for CNC lathes. This tool image shows the current tool position, valid for the processed program section. Normally, the graphic image moves along the contour, without leaving any traces. A variation is that the tool will remain at the contour change points only, but nowhere else. This is called static display and is very important for some machining operations. Premium CAM software also allows to design a customized tool shape, including the tool holder and use it on the screen to simulate actual tool path. Shaded 3D tools add even more realism to program viewing.

Also important is the representation of the tool path for lathe tools. Many cutting tools for a CNC lathe have a back angle. A high quality software should also evaluate the tool back angle in its calculations and in the display.

◆ CAD Interface

A stand alone CNC programming system does not need a CAD software for the geometry definitions. It can create its own. Yet, in a any CAD/CAM system is important to have the *option* of importing part geometry *from* a CAD system. Even if a company does not need CAD, it should be prepared to accept its files, perhaps from customers or company branch offices.

Needless to say, if a CAD software is not available, the computer cannot not accept the drawing files generated by such a software. These files are proprietary and their structure is not a matter of public access. Therefore, there must be another way to interchange drawing files. There is another way - *use a different file format.*

File Exchange Formats

The need to exchange design files between different software systems has always been a prime requirement. There are many competing formats of a neutral file format. The oldest of them is called *IGES (Initial Graphics Exchange Specification)*, originally developed to transfer complex design files from one software to another. Another format that is also used, is the *DXF* format by Autodesk™.

The *DXF (Drawing eXchange Format* or *Data eXchange Format)* is considered by many to be *the* standard of drawing file exchange between micro computers. It has been developed by Autodesk™, Inc., the developers of the popular AutoCAD™, the most widely used PC based CAD in the world. DXF format is suitable only for common geometric elements, such as points, lines, arcs and a few others.

The CNC software should also support an interface between the neutral files generated by a CAD system. Depending on the nature of a particular programming application, the DXF interface may be needed for simpler jobs, and IGES for more complex geometries. High quality CNC software offers at least these two formats, usually many more. Keep in mind that the format and structure of the translators, such as DXF or IGES, is not in the hands of the CNC software developer, therefore it is a subject to change.

SUPPORT AND MANAGEMENT

Hardware and software for CNC programming work can be costly. It can represent a significant investment of money *and* people and can become a total failure if it is not used properly. A failure is not the actual loss of the hardware and software cost. The real and heaviest loss is in the increased productivity, speed and quality that was expected but never materialized. The loss is also in the confidence the company employees put into the technology. These losses can be high. To prevent such prospects, keep three key elements in mind when planning a CNC programming system:

- ❏ High quality training program for long term skills
- ❏ System management philosophy and strategies
- ❏ Technical support for hardware and software

No single item in the list is any more important than the others - they are *all equally* important.

◆ Training

Training should be planned, thorough, and professional. Many successful programs apply three levels of training. Some companies do not place enough emphasis on training, despite many studies and examples proving that good quality training does work. The lack of time and perceived high costs are often used as excuses. Training is a necessary investment for any company that wants to be competitive.

Training level 1

The *first level* of training should be aimed at the person with *none* or *very little* computer experience. It should introduce the CNC software to the programmer who programs manually. It should be an overall training, mainly general in nature, with the emphasis on the system features and capabilities - as they relate to the company where the software is installed. The typical general approach should be balanced by explaining the philosophy behind the software design, and the structure of menus and commands. It is very important to show the student what the software can do in skilled hands. The first level should be done when the software is purchased. The objective is to give the programmer enough tools to *play* with the software, to grow into it. A simple way to achieve this goal is to try out simple projects while still programming manually the important jobs.

Training level 2

The *second level* is most beneficial two or three weeks *after* the first level completion. It is the most critical level of the three. It should include a *systematic approach* to all software features, with special emphasis on features relating to the machining operations used locally. This training level eliminates manual programming, and marks the beginning of a new era. Supervisors should evaluate the complexity of the first few jobs to be programmed and select, if possible, the less difficult jobs to build a little confidence.

Training level 3

The *third level* is usually done 2-3 months later. It covers problems, questions, difficulties and concerns, introduces tips, shortcuts, etc. The purpose of this level is to create a long term confidence. At this stage, the programmer has many questions. Professional instructor can answer all questions, weed out bad habits, and offer further guidance.

◆ System Management

A reliable operation of all system elements is crucial to the success of CNC software. Use of any software requires good organization, it needs strategies, it needs focus, and it definitely needs a professional management. System management establishes standards and procedures for CNC and related operations. Concerns about people selection, data backup methods, confidentiality and security, work environment quality , etc., are not confined to a single discipline and should be important in the overall company culture.

◆ Technical Support

Technical support is an important part of the system management. A service contract or a support package can be usually negotiated with the vendor, covering installation, hardware, update policies, new developments, etc. An important part of technical support is the speed and reliability of handling *emergency* situations. If a hard disk fails - and a data back up *does* exist - what can be done? The CNC shop is waiting for the critical job, while the programmer cannot send program data to the machine, because an inexpensive hard disk failed. Support should cover both the hardware and software. All support promised by the vendor should always be written down. Know *exactly* what the bill is for. *If something isn't in the contract, it usually isn't available.*

THE END AND THE BEGINNING

What the future of CNC technology holds is always hard to predict. There are many indications where the technology will be going. System controls with more computing power, more standardized approach to programming, more solid modeling, more 3D, better storage methods, etc., are in the works. Changes are also inevitable in work skills.

Stand alone CNC machines will always be needed. On the CNC machining centers, there will be much more emphasis on faster machining rates. On the CNC lathes, the natural way of development would be to adapt the tool indexing techniques of the machining centers. This would increase the number of cutting tools available and keep inactive tools away from the machining area. Also watch for features that eliminate secondary operations, such as complex milling features on lathes and built-in part indexing.

Predictions for computers are difficult at best, except that their power will increase. Hardware has developed at a higher rate than software and this will not change soon. CNC software is no exception. The winner of the competitive race will be the one that can combine hardware, software and people, makes a product for a reasonable price and markets it across the world. The protectionist economy does not work, and trading will not be confined to several 'local' blocks - it will be a part of the true global economy. Before too many personal opinions force their way out, it is time to end and say *"Learn, work, and then learn again"*.

REFERENCE TABLES

Decimal Equivalents

The following chart lists fractional, wire gauge (number), letter and metric (mm) values for given decimal equivalents in inches.

Decimal inch	Fraction	Number / Letter	Metric (mm)
.0059		97	0.15
.0063		96	0.16
.0067		95	0.17
.0071		94	0.18
.0075		93	0.19
.0079		92	0.20
.0083		91	0.21
.0087		90	0.22
.0091		89	0.23
.0095		88	0.24
.0100		87	0.25
.0105		86	
.0110		85	0.28
.0115		84	
.0118			0.30
.0120		83	
.0125		82	
.0126			0.32
.0130		81	
.0135		80	
.0138			0.35
.0145		79	
.0150			0.38
.0156	1/64		
.0157			0.40
.0160		78	
.0177			0.45
.0180		77	
.0197			0.50
.0200		76	
.0210		75	
.0217			0.55
.0225		74	
.0236			0.60
.0240		73	
.0250		72	
.0256			0.65
.0260		71	
.0276			0.70
.0280		70	
.0292		69	
.0295			0.75
.0310		68	
.0313	1/32		
.0315			0.80
.0320		67	
.0330		66	

Decimal inch	Fraction	Number / Letter	Metric (mm)
.0335			0.85
.0350		65	
.0354			0.90
.0360		64	
.0370		63	
.0374			0.95
.0380		62	
.0390		61	
.0394			1.00
.0400		60	
.0410		59	
.0413			1.05
.0420		58	
.0430		57	
.0433			1.10
.0453			1.15
.0465		56	
.0469	3/64		
.0472			1.20
.0492			1.25
.0512			1.30
.0520		55	
.0531			1.35
.0550		54	
.0551			1.40
.0571			1.45
.0591			1.50
.0595		53	
.0610			1.55
.0625	1/16		
.0630			1.60
.0635		52	
.0650			1.65
.0669			1.70
.0670		51	
.0689			1.75
.0700		50	
.0709			1.80
.0728			1.85
.0730		49	
.0748			1.90
.0760		48	
.0768			1.95
.0781	5/64		
.0785		47	
.0787			2.00
.0807			2.05
.0810		46	
.0820		45	
.0827			2.10
.0846			2.15
.0860		44	
.0866			2.20

Decimal inch	Fraction	Number / Letter	Metric (mm)
.0886			2.25
.0890		43	
.0906			2.30
.0925			2.35
.0935		42	
.0938	3/32		
.0945			2.40
.0960		41	
.0965			2.45
.0980		40	
.0984			2.50
.0995		39	
.1015		38	
.1024			2.60
.1040		37	
.1063			2.70
.1065		36	
.1083			2.75
.1094	7/64		
.1100		35	
.1102			2.80
.1110		34	
.1130		33	
.1142			2.90
.1160		32	
.1181			3.00
.1200		31	
.1220			3.10
.1250	1/8		
.1260			3.20
.1280			3.25
.1285		30	
.1299			3.30
.1339			3.40
.1360		29	
.1378			3.50
.1405		28	
.1406	9/64		
.1417			3.60
.1440		27	
.1457			3.70
.1470		26	
.1476			3.75
.1495		25	
.1496			3.80
.1520		24	
.1535			3.90
.1540		23	
.1562	5/32		
.1570		22	
.1575			4.00
.1590		21	
.1610		20	
.1614			4.10
.1654			4.20
.1660		19	
.1673			4.25
.1693			4.30
.1695		18	
.1719	11/64		
.1730		17	
.1732			4.40
.1770		16	

Decimal inch	Fraction	Number / Letter	Metric (mm)
.1772			4.50
.1800		15	
.1811			4.60
.1820		14	
.1850		13	4.70
.1870			4.75
.1875	3/16		
.1890		12	4.80
.1910		11	
.1929			4.90
.1935		10	
.1960		9	
.1969			5.00
.1990		8	
.2008			5.10
.2010		7	
.2031	13/64		
.2040		6	
.2047			5.20
.2055		5	
.2067			5.25
.2087			5.30
.2090		4	
.2126			5.40
.2130		3	
.2165			5.50
.2188	7/32		
.2205			5.60
.2210		2	
.2244			5.70
.2264			5.75
.2280		1	
.2283			5.80
.2323			5.90
.2340		A	
.2344	15/64		
.2362			6.00
.2380		B	
.2402			6.10
.2420		C	
.2441			6.20
.2460		D	
.2461			6.25
.2480			6.30
.2500	1/4	E	
.2520			6.40
.2559			6.50
.2570		F	
.2598			6.60
.2610		G	
.2638			6.70
.2556	17/64		
.2657			6.75
.2660		H	
.2677			6.80
.2717			6.90
.2720		I	
.2756			7.00
.2770		J	
.2795			7.10
.2810		K	
.2812	9/32		
.2835			7.20

Decimal inch	Fraction	Number / Letter	Metric (mm)
.2854			7.25
.2874			7.30
.2900		L	
.2913			7.40
.2950		M	
.2953			7.50
.2969	19/64		
.2992			7.60
.3020		N	
.3031			7.70
.3051			7.75
.3071			7.80
.3110			7.90
.3125	5/16		
.3150			8.00
.3160		O	
.3189			8.10
.3228			8.20
.3230		P	
.3248			8.25
.3268			8.30
.3281	21/64		
.3307			8.40
.3320		Q	
.3346			8.50
.3386			8.60
.3390		R	
.3425			8.70
.3438	11/32		
.3445			8.75
.3465			8.80
.3480		S	
.3504			8.90
.3543			9.00
.3580		T	
.3583			9.10
.3594	23/64		
.3622			9.20
.3642			9.25
.3661			9.30
.3680		U	
.3701			9.40
.3740			9.50
.3750	3/8		
.3770		V	
.3780			9.60
.3819			9.70
.3839			9.75
.3858			9.80
.3860		W	
.3898			9.90
.3906	25/64		
.3937			10.00
.3970		X	
.4040		Y	
.4062	13/32		
.4130		Z	
.4134			10.50
.4219	27/64		
.4331			11.00
.4375	7/16		
.4528			11.50
.4531	29/64		

Decimal inch	Fraction	Number / Letter	Metric (mm)
.4688	15/32		
.4724			12.00
.4844	31/64		
.4921			12.50
.5000	½		12.70
.5118			13.00
.5156	33/64		
.5312	17/32		
.5315			13.50
.5469	35/64		
.5512			14.00
.5625	9/16		
.5709			14.50
.5781	37/64		
.5906			15.00
.5938	19/32		
.6094	39/64		
.6102			15.50
.6250	5/8		
.6299			16.00
.6406	41/64		
.6496			16.50
.6562	21/32		
.6693			17.00
.6719	43/64		
.6875	11/16		
.6890			17.50
.7031	45/64		
.7087			18.00
.7188	23/32		
.7283			18.50
.7344	47/64		
.7480			19.00
.7500	3/4		
.7656	49/64		
.7677			19.50
.7812	25/32		
.7874			20.00
.7969	51/64		
.8071			20.50
.8125	13/16		
.8268			21.00
.8281	53/64		
.8438	27/32		
.8465			21.50
.8594	55/64		
.8661			22.00
.8750	7/8		
.8858			22.50
.8906	57/64		
.9055			23.00
.9062	29/32		
.9219	59/64		
.9252			23.50
.9375	15/16		
.9449			24.00
.9531	61/64		
.9646			24.50
.9688	31/32		
.9843			25.00
.9844	63/64		
1.0000	1		25.40

All tap drill sizes in the following tables are based on the approximate full thread depth of 72-77% of nominal.

English Threads - UNC/UNF

Thread-TPI	Tap Drill Size	Inch equivalent	Metric alternative
#0-80	3/64	.0469	
1/16-64	3/64	.0469	
#1-64	#53	.0595	
#1-72	#53	.0595	
#2-56	#50	.0700	
#2-64		.0709	1.80
3/32-48	#49	.0730	
#3-48	#47	.0785	
#3-56	#45	.0820	
#4-32	#45	.0820	
#4-36	#44	.0860	
#4-40	#43	.0890	
#4-48	#42	.0935	
#5-40	#39	.0995	
#5-44	#37	.1040	
1/8-40	#38	.1015	
#6-32	#36	.1065	
#6-36	#34	.1110	
#6-40	#33	.1130	
5/32-32	1/8	.1250	
5/32-36	#30	.1285	
#8-32	#29	.1360	
#8-36	#29	.1360	
#8-40	#28	.1405	
3/16-24	#26	.1470	
3/16-32	#22	.1570	
#10-24	#25	.1495	
#10-28	#23	.1540	
#10-30	#22	.1570	
#10-32	#21	.1590	
#12-24	#16	.1770	
#12-28	#14	.1820	
#12-32	#13	.1850	4.70
7/32-24	#16	.1770	
7/32-32	#12	.1890	4.80
#14-20	#10	.1935	
#14-24	#7	.2010	5.10
1/4-20	#7	.2010	5.10
1/4-28	#4	.2090	
1/4-32	7/32	.2188	5.50
5/16-18	F	.2570	6.50
5/16-20	17/64	.2656	
5/16-24	I	.2720	6.90
5/16-32	9/32	.2813	7.10
3/8-16	5/16	.3125	8.00
3/8-20	21/64	.3281	
3/8-24	Q	.3320	8.50
3/8-32	11/32	.3438	
7/16-14	U	.3680	9.40
7/16-20	25/64	.3906	9.90
7/16-24	X	.3970	10.00
7/16-28	Y	.4040	
1/2-13	27/64	.4219	
1/2-20	29/64	.4531	11.50
1/2-28	15/32	.4688	
9/16-12	31/64	.4844	

Thread-TPI	Tap Drill Size	Inch equivalent	Metric alternative
9/16-18	33/64	.5156	13.00
9/16-24	33/64	.5156	13.00
5/8-11	17/32	.5313	13.50
5/8-12	35/64	.5469	
5/8-18	37/64	.5781	
5/8-24	37/64	.5781	
11/16-12	39/64	.6094	
11/16-16	5/8	.6250	
11/16-24	41/64	.6406	
3/4-10	21/32	.6563	16.50
3/4-12	43/64	.6719	17.00
3/4-16	11/16	.6875	17.50
3/4-20	45/64	.7031	17.50
3/4-28	23/32	.7188	
13/16-12	47/64	.7344	
13/16-16	3/4	.7500	
7/8-9	49/64	.7656	19.50
7/8-12	51/64	.7969	20.00
7/8-14	13/16	.8125	
7/8-16	13/16	.8125	
7/8-20	53/64	.8281	
15/16-12	55/64	.8594	
15/16-16	7/8	.8750	
15/16-20	57/64	.8906	
1-8	7/8	.8750	
1-12	59/64	.9219	
1-14	15/16	.9375	
1-20	61/64	.9531	
1 1/16-12	63/64	.9844	
1 1/16-16	1.0	1.0000	

Straight Pipe Taps NPS

Tap Size	Tap Drill ⌀ (in.)	Metric alternative
1/8-27	S	8.80
1/4-18	29/64	11.50
3/8-18	19/32	15.00
1/2-14	47/64	18.50
3/4-14	15/16	23.75
1-11 1/2	1 3/16	30.25
1 1/4-11 1/2	1 33/64	38.50
1 1/2-11 1/2	1 3/4	44.50
2-11 1/2	2 7/32	56.00

Pipe Size	TPI	Tap Drill	Decimal Size
1/16	27	1/4	.2500
1/8	27	11/32	.3438
1/4	18	7/16	.4375
3/8	18	37/64	.5781
1/2	14	23/32	.7188
3/4	14	59/64	.9219
1.0	11-1/2	1-5/32	1.1563
1-1/4	11-1/2	1-1/2	1.5000
1-1/2	11-1/2	1-3/4	1.7500
2.0	11-1/2	2-7/32	2.2188

Taper Pipe Taps NPT

Tap Size	Tap Drill ⌀ (in.)	Metric alternative
1/16-27	D	6.30
1/8-27	R	8.70
1/4-18	7/16	11.10
3/8-18	37/64	14.50
½-14	45/64	18.00
3/4-14	59/64	23.25
1-11 ½	1 5/32	29.00
1 1/4-11 ½	1 ½	38.00
1 ½-11 ½	1 47/64	44.00
2-11 ½	2 7/32	56.00
2 ½-8	2 5/8	67.00
3-8	3 1/4	82.50

Pipe Size	TPI	Drilled Only		Taper Reamed	
		Tap Drill	Dec. Size	Tap Drill	Dec. Size
1/16	27	D	.2460	15/64	.2344
1/8	27	Q	.3320	21/64	.3281
1/4	18	7/16	.4375	27/64	.4219
3/8	18	37/64	.5781	9/16	.5625
½	14	45/64	.7031	11/16	.6875
3/4	14	29/32	.9062	57/64	.8906
1.0	11-1/2	1-9/64	1.1406	1-1/8	1.1250
1-1/4	11-1/2	1-31/64	1.4844	1-15/32	1.4688
1-1/2	11-1/2	1-47/64	1.7344	1-23/32	1.7188
2.0	11-1/2	2-13/64	2.2031	2-3/16	2.1875

Metric Coarse Threads

Nominal ⌀ x Pitch (mm)	Tap Drill ⌀ (mm)	Inch equivalent
M1 x 0.25	0.75	.0295
M1.2 x 0.25	0.95	.0374
M1.4 x 0.3	1.10	.0433
M1.5 x 0.35	1.15	.0453
M1.6 x 0.35	1.25	.0492
M1.8 x 0.35	1.45	.0571
M2 x 0.4	1.60	.0630
M2.2 x 0.45	1.75	.0689
M2.5 x 0.45	2.05	.0807
M3 x 0.5	2.50	.0984
M3.5 x 0.6	2.90	.1142
M4 x 0.7	3.30	.1299
M4.5 x 0.75	3.75	.1476
M5 x 0.8	4.20	.1654
M6 x 1	5.00	.1969
M7 x 1	6.00	.2362
M8 x 1.25	6.75	.2657
M9 x 1.25	7.75	.3051
M10 x 1.5	8.50	.3346
M11 x 1.5	9.50	.3740
M12 x 1.75	10.20	.3937
M14 x 2	12.00	.4724
M16 x 2	14.00	.5512
M18 x 2.5	15.50	.6102
M20 x 2.5	17.50	.6890
M22 x 2.5	19.50	.7677
M24 x 3	21.00	.8268
M27 x 3	24.00	.9449
M30 x 3.5	26.50	1.0433

Metric Fine Threads

Nominal ⌀ x Pitch (mm)	Tap Drill ⌀ (mm)	Inch equivalent
M3 x 0.35	2.65	.1043
M3.5 x 0.35	3.15	.2283
M4 x 0.5	3.50	.1378
M4.5 x 0.5	4.00	.1575
M5 x 0.5	4.50	.1772
M5.5 x 0.5	5.00	.1969
M6 x 0.75	5.25	.2067
M7 x 0.75	6.25	.2461
M8 x 1	7.00	.2756
M9 x 1	8.00	.3150
M10 x 0.75	9.25	.3642
M10 x 1	9.00	.3543
M10 x 1.25	8.75	.3445
M11 x 1	10.00	.3937
M12 x 1	11.00	.4331
M12 x 1.25	10.75	.4232
M12 x 1.5	10.50	.4134
M13 x 1.5	11.50	.4528
M13 x 1.75	11.25	.4429
M14 x 1.25	12.75	.5020
M14 x 1.5	12.50	.4921
M15 x 1.5	13.50	.5315
M16 x 1	15.00	.5906
M16 x 1.5	14.50	.5709
M17 x 1.5	15.50	.6102
M18 x 1	17.00	.6693
M18 x 1.5	16.50	.6496
M18 x 2	16.00	.6299
M20 x 1	19.00	.7480
M20 x 1.5	18.50	.7283
M20 x 2	18.00	.7087
M22 x 1	21.00	.8268
M22 x 1.5	20.50	.8071
M22 x 2	20.00	.7874
M24 x 1	23.00	.9055
M24 x 1.5	22.50	.8858
M24 x 2	22.00	.8661
M25 x 1.5	23.50	.9252
M27 x 2	25.00	.9843
M28 x 2	26.00	1.0236
M30 x 2	28.00	1.1024
M30 x 3	27.00	1.0630

Index

E

F

G

N

O

P

Q

R

S

T